ADME AND TRANSLATIONAL PHARMACOKINETICS/ PHARMACODYNAMICS OF THERAPEUTIC PROTEINS

ADME AND TRANSLATIONAL PHARMACOKINETICS/ PHARMACODYNAMICS OF THERAPEUTIC PROTEINS

Applications in Drug Discovery and Development

Edited by

HONGHUI ZHOU
FRANK-PETER THEIL

WILEY

Published by John Wiley & Sons, Inc., Hoboken, New Jersey
Published simultaneously in Canada

Library of Congress Cataloging-in-Publication Data:

ADME and translational pharmacokinetics/pharmacodynamics of therapeutic proteins : applications in drug discovery and development / edited by Honghui Zhou, Frank-Peter Theil.
p. ; cm.
Includes bibliographical references and index.
ISBN 978-1-118-89864-2 (hardback)
I. Zhou, Honghui, editor. II. Theil, Frank-Peter, editor.
[DNLM: 1. Antibodies, Monoclonal–therapeutic use. 2. Protein Engineering. 3. Proteins–pharmacokinetics. 4. Drug Discovery. QU 55.95]
RM301.5
615.7–dc23

2015024770

Set in 10/12pt Times by SPi Global, Pondicherry, India

Printed in the United States of America

CONTENTS

LIST OF CONTRIBUTORS

Omoniyi J. Adedokun, Janssen Research and Development, LLC, Spring House, PA, USA

Bo An, Department of Pharmaceutical Sciences, State University of New York at Buffalo, Buffalo, NY, USA

Andreas Baumann, Bayer Pharma Aktiengesellschaft, Berlin, Germany

C. Andrew Boswell, Genentech Research and Early Development, South San Francisco, CA, USA

David Canter, Keck School of Medicine of the University of Southern California, Los Angeles, CA, USA

Yanguang Cao, State University of New York at Buffalo, Buffalo, NY, USA; The University of North Carolina at Chapel Hill, Chapel Hill, NC, USA

Alan L. Epstein, Keck School of Medicine of the University of Southern California, Los Angeles, CA, USA

Thierry Flandre, Novartis Pharma AG, Basel, Switzerland

Gert Fricker, Ruprecht-Karls University Heidelberg, Heidelberg, Germany

Ludivine Fronton, Roche Pharmaceutical Research and Early Development, Basel, Switzerland

Saskia Fuhrmann, Universität Potsdam Institute of Biochemistry and Biology, Potsdam, Germany; Freie Universität Berlin, Berlin, Germany

María J. Garrido, University of Navarra, Pamplona, Spain

Megan Gibbs, Clinical Pharmacology, Modeling and Simulation, Amgen Inc., Thousand Oaks, CA, USA

José David Gómez-Mantilla, University of Navarra, Pamplona, Spain

Chao Han, Janssen Research & Development, LLC, Spring House, PA, USA

Tae H. Han, Stemcentrx, Inc., South San Francisco, CA, USA

Kenta Haraya, Chugai Pharmaceutical Co., Ltd., Gotemba City, Shizuoka Prefecture, Japan

Justin T. Hoffman, Pfizer Oncology, San Diego, CA, USA

Peisheng Hu, Keck School of Medicine of the University of Southern California, Los Angeles, CA, USA

Wilhelm Huisinga, Universität Potsdam, Potsdam, Germany

Susan Hurst, Pfizer Worldwide Research and Development, Groton, CT, USA

Tomoyuki Igawa, Chugai Pharmaceutical Co., Ltd., Gotemba City, Shizuoka Prefecture, Japan

Yuki Iwayanagi, Chugai Pharmaceutical Co., Ltd., Gotemba City, Shizuoka Prefecture, Japan

Julie K. Jang, Keck School of Medicine of the University of Southern California, Los Angeles, CA, USA

Stewart Jones, AstraZeneca, Cambridge, UK

Sean B. Joseph, Genentech Research and Early Development, South San Francisco, CA, USA

William J. Jusko, State University of New York at Buffalo, Buffalo, NY, USA

Jitendra Kanodia, Genentech Research and Early Development, South San Francisco, CA, USA

Leslie A. Khawli, Keck School of Medicine of the University of Southern California, Los Angeles, CA, USA

Ben-Fillippo Krippendorff, Roche Pharmaceutical Research and Early Development, Basel, Switzerland

Harald Kropshofer, Roche Pharmaceutical Research and Early Development, Basel, Switzerland

Anne Kümmel, Novartis Pharmaceuticals AG, Basel, Switzerland

Jiunn H. Lin, 3D BioOptimal Co Ltd., Suzhou, China

Peter Lloyd, KinDyn Consulting Ltd., Horsham, UK

Philip J. Lowe, Novartis Pharmaceuticals AG, Basel, Switzerland

Anne Mahringer, Ruprecht-Karls University Heidelberg, Heidelberg, Germany

Soichiro Matsushima, Novartis Pharmaceuticals K.K., Tokyo, Japan

Christina Lourdes Mayer, Janssen Research & Development, LLC, Spring House, PA, USA

Bernd Meibohm, University of Tennessee Health Science Center, Memphis, TN, USA

Guy M.L. Meno-Tetang, UCB Celltech, Slough, Berkshire, UK

Kourosh Parivar, Pfizer Oncology, San Diego, CA, USA

Andrew G. Popplewell, UCB Celltech, Slough, Berkshire, UK

Saileta Prabhu, Genentech Research and Early Development, South San Francisco, CA, USA

Thomayant Prueksaritanont, Merck Research Laboratories, West Point, PA, USA

Jun Qu, Department of Pharmaceutical Sciences, State University of New York at Buffalo, Buffalo, NY, USA

Chetan Rathi, University of Tennessee Health Science Center, Memphis, TN, USA

Wolfgang F. Richter, Roche Pharmaceutical Research and Early Development, Basel, Switzerland

Malcolm Rowland, University of Manchester, Manchester, UK; University of California, San Francisco, CA, USA

Elizabeth Y. Shang, US Food and Drug Administration, Silver Spring, MD, USA

Vikram Sinha, US Food and Drug Administration, Silver Spring, MD, USA

Andrej Skerjanec, Sandoz Biosimilars, Novartis, Basel, Switzerland

Tatsuhiko Tachibana, Chugai Pharmaceutical Co., Ltd., Gotemba City, Shizuoka Prefecture, Japan

Sarah Taplin, Novartis Pharma AG, Basel, Switzerland

Greg M. Thurber, University of Michigan, Ann Arbor, MI, USA

Jay Tibbitts, UCB Celltech, Slough, UK

Iñaki F. Trocóniz, University of Navarra, Pamplona, Spain

Christina Vasalou, Novartis Institutes for Biomedical Research, Cambridge, MA, USA

Diane D. Wang, Pfizer Oncology, San Diego, CA, USA

Weirong Wang, Janssen Research and Development, LLC, Spring House, PA, USA

Yaning Wang, US Food and Drug Administration, Silver Spring, MD, USA

Zhenhua Xu, Janssen Research and Development, LLC, Spring House, PA, USA

Donghua Yin, Pfizer Worldwide Research and Development, San Diego, CA, USA

Theresa Yuraszeck, Clinical Pharmacology, Modeling and Simulation, Amgen Inc., Thousand Oaks, CA, USA

Ming Zhang, Department of Pharmaceutical Sciences, State University of New York at Buffalo, Buffalo, NY, USA

Baiteng Zhao, Seattle Genetics, Inc., Bothell, WA, USA

Liang Zhao, US Food and Drug Administration, Silver Spring, MD, USA

Ping Zhao, US Food and Drug Administration, Silver Spring, MD, USA

Songmao Zheng, Janssen Research & Development, LLC, Spring House, PA, USA

Honghui Zhou, Janssen Research & Development, LLC, Spring House, PA, USA

FOREWORD

The use of biologically derived sources for the treatment or amelioration of diseases and conditions is as old as the existence of the human race. But most extracts and preparations from plants and animals were un- or ill-defined mixtures, without a clear understanding of the active ingredients, if any. The past 50 years has seen a transformational change with the development and availability of so many new specific and potent drugs to address a myriad of previously unmet medical needs. Although the first few decades of this period, from the 1960s to 1980s, was the age of small molecular entities, the last two decades can be said to be the age of the emergence of the large drug molecule, comprising approximately 35% of all new drugs approved during this latter period. This successful development of a plethora of large (often recombinant) molecules is one of the fruits of the preceding explosion in molecular biology and biotechnology; hence the term therapeutic biologics. They span a dizzying array of compounds of varying molecule size and complexity, from relatively simple molecules, such as insulin, to antibodies, to vaccines, and beyond. What they share in common, and in contrast to most small molecular weight drugs, are the building blocks, namely amino acids. Also because of lability of many protein drugs in the gastrointestinal tract, they often need to be administered parenterally to be therapeutically effective, in contrast to small molecular weight drugs for which the oral route is the most common.

This book is about the absorption, distribution, metabolism, and excretion (ADME) and pharmacokinetics/pharmacodynamics (PK/PD) of protein drugs, in particular, but not exclusively, monoclonal antibodies, primarily because so many of the currently approved protein drugs, and likely future ones, are of this class. Unlike many small molecules, which access all sites within the body, including intracellular targets, currently most protein drugs tend to be restricted to extracellular targets, because of their relatively large size and polarity. Even so, there are issues regarding the movement of protein drugs from the vascular to target sites within the tissue interstitial space. The need for this book exists because it has become clear that although there are some similarities, many of the processes that operate or apply to small molecular weight drugs and guide their development do not apply, or are less explicable, to protein drugs, about which our understanding is at an early stage. Still, there are lessons to be learned from our experience with small molecular entities.

Before the 1950s, the development of small molecular weight drugs was based almost exclusively on final outcomes, efficacy, and safety, with virtually no concern as to what happened to a drug within the body upon its administration. The thalidomide catastrophe in the 1960s and the life-threatening terfenadine–ketoconazole interaction in the early 1990s were among landmark events that focused (regulatory) attention on the need to better understand the ADME processes controlling the fate of such drugs within the body and effects on the body. Also, linking PK/PD was seen as the way forward to quantitatively map the dose–response–time surface within the patient population to better inform the design and development of optimal dosage regimens.

Similarly, because early bioengineered protein drugs were purified forms of ones already found in the body, there was virtually no attention on, or a considered need to study, their ADME. Also, when applied, their PK/PD was invariably described empirically. These views are rapidly disappearing for various reasons. We became aware that some protein drugs exhibit "strange" phenomena, such as target-mediated disposition, and that many were primarily eliminated by tissues or organs other than the kidneys or liver (the predominant organs for small molecules). We also started to manipulate natural proteins to produce a whole array of modified proteins to improve their PK or efficacy, or linking antibodies to small potent drug molecules to facilitate specific targeting (the magic bullet). These and other events and factors have driven the impetus for a better understanding of their ADME and PK/PD, which has now emerged to the point that it is profoundly influencing the design and development of new therapeutic proteins. It is in this context that this timely book on protein drugs should be viewed. Collected together are a unique group of authors, with experience in all aspects of ADME and PK/PD of protein drugs, from characterization of the protein itself, to development of analytical methods, to their absorption and processing within the body, including the role of specific mechanisms for their distribution and stabilization there, to effects produced, and to the application of modeling and simulation, at the academic, industrial, and regulatory levels. This book will be of great help to those newly entering the field of therapeutic proteins, as well as those already working in it, and should act as a stimulus to undertake further research to better understand this important and ever-growing group of therapeutic agents. I am also sure that it will be the first of many such books, extending beyond monoclonal antibodies to the vast array of "nonantibody" protein drugs.

In my lifetime, I have seen mechanistic and quantitative understanding of the PK and PD of small molecular weight drugs advance to a stage that physiologically (systems) based modeling, coupled with *in vitro* and *in silico* methodologies, has now become an integral part of their discovery and development, improving efficiency and predictability. I can envisage the day that such approaches will routinely extend to therapeutic proteins for the betterment of patients.

Malcolm Rowland
Professor Emeritus, University of Manchester,
Manchester, United Kingdom; Adjunct Professor,
University of California,
San Francisco, CA

1

ADME FOR THERAPEUTIC BIOLOGICS: WHAT CAN WE LEVERAGE FROM GREAT WEALTH OF ADME KNOWLEDGE AND RESEARCH FOR SMALL MOLECULES

WEIRONG WANG[1] AND THOMAYANT PRUEKSARITANONT[2]

[1] *Janssen Research and Development, LLC, Spring House, PA, USA*

[2] *Merck Research Laboratories, West Point, PA, USA*

1.1 INTRODUCTION

Over the past decade, there has been increased investment to the development of biotechnologically derived drug products or biologics (including peptides, proteins, and monoclonal antibodies, mAbs, aggregately referred as large molecule (LM) drugs) in pharmaceutical companies [1, 2]. These are attributable to the reported therapeutic success of this modality thus far, together with the rapid advancement and breakthroughs in the fields of recombinant DNA biotechnology and molecular biology. However, reports on mechanistic investigation of absorption, distribution, metabolism, and excretion (ADME) processes for LMs are sparse and our current understanding of the associated mechanisms and key determinants of pharmacokinetic (PK) properties is scant [3]. Conceivably, these are related to the fact that the biopharmaceutical industry is still at an early stage, relative to the traditional pharmaceutical counterpart; the first approved LM drug product was in 1980s [4], several decades after many small molecule (SM) drugs were on the market. In addition, unlike the discovery and development of SM drugs, where the sciences and the functional role of drug metabolism and pharmacokinetics (DMPK) in studying and understanding ADME processes have been well recognized as an indispensable and integral discipline spanning from early discovery to development and postmarketing spaces [5], the function of DMPK in support of LM drug development is somewhat limited to mostly *in vivo* PK and/or pharmacokinetics/pharmacodynamics (PK/PD) studies, typically after candidate selection and primarily in the clinical space. Despite the intrinsic difference between SM and LM drugs, it should be of particular interest to appraise the relevance and applicability of what we have learned over the past few decades from the discovery and development of SM drugs to the same process of LMs. Thus, in this chapter, a brief historical perspective is presented on how the roles of DMPK and the key enablers for studying the ADME processes of SM drugs and their underlying mechanisms have evolved over time in order to influence internal de-risking strategy and decisions. External factors, such as changing regulatory environments and evolving LM discovery and development landscape, are briefly reviewed. Also presented is an overview of a DMPK concept analogy between SMs and LMs, as well as case examples to demonstrate the applicability of SM DMPK knowledge and experiences to LM drug discovery and development.

1.2 SM DRUG DISCOVERY AND DEVELOPMENT: HISTORICAL PERSPECTIVE

1.2.1 Evolving Role of DMPK: Paradigm Shift

It has long been well recognized that the drug discovery and development process is very expensive, largely due to a high development attrition rate and prolonged development time to meet the requirement for more extensive and

ADME and Translational Pharmacokinetics/Pharmacodynamics of Therapeutic Proteins: Applications in Drug Discovery and Development, First Edition. Edited by Honghui Zhou and Frank-Peter Theil.

complex clinical trials [1, 6–8]. In 1990s, poor human PK and bioavailability were the most significant cause of attrition for SM drugs, accounting for approximately 40% of all attrition in development. This number was dramatically reduced to approximately 8% by 2000 [7]. Such a drastic difference has been attributable primarily to a *Paradigm shift* in the roles of DMPK from little involvement decades before 1990 to active participation in SM drug early discovery starting in late 1980s [5]. Previously, compounds were selected mainly based on *in vitro* potency and *in vivo* efficacy in animal studies, with little attention being paid to the exposure or PK as an important measure connecting pharmacodynamics (PD)/efficacy/safety profiles, or consideration to commonly observed differences in these profiles between animals and humans. The integration of DMPK support as a key component of the overall drug discovery process helped to better understand ADME properties and filled these gaps, thus enabling proper data interpretations and rationale-based predictions of DMPK-related properties in humans [9–13]. As a result, potential liabilities of new chemical entities *in humans* were dialed out as early as possible, leading to increased likelihood for preclinical candidates to be developed successfully as therapeutic agents.

1.2.2 Key Enablers to Successful DMPK Support

The aforementioned successful DMPK support would not have been possible without numerous advances over the past few decades in drug metabolism sciences and technologies, which have provided powerful tools to enable DMPK scientists to shape SM drug metabolism research. Of special note are two key enablers, signifying game changers within the time period of interest (late-1980s to late-1990s): (i) rapid advancement of cytochrome P450 (CYP) science and (ii) availability of liquid chromatography–mass spectrometry (LC–MS). As will be described in later sections, these elements and associated wealth of information generated over the last few decades can be leveraged and applied to support LM drug development.

The CYP enzymes play central roles in the metabolism of SMs; it is estimated that more than 70% of marketed SM drugs were eliminated primarily by CYPs [13]. CYP enzymes were discovered in 1958, and research on their structure, function, regulation, and tissue expression levels, as well as their role in drug metabolism, was rapidly expanded in the 1980–1990s [14–16]. Such rapid advancement provided fundamental concepts and important tools that helped leverage preclinical/*in vitro* results as a bridge to clinical outcomes, consequently enabling one to predict, understand, and manage clinical findings, particularly with respect to human clearance and PK variability due to factors such as CYP-mediated drug–drug interaction (DDI) or CYP polymorphism [13, 16–18]. Specifically, for compounds with CYPs as the major or sole contributor to their metabolism, human metabolic clearance can be reasonably predicted based simply on *in vitro* metabolism studies with recombinant CYP isoforms, corrected for relative expression levels of each isoform in tissues [19]. In addition, the knowledge of CYP substrate specificity, multiplicity, and responses to factors, such as inducers and inhibitors, has provided a means to quantitatively predict, based on *in vitro* studies with specific CYP marker substrates or inhibitors/inducers, the magnitude of DDI, thus enabling a selection of candidates at discovery stage that do not bear considerable liability to serious clinical DDIs, either as perpetrators or victims [16–18, 20]. The DDI prediction results have also been used (and accepted by regulatory agencies) to inform inclusion and exclusion criteria for clinical programs, decide whether a clinical DDI study is needed, and inform product labeling with respect to dosage adjustment and warning/contraindication when used with other medications [21, 22]. Collectively, advances in understanding CYPs, the primary determinant for clearance mechanism of majority of SM drugs, has helped reduce drug development failure rate due to undesirable human PK properties.

In the area of tools and technologies, the successful coupling of high performance liquid chromatography with mass spectrometry (MS) has provided unprecedented sensitivity, selectivity, and high throughput that has facilitated the rapid assessment of ADME properties and the multiplicity of their governing factors for SM candidates in animals and humans [23–26]. Capitalizing on chromatographic separation and mass selectivity, the LC–MS technology enables the quantitation of coeluting or overlapping analytes, which otherwise would be constrained by chromatographic resolution. A dramatic outcome of this feature is the various *in vivo* and *in vitro* cassette studies in which more than one compounds were administered or incubated for the screening of DMPK properties, including metabolic stability, DDI liability, and plasma protein binding [23–25]. Along with the accelerated method development similarly attributed to the extraordinary selectivity and sensitivity of LC–MS, this practice has tremendously facilitated the speed and throughput of analyses of samples of low concentrations or of small volumes. Likewise, LC–MS technology has reshaped the business of metabolite characterization, allowing rapid detection and identification of major metabolites of drug candidates so that the result can be fed back into the cycle in time to influence the synthetic chemistry effort. Together, this powerful technology has enabled informed decisions to be made rapidly on a large number of candidates, each available in a small quantity, during the discovery stage. It has also enabled other in-depth mechanistic investigations into the governing factors of ADME processes, as well as detailed and accurate characterization of ADME properties of development candidates required for risk mitigation and regulatory submission [5, 10, 26]. With the recent advent of new chromatographic techniques, such as ultraperformance liquid chromatography, and more sophisticated MS, such as high resolution MS [27], this technology will continue to be the most powerful tool for drug discovery and development for SMs, and potentially for LMs alike.

1.2.3 Regulatory Considerations

Successful development of a drug candidate requires the right set of high quality data to help inform decisions not only internally, but also decisions by regulatory authorities. In-depth industry analysis by PhRMA has attributed much of the increasing R&D costs to the extending development times in clinical phases (10–15 years), greatly influenced by the increased regulatory demands in today's low risk, low tolerance environment, and stemmed primarily from the withdrawal of several prominent prescription drugs from the market over the past decades for safety reasons. Of special note was the withdrawal of the drugs from the U.S. market in 1990s, half of which due to serious and unmanageable safety issues as a result of PK and/or PD DDIs. These occurrences prompted the FDA to publish guidance documents for industry to encourage the characterization of DDI potential for a new molecular entity early in the drug development process [21]. The first two guidance documents: one on *in vitro* DDI, published in 1997, and the other on *in vivo* DDI, published in 1999, focused on metabolic DDI due to CYPs, and was based primarily on considerable advances in our understanding of roles of the CYP family at the time. In the latest draft DDI guidance recently issued [22], there are recommendations to conduct many additional drug transporters, and drug interaction studies for LMs have been included for the first time. Given the current status and understanding of drug transporter sciences relative to the CYPs [28], the inclusion of drug transporters in the latest guidance suggested that the FDA has become more proactive in embracing evolving sciences in their decision making. Likewise, much less is known about LM drugs in their DMPK properties and underlying DDI mechanisms in comparison with SM drugs. Consistent with this, the time span between the first approved LM drug in 1986 and the anticipated DDI guidance is much shorter than the corresponding time span of many decades for SM drugs. This apparently speedy process for LMs may be attributable to the decision of the 2003 FDA to transfer the regulatory responsibility from the Center for Biologics Evaluation and Research (CBER) to Center for Drug Evaluation and Research (CDER), who has been overseeing the regulatory approval of SM drugs and has provided more comprehensive information on ADME properties and associated DDI implications. It is conceivable that there will be increasing regulatory demands for other DMPK-related information for LMs in the near future. In fact, the CDER Science Prioritization and Review Committee has recently highlighted several relevant LM DMPK aspects warranting additional research and further understanding [29], suggesting that the most relevant factors that affect the PK/PD determinants of LMs, such as a variety of specific receptors that can influence protein $t_{1/2}$ and distribution (e.g., delivery of therapeutic enzymes to the correct cellular compartment), should be identified.

1.3 LM DRUG DISCOVERY AND DEVELOPMENT

1.3.1 Role of DMPK: Current State

From a DMPK perspective, the current state for an LM support paradigm and ADME knowledge is similar to where we were with SM drugs a few decades ago. First, DMPK is involved primarily in the development space after a preclinical LM candidate has already been selected, and much less at the early drug discovery stage of the optimization and selection of LM candidates. This conventional mindset, widespread in many biotech and pharmaceutical firms, resembles what was practiced for SMs before the 1990s, and may stem from a wide belief that PK of biologics, especially mAbs, is well behaved/predictable and that this property is not known to be a major success-limiting factor, based on a historical record of relatively low attrition rate for LMs versus SMs. However, the view that PK of all mAbs is well behaved and the deviation of typical mAb PK properties is due primarily to their intended target binding has recently been challenged. For example, a specific off-target interaction of an anti-FGFR4 mAb candidate has been identified as the cause for its rapid clearance, poor target tissue biodistribution, and limited efficacy [30]. The authors concluded that screens typically developed to identify general nonspecific interactions are likely to miss the rare but highly specific off-target binding observed in this study. Similarly, we found that several of our early mAb candidates displayed much shorter half-life ($t_{1/2}$) than anticipated [31]. This less than desirable DMPK property was recognized after DMPK involvement following candidate selection. Some of the candidates were eventually terminated due to the poor PK behavior and safety concerns. In addition, an examination of the clinical PK of approved mAbs clearly showed that mAbs can exhibit different PK at their saturated dose [32]. It is also notable that the relatively low attrition rate of LMs that is often referred to may not be replicated going forward, considering the increasingly competitive LM landscape and an evolving LM pipeline enriched with a variety of new and untested engineering technology platforms [33].

Furthermore, current DMPK approaches for LM support in preclinical development is usually limited to *in vivo* PK studies in laboratory animals, including mice, rats, dogs, and monkeys. In the case of mAbs, it has been widely accepted that nonhuman primate (NHP) is a representative animal model for human PK, and human PK prediction is typically performed using an empirical allometric scaling approach heavily dependent on this single species [34]. However, recent publications suggest potential issues with this approach. Vugmeyster et al. have shown that an anti-amyloid beta Ab2, a humanized mAb against amino acids three to six of primate amyloid beta, exhibited faster clearance, with a much shorter $t_{1/2}$ of less than 2.5 days, compared to approximately 13 days for a control antibody (no affinity to the target) in monkeys [35]. Additional mechanistic studies revealed that the fast elimination of Ab2 was linked to off-target binding to fibrinogen specific to monkeys and not humans, and thus provided a basis for a projected much slower elimination of Ab2 in humans. The prediction was later proven in a clinical trial [36]. Clearly, without appropriate DMPK input and mechanistic insights, this compound would have been precluded from further development. There are few other examples along this line (Merck internal database; Dr. FP. Theil, personal communication), where NHP PK failed to inform human PK correctly, due to either under- or overprediction. Collectively, these cases underscore our currently limited knowledge about the ADME processes of LMs and their determinants, which are even less adequate than those we knew for SMs in the 1980s.

One of major barriers limiting our understanding in ADME properties of LMs may be related to lack of appropriate analytical tools. It is well known that the structural complexity of LMs has posed formidable bioanalytical challenges. The commonly used bioanalytical methods for the determination of LMs in biological fluids are ligand-binding assays that are immunological in nature. These assays usually have an associated degree of nonspecificity. For example, multiple forms of mAb and ligand can exist *in vivo*, including free mAb, free ligand, and mono- and/or bivalent complexes of mAb and ligand. Given the complexity of the dynamic binding equilibrium occurring in the body after dosing, and multiple sources of perturbation of the equilibrium during bioanalysis, *ex vivo* quantification of the forms of interest (free, bound, or total mAb and ligand) may differ from the actual ones *in vivo* [37]. Several other possible weaknesses that may result in erroneous characterization of drug disposition have also been identified and recognized by regulatory agencies. These shortcomings, which include interferences from structurally related compounds such as endogenous proteins, degraded or catabolic products that are immunoreactive but may or may not be active or may elicit activity with different potencies, will certainly complicate data interpretations and hamper in-depth understanding of underlying mechanisms [38, 39]. LC–MS is emerging as a highly useful complementary tool for qualitative and quantitative applications to LMs [40–42]. However, the routine use of LC–MS is still hampered by the relatively time-consuming development process due to complex sample preparations, such as immunocapture and enzyme digestion of LMs, and limited sensitivity as compared to a typical immunoassay [42–44].

1.3.2 SM/LM DMPK Analogy

On a high level, PK/PD models and concepts are generally similar between SMs and LMs. In other words, PD is linked to PK (or specifically drug concentrations at biophase (C_e), which is related to systemic concentrations (C_p)), following certain relationships defined by molecular mechanisms of action of a drug, irrespective of its modality [45]. Similarly, PK is a collective depiction of ADME processes for both SMs and LMs. However, at the next level down, including ADME processes and associated underlying determinants, there are differences between the two modalities. For SMs, the ADME processes are relatively well studied and are mainly governed by (i) specific characteristics of a compound, including its physicochemical properties and ability to interact with transporters, drug-metabolizing enzymes, and binding proteins and (ii) physiological factors that govern the exposure of the compound to those proteins, such as distribution, tissue localization, and organ blood flow [46]. Not only have the nature of these interactions and their governing factors been mostly characterized, appropriate tools required for the studies have also been largely available. As illustrated in Figure 1.1 following a typical oral administration, an SM drug is absorbed either via passive diffusion and/or active transport, and then subjected to first-pass metabolism in the intestine and/or liver, before reaching systemic circulation for distribution to tissues and other organs of elimination, including kidney. Systemic bioavailability (F), a PK parameter central to efficacy and safety of a drug candidate, is a product of these processes.

In general, the ADME processes for LMs are much less characterized, as compared to SMs, even though their ADME processes are similar in concept. Unlike SMs, oral administration is precluded by molecular size, hydrophilicity, and gastric degradation of LMs. LMs are administered intravenously, intramuscularly, or subcutaneously (SC). As illustrated in Figure 1.1, following an SC administration, an LM is absorbed and potentially subjected to metabolism/catabolism at the injection site as well as during transport through the lymphatic system before reaching blood circulation. This is based largely on limited studies in sheep [47], and more recently in rats and dogs [48]. Analogously to SM drugs, systemic bioavailability, F, is a product of these processes. But unlike SMs, there are no established methods to measure the extent of absorption or presystemic catabolism. There is also little knowledge on the factors that can impact these parameters in animals or humans [3, 48]. Not surprisingly, it remains a challenge to extrapolate the SC absorption results in preclinical species into humans for LMs.

The majority of SMs enter tissues by passive diffusion, and the key determinant of tissue distribution includes nonselective binding to tissue proteins. Many SM drugs have also been reported to enter tissues via active transport, and the transporters involved have been identified. Similar to

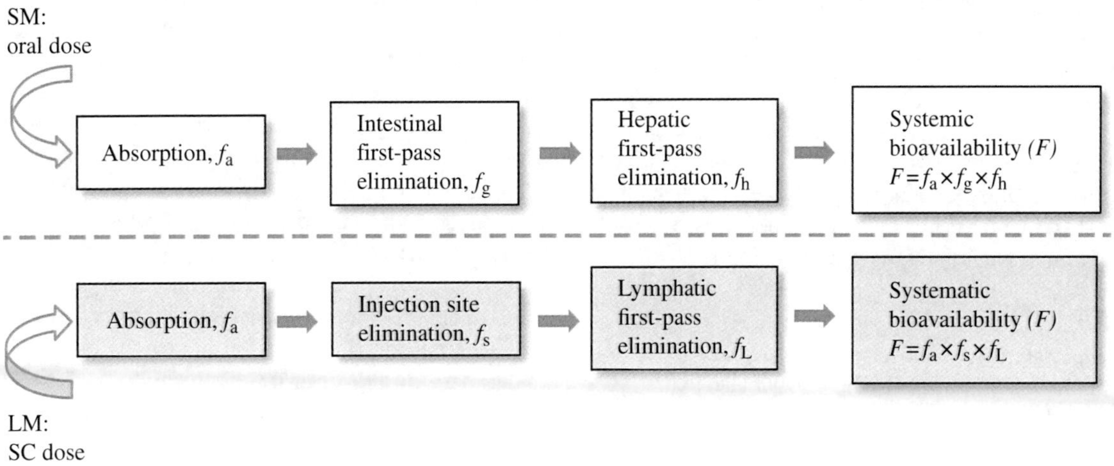

FIGURE 1.1 The ADME concept analogy between SMs and LMs. f_a = fraction absorbed; f_g = fraction-escaped gut first-pass elimination; f_h = fraction-escaped hepatic first-pass elimination; f_S = fraction-escaped degradation/catabolism at the injection site; f_L = fraction-escaped degradation/catabolism in the lymphatic system.

SMs, once entering the blood circulation, the LMs must cross the vascular wall to reach the site of action in target tissue(s) in order to exert their pharmacological activity before being eliminated via metabolism or other elimination pathways. Because of their molecular size, distribution of LMs into tissues is generally slow, and via the so-called convective transport through pores on capillary walls, as well as transcytosis from circulation to the extracellular space [49, 50]. Unlike SMs, distribution of LMs is usually limited to extracellular fluids due to their size and hydrophilicity. This fundamental difference between the two modalities is in line with the location of their respective biological targets. Namely, the targets are either soluble or on the cell surface for LM drugs, which are in contrast to the intracellular location for most SM drugs.

Presence of target in peripheral tissues can significantly change the tissue distribution of LMs, and leads to potential disconnect between plasma and tissue levels for LM. Given the importance of understanding the distribution to target tissue for LMs and the availability of related tools, tissue distribution studies have been more frequently conducted (vs other ADME-related types of studies) for LMs [51]. However, despite the wealth of LM (especially mAbs) tissue distribution data that had been accumulated [51, 52], in-depth mechanistic studies on these processes are scarce, and it currently is still unclear which transport pathway, convective transport or transcytosis, would be quantitatively more important in terms of extravasation of protein drugs from blood circulation [3].

As described earlier, SMs are eliminated from the body predominantly via metabolism with CYPs as the major metabolizing enzymes. Direct excretion into bile and/or urine is the other major elimination pathway for a variety of compounds that escape metabolism. For LM, common mechanisms of elimination include filtration (e.g., into urine), secretion (e.g., into the bile), and biotransformation (e.g., metabolism or catabolism). Smaller size LMs are subjected to elimination via kidney. Contrary to SMs, LMs are typically not subjected to metabolism by CYP enzymes, but generally believed to be catabolized to peptides and amino acids via proteolysis throughout the body, either extracellularly or intracellularly following fluid-based pinocytosis/receptor-mediated endocytosis. Subsequent to pinocytosis/endocytosis, the LMs usually are catabolized inside the cells. A notable exception for this process is with mAbs and endogenous immunoglobulin Gs/albumin, which are protected from degradation by binding to Fc receptor of the neonatal (FcRn) (at acidic pH) with subsequent dissociation (at neutral pH) to recycle back into circulation. As a result, these molecules display a distinct key feature of relatively long elimination half-lives usually in weeks [3, 49]. Although the impact of the FcRn salvage pathway on IgG PK has been established since the early 1990s [53–55], our understanding of the relevant molecular mechanisms and implications is still limited. For examples, until our recent publication showing that Fab domain may also impact the FcRn interaction [31], it had been commonly assumed that IgGs with the same Fc sequences would bind to FcRn equally and be protected by FcRn similarly. We have shown that mAbs with wild-type human Fc sequences interacted with FcRn with considerable differences in both binding at acidic pH and dissociation at neutral pH, thus exhibiting a wide range of $t_{1/2}$ and clearance [31]. Based on these results, we have implemented *in vitro* FcRn binding/dissociation assays, and *in vivo* human FcRn mouse studies, as useful screening and funneling tools for PK assessment of mAbs with wild-type Fc sequences. Fortunately, we have witnessed a rapid rise in FcRn-related researches over the past few years [56–61]. This increasing trend, which is reminiscent to what happened with CYPs decades ago, is an important step toward full integration of DMPK to IgG drug discovery and development.

Another unique elimination mechanism for LMs is target-mediated drug disposition (TMDD) [62]. Even though TMDD was first described for conventional SM drugs by Dr. Gerhard Levy in 1994 [63], the PK of conventional SM drugs is usually independent of their targets because the fraction of SMs involved in the target binding is usually negligible. In contrast, TMDD is common for LM drugs, especially mAbs, due to their relatively low nonspecific systemic clearance and extremely high target-binding affinity. The target–drug interaction and subsequent degradation thus contribute significantly not only to the PD, but also to the PK of LM drugs [62]. A resulting key feature of TMDD is non-linearity in PK with higher clearance observed at lower doses. In addition, at a given dose, PK of LMs with TMDD can also be altered with changes in PD reflective of target expression-level alterations. As is the case for their absorption, distribution, and metabolism, the underlying mechanisms and factors influencing elimination of LMs have not been extensively investigated, especially as compared to SMs.

1.3.3 Leveraging SM Experience: Case Examples

Given all of the considerations above, there is a need to advance LM ADME sciences and develop enabling tools/technologies for ADME studies of LMs, similar to the two fundamental elements vital to the successful SM discovery and development. Equally importantly, realization of these two elements requires more active and timely participation of DMPK scientists over the entire continuum of LM drug discovery and development. In this section, we present examples to illustrate how the same principles and knowledge gained from SMs can be applied to LMs, exemplifying the impact of early and better understanding in ADME processes in the discovery and early preclinical development spaces.

1.3.3.1 Example 1: LM-SM DDIs—Leveraging Knowledge on CYPs Recently, CYP-mediated DDI observed when LMs were coadministered with SMs has been a subject of increasing interest for LM drugs across industry and regulatory scientists [64–66]. These DDIs typically involves LMs that target cytokines and/or treat inflammatory diseases, both of which can impact CYPs [67]. Along the same line with SM–SM DDIs, but with an added consideration of altered levels of endogenous cytokines in disease settings, the CYP knowledge and tools could potentially be applied to explore the utility of *in vitro* CYP studies to quantitatively predict the LM–SM DDI risk. Indeed, as a first step toward the prediction, a model has been recently developed using *in vitro* CYP suppression data with interleukin-6 (IL-6) from hepatocytes to simulate the disease–drug interactions reported in clinical studies with sensitive CYP3A SM substrates [68]. The results were encouraging and clearly highlighted the complexity associated with underlying pathological factors. This is not surprising given the nature of interactions that are primarily disease state dependent and/or drug target dependent, which are different from and more complicated than typical DDIs observed between SMs. Nevertheless, by leveraging the existing knowledge on CYPs, and particularly their regulation factors, it has been possible to provide insights into the underlying mechanisms for the observed DDIs [64, 65, 67]. More importantly, the wealth of information on CYPs has enabled a consensus framework to be developed in a relatively short time among industry and regulatory agencies that entails a general approach for LM–SM DDI assessment during drug development [67]. There remains, however, a need for additional research in disease biology and physiologically relevant *in vitro* systems to facilitate *in vitro–in vivo* extrapolations of the impact of LMs or diseases on CYPs, and eventually successful prediction of LM–SM DDIs.

1.3.3.2 Example 2: LC–MS to Characterize In Vivo Transformation of mAb—Key Enabler in Candidate Selection In this example, we show that LC–MS can provide invaluable information to aid in the understanding of LM disposition important to candidate selection, similar to its role in SM drug discovery support. Therapeutic proteins are subjected to transformation mechanisms such as deamidation, oxidation, and isomerization. These processes usually result in relatively small structural changes in the parent drugs. Such small structural changes may be difficult for a conventional immunoassay to differentiate, but they can still affect biological activity, PK, and immunogenicity of a therapeutic protein [69]. LC–MS is commonly used to detect Asp isomerization in proteins during stability testing at relatively high protein concentrations (mg/mL levels), but not in plasma from *in vivo* studies, due in part to the difficulties in sample analysis resulting from the complex matrix and requirement for high sensitivity.

The first demonstration of *in vivo* Asp isomerization with significant impact on the function of a model mAb (mAb X) has been recently shown [33]. In this case, liquid chromatography with high resolution mass spectrometry (LC-HRMS) provided qualitative and quantitative information on the structurally modified products of therapeutic proteins in biological matrices. It was found that this mAb completely lost its target-binding ability due to isomerization of a single Asp in the complementary determining region (CDR) (isoAsp-mAb X) following an accelerated stability test at 40 °C over 3 months. This raised a question with respect to the *in vivo* relevance of this *in vitro* occurrence and the developability of this mAb. For this, an LC–MS assay was needed since the immunoassay used for the PK evaluation of mAb X was incapable of distinguishing the parent compound from its inactive isomer. Coupled with immunocapture, using biotinylated mouse antihuman IgG (Fc) antibody to enrich analytes and following trypsin digestion of mAb X, a unique 43-amino acid peptide that contains the Asp of interest (Pep A) and isoPep A (surrogates for the parent and isoAsp-mAb X, respectively) was separated and detected by LC–HRMS. The isoAsp-mAb X/parent ratio was found in mouse serum with an increase in the absolute levels of isoAsp-mAb X of approximately 45% from Day 2 to Day 28. The result from this work provided direct evidence of Asp isomerization *in vivo* and thus disqualified mAb X from further development consideration.

1.3.3.3 Example 3: Mechanism-Based Human PK Prediction For SM drugs, knowledge gained over recent decades has established a foundation for a "bottom-up" physiologically-based PK (PBPK) modeling approach to integrate drug-specific parameters obtained *in vitro* using human tissues or, for the majority of SMs, the major PK determinant CYP systems to predict and provide mechanistic insights into the PK properties in humans under various intrinsic and extrinsic factors [19, 70]. For LM drugs, owing to limited understanding of the ADME determinants and especially with regard to scalability from *in vitro* systems to *in vivo* and/or from preclinical species to humans, a fully bottom-up PBPK-based model has either yet to be completely validated or widely applied for prospective human PK prediction. Nevertheless, there has been some recent progress of PBPK models with mAbs, with potential translational utility to human situations [71–74]. For example, Abuqayyas and Balthasar have recently developed a PBPK model with TMDD components in the tumor compartment to predict the disposition of mAbs *a priori* in plasma and in tissues, including tumors that express target antigens in mice [74]. In addition to FcRn–IgG interaction, the model structure included the following determinants: antibody–target-binding affinity, target expression levels, rates of internalization of mAb–target complexes, plasma and lymphatic flow rates, and the tumor vascular volume. Two mAbs, one with high tumor target antigen-binding affinity and one nonbinding control, were examined. The exposure of both mAbs in plasma, tumor, and other tissues was predicted reasonably well in the xenograft-bearing SCID mice [74]. It is anticipated that further development of LM PBPK models, coupled with better understanding in target–LM interaction biology and improved experimental methods to characterize target expression and dynamics, will eventually allow *a priori* prediction of LM plasma and tissue disposition in humans. A dedicated chapter (by Yanguang Cao and William Jusko) in this book focusing on PBPK for therapeutic mAbs provides more details about how to apply this technique during drug development.

Currently, human PK for LMs not subject to nonlinear or species-specific clearance mechanisms is predicted reasonable well from preclinical PK with the principle of allometry [34, 75–77]. This is because certain general elimination processes of LMs are governed primarily by physiological parameters, which can be scaled between species in a

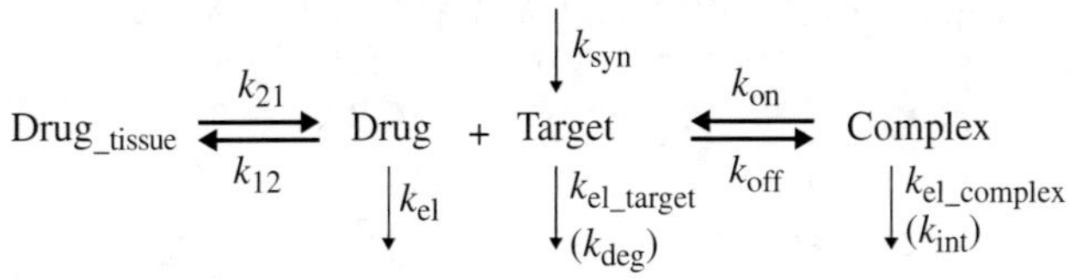

FIGURE 1.2 Schematic representation of a TMDD model for description of the interaction between a drug and its target. k_{12} = transfer rate constant of drug from central to peripheral (tissue) compartment; k_{21} = transfer rate constant of drug from peripheral (tissue) to central compartment; k_{el} = elimination rate constant of drug; k_{syn} = synthesis rate of target; k_{el_target} (k_{deg}) = elimination rate constant of target; k_{on} = association rate constant; k_{off} = dissociation rate constant; $k_{el\_complex}$ (k_{int}) = elimination rate constant of drug/target complex.

compound-independent manner. However, for LMs subject to nonlinear clearance, a more mechanism-based modeling approach incorporating the impact of targets on PK is needed for human PK prediction. A nice example was reported by Luu et al. recently on how mechanistic modeling can be used to predict human PK of a mAb exhibiting TMDD [78]. PF-03446962 is a human mAb against ALK1 (activin receptor-like kinase 1) that exhibited nonlinear PK, a hallmark of TMDD, in monkeys. A TMDD model as depicted in Figure 1.2 was used to capture the plasma PK profiles of PF-03446962 following single and multiple doses. The mechanism-related parameters, such as k_{on}, k_{off}, k_{deg}, and k_{int} rates were experimentally determined for both monkeys and humans. Together with allometric scaling of monkey PK parameters (e.g., k_{el}, k_{12}, and k_{21}), the model successfully predicted the plasma PK profile of PF-03446962 in humans.

1.3.3.4 Example 4: PK/PD Modeling for LMs A mechanistic PK/PD modeling approach has been increasingly used to help define and better understand systemic exposure–effect (efficacy or safety) relationship, a key element to successful SM drug development. This approach can similarly be applied to the development of LM drugs. In fact, of all DMPK aspects of LMs, this is the area that has been relatively well developed and received great attention from DMPK scientists supporting LM drug development. As described earlier, unlike most of SM drugs, the PK and PD of LM drugs are often interrelated. For these molecules, TMDD model has not only been an important tool to characterize the PK of LMs, it has also been incorporated widely into the PK/PD modeling of LMs to characterize PD effect and dose–response relationship for LMs, delineating the impact of target engagement (TE) on downstream pharmacological effects.

One such example was presented by Ng et al., for TRX1, an anti-CD4 mAb [79]. TRX1 exhibited typical target-mediated nonlinear PK characteristics in humans. Binding of TRX1 to CD4 receptors on circulating T cells leads to down-modulating the CD4 receptors in a dose- and concentration-dependent manner, which in turn changes the extent of target impact on TRX1 PK. Therefore, a receptor-binding-based PK/PD model as depicted in Figure 1.2 was also used to describe the PK and PD (CD4 target binding) of TRX1. Serum TRX1 concentration and total and free CD4 levels were measured and fitted into the model simultaneously to account for the effect of PD on PK. This mechanism-based PK/PD model was later used to simulate PK/PD-time profiles after different dosing regimens to help guide the dose selection in future clinical studies.

For LM drugs against soluble targets, the impact of target binding on drug PK may not be as apparent, depending on whether the elimination rate for drug–target complex is similar to that of the free drug [80, 81]. Nevertheless, understanding the interplay between drug and target is essential in determining the dosing regimen for LMs. Following LM treatment, there is often a rapid accumulation of drug/target complex due to dramatic differences in the elimination rates of free target and LM drug/target complex [81]. Dissociation of the accumulated LM drug/target complex will result in the return of free target to baseline while free drug levels are still orders of magnitude higher than the free target levels. As shown by Wang et al., following treatment of siltuximab, an anti-IL-6 mAb, in cynomolgus monkeys, total IL-6 levels reached 10,000–100,000-fold above the IL-6 baseline, and free IL-6 returns to baseline when siltuximab levels were $>10^6$-fold higher than the IL-6 baseline (also 100–1000-fold higher than the highest total IL-6 levels [81]). A quantitative PK/TE model that takes into account the production rate of IL-6, elimination rates of IL-6 and siltuximab/IL-6 complex, equilibrium dissociation constant between siltuximab and IL-6, as well as the PK characteristics of siltuximab was established via simultaneous fitting of total siltuximab, total IL-6, and free IL-6 concentration profiles [81]. The model provided estimation of all model parameters and was used successfully to predict the free IL-6 profiles at higher siltuximab doses, where the accurate determination of free IL-6 concentration became technically too difficult. This kind of integrated PK/TE/PD modeling approach provided a framework for prediction of efficacious dose levels and duration of action for mAbs against soluble ligands with rapid turnover.

1.4 CONCLUSIONS

Over the past few decades, a better understanding of ADME processes, brought about by participation of DMPK scientists in early discovery through late development, has been crucial to enhancing the possibility of success of SM drugs. We attribute the success of DMPK involvement to the combination of substantial progresses in the drug metabolism sciences, particularly in the area of CYPs, the major enzymes responsible for clearance mechanisms of a large number of SMs, with the availability of powerful tools, notably the LC–MS technology. Compared to SMs, the role of DMPK in supporting LM drug discovery and development is far behind and should be increased and expanded to cover the entire process. This point of view is underpinned by a number of factors, including the evolving and competitive biotechnology landscape, and imminent/growing regulatory pressure. A few case examples are presented to illustrate the relevance and transferability of strategies and experiences of DMPK support for SM drugs to LM drugs. A similar path used for SM drug discovery and development, especially with respect to establishing mechanistic understanding in ADME properties and associated determinants, as well as developing necessary tools and technology, can be followed in the endeavors to increase the possibility of success of a safe and effective LM candidate.

REFERENCES

[1] DiMasi JA, Feldman L, Seckler A, Wilson A. Trends in risks associated with new drug development: success rates for investigational drugs. Clin Pharmacol Ther 2010;87:272–277.

[2] Kaitin KI, DiMasi MA. Pharmaceutical innovation in the 21st century: new drug approvals in the first decade, 2000–2009. Clin Pharmacol Ther 2011;89:183–188.

[3] Lin JH. Pharmacokinetics of biotech drugs: peptides, proteins and monoclonal antibodies. Curr Drug Metab 2009;10: 661–691.

[4] Waldmann TA. Immunotherapy: past present and future. Nat Med 2003;9:269–277.

[5] Baillie TA. Metabolism and toxicity of drugs. Two decades of progress in industrial drug metabolism. Chem Res Toxicol 2008;21:129–137.

[6] Paul SM, Mytelka DS, Dunwiddie CT, Persinger CC, Munos BH, Lindborg SR, Schacht AL. How to improve R&D productivity: the pharmaceutical industry's grand challenge. Nat Rev Drug Discov 2010;9:203–214.

[7] Kola I, Landis J. Can the pharmaceutical industry reduce attrition rates? Nat Rev Drug Discov 2004;3:711–715.

[8] Sung NS, Crowley WF Jr, Genel M, Salber P, Sandy L, Sherwood LM, et al. Central challenges facing the national clinical research enterprise. JAMA 2003;289:1278–1287.

[9] MacCoss M, Baillie TA. Organic chemistry in drug discovery. Science 2004;303:1810–1813.

[10] Lin JH, Lu AYH. Role of pharmacokinetics and metabolism in drug discovery and development. Pharmacol Rev 1997;49:403–449.

[11] Lin JH. Species similarities and differences in pharmacokinetics. Drug Metab Dispos 1995;23:1008–1021.

[12] Tang C, Prueksaritanont T. Use of *in vivo* animal models to assess pharmacokinetic drug-drug Interactions. Pharm Res 2010;27:1772–1787.

[13] Wienkers LC, Heath TG. Predicting *in vivo* drug interactions from *in vitro* drug discovery data. Nat Rev Drug Discov 2004;4:825–833.

[14] Guengerich FP. Cytochrome P450s and other enzymes in drug metabolism and toxicity. AAPS J 2006;8:E101–E111.

[15] Estabrook RW. A passion for P450s (remembrances of the early history of research on cytochrome P450). Drug Metab Dispos 2003;31:1461–1473.

[16] Lin JH, Lu AYH. Inhibition and induction of cytochrome P450 and the clinical implications. Clin Pharmacokinet 1998; 5:361–390.

[17] Prueksaritanont T, Ma B, Tang C, Meng Y, Assang C, Lu P, et al. Metabolic interactions between mibefradil and HMG-CoA reductase inhibitors: an *in vitro* investigation with human liver preparations. Br J Clin Pharmacol 1999;47:291–298.

[18] Ingelman-Sundberg M. Implications of polymorphic cytochrome P450-dependent drug metabolism for drug development. Drug Metab Dispos 2001;29:570–573.

[19] Shiran MR, Proctor NJ, Howgate EM, Rowland-Yeo K, Tucker GT, Rostami-Hodjegan A. Prediction of metabolic drug clearance in humans: *in vitro-in vivo* extrapolation vs allometric scaling. Xenobiotica 2006;36:567–580.

[20] Zhou SF, Liu JP, Chowbay B. Polymorphism of human cytochrome P450 enzymes and its clinical impact. Drug Metab Rev 2009;41:89–295.

[21] Huang SM, Strong JM, Zhang L, Reynolds KS, Nallani S, Temple R, et al. New era in drug interaction evaluation: US Food and Drug Administration update on CYP enzymes, transporters, and the guidance process. J Clin Pharmacol 2008;48:662–670.

[22] U.S. Department of Health and Human Services, Food and Drug Administration, Center for Drug Evaluation and Research (CDER). 2012. Guidance for industry drug interaction studies —study design, data analysis, implications for dosing, and labeling recommendations, draft duidance; 2012 February.

[23] Hopfgartner G, Bourgogne E. Quantitative high-throughput analysis of drugs in biological matrices by mass spectrometry. Mass Spectrom Rev 2003;22:195–214.

[24] Korfmacher WA. Principles and applications of LC-MS in new drug discovery. Drug Discov Today 2005;10:1357–1367.

[25] Castro-Perez J, Plumb R, Granger JH, Beattie L, Joncour K, Wright A. Increasing throughput and information content for *in vitro* drug metabolism experiments using ultra-performance liquid chromatography coupled to a quadrupole time-of-flight mass spectrometer. Rapid Commun Mass Spectrom 2005;19:843–848.

[26] Korfmacher WA. Use of mass spectrometry for drug metabolism studies. Curr Drug Metab 2006;7:455–563.

[27] Plumb R, Castro-Perez J, Granger J, Beattie I, Joncour K, Wright A. Ultra-performance liquid chromatography coupled to quadrupole-orthogonal time-of-flight mass spectrometry. Rapid Commun Mass Spectrom 2004;18:2331–2337.

[28] Prueksaritanont T, Chu X, Gibson C, Cui D, Yee KL, Ballard J, Cabalu T, Hochman J. Drug–drug interaction studies: regulatory guidance and an industry perspective. AAPS J 2013;15:629–645.

[29] U.S. Department of Health and Human Services, Food and Drug Administration, Center for Drug Evaluation and Research (CDER), The CDER Science Prioritization and Review Committee (SPaRC). 2011. Identifying CDER's science and research needs report; 2011 July.

[30] Bumbaca D, Wong A, Drake E, Reyes AE II, Lin BC, Stephan JP, Desnoyers L, Shen BQ, Dennis MS. Highly specific off-target binding identified and eliminated during the humanization of an antibody against FGF receptor 4. MAbs 2011;3:376–386.

[31] Wang W, Lu P, Fang Y, Hamuro L, Pittman T, Carr B, et al. Monoclonal antibodies with identical Fc sequences can bind to FcRn differentially with pharmacokinetic consequences. Drug Metab Dispos 2011;39:1469–1477.

[32] Keizer RJ, Huitema AD, Schellens JH, Beijnen JH. Clinical pharmacokinetics of therapeutic monoclonal antibodies. Clin Pharmacokinet 2010;49:493–507.

[33] Prueksaritanont T, Tang C. ADME of biologics-what have we learned from small molecules? AAPS J 2012;14:410–419.

[34] Wang W, Prueksaritanont T. Prediction of human clearance of therapeutic proteins: simple allometric scaling method revisited. Biopharm Drug Dispos 2010;31:253–263.

[35] Vugmeyster Y, Szklut P, Wensel D, Ross J, Xu X, Awwad M, et al. Complex pharmacokinetics of a humanized antibody against human amyloid beta peptide, anti-abeta ab2, in nonclinical species. Pharm Res 2011;28:1696–1706.

[36] Black RS, Sperling RA, Safirstein B, Motter RN, Pallay A, Nichols A, Grundman M. A single ascending dose study of bapineuzumab in patients with Alzheimer disease. Alzheimer Dis Assoc Disord 2010;24:198–203.

[37] Lee JW, Kelley M, King LE, Yang J, Salimi-Moosavi H, Tang MT, et al. Bioanalytical approaches to quantify "total" and "free" therapeutic antibodies and their targets: technical challenges and PK/PD applications over the course of drug development. AAPS J 2011;13:99–110.

[38] European Medicines Agency (EMA), Committee for Medicinal Products for Human Use (CHMP). Draft guideline on the clinical investigation of the pharmacokinetics of therapeutic proteins; 2005 July.

[39] Tang C, Prueksaritanont T. Theoretical analysis of interplay of therapeutic protein rug and circulating soluble target: temporal profiles of 'free' and 'total' drug and target. Pharm Res 2011;28:2447–2457.

[40] Ezan E, Dubois M, Becher F. Bioanalysis of recombinant proteins and antibodies by mass spectrometry. Analyst 2009; 134:825–834.

[41] Hall MP, Gegg C, Walker K, Spahr C, Ortiz R, Patel V, et al. Ligand-binding mass spectrometry to study biotransformation of fusion protein drugs and guide immunoassay development: strategic approach and application to peptibodies targeting the thrombopoietin receptor. AAPS J 2010;12:576–585.

[42] Campbell Larry J, Yves Le Blanc JC. Peptide and protein drug analysis by MS: challenges and opportunities for the discovery environment. Bioanalysis 2011;3:645–657.

[43] Zheng JN, Mehl J, Zhu YX, Xin BM, Olah T. Application and challenges in using LC-MS assays for absolute quantitative analysis of therapeutic proteins in drug discovery. Bioanalysis 2014;6:859–879.

[44] Onami I, Ayabe M, Murao N, Ishigai M. A versatile method for protein-based antigen bioanalysis in non-clinical pharmacokinetics studies of a human monoclonal antibody drug by an immunoaffinity liquid chromatography-tandem mass spectrometry. J Chromatogr 2014;1334:64–71.

[45] Mager DE, Wyska E, Jusko WJ. Diversity of mechanism-based pharmacodynamic models. Drug Metab Dispos 2003;31:510–518.

[46] Prueksaritanont T, Lin JH, Baillie TA. Complicating factors in safety testing of drug metabolites: kinetic differences between generated and preformed metabolites. Toxicol Appl Pharmacol 2006;217:143–152.

[47] Porter CJ, Charman SA. Lymphatic transport of proteins after subcutaneous administration. J Pharm Sci 2000;89:297–310.

[48] Wang W, Chen N, Shen X, Cunningham P, Fauty S, Michel K, et al. Lymphatic transport and catabolism of therapeutic proteins following subcutaneous administration to rats and dogs. Drug Metab Dispos 2012;40:952–962.

[49] Lobo ED, Hansen RJ, Balthasar JP. Antibody pharmacokinetics and pharmacodynamics. J Pharm Sci 2004;93: 2645–2668.

[50] Wang W, Wang EQ, Balthasar JP. Monoclonal antibody pharmacokinetics and pharmacodynamics. Clin Pharmacol Ther 2008;84:548–558.

[51] Vugmeyster Y, Defranco D, Szklut P, Wang Q, Xu X. Biodistribution of [(125)I]-labeled therapeutic proteins: application in protein drug development beyond oncology. J Pharm Sci 2010;99:1028–1045.

[52] Boswell CA, Bumbaca D, Fielder PJ, Khawli LA. Compartmental tissue distribution of antibody therapeutics: experimental approaches and interpretations. AAPS J 2012; 14:612–618.

[53] Ghetie V, Hubbard JG, Kim JK, Tsen MF, Lee YF, Ward ES. Abnormally short serum half-lives of IgG in beta 2-microglobulin-deficient mice. Eur J Immunol 1996;26:690–696.

[54] Israel EJ, Wilsker DF, Hayes KC, Schoenfeld D, Simister NE. Increased clearance of IgG in mice that lack beta(2)-microglobulin: possible protective role of FcRn. Immunology 1996;89:573–578.

[55] Junghans RP, Anderson CL. The protection receptor for IgG catabolism is the beta(2)-microglobulin-containing neonatal intestinal transport receptor. Proc Natl Acad Sci USA 1996;93:5512–5516.

[56] Garg A, Balthasar JP. Investigation of the influence of FcRn on the distribution of IgG to the brain. AAPS J 2009;11:553–557.

[57] Zalevsky J, Chamberlain AK, Horton HM, Karki S, Leung IW, Sproule TJ, Lazar GA, Roopenian DC, Desjarlais JR. Enhanced antibody half-life improves *in vivo* activity. Nat Biotechnol 2010;28:157–159.

[58] Wang W, Vlasak J, Li Y, Pristatsky P, Fang Y, Pittman T, et al. Impact of methionine oxidation in human IgG1 Fc on serum half-life of monoclonal antibodies. Mol Immunol 2011;48:860–866.

[59] Deng R, Meng YG, Hoyte K, Lutman J, Lu Y, Iyer S, DeForge LE, Theil FP, Fielder PJ, Prabhu S. Subcutaneous bioavailability of therapeutic antibodies as a function of FcRn binding affinity in mice. MAbs 2012;4:101–109.

[60] Yip V, Palma E, Tesar DB, Mundo EE, Bumbaca D, Torres EK, Reyes NA, Shen BQ, Fielder PJ, Prabhu S, Khawli LA, Boswell CA. Quantitative cumulative biodistribution of antibodies in mice: effect of modulating binding affinity to the neonatal Fc receptor. MAbs 2014;6:689–696.

[61] Datta-Mannan A, Wroblewski VJ. Application of FcRn binding assays to guide mAb development. Drug Metab Dispos 2014;42:1867–1872.

[62] Mager DE, Jusko WJ. General pharmacokinetic model for drugs exhibiting target-mediated drug disposition. J Pharmacokinet Pharmacodyn 2001;28:507–532.

[63] Levy G. Pharmacologic target-mediated drug disposition. Clin Pharmacol Ther 1994;56:248–252.

[64] Zhou H, Mascelli MA. Mechanisms of monoclonal antibody-drug interactions. Annu Rev Pharmacol Toxicol 2011;51:359–372.

[65] Lee JI, Zhang L, Men AY, Kenna LA, Huang SM. CYP-mediated therapeutic protein-drug interactions clinical findings, proposed mechanisms and regulatory implications. Clin Pharmacokinet 2010;49:295–310.

[66] Girish S, Martin SW, Peterson MC, Zhang LK, Zhao H, Balthasar J, et al. AAPS workshop report: strategies to address therapeutic protein-drug interactions during clinical development. AAPS J 2011;13:405–416.

[67] Evers R, Dallas S, Dickmann LJ, Fahmi OA, Kenny JR, Kraynov E, Nguyen T, Patel AH, Slatter JG, Zhang L. Critical review of preclinical approaches to investigate cytochrome P450–mediated therapeutic protein drug-drug interactions and recommendations for best practices: a white paper. Drug Metab Dispos 2013;41:1598–1609.

[68] Machavaram KK, Almond LM, Rostami-Hodjegan A, Gardner I, Jamei M, Tay S, Wong S, Joshi A, Kenny JR. A physiologically based pharmacokinetic modeling approach to predict disease–drug interactions: suppression of CYP3A by IL-6. Clin Pharmacol Ther 2013;94:260–268.

[69] Zhong X, Wright JF. Biological insights into therapeutic protein modifications throughout trafficking and their biopharmaceutical applications. Int J Cell Biol 2013;2013:273086.

[70] Rowland M, Peck C, Tucker G. Physiologically-based pharmacokinetics in drug development and regulatory science. Annu Rev Pharmacol Toxicol 2011;51:45–73.

[71] Garg A, Balthasar JP. Physiologically-based pharmacokinetic (PBPK) model to predict IgG tissue kinetics in wild-type and FcRn-knockout mice. J Pharmacokinet Pharmacodyn 2007;34(5):687–709.

[72] Chen Y, Balthasar JP. Evaluation of a catenary PBPK model for predicting the *in vivo* disposition of mAbs engineered for high-affinity binding to FcRn. AAPS J 2012;14:850–859.

[73] Shah DK, Betts AM. Towards a platform PBPK model to characterize the plasma and tissue disposition of monoclonal antibodies in preclinical species and human. J Pharmacokinet Pharmacodyn 2012;39:67–86.

[74] Abuqayyas L, Balthasar JP. Application of PBPK modeling to predict monoclonal antibody disposition in plasma and tissues in mouse models of human colorectal cancer. J Pharmacokinet Pharmacodyn 2012;39:683–710.

[75] Mahmood I. Interspecies scaling of protein drugs: prediction of clearance from animals to humans. J Pharm Sci 2004; 93:177–185.

[76] Ling J, Zhou H, Jiao Q, Davis HM. Interspecies scaling of therapeutic monoclonal antibodies: initial look. J Clin Pharmacol 2009;49:1382–1402.

[77] Deng R, Iyer S, Theil FP, Mortensen DL, Fielder PJ, Prabhu S. Projecting human pharmacokinetics of therapeutic antibodies from nonclinical data: what have we learned? MAbs 2011;3:61–66.

[78] Luu KT, Bergqvist S, Chen E, Hu-Lowe D, Kraynov E. A model-based approach to predicting the human pharmacokinetics of a monoclonal antibody exhibiting target-mediated drug disposition. J Pharmacol Exp Ther 2012;341:702–708.

[79] Ng CM, Stefanich E, Anand BS, Fielder PJ, Vaickus L. Pharmacokinetics/pharmacodynamics of nondepleting anti-CD4 monoclonal antibody (TRX1) in healthy human volunteers. Pharm Res 2006;23:95–103.

[80] Meno-Tetang GM, Lowe PJ. On the prediction of the human response: a recycled mechanistic pharmacokinetic/pharmacodynamic approach. Basic Clin Pharmacol Toxicol 2005;96:182–192.

[81] Wang W, Wang X, Doddareddy R, Fink D, McIntosh T, Davis HM, Zhou H. Mechanistic pharmacokinetic/target engagement/pharmacodynamic (PK/TE/PD) modeling in deciphering interplay between a monoclonal antibody and its soluble target in cynomolgus monkeys. AAPS J 2014;16:129–139.

2

PROTEIN ENGINEERING: APPLICATIONS TO THERAPEUTIC PROTEINS AND ANTIBODIES

ANDREW G. POPPLEWELL
UCB Celltech, Slough, Berkshire, UK

2.1 INTRODUCTION

The development of recombinant DNA methodology and genetic engineering from the late 1970s onwards opened up a range of possibilities for making specific changes to primary protein sequences. Together with advances in protein structure determination and molecular modeling techniques, these developments combined to create the discipline of protein engineering. As such techniques grew and became more widespread, protein engineering entered into mainstream use and led directly to the success and expansion of therapeutic biologics. The ability to tailor the duration of activity of recombinant insulin [1], to increase serum retention times by making precisely joined receptor–Fc fusion proteins [2, 3], and to change antibody scaffolds to reduce immunogenic potential [4] are all examples of the successful application of protein engineering to therapeutic proteins, creating solutions that would not be possible with the natural protein sequence.

To make changes in protein sequences, one must understand the contribution of those sequences to both the structure and function of the protein molecule. In recent years, new DNA-sequencing technology has led to a great increase in the size of databases of protein sequence. Although not quite so rapidly, more three-dimensional high resolution protein structures have been solved through X-ray crystallography and nuclear magnetic resonance (NMR) techniques, and developments in structure prediction and modeling are also progressing at speed. Thus, our understanding of the relationship between sequence, structure, and function is constantly evolving to expand the knowledge base of the protein engineer, which allows us to optimize therapeutic proteins with regard to their behavior *in vivo* to deliver therapeutic benefits and limit safety issues. However, being able to change or tailor a protein sequence brings also associated risks, making modifications for the benefit of one property might adversely influence another characteristic, as the full consequences of sequence change are rarely if ever entirely predictable. Thus, alongside protein engineering technology, the development of protein characterization methodology has been a requirement to ensure that the engineered molecules are fit-for-purpose as therapeutics.

This chapter serves as an introduction to protein engineering, briefly explaining the basic concepts and methods before exemplifying with some success stories of applications to therapeutic proteins. The examples are not comprehensive and are somewhat biased toward therapeutic antibodies, in part because of the author's own area of expertise and also because antibodies are the leading class of biological drugs, and today make up 5 of the top 10 best-selling therapeutics on the market [5]. Protein engineering has been pivotal to the success of this drug class and has impacted throughout the therapeutic antibody development pathway, from antibody discovery right through to manufacture.

2.2 METHODS OF PROTEIN ENGINEERING

2.2.1 General Techniques

The discipline of protein engineering is underpinned by a relatively small number of practical methodologies. As mentioned above, the basic tools of the protein engineer are those

ADME and Translational Pharmacokinetics/Pharmacodynamics of Therapeutic Proteins: Applications in Drug Discovery and Development, First Edition. Edited by Honghui Zhou and Frank-Peter Theil.

of the molecular biologist, with DNA sequencing, plasmid preparation, subcloning by restriction/ligation, and the polymerase chain reaction (PCR) providing the foundation. Molecular cloning using DNA restriction/modification enzymes enables the genes for proteins to be manipulated and expressed in recombinant systems, and DNA sequencing is used to confirm that coding sequences are as expected. Informatics skills are also becoming more useful as there is a need to access sequence and structure databases [6], to align sequences, and to visualize structures using molecular display tools.

2.2.2 Introducing Specific, Directed Sequence Changes

The simplest sequence changes available to a protein engineer are the directed mutation of specific residues to other amino acids, by making changes to the DNA encoding the recombinant protein. This site-directed mutagenesis is usually achieved by synthesizing oligonucleotide primers encoding mismatches for specific change(s), and using these to prime replication of the gene using an appropriate DNA polymerase. The mutated gene is then subcloned into a vector using standard molecular cloning techniques. Before the advent of PCR, the method of Kunkel was widely used to ensure that the newly synthesized mutated gene could be selected in favor of the original, nonmutated template [7]. Another method was to use double-stranded oligonucleotide cassettes, made by annealing complementary mutation-containing oligonucleotides and ligating them between restriction sites, either naturally occurring or introduced "silently" into the sequence through a previous round of mutagenesis [8, 9]. Although efficient as long as sites for high-fidelity restriction enzymes could be used, this method was limited by the length of oligonucleotide that could be accurately synthesized and purified, usually in the range of 60–80 bases. The introduction of PCR [10] led to rounds of sequence-specific amplification being used to generate large amounts of mismatched, primer-encoded, altered fragments, which could be cloned through a variety of methods. In its simplest form, forward and reverse primers are used, one of which contains a sequence-specific mismatch. Reaction cycles then give an exponential increase in the mutated sequence to "flood-out" the original template, and the amplified fragment is then cloned. Again, this method is dependent on the presence of a suitable restriction site close to the mismatch mutation. To overcome this, several primers can be used and a multistep PCR strategy employed using a technique known as mutagenesis by overlap extension [11]. Complementary oligonucleotide primers are used to prime the synthesis of two DNA fragments having overlapping ends. These fragments are combined in a subsequent "fusion" reaction in which the overlapping ends anneal, allowing the 3′ overlap of each strand to serve as a primer for the 3′ extension of the complementary strand. The resulting fusion product is amplified further by PCR using flanking primers.

The availability of high-fidelity, proof-reading, thermo-stable polymerases has led to the development of protocols using whole-plasmid PCR for site-directed mutagenesis, negating the need for restriction/ligation subcloning of amplified fragments [12]. Today, a variety of commercial kits are available offering rapid, site-specific mutagenesis of double-stranded plasmid DNA.

2.2.3 Fragment Fusion

In addition to single amino acid replacements, there are occasions when domain deletion or domain (or whole protein) fusion are desired. Again PCR methods are widely used to achieve this, the splicing by overlap extension method being a particularly useful technique to link genes without using restriction enzymes [13]. More recently, other gene assembly methods have been developed [14–16]. The Gibson Assembly technique provides an efficient means for joining any number of overlapping fragments, often generated by PCR, in a single tube. It employs three enzymatic activities: a 5′ exonuclease, the 3′ extension activity of a DNA polymerase, and DNA ligase activity. The 5′ exonuclease activity chews back the 5′ end sequences and exposes the complementary sequence for annealing. The polymerase fills the gaps on the annealed regions, and the DNA ligase then seals the break and covalently links the DNA fragments together.

2.2.4 Gene Synthesis

There are also occasions when a large number of sequence changes are desired, or when a number of short amino acid sections are required to be introduced into another protein scaffold, an example being antibody humanization by complementarity-determining region (CDR) grafting (see Section 2.4.1). In such instances, gene synthesis is a viable option. Gene synthesis can also be used if codon optimization is required to aid recombinant protein expression. Initially, gene synthesis involved synthesizing and annealing complementary oligonucleotides, then ligating through overlapping cohesive ends [17]. Once again, PCR revolutionized the process enabling savings in expense and time, as single-stranded oligonucleotides could be annealed and gaps filled with thermo-stable polymerase before amplification of full length products. Today, most gene synthesis processes can be outsourced to companies specializing in the technique, who can also optimize codons, minimize unwanted secondary structure, and remove cryptic splice sites.

2.2.5 Molecular "Evolution" through Display and Selection

In addition to making precise mutations to a protein to achieve desired changes, an alternative approach is to make a large number of random or "semi-random" changes (a library) and to select the variants with desired properties (such as improved binding to a ligand). Sequence changes can be introduced through error-prone PCR [18] or through a technique called DNA shuffling [19]. In this approach, the sequences of closely related proteins are randomly cut with DNase I and then denatured and reannealed, with gaps filled by PCR. Oligonucleotides can be added to the gene fragments to increase diversity. In this way, recombination-like crossovers can be produced in a given sequence.

Display technologies, first described by Smith in the form of phage display of peptides through their incorporation as a fusion into the sequence of the phage gene III protein [20], have become increasingly important. The key to the technology is the physical link between the genotype, the DNA within the phage particle, and the phenotype, the property conferred by the peptide or protein displayed on the surface of the phage particle. This linkage facilitates sequence determination of the displayed molecule following selection of the phage through the specific binding property. Phage display is often used to display antibody fragments, usually single-chain variable fragment (scFv) or fragment antigen binding (Fab) (see Fig. 2.1), to select molecules with a variety of properties, including

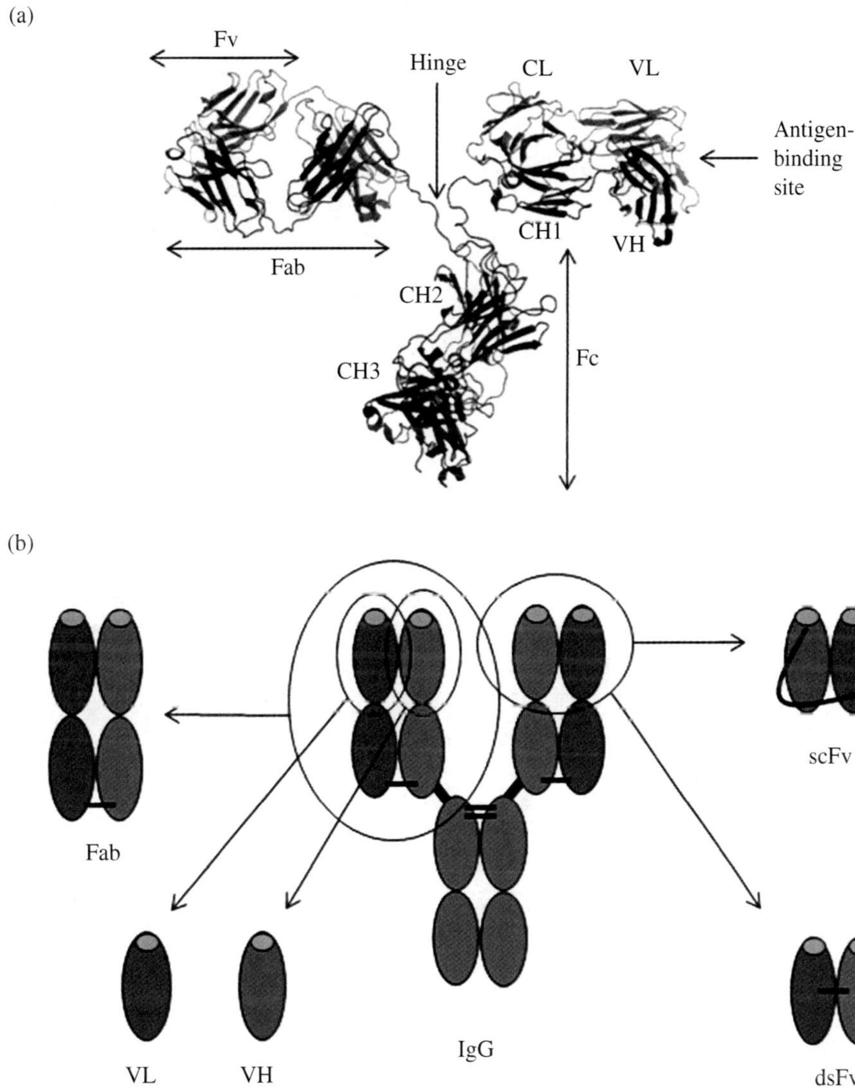

FIGURE 2.1 IgG structure and major antibody fragments. (a) Ribbon diagram of a murine IgG2a molecule (carbohydrate not shown) from its crystal structure (Protein Data Bank (PDB) 1igt [21]). The heavy chain is colored grey, the light chain red and the CDRs yellow. Structural features are highlighted (see text for details). (b) Schematic of IgG molecule showing derivation of the major antibody fragments. Color scheme is as for (a). The fragments shown are the Fab (fragment antigen binding), the scFv (single-chain variable fragment), the dsFv (disulfide-linked variable fragment), and the VH/VL (variable heavy/variable light, single domains). (*See insert for color representation of this figure.*)

improved antigen binding, cross-reactivity between species or increased stability. This selection for improvements enables the use of random sequence modifications to introduce variation into the library, such as those generated by error-prone PCR [22]. Alternatively, a more directed approach can be used to vary the sequence at specific positions only, by use of oligonucleotide synthesis using degenerate codons. For example, the triplet codon NNN, where N represents any of the four possible nucleotide bases, encodes all possible amino acids but also includes the three stop codons, TAG, TAA, and TGA. By using the triplet NNK at positions to be changed, where K represents G or T, the stop codons TAA and TGA are avoided while triplets encoding all 20 amino acids are still represented. However, NNK encodes an unequal distribution of amino acids, introducing bias to the library particularly when a large number of positions are altered. An alternative approach is the use of specialized trinucleotide phosphoramidites in oligonucleotide synthesis, which can be used to introduce any desired codons at equal frequency (or unequally if desired) [23, 24].

Phage display was first applied to antibody fragments (scFv) by McCafferty et al. [25]. Other display methods have also been developed including bacterial display, yeast display, mammalian cell display, and *in vitro* display [26–30]. Such techniques are now widely applied and, through the use of innovative "panning" and selection strategies, molecular evolution and display technologies have become extremely useful and powerful techniques in protein engineering.

2.3 APPLICATIONS OF PROTEIN ENGINEERING TO NON-ANTIBODY THERAPEUTIC PROTEINS

The focus of this chapter is the application of protein engineering to therapeutic antibodies, but it is also worth mentioning the contribution to other non-antibody approved biologics. For example, following the approval of recombinant insulin in the early 1980s, protein engineering created insulin analogs to modulate the pharmacokinetic (PK) and pharmacodynamic properties of the insulin molecule. The structural information from X-ray crystallography led to the rational design of fast-acting and slow-acting forms of the molecule, which have been tremendously useful in providing treatment regimens to patients with type I diabetes [1]. Fast-acting forms include insulin lispro, engineered by point mutation of two adjacent residues of the beta chain (proline (Pro-28) to lysine (Lys) and Lys-29 to Pro) to de-stabilize dimerization of the B-subunit facilitating rapid capillary absorption of active monomer [31, 32]. Slow-acting forms include insulin glargine, which has a substitution of glycine (Gly) for asparagine (Asn) and two arginine residues (Arg) added to the carboxy terminus of the B-chain [33]. The additional Arg amino acids raise the isoelectric point (pI) from 5.4 to 6.7, creating a molecule that is soluble at the formulation pH (pH 4) but that comes out of solution as it passes through its pI upon injection to physiologic pH (pH 7.4). This gives a slow absorption from the injection site [34]. The Asn to Gly substitution prevents deamidation of the acid-sensitive asparagine at acidic pH (see Section 2.4.2).

Another notable example of protein engineering in biological therapeutics is the fusion of the Fc region of the antibody molecule (see Fig. 2.1) to another non-antibody molecule; examples include the extracellular portion of a membrane bound receptor as in etanercept [35], or a peptide as in romiplostim [36]. These Fc fusions confer long serum half-life to the partner protein due to Fc-mediated salvage (via the neonatal Fc receptor (FcRn)) and can confer other advantages such as improving stability or solubility and facilitating purification.

Other engineering approaches have been employed to extend the serum half-life of proteins, including conjugation of polyethylene glycol (PEG) [37] and glycoengineering. Darbepoetin alfa is an example of the latter, a modified form of recombinant erythropoietin-α engineered to introduce two additional glycosylation sites [38]. Its consequential extended *in vivo* half-life permits less frequent dosing.

2.4 APPLICATIONS OF PROTEIN ENGINEERING TO THERAPEUTIC ANTIBODIES

Antibodies and antibody-based drugs are the largest class of protein therapeutics and have benefited hugely from protein engineering interventions and advances, some of which are outlined in this section. These advances have relied on detailed knowledge of antibody structure, through techniques including high resolution X-ray crystallography, and an increasing understanding of the relationship between this structure and biological activity. Figure 2.1 shows the structure of the IgG antibody molecule. It consists of two heavy chains and two light chains, covalently linked by disulfide bonds. Two Fab arms are linked through a flexible hinge to the Fc (fragment crystallizable). Each chain is composed of constant (C) and variable (V) regions. Each V-region consists of frameworks onto which are attached the CDRs; these are structural loops (three from variable heavy (VH) and three from variable light (VL)), which are largely responsible for antigen binding. The Fc region confers the effector function of the molecule through its interactions with Fc receptors on cells. These functions include ADCC (antibody-dependent cellular cytotoxicity), CDC (complement-dependent cytotoxicity), and ADCP (antibody-dependent cellular phagocytosis), and the Fc also mediates catabolic salvage through FcRn. Protein engineering of the molecule has taken many forms and some of these are outlined below, grouped under the desired functional outcome of the engineering.

2.4.1 Reduction of Immunogenicity

Antibodies destined for therapeutic use are often raised following immunization of animals, usually mice or rats, and need to be engineered to prevent an immune response in patients against the non-human sequences. Initially, this was achieved through chimerization, replacement of the constant regions of the rodent antibody with those from a human antibody [39, 40]. The therapeutic antibodies infliximab and rituximab are examples of chimeric antibodies that have been hugely successful as therapeutics. However, chimeric antibodies still have non-human V-regions that have the potential to elicit a human anti-chimeric antibody (HACA) response. In reality, human and rodent V-regions vary in their degree of primary sequence identity and HACA is not always an issue [41]. Methods to reduce the nonhuman content still further by V-region humanization were first developed in the mid-1980s using the technique of CDR grafting [42]. In this technique, DNA encoding the CDRs is fused onto DNA encoding the V-region framework sequence of a human antibody. Usually, it is also necessary to include a small number of non-CDR residues from the parental antibody in the final humanized sequence, in order to retain the full affinity of the original antibody in the engineered form. In practice, several variants are usually generated and are tested empirically for activity. A number of methods are available to determine these residues [43–45]. Rather than incorporate non-human framework residues in the final variant to recover full activity, some groups will retain fully germ-line frameworks but will subject CDRs to *in vitro* affinity maturation to enhance activity (see Section 2.4.5). Other methods of humanization have been developed; some are variations of this grafting method while others rely on phage display to select active variants [4]. The technique of "guided selection" is a phage display method that was used to create the therapeutic antibody adalimumab from its hybridoma precursor D2E7 [46]. Phage display is also used for antibody discovery to generate human antibodies directly from naïve human libraries [47]. Another method used to generate human antibodies directly is to use mouse strains in which the mouse immunoglobulin loci have been replaced with human sequences [48–50]. Antibodies generated from either the phage or the transgenic route are often subjected to further engineering, including *in vitro* affinity maturation and "germ-lining" to replace unusual residues in the framework regions with those present in the pre-mutated germ-line genes to reduce the potential for immunogenicity. Immunogenicity has multifactorial causes and is influenced by, for example, dose, route of delivery, presence of protein aggregates, use of concomitant medication, genetic background, as well as sequence. In addition to reducing the non-human primary sequence, some groups have looked to de-immunize protein sequences by engineering to remove T-cell epitopes, identified by either *in silico* prediction or *in vitro* methods including T-cell stimulation assays [51, 52].

2.4.2 Improving Stability and Biophysical Properties

Therapeutic proteins and antibodies need to be robust and stable, and protein engineering can be used to improve these characteristics. An example is the human IgG4 molecule. This can be used therapeutically when an inactive isotype is required (see Section 2.4.3) but it has an inherent tendency to dissociate into "half-molecule" through reduction of its single hinge disulfide, a phenomenon not seen with the other human immunoglobulin isotypes. A process known as "Fab-arm exchange" occurs, in which the Fab arms from different antibody molecules can pair together to form natural bispecific antibodies [53, 54]. By using protein engineering to mutate a serine (Ser) residue in the gamma-4 hinge to Pro, Angal et al. were able to create an IgG4 mutant with a more rigid and stable IgG1-like hinge, which no longer undergoes the half-molecule formation nor the consequent Fab-arm exchange [55]. Gemtuzumab ozogamicin is an example of a recombinant antibody bearing this engineered and stabilized IgG4 hinge [56].

Therapeutic antibodies also need to be resistant to degradation during production and storage, and need to be formulated as a homogeneous preparation that conforms within strict parameters [57]. Table 2.1 summarizes the major degradation and unwanted post-translational modifications occurring within the usual range of pH exposure encountered during expression, purification, and storage. Normally, this will be in the pH range 4.0–7.5, although viral inactivation steps often include an extended holding period at lower pH. Reactions are likely to be accelerated at elevated temperatures. To some extent, degradation-susceptible sites can be predicted and engineered out of the sequence, provided the changes introduced do not adversely affect affinity or create additional liabilities, including increased immunogenic potential and reduced biological activity. As an example, the Asn–Gly dipeptide motif is particularly susceptible to non-enzymatic deamidation, resulting in the elimination of the amide group of the Asn residue giving aspartate (Asp) via a succinimide intermediate [58]. Deamidation results in an increase in negative charge creating product heterogeneity and could have a deleterious effect on protein structure and function. Rates of deamidation are influenced not only by factors including pH, temperature, and ionic strength, but also by both the primary sequence context and secondary structure around the affected residue. When the sequence occurs in a solvent-exposed part of the molecule such as within a CDR, deamidation is often more likely. Hence, as well as sequence scrutiny, knowledge of the structural environment of the particular sequence can be informative. Using site-directed mutagenesis, the protein engineer may be able to remove such sites, either by changing the Asn residue or by replacing the Gly residue with a more hydrophobic one less likely to induce Asn deamidation. Another way around this issue is to use a lyophilized formulation so that the deamidation reaction cannot occur

TABLE 2.1 Major Antibody Degradation and Post-translational Modifications with Possible Protein Engineering Interventions

Reaction	Susceptible Sequence	Suggested Mutagenesis Options	Comments
V-region N-linked glycosylation	Asn-X-Ser/Thr where X is any amino acid except Pro	Ala-X-Ser/Thr Asn-X-Ala Gln-X-Ser/Thr	Other changes can be considered Cys at Ser/Thr position can occasionally be glycosylated
Deamidation	Asn-X [or Gln-X] where X is a small, polar residue, particularly Gly (or Ser)	Ala-X Ser-X Asn–Ala Asn–Val Asn–Leu	Asn more susceptible than Gln Very much context dependent, as secondary structure can have significant influence on occurrence/rate Also influenced by temperature, pH, and ionic strength
Asp isomerization	Asp-X where X is a small, polar residue, particularly Gly (or Ser)	Glu-X Asp–Ala Asp–Val	Very much context dependent, as secondary structure can have significant influence on occurrence/rate
Acid hydrolysis	Asp–Pro	Ala–Pro Ser–Pro Glu–Pro	Replacement of Pro is possible but could affect secondary structure
Side-chain oxidation	Trp, Met	Ala Ser	His and Tyr are also susceptible residues under more extreme conditions Many other substitutions are possible
N-terminal pyroglutamate formation	Gln	Glu	Pyroglutamate blocks N-terminal protein sequencing Results in product heterogeneity
C-terminal lysine clipping	Lys	Delete Lys	Results in product heterogeneity
Unpaired cysteine reactivity	Cys	Ser Ala	Local environment can influence the choice of replacement residue

upon storage, but this has implications for patient convenience and compliance.

Asp residues can isomerize to iso-Asp especially if the sequence is Asp–Gly [59]. This does not change overall charge but iso-Asp has the potential to be immunogenic. Side-chain oxidation can occur for reactive amino acids including methionine (Met) and tryptophan (Trp) residues in exposed positions, although under mild conditions this is not usually an issue [60]. Similarly, free cysteine (Cys) sulfydryl can be oxidized to form Cys disulfide bonds, depending on the local environment. The presence of single unpaired Cys residues in CDRs is usually undesirable and replacement can be considered (often with Ser). N-terminal glutamine (Gln) residues, frequently found on heavy chains, are liable to cyclize to form pyroglutamate, by the free NH_2 group reacting with the side chain [57]. This occurs naturally in antibodies but the reaction may not go to completion and any heterogeneity in preparations of recombinant antibodies is undesirable. A potential solution is to replace the Gln with glutamate (Glu), also a common N-terminal residue on antibody heavy chains. Clipping of the C-terminal lysine (Lys) residues in both human IgG1 and IgG4 can also occur and gives rise to a difference in charge [57]. Again an option to avoid this is to delete the C-terminal residue by protein engineering.

Another predictable and measurable potential sequence liability is the occurrence of an N-linked glycosylation site within a V-region. These are present in about 10–20% of all V-regions (depending on the species), and it is usual to try to remove these to ensure product homogeneity. If the V-region site is present in a framework region, it can usually be mutated out as part of the humanization process and reverted back to the human germ-line sequence, but glycosylation sites within CDRs can be more problematical. The consensus site is asparagine-X-threonine/serine (Asn-X-Thr/Ser), where X can be any amino acid other than Pro, so mutation of either the Asn or the Thr/Ser residue can be used to successfully ablate the site. Occasionally, V-region carbohydrate moieties can be involved in antigen binding, so affinity will need to be assessed after mutation. During the humanization of the antibody 5/44, which became the antibody element of inotuzumab ozogamicin, a glycosylation site within heavy-chain CDR2 was ablated by changing the Thr to Ala [61].

Finally, acid hydrolysis of the Asp–Pro dipeptide bond is a well-known phenomenon and may cause cleavage of the V-region polypeptide chain, but again this is unlikely under "standard" conditions. As with many of these potential sequence liability "hot-spots," protein engineering offers the potential to change the sequence to mitigate the risk.

In terms of using protein engineering to improve biophysical properties such as increased thermal stability, improved solubility, and reduced propensity to aggregate, the field is awaiting improvements in the prediction of such properties from primary sequence. However, some steps have been taken including CDR-grafting strategies onto particularly stable frameworks [62], introducing additional disulfide bonds between VH and VL regions [63], and using library technologies and high throughput screening to select for thermo-stable antibody fragments [64, 65]. Analysis of V-region structures, particularly of CDR conformations and their interactions with framework residues, permits a rational approach to be pursued for stability improvement and for the elimination of hydrophobic patches that could drive solubility issues.

2.4.3 Tailoring Mechanism of Action

There are many ways in which protein engineering has been used to tailor the mechanism of action of a therapeutic antibody, and a major example is in format engineering. A wide variety of formats are available to the antibody engineer. Starting with the "natural" human IgG formats, it is straightforward from the engineering perspective to change the isotype simply by subcloning the VH gene into vectors bearing the heavy-chain isotype of choice. For an "active" mechanism of action such as ADCC or CDC, the IgG1 isotype is used (e.g., trastuzumab). For simple ligand/receptor blockade or when cell killing needs to be avoided, an inactive isotype is preferred, either IgG2 (e.g., denosumab) or IgG4 (e.g., natalizumab).

Protein engineering offers further options to the natural human isotypes for reducing or increasing Fc effector function reactivity. For example, IgG backbones have been generated with even less activity and greater stability than the wild-type IgG4 and IgG2 [66–68]. Eculizumab is a hybrid of IgG2 and IgG4, with a gamma-2 CH1 and hinge and a gamma-4 CH2 and CH3 [69]. Otelixizumab, in trials for type I diabetes, has an IgG1 bearing an Asn to Ala change, which ablates the constant region glycosylation site to reduce the activity of the gamma-1 isotype [70]. Teplizumab carries two mutations in the upper hinge region of gamma-1 CH2, Leucine (Leu)-234 to Ala and Leu-235 to Ala (EU numbering [71]), which reduce effector function by inhibiting FcR binding.

Point mutations have also been made to the gamma-1 Fc region in order to enhance activity and create more potent cell-killing molecules. These include the Ser-239 to Asp, Isoleucine (Ile)-332 to Glu changes that offer significant improvements in ADCC [72], plus the Ser-267 to Glu, Histidine (His)-268 to Ser, Ser-324 to Thr combination to improve CDC [73]. In addition, since glycosylation at Asn-297 is critical for Fc structure and effector function, glycoengineering has been undertaken to enhance activity [74].

Because of its modular nature, the IgG molecule lends itself to deletion and modification to form fragments (Fig. 2.1), which have shown success in the clinic and offer even greater future potential. The simplest is the Fab fragment that offers monovalent antigen binding without any Fc-mediated effector function. It also lacks the ability to interact with FcRn and therefore exhibits rapid clearance from the plasma. Ranibizumab and abciximab are examples of approved therapeutic Fabs. Smaller fragments include scFv, which consists of just the VL and VH domains joined by a flexible linker peptide, and the disulfide-linked variable fragment (dsFv), in which an additional disulfide bond has been introduced to increase stability [75–77]. Single binding domains have also been used, isolated VL or VH domains representing the smallest antigen-binding units from a natural IgG [78]. Such small fragments share a simple structure that opens up the possibility of microbial expression, although the smaller the fragment, the more exposure of regions at domain interfaces in the intact molecule, giving increased potential for aggregation-related liabilities. The domain structure of the antibody molecule also lends itself to other engineered formats designed to achieve new modalities, such as bispecificity. Several approaches have been taken to promote heavy-chain heterodimer formation, including a "knobs into holes" method [79] and a method exploiting the phenomenon of Fab-arm exchange [80]. Other approaches aim to combine IgG with additional smaller antigen-binding fragments fused to the C-terminus of heavy or light chain [81]. It is beyond the scope of this chapter to describe in full this plethora of bispecific strategies, but they do serve to highlight the power of protein engineering techniques to create Ig-based novel molecules to achieve new therapeutic modalities.

2.4.4 Influencing Distribution and PK

Protein engineering of antibody sequences can be used to influence distribution and PK. The Fc domain confers antibody half-life through its interaction with FcRn. Circulating IgG is taken into cells by pinocytosis but within the acidic endosome it is able to bind FcRn through a pH-dependent interaction. It therefore escapes lysosomal degradation and is recycled to the plasma membrane and back into the circulation following release at neutral pH. Mutation of residues in the CH2/CH3 interface region have been shown to increase the affinity of the FcRn interaction [82], leading to modest increases in serum half-life. Similarly, deletion of the CH2 and CH3 domains as in some fragments removes the interaction with FcRn giving a much shorter serum retention time. The reduction in size of certain antibody fragments including Fab and smaller molecules, coupled to the absence of the FcRn interaction site, gives a very rapid clearance from blood by the kidneys but also gives much more rapid distribution to the tissues. For instances where monovalent antibody fragments are preferred but where long half-life is necessary, an approach other than to use the Fc region can be applied. Some groups have achieved this using the albumin molecule, an abundant serum protein that is also rescued and recycled by FcRn. This could be achieved by direct fusion of the Fab or scFv fragment to albumin [83], or by creation of a bispecific antibody fragment with one "arm," an antibody-binding site directed against albumin such that it does not interfere with its FcRn interaction site [84, 85]. In this way, the molecule can "piggy-back" to albumin to achieve antibody-like serum retention without any of the properties of the Fc. Another solution is to use PEGylation. Certolizumab pegol consists of a Fab′ fragment engineered to have a single free cysteine in its hinge region to which a PEG moiety is site-specifically conjugated. The PEG moiety is thought to create a large hydrodynamic water shell around the Fab fragment, which prevents kidney filtration and may reduce its tendency to diffuse out of normal blood but results in efficient tissue penetration at sites of inflammation [86]. Other non-Fc-mediated means of extending the half-life of antibody fragments are under development.

The pH sensitivity of antigen binding can play a role in modulating target-mediated clearance, and changes to pI can influence non-specific antibody clearance. Such changes are made by site-directed mutagenesis to introduce or remove charged amino acids to the IgG surface. These themes are discussed in the next chapter.

2.4.5 Improving Ligand/Receptor Interaction

Protein engineering techniques are often applied to improve or modulate the affinity of an antibody for its antigen. This is particularly relevant for antibodies generated by phage display or using transgenic mice with human immunoglobulin genes, as these technologies tend not to generate antibodies with the same affinity as immunization-based techniques with wild-type animals. Methods of "*in vitro* affinity maturation" tend to fall into two categories: display-based methods and structure-based methods.

Display-based methods rely on building libraries of variants of the original molecule bearing changes at positions likely to be involved in the antigen interaction. Panning methods under high-stringency binding conditions are then employed to select those variants with improved affinity. Ranibizumab is an example of a therapeutic antibody derived by affinity maturation of a precursor molecule [87]. As well as phage display, a number of other display methodologies have been used for affinity maturation including yeast display, mammalian cell display, bacterial display, and *in vitro* display. Diversity can be introduced randomly across the V-gene but is usually targeted to specific locations: either the individual CDRs or a subset of residues within the CDRs (e.g., predicted antigen "contact" residues). Because of the size limits of library construction, it is extremely challenging to randomize an entire binding site simultaneously. One approach to overcome this is to focus on CDR-H3 that often forms the majority of antigen contacts, or to proceed sequentially CDR by CDR, optimizing one at a time and using the new sequence as the start point for subsequent library generation for the next CDR [88].

Structure-based methods rely on precise structural information of the antibody–antigen interaction, usually from high-resolution X-ray crystallography or from homology models when structures are not available, from which a relatively small number of rational and highly specific mutations in the antibody binding site can be constructed [89]. A combination of structure-based prediction and a small, focused display library to interrogate these predicted positions can also be a productive approach.

2.5 FUTURE PERSPECTIVES

Protein engineering has clearly been a major driver in the success of therapeutic proteins and in particular therapeutic antibodies. In the next few years, we are likely to see many of the engineering advances discussed in this chapter, including Fc engineering and novel bispecific antibody formats, reach the market as approved medicines. There are also numerous antibody alternatives being developed, novel scaffolds engineered to have antibody-like binding properties but without any semblance of Fc activity. It will be interesting to see whether these molecules, which can have advantages in terms of expression, stability, and reduced size (and hence perhaps improved tissue penetration), can find a therapeutic niche.

As more antibody sequences are analyzed on the back of Next Generation Sequencing technologies, as more antibody structures are solved, and as molecular modeling and sequence/structure prediction software improves, the impact of protein engineering can only increase further. There is also the prospect that *in silico* prediction of properties from protein sequence, including propensity to aggregate, efficiency of expression, and immunogenicity, will provide further protein engineering opportunities. One exciting application beginning to gain ground is *de novo* antibody design, the ability to design a human antibody sequence to any given epitope on an antigen and to "build-in" the desired activity such as ligand blockade. The ability to do this accurately and reproducibly may require a combination of structure-based rational design and directed evolution through display technologies. This would be particularly impactful for hard-to-immunize targets.

There is still a major drive to produce medicines with improved efficacy, greater safety, reduced immunogenicity, and improved stability. Optimizing drug disposition by improved *in vivo* ADME behavior remains an important focus of protein engineering to obtain protein structures, which provide functional behavior resulting into beneficial efficacy and limiting safety issues. The following chapter illustrates some specific protein engineering techniques with special application toward improving drug disposition. Predicting the future triumphs of protein engineering is difficult, but it is sure to continue to drive therapeutic proteins to further success and to bring innovative and effective treatments to patients.

REFERENCES

[1] Berenson DF, Weiss AR, Wan ZL, Weiss MA. Insulin analogs for the treatment of diabetes mellitus: therapeutic applications of protein engineering. Ann NY Acad Sci 2011;1243:E40–E54.

[2] Beck A, Reichert JM. Therapeutic Fc-fusion proteins and peptides as successful alternatives to antibodies. MAbs 2011;3:415–416.

[3] Czajkowsky DM, Hu J, Shao Z, Pleass RJ. Fc-fusion proteins: new developments and future prospects. *EMBO Mol Med* 2012;10:1015–1028.

[4] Almagro JC, Fransson J. Humanization of antibodies. Front Biosci 2008;13:1619–1633.

[5] Palmer E. 2014. The 10 best selling drugs of 2013. *Fierce pharma*. Available at http://www.fiercepharma.com/special-reports/10-best-selling-drugs-2013. Accessed 2015 Jun 3.

[6] Berman HM, Westbrook J, Feng Z, Gilliland G, Bhat TN, Weissig H, Shindyalov IN, Bourne PE. The Protein Data Bank. Nucleic Acids Res 2000;28:235–242.

[7] Kunkel TA. Rapid and efficient site-directed mutagenesis without phenotypic selection. Proc Natl Acad Sci U S A 1985;82:488–492.

[8] Wells JA, Vasser M, Powers DB. Cassette mutagenesis: an efficient method for generation of multiple mutations at defined sites. Gene 1985;34:315–323.

[9] Popplewell AG, Gore MG, Scawen M, Atkinson T. Synthesis and mutagenesis of an IgG-binding protein based upon protein A of Staphylococcus aureus. Protein Eng 1991;4:963–970.

[10] Mullis KB, Falooona FA. Specific synthesis of DNA *in vitro* via a polymerase-catalysed chain reaction. Methods Enzymol 1987;155:335–350.

[11] Hunt HD, Horton RM, Pullen JK, Pease LR. Site-directed mutagenesis by overlap extension using the polymerase chain reaction. Gene 1989;77:51–59.

[12] Zheng L, Baumann U, Reymond JL. An efficient one-step site-directed mutagenesis protocol. Nucleic Acids Res 2004;32:e115.

[13] Horton RM, Hunt HD, Ho SN, Pullen JK, Pease LR. Engineering hybrid genes without the use of restriction enzymes: gene splicing by overlap extension. Gene 1989;77:61–68.

[14] Engler C, Kandzia R, Marilllonnet S. One pot, one step, precision cloning method with high throughput capability. PLoS One 2008;3:e3647.

[15] Gibson DG, Young L, Chuang RY, Venter JC, Hutchison CA III, Smith HO. Enzymatic assembly of DNA molecules up to several hundred kilobases. Nat Methods 2009;6:343–345.

[16] Gibson DG. Enzymatic assembly of overlapping DNA fragments. Methods Enzymol 2011;498:349–361.

[17] Edge MD, Green AR, Heathcliffe GR, Meacock PA, Schuch W, Scanlon DB, Atkinson TC, Newton CR, Markham AF. Total synthesis of a human leukocyte interferon gene. Nature 1981;292:756–762.

[18] Leung DW, Chen E, Goeddel DV. A method for random mutagenesis of a defined DNA segment using a modified polymerase chain reaction. Technique 1989;1:11–15.

[19] Stemmer WPC. DNA shuffling by random fragmentation and reassembly: *in vitro* recombination for molecular evolution. Proc Natl Acad Sci USA 1994;91: 10747–10751.

[20] Smith GP. Filamentous fusion phage: novel expression vectors that display cloned antigens on the virion surface. Science 1985;228:1315–1317.

[21] Harris LJ, Larson SB, Hasel KW, Day J, Greenwood A, McPherson A. The three-dimensional structure of an intact monoclonal antibody for canine lymphoma. Nature 1992; 360:369–372.

[22] Hawkins RE, Russell SJ, Winter G. Selection of phage antibodies by binding affinity. Mimicking affinity maturation. J Mol Biol 1992;226:889–896.

[23] Sondek J, Shortle D. A general strategy for random insertion and substitution mutagenesis: substoichiometric coupling of trinucleotide phosphoramidites. Proc Natl Acad Sci U S A 1992;89:3581–3585.

[24] Yanez J, Arguello M, Osunu J, Soberon X, Gaytan P. Combinatorial codon-based amino acid substitutions. Nucleic Acids Res 2004;32:e158.

[25] McCafferty J, Griffiths AD, Winter GH, Chiswell DJ. Phage antibodies: filamentous phage displaying antibody variable domains. Nature 1990;348:552–554.

[26] Boder ET, Wittrup KD. Yeast surface display for screening combinatorial polypeptide libraries. Nat Biotechnol 1997;15: 553–557.

[27] Gai SA, Wittrup KD. Yeast surface display for protein engineering and characterization. Curr Opin Struct Biol 2007;17:467–473.

[28] Harvey BR, Georgiou G, Hayhurst A, Jeong KJ, Iverson BL, Rogers GK. Anchored periplasmic expression, a versatile technology for the isolation of high-affinity antibodies from Escherichia coli-expressed libraries. Proc Natl Acad Sci U S A 2004;101:9193–9198.

[29] Smith ES, Zauderer M. Antibody library display on a mammalian virus vector: combining the advantages of both phage and yeast display into one technology. Curr Drug Discov Technol 2014;11:48–55.

[30] Hanes J, Pluckthun A. *In vitro* selection and evolution of functional proteins by using ribosome display. Proc Natl Acad Sci U S A 1997;94:4937–4942.

[31] Brems DM, Alter LA, Beckage MJ, Chance RE, DiMarchi RD, Green LK, Long HB, Pekar AH, Shields JE, Frank BH. Altering the association properties of insulin by amino acid replacement. Prot Eng 1992;5:527–533.

[32] DiMarchi RD, Chance RE, Long HB, Shields JE, Slieker LJ. Preparation of an insulin with improved pharmacokinetics relative to human insulin through consideration of structural

homology with insulin-like growth factor 1. Horm Res 1994;41(2 Suppl):93–96.

[33] Markussen J, Diers I, Hougaard P, Langkjaer L, Norris K, Snel L, Sørensen AR, Sørensen E, Voigt HO. Soluble prolonged-acting insulin derivatives. Degree of protraction, crystallizability and chemical stability of insulins substituted at positions A21, B13, B23, B27 and B30. Protein Eng 1988;2:157–166.

[34] Heinemann LR, Linkeschova R, Rave K, Hompesch B, Sedlak M, Heise T. Time action profile of the long-acting insulin analog insulin glargine (HOE901) in comparison with those of NPH insulin and placebo. Diabetes Care 2000;23:644–649.

[35] Garrison L, McDonnell ND. Etanercept: therapeutic use in patients with rheumatoid arthritis. Ann Rheum Dis 1999;58: 165–169.

[36] Molineux G, Newland A. Development of romiplostim for the treatment of patients with chronic immune thrombocytopenia: from bench to bedside. Br J Haematol 2010;150:9–20.

[37] Jevsevar S, Kunstelj M, Porekar VG. PEGylation of therapeutic proteins. Biotechnol J 2010;5:113–128.

[38] Sinclair AM, Elliott S. Glycoengineering: the effect of glycosylation on the properties of therapeutic proteins. J Pharm Sci 2005;94:1626–1635.

[39] Morrison SL, Johnson MJ, Herzenberg LA, Oi VT. Chimeric human antibody molecules: mouse antigen-binding domains with human constant region domains. Proc Natl Acad Sci U S A 1984;81:6851–6855.

[40] Boulianne GL, Hozumi N, Shulman MJ. Production of functional chimaeric mouse/human antibody. Nature 1984;312:643–646.

[41] Clark M. Antibody humanization: a case of the 'Emperor's new clothes'? Immunol Today 2000;21:397–402.

[42] Jones PT, Dear PH, Foote J, Neuberger MS, Winter G. Replacing the complementarity-determining regions in a human antibody with those from a mouse. Nature 1986;321:522–525.

[43] Adair JR, Athwal DS, Emtage JS. Humanised antibodies. WO91/09967. 1991.

[44] Queen CL, Schneider WP, Landolfi NF, Coelingh KL. Improved humanized immunoglobulins. WO92/11018. 1992.

[45] Carter PJ, Presta LG. Method for making humanised antibodies. WO94/04679. 1994.

[46] Osbourn J, Groves M, Vaughan T. From rodent reagents to human therapeutics using antibody guided selection. Methods 2005;36:61–68.

[47] Marks JD, Hoogenboom HR, Bonnert TP, McCafferty J, Griffiths AD, Winter G. By-passing immunization. Human antibodies from V-gene libraries displayed on phage. J Mol Biol 1991;222:581–597.

[48] Bruggemann M, Caskey HM, Teale C, Waldmann H, Williams GT, Surani MA, Neuberger MS. A repertoire of monoclonal antibodies with human heavy chains from transgenic mice. Proc Natl Acad Sci U S A 1989;86:6709–6713.

[49] Lonberg N, Taylor LD, Harding FA, Trounstine M, Higgins KM, Schramm SR, Kuo CC, Mashayekh R, Wymore K, McCabe JG, Munoz-O'Regan D, O'Donnell SL, Lapachet ESG, Bengoechea T, Fishwild DM, Carmack CE, Kay RM, Huszar D. Antigen-specific human antibodies from mice comprising four distinct genetic modifications. Nature 1994;368:856–859.

[50] Green LL, Hardy MC, Maynard-Currie CE, Tsuda H, Louie DM, Mendez MJ, Abderrahim H, Noguchi M, Smith DH, Zeng Y, David NE, Sasai H, Garza D, Brenner DG, Hales JF, McGuinness RP, Capon DJ, Klapholz S, Jakobovits A. Antigen-specific human monoclonal antibodies from mice engineered with human Ig heavy and light chain YACs. Nat Genet 1994;7:13–21.

[51] De Groot AS, Moise L. Prediction of immunogenicity for therapeutic proteins: state of the art. Curr Opin Drug Discuss Devel 2007;10:332–340.

[52] Holgate RG, Baker MP. Circumventing immunogenicity in the development of therapeutic antibodies. IDrugs 2009;12: 233–237.

[53] Van der Neut Kolfschoten M, Schuurman J, Losen M, Bleeker WK, Martínez-Martínez P, Vermeulen E, den Bleker TH, Wiegman L, Vink T, Aarden LA, De Baets MH, van de Winkel JG, Aalberse RC, Parren PW. Anti-inflammatory activity of human IgG4 antibodies by dynamic Fab arm exchange. Science 2007;317:1554–1557.

[54] Labrijn AF, Buijsse AO, van den Bremer ET, Verwilligen AY, Bleeker WK, Thorpe SJ, Killestein J, Polman CH, Aalberse RC, Schuurman J, van de Winkel JG, Parren PW. Therapeutic IgG4 antibodies engage in Fab arm exchange with endogenous human IgG4 *in vivo*. Nat Biotechnol 2009;27:767–771.

[55] Angal S, King DJ, Bodmer MW, Turner A, Lawson AD, Roberts G, Pedley B, Adair JR. A single amino acid substitution abolishes the heterogeneity of chimeric mouse/human (IgG4) antibody. Mol Immunol 1993;30:105–108.

[56] Hamann PR, Hinman LM, Hollander I, Beyer CF, Lindh D, Holcomb R, Hallett W, Tsou HR, Upeslacis J, Shochat D, Mountain A, Flowers DA, Bernstein I. Gemtuzumab ozogamicin, a potent and selective anti-CD33 antibody-calicheamicin conjugate for treatment of acute myeloid leukemia. Bioconjug Chem 2002;13:47–58.

[57] Liu H, Gaza-Bulseco G, Faldu D, Chumsae C, Sun J. Heterogeneity of monoclonal antibodies. J Pharm Sci 2008; 97:2426–2447.

[58] Robinson NE. Protein deamidation. Proc Natl Acad Sci U S A 2002;99:5283–5288.

[59] Geiger T, Clarke S. Deamidation, isomerization and racemization at asparaginyl and aspartyl residues in peptides. Succinimide-linked reactions that contribute to protein degradation. J Biol Chem 1987;262:785–794.

[60] Ji JA, Zhang B, Cheng W, Wang YJ. Methionine, tryptophan and histidine oxidation in a model protein, PTH: mechanism s and stabilization. J Pharm Sci 2009;98:4485–4500.

[61] Dijoseph JF, Popplewell A, Tickle S, Ladyman H, Lawson A, Kunz A, Khandke K, Armellino DC, Boghaert ER, Hamann P, Zinkewich-Peotti K, Stephens S, Weir N, Damle NK. Antibody-targeted chemotherapy of B-cell lymphoma using calicheamicin conjugated to murine or humanized antibody against CD22. Cancer Immunol Immunother 2005;54:11–24.

[62] Ewert S, Honegger A, Pluckthun A. Stability improvements of antibodies for extracellular and intracellular applications: CDR grafting to stable frameworks and structure-based framework engineering. Methods 2004;34:184–199.

[63] Glockshuber R, Malia M, Pfitzinger I, Pluckthun A. A comparison of strategies to stabilize immunoglobulin Fv fragments. Biochemistry 1990;29:1362–1367.

[64] Miller BR, Glaser SM, Demarest SJ. Rapid screening platform for stabilization of scFvs in Escherichia coli. Methods Mol Biol 2009;525:279–289.

[65] Miller BR, Demarest SJ, Lugovskoy A, Huang F, Wu X, Snyder WB, Croner LJ, Wang N, Amatucci A, Michaelson JS, Glaser SM. Stability engineering of scFvs for the development of bispecific and multivalent antibodies. Protein Eng Des Sel 2010;23:549–557.

[66] An Z, Forrest G, Moore R, Cukan M, Haytko P, Huang L, Vitelli S, Zhao JZ, Lu P, Hua J, Gibson CR, Harvey BR, Montgomery D, Zaller D, Wang F, Strohl W. IgG2m4, an engineered antibody isotype with reduced Fc function. MAbs 2009;1:572–579.

[67] Alegre ML, Collins AM, Pulito VL, Brosius RA, Olson WC, Zivin RA, Knowles R, Thistlethwaite JR, Jolliffe LK, Bluestone JA. Effect of a single amino acid mutation on the activating and immunosuppressive properties of a 'humanized' OKT3 monoclonal antibody. J Immunol 1992;148:3461–3468.

[68] Armour KL, Clark MR, Ag H, Williamson LM. Recombinant human IgG molecules lacking Fcgamma receptor 1 binding and monocyte triggering activities. Eur J Immunol 1999;29:2613–2614.

[69] Rother RP, Rollins SA, Mojcik CF, Brodsky RA, Bell L. Discovery and development of the complement inhibitor eculizumab for the treatment of paroxysmal nocturnal hemoglobinuria. Nat Biotechnol 2007;25:1256–1264.

[70] Tao MH, Morrison SL. Studies of aglycosylated chimeric mouse-human IgG. Role of carbohydrate in the structure and effector functions mediated by the human IgG constant region. J Immunol 1989;143:2595–2601.

[71] Edelman GM, Cunningham BA, Gall WE, Gottlieb PD, Rutishauser U, Waxdal MJ. The covalent structure of an entire gammaG immunoglobulin molecule. Proc Natl Acad Sci U S A 1969;63:78–85.

[72] Lazar GA, Dang W, Karki S, Vafa O, Peng JS, Hyun L, Chan C, Chung HS, Eivazi A, Yoder SC, Vielmetter J, Carmichael DF, Hayes RJ, Dahiyat BI. Engineered antibody Fc variants with enhanced effector function. Proc Natl Acad Sci U S A 2006;103:4005–4010.

[73] Moore GL, Chen H, Karki S, Lazar GA. Engineered Fc variant antibodies with enhanced ability to recruit complement and mediate effector functions. MAbs 2010;2: 320–328.

[74] Raju TS. Terminal sugars of Fc glycans influence antibody effector functions of IgGs. Curr Opin Immunol 2008;20: 471–478.

[75] Bird RE, Hardman KD, Jacobson JW, Johnson S, Kaufman BM, Lee SM, Lee T, Pope SH, Riordan GS, Whitlow M. Single-chain antigen-binding proteins. Science 1988;242: 423–426.

[76] Huston JS, Levinson D, Mudgett-Hunter M, Tai MS, Novotný J, Margolies MN, Ridge RJ, Bruccoleri RE, Haber E, Crea R, Oppermann H. Protein engineering of antibody binding sites: recovery of specific activity in an anti-digoxin single-chain Fv analogue produced in Escherichia coli. Proc Natl Acad Sci U S A 1988;85:5879–5883.

[77] Brinkmann U, Reiter Y, Jung SH, Lee B, Pastan I. A recombinant immunotoxin containing a disulphide-stabilized Fv fragment. Proc Natl Acad Sci U S A 1993;90:7538–7542.

[78] Holt LJ, Herring C, Jespers LS, Woolven BP, Tomlinson IM. Domain antibodies: proteins for therapy. Trends Biotechnol 2003;21:484–490.

[79] Ridgway JB, Presta LG, Carter P. Knobs-into-holes engineering of antibody CH3 domains for heavy chain dimerization. Protein Eng 1996;9:617–621.

[80] Labrijn AF, Meesters JI, de Goeij BE, van den Bremer ET, Neijssen J, van Kampen MD, Strumane K, Verploegen S, Kundu A, Gramer MJ, van Berkel PH, van de Winkel JG, Schuurman J, Parren PW. Efficient generation of stable bispecific IgG1 by controlled Fab-arm exchange. Proc Natl Acad Sci U S A 2013;110:5145–5150.

[81] Coloma MJ, Morrison SL. Design and production of novel tetravalent bispecific antibodies. Nat Biotechnol 1997; 15:159–163.

[82] Dall'Acqua WF, Woods RM, Ward ES, Palaszynski SR, Patel NK, Brewah YA, Wu H, Kiener PA, Langermann S. Increasing the affinity of a human IgG1 for the neonatal Fc receptor: biological consequences. J Immunol 2002;169:5171–5180.

[83] Smith BJ, Popplewell A, Athwal D, Chapman AP, Heywood S, West SM, Carrington B, Nesbitt A, Lawson AD, Antoniw P, Eddelston A, Suitters A. Prolonged *in vivo* residence times of antibody fragments associated with albumin. Bioconjug Chem 2001;12:750–756.

[84] Tijink BM, Laeremans T, Budde M, Stigter-van Walsum M, Dreier T, de Haard HJ, Leemans CR, van Dongen GA. Improved tumour targeting of anti-epidermal growth factor receptor Nanobodies through albumin binding: taking advantage of modular nanobody technology. Mol Cancer Ther 2008;7:2288–2297.

[85] Holt LJ, Basran A, Jones K, Chorlton J, Jespers LS, Brewis ND, Tomlinson IM. Anti-serum albumin domain antibodies for extending the half-lives of short lived drugs. Protein Eng Des Sel 2008;21:283–288.

[86] Palframan R, Airey M, Moore A, Vugler A, Nesbitt A. Use of biofluorescence imaging to compare the distribution of certolizumab pegol, adalimumab and infliximab in the inflamed paws of mice with collagen-induced arthritis. J Immunol Methods 2009;348:36–41.

[87] Ferrara N, Damico L, Shams N, Lowman H, Kim R. Development of ranibizumab, an anti-vascular endothelial growth factor antigen binding fragment, as therapy for neovascular age-related macular degeneration. Retina 2006;26:859–870.

[88] Yang WP, Green K, Pinz-Sweeney S, Briones AT, Burton DR, Barbas CF III. CDR walking mutagenesis for the affinity maturation of a potent human anti-HIV-1 antibody into the picomolar range. J Mol Biol 1995;254:392–403.

[89] Kiyoshi M, Caaveiro JM, Miura E, Nagatoishi S, Nakakido M, Soga S, Shirai H, Kawabata S, Tsumoto K. Affinity improvement of a therapeutic antibody by structure-based computational design: generation of electrostatic interactions in the transition state stabilizes the antibody-antigen complex. PLoS One 2014;9:e87099.

3

THERAPEUTIC ANTIBODIES—PROTEIN ENGINEERING TO INFLUENCE ADME, PK, AND EFFICACY

Tatsuhiko Tachibana, Kenta Haraya, Yuki Iwayanagi and Tomoyuki Igawa
Chugai Pharmaceutical Co., Ltd., Gotemba City, Shizuoka Prefecture, Japan

3.1 INTRODUCTION

In recent years, research focused on pharmacokinetic properties of therapeutic antibodies has been on the rise as the market share of therapeutic antibodies continues to increase. It is currently known that there are two major mechanisms that determine antibody pharmacokinetics: nonspecific elimination and target-dependent elimination [1, 2]. Nonspecific elimination is a common mechanism for antibodies targeting both membrane-bound and soluble antigens, whereas target-dependent elimination is commonly observed if the target of the antibody is a membrane-bound antigen. In mammalian species, antibodies can be taken up into the cell by fluid-phase micropinocytosis and either recycled back to the blood by neonatal Fc receptor (FcRn) or degraded in the lysosome [3]. Renal elimination of antibodies (MW 150 kDa) is minimal because glomerular filtration rate is significant only for proteins with an MW below 50 kDa. To address the phenomenon of nonspecific clearance, researchers have mainly focused on two general approaches: Fc engineering to improve FcRn-mediated recycling [4–8] and reducing the nonspecific uptake of antibodies by cells [9, 10]. In Sections 3.2 and 3.3, we will be focusing on protein engineering strategies to reduce nonspecific uptake. One of those strategies tries to reduce the isoelectric point (pI) of the antibody by protein engineering, which can be an effective way to reduce nonspecific cellular uptake as well as improve various pharmacokinetic properties of the antibody. The putative mechanism by which pI reduction decreases the antibody's elimination rate is that lower-pI antibodies have more negative charges on their surface, which repels the negative charges (i.e., sialic acids and glycosaminoglycans) on the cell membrane where micropinocytosis takes place. Conversely, an antibody with a higher *pI* value has more positive charges on its surface and exhibits faster cellular uptake due to its interaction with negative charges on the cell membrane. It is believed that the negatively charged extracellular matrix (ECM) also contributes to the pharmacokinetic profile of antibodies and other proteins [11].

In the case of antibodies targeting membrane-bound antigens, circulating antibodies bind to the membrane-type antigen, are internalized into the cell, and are then degraded in the lysosome if they cannot escape the lysosomal clearance pathway. Generally, this target-mediated elimination is a saturable process; if the antibody dose is high enough, the antibody amount will be in excess of that of the cell surface target antigen, which will saturate all antigens in the body. As a consequence of target saturation, the contribution of target-mediated clearance to total antibody clearance will be low. Conversely, if the antibody dose is low, target-mediated clearance will account for a larger portion of the total clearance. Some factors that can affect target-mediated antibody clearance include the expression pattern and expression level of the antigen in the body, the internalization rate of the antigen, and the affinity of the antibody. Recently, we have developed a technology to reduce target-mediated clearance by adding a pH-dependent antigen binding property to the therapeutic antibody, thereby increasing its chances of being recycled [12]. This technology will be discussed in more detail in Section 3.4.

For antibodies targeting soluble antigens, conventional antibodies still face significant drawbacks even if their

ADME and Translational Pharmacokinetics/Pharmacodynamics of Therapeutic Proteins: Applications in Drug Discovery and Development,
First Edition. Edited by Honghui Zhou and Frank-Peter Theil.

pharmacokinetic profile is in the acceptable range. In many cases, the target antigen accumulates in plasma after the therapeutic antibody is administered because of reduced antigen clearance after forming a complex with the antibody. In some cases, higher concentrations of the therapeutic antibody are needed due to this accumulation of antigen, which may require substantially higher dose levels. Recently, we have also developed a technology to reduce antigen accumulation by optimized pH-dependent antigen binding and increased affinity to FcRn or Fcγ receptors (FcγRs) [13]. This technology will be described in Section 3.5.

3.2 RELATIONSHIP BETWEEN pI AND PHARMACOKINETICS

3.2.1 pI and Clearance

We have previously reported that the plasma half-life of antibodies can be prolonged by substituting amino acids in the variable region of the antibody to decrease pI without impairing the target affinity [9]. We also reported that to engineering the antibody to have higher pI resulted in a shorter plasma half-life. Figure 3.1 shows the relationship between antibody pI and plasma clearance in mice for some antibodies. Monoclonal antibodies (mAbs) A1, B1, C1, and D1 are all humanized mAbs with IgG4 heavy chains and kappa light chains. pI-increased A1 variants, mAbs A2 and A3, were obtained by introducing mutations in the variable region. mAbs B2 and B3 are pI-decreased variants of B1. These antibodies do not cross-react with mouse antigen; therefore, the clearance observed can be considered primarily driven by nonspecific elimination. The results show that the pharmacokinetic profile of a therapeutic antibody can be improved by mutagenesis of its variable region without changing its constant region indicating that nonspecific clearance of higher-pI mAbs is increased by suboptimal variable region structures composed of positively charged amino acids.

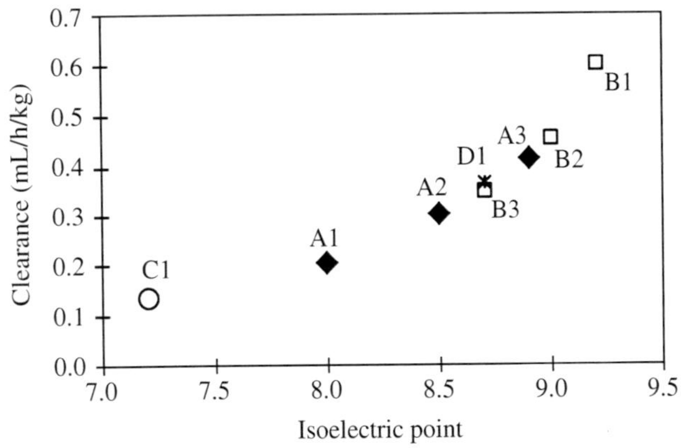

FIGURE 3.1 Correlation between pI and clearance—humanized IgG4 mAbs A1, A2, A3 (closed diamonds), B1, B2, B3 (open square), C1 (open circles), and D1 (asterisk) were intravenously injected into normal mice. Clearance values were determined from plasma concentration–time profiles and plotted against their *pI* values.

There are some points to be considered when modifying an antibody's pI. Introducing mutations in the constant region is not recommended because it may cause a higher risk of immunogenicity. One option is to change the framework sequence to another germ-line sequence that has the appropriate pI but changing the framework sequence often results in impaired affinity. To minimize the risk of immunogenicity, it is preferable to introduce mutations into the complementarity determining region (CDR) rather than the framework region. It is also important to introduce a mutation at an amino acid that is located on the surface of the antibody because the antibody's surface charge greatly affects its interactions with the cell surface. When reduced pI is desired, changing asparagine (Asn) to aspartic acid (Asp) or glutamine (Gln) to glutamic acid (Glu) can be a good first option because of the similarity in molecular size. If increased pI is needed, the reverse (Asp to Asn or Glu to Gln) can be applied. Otherwise, arginine and lysine can be introduced to increase pI or arginine and lysine can be replaced with another amino acid to reduce pI. In all cases, it should be confirmed that mutagenesis does not cause impaired affinity or poor physicochemical properties.

Chemical modification is another method to change the pI of a protein. It is reported that both cationization and anionization of antibodies resulted in increased clearance [14–16]. One possible reason for this is that these chemical modifications inhibited binding of the antibody to FcRn, and therefore the resulting antibody with reduced affinity to FcRn could not be recycled effectively.

3.2.2 pI and Distribution

To elucidate the link between pI and tissue distribution, steady-state distribution volume (V_{dss}) and distribution volumes of the central compartment (V_{d1}) and peripheral compartment (V_{d2}) of the previously mentioned mAbs A1, B1, C1, and D1 were calculated by noncompartmental analysis and two-compartment analysis. The relationship between pI and distribution volume (V_{dss}, V_{d1}, V_{d2}) is shown in Figure 3.2. Antibodies with higher pI have larger V_{d2} whereas all antibodies have comparable V_{d1} that is close to the plasma volume in the body. As a result, antibodies with higher pI have larger V_{dss}, which represents the sum of V_{d1} and V_{d2}. These results suggest that all antibodies can be quickly distributed in plasma regardless of pI and then gradually distributed to the tissues, but negatively charged antibodies with low pI are distributed to the tissues more slowly due to electrostatic repulsion with the negative surface charges of endothelial and epithelial cells in the vasculature and other tissues.

As mentioned above, the effect of pI on tissue distribution was observed from the analysis of plasma concentration–time

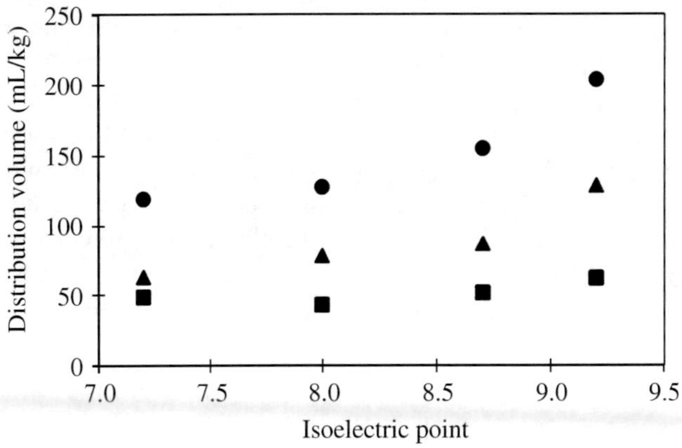

FIGURE 3.2 Correlation between pI and distribution volume—humanized IgG4 mAbs A1, B1, C1, and D1 were intravenously injected into normal mice. Distribution volume of the central compartment (closed square) and the peripheral compartment (closed triangle) and steady-state distribution volume (closed circle) were plotted against their *pI* values.

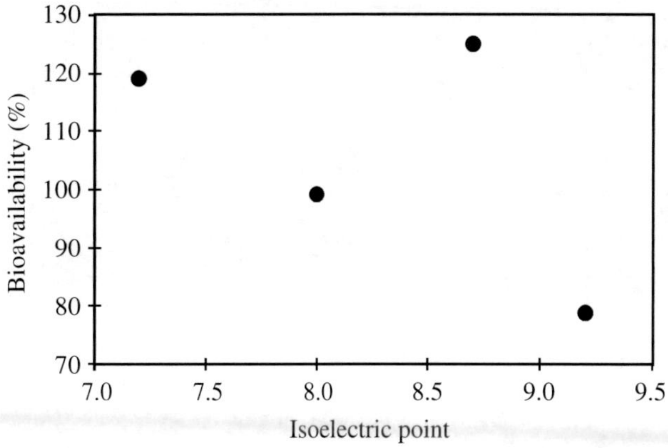

FIGURE 3.3 Relationship between pI and bioavailability—humanized IgG4 mAbs A1, B1, C1, and D1 were subcutaneously injected into normal mice. Bioavailability values were plotted against their *pI* values.

profiles. There is also a report [17] that proved the relationship between pI and distribution volume by directly measuring antibody concentrations in tissues. In this report, the plasma concentration-based tissue distribution volumes in Hindlimb skin (excluding capillary volume) of a positively charged antibody ($pI = 8.7$) and a negatively charged antibody ($pI = 6.6$) were 0.150 mL/g tissue and 0.088 mL/g tissue, respectively. Based on these results, the positively charged antibody can be distributed 1.7-fold more than the negatively charged antibody. They also measured the interstitial fluid concentration of the antibodies and determined the relative excluded volume fraction in the interstitial fluid space. The relative excluded volume fractions of the positively charged antibody ($pI = 8.7$) and the negatively charged antibody ($pI = 6.6$) in Hindlimb skin were 0.37 and 0.65, respectively. It was discussed in that report that highly negatively charged hyaluronan repels the negatively charged antibody and hinders its distribution into the interstitial fluid space.

In the case of chemical modification, it has been reported that cationization of proteins results in increased distribution [14] and anionization results in decreased distribution [18]. The distribution of antibody into tumor tissue is also reported to be impaired by decreasing the pI of the antibody [15, 19]. These results suggest that improved distribution to the target tissue by increased pI is an attractive approach to achieve better efficacy; however, increased pI often results in a poor plasma pharmacokinetic profile and thus cautious optimization of pI is needed.

3.2.3 pI and SC Absorption

To investigate the link between pI and the bioavailability (F) of subcutaneously (SC) injected antibodies, the previously mentioned mAbs A1, B1, C1, and D1 were injected SC into mice and F was calculated. mAb B1, which has the highest *pI* (9.2) among the four antibodies, exhibited the lowest F (79%) but the relationship between pI and F was not so clear because the other three antibodies showed an apparent complete F (Figure 3.3).

There is a report that describes antibodies with various pIs that were intravenously and SC injected into minipigs [20]. Antibodies with higher pI tend to have larger clearance and lower F in minipigs. A negative correlation between pI and F in humans was also reported. To avoid overestimation when predicting F in humans, it should be noted that F in humans tends to be lower than that in minipigs and monkeys [20, 21].

3.2.4 pI and FcRn Function

It is considered difficult to directly evaluate the effect of antibody pI on the efficiency of recycling by FcRn. Instead, we injected mAbs A1, A3, B1, and C1 into beta 2 microglobulin knockout (β2mKO) mice, which lack functional FcRn and compared their clearance ($CL_{\beta 2mKO}$) with that of wild-type mice (CL_{WT}). It was observed that antibodies with higher pI have larger clearance even in β2mKO mice; therefore, the difference in pI greatly affected the uptake rate by pinocytosis. Clearance ratios ($CL_{\beta 2mKO}/CL_{WT}$), which can be considered as recycling efficiency by FcRn, were comparable among the four antibodies (Figure 3.4). Thus, it is suggested that pI has a minimal effect on functional FcRn.

3.3 NONSPECIFIC/SPECIFIC OFF-TARGET BINDING

3.3.1 Nonspecific Binding and Clearance

It is thought that the reason for the high clearance of high pI antibodies is because the negative charge on the cell surface

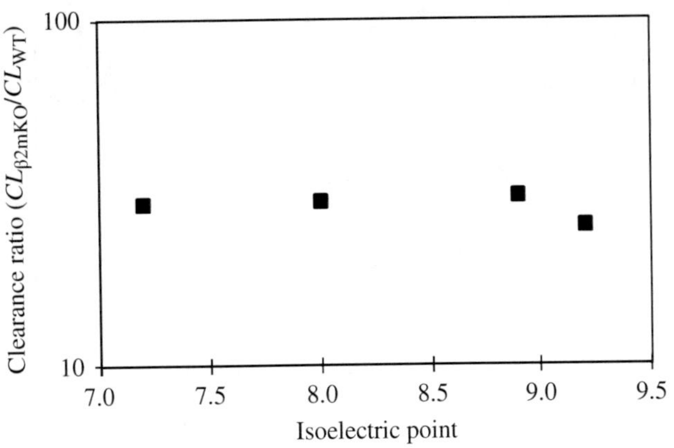

FIGURE 3.4 Relationship between pI and clearance ratio between β2mKO mice and normal mice ($CL_{\beta2mKO}/CL_{WT}$)—humanized IgG4 mAbs A1, A3, B1, and C1 were intravenously injected into β2mKO mice. The clearance ratio between β2mKO mice and normal mice was plotted against their *pI* values. Similar recycling efficiency was observed among the four antibodies.

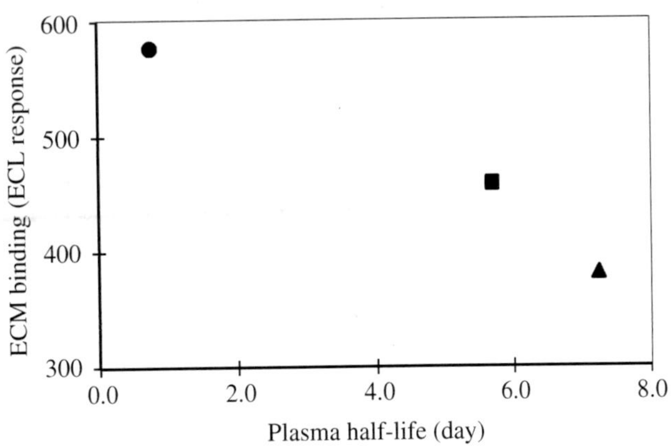

FIGURE 3.5 Correlation between ECM binding and plasma half-life—humanized mAbs C7 (triangle), C8 (square), and C9 (circle) were intravenously injected into human FcRn-transgenic mice. The plasma half-life was plotted against their ECM binding property. Longer half-life was achieved by reducing ECM binding.

attracts the positively charged antibody, which is then internalized more rapidly into the cell. There are some reported methods to directly measure the interaction between antibodies and components in the body. Evaluation of ECM binding is one of these methods. mAb C9 is an antibody that has a very high binding property to ECM and very short plasma half-life in mice [22]. mAbs C7 and C8, which have reduced ECM-binding property, were created by mutagenesis of mAb C9. To investigate the link between ECM binding and antibody pharmacokinetics, these antibodies were injected into human FcRn-transgenic mice. It was revealed that plasma half-life can be greatly improved by reducing ECM binding (Figure 3.5).

There is a report that argues that antibodies with high clearance can be identified by evaluating binding to baculovirus particles [10]. The authors of this report claim that baculovirus mimics the membrane of infected cells and their proposed method has the advantage that any proteins homologous to the antigen are not expressed on the surface of baculovirus, so target-related binding can be eliminated in the assay. There is also a report wherein heparin binding of antibodies was evaluated [23]; however, the relationship between heparin binding and antibody clearance remains uncertain.

3.3.2 Specific Off-Target Binding and Clearance

Even though antibodies are believed to have high specificity to their target antigen, there are some cases wherein specific off-target binding of the antibody results in a poor pharmacokinetic profile in some species. There are cases wherein the off-target molecule is identified and others where it is unknown. If the molecule that causes the off-target binding is a soluble molecule found in plasma, it can be identified relatively easily and the antibody can be optimized to have reduced affinity to the off-target molecule found in preclinical species and human [24]. However, if the off-target molecule is unknown, there remains a risk of showing a poor pharmacokinetic profile in humans even if the antibody could be optimized to have improved pharmacokinetics in the preclinical species [25].

3.4 pH-DEPENDENT ANTIGEN BINDING TO REDUCE TARGET-MEDIATED ELIMINATION

3.4.1 Concept of Recycling Antibody

Generally, proteins on the cell membrane (mainly receptors) are internalized into the cell and degraded in the lysosome by constant turnover. Also, ligand proteins of receptors are internalized together with the receptor upon binding. It has been reported that erythropoietin (EPO), which is a protein hormone for erythropoiesis control, is internalized with the EPO receptor (EPOR) as a complex after binding to EPOR and then subsequently degraded in the lysosome [26]. Thus, after administration of recombinant EPO *in vivo*, dose-dependent nonlinear pharmacokinetics via EPOR-mediated clearance was observed [27]. Such receptor-mediated clearance was also observed for granulocyte-colony stimulating factor (G-CSF) [28], thrombopoietin [29], and interferon-β [30]. Due to this rapid elimination by receptor-mediated clearance, high and frequent administration is required for treatment. Since some therapeutic mAbs target receptors, target antigen-mediated internalization may also interfere with mAb pharmacokinetics.

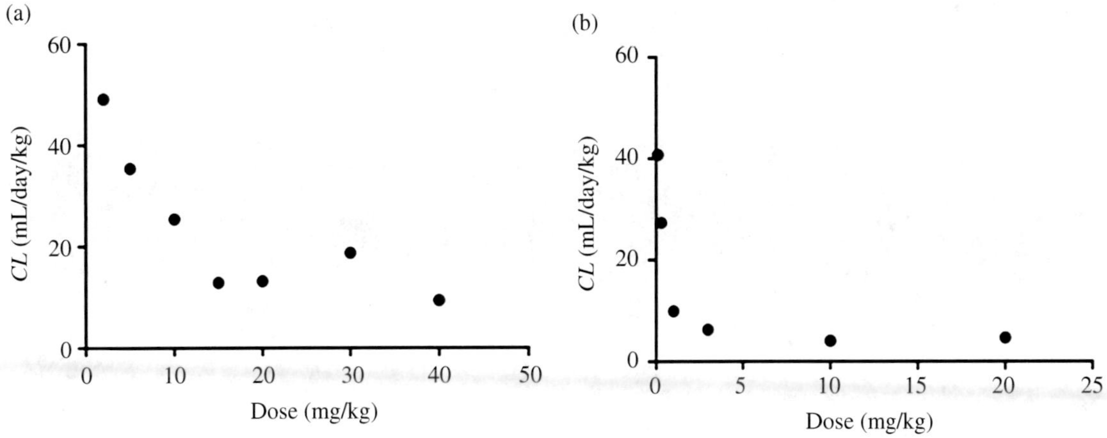

FIGURE 3.6 Relationship between dose and clearance in an anti-NRP-1 antibody (a) and an anti-IFNαR antibody (b) in humans.

Nonlinear pharmacokinetics has been observed frequently during clinical and preclinical development of mAbs. For membrane-bound antigens such as IL-6R [31], EGFR (epidermal growth factor receptor) [32], and IL 17R [33], mAbs have shown a nonlinear dose–exposure relationship due to the saturation of target antigen-mediated clearance. The relationships between dose and clearance for an anti-neuropilin-1 (NRP-1) antibody [34] and an anti-interferon α receptor (IFNαR) antibody [35] are shown in Figure 3.6. Although large clearance was observed at low doses, clearance decreased with increasing dose due to saturation of target antigen-mediated clearance. Because membrane-bound antigen is internalized into the cell, complexes of mAbs and membrane-bound antigen can also be internalized, transferred to the endosome, and degraded in the lysosome (Figure 3.7a). Therefore, rapid elimination was observed at low plasma mAb concentration and high dose was required to maintain a sufficiently high plasma concentration of these mAbs. Minimization of target antigen-mediated clearance may prolong mAb pharmacokinetics and reduce injection dose or frequency.

It is well known that mAbs are rescued from the endosome to the plasma via the FcRn recycling system [37]. Because FcRn–IgG binding is pH dependent, IgG binds to FcRn only at acidic pH (i.e., in the endosome). After recycling from endosome to plasma, IgG is released from FcRn due to this pH-dependent binding property. Therefore, mAbs have a long half-life *in vivo* compared to small-molecule drugs and other protein drugs (EPO, G-CSF, blood coagulation factor, etc.). On the other hand, conventional mAbs have no pH dependency in antigen binding between plasma (pH 7.4) and endosome (pH 6.0). Hence, after internalization of complexes of mAbs and membrane-bound antigen, mAbs cannot dissociate from the membrane-bound antigen in the endosome. As a result, these complexes are transferred to the lysosome and degraded without recycling by FcRn (Figure 3.7a). If mAbs can dissociate from their membrane-bound antigen in the endosome, they can be recycled efficiently from endosome to plasma by FcRn (Figure 3.7b).

3.4.2 pH Dependency and Target-Mediated Elimination

There are many environmental differences between the plasma and the endosome, such as pH [38], calcium ion concentration [39], and protein expression [40]. Among them, pH is the most well known; utilizing this pH difference between plasma and endosome, a number of drug delivery systems for drug release in the endosome have been reported, especially for cancer treatment [41, 42].

Histidine has got a pK_a of approximately 6.0. Therefore, at pH 7.4 (in blood), histidine is uncharged. However, below pH 6.0 (in endosome), histidine is protonated and has a positive charge. Due to this pH-dependent change in structure, histidine has been applied to various types of protein engineering. In one study, Sarkar et al. applied histidine substitutions to G-CSF to increase its half-life [43]. It has been reported that G-CSF was rapidly eliminated via G-CSF receptor (G-CSFR)-mediated internalization [28]. In this study, histidine-mutated G-CSF showed pH-dependent binding at pH 7.4 and 5.5 to G-CSFR. Moreover, histidine-mutated G-CSF had improved half-life in the cell perhaps due to an improvement in endosomal sorting and recycling by dissociation from G-CSFR at endosomal acidic pH.

The first report of mAb engineering using histidine to introduce pH-dependent antigen binding was from Igawa et al. [12]. Tocilizumab, an anti-IL-6R antibody, showed IL-6R-mediated nonlinear pharmacokinetics in human IL-6R-transgenic mice [12], cynomolgus monkeys [12], and humans [31]. To minimize this IL-6R-mediated clearance, an anti-IL-6R antibody with pH-dependent antigen binding property, named recycling tocilizumab, was

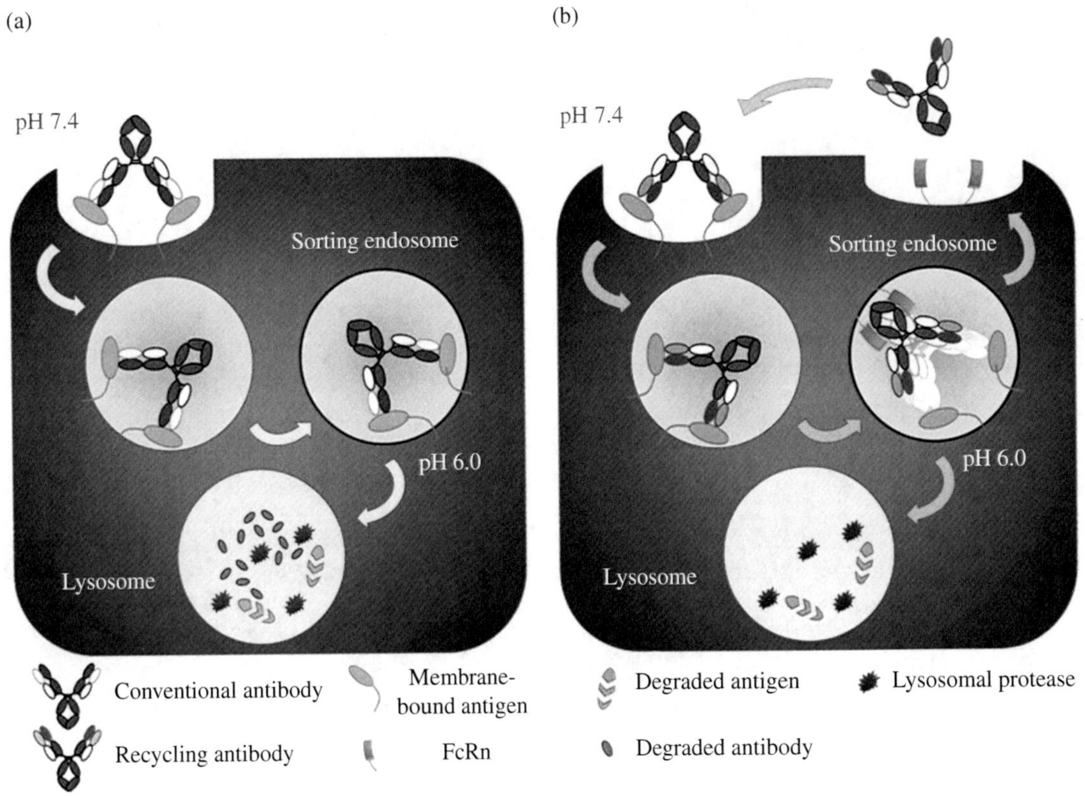

FIGURE 3.7 Mechanism of antibody clearance in the case of membrane-bound antigen for (a) conventional antibody and (b) pH-dependent antigen-binding antibody. Reprinted from Igawa et al. [36], with permission from Elsevier. (*See insert for color representation of this figure.*)

generated using the histidine mutagenesis approach and showed prolonged pharmacokinetic profile in cynomolgus monkeys compared with the control antibody without pH-dependent antigen binding [12]. C-reactive protein, which is a pharmacodynamic marker of IL-6R neutralization, was also suppressed for a longer duration by recycling tocilizumab. Importantly, although recycling tocilizumab showed comparable binding affinity against IL-6R at pH 7.4, only the recycling antibody showed rapid dissociation against IL-6R at pH 6.0. These results suggest that recycling tocilizumab can be dissociated from IL-6R in the acidic endosome and recycled back to the blood by FcRn. Furthermore, recycling tocilizumab demonstrated significantly improved pharmacokinetics compared to tocilizumab in a Phase I clinical study [36], thus validating the efficacy of pH-dependent antigen binding in humans.

Thereafter, a number of pH-dependent antigen-binding antibodies have been reported. Chaparro-Riggers et al. reported a pH-dependent antibody against proprotein convertase subtilisin kexin 9 (PCSK9) [44], which is currently an attractive target for hyperlipidemia treatment. Although PCSK9 is a soluble antigen, an anti-PCSK9 antibody without pH-dependent property showed nonlinear pharmacokinetics in mice and cynomolgus monkeys via an unknown PCSK9-mediated mechanism. Using the histidine mutagenesis approach, a pH-dependent anti-PCSK9 antibody was generated. Histidine-mutated variant J17 showed a 7.7-fold difference in dissociation rate between pH 6.0 and 7.4 against human PCSK9 and a 12-fold difference against cynomolgus PCSK9. Moreover, J17 showed significantly improved half-life in mice and cynomolgus monkeys compared to the control antibody without pH dependency.

Traxlmayr et al. reported a pH-dependent antibody against human epidermal growth factor receptor 2 (Her2) [45], a validated target antigen for cancer treatment. Three anti-Her2 antibodies (trastuzumab, pertuzumab, and trastuzumab-DM1) have already been approved. Nonlinear pharmacokinetics via Her2 internalization in humans was observed for these antibodies [46–48]. The pH-dependent antibodies against Her2 were generated by using the histidine mutagenesis approach and showed rapid dissociation at pH 6.0 against Her2 in BIAcore assays. The authors also investigated pH-dependent binding of these antibodies to Her2-positive cells at pH 7.4 and 6.0. Significantly weaker binding to Her2 on the cell surface at 6.0 was observed only for pH-dependent antibodies. Importantly, the authors concluded that not only histidine but also amino acids next to histidine affect pH dependency. Although an *in vivo* study was not conducted to evaluate the improvement of pharmacokinetics in this paper, future reports with *in vivo* data are expected.

The effectiveness and advantages of pH-dependent antigen binding have been shown for a number of target antigens. Reducing the dosing frequency can decrease the number of hospital visits for patients. Moreover, lowering the dose may allow the application of SC dosing that cannot be achieved by conventional mAbs, and SC dosing enables patients to do self-injection at home. Thus, this technology can give patients substantial benefits by minimizing target antigen-mediated clearance.

3.5 SOLUBLE ANTIGEN SWEEPING

3.5.1 Concept of Sweeping Antibody

In the case of recycling antibodies targeting soluble antigens, antibody-bound antigen is cleared from the plasma at the speed of nonspecific intracellular uptake of the antigen–antibody complex (Figure 3.8). The internalized recycling antibody itself, after dissociating from the antigen within the acidic endosome, is recycled back to the plasma by FcRn and this unbound, returned antibody can bind to another antigen (recycling of an antibody). This enables the recycling antibody to neutralize more than two antigen molecules, in sharp contrast with the conventional non-pH-dependent antibody that can only neutralize one to two antigen molecules [12].

Based on this pH-dependent recycling antibody, sweeping antibody technology was developed to increase the antibody recycling rate and achieve more efficient antigen clearance from plasma [13]. The concept behind sweeping antibodies is described in Figure 3.8. As discussed previously, the antibody recycling rate and antigen clearance rate of a recycling antibody is determined by the intracellular uptake rate of the antigen–antibody complex. Therefore, we tried to enhance the efficacy of recycling antibodies by increasing the cellular uptake rate of the antigen–antibody complex. To evaluate our sweeping antibodies, we used a steady-state model in which soluble antigen is continuously infused by a pump implanted under the skin on the backs of mice to mimic the endogenous situation [13]. We monitored the plasma concentration profiles of both antigen and antibody because the former reflects the extent and the latter reflects the duration of the antigen clearance effect of the sweeping antibody.

There are many ways to enhance the intracellular uptake rate of antigen–antibody complexes, including enhancing the nonspecific uptake rate as mentioned in Sections 3.2 and 3.3. We have established sweeping antibody technologies targeting soluble monomeric antigens (with only one antibody binding site) that can be quickly taken up into the cell in an FcRn- or FcγRIIb-dependent manner, and each antibody was designated as an "FcRn sweeping antibody" or "FcγRIIb sweeping antibody," respectively. In the following sections, the concepts of FcRn sweeping antibodies and FcγRIIb sweeping antibodies will be introduced.

3.5.2 FcRn-Mediated Sweeping

FcRn has an innate function as a recycling receptor for IgG from the endosome to the plasma. IgG plasma concentration can be as high as 10–20 mg/mL in humans [49], so the expression level of FcRn should be high enough to enhance the intracellular uptake rate of antigen–antibody complexes significantly. Moreover, considering its function, most antibodies internalized by FcRn are expected to be recycled back to the plasma, so FcRn is a promising candidate for a sweeping receptor.

As discussed in Section 3.4, when endogenous IgG is transferred to the acidic endosome, FcRn binds to IgG and recycles it back to the cell surface [37]. Thus, native IgG does not bind to FcRn at neutral pH (pH 7.4) in the plasma but binds to FcRn at acidic pH (pH 5.8–6.0) within the endosome [50]. On the other hand, in the case of an FcRn sweeping antibody, the Fc portion of a recycling antibody is

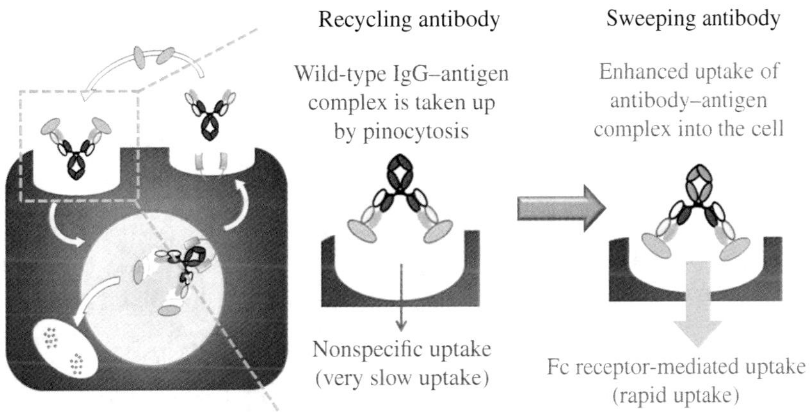

FIGURE 3.8 Mechanism of recycling and sweeping antibodies to clear antigens from plasma. (*See insert for color representation of this figure.*)

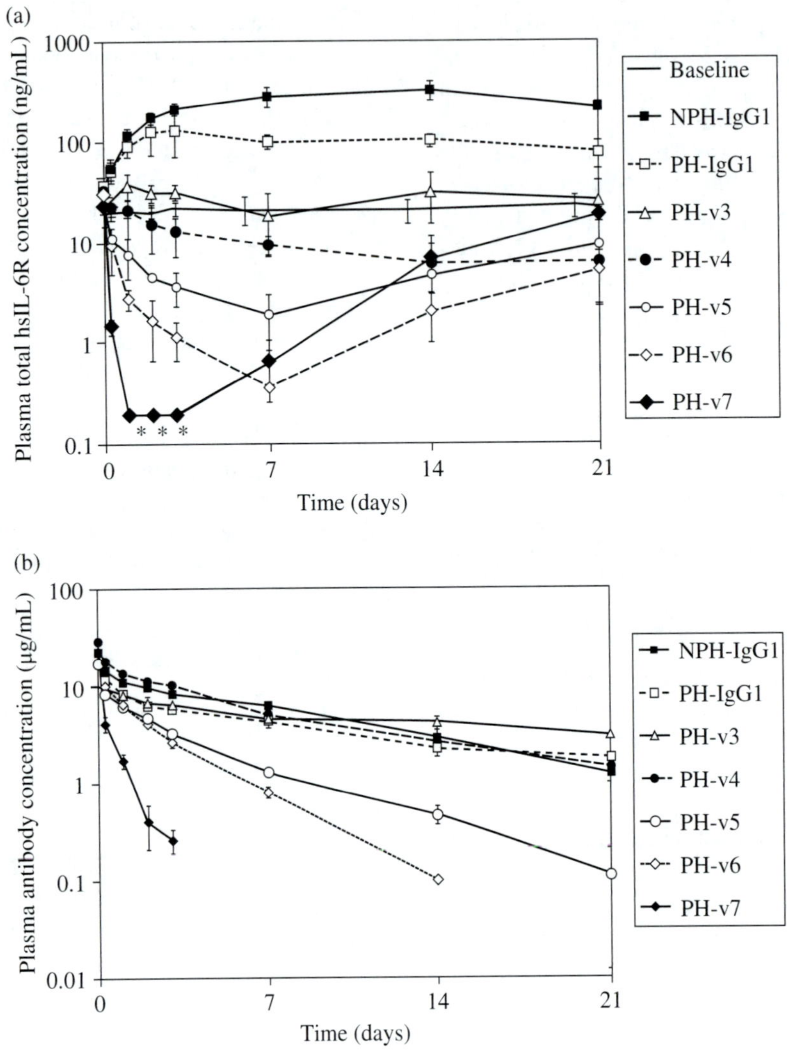

FIGURE 3.9 Effect of FcRn-binding affinity at pH 7.4 on antigen and antibody PK profile in human FcRn-transgenic mice. IL-6R (a) and antibody (b) concentration profiles are shown.

engineered to bind to FcRn at both neutral pH and acidic pH [13], so that the antigen–antibody complex can bind to FcRn at the cell surface and be internalized in an FcRn-dependent manner. Figure 3.9a,b shows the effect of FcRn-binding affinity at pH7.0 on antigen and antibody PK profiles in human FcRn-transgenic mice (B6.mFcRn−/−. hFcRn Tg line 32+/+ mice; Jackson Laboratories) [13]. Several Fc variants with different binding affinities to FcRn at neutral pH (Table 3.1) were evaluated using the steady-state model and the results indicate that the stronger Fc binding to FcRn becomes, the more effectively the sweeping antibody eliminates antigen from circulation. Although accelerating internalization by FcRn also tends to increase antibody clearance, this rise in antibody clearance is much smaller than the rise in antigen clearance. This means that after FcRn-mediated internalization of antigen–antibody complexes, the sweeping antibody dissociates from the antigen within the acidic endosome and the dissociated antigen is degraded in the lysosome, as observed in recycling antibodies whose antigen–antibody complexes are taken up into the cell in a nonspecific manner, whereas the sweeping antibody itself is recycled back to the plasma with a rather high recycling rate.

Because of this high recycling rate, the FcRn sweeping antibody with moderate FcRn-binding affinity at neutral pH provides moderate but long-acting antigen sweeping. The results of the FcRn sweeping antibody PH-v4, whose FcRn affinity is around 120 nM, indicate that by accelerating the uptake rate of the antigen–antibody complex by FcRn, plasma antigen concentration can be reduced by about 15-fold lower than that of the recycling antibody (PH-IgG1) and 30-fold lower than that of the conventional antibody (NPH-IgG1) while maintaining a comparable antibody PK profile. This demonstrates that the PH-v4 type FcRn sweeping antibody provides more than 30-fold reduction of dosage over a conventional antibody even with infinite affinity, a level that can never be achieved with a conventional antibody. On the other hand, the FcRn sweeping antibody PH-v5, whose FcRn-binding affinity is below 80 nM, provides short-term but extensive reduction of plasma antigen concentration. PH-v6 was able to make antigen concentration about 1000-fold lower than the conventional antibody (NPH-IgG1) with only a fourfold increase in antibody clearance. Thus, this type of sweeping antibody could antagonize a huge amount of antigen by reducing plasma antigen concentration below the baseline level, whereas a conventional antibody, even with infinite affinity, would be completely ineffective in such a situation.

These results in human FcRn-transgenic mice indicate that by enhancing the FcRn-binding affinity of recycling antibodies at neutral pH, we can eliminate soluble monomeric antigens from the plasma much more effectively than with a recycling antibody. In addition, since changing the binding affinity to FcRn generates antibodies with different extents and durations of antigen sweeping, the antigen-sweeping profile can be readily customized to enable the FcRn sweeping antibody to be broadly applied to various antigens.

TABLE 3.1 Binding Affinity to FcRn at Neutral pH

	K_D (nM) at pH 7.0		
Fc Variant	Mouse FcRn	Human FcRn	Mutations
IgG1	3918	88000	–
v1	52	NT	I332V/N434Y
v2	NT	155	M252W/N434W
v3	NT	288	M252Y/N434Y
v4	NT	120	M252Y/N286E/N434Y
v5	NT	77	M252Y/T307Q/Q311A/N434Y
v6	NT	35	M252Y/V308P/N434Y
v7	NT	4	S239K/M252Y/D270F/N286E/T307Q/V308P/Q311A/M428I/N434Y
v0	No binding	No binding	I253A

3.5.3 FcγRIIb-Mediated Sweeping

FcγRs is one of the major IgG receptor classes and Fc engineering to modulate Fc–FcγR interaction has been successfully applied to enhance the potency of antibody therapeutics targeting membrane-bound antigens [51, 52]. In addition, it has long been said that a large- to mid-size polyclonal antigen–antibody immune complex is internalized and cleared by hepatic FcγR via multivalent binding of the Fc to FcγR, while monomeric antigen–antibody complexes containing a single Fc (1:1 or 1:2 complex formed by mAb and monomeric antigen) is not internalized by FcγR because the monovalent interaction between one Fc and FcγR is weak [53–57]. Therefore, we expect that by enhancing Fc binding affinity to FcγR at neutral pH, the intracellular uptake rate of monomeric antigen–antigen complexes can be accelerated, which would lead to enhanced antigen sweeping of recycling antibodies.

There are four types of FcγRs in mice and, among them, the highly homologous mice FcγRII and FcγRIII have been thought to eliminate immune complexes from the plasma [54, 55]. Fc engineering to enhance binding affinity to both mouse FcγRII and FcγRIII (Table 3.2) was applied to recycling antibodies to examine whether these receptors could increase the intracellular uptake of antigen–antibody complexes in a C57/BL normal mouse steady-state model. In addition, in order to separate the contribution of each receptor, we evaluated the FcγRII- and FcγRIII-enhanced sweeping antibody (PH-mIgG1-Fx) in mFcγRII$^{-/-}$ mice and mFcγRIII$^{-/-}$ mice. We observed that the sweeping antibody with enhanced affinity to both mFcγRII and mFcγRIII could accelerate antigen clearance more than the recycling antibody with silenced mFcγR binding (PH-mIgG1-mFcγR(−)) in both normal mice and mFcγRIII$^{-/-}$ mice, but this accelerated clearance was drastically diminished in mFcγRII$^{-/-}$ mice (Figure 3.10). This demonstrates that mouse FcγRII has a major role in the internalization of antigen–antibody complexes containing one antibody in mice and it can be used as an alternative sweeping receptor instead of FcRn to enhance the antigen sweeping effect of recycling antibodies. As in the case of the FcRn sweeping antibody, we can also customize the antigen sweeping profile of the FcγRII sweeping antibody by modulating the Fc binding affinity to FcγRII. In addition, the profile of the FcγRII sweeping antibody (PH-mIgG1-Fx) in mFcγRIII$^{-/-}$ mice is almost the same as that

TABLE 3.2 Binding Affinity to FcγR

	K_D (nM) at pH 7.4					
Fc Variant	Mouse FcgRI	Mouse FcγRII	Mouse FcγRIII	Mouse FcγRIV	Human FcγRIIb	Mutations
mIgG1	N.D.	110	210	N.D.	NT	
mIgG1-FcγR(−)	N.D.	N.D.	N.D.	N.D.	NT	P235K/S239K
mIgG1-Fx	N.D.	1.2	3.6	N.D.	NT	S239D/A327D

Modified from Igawa et al. [13].
N.D. not determined and NT, not tested.

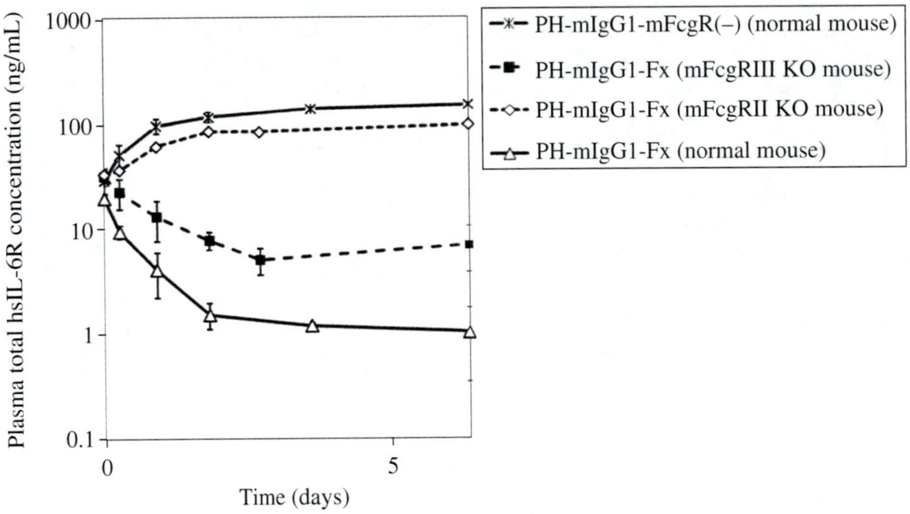

FIGURE 3.10 Contribution of FcγRII and FcγRIII in FcγR-mediated sweeping.

of the FcRn sweeping antibody (PH-v4) in hFcRn-transgenic mice; thus, the FcγRII sweeping antibody could reduce plasma antigen concentration by as much as the FcRn sweeping antibody with comparable antibody half-life. This means that, as with the FcRn sweeping antibody, the majority of the FcγRII sweeping antibody internalized by FcγRII could be returned back to the plasma. Although further investigation is needed on how antibodies taken up into the cell in an FcγRII-dependent manner could be recycled back to the cell surface, it has been shown by *in vitro* experiments that hFcγRIIb has recycling capabilities and that an immune complex internalized by hFcγRIIb is constitutively recycled back to the cell surface after internalization [58, 59]. Moreover, although some studies have shown that FcγR does not contribute to the elimination of wild-type antibody itself [60], our studies using the recycling antibody PH-IgG1 have revealed that FcγR contributes to the cellular uptake of wild-type IgG1. This is consistent with the fact that wild-type mIgG1can bind to mFcγRII and mFcγRIII at neutral pH (Table 3.2). Thus, it can be said that the FcγRII sweeping antibody accelerates antigen clearance from the circulation by enhancing this natural IgG1 uptake pathway through increased Fc binding affinity to FcγRII.

In addition, we generated a transgenic mouse that expresses human FcγRIIb, which is the human homolog of mouse FcγRII, and confirmed that a sweeping antibody with selectively enhanced hFcγRIIb affinity could also accelerate antigen elimination in hFcγRIIb-transgenic mice. Thus, we believe that by increasing the hFcγRIIb binding activity of recycling antibodies, we can sweep soluble antigens from the circulation in humans.

At present, we have established two effective sweeping antibody technologies using either FcRn or FcγRIIb as a sweeping receptor. We can choose the appropriate strategy depending on the situation and the target antigen profile (function, expression pattern, etc.), which makes sweeping antibodies applicable to a wide range of antigens.

3.6 FUTURE PERSPECTIVES

There is severe competition in the development of therapeutic antibodies. There are over 400 antibodies in the market or in clinical study and 177 antibodies are competing against 34 major antigens (Surveyed in July 2012). This means that by using only conventional technologies, companies cannot achieve differentiation and can only access a limited number of target antigens. Therefore, innovative antibody engineering technologies such as those introduced in this chapter will play more and more important roles in antibody drug discovery. With these technologies, a best-in-class strategy can be taken for clinically validated targets; for example, developing SC injectable formulations or drugs with a longer effective duration will be possible by applying these technologies. Moreover, expansion of target space is also feasible with these innovative technologies. For example, sweeping antibodies can antagonize high concentration antigens that cannot be handled by realistic doses of conventional antibodies, allowing a first-in-class strategy to be taken. Thus, various antibody engineering technologies will be applied to more antibody therapeutics than ever before and may be used in a clinical setting in the near future.

REFERENCES

[1] Ling J, Zhou H, Jiao Q, Davis HM. Interspecies scaling of therapeutic monoclonal antibodies: initial look. J Clin Pharmacol 2009;49(12):1382–1402.

[2] Tabrizi MA, Tseng CM, Roskos LK. Elimination mechanisms of therapeutic monoclonal antibodies. Drug Discov Today 2006;11(1–2):81–88.

[3] Junghans RP, Anderson CL. The protection receptor for IgG catabolism is the beta2-microglobulin-containing neonatal intestinal transport receptor. Proc Natl Acad Sci USA 1996; 93(11):5512–5516.

[4] Hinton PR, Johlfs MG, Xiong JM, Hanestad K, Ong KC, Bullock C, Keller S, Tang MT, Tso JY, Vásquez M, Tsurushita N. Engineered human IgG antibodies with longer serum half-lives in primates. J Biol Chem 2004;279(8):6213–6216.

[5] Oganesyan V, Damschroder MM, Woods RM, Cook KE, Wu H, Dall'acqua WF. Structural characterization of a human Fc fragment engineered for extended serum half-life. Mol Immunol 2009;46(8–9):1750–1755.

[6] Deng R, Loyet KM, Lien S, Iyer S, DeForge LE, Theil FP, Lowman HB, Fielder PJ, Prabhu S. Pharmacokinetics of humanized monoclonal anti-tumor necrosis factor-{alpha} antibody and its neonatal Fc receptor variants in mice and cynomolgus monkeys. Drug Metab Dispos 2010;38(4):600–605.

[7] Yeung YA, Leabman MK, Marvin JS, Qiu J, Adams CW, Lien S, Starovasnik MA, Lowman HB. Engineering human IgG1 affinity to human neonatal Fc receptor: impact of affinity improvement on pharmacokinetics in primates. J Immunol 2009;182(12):7663–7671.

[8] Robbie GJ, Criste R, Dall'acqua WF, Jensen K, Patel NK, Losonsky GA, Griffin MP. A novel investigational Fc-modified humanized monoclonal antibody, motavizumab-YTE, has an extended half-life in healthy adults. Antimicrob Agents Chemother 2013;57(12):6147–6153.

[9] Igawa T, Tsunoda H, Tachibana T, Maeda A, Mimoto F, Moriyama C, Nanami M, Sekimori Y, Nabuchi Y, Aso Y, Hattori K. Reduced elimination of IgG antibodies by engineering the variable region. Protein Eng Des Sel 2010;23(5): 385–392.

[10] Hotzel I, Theil FP, Bernstein LJ, Prabhu S, Deng R, Quintana L, Lutman J, Sibia R, Chan P, Bumbaca D, Fielder P, Carter PJ, Kelley RF. A strategy for risk mitigation of antibodies with fast clearance. MAbs 2012;4(6):753–760.

[11] Holash J, Davis S, Papadopoulos N, Croll SD, Ho L, Russell M, Boland P, Leidich R, Hylton D, Burova E, Ioffe E, Huang T, Radziejewski C, Bailey K, Fandl JP, Daly T, Wiegand SJ, Yancopoulos GD, Rudge JS. VEGF-Trap: a VEGF blocker with potent antitumor effects. Proc Natl Acad Sci U S A 2002;99(17):11393–11398.

[12] Igawa T, Ishii S, Tachibana T, Maeda A, Higuchi Y, Shimaoka S, Moriyama C, Watanabe T, Takubo R, Doi Y, Wakabayashi T, Hayasaka A, Kadono S, Miyazaki T, Haraya K, Sekimori Y, Kojima T, Nabuchi Y, Aso Y, Kawabe Y, Hattori K. Antibody recycling by engineered pH-dependent antigen binding improves the duration of antigen neutralization. Nat Biotechnol 2010;28(11):1203–1207.

[13] Igawa T, Maeda A, Haraya K, Tachibana T, Iwayanagi Y, Mimoto F, Higuchi Y, Ishii S, Tamba S, Hironiwa N, Nagano K, Wakabayashi T, Tsunoda H, Hattori K. Engineered monoclonal antibody with novel antigen-sweeping activity *in vivo*. PLoS One 2013;8(5):e63236.

[14] Bickel U, Lee VMY, Pardridge WM. Pharmacokinetic differences between ^{111}In- and ^{125}I-Labeled cationized monoclonal antibody against β-Amyloid in mouse and dog. Drug Deliv 1995;2(2):128–135.

[15] Dellian M, Yuan F, Trubetskoy VS, Torchilin VP, Jain RK. Vascular permeability in a human tumour xenograft: molecular charge dependence. Br J Cancer 2000;82(9):1513–1518.

[16] Yamasaki Y, Sumimoto K, Nishikawa M, Yamashita F, Yamaoka K, Hashida M, Takakura Y. Pharmacokinetic analysis of *in vivo* disposition of succinylated proteins targeted to liver nonparenchymal cells via scavenger receptors: importance of molecular size and negative charge density for *in vivo* recognition by receptors. J Pharmacol Exp Ther 2002;301(2): 467–477.

[17] Wiig H, Tenstad O. Interstitial exclusion of positively and negatively charged IgG in rat skin and muscle. Am J Physiol Heart Circ Physiol 2001;280 (4):H1505–H1512.

[18] ten Kate CI, Fischman AJ, Rubin RH, Fucello AJ, Riexinger D, Wilkinson RA, Du L, Khaw BA, Strauss HW. Effect of isoelectric point on biodistribution and inflammation: imaging with indium-111-labelled IgG. Eur J Nucl Med 1990;17(6–8):305–309.

[19] Shin IS, Lee SM, Kim HS, Yao Z, Regino C, Sato N, Cheng KT, Hassan R, Campo MF, Albone EF, Choyke PL, Pastan I, Paik CH. Effect of chelator conjugation level and injection dose on tumor and organ uptake of 111In-labeled MORAb-009, an anti-mesothelin antibody. Nucl Med Biol 2011;38(8):1119–1127.

[20] Zheng Y, Tesar DB, Benincosa L, Birnböck H, Boswell CA, Bumbaca D, Cowan KJ, Danilenko DM, Daugherty AL, Fielder PJ, Grimm HP, Joshi A, Justies N, Kolaitis G, Lewin-Koh N, Li J, McVay S, O'Mahony J, Otteneder M, Pantze M, Putnam WS, Qiu ZJ, Ruppel J, Singer T, Stauch O, Theil FP, Visich J, Yang J, Ying Y, Khawli LA, Richter WF. Minipig as a potential translatable model for monoclonal antibody pharmacokinetics after intravenous and subcutaneous administration. MAbs 2012;4(2):243–255.

[21] Richter WF, Bhansali SG, Morris ME. Mechanistic determinants of biotherapeutics absorption following SC administration. AAPS J 2012;14(3):559–570.

[22] Igawa T, Nishida Y. Method for improving physical properties of antibody. US patent 2014/0080153 A1. 2014.

[23] Datta-Mannan A, Chow CK, Dickinson C, Driver D, Lu J, Witcher DR, Wroblewski VJ. FcRn affinity-pharmacokinetic relationship of five human IgG4 antibodies engineered for improved *in vitro* FcRn binding properties in cynomolgus monkeys. Drug Metab Dispos 2012;40(8):1545–1555.

[24] Bumbaca D, Wong A, Drake E, Reyes AE II, Lin BC, Stephan JP, Desnoyers L, Shen BQ, Dennis MS. Highly specific off-target binding identified and eliminated during the humanization of an antibody against FGF receptor 4. MAbs 2011;3(4):376–386.

[25] Xin Y, Bai S, Damico-Beyer LA, Jin D, Liang WC, Wu Y, Theil FP, Joshi A, Lu Y, Lowe J, Maia M, Brachmann RK, Xiang H. Anti-neuropilin-1 (MNRP1685A): unexpected pharmacokinetic differences across species, from preclinical models to humans. Pharm Res 2012;29(9):2512–2521.

[26] Walrafen P, Verdier F, Kadri Z, Chrétien S, Lacombe C, Mayeux P. Both proteasomes and lysosomes degrade the activated erythropoietin receptor. Blood 2005;105(2):600–608.

[27] Kato M, Kamiyama H, Okazaki A, Kumaki K, Kato Y, Sugiyama Y. Mechanism for the nonlinear pharmacokinetics of erythropoietin in rats. J Pharmacol Exp Ther 1997;283(2): 520–527.

[28] Kuwabara T, Uchimura T, Kobayashi H, Kobayashi S, Sugiyama Y. Receptor-mediated clearance of G-CSF derivative nartograstim in bone marrow of rats. Am J Physiol 1995;269(1, Pt 1):E1–E9.

[29] Jin F, Krzyzanski W. Pharmacokinetic model of target-mediated disposition of thrombopoietin. AAPS PharmSci 2004;6(1):E9.

[30] Mager DE, Neuteboom B, Efthymiopoulos C, Munafo A, Jusko WJ. Receptor-mediated pharmacokinetics and pharmacodynamics of interferon-beta1a in monkeys. J Pharmacol Exp Ther 2003;306(1):262–270.

[31] Frey N, Grange S, Woodworth T. Population pharmacokinetic analysis of tocilizumab in patients with rheumatoid arthritis. J Clin Pharmacol 2010;50(7):754–766.

[32] Weiner LM, Belldegrun AS, Crawford J, Tolcher AW, Lockbaum P, Arends RH, Navale L, Amado RG, Schwab G, Figlin RA. Dose and schedule study of panitumumab monotherapy in patients with advanced solid malignancies. Clin Cancer Res 2008;14(2):502–508.

[33] Martin DA, Churchill M, Flores-Suarez L, Cardiel MH, Wallace D, Martin R, Phillips K, Kaine JL, Dong H, Salinger D, Stevens E, Russell CB, Chung JB. A phase Ib multiple ascending dose study evaluating safety, pharmacokinetics, and early clinical response of brodalumab, a human anti-IL-17R antibody, in methotrexate-resistant rheumatoid arthritis. Arthritis Res Ther 2013;15(5):R164.

[34] Weekes CD, Beeram M, Tolcher AW, Papadopoulos KP, Gore L, Hegde P, Xin Y, Yu R, Shih LM, Xiang H, Brachmann RK, Patnaik A. A phase I study of the human monoclonal anti-NRP1 antibody MNRP1685A in patients with advanced solid tumors. Invest New Drugs 2014;32(4):653–660.

[35] Goldberg A, Geppert T, Schiopu E, Frech T, Hsu V, Simms RW, Peng SL, Yao Y, Elgeioushi N, Chang L, Wang B, Yoo S. Dose-escalation of human anti-interferon-alpha receptor monoclonal antibody MEDI-546 in subjects with systemic sclerosis: a phase 1, multicenter, open label study. Arthritis Res Ther 2014;16(1):R57.

[36] Igawa T, Mimoto F, Hattori K. pH-dependent antigen-binding antibodies as a novel therapeutic modality. Biochim Biophys Acta 2014;1844(11):1943–1950.

[37] Roopenian DC, Akilesh S. FcRn: the neonatal Fc receptor comes of age. Nat Rev Immunol 2007;7(9):715–725.

[38] Geisow MJ, Evans WH. pH in the endosome. Measurements during pinocytosis and receptor-mediated endocytosis. Exp Cell Res 1984;150(1):36–46.

[39] Gerasimenko JV, Tepikin AV, Petersen OH, Gerasimenko OV. Calcium uptake via endocytosis with rapid release from acidifying endosomes. Curr Biol 1998;8(24):1335–1338.

[40] Feng Y, Press B, Wandinger-Ness A. Rab 7: an important regulator of late endocytic membrane traffic. J Cell Biol 1995;131(6, Pt 1):1435–1452.

[41] Zhang H, Wang C, Chen B, Wang X. Daunorubicin TiO2 nanocomposites as a "smart" pH-responsive drug delivery system. Int J Nanomedicine 2012;7:235–242.

[42] Nishimura Y, Takeda K, Ezawa R, Ishii J, Ogino C, Kondo A. A display of pH-sensitive fusogenic GALA peptide facilitates endosomal escape from a Bio-nanocapsule via an endocytic uptake pathway. J Nanobiotechnol 2014;12:11.

[43] Sarkar CA, Lowenhaupt K, Horan T, Boone TC, Tidor B, Lauffenburger DA. Rational cytokine design for increased lifetime and enhanced potency using pH-activated "histidine switching". Nat Biotechnol 2002;20(9):908–913.

[44] Chaparro-Riggers J, Liang H, DeVay RM, Bai L, Sutton JE, Chen W, Geng T, Lindquist K, Casas MG, Boustany LM, Brown CL, Chabot J, Gomes B, Garzone P, Rossi A, Strop P, Shelton D, Pons J, Rajpal A. Increasing serum half-life and extending cholesterol lowering *in vivo* by engineering antibody with pH-sensitive binding to PCSK9. J Biol Chem 2012;287(14):11090–11097.

[45] Traxlmayr MW, Lobner E, Hasenhindl C, Stadlmayr G, Oostenbrink C, Rüker F, Obinger C. Construction of pH-sensitive Her2-binding IgG1-Fc by directed evolution. Biotechnol J 2014;9(8):1013–1022.

[46] Cosson VF, Ng VW, Lehle M, Lum BL. Population pharmacokinetics and exposure-response analyses of trastuzumab in patients with advanced gastric or gastroesophageal junction cancer. Cancer Chemother Pharmacol 2014;73(4): 737–747.

[47] Agus DB, Gordon MS, Taylor C, Natale RB, Karlan B, Mendelson DS, Press MF, Allison DE, Sliwkowski MX, Lieberman G, Kelsey SM, Fyfe G. Phase I clinical study of pertuzumab, a novel HER dimerization inhibitor, in patients with advanced cancer. J Clin Oncol 2005;23(11): 2534–2543.

[48] Krop IE, Beeram M, Modi S, Jones SF, Holden SN, Yu W, Girish S, Tibbitts J, Yi JH, Sliwkowski MX, Jacobson F, Lutzker SG, Burris HA. Phase I study of trastuzumab-DM1, an HER2 antibody-drug conjugate, given every 3 weeks to patients with HER2-positive metastatic breast cancer. J Clin Oncol 2010;28(16):2698–2704.

[49] Waldmann TA, Strober W. Metabolism of immunoglobulins. Prog Allergy 1969;13:1–110.

[50] Hinton PR, Xiong JM, Johlfs MG, Tang MT, Keller S, Tsurushita N. An engineered human IgG1 antibody with longer serum half-life. J Immunol 2006;176(1):346–356.

[51] Clynes R, Takechi Y, Moroi Y, Houghton A, Ravetch JV. Fc receptors are required in passive and active immunity to melanoma. Proc Natl Acad Sci U S A 1998;95(2):652–656.

[52] Clynes RA, Towers TL, Presta LG, Ravetch JV. Inhibitory Fc receptors modulate *in vivo* cytotoxicity against tumor targets. Nat Med 2000;6(4):443–446.

[53] Arend WP, Sturge JC. Composition and biologic properties of soluble IgG-anti-IgG immune complexes: effects of variations in the specificity of rabbit antibodies to different structural components of human IgG. J Immunol 1979;123(1): 447–454.

[54] Benacerraf B, Sebestyen M, Cooper NS. The clearance of antigen antibody complexes from the blood by the reticuloendothelial system. J Immunol 1959;82(2):131–137.

[55] Ganesan LP, Kim J, Wu Y, Mohanty S, Phillips GS, Birmingham DJ, Robinson JM, Anderson CL. FcγRIIb on liver sinusoidal endothelium clears small immune complexes. J Immunol 2012;189(10):4981–4988.

[56] Kurlander RJ, Ellison DM, Hall J. The blockade of Fc receptor-mediated clearance of immune complexes *in vivo* by

a monoclonal antibody (2.4G2) directed against Fc receptors on murine leukocytes. J Immunol 1984;133(2):855–862.

[57] Lukehart SA, Tam MR, Hom J, Baker-Zander SA, Holmes KK, Nowinski RC. Characterization of monoclonal antibodies to Treponema pallidum. J Immunol 1985;134(1):585–592.

[58] Bergtold A, Desai DD, Gavhane A, Clynes R. Cell surface recycling of internalized antigen permits dendritic cell priming of B cells. Immunity 2005;23(5):503–514.

[59] Zhang CY, Booth JW. Divergent intracellular sorting of Fc{gamma}RIIA and Fc{gamma}RIIB2. J Biol Chem 2010; 285(44):34250–34258.

[60] Leabman MK, Meng YG, Kelley RF, DeForge LE, Cowan KJ, Iyer S. Effects of altered FcγR binding on antibody pharmacokinetics in cynomolgus monkeys. MAbs 2013;5(6):896–903.

4

ADME FOR THERAPEUTIC BIOLOGICS: ANTIBODY-DERIVED PROTEINS AND PROTEINS WITH NOVEL SCAFFOLDS*

CHETAN RATHI AND BERND MEIBOHM

University of Tennessee Health Science Center, Memphis, TN, USA

4.1 INTRODUCTION

Therapeutic monoclonal antibodies (mAbs) are the largest and most rapidly growing class of therapeutic proteins [1]. More than 40 mAbs have already received approval for various medical conditions in the United States. In 2012, mAbs were the top selling class of biologics with sales of $24.6 billion in the United States, which was an increase of 18.3% over 2011 compared to only 2.5% rise in overall pharmaceutical sales [2]. The success of mAb-based therapeutics can be attributed to the numerous advantages that this class of biologics carries including the ability to create therapeutics with high affinity and selectivity for a wide range of targets [3], the potential to modulate the native properties of mAbs for customized therapeutic applications [4], and the usually high tolerability and limited off-target toxicity. These factors have also contributed to lower attrition rates for mAbs during the drug development process as 17% of the humanized/human mAbs that entered clinical trials achieved market authorization [3, 5].

Despite their many favorable properties, however, there are also multiple challenges associated with the development of mAbs as therapeutics. Their large size (~150 kDa) hinders tissue penetration and delivery to intracellular targets, which, for example, limits the efficacy in many solid tumor indications [6]. Immunogenicity has led to the formation of anti-drug antibodies (ADA) for numerous mAbs, which was subsequently responsible for a reduced systemic exposure, loss of efficacy, and/or adverse events [7].

Recent advances in protein engineering have opened up new avenues of modulating the desired properties of therapeutic proteins. This includes not only the optimization of existing protein therapeutics by approaches such as glycoengineering [1], but also the creation of new scaffolds such as antibody–drug conjugates (ADCs), antibody fragments, antibody-based fusion proteins, nanobodies, and bispecifics [8].

ADCs and bipecifics are among the most promising classes of antibody-based therapeutics with a large number of candidates in various stages of preclinical and clinical development. Pharmacokinetic (PK) characterization is one of the most important determinants toward rational development of these novel scaffolds. In this review, we discuss the ADME properties of ADCs and bispecifics and its role in guiding the development of next-generation therapeutic biologics.

4.2 ANTIBODY–DRUG CONJUGATES

ADCs are mAbs that are chemically bonded to cytotoxic drugs (Fig. 4.1). Cytotoxic drugs are highly potent but they can cause substantial toxicity due to their relative nonselectivity. mAbs, on the other hand, have high affinity and specificity for their target but have limited potency based on their conventional mechanisms of action that include ADCC (antibody-dependent

* Part of this chapter has previously been published in *Journal of Clinical Pharmacology* 2015;55 Suppl 3:S21–8.

ADME and Translational Pharmacokinetics/Pharmacodynamics of Therapeutic Proteins: Applications in Drug Discovery and Development, First Edition. Edited by Honghui Zhou and Frank-Peter Theil.

cell-mediated cytotoxicity), CDC (complement-dependent cytotoxicity), or disruption of signal transduction [8]. The concept behind the design of ADCs is to combine the strengths of both cytotoxic drug and mAb with the help of a linker that is controlling the release of drug upon ADC internalization in target cells. This helps in widening the narrow therapeutic index of the cytotoxic molecule and may also overcome drug-resistance mechanism(s) often limiting its therapeutic potential (Fig. 4.2) [9, 10]. For example, clinical development of maytansine was halted in the 1980s because of the lack of therapeutic window; however, it has been successfully used as a conjugated drug in T-DM1, which received approval in 2013 for the treatment of human epidermal growth factor receptor 2 (HER2+) metastatic breast cancer [11].

Gemtuzumab ozogamicin (Mylotarg®) was the first approved ADC for the treatment of acute myeloid leukemia (AML) in the clinic [12]. It was withdrawn from marketing in 2010, however, for its failure to improve response rates in clinical trials. One of the challenges associated with ADC is that the total dose that can be practically delivered to the patients was restricted by pricing constraints [13]. In addition, conjugating a higher number of drug molecules per antibody molecule was not fruitful in increasing potency [14]. These challenges were addressed during the design of the second generation of ADCs that incorporated more potent cytotoxic drugs and exhibited improved linker stability [15]. The second-generation ADCs that have so far received approval by FDA are Adcetris® (brentuximab vedotin) in 2011 for the treatment of Hodgkin's lymphoma and anaplastic large-cell lymphoma, and Kadcyla® (ado-trastuzumab emtansine, T-DM1) approved in 2013 (Table 4.1).

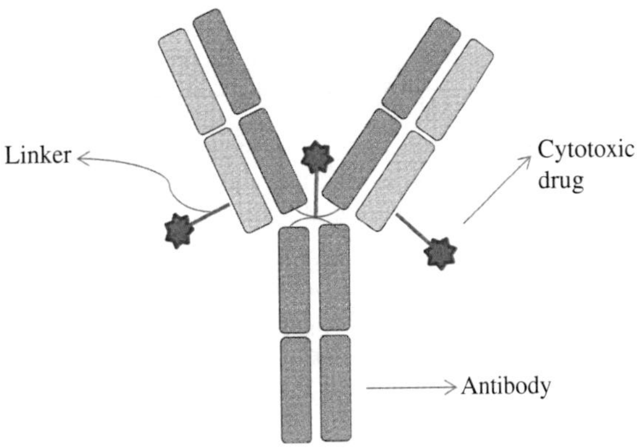

FIGURE 4.1 Schematic of ADC.

4.2.1 Components of ADCs

An ADC comprises three components: (i) antibody, (ii) linker, and (iii) cytotoxic drug.

4.2.1.1 Antibody The role of the antibody in the context of an ADC is mainly for targeted delivery of cytotoxic molecules to tumor cells enriched with cell surface antigens that are recognized by the antibody. An ideal antibody for an ADC is one which has high specificity and affinity, and is human or humanized to reduce the chances of immunogenicity and mimic the long half-life of endogenous immunoglobulin G (IgG). Unlike the therapeutic mAbs that are designed to suppress specific tumor pathway, the antibody component of ADCs may or may not have biological activity. Their primary role is to provide high specificity for antigens that are expressed either only on tumor cells or at significantly higher levels in tumors [16, 17]. The overall PK characteristics of an ADC are driven by its antibody component and are expected to be similar to those of IgG molecules. Deviation in their PK behavior may be associated with conjugation of the cytotoxic drug, which is discussed in more detail in the following sections.

4.2.1.2 Linker The linker plays an important role in the stability of ADCs. An ideal linker is one that is stable in systemic circulation to minimize release and toxicity of the cytotoxic drug only upon internalization into the target cell after targeted-mediated cellular uptake or nonspecific pinocytosis. The linkers can be broadly classified into two main categories: (i) cleavable linkers and (ii) noncleavable. Cleavable

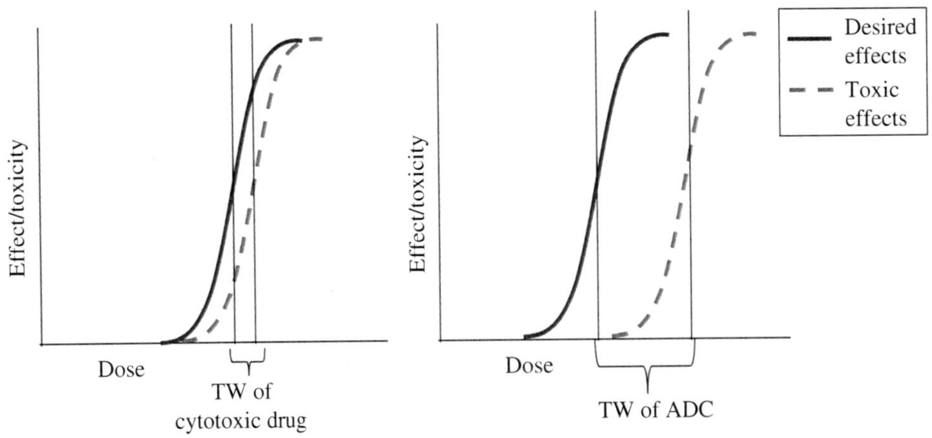

FIGURE 4.2 Comparison of therapeutic window (TW) between cytotoxic drug and ADC.

TABLE 4.1 Overview of FDA Approved ADC

Component	Kadcyla® (Trastuzumab Emtansine)	Adcetris® (Brentuximab Vedotin)
Antibody	Trastuzumab (anti-HER2)	Brentuximab (anti-CD30)
Cytotoxic drug	Mertansine (DM1)	Monomethyl auristatin E (MMAE)
Linker	MCC (thioether-noncleavable)	MC-vc-PAB (protease cleavable)

linkers release the cytotoxic drug using various mechanisms within the cells that includes low pH in lysosomes (acid labile linkers), glutathione levels (disulfide linkers), and lysosomal proteases (protease-cleavable linkers). On the other hand, noncleavable linkers (thioether linkers) require degradation of the mAb backbone to release the drug [18].

4.2.1.3 Cytotoxic Drug Cytotoxic drugs used in ADCs are highly potent antitumor agents effective even in picomolar concentrations. For example, DM1 derivatives are 25–4000-fold more potent compared to the currently used conventional chemotherapeutic agents [19–22]. Cytotoxic agents can be classified into two main categories based on their mechanisms of action that include microtubule inhibitors (auristatins and maytansioids) and DNA-damaging agents (calicheamicin, duocarmycins, anthracyclines, and pyrrolobenzodiazepine dimers) [18]. In addition to their inherent biological activity, physicochemical characteristics such as permeability along with their metabolism and transporter-substrate attributes govern bystander effects and resistance development. During the synthesis of ADCs, small molecule cytotoxic drugs are conjugated to the mAb that produces a heterogeneous mixture of ADC species differing mainly in two aspects: (i) different numbers of drug molecules linked to the mAb (drug–antibody ratio: DAR) and (ii) different locations of drug linkage on the mAbs [14, 23]. To characterize the DAR species, novel bioanalytical techniques were developed such as affinity capture capillary liquid chromatography–mass spectrometry (LC–MS) and affinity capture hydrophobic interaction chromatography [24, 25]. In addition to the manufacturing process, heterogeneity is also partially attributed to deconjugation of the small molecule component from the ADC due to enzymatic or chemical processes taking place in the biological system [26].

4.2.2 Types of ADC Analytes and Their PK Interpretation

The PK of ADCs is usually characterized by total antibody (conjugated and unconjugated antibody), conjugated antibody, conjugated drug, unconjugated antibody, and unconjugated drug. Quantitative assessments of these analytes are necessary for fully understanding ADC PK (Fig. 4.3).

The total antibody concentration consists of both the conjugated and unconjugated forms of the ADC, which are usually analyzed using an enzyme-linked immunosorbent assay (ELISA). The PK of total antibody provides insights into the antibody-associated features of ADCs and their *in vivo* stability, which is critical during the optimization stage of ADC development [11].

The concentration of the conjugated antibodies (which includes a heterogeneous mixture of species with a varying DAR) is generally measured using ELISA. Its systemic exposure is associated with efficacy and is relevant for most of the ADC PK analyses. The concentration levels of conjugated antibody decline more rapidly than total antibody concentration because two processes regulate the decline of conjugated antibody, elimination of intact ADC and complete deconjugation of cytotoxic drug to form "naked" antibody, while total antibody concentrations are solely determined by elimination of ADC and unconjugated Ab [11]. The rate of loss of drug from conjugated antibody determines the separation between the PK profiles of conjugated antibody and total antibody. The faster the rate of loss of drug from ADC, the wider will be the separation between the PK profiles of conjugated antibody and total antibody. This comparative analysis can be performed for optimizing the relative linker stability, which in turn is dependent on the type and site of conjugation.

PK interpretation of this analyte is complicated by the conjugated assay formats that may have varying sensitivities for different DAR species [27]. These DAR species are associated with different potencies. This makes the dose–response assessment challenging even for the ADCs with similar PK properties as they may have different DAR distributions resulting in different pharmacological activities [27]. Careful planning of the assay strategy is required for accurate and consistent measurement of conjugated ADCs, which is essential for establishing the concentration–effect relationship.

Measurement of conjugated drug can be substituted for conjugated antibody as a surrogate for ADC PK. It represents the total amount of drug covalently bound to the antibody and is typically measured by affinity capture LC–MS in which the ADC is first extracted out from plasma followed by LC–MS/MS quantification of the conjugated drug [28]. Although the interpretation of conjugated drug is similar to conjugated antibody, the conjugated drug assay is not conclusive regarding the antibody concentration, which was linked to the drug due to the heterogeneity of the molecular species [11, 29]. For example, a conjugated drug concentration of 50 nM ADC with average DAR of 1 would be equivalent to 10 nM ADC with average DAR of 5, but they may differ in their pharmacological activity. The representation of concentrations in molar units is useful for improving the interpretation of PK profiles of ADC species having a wide range of molecular weights. The difference in the molar concentrations of

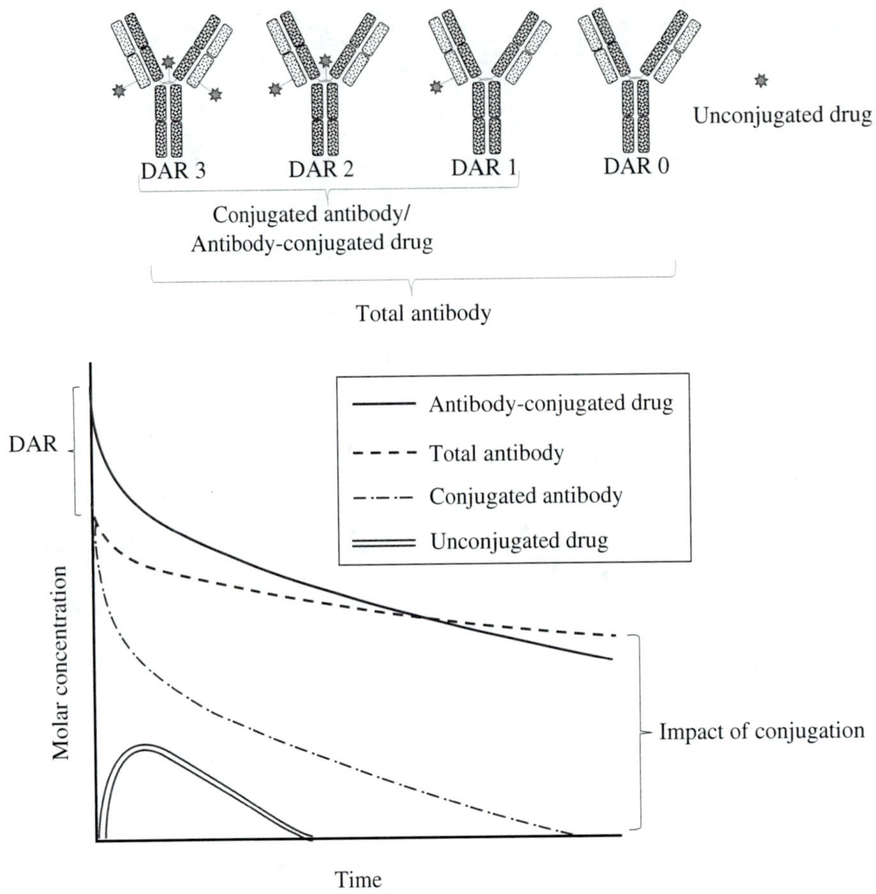

FIGURE 4.3 Typical PK profile of different analytes of ADC.

conjugated drug and total antibody at the time of dosing represents the initial average DAR [11].

Unconjugated cytotoxic drug released from the ADC may be related to adverse effect or reduction in efficacy. Highly sensitive LC–MS or ELISA methods are used to measure these usually very low drug concentrations with accuracy and precision [30, 31]. Additional investigations are also needed for identifying and quantifying any pharmacologically active metabolites or related products of cytotoxic drug in order to provide complete toxicity assessment and determine the potential for drug–drug interactions (DDIs).

Unconjugated antibody formed after deconjugation of drug from the ADC usually has limited or no biological activity, and hence is rarely measured by analytical techniques. When needed, this concentration can be calculated from the difference in concentration between total antibody and conjugated antibody [32, 33].

4.2.3 PK of ADC

Intravenous dosing has been used for administration of all ADCs that are either on the market or in the clinical development stage. Both brentuximab vedotin and ado-trastuzumab emtansine attain C_{max} levels near the end of infusion and thereafter undergo multiexponential decrease in concentration levels [15]. Subcutaneous (SC) administration has also been explored as an alternative dosing strategy for ADCs. In addition, intraperitoneal (IP) administration is preferred in preclinical studies for dosing convenience [34, 35]. The mechanism of absorption through IP or SC is not yet been investigated; however, lymphatic drainage from the absorption site into the circulation seems to be the probable route [36].

The distribution of ADC is quite often similar to that of the unconjugated mAb because it forms the backbone of the ADC. The high molecular mass and the hydrophilicity/polarity of ADC limit its distribution, with concentrations in the interstitial space approximately 10-fold lower than in serum. The permeability of ADC across the cell barriers or tissues takes place by transcellular or paracellular transport, involving the processes of diffusion, convection, and cellular uptake. The distribution of ADCs can also be affected by the expression of the target antigen and the binding affinity or avidity of the ADC to the target [37].

Conjugation is an important determinant of the PK behavior of ADCs. The location of conjugation on the ADC

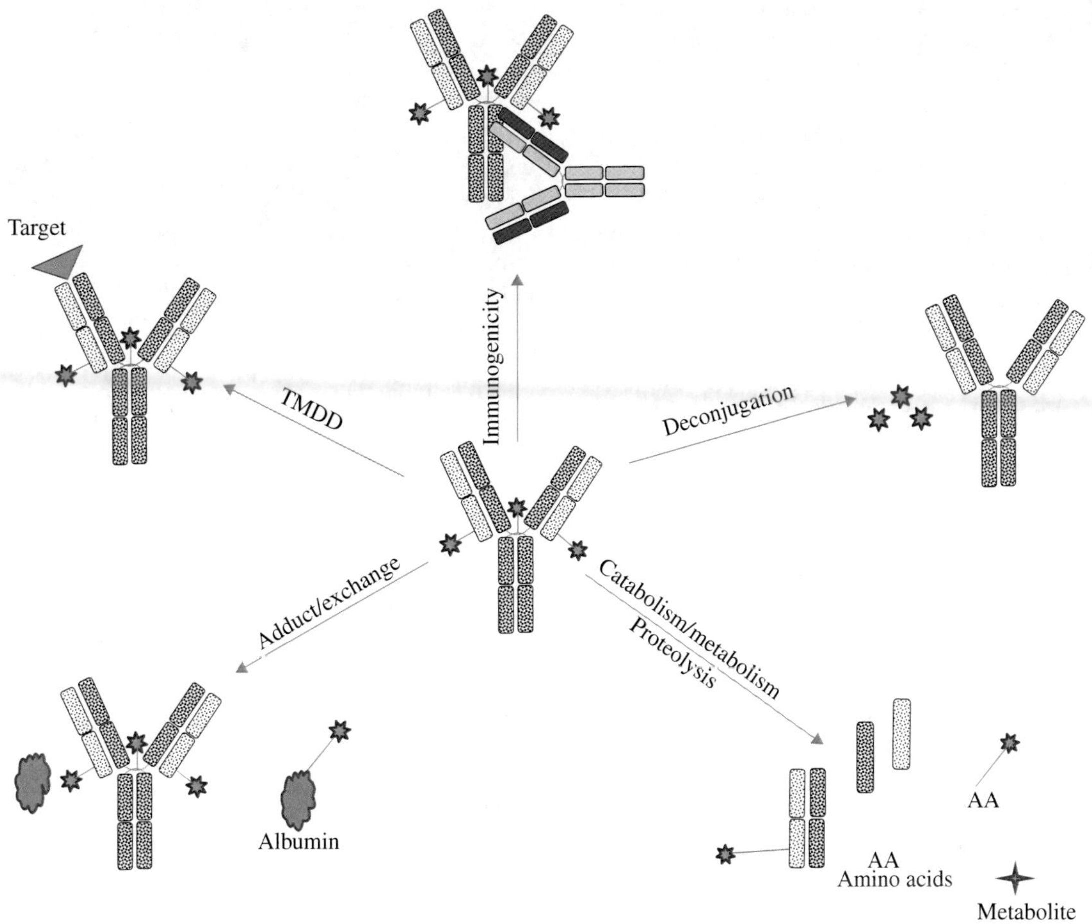

FIGURE 4.4 *In vivo* fate of ADC. (Adapted from King [38].)

can influence its catabolism (Fig. 4.4). For example, the PK of ADCs conjugated at two different sites (C16-HC and C16-LC) was found to be different in rats, although both had similar binding affinity to neonatal Fc receptor (FcRn). C16-LC had a similar PK profile as naked C16 mAb. C16-HC, on the other hand, showed dramatically different PK profiles from the naked C16 mAb in rat with rapid drug release [39]. This may be because the conjugation site interferes with the process of FcRn-mediated recycling [40], Fcγ interaction, receptor-mediated endocytosis [41], water accessibility, and surface charge structure that ultimately modulate the rate of catabolism.

The degree of conjugation also impacts the PK properties and efficacy. ADCs containing a heterogeneous mixture of molecular species having different DARs could potentially have distinct potency and PK [14, 42]. Hamblett et al. [14] synthesized ADC with anti-CD30 antibody conjugated to monomethyl auristatin E (MMAE) using the maleimidocaproyl-valine-citrulline-p-aminobenzoyloxycarbonyl (MC-vc-PAB) linker with three different drug loads: DAR-2, DAR-4, and DAR-8. The ADC with a higher DAR (DAR-8) had faster clearance and was less tolerated in mice compared to ADCs with a lower DAR (DAR-2 and DAR-4). In addition, ADC with DAR-4 was equally efficacious as DAR-8 at the same total antibody doses but half the amount of MMAE in DAR-4. This study also indicates that an improvement in *in vitro* potency with an increase in DAR may not necessarily translate into *in vivo* efficacy [18].

To minimize heterogeneity, new strategies have been tried for producing ADC with a uniform DAR [43–46]. The new site-specific conjugation strategies include the use of engineered cysteine, modified amino acids, and enzymatic conjugation by glucotransferases and transglutaminases. They have generated ADCs with defined DAR, which have improved stability and a better therapeutic index compared to those using conventional conjugation sites [40, 45–47]. However, it is still challenging to control the heterogeneity associated with the *in vivo* process.

The degree of conjugation may also influence the tissue distribution of the antibody, which is evident from the increased hepatic uptake of ADCs compared to naked mAbs [48–50]. For example, MMAE conjugated to anti-STEAP1 had significantly higher liver uptake than the unconjugated anti-STEAP1. It was hypothesized that higher hydrophobicity of the ADC facilitated greater reticuloendothelial system clearance and consequently led to elevated levels of ADC in the liver as compared to the unconjugated mAb [49]. To further explore this hypothesis, a separate study was conducted to compare plasma clearance of three ADCs with the same DAR (DAR-8) but different linkers: (i) MC-vc-MMAF (maleimidocaproyl-valine-citrulline-monomethyl auristatin F) linker (more hydrophobic), (ii) MC-MMAF linker (less hydrophobic), and (iii) AT–Glu-MDpr (a novel auristatin T (AT)-based drug linker to minimize hydrophobicity). The plasma clearance was in the following order: MC-vc-MMAF linker>MC-MMAF linker>AT–Glu-MDpr. In addition, the ADC with the AT–Glu-MDpr linker had comparable hepatic uptake as its parent antibody in a perfused liver system. It was also found that the antibody with reduced interchain disulfide bonds had clearance similar to that of naked antibody. These observations indicate that it is the intrinsic hydrophobicity of the linker rather than the destabilization of the antibody structure by reduced disulfide bonds, which contributes toward increased plasma clearance and hepatic uptake of ADCs [18, 51].

The catabolic profile of ADCs is also influenced by the type of linker. Cleavable linkers release cytotoxic drug and this process is referred to as "deconjugation," while the noncleavable linkers release drug attached to an amino-acid: for example, maytansinoid ADCs with noncleavable linker MCC-DM1 (succinimidyl-4-(*N*-maleimidomethyl)cyclohexane-1-carboxylate (MCC)) was degraded to a single major catabolite, lysine-MCC-DM1, whereas ADCs with cleavable linkers SPP-DM1 (*N*-succinimidyl-4-(2-pyridyldithio)pentanoate (SPP)) were degraded into multiple catabolites including DM1, lysine-SPP-DM1, *S*-methyl-DM1, *S*-methyl-DM1 sulfoxide, and *S*-methyl-DM1-sulfone [26]. Some of these catabolites could contribute to antitumor activity by "bystander effect," whereby cell membrane permeable ADC catabolites produced in one cell diffuse to neighboring cells and exhibit their effect [26, 52].

Many ADC linkers also utilize maleimide chemistry for binding the cytotoxic component to the cysteine residues on the mAb with the help of thiosuccinimide linkers. These linkers are unstable in plasma and allow maleimide exchange of the linker drug with reactive thiols in albumin, free cysteine, or reduced glutathione [18, 53]. To overcome this limitation, a new class of linkers was developed with a basic amino group incorporated adjacent to the maleimide that prevented maleimide exchange by catalyzing succinimide ring hydrolysis and hence conferring more stability to the ADC [54].

Usually the cytotoxic agent in the ADC acts intracellularly and therefore the rate of target internalization and rate of target replenishment on cell surface are critical determinants for the sustained accumulation of active drug inside the tumor [55]. Target shedding from the cell surface into the circulation may alter the distribution of cytotoxic drug/conjugated antibody by increasing the liver uptake of target–ADC complex, which may lead to hepatotoxicity [56]. A multiple-dose study of ado-trastuzumab emtansine at 3.6 mg/kg every 3 weeks resulted in no accumulation whereas increased concentrations of gemtuzumab ozogamicin were observed after the second dose, which could be explained by a decrease in clearance by CD33-positive blast cells following the first dose [57, 58]. The cytotoxic drug upon release from the ADC has typical elimination characteristics of a small molecule drug such as hepatic metabolism and renal and biliary excretion, mediated by CYP and non-CYP enzymes, and transporters. For example, the elimination of DM-1 in rats after administration of T-DM1 was mainly through the fecal/biliary route with 80% radioactivity recovered in the feces and 50% in the bile [53]. The rate of formation of the cytotoxic drug from the ADC is quite often slower than the clearance of the cytotoxic drug itself, resulting in formation rate-limited PK [59]. This may explain the relatively low systemic exposure (100–1000-fold lower) observed for the cytotoxic drug component as compared to that of the ADC [60, 61]. For example, the mean maximum free concentrations for MMAE and brentuximab vedotin were 5–7 ng/mL and 32.0–45.0 μg/mL, respectively, after every 3-week administration of 1.8–2.7 mg/kg of brentuximab vedotin [61]. A biodistribution study of anti-CD70 antibody h1F6 conjugated to auristatin conducted in mice showed similar exposure for auristatin and mAb in all tissues except liver, where concentrations of auristatin were higher than those of mAb. In addition, the distribution of auristatin in tumor was several folds higher than that in normal tissue [48]. Cytotoxic small molecules may have limited penetration into cells or tissues due to their physicochemical properties. However, on combination with the mAb in the ADC format, small molecules are able to reach those areas that were inaccessible before.

As ADCs have a relatively narrow therapeutic index, it is necessary to evaluate the impact of combination therapy on the PK of ADCs and unconjugated cytotoxic drug and also assess the effect of ADC on the PK of coadministered therapy. The potent cytotoxic drugs are frequently substrates of CYP enzymes and efflux transporters such as P-gp; for example, both DM1 and MMAE get primarily metabolized by CYP3A4 and are also substrates of P-gp [22, 62]. These small molecule drugs could potentially be "victims" of DDI when coadministered with other CYP inhibitors or inducers. They are unlikely, however, to act as a "perpetrator" as clinically observed concentrations for cytotoxic drugs are several orders of magnitude lower than the IC_{50} required for inhibition or induction. Some of these theoretical risk assessments have also been tested by dedicated DDI clinical studies. It has been observed that when T-DM1 is combined with taxanes, the PK properties remain unchanged because taxanes and T-DM1 neither inhibit nor induce CYP at clinically relevant concentrations [22]. Brentuximab vedotin was found to have no influence on the PK of midazolam, a CYP3A4 substrate. In the same study, the exposure of unconjugated MMAE increased by approximately 34% when brentuximab vedotin was coadministered with ketoconazole (a potent CYP3A4 inhibitor) and decreased by approximately 46% when combined with rifampin (a potent CYP3A4 inducer). These situations require close monitoring in patients for MMAE associated adverse events. No dose adjustments, however, have been recommended because of the lack of a defined relationship between MMAE exposure and clinical response or probability of neutropenia, peripheral neuropathy, or thrombocytopenia [22].

4.2.4 Immunogenicity of ADC

ADCs, like other biologics, pose a theoretical risk of eliciting undesired immune responses. The potential factors responsible for immunogenicity of ADCs are (i) structural distortion of the tertiary structure of the mAb, (ii) heterogeneity of the ADC product, and (iii) extrinsic (patient related) factors. In the clinic, however, the incidence rate of immunogenicity has been found to be low, between 0% and 5%. Both, T-DM1 and brentuximab vedotin reported minor incidence rates (<5%) of immunogenicity in patients exposed to a repeated number of dosing cycles. A fair comparison of immunogenicity incidence rate between products, however, is challenging because of the semiquantitative nature of ADA measurement methods.

4.2.5 Exposure–Response of ADCs

Integration of PK with efficacy and toxicity data allows for quantitative assessment of exposure–response (E–R) relationships, which can be used for providing recommendations regarding dosing regimens as well as translation of PK/pharmacodynamic (PK/PD) and dosing strategies from preclinical to clinical species. Characterization of the E–R relationship becomes particularly important for ADCs because of their typically narrow therapeutic index and hence the need for optimizing the dosing regimen. In case of Adcetris®, a well-defined E–R relationship could be established between average steady-state concentration (C_{ss}) trough of ADC and probability of overall response rate (ORR) with ORR increasing with increasing average C_{ss} trough of ADC. In some cases, however, it may be difficult to define an E–R relationship. This may be because of a number of challenges including (i) the presence of multiple active species; (ii) variability in the target expression, affinity, or potency between preclinical and clinical species; and (iii) bioanalytical limitations in identifying all the active species. For example, in case of Mylotarg®, no E–R was obtained between maximum response and exposure of total antibody. It is essential to have an in-depth understanding of the relevant analytes for proper assessment of E–R relationships.

4.2.6 Dose-Dependent PK of ADCs

Dose escalation studies of ADC give us some insights into potential dose-dependent changes in their PK. The dose-clearance pattern was found to be nonlinear for AVE9633, BT062, MLN2704, and ado-trastuzumab emtansine where the clearance decreased with increasing dose indicating target-mediated drug disposition (TMDD). However, the therapeutic window for these ADCs was within the linear range. On the other hand, clearance was found to be constant with increasing dose levels for huC242-DM1, IMGN901, and PSMA-ADC. The unconjugated drug–plasma exposure was proportional to the conjugated antibody exposure for brentuximab vedotin and PSMA-ADC whereas no relationship could be derived for ado-trastuzumab emtansine and SAR3419 [58]. The limited information available from these studies indicates that the relationships observed were irrespective of ADC target location, that is, hematological cancers or solid tumors and the type of linker, that is, "stable" or "cleavable" [58].

4.3 BISPECIFICS

Bispecific antibody constructs (in the following named "bispecifics") are one of the modalities that have received much attention during recent years. Bispecifics are characterized by the ability to simultaneously bind two targets [63]. There are three main dual-targeting strategies that are pursued by bispecifics:

1. **Simultaneous Inhibition of Two Targets.** This approach can include the binding of two different receptors, two different ligands, or a combination thereof. It may be useful in the case of disease processes that involve redundant or synergistic regulation by different receptor systems in a complex network of signaling pathways and where it is desirable to modulate two receptor systems concurrently. Such a bispecific would not only provide an alternative to combination therapy for improved therapeutic efficacy but also potentially make the drug development and regulatory approval less complicated and cheaper compared to single-agent combination therapy [64].
2. **Retargeting.** Bispecifics may also be advantageous when an effector cell type needs to be recruited to a target. In this case, the bispecific binds with one arm to a cell surface receptor specific for a target cell, for example, a tumor cell, while the other arm recruits an immune effector cell by binding to an antigen characteristic for that cell type, with the intent to elicit ADCC [65].
3. **Increased Specificity.** Simultaneously targeting two different receptors on the target cell may not only overcome redundant signaling pathways but also increase specificity for cells that express both antigens.

These general approaches for bispecifics have also been combined with delivering payloads, such as enzymes for prodrug activation, cytokines, radionuclides, and toxins to augment efficacy [63].

4.3.1 Bispecific Antibody Formats

The structural features of bispecifics are the major determinants for their clinical PK and PD behavior. Bispecifics are available in a wide variety of formats with sheer limitless scaffold variations, with a limited selection presented in Figure 4.5. Some of the first attempts to develop bispecifics consisted of full-length antibodies developed by chemical crosslinking or fusion of two hybridoma cell lines resulting in the formation of a hybrid hybridoma or quadroma cell line. A random assembly of the two versions of heavy and light chains, however, made it difficult to consistently manufacture homogeneous batches on a large scale. Catumaxomab, the first bispecific to receive market authorization in Europe in 2009, overcame this challenge by minimizing random association through species-specific heavy/light chain pairing by fusing mouse IgG2a and rat IgG2b cell lines [66]. Chemical crosslinking as an alternative approach for the generation of bispecifics has been used for full-length antibodies as well as F(ab′) fragments, the latter involves covalent conjugation of two different mAbs followed by enzymatic digestion [65].

Although bispecifics from full-length antibodies successfully demonstrated biological efficacy, they were also associated with numerous limitations including difficulties in large-scale production, immunogenicity, and Fc-receptor-mediated side effects such as cytokine release, thrombocytopenia, and leukopenia [66–68]. Some of these weaknesses were overcome by advancements in recombinant DNA technology that provided a plethora of recombinant bispecific constructs ranging from variants of whole IgG-like molecules to formats based on single-chain variable fragments (scFvs).

Dual-variable domain-immunoglobulin (DVD-Ig) is an example for an IgG-like molecule having bispecificity and tetravalency. The V_L and V_H domains of an antibody toward the first antigen are covalently linked via short linkers to the N-terminal end of the respective light and heavy chains of a mAb against the second antigen [69]. DVD-Ig can be engineered from any two mAbs of distinct specificities while retaining the bioactivity of the parent mAbs. The versatility of DVD-Ig is exhibited by their numerous features that include (i) the ability to combine mAbs of different origins (human, humanized, chimeric, or murine) and isotypes (κ and λ), (ii) the ability to target proteins of different sizes, and (iii) the ease of production and purification [70].

Bispecific modalities based on scFv are designs containing minimal binding domains that are vital for antigen recognition and binding. Of particular interest are the tandem scFv (~55–60 kDa) and bispecific diabodies (~60 kDa). Other recombinant formats include single-chain diabodies and tandem diabodies [66, 71].

Bispecific T-cell engager (BiTE) is a type of tandem scFv formed by covalent binding of two scFv molecules with a flexible peptide linker. Blinatumomab is a BiTE that is formed by genetically engineered murine scFv lacking the Fc domain

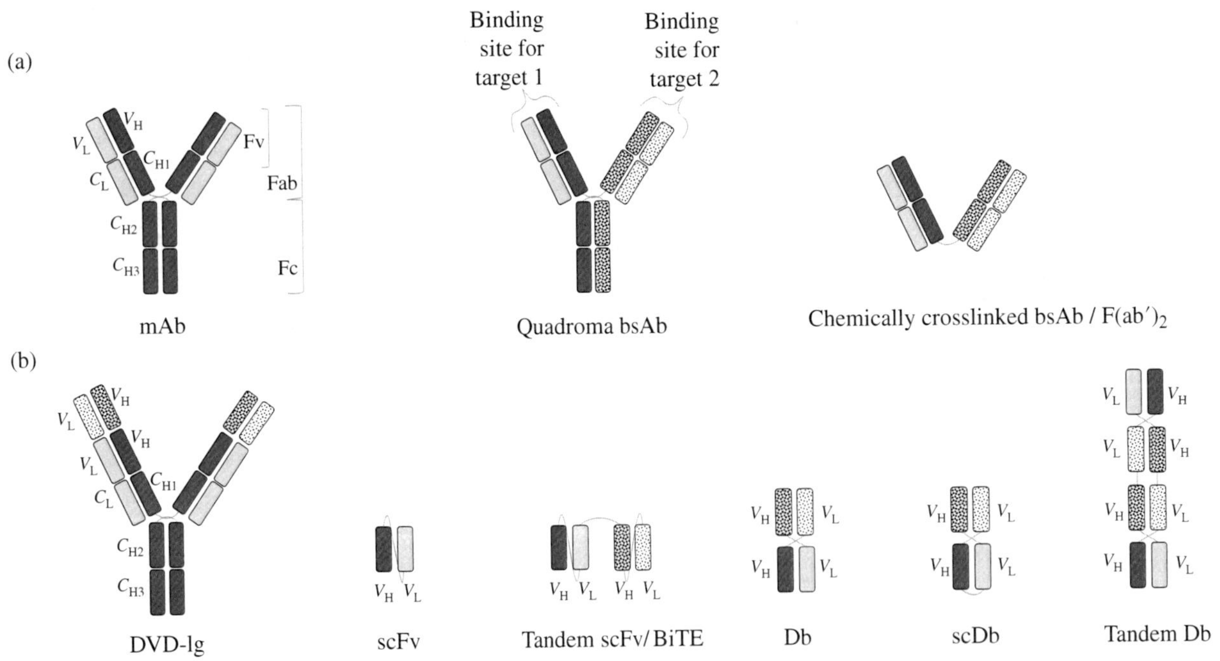

FIGURE 4.5 (a) Monoclonal antibody (mAb) and "first-generation" bispecific antibodies (bsAbs). (b) Recombinant bispecific antibodies: dual variable domain-immunoglobulin (DVD-Ig); single-chain variable fragment (scFv); tandem scFv/bispecific T-cell engager (BiTE); diabody (Db); single-chain Db; tandem Db. C_{Hn}, constant heavy chain; C_L, constant light chain; V_H, variable heavy chain; V_L, variable light chain; Fab, antigen-binding fraction; Fc, crystallizable fraction; F_v, variable fraction; connecting lines represents linker domain.

and contains two distinct variable regions directed toward T-cell-specific CD3 and B-cell-specific CD19 connected with a flexible five-residue peptide linker [72]. Blinatumomab retargets cytotoxic T-lymphocytes against the CD-19-expressing B-cells for the treatment of B-lineage acute lymphoblastic leukemia (ALL).

In contrast to tandem scFv, bispecific diabody constructs are also heterodimeric molecules consisting of the variable domains of two different antibodies, but are formed by noncovalent association. A very short peptide linker prevents intrachain interactions (Fig. 4.5) [71]. Noncovalent interactions between the V_H and V_L chains, however, can lead to instability and consequently aggregation of proteins. This limitation was partly overcome by introducing intrachain disulfide bonds or generating single-chain diabodies. The length of the peptide linker can be further extended to generate tandem diabodies that have improved stability and affinity [63, 73, 74].

4.3.2 PK of Bispecific Constructs

The disposition of bispecifics based on full-length IgG molecules is similar to that of intact monospecific mAbs [75, 76]. Similar to classic mAbs, bispecifics have negligible oral bioavailability due to low permeability through the gut wall and high gastrointestinal protease activity. Thus, parenteral routes of administration, particularly intravenous and SC administration, are the most frequently utilized dosing pathways for bispecifics. Direct delivery to the site of action has also been described for attaining high local concentrations.

The distribution of bispecifics is governed by their molecular weight, physicochemical properties including charge, and other structural features such as binding to Fc receptors. A biexponential PK profile is usually observed after intravenous administration of IgG-like bispecifics, with a volume of distribution equal to or slightly larger than the plasma volume (3–8 L) representing the vascular space and to a lesser degree the interstitial space [72, 77]. The movement of bispecifics from the systemic circulation to the interstitial space is, similar to other large proteins, predominantly facilitated by convective transport rather than diffusion, thereby following the unidirectional fluid flux from the vascular space into the interstitial space. The subsequent removal from the interstitial space back into the vascular space is regulated by lymphatic drainage [78, 79]. For example, PK studies of an investigational DVD-Ig, anti-IL-12/IL-18 1D4.1-325, in rats exhibited a biexponential plasma concentration profile similar to that observed for conventional mAbs. The observed PK profile based on measurements from two different methods that included IL-12-specific and IL-18-specific capture ELISA assays were very similar indicating that the two variable domains were stable *in vivo*. In addition, the DVD-Ig also seemed to have tissue penetration comparable to normal IgG based on their similarities in the volume of distribution [70]. Similarly, the PK of a bispecific IgG1 against epidermal growth factor receptor (EGFR) and CD20 was also found to be comparable to classic IgG1 [80].

As many antibody-based drugs have targets outside the vascular space, tissue penetration becomes a critical factor for therapeutic efficacy, especially in the case of solid tumors. To complicate the situation further, the environment in solid tumors is different from normal tissues and is usually characterized by poorly organized vasculature resulting in sluggish blood flow with unstable rheology. The additional lack of functional lymphatic drainage leads to high interstitial fluid pressure that restricts convective extravasation, the main mechanism for tissue access of large proteins [78, 79]. Consequently, typical mAbs (150 kDa) poorly diffuse into solid tumors [81]. Another important determinant for tumor penetration is the binding affinity to the antigen, with moderate binding affinity being preferred over high affinity for deeper tumor penetration [82]. Compared to full-length mAbs, smaller bispecific formats (55–60 kDa) such as $F(ab')_2$, tandem scFv, and diabodies seem to have more uniform distribution and deeper penetration into solid tumors [83]. In addition to their size, multivalent bispecifics have higher avidity that further promotes prolonged target retention [83]. These characteristics suggest smaller formats of bispecifics to be more advantageous for the treatment of solid tumors than regular mAbs.

Bispecifics that are designed for the retargeting strategy have a moiety that specifically binds to immune cells of the body. After binding to the immune cells in the blood stream, tissue penetration of bispecifics may be restricted due to the size and charge of the resulting complex. Having low affinity to the immune cell antigen and maintaining the bispecific plasma concentration below the dissociation constant for the immune cell binding moiety are therefore considered favorable for efficient extravasation without being trapped by circulating immune cells [84].

Similar to other protein therapeutics, bispecifics are eliminated by catabolism resulting in peptides and amino acids that are reutilized for de novo protein synthesis. This nonspecific proteolytic degradation can be carried out ubiquitously throughout the body, in particular by endothelial cells and cells of the reticuloendothelial system [85, 86]. Intracellular uptake is hereby a prerequisite and is usually facilitated by pinocytosis or similar endocytotic processes. In addition to this nonspecific clearance pathway, bispecifics may also undergo classical TMDD after binding to one of their targets, where the target binding results in intracellular uptake and subsequent lysosomal degradation [87]. Since this process can be saturated at therapeutic concentrations and may constitute a major elimination pathway for some bispecifics, nonlinear PK behavior with higher clearance and shorter half-life at low doses and lower clearance and longer half-life at high doses is often observed. The target-mediated clearance of bispecifics is usually driven by one of the targets. For example, target-mediated clearance of MEHD7945A (anti-HER3/EGFR) in humans was comparable to the anti-EGFR mAb (e.g., panitumumab), suggesting a minor role played by HER3 in target-mediated clearance. This may be attributed to the differences in expression and turnover rate of targets and binding affinity between target and ligand.

For normal mAbs and full IgG bispecifics, interaction with the FcRns provides a well-described mechanism that recycles them after intracellular uptake by preventing them from lysosomal degradation [88]. Similar to endogenous IgG and monospecific mAbs, this salvage pathway also prolongs the elimination of IgG-type bispecifics. Bispecific fragments such as F(ab′)$_2$, however, lack the FcRn-binding site on the Fc domain. As they cannot interact with the FcRn salvage pathway, their half-lives are substantially shorter than for full-size mAbs. For example, the F(ab′)$_2$ molecule H22xKi-4, with a molecular weight of 104 kDa, only exhibited a half-life of 11.1 h in patients with Hodgkin lymphoma [75], compared to the 2–3 weeks for most mAbs [89, 90].

In addition to FcRn binding, the Fc domain of IgG-type bispecifics can also bind to Fcγ receptors on immune cells. This interaction contributes toward the effector function such as ADCC of mAbs, including IgG-type bispecifics, and also influences their PK behavior as additional elimination pathway. Modification in the glycan chain and advanced protein engineering techniques have been utilized to modify the binding affinity of the Fc domain of single-targeting and bispecific mAbs toward the FcRn and Fcγ receptors for modulating their PK and PD behavior [91, 92].

The charge on bispecifics can also have a substantial impact on their disposition. Modifications of the isoelectric point of an antibody by one or more units can result in significant differences in its PK [93]. For example, an investigational IgG-type bispecific antibody for the treatment of hemophilia was found to have unexpectedly higher clearance in mice. This was partially attributed to the large positive charge cluster in the variable region that might have increased the nonspecific binding to the extracellular matrix and subsequently increased its clearance. To overcome this limitation, a single Tyr30Glu mutation was carried out that markedly neutralized the charge cluster and increased the plasma half-life without any compromise in pharmacological activity [94].

While elimination by renal metabolism or even renal excretion is not relevant for intact mAbs, including IgG-type bispecifics, smaller bispecific constructs such as tandem scFv and Db (diabody) with molecular weights less than 60 kDa are below the renal filtration cutoff, and can undergo glomerular filtration and subsequent catabolism by intracellular uptake and lysosomal degradation, resulting in short elimination half-lives [95]. This challenge has been successfully overcome by molecular modifications including pegylation, N-glycosylation, or fusion with human albumin (covalently or using albumin-binding domains) [63, 71], thereby increasing the hydrodynamic volume and preventing glomerular filtration. These strategies are useful when Fc-mediated effector functions are not required or even detrimental. For example, a novel bispecific tandem scFv named MM-111 targeting ErbB2/ErbB3 was expected to have a short half-life of 5 h in mice typically observed for tandem svFc. Incorporation of human serum albumin as a linker between the two scFv extended the half-life of MM-111 to 16–20 h in mice and up to 99 h in cynomolgus monkeys [96]. But these half-life extending modifications may reduce the pharmacological activity because of steric hindrance preventing the effective interaction between the target antigen and the epitope [71, 97]. For example, N-glycosylation of a bispecific diabody increased the exposure by two- to threefold, but reduced the bioactivity three- to fivefold compared to the unmodified molecule [98].

4.3.3 Immunogenicity of Bispecific Constructs

Immunogenicity can also have a significant influence on the PK and PD of any therapeutic protein, including bispecific constructs. Formation of ADA that bind to the therapeutic protein can either have neutralizing or nonneutralization effects with regard to target interaction, but even nonneutralizing ADA may reduce the therapeutic activity by triggering an additional clearance pathway for the therapeutic protein through immune complex formation and subsequent degradation [7]. Compared to the monospecific mAbs and IgG-type bispecifics, however, scFv-based platforms such as tandem scFv and diabodies are generally less immunogenic because of the absence of an Fc domain in the molecule [71, 74]. Bispecific diabodies are even less likely to prompt immune reactions due to their compact size [99]. Thus, small bispecific constructs may have a substantial clinical advantage compared to classic mAbs with regard to immunogenicity.

4.3.4 Examples of Bispecific Therapeutics—Oncology Indications

4.3.4.1 Catumaxomab Catumaxomab is an IgG-type bispecific antibody that is produced by a rat/mouse hybrid quadroma cell line. It is called a "triomab" as it has trifunctionality with an anti-epithelial cell adhesion molecule (EpCAM) domain for EpCAM antigen expressed on tumor cells, an anti-CD3 domain directed toward T-lymphocytes, and an intact Fc region that binds and activates Fcγ-receptor positive cells such as CD64+ accessory cells and CD16+ natural killer cells [100]. With the help of its trifunctionality, catumaxomab redirects cells of the innate and adaptive immune system to EpCAM antigen-expressing tumor cells that ultimately results in cell-mediated tumor cell killing [65]. Catumaxomab received in 2009 marketing authorization by the European Medicine Agency (EMA) for the treatment of malignant ascites in patients with EpCAM-positive carcinoma where standard therapy is not available or no longer feasible. It is administered locally via intraperitoneal infusion to obtain high concentrations at the site of action, that is, the tumors in the peritoneal cavity. Systemic absorption was found to be low (<1%) and high interindividual variation was observed in the local ascites fluid and systemic concentrations. The peak plasma concentrations of catumaxomab ranged from 0 to 2290 pg/mL [101]. Tumor load and effector cells were expected to impact PK, which could be confirmed in an animal model: binding to immune effector cells and tumor cells sequestered catumaxomab in a dose-dependent manner to the peritoneal cavity, resulting in decreasing systemic bioavailability with increasing tumor load and immune effector cell availability [101].

The observed mean elimination half-life was 2.1 days [101]. This substantially shorter half-life compared to human IgG molecules can readily be explained by the substantially reduced binding affinity of rat and murine Fc fragments to the human FcRn compared to human Fc fragments [102], thereby resulting in a substantially less efficient FcRn-mediated recycling of catumaxomab compared to humanized or human mAb. As a rat/murine chimeric antibody, catumaxomab is expected to have intrinsic immunogenicity upon repeated administration. This was indeed observed in a dose escalation study of catumaxomab with intraperitoneal administration, where ADA developed after 10–16 days during multiple-dose treatment [101].

4.3.4.2 Blinatumomab Blinatumomab is a tandem scFv using the BiTE format. It has recently received FDA approval after obtaining breakthrough therapy designation for the treatment of Philadelphia chromosome-negative relapsed/refractory B-lineage ALL in adult patients, specifically for eradication of minimal residual disease (MRD) [103, 104]. As a single agent, it was able to achieve complete and durable molecular remission in approximately 70% of the patients with MRD-ALL and relapsed ALL. Blinatumomab is a genetically engineered murine tandem scFv that lacks the antibody Fc domains and contains two distinct variable regions for binding T-cell-specific CD3 and B-cell-specific CD19. By targeting cytotoxic T-lymphocytes against the CD-19-expressing B-cells, the T-cells become activated within minutes and induce perforin-mediated death to the targeted B-cells. In contrast to CD20, the target of current antibody-based immunotherapy, CD19, is expressed not only on regular B-cells but also on the earliest B-precursor lymphocytes that are malignantly transformed in ALL [103].

Blinatumomab has a volume of distribution of $1.61 \pm 0.74\,L/m^2$ that is similar to that of mAbs. It has a short serum half-life of 1.25 ± 0.63 h and a relatively high clearance of approximately 2 L/h, which are the consequence of a lack of FcRn-mediated recycling due to the lack of the Fc domain and its low molecular weight (55 kDa) that allows for glomerular filtration and subsequent renal metabolism [72]. In bilaterally nephrectomized mice, systemic exposure and half-life were found to be higher compared with the control group. PK data in patients with mild-to-moderate renal impairment, however, did not show a clinically meaningful difference in blinatumomab exposure [105].

As a consequence of the short elimination half-life, blinatumomab is administered as a 4-week continuous intravenous infusion at a dose of $15\,\mu g/m^2/24\,h$, resulting in a steady-state serum concentration of 731 ± 163 pg/mL [105, 106]. Shorter infusion times had also been explored but seemed to result in higher incidences of adverse effects including neurologic symptoms and cytokine-release syndrome [107].

Although blinatumomab was derived from murine sources, only a negligible number of patients (<1%) seem to develop human antimouse antibodies during therapy [103, 106, 108]. Absence of the Fc region along with B-cell depletion is assumed to be the critical component for this low immunogenicity [71].

4.3.5 Examples of Bispecific Therapeutics—CNS Indications

The application of mAbs in central nervous system-related indications such as neurodegenerative diseases has so far been hampered by their usually extremely low brain penetration through the blood–brain barrier, with IgG concentrations in cerebrospinal fluid around 0.1% of serum concentrations [109]. Bispecific constructs have been suggested to facilitate mAb transport across the blood–brain barrier. This has been exemplified by a bispecific antibody that blocks with one arm the activity of β-secretase (BACE1) to reduce the production of amyloid-β in Alzheimer's disease [110]. The second arm of this bispecific is targeted toward the transferrin receptor (TfR) to facilitate receptor-mediated transcytosis. The TfR is highly expressed in brain endothelial cells and has sufficient capacity to transfer therapeutic IgG [111–113]. It was observed that anti-TfR antibodies that bind with high affinity to TfR remain bound within the blood–brain barrier whereas lower-affinity anti-TfR antibody variants entered the brain. Based on this knowledge, a bispecific mAb was designed with one arm comprising a low affinity anti-TfR antibody and the other arm comprising the high affinity anti-BACE1 antibody. This bispecific antibody achieved higher brain penetration and was substantially more effective in reducing amyloid-β in mice compared to the monospecific mAb [114]. These results suggest that bispecific mAbs utilizing TfR-mediated transcytosis may provide a platform technology for increasing brain penetration for antibody-based therapeutics.

4.3.6 Examples of Bispecific Therapeutics—Ocular Indications

Local therapy by intravitreal administration of bispecifics is being considered as a potential treatment option for ocular diseases such as age-related macular degeneration and diabetic macular edema. A tetravalent scFv-Fc-scFv bispecific construct simultaneously targeting platelet-derived growth factor receptor-beta (PDGFRβ) and vascular-endothelial growth factor (VEGF) has shown enhanced activity over the monospecific mAb illustrating the usefulness of blocking both angiogenic pathways with a single agent [115]. This construct exhibited a relative long half-life of 19 days in mice, most likely due to the intact Fc domain allowing for interaction with the FcRn salvage pathway. As long residence times in the systemic circulation are usually undesired for local therapy after intravitreal administration, bispecific formats without the Fc fragment, such as $F(ab')_2$, may be preferable for local ophthalmologic indications. Additional benefits would include lack of any unwanted activation of ADCC for improved safety [116]. Smaller bispecific constructs, however, typically have shorter intravitreal half-lives because of faster diffusion into the plasma and thus have to be administered more frequently. A compromise may be a recently described, full-length bispecific mAb with one arm binding to VEGF while the other arm binding to angiopoietin-2 [117]. The Fc domain has been engineered incorporating mutations for precluding FcRn binding. The ocular PK of this bispecific in mice indicated a longer intravitreal half-life compared to smaller antibody fragments such as $F(ab')_2$, but higher plasma clearance compared to unmodified IgG for reduced systemic exposure [117].

4.4 CONCLUSIONS

The recent decade has witnessed a surge of interest in ADCs and bispecifics for a variety of indications, and there will be a growing number of these biotherapeutics entering clinical studies and applied pharmacotherapy in the next years. Although a plethora of ADCs and creative bispecific formats have emerged of which only a few scaffolds have been presented in this review, their success will ultimately depend on their improvement of the risk–benefit ratio compared to traditional mAbs, upon their drug-like properties in terms of aggregation and stability, and upon their large-scale manufacturing cost [118]. From the clinical pharmacology perspective, it will be critical that these novel scaffolds can strike a balance between access to the site of action, low immunogenicity, and a sufficiently long half-life for generating and maintaining the desired therapeutic effects. As additional clinical data are unveiled in the next years, it will shed more light on the clinical pharmacology of ADCs and bispecifics and which scaffolds are more appropriate for a particular therapeutic indication.

REFERENCES

[1] Meibohm B. Protein engineering for improved pharmacologic characteristics of established monoclonal antibody-based therapeutics. Clin Pharmacokinet 2014;53:863–864.

[2] Aggarwal RS. What's fueling the biotech engine-2012 to 2013. Nat Biotechnol 2014;32:32–39.

[3] Carter PJ. Introduction to current and future protein therapeutics: a protein engineering perspective. Exp Cell Res 2011;317:1261–1269.

[4] Chan AC, Carter PJ. Therapeutic antibodies for autoimmunity and inflammation. Nat Rev Immunol 2010;10:301–316.

[5] Reichert JM. Monoclonal antibodies as innovative therapeutics. Curr Pharm Biotechnol 2008;9:423–430.

[6] Thurber GM, Schmidt MM, Wittrup KD. Factors determining antibody distribution in tumors. Trends Pharmacol Sci 2008;29:57–61.

[7] Chirmule N, Jawa V, Meibohm B. Immunogenicity to therapeutic proteins: impact on PK/PD and efficacy. AAPS J 2012;14:296–302.

[8] Wu AM, Senter PD. Arming antibodies: prospects and challenges for immunoconjugates. Nat Biotechnol 2005;23:1137–1146.

[9] Kustova Y, Espey MG, Sung EG, Morse D, Sei Y, Basile AS. Evidence of neuronal degeneration in C57B1/6 mice infected with the LP-BM5 leukemia retrovirus mixture. Mol Chem Neuropathol 1998;35:39–59.

[10] Engelman JA, Zejnullahu K, Mitsudomi T, Song Y, Hyland C, Park JO, Lindeman N, Gale CM, Zhao X, Christensen J, Kosaka T, Holmes AJ, Rogers AM, Cappuzzo F, Mok T, Lee C, Johnson BE, Cantley LC, Jänne PA. MET amplification leads to gefitinib resistance in lung cancer by activating ERBB3 signaling. Science 2007;316:1039–1043.

[11] Lin K, Tibbitts J. Pharmacokinetic considerations for antibody drug conjugates. Pharm Res 2012;29:2354–2366.

[12] Bross PF, Beitz J, Chen G, Chen XH, Duffy E, Kieffer L, Roy S, Sridhara R, Rahman A, Williams G, Pazdur R. Approval summary: gemtuzumab ozogamicin in relapsed acute myeloid leukemia. Clin Cancer Res 2001;7:1490–1496.

[13] Casi G, Neri D. Antibody-drug conjugates: basic concepts, examples and future perspectives. J Control Release 2012;161:422–428.

[14] Hamblett KJ, Senter PD, Chace DF, Sun MM, Lenox J, Cerveny CG, Kissler KM, Bernhardt SX, Kopcha AK, Zabinski RF, Meyer DL, Francisco JA. Effects of drug loading on the antitumor activity of a monoclonal antibody drug conjugate. Clin Cancer Res 2004;10:7063–7070.

[15] Han TH, Zhao B. ADME considerations for the development of antibody-drug conjugates. Drug Metab Dispos 2014;42: 1914–1920.

[16] Silver DA, Pellicer I, Fair WR, Heston WD, Cordon-Cardo C. Prostate-specific membrane antigen expression in normal and malignant human tissues. Clin Cancer Res 1997;3:81–85.

[17] Bander NH. Antibody-drug conjugate target selection: critical factors. Methods Mol Biol 2013;1045:29–40.

[18] Kamath AV, Iyer S. Preclinical pharmacokinetic considerations for the development of antibody drug conjugates. *Pharm Res* 2014[Epub ahead of print]. DOI: 10.1007/s11095-014-1584-z.

[19] Teicher BA, Doroshow JH. The promise of antibody-drug conjugates. N Engl J Med 2012;367:1847–1848.

[20] Kovtun YV, Audette CA, Mayo MF, Jones GE, Doherty H, Maloney EK, Erickson HK, Sun X, Wilhelm S, Ab O, Lai KC, Widdison WC, Kellogg B, Johnson H, Pinkas J, Lutz RJ, Singh R, Goldmacher VS, Chari RV. Antibody-maytansinoid

conjugates designed to bypass multidrug resistance. Cancer Res 2010;70:2528–2537.

[21] Junttila TT, Li G, Parsons K, Phillips GL, Sliwkowski MX. Trastuzumab-DM1 (T-DM1) retains all the mechanisms of action of trastuzumab and efficiently inhibits growth of lapatinib insensitive breast cancer. Breast Cancer Res Treat 2011;128:347–356.

[22] Lu D, Sahasranaman S, Zhang Y, Girish S. Strategies to address drug interaction potential for antibody-drug conjugates in clinical development. Bioanalysis 2013;5:1115–1130.

[23] Singh R, Erickson HK. Antibody-cytotoxic agent conjugates: preparation and characterization. Methods Mol Biol 2009;525:445–467, xiv.

[24] Xu K, Liu L, Saad OM, Baudys J, Williams L, Leipold D, Shen B, Raab H, Junutula JR, Kim A, Kaur S. Characterization of intact antibody-drug conjugates from plasma/serum *in vivo* by affinity capture capillary liquid chromatography-mass spectrometry. Anal Biochem 2011;412:56–66.

[25] Hengel SM, Sanderson R, Valliere-Douglass J, Nicholas N, Leiske C, Alley SC. Measurement of *in vivo* drug load distribution of cysteine-linked antibody-drug conjugates using microscale liquid chromatography mass spectrometry. Anal Chem 2014;86:3420–3425.

[26] Sun X, Widdison W, Mayo M, Wilhelm S, Leece B, Chari R, Singh R, Erickson H. Design of antibody-maytansinoid conjugates allows for efficient detoxification via liver metabolism. Bioconjug Chem 2011;22:728–735.

[27] Stephan JP, Chan P, Lee C, Nelson C, Elliott JM, Bechtel C, Raab H, Xie D, Akutagawa J, Baudys J, Saad O, Prabhu S, Wong WL, Vandlen R, Jacobson F, Ebens A. Anti-CD22-MCC-DM1 and MC-MMAF conjugates: impact of assay format on pharmacokinetic parameters determination. Bioconjug Chem 2008;19:1673–1683.

[28] Xu K, Liu L, Dere R, Mai E, Erickson R, Hendricks A, Lin K, Junutula JR, Kaur S. Characterization of the drug-to-antibody ratio distribution for antibody-drug conjugates in plasma/serum. Bioanalysis 2013;5:1057–1071.

[29] Sanderson RJ, Hering MA, James SF, Sun MM, Doronina SO, Siadak AW, Senter PD, Wahl AF. *In vivo* drug-linker stability of an anti-CD30 dipeptide-linked auristatin immunoconjugate. Clin Cancer Res 2005;11:843–852.

[30] Xie H, Audette C, Hoffee M, Lambert JM, Blattler WA. Pharmacokinetics and biodistribution of the antitumor immunoconjugate, cantuzumab mertansine (huC242-DM1), and its two components in mice. J Pharmacol Exp Ther 2004;308:1073–1082.

[31] Wang L, Amphlett G, Blattler WA, Lambert JM, Zhang W. Structural characterization of the maytansinoid-monoclonal antibody immunoconjugate, huN901-DM1, by mass spectrometry. Protein Sci 2005;14:2436–2446.

[32] Tijink BM, Buter J, de Bree R, Giaccone G, Lang MS, Staab A, Leemans CR, van Dongen GA. A phase I dose escalation study with anti-CD44v6 bivatuzumab mertansine in patients with incurable squamous cell carcinoma of the head and neck or esophagus. Clin Cancer Res 2006;12:6064–6072.

[33] Rupp U, Schoendorf-Holland E, Eichbaum M, Schuetz F, Lauschner I, Schmidt P, Staab A, Hanft G, Huober J, Sinn HP, Sohn C, Schneeweiss A. Safety and pharmacokinetics of bivatuzumab mertansine in patients with CD44v6-positive metastatic breast cancer: final results of a phase I study. Anticancer Drugs 2007;18:477–485.

[34] Sharkey RM, Karacay H, Govindan SV, Goldenberg DM. Combination radioimmunotherapy and chemoimmunotherapy involving different or the same targets improves therapy of human pancreatic carcinoma xenograft models. Mol Cancer Ther 2011;10:1072–1081.

[35] Sharkey RM, Govindan SV, Cardillo TM, Goldenberg DM. Epratuzumab-SN-38: a new antibody-drug conjugate for the therapy of hematologic malignancies. Mol Cancer Ther 2012;11:224–234.

[36] Zhao L, Ji P, Li Z, Roy P, Sahajwalla CG. The antibody drug absorption following subcutaneous or intramuscular administration and its mathematical description by coupling physiologically based absorption process with the conventional compartment pharmacokinetic model. J Clin Pharmacol 2013;53:314–325.

[37] Rhoden JJ, Wittrup KD. Dose dependence of intratumoral perivascular distribution of monoclonal antibodies. J Pharm Sci 2012;101:860–867.

[38] King L. AAPS Annual Meeting November 5, 2014.

[39] Strop P, Liu SH, Dorywalska M, Delaria K, Dushin RG, Tran TT, Ho WH, Farias S, Casas MG, Abdiche Y, Zhou D, Chandrasekaran R, Samain C, Loo C, Rossi A, Rickert M, Krimm S, Wong T, Chin SM, Yu J, Dilley J, Chaparro-Riggers J, Filzen GF, O'Donnell CJ, Wang F, Myers JS, Pons J, Shelton DL, Rajpal A. Location matters: site of conjugation modulates stability and pharmacokinetics of antibody drug conjugates. Chem Biol 2013;20:161–167.

[40] Junutula JR, Raab H, Clark S, Bhakta S, Leipold DD, Weir S, Chen Y, Simpson M, Tsai SP, Dennis MS, Lu Y, Meng YG, Ng C, Yang J, Lee CC, Duenas E, Gorrell J, Katta V, Kim A, McDorman K, Flagella K, Venook R, Ross S, Spencer SD, Lee Wong W, Lowman HB, Vandlen R, Sliwkowski MX, Scheller RH, Polakis P, Mallet W. Site-specific conjugation of a cytotoxic drug to an antibody improves the therapeutic index. Nat Biotechnol 2008;26:925–932.

[41] Ritchie M, Tchistiakova L, Scott N. Implications of receptor-mediated endocytosis and intracellular trafficking dynamics in the development of antibody drug conjugates. MAbs 2013;5:13–21.

[42] Okeley NM, Alley SC, Senter PD. Advancing antibody drug conjugation: from the laboratory to a clinically approved anti-cancer drug. Hematol Oncol Clin North Am 2014;28:13–25.

[43] McDonagh CF, Kim KM, Turcott E, Brown LL, Westendorf L, Feist T, Sussman D, Stone I, Anderson M, Miyamoto J, Lyon R, Alley SC, Gerber HP, Carter PJ. Engineered anti-CD70 antibody-drug conjugate with increased therapeutic index. Mol Cancer Ther 2008;7:2913–2923.

[44] Kung Sutherland MS, Walter RB, Jeffrey SC, Burke PJ, Yu C, Kostner H, Stone I, Ryan MC, Sussman D, Lyon RP, Zeng W, Harrington KH, Klussman K, Westendorf L, Meyer D, Bernstein ID, Senter PD, Benjamin DR, Drachman JG, McEarchern JA. SGN-CD33A: a novel CD33-targeting antibody-drug conjugate using a pyrrolobenzodiazepine dimer is active in models of drug-resistant AML. Blood 2013;122:1455–1463.

[45] Junutula JR, Bhakta S, Raab H, Ervin KE, Eigenbrot C, Vandlen R, Scheller RH, Lowman HB. Rapid identification of reactive cysteine residues for site-specific labeling of antibody-Fabs. J Immunol Methods 2008;332:41–52.

[46] Dornan D, Bennett F, Chen Y, Dennis M, Eaton D, Elkins K, French D, Go MA, Jack A, Junutula JR, Koeppen H, Lau J, McBride J, Rawstron A, Shi X, Yu N, Yu SF, Yue P, Zheng B, Ebens A, Polson AG. Therapeutic potential of an anti-CD79b antibody-drug conjugate, anti-CD79b-vc-MMAE, for the treatment of non-Hodgkin lymphoma. Blood 2009;114: 2721–2729.

[47] Jackson D, Atkinson J, Guevara CI, Zhang C, Kery V, Moon SJ, Virata C, Yang P, Lowe C, Pinkstaff J, Cho H, Knudsen N, Manibusan A, Tian F, Sun Y, Lu Y, Sellers A, Jia XC, Joseph I, Anand B, Morrison K, Pereira DS, Stover D. *In vitro* and *in vivo* evaluation of cysteine and site specific conjugated herceptin antibody-drug conjugates. PLoS One 2014;9:e83865.

[48] Alley SC, Zhang X, Okeley NM, Anderson M, Law CL, Senter PD, Benjamin DR. The pharmacologic basis for antibody-auristatin conjugate activity. J Pharmacol Exp Ther 2009;330:932–938.

[49] Boswell CA, Mundo EE, Zhang C, Bumbaca D, Valle NR, Kozak KR, Fourie A, Chuh J, Koppada N, Saad O, Gill H, Shen BQ, Rubinfeld B, Tibbitts J, Kaur S, Theil FP, Fielder PJ, Khawli LA, Lin K. Impact of drug conjugation on pharmacokinetics and tissue distribution of anti-STEAP1 antibody-drug conjugates in rats. Bioconjug Chem 2011;22:1994–2004.

[50] Herbertson RA, Tebbutt NC, Lee FT, MacFarlane DJ, Chappell B, Micallef N, Lee ST, Saunder T, Hopkins W, Smyth FE, Wyld DK, Bellen J, Sonnichsen DS, Brechbiel MW, Murone C, Scott AM. Phase I biodistribution and pharmacokinetic study of Lewis Y-targeting immunoconjugate CMD-193 in patients with advanced epithelial cancers. Clin Cancer Res 2009;15:6709–6715.

[51] Doronina SO, Setter JR, Bovee TD, Anderson ME, Jonas M, Daniho S, Kostner H, Senter PD, Lyon RP. Elucidating the role of drug-linker hydrophobicity in the disposition of antibody-drug conjugates. Cancer Res 2014;74:4470.

[52] Erickson HK, Lewis Phillips GD, Leipold DD, Provenzano CA, Mai E, Johnson HA, Gunter B, Audette CA, Gupta M, Pinkas J, Tibbitts J. The effect of different linkers on target cell catabolism and pharmacokinetics/pharmacodynamics of trastuzumab maytansinoid conjugates. Mol Cancer Ther 2012;11:1133–1142.

[53] Shen BQ, Bumbaca D, Saad O, Yue Q, Pastuskovas CV, Khojasteh SC, Tibbitts J, Kaur S, Wang B, Chu YW, LoRusso PM, Girish S. Catabolic fate and pharmacokinetic characterization of trastuzumab emtansine (T-DM1): an emphasis on preclinical and clinical catabolism. Curr Drug Metab 2012;13:901–910.

[54] Lyon RP, Setter JR, Bovee TD, Doronina SO, Hunter JH, Anderson ME, Balasubramanian CL, Duniho SM, Leiske CI, Li F, Senter PD. Self-hydrolyzing maleimides improve the stability and pharmacological properties of antibody-drug conjugates. Nat Biotechnol 2014;32:1059–1062.

[55] Sievers EL, Senter PD. Antibody-drug conjugates in cancer therapy. Annu Rev Med 2013;64:15–29.

[56] Lovdal T, Andersen E, Brech A, Berg T. Fc receptor mediated endocytosis of small soluble immunoglobulin G immune complexes in Kupffer and endothelial cells from rat liver. J Cell Sci 2000;113 (Part 18):3255–3266.

[57] Dowell JA, Korth-Bradley J, Liu H, King SP, Berger MS. Pharmacokinetics of gemtuzumab ozogamicin, an antibody-targeted chemotherapy agent for the treatment of patients with acute myeloid leukemia in first relapse. J Clin Pharmacol 2001;41:1206–1214.

[58] Deslandes A. Comparative clinical pharmacokinetics of antibody-drug conjugates in first-in-human Phase 1 studies. MAbs 2014;6:859–870.

[59] Sapra P, Damelin M, Dijoseph J, Marquette K, Geles KG, Golas J, Dougher M, Narayanan B, Giannakou A, Khandke K, Dushin R, Ernstoff E, Lucas J, Leal M, Hu G, O'Donnell CJ, Tchistiakova L, Abraham RT, Gerber HP. Long-term tumor regression induced by an antibody-drug conjugate that targets 5T4, an oncofetal antigen expressed on tumor-initiating cells. Mol Cancer Ther 2013;12:38–47.

[60] Girish S, Gupta M, Wang B, Lu D, Krop IE, Vogel CL, Burris Iii HA, LoRusso PM, Yi JH, Saad O, Tong B, Chu YW, Holden S, Joshi A. Clinical pharmacology of trastuzumab emtansine (T-DM1): an antibody-drug conjugate in development for the treatment of HER2-positive cancer. Cancer Chemother Pharmacol 2012;69:1229–1240.

[61] Younes A, Bartlett NL, Leonard JP, Kennedy DA, Lynch CM, Sievers EL, Forero-Torres A. Brentuximab vedotin (SGN-35) for relapsed CD30-positive lymphomas. N Engl J Med 2010;363:1812–1821.

[62] Wong S, Bumbaca D, Yue Q, Halladay J, Kenny JR, Salphati L, Saad O, Tibbitts J, Khojasteh C, Girish S, Shen BQ. Abstract A136: nonclinical disposition, metabolism, and *in vitro* drug-drug interaction assessment of DM1, a component of trastuzumab emtansine (T-DM1). Mol Cancer Ther 2011;10:A136.

[63] Kontermann RE. Dual targeting strategies with bispecific antibodies. MAbs 2012;4:182–197.

[64] May C, Sapra P, Gerber HP. Advances in bispecific biotherapeutics for the treatment of cancer. Biochem Pharmacol 2012;84:1105–1112.

[65] Hollander N. Bispecific antibodies for cancer therapy. Immunotherapy 2009;1:211–222.

[66] Byrne H, Conroy PJ, Whisstock JC, O'Kennedy RJ. A tale of two specificities: bispecific antibodies for therapeutic and diagnostic applications. Trends Biotechnol 2013;31:621–632.

[67] Ott MG, Marme F, Moldenhauer G, Lindhofer H, Hennig M, Spannagl R, Essing MM, Linke R, Seimetz D. Humoral response to catumaxomab correlates with clinical outcome: results of the pivotal phase II/III study in patients with malignant ascites. Int J Cancer 2012;130:2195–2203.

[68] Heiss MM, Murawa P, Koralewski P, Kutarska E, Kolesnik OO, Ivanchenko VV, Dudnichenko AS, Aleknaviciene B, Razbadauskas A, Gore M, Ganea-Motan E, Ciuleanu T, Wimberger P, Schmittel A, Schmalfeldt B, Burges A, Bokemeyer C, Lindhofer H, Lahr A, Parsons SL. The trifunctional antibody catumaxomab for the treatment of malignant ascites due to epithelial cancer: results of a prospective randomized phase II/III trial. Int J Cancer 2010;127:2209–2221.

[69] Jin P, Zhu Z. The design and engineering of IgG-like bispecific antibodies. In: Kontermann RE, editor. *Bispecific Antibodies*. Berlin, Heidelberg: Springer; 2011. pp. 151–169.

[70] Wu C, Ying H, Grinnell C, Bryant S, Miller R, Clabbers A, Bose S, McCarthy D, Zhu RR, Santora L, Davis-Taber R, Kunes Y, Fung E, Schwartz A, Sakorafas P, Gu J, Tarcsa E, Murtaza A, Ghayur T. Simultaneous targeting of multiple disease mediators by a dual-variable-domain immunoglobulin. Nat Biotechnol 2007;25:1290–1297.

[71] Lameris R, de Bruin RC, Schneiders FL, van Bergen En Henegouwen PM, Verheul HM, de Gruijl TD, van der Vliet HJ. Bispecific antibody platforms for cancer immunotherapy. Crit Rev Oncol Hematol 2014;92 (3):153–165.

[72] Portell CA, Wenzell CM, Advani AS. Clinical and pharmacologic aspects of blinatumomab in the treatment of B-cell acute lymphoblastic leukemia. Clin Pharmacol 2013;5:5–11.

[73] Baum V, Buhler P, Gierschner D, Herchenbach D, Fiala GJ, Schamel WW, Wolf P, Elsässer-Beile U. Antitumor activities of PSMAxCD3 diabodies by redirected T-cell lysis of prostate cancer cells. Immunotherapy 2013;5:27–38.

[74] Kontermann RE. Recombinant bispecific antibodies for cancer therapy. Acta Pharmacol Sin 2005;26:1–9.

[75] Borchmann P, Schnell R, Fuss I, Manzke O, Davis T, Lewis LD, Behnke D, Wickenhauser C, Schiller P, Diehl V, Engert A. Phase 1 trial of the novel bispecific molecule H22xKi-4 in patients with refractory Hodgkin lymphoma. Blood 2002;100:3101–3107.

[76] Meibohm B. Pharmacokinetics and pharmacodynamics of peptide and protien therapeutics. In: Crommelin DJA, Sindelar RD, Meibohm B, editors. *Pharmaceutical Biotechnology: Fundamentals and Applications*. 4th ed. New York: Springer; 2013. pp. 101–132.

[77] Tang L, Persky AM, Hochhaus G, Meibohm B. Pharmacokinetic aspects of biotechnology products. J Pharm Sci 2004;93:2184–2204.

[78] Meibohm B. Pharmacokinetics and half-life of protein therapeutics. In: Kontermann RE, editor. *Therapeutic Proteins: Strategies to Modulate their Plasma Half-lives*. Wiley-VCH; 2012. pp. 23–38.

[79] Rippe B, Haraldsson B. Transport of macromolecules across microvascular walls: the two-pore theory. Physiol Rev 1994;74:163–219.

[80] Labrijn AF, Meesters JI, de Goeij BE, van den Bremer ET, Neijssen J, van Kampen MD, Strumane K, Verploegen S, Kundu A, Gramer MJ, van Berkel PH, van de Winkel JG, Schuurman J, Parren PW. Efficient generation of stable bispecific IgG1 by controlled Fab-arm exchange. Proc Natl Acad Sci U S A 2013;110:5145–5150.

[81] Chames P, Van Regenmortel M, Weiss E, Baty D. Therapeutic antibodies: successes, limitations and hopes for the future. Br J Pharmacol 2009;157:220–233.

[82] Fujimori K, Covell DG, Fletcher JE, Weinstein JN. A modeling analysis of monoclonal antibody percolation through tumors: a binding-site barrier. J Nucl Med 1990;31:1191–1198.

[83] Holliger P, Hudson PJ. Engineered antibody fragments and the rise of single domains. Nat Biotechnol 2005;23:1126–1136.

[84] List T, Neri D. Biodistribution studies with tumor-targeting bispecific antibodies reveal selective accumulation at the tumor site. MAbs 2012;4:775–783.

[85] Tang L, Meibohm B. Pharmacokinetics of peptides and proteins. In: Meibohm B, editor. *Pharmacokinetics and Pharmacodynamics of Biotech Drugs*. Weinheim: Wiley-VCH; 2006. pp. 17–43.

[86] Tabrizi MA, Tseng CM, Roskos LK. Elimination mechanisms of therapeutic monoclonal antibodies. Drug Discov Today 2006;11:81–88.

[87] Mager DE. Target-mediated drug disposition and dynamics. Biochem Pharmacol 2006;72:1–10.

[88] Roopenian DC, Akilesh S. FcRn: the neonatal Fc receptor comes of age. Nat Rev Immunol 2007;7:715–725.

[89] Diao L, Meibohm B. Pharmacokinetics and pharmacokinetic-pharmacodynamic correlations of therapeutic peptides. Clin Pharmacokinet 2013;52:855–868.

[90] Dirks NL, Meibohm B. Population pharmacokinetics of therapeutic monoclonal antibodies. Clin Pharmacokinet 2010;49:633–659.

[91] Yamane-Ohnuki N, Satoh M. Production of therapeutic antibodies with controlled fucosylation. MAbs 2009;1:230–236.

[92] Yip V, Palma E, Tesar DB, Mundo EE, Bumbaca D, Torres EK, Reyes NA, Shen BQ, Fielder PJ, Prabhu S, Khawli LA, Boswell CA. Quantitative cumulative biodistribution of antibodies in mice: effect of modulating binding affinity to the neonatal Fc receptor. MAbs 2014;6:689–696.

[93] Boswell CA, Tesar DB, Mukhyala K, Theil FP, Fielder PJ, Khawli LA. Effects of charge on antibody tissue distribution and pharmacokinetics. Bioconjug Chem 2010;21:2153–2163.

[94] Sampei Z, Igawa T, Soeda T, Okuyama-Nishida Y, Moriyama C, Wakabayashi T, Tanaka E, Muto A, Kojima T, Kitazawa T, Yoshihashi K, Harada A, Funaki M, Haraya K, Tachibana T, Suzuki S, Esaki K, Nabuchi Y, Hattori K. Identification and multidimensional optimization of an asymmetric bispecific IgG antibody mimicking the function of factor VIII cofactor activity. PLoS One 2013;8:e57479.

[95] Meibohm B, Zhou H. Characterizing the impact of renal impairment on the clinical pharmacology of biologics. J Clin Pharmacol 2012;52:54S–62S.

[96] McDonagh CF, Huhalov A, Harms BD, Adams S, Paragas V, Oyama S, Zhang B, Luus L, Overland R, Nguyen S, Gu J, Kohli N, Wallace M, Feldhaus MJ, Kudla AJ, Schoeberl B, Nielsen UB. Antitumor activity of a novel bispecific antibody that targets the ErbB2/ErbB3 oncogenic unit and inhibits heregulin-induced activation of ErbB3. Mol Cancer Ther 2012;11:582–593.

[97] Stork R, Campigna E, Robert B, Muller D, Kontermann RE. Biodistribution of a bispecific single-chain diabody and its half-life extended derivatives. J Biol Chem 2009;284:25612–25619.

[98] Stork R, Zettlitz KA, Muller D, Rether M, Hanisch FG, Kontermann RE. N-glycosylation as novel strategy to improve pharmacokinetic properties of bispecific single-chain diabodies. J Biol Chem 2008;283:7804–7812.

[99] Buhler P, Wolf P, Gierschner D, Schaber I, Katzenwadel A, Schultze-Seemann W, Wetterauer U, Tacke M, Swamy M,

Schamel WW, Elsässer-Beile U. A bispecific diabody directed against prostate-specific membrane antigen and CD3 induces T-cell mediated lysis of prostate cancer cells. Cancer Immunol Immunother 2008;57:43–52.

[100] Seimetz D, Lindhofer H, Bokemeyer C. Development and approval of the trifunctional antibody catumaxomab (anti-EpCAM x anti-CD3) as a targeted cancer immunotherapy. Cancer Treat Rev 2010;36:458–467.

[101] Ruf P, Kluge M, Jager M, Burges A, Volovat C, Heiss MM, Hess J, Wimberger P, Brandt B, Lindhofer H. Pharmacokinetics, immunogenicity and bioactivity of the therapeutic antibody catumaxomab intraperitoneally administered to cancer patients. Br J Clin Pharmacol 2010;69:617–625.

[102] Ober RJ, Radu CG, Ghetie V, Ward ES. Differences in promiscuity for antibody-FcRn interactions across species: implications for therapeutic antibodies. Int Immunol 2001;13:1551–1559.

[103] Topp MS, Kufer P, Gokbuget N, Goebeler M, Klinger M, Neumann S, Horst HA, Raff T, Viardot A, Schmid M, Stelljes M, Schaich M, Degenhard E, Köhne-Volland R, Brüggemann M, Ottmann O, Pfeifer H, Burmeister T, Nagorsen D, Schmidt M, Lutterbuese R, Reinhardt C, Baeuerle PA, Kneba M, Einsele H, Riethmüller G, Hoelzer D, Zugmaier G, Bargou RC. Targeted therapy with the T-cell-engaging antibody blinatumomab of chemotherapy-refractory minimal residual disease in B-lineage acute lymphoblastic leukemia patients results in high response rate and prolonged leukemia-free survival. J Clin Oncol 2011;29:2493–2498.

[104] Viardot A, Goebeler M, Pfreundschuh M, Adrian N, Libicher M, Degenhard E, Stieglmaier J, Zhang A, Nagorsen D, Bargou RC. Open-label phase 2 study of the bispecific t-cell engager (BiTE®) blinatumomab in patients with relapsed/refractory diffuse large b-cell lymphoma. Blood 2013;122:1811.

[105] Amgen Inc.. *Background Information for the Pediatric Subcommittee of the Oncologic Drugs Advisory Committee Meeting – 04 October 2012*. US Food and Drug Administration; 2012.

[106] Klinger M, Brandl C, Zugmaier G, Hijazi Y, Bargou RC, Topp MS, Gökbuget N, Neumann S, Goebeler M, Viardot A, Stelljes M, Brüggemann M, Hoelzer D, Degenhard E, Nagorsen D, Baeuerle PA, Wolf A, Kufer P. Immunopharmacologic response of patients with B-lineage acute lymphoblastic leukemia to continuous infusion of T cell-engaging CD19/CD3-bispecific BiTE antibody blinatumomab. Blood 2012;119:6226–6233.

[107] Nagorsen D, Kufer P, Baeuerle PA, Bargou R. Blinatumomab: a historical perspective. Pharmacol Ther 2012;136:334–342.

[108] Bargou R, Leo E, Zugmaier G, Klinger M, Goebeler M, Knop S, Noppeney R, Viardot A, Hess G, Schuler M, Einsele H, Brandl C, Wolf A, Kirchinger P, Klappers P, Schmidt M, Riethmüller G, Reinhardt C, Baeuerle PA, Kufer P. Tumor regression in cancer patients by very low doses of a T cell-engaging antibody. Science 2008;321:974–977.

[109] Wurster U, Haas J. Passage of intravenous immunoglobulin and interaction with the CNS. J Neurol Neurosurg Psychiatry 1994;57 (Suppl):21–25.

[110] Atwal JK, Chen Y, Chiu C, Mortensen DL, Meilandt WJ, Liu Y, Heise CE, Hoyte K, Luk W, Lu Y, Peng K, Wu P, Rouge L, Zhang Y, Lazarus RA, Scearce-Levie K, Wang W, Wu Y, Tessier-Lavigne M, Watts RJ. A therapeutic antibody targeting BACE1 inhibits amyloid-beta production *in vivo*. Sci Transl Med 2011;3:84ra43.

[111] Kissel K, Hamm S, Schulz M, Vecchi A, Garlanda C, Engelhardt B. Immunohistochemical localization of the murine transferrin receptor (TfR) on blood-tissue barriers using a novel anti-TfR monoclonal antibody. Histochem Cell Biol 1998;110:63–72.

[112] Jones AR, Shusta EV. Blood-brain barrier transport of therapeutics via receptor-mediation. Pharm Res 2007;24:1759–1771.

[113] Fishman JB, Rubin JB, Handrahan JV, Connor JR, Fine RE. Receptor-mediated transcytosis of transferrin across the blood-brain barrier. J Neurosci Res 1987;18:299–304.

[114] Yu YJ, Zhang Y, Kenrick M, Hoyte K, Luk W, Lu Y, Atwal J, Elliott JM, Prabhu S, Watts RJ, Dennis MS. Boosting brain uptake of a therapeutic antibody by reducing its affinity for a transcytosis target. Sci Transl Med 2011;3:84ra44.

[115] Mabry R, Gilbertson DG, Frank A, Vu T, Ardourel D, Ostrander C, Stevens B, Julien S, Franke S, Meengs B, Brody J, Presnell S, Hamacher NB, Lantry M, Wolf A, Bukowski T, Rosler R, Yen C, Anderson-Haley M, Brasel K, Pan Q, Franklin H, Thompson P, Dodds M, Underwood S, Peterson S, Sivakumar PV, Snavely M. A dual-targeting PDGFRbeta/VEGF-A molecule assembled from stable antibody fragments demonstrates anti-angiogenic activity *in vitro* and *in vivo*. MAbs 2010;2:20–34.

[116] Le Couter J, Scheer JM. Bispecific therapeutics for ophthalmic indications: target selection and the optimal molecular format. Expert Rev Ophthalmol 2014;9:217–225.

[117] Duerr H, Herting F, Klein C, Regula JT, Rueth M, and Stubenrauch KG Patent application: bispecific Anti-VEGF/Anti-ANG-2 antibodies and their use in the treatment of ocular vascular diseases. US20140017244 A1. editor: US Patent and Trademark Office. 2014.

[118] Beck A, Wurch T, Bailly C, Corvaia N. Strategies and challenges for the next generation of therapeutic antibodies. Nat Rev Immunol 2010;10:345–352.

5

OVERVIEW OF ADME AND PK/PD OF ADCs

BAITENG ZHAO[1] AND TAE H. HAN[2]

[1]*Seattle Genetics, Inc., Bothell, WA, USA*

[2]*Stemcentrx, Inc., South San Francisco, CA, USA*

5.1 INTRODUCTION TO ADC

Antibody–drug conjugates (ADCs) are an important class of biotherapeutics for the treatment of cancers and potentially other diseases. The concept of ADC technology is to selectively deliver a potent small molecule drug (payload or abbreviated as drug) to the intended tissues or cells using the specificity of an antibody to its target, such as a cell surface antigen, while sparing the other tissues or cells with minimal or no expression of such target [1]. Although the first ADC was approved for marketing more than 14 years ago and there are currently two ADCs on the U.S. market, the concept of ADC-like therapeutics has actually been postulated more than 100 years ago by the German physician Paul Ehrlich, who reasoned that if a compound could be made that selectively targeted a disease-causing organism, then a toxin for that organism could be delivered along with the agent of selectivity to kill only the organism targeted, hence the "magic bullet" [2]. Such concept was first fulfilled in May 2000 with the accelerated approval of gemtuzumab ozogamicin (GO, Mylotarg™) [3], an ADC for patients with relapsed acute myelogenous leukemia (AML). Although GO was voluntarily discontinued in the U.S. market in June 2010 for failing to demonstrate adequate clinical benefit to patients enrolled in a late-stage clinical trial, two new ADCs have since been approved for marketing worldwide—brentuximab vedotin (ADCETRIS™) in August 2011 and ado-trastuzumab emtansine (T-DM1, KADCYLA™) in February 2013. These approvals represent the clinical validation of the ADC concept and more ADCs may follow in the near future.

As shown in Figure 5.1, the key components of an ADC are a monoclonal antibody (mAb) or an antibody derivative for a specific cell surface target, a potent small molecule drug that interferes a critical cellular process, and a linker that covalently conjugates the mAb and payload. The primary mechanism of action (MOA) of an ADC is through the specific binding of ADC on a cell surface target, internalization of the binding complex, release of the small molecule drug or derivative inside the cell, and interference of a cellular process that induces cell death, as shown in Figure 5.2. The goal is to minimize the payload exposure on normal tissues while maximizing the exposure on the intended tissues, such as a tumor. Given this MOA, it is understandable that the critical parameters for ADC development may include differential target-expression profiles on the intended versus unintended cells or tissues, drug-linker stability in circulation, accessibility of the site of action, internalization and selective intracellular release of small molecule drug via ADC binding to its target, drug-to-antibody ratio (DAR), and the potency of the released payload. Since the eventual driver of the ADC pharmacological effects is the intracellular released drug concentrations, many other factors can also impact the effectiveness of an ADC, such as the properties of the ADCs and the released payloads related to their absorption, distribution, metabolism/catabolism, and elimination (ADME). Many of the advancements in ADC technology that have been and will continuously to be made revolve around improving the pharmacokinetic (PK) and ADME characteristics of ADCs. Thus, it is pivotal to characterize the ADME of ADCs and understand the impact to the optimal dose and schedule for clinical applications.

ADME and Translational Pharmacokinetics/Pharmacodynamics of Therapeutic Proteins: Applications in Drug Discovery and Development,
First Edition. Edited by Honghui Zhou and Frank-Peter Theil.

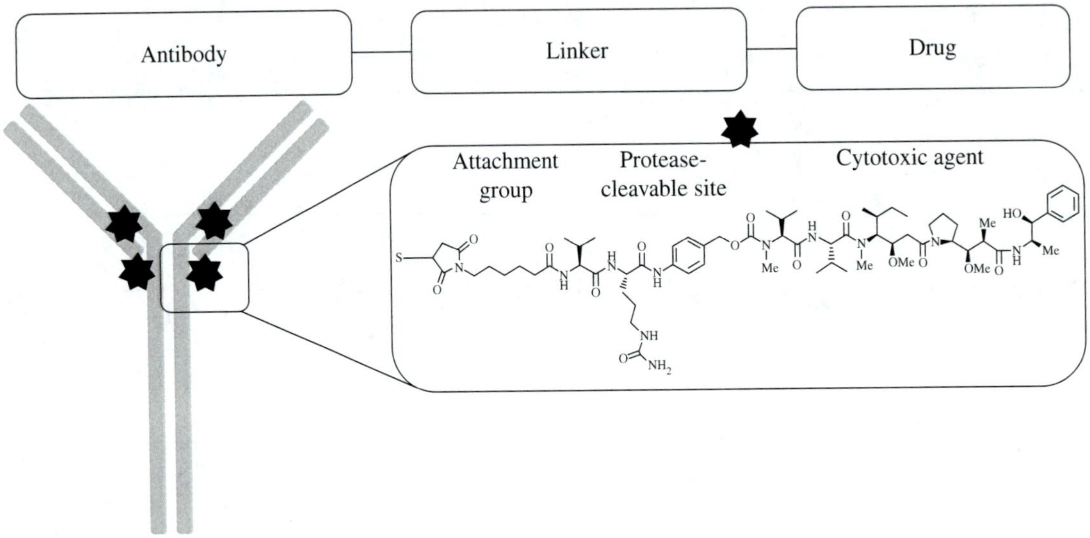

FIGURE 5.1 Typical components of ADCs. (From Han and Zhao [4].)

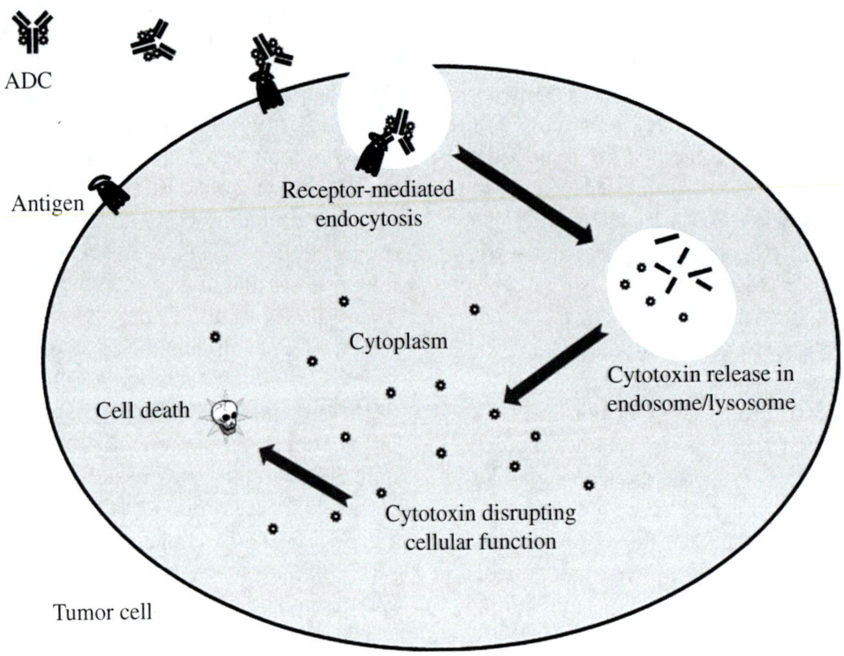

FIGURE 5.2 Mechanism of action of ADCs.

ADC technology can be applied to many therapeutic areas, and currently most of the ADC development has been focused on oncology, so this chapter will primarily discuss the ADME properties for ADCs in oncology therapy and associated PK and PK/PD (pharmacokinetic/pharmacodynamic) modeling of ADCs. Table 5.1 provides the ADCs that are currently in various clinical development stages with their targets, lead indications, and conjugated small molecule chemotherapy types. The MOA of the drugs once released inside the cells is shown in Table 5.2.

5.2 ABSORPTION

Various dosing routes have been employed for mAb-based therapeutics, such as intravenous (IV), subcutaneous (SC), and intramuscular (IM). The SC and IM administrations are usually considered more convenient for patients and do not require hospital visits; however, ADCs have thus far been only given via IV administration. One of the primary reasons is the potential for injection-site toxicity associated with ADCs. The payloads of current ADCs are typically

TABLE 5.1 ADCs in Clinical Development as of July 2014

ADC	Tumor	Target	Chemotherapy Type	Latest Phase of Development
Brentuximab vedotin	HL, ALCL	CD30	Auristatin	Approved
Ado-trastuzumab emtansine	mBC	HER2	Maytansine	Approved
Inotuzumab ozogamicin (CMC-544)	NHL, ALL	CD22	Calicheamicin	III
BT-062	MM	CD138	Maytansine	II
Glembatumumab vedotin (CDX-011)	mBC, melanoma	GPNMB	Auristatin	II
PSMA ADC	Prostate cancer	PMSA	Auristatin	II
Pinatuzumab vedotin (RG7593)	NHL	CD22	Auristatin	II
Polatuzumab vedotin (RG7596)	NHL	CD79B	Auristatin	II
hLL1-DOX	MM	CD74	Doxorubicin	II
SAR3419	DLBCL	CD19	Maytansine	II
IMMU-132	Solid tumors	TACSTD2 (TROP-2)	SN-38	II
Labetuzumab-SN-38 (IMMU-130)	mCRC	CEACAM5	SN-38	II
ABT-414	Solid tumors	EDFR	Auristatin	II
RG7599	Ovarian, NSCLC	NaPi2b	Auristatin	II
SGN-LIV1A	MBC	LIV-1	Auristatin	I
ASG-22ME	Solid tumors	Nectin-4	Auristatin	I
ASG-15ME	Bladder cancer	SLITRK6	Auristatin	I
AGS-16M8F	RCC	AGS-16	Auristatin	I
BAY-94-9343	Solid tumors	Mesothelin	Maytansine	I
BIIB015	Solid tumors	Cripto	Maytansine	I
IMGN529	NHL, CLL	CD37	Maytansine	I
IMGN853	Solid tumors	Folate receptor-1	Maytansine	I
IMGN289	Solid tumors	EGFR	Maytansine	I
RG7450	Prostate cancer	STEAP1	Auristatin	I
RG7458	Ovarian cancer	MUC16	Auristatin	I
RG7636	Melanoma	ETBR	Auristatin	I
SAR566658	Solid tumors	CA6	Maytansine	I
MLN0264	Colorectal cancer	GCC	Auristatin	I
AMG 595	Glioma	EGFRvIII	Maytansine	I
AMG 172	RCC	CD70	Maytansine	I
SGN-CD19A	NHL, ALL	CD19	Auristatin	I
SGN-CD33A	AML	CD33	PBD dimer	I
PF-0626350	Solid tumors	5T4	Auristatin	I
SC16LD6.5	SCLC	Undisclosed	DNA-damaging agent	I
SGN-CD70A	NHL, RCC	CD70	Auristatin	I

ALCL, anaplastic large-cell lymphoma; ALL, acute lymphocytic leukemia; CEA, carcinoembryonic antigen; CLL, chronic lymphocytic leukemia; DLBCL, diffuse large B-cell lymphoma; EGFR, epidermal growth factor receptor; ETBR, endothelin B receptor; GCC, guanylyl cyclase C; GPNMB, glycoprotein NMB; HL, Hodgkin's lymphoma; MBC, metastatic breast cancer; mCRC, metastatic colorectal cancer; MM, multiple myeloma; NHL, non-Hodgkin's lymphoma; NSCLC, non-small-cell lung cancer; PBD, pyrrolobenzodiazepine; PSMA, prostate-specific membrane antigen; RCC, renal cell carcinoma; SCLC, small-cell lung cancer; STEAP1, six-transmembrane epithelial antigen of prostate 1; and TACSTD2, tumor-associated calcium signal transducer 2.

TABLE 5.2 Mechanisms of Action of the Unconjugated Small Molecule Drug

Mechanism of Action	Small Molecule Cytotoxic Agent
Microtubule disruption	Auristatin, maytansine
DNA damage	PBD dimer, calicheamicin, doxorubicin
Topoisomerase I inhibition	SN-38

very potent small molecule cytotoxic agents. Even though ADCs are designed to be effectively inert before reaching and binding to their targets, active payloads may be released through the nonspecific release of drug and local nonspecific catabolism of the ADCs. The future design criteria for ADCs may include increased stability in nontarget tissues and circulation, and the released payloads to remain inactive in normal cells and become activated only in tumor cells. In turn, these advancements may allow alternative routes of administration (e.g., SC or IM). Until then, IV administration will continue to be the route of choice for ADCs.

5.3 DISTRIBUTION

The biodistribution characteristics of ADCs are expected to be similar to those of unmodified or "naked" antibodies given the similarity in size, molecular structure, and binding kinetics. This is despite the conjugation of one or more small molecule payloads that can alter the local physiochemical properties of the antibody. Limited distribution in vasculature and peripheral tissues and recirculation through lymphatic systems are reasonable assumptions for ADC PK.

Similar to naked antibodies, ADC distribution can be affected by tissue target expression and other factors affecting antibody biodistribution. Tissues that express the ADC target, such as the tumor in xenograft mouse studies, are typically seen with the highest ADC concentrations [5–7]. Alteration of the physiochemical properties and modified binding affinity to neonatal Fc receptor (FcRn) may also affect tissue distribution of an ADC [8–10]. Such topics have been covered extensively in other chapters and are not further discussed here.

Unique to ADCs, internalization by the target-expressing cells is one of the necessary steps in the MOA of ADCs in order to bring drug into the cells. This process may be affected by drug conjugation. A study with anti-CD20 ADCs showed that ADCs could be internalized by the CD20-positive cells more rapidly and efficiently than the naked antibody, and different conjugated drugs appeared to also affect the internalization rate [11]. These findings warrant a closer evaluation of some cell surface targets that are often deemed not suitable for ADC development due to minimal or slow internalization based on naked antibody data. Testing of multiple ADCs with alternative drug linkers should be considered to validate a target.

The biodistribution of the small molecule drugs, on the contrary, is fundamentally altered when administered as ADC comparing to when dosed as unconjugated small molecule itself. The drugs given in either form are typically widely distributed and their volume of distributions is usually much greater than the sum of plasma and extracellular fluid volume. The extent of distribution is largely dictated by its physicochemical properties; however, when conjugated to an ADC, the distribution of the drug into a particular tissue can be affected dramatically and reflect the biodistribution kinetics of the ADC. This modified biodistribution is especially evident for small molecule drugs with poor cell permeability since an ADC can deliver the drug into a cell through target binding and internalization. The ADC-dependent drug biodistribution is indeed one of the key attributes of ADC technology to improve the therapeutic index of the drug itself. Better understanding and leveraging this process should help the development of current and future ADCs. It is important to point out that since the biodistribution of drug can be affected by conjugation, biodistribution studies using the small molecule drug alone may not reflect the tissue distribution of the released drug from an ADC.

Another factor for consideration of the small molecule drug biodistribution is the so-called bystander effect. Ideally, the ADC target is highly and uniformly expressed in the tumor so that ADC is internalized and a sufficient amount of the drug is brought into all tumor cells. In reality, the expression of the tumor cell surface target can be highly heterogeneous, resulting in the coexistence of high, low, or negative target-expressing cells. The cell permeability of a drug could impact its ability to kill the surrounding cells once released by the target-expressing cells. It was shown that an antit-issue factor ADC conjugated to the highly permeable monomethyl auristatin E (MMAE) can induce complete tumor regression in patient-derived xenograft models even when tissue factor was expressed in only 25–50% of the tumor cells, thus the "bystander effect" [12]. On the other hand, it is likely that a less permeable drug will accumulate inside target-expressing cells better for greater antitumor effect while limiting exposure to the nontarget-expressing normal cells. These biodistribution properties of the small molecule drugs should be taken into account when evaluating potential drug linkers in preclinical ADC development in conjunction with the target-expression characteristics of the diseases.

Similar to other small molecule drugs, the drugs released from ADCs can be the substrate of uptake and/or efflux transporters, which could result in an increase or decrease in the intracellular released drug exposures in tumor and/or normal tissues [13]. This may have implication on both the antitumor effect and toxicities of an ADC. Preclinical and clinical characterization and evaluations should therefore be considered for the released drugs. All three cytotoxins (calicheamicin, MMAE, and derivative of maytansine 1 (DM1)) used in the approved ADCs have been implicated as a substrate of P-gp [14, 15].

In summary, the biodistribution of ADC and released drug is of particular importance in determining the antitumor effect and the toxicities of an ADC and proper evaluations should be implemented during ADC development.

5.4 METABOLISM/CATABOLISM

Understanding the catabolism of ADCs and the metabolism of the small molecule drug is important to identify the potential for clinical drug–drug interactions (DDIs), the effect of organ impairment, and to design ADC constructs with properties that result in a more effective therapeutic. Owing to having antibody and small molecule characteristics, the catabolism and metabolism of an ADC are complex to characterize and leverage for optimization. Furthermore, with the design space starting with the three basic elements (mAb, linker, and drug) and the weak *in vitro* to *in vivo* correlation, each permutation geometrically increases the design space and efforts required to identify an optimal ADC therapeutic.

Recently, efforts have also been put forth to understand the catabolism of protein therapeutics and potential DDIs with protein therapeutics, [16–19], but those efforts remain early at this point. Like unconjugated antibodies, the antibody portion of an ADC is expected to be catabolized through proteolytic degradation to constituent amino acids and reused for the production of other proteins. Because the small molecule payload of an ADC may alter the distribution of an ADC compared to the unconjugated mAb, the tissues that catabolize an ADC and yield released drug may differ. Analytical techniques, including imaging-mass spectrometry or dual-labeled ADCs, may yield insight into the tissues that result in drug release from an ADC [5, 20, 21].

Drug conjugation can also impact the antibody conformational stability. It has been shown that upon conjugation, an ADC can maintain similar secondary and tertiary structure as an unconjugated mAb; however, certain ADCs are more easily destabilized at higher temperatures [22, 23]. Conjugation may also introduce more hydrophobic regions on the mAb, which then results in lower colloidal stability and present a greater aggregation potential. These aspects may explain, in part, the observations that ADCs with higher DAR (6 or 8) tend to have a higher potential to form high molecular weight species [24]. Furthermore, ADCs generally have shorter terminal half-lives compared to those of the corresponding naked mAb, partially owing to loss of drug. For example, the average serum half-life of brentuximab vedotin was 4–6 days in patients, shorter than that of the unconjugated mAb, SGN-30, which had a terminal half-life of 1–3 weeks [25, 26]. Secondly, the number of conjugates on an ADC may affect ADC catabolism, with higher DAR leading to shorter serum half-life and lower exposure [27].

Finally, the specific mechanism of uptake into the cell, whether it be FcRn- and Fc-gamma-mediated uptake or receptor-mediated endocytosis, may alter the catabolism of an ADC. Uptake through those mechanisms may result in altered drug release [28]. To maximize the ADC exposure for a given target at a given dose, these and other factors should be taken into consideration for the lead selection of an ADC construct.

The small molecule payload of an ADC, like other small molecule drugs, is subject to biotransformation. They may be metabolized by cytochrome P450 (CYP) and other enzymes and subject to potential DDI from CYP inhibitors or inducers [29].

The unconjugated drugs from both brentuximab vedotin and T-DM1 were primarily metabolized by CYP3A enzymes based on *in vitro* data [30, 31]. Following the administration of T-DM1 in patients, several DM1-containing catabolites were identified in plasma suggesting the potential for multiple pathways for T-DM1 degradation [32]. The metabolism of maytansinoids has been studied in detail and suggests the low potential for DDIs, similar to MMAE [33–35]. Multiple MMAE metabolites following brentuximab vedotin administration in patients were identified in feces and urine; however, they were at very low levels and only detectable in highly concentrated samples [30]. Both MMAE and lys-MCC-DM1 (and related DM1 species) have been shown to be substrates of CYP enzymes and alter the activity of CYP enzymes, *in vitro*, but not in humans for MMAE.

The potential for clinically meaningful DDIs was evaluated both *in vitro* and in clinical trials for brentuximab vedotin and T-DM1. Both programs focused primarily on the potential for CYP-mediated DDIs with a dedicated clinical evaluation for brentuximab vedotin. MMAE, the small molecule drug of brentuximab vedotin and a number of other ADCs, was found to be a substrate of CYP3A, but neither an inhibitor nor inducer of CYP3A [29]. This was consistent with the preclinical data for MMAE and its potential for DDIs. A population PK analysis was performed and specific examples of coadministered drugs were evaluated to determine the potential for DDIs associated with DM1, the small molecule payload of T-DM1 [31]. No clear conclusion was drawn from this analysis.

5.5 DRUG-LINKER STABILITY

Drug-linker stability is a key attribute required for the development of a successful ADC and has been a primary focus in ADC technology [1, 36]. Premature release of the conjugated drug prior to reaching its intended target may cause unwanted systemic toxicity and reduce the amount of drug delivered to the target tissues. Early drug linkers, such as those employed by the cBR96-doxorubicin immunoconjugate and GO (hydrazone), did not result in clinically viable ADCs. Drug-linker stability, in addition to a number of other factors, was identified as contributing to the lack of clinical success [37]. New ADC design principles were constructed based on this experience and the current generation of drug linkers was developed to overcome the initial design flaws. In order to confer plasma stability while maintaining the ability to release the drug in the target cells, protease-cleavable and noncleavable drug linkers have been developed and are in use.

Current ADC technology employs plasma stable linker chemistry that is designed to be either enzymatically cleavable, such as the dipeptide linker used in brentuximab vedotin, or noncleavable, such as the thioether linker in

T-DM1 [38, 39]. The cleavable linkers release drug by endosomal and lysosomal proteases without requiring the degradation of the mAb component [38], whereas noncleavable linkers require degradation of the mAb backbones to release the drug [40]. In addition, the type of drug linkers may also play a role in the circulation stability of an ADC. When trastuzumab was conjugated to DM1 using either a thioether or disulfide linker, the resulting ADCs had different plasma clearance [41].

A number of methods exist to attach a drug linker to an antibody. The current generation of ADCs typically employs maleimide chemistry to attach drug linkers to cysteine or lysine residues on antibodies. The potential for maleimide transfer to other nucleophiles or nucleophilic groups exists through a maleimide elimination reaction [42]. This transfer has been observed for ADCs; however, the circulating concentrations of these species are appreciably lower than those of ADC. Other mechanisms may compete with maleimide transfer, such as proteolytic cleavage in cells or hydrolysis of the linker. It has been found that hydrolysis of the succinimide ring in the linker can reduce maleimide transfer and result in more stable conjugation. Lyon et al. engineered a new class of drug linkers with a basic amino group incorporated adjacent to the maleimide that can catalyze the ring hydrolysis at neutral pH and room temperature. As a result, ADCs employing this chemistry are highly stable compared to traditional drug-linker formats [43]. In addition to using alternative chemistry, selection of the antibody conjugation sites through site-specific mutations can also confer stability or instability to the ADC construct. Conjugation to certain mutated cysteine or nonnatural amino acid sites appears to result in more stable ADCs than those using the conventional conjugation sites. Improved efficacy and tolerability in animal models have been shown with these site-specific conjugated ADCs [42, 44, 45].

5.6 ELIMINATION

Characterizing the route of elimination for an ADC and small molecule drug is a key to understanding the impact of certain patient-related factors, such as organ impairment, and the potential for drug interactions. An ADC is eliminated through two processes: catabolism through proteolytic degradation in tumor and other tissues to constituent amino acids of the mAb portion or deconjugation of the drug to form the naked mAb. Following drug release, the small molecule drug is excreted by either renal or hepatic pathways, or both [29, 32].

In the case of T-DM1, an integrated population PK analysis suggested that deconjugation accounted for slightly more than half of the ADC clearance while catabolism accounts for the rest of the ADC clearance [46]. For brentuximab vedotin, only MMAE was identified as the species released upon catabolism of the ADC [47]. Based on population PK analysis, patient body size was identified as a statistically significant covariate for the clearance of both brentuximab vedotin and T-DM1. Body weight-based dosing is considered appropriate for both ADCs [30, 31].

For the two primary drugs used in the construct of an ADC, MMAE and DM1, the majority of those drugs are excreted by the liver through the biliary pathway in rats and also likely in humans. Following T-DM1 administration about 80% of radioactivity was recovered in the feces and 50% was recovered in the bile in rats [32]. A clinical excretion study of brentuximab vedotin showed that, in a 1-week period, the primary excretion route of MMAE is via feces, which account for approximately 72% of the recovered MMAE. The rest of MMAE was recovered in urine [29]. Multiple MMAE metabolites following brentuximab vedotin administration in patients were identified in feces and urine; however, they were at very low levels and only detectable in highly concentrated samples [30]. These data suggest that, for DM1 and MMAE, the liver is the major elimination organ and ADCs releasing these drug species should be considered for the evaluation of relationship of hepatic function and drug exposure during clinical development.

5.7 CLINICAL PK

Many ADCs that are currently being evaluated in humans are a heterogeneous mixture consisting of various fractions of ADCs with different DARs and often a small amount of unconjugated antibody. Ideally, each individual species is monitored in the biological matrices due to its potentially different PK and PD (pharmacodynamic) and relationship to antitumor activity and safety; however, there are technical challenges to do so and current practices are to monitor three major analytes: the ADC that accounts for the conjugated mAb (or alternatively, antibody-conjugated cytotoxin), total antibody (TAb) that consists of both conjugated and unconjugated mAb, and unconjugated drug. Additional drug species, such as the metabolites of the release drug or catabolism products, may be evaluated if warranted based on preclinical and clinical data.

The typical PK profiles of ADC and TAb are generally similar to those of other antibody-based therapeutics. Maximum blood concentrations are achieved at the end of infusion following IV administration and a multiexponential decline in concentration with terminal half-lives in the range of days is typically observed. The volumes of distribution are usually close to the plasma and interstitial fluid volume. TAb concentrations are theoretically higher than those of the ADC and may appear to decline more slowly. Given the similarity of PK profiles between TAb and ADC, evaluation of both may not always be necessary in late-stage studies and one of them may be dropped or reduced if justified. The

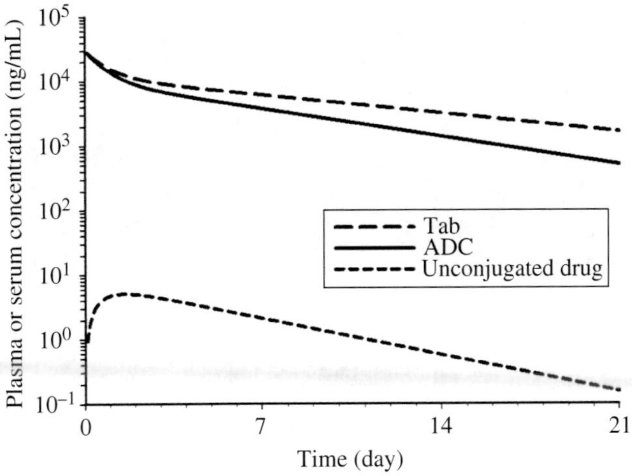

FIGURE 5.3 Typical PK profiles of the measured ADC analytes. (From Han and Zhao [4].)

released small molecule drug usually displays formation-limited kinetics with an apparent terminal half-life similar to that of the ADC, although this does not necessarily have to be the case. The difference in exposure between ADCs and released drug is very large, often in the range of 100–1000-fold when compared using the same molar concentration unit [26, 48]. The typical PK profiles of the abovementioned analytes and their relationship are shown in Figure 5.3.

As discussed earlier, the potential impact of DDI and organ impairment on ADC PK should be considered for evaluation given the high likelihood of concomitant use of other medicines and relatively severe conditions in the patient populations. The DDI potential for ADCs is potentially twofold—one related to the antibody and target properties of ADC, much like other therapeutic proteins, and the other related to the released drug, similar to other classic small molecule drugs [17, 49–51]. The DDI related to the former has been discussed in more detail and is not further expanded here. The latter can be found with extensive coverage in related literature. With only two approved ADCs on the U.S. market and many of the ADCs in clinical development employing the same or similar chemotherapy types as shown in Table 5.1, very limited preclinical and clinical data are available. For brentuximab vedotin, a clinical DDI study was conducted and the data suggested that brentuximab vedotin did not affect midazolam exposures and unconjugated MMAE exposures were 31–46% lower with rifampin and 34–73% higher with ketoconazole [29]. As expected, brentuximab vedotin ADC exposures were unaffected by concomitant rifampin or ketoconazole. Since the small molecule drug is typically eliminated through liver and/or kidney, its exposures might be elevated in patients with organ impairments. A clinical excretion study with brentuximab vedotin showed that MMAE was excreted primarily via feces with the rest via urine when dosed as ADC [29]. Results are not yet available from the clinical studies in patients with hepatic impairment for both brentuximab vedotin and T-DM1, although both are currently under investigation. Preliminary data suggested that mild or moderate renal impairment did not meaningfully affect the brentuximab vedotin ADC PK and severe renal impairment (creatinine clearance <30 mL/min) may be associated with an increase in the unconjugated MMAE exposures [52].

In summary, the clinical PK of ADCs have been evaluated for multiple analytes due to the complex nature of ADC construct and all the common considerations in PK, DDI, and organ impairment for therapeutic proteins and small molecule drugs should be applied to ADC drug development.

5.8 PK AND PK/PD MODELING FOR ADCs

An exposure–response relationship exists for all therapeutic proteins and small molecule drugs. During drug development, it is important to determine this relationship in order to guide first-in-human study designs based on preclinical data, and to identify the optimal dose regimens in the target diseases and patient populations. ADCs are no exception to these considerations and, on the contrary, given their complex PK and MOA, a more mechanistic approach is warranted in order to fully evaluate and identify the critical factors in understanding ADC PK/PD relationship.

First, it is critical to comprehend the complex PK characteristics ADCs display. As discussed earlier, most of the current ADCs in clinical development are a defined heterogeneous mixture of ADCs with different DARs. Once administered, the relative proportion of each DAR species changes due to deconjugation process and differential clearance, resulting in a decrease in average DAR over time [27, 53]. These changes may impact the PK of various measured analytes and, more importantly, the amount of small molecule payload delivered to the target-expressing cells at a given time. Ideally, these processes should be taken into account in a mechanistic PK or PK/PD model, especially for ADCs with longer circulating half-lives, as the effect could occur at later times. However, there are challenges in model implementation since individual DAR species are not often measured in the biological samples. For ADCs with short circulating half-lives, either due to target-mediated drug disposition (TMDD) or other reasons, an assumption of minimal deconjugation and differential clearance might be reasonable for the short duration of measurable drug exposure.

Multiple modeling efforts have been attempted to incorporate the deconjugation process and/or nonlinear clearance in ADC PK models. Chudasama et al. developed a semi-mechanistic population PK model to characterize T-DM1 concentration–time profiles in patients [54]. In their model, a series of transit compartments were used to describe the deconjugation process from higher to lower DARs and all

DAR species were assumed to have the same disposition parameters. Using rat and monkey PK data, where the concentrations of different T-DM1 DAR species were measured, a similar model suggested that the higher DAR species (≥3) deconjugated faster than those lower DAR species and were the slowest when DAR = 1 [55]. Model simulations suggest when dosed once every 3 weeks, the average DAR is close to 1 at steady-state Ctrough. Following the same approach and using an MMAE-conjugated ADC, Lu et al. found that a Weibull model best described the deconjugation rate constant change with DAR when fitting to monkey PK data [56]. Given the nonlinear clearance observed with some ADCs, it is expected that the TMDD model commonly used for mAb may be extended to ADC. A full mechanistic PK model encompassing all major processes of ADC distribution, binding, and elimination of different DAR species was proposed and multiple approximations were applied to simplified the model based on different assumptions [57]. Such models were able to describe the simulated T-DM1 PK data. It is conceivable that as ADC doses continue to move lower with more potent small molecule payloads or in a disease where the ADC targets are abundant and/or well perfused, nonlinear ADC clearance due to TMDD will be more frequently observed and such models more commonly applied.

Since ADCs are administered directly into the circulation, multiple processing steps are involved for an ADC to kill targeted cells [36]. The key processes are (i) distribution to the site of action (e.g., tumor for an oncology application); (ii) binding to the specific antigen target on the cell surface; (iii) internalization of the antigen target–ADC complex into cells; (iv) intracellular release of the potent small molecule payloads through degradation of the ADC and/or cleavage of the drug linker; and (v) interference of a critical cellular machinery to cause cell death. The ultimate driver of the pharmacologic effects of ADCs is the intracellular unconjugated drug concentrations. The antibody portion of an ADC can also confer pharmacological activities, but because of the intrinsically high potency of an ADC, the concentrations of circulation ADC typically are lower than those needed to engage antibody-mediated pharmacology. Given the multiple steps between ADC administration and small molecule drug release, the plasma drug exposure may not be a good surrogate of the target tissue exposure for ADCs. Therefore, a mechanistic approach is needed to account for the key steps as discussed above and to understand the PK/PD relationship of an ADC and its antitumor effect.

Based on the MOA, the following factors should be considered in mechanistic PK/PD models for ADCs:

- The DAR and stability of the ADC drug linker in circulation
- The accessibility of the sites of action to ADC, for example, leukemia versus lymphoma versus solid tumors
- The size of the tumor or disease burden
- The abundance and production rate (or turnover rate) of the cell surface target
- The binding affinity (or k_{on} and k_{off}) of the ADC to its target and, upon binding, the rate of internalization and release of the small molecule drug
- The rate associated with the drug diffusion into and out of cells and tissues, binding and disassociation to its intracellular target
- The distance of the target-expressing cells from microvasculature
- The change in disease burden following treatment and the resulting altered total target amount.

In a PK/PD model developed using published brentuximab vedotin preclinical and clinical data, Shah and colleagues have demonstrated that multiscale PK and PD data are needed to successfully build a mechanistic PK/PD model for an ADC and the model can be used to describe all *in vitro* and *in vivo* data and predict clinical responses [58]. Some of the key features of the mechanistic ADC PK/PD model include simultaneous characterization of the blood PK of both ADC and the released small molecule payload, deconjugation of the payload from ADC in circulation and tissues, incorporation of both surface and vascular exchange for ADC and payload distribution between central and tumor compartments, receptor binding and internalization of ADC into cells and intracellular payload release, and tumor intracellular payload concentration-driven cell kill and tumor shrinkage. In a subsequent publication, Shah further validated the mechanistic PK/PD model for its ability to *a priori* predict tumor concentrations of ADC and the released payload [59]. Model simulations and sensitivity analyses suggest that the deconjugation process and tumor size may contribute to the released payload exposure in blood and tumor. Given the nonlinearity involved in many of the ADC PK and PD processes, it was of no surprise for the analyses to reveal that the sensitivity of certain parameters is dose dependent. Additional improvement to the PK/PD model could potentially be made by including the more mechanistic deconjugation and clearance processes as discussed above on ADC PK model development. Nonetheless, the mechanistic PK/PD model developed provides a useful tool for translating preclinical experimental data to clinical predictions and for evaluating alternative ADCs during preclinical lead selection and optimization. More details in modeling methods and strategies are discussed in a separate chapter.

5.9 SUMMARY

ADC technologies utilize the specificity that antibodies bring and the potency of the small molecule drug to deliver a highly active agent to target-expression cells and minimize

TABLE 5.3 Attributes of Small Molecule Drugs (SMDs), Antibodies (mAb), and ADCs. ADCs are a class of complex therapeutic modality composed of a monoclonal antibody and a small molecule drug that are chemically conjugated together. They carry many of the same attributes of therapeutic antibodies and small molecule drugs that require careful characterization and optimization during preclinical and clinical development

	SMD	mAb	ADC
MW (Da)	Typically <1 K	~150 K	~150 K
Dosing route	Typically orally	Subcutaneous or intravenous	Intravenous
Distribution	High V_d and well distributed into tissues; potential substrate of transporters	V_d close to plasma volume; limited tissue distribution	ADC similar to mAb; unconjugated drug similar to SMD
Metabolism	Mainly phase I and II metabolism	Catabolism	Both
Excretion	Mainly biliary and renal excretion	Mostly recycled by body	Both
Circulation half-life	Typically short (h)	Long (d)	Long for both ADC and unconjugated drug (d) when dosed as ADC
PK linearity	Usually linear at low dose and nonlinear at high dose	Usually linear at high dose and nonlinear at low dose	Similar to mAb
PK analytes	Drug and metabolites	Antibody	ADC or conjugated drug, total antibody, and unconjugated drug
Analytical methods	Typically LC-MS/MS	Typically ligand binding assays	Both
Immunogenicity	No	Yes	Yes
API heterogeneity	Uniform single entity	Generally uniform	Mixture of ADCs with different DARs

From Han and Zhao [4].

API, active pharmaceutical ingredient; LC-MS/MS, liquid chromatography with tandem mass spectrometry; MW, molecular weight; V_d, volume of distribution.

systemic exposure to this agent. Successes in marketing approval and current clinical development have demonstrated the unique place of ADCs in the treatment of cancers and potentially other diseases. ADCs are a complex drug modality and careful considerations and evaluations of many of the ADME and PK/PD properties are required to fully achieve their potential and realize the promises ADCs can bring. A number of these factors have been discussed and a summary of some of the attributes for ADCs in comparison to mAb and small molecule drug is provided in Table 5.3 for reference.

REFERENCES

[1] Carter PJ, Senter PD. Antibody-drug conjugates for cancer therapy. Cancer J 2008;14:154–169.

[2] Strebhardt K, Ullrich A. Paul Ehrlich's magic bullet concept: 100 years of progress. Nat Rev Cancer 2008;8:473–480.

[3] Bross PF, Beitz J, Chen G, Chen XH, Duffy E, Kieffer L, Roy S, Sridhara R, Rahman A, Williams G, Pazdur R. Approval summary: gemtuzumab ozogamicin in relapsed acute myeloid leukemia. Clin Cancer Res 2001;7:1490–1496.

[4] Han TH, Zhao B. Absorption, distribution, metabolism, and excretion considerations for the development of antibody-drug conjugates. Drug Metab Dispos 2014;42:1914–1920.

[5] Alley SC, Zhang X, Okeley NM, Anderson M, Law CL, Senter PD, Benjamin DR. The pharmacologic basis for antibody-auristatin conjugate activity. J Pharmacol Exp Ther 2009;330:932–938.

[6] Boswell CA, Mundo EE, Firestein R, Zhang C, Mao W, Gill H, Young C, Ljumanovic N, Stainton S, Ulufatu S, Fourie A, Kozak KR, Fuji R, Polakis P, Khawli LA, Lin K. An integrated approach to identify normal tissue expression of targets for antibody-drug conjugates: case study of TENB2. Br J Pharmacol 2013;168:445–457.

[7] Boswell CA, Mundo EE, Zhang C, Stainton SL, Yu SF, Lacap JA, Mao W, Kozak KR, Fourie A, Polakis P, Khawli LA, Lin K. Differential effects of predosing on tumor and tissue uptake of an 111In-labeled anti-TENB2 antibody-drug conjugate. J Nucl Med 2012;53:1454–1461.

[8] Boswell CA, Tesar DB, Mukhyala K, Theil FP, Fielder PJ, Khawli LA. Effects of charge on antibody tissue distribution and pharmacokinetics. Bioconjug Chem 2010;21:2153–2163.

[9] Boylan NJ, Zhou W, Proos RJ, Tolbert TJ, Wolfe JL, Laurence JS. Conjugation site heterogeneity causes variable electrostatic properties in Fc conjugates. Bioconjug Chem 2013;24:1008–1016.

[10] Yip V, Palma E, Tesar DB, Mundo EE, Bumbaca D, Torres EK, Reyes NA, Shen BQ, Fielder PJ, Prabhu S, Khawli LA, Boswell CA. Quantitative cumulative biodistribution of antibodies in mice: effect of modulating binding affinity to the neonatal Fc receptor. MAbs 2014;6:689–696.

[11] Law CL, Cerveny CG, Gordon KA, Klussman K, Mixan BJ, Chace DF, Meyer DL, Doronina SO, Siegall CB, Francisco JA, Senter PD, Wahl AF. Efficient elimination of B-lineage lymphomas by anti-CD20-auristatin conjugates. Clin Cancer Res 2004;10:7842–7851.

[12] Breij EC, De Goeij BE, Verploegen S, Schuurhuis DH, Amirkhosravi A, Francis J, Miller VB, Houtkamp M, Bleeker WK, Satijn D, Parren PW. An antibody-drug conjugate that targets tissue factor exhibits potent therapeutic activity against a broad range of solid tumors. Cancer Res 2014;74:1214–1226.

[13] Chu X, Korzekwa K, Elsby R, Fenner K, Galetin A, Lai Y, Matsson P, Moss A, Nagar S, Rosania GR, Bai JP, Polli JW, Sugiyama Y, Brouwer KL, International Transporter C. Intracellular drug concentrations and transporters: measurement, modeling, and implications for the liver. Clin Pharmacol Ther 2013;94:126–141.

[14] Genentech. *KADCYLA™ Prescribing Information*. South San Francisco, CA: Genentech, Inc.; 2013.

[15] Seattle Genetics Inc.. *ADCETRIS™ Prescribing Information*. Bothell, WA: Seattle Genetics, Inc.; 2011.

[16] Dickmann LJ, Patel SK, Wienkers LC, Slatter JG. Effects of interleukin 1beta (IL-1beta) and IL-1beta/interleukin 6 (IL-6) combinations on drug metabolizing enzymes in human hepatocyte culture. Curr Drug Metab 2012; 13:930–937.

[17] Evers R, Dallas S, Dickmann LJ, Fahmi OA, Kenny JR, Kraynov E, Nguyen T, Patel AH, Slatter JG, Zhang L. Critical review of preclinical approaches to investigate cytochrome p450-mediated therapeutic protein drug-drug interactions and recommendations for best practices: a white paper. Drug Metab Dispos 2013;41:1598–1609.

[18] Kenny JR, Liu MM, Chow AT, Earp JC, Evers R, Slatter JG, Wang DD, Zhang L, Zhou H. Therapeutic protein drug-drug interactions: navigating the knowledge gaps-highlights from the 2012 AAPS NBC Roundtable and IQ Consortium/FDA workshop. AAPS J 2013;15:933–940.

[19] Slatter JG, Wienkers LC, Dickmann LJ. Drug interactions of cytokines and anticytokine therapeutic proteins. Pharm Sci Encycl 2013;13:215–237.

[20] Bonnel D, Legouffe R, Willand N, Baulard A, Hamm G, Deprez B, Stauber J. MALDI imaging techniques dedicated to drug-distribution studies. Bioanalysis 2011;3:1399–1406.

[21] Hengel SM, Sanderson R, Valliere-Douglass J, Nicholas N, Leiske C, Alley SC. Measurement of *in vivo* drug load distribution of cysteine-linked antibody-drug conjugates using microscale liquid chromatography mass spectrometry. Anal Chem 2014;86:3420–3425.

[22] Adem YT, Schwarz KA, Duenas E, Patapoff TW, Galush WJ, Esue O. Auristatin antibody drug conjugate physical instability and the role of drug payload. Bioconjug Chem 2014;25(4):656–664.

[23] Guo J, Kumar S, Prashad A, Starkey J, Singh SK. Assessment of physical stability of an antibody drug conjugate by higher order structure analysis: impact of thiol-maleimide chemistry. Pharm Res 2014;31:1710–1723.

[24] Beckley NS, Lazzareschi KP, Chih HW, Sharma VK, Flores HL. Investigation into temperature-induced aggregation of an antibody drug conjugate. Bioconjug Chem 2013;24: 1674–1683.

[25] Bartlett NL, Younes A, Carabasi MH, Forero A, Rosenblatt JD, Leonard JP, Bernstein SH, Bociek RG, Lorenz JM, Hart BW, Barton J. A phase 1 multidose study of SGN-30 immunotherapy in patients with refractory or recurrent CD30+ hematologic malignancies. Blood 2008;111:1848–1854.

[26] Younes A, Bartlett NL, Leonard JP, Kennedy DA, Lynch CM, Sievers EL, Forero-Torres A. Brentuximab vedotin (SGN-35) for relapsed CD30-positive lymphomas. N Engl J Med 2010;363:1812–1821.

[27] Hamblett KJ, Senter PD, Chace DF, Sun MM, Lenox J, Cerveny CG, Kissler KM, Bernhardt SX, Kopcha AK, Zabinski RF, Meyer DL, Francisco JA. Effects of drug loading on the antitumor activity of a monoclonal antibody drug conjugate. Clin Cancer Res 2004;10:7063–7070.

[28] Tjelle TE, Brech A, Juvet LK, Griffiths G, Berg T. Isolation and characterization of early endosomes, late endosomes and terminal lysosomes: their role in protein degradation. J Cell Sci 1996;109(Pt. 12):2905–2914.

[29] Han TH, Gopal AK, Ramchandren R, Goy A, Chen R, Matous JV, Cooper M, Grove LE, Alley SC, Lynch CM, O'Connor OA. CYP3A-mediated drug-drug interaction potential and excretion of brentuximab vedotin, an antibody-drug conjugate, in patients with CD30-positive hematologic malignancies. J Clin Pharmacol 2013;53:866–877.

[30] FDA Center for Drug Evaluation and Research US (2011) Clinical pharmacology and biopharmaceutics review(s). Application Number: 125388Orig1s000.

[31] FDA Center for Drug Evaluation and Research US (2013) Clinical pharmacology and biopharmaceutics review(s). Application Number: 125427Orig1s000.

[32] Shen BQ, Bumbaca D, Saad O, Yue Q, Pastuskovas CV, Khojasteh SC, Tibbitts J, Kaur S, Wang B, Chu YW, Lorusso PM, Girish S. Catabolic fate and pharmacokinetic characterization of trastuzumab emtansine (T-DM1): an emphasis on preclinical and clinical catabolism. Curr Drug Metab 2012a;13:901–910.

[33] Davis JA, Rock DA, Wienkers LC, Pearson JT. *In vitro* characterization of the drug-drug interaction potential of catabolites of antibody-maytansinoid conjugates. Drug Metab Dispos 2012;40:1927–1934.

[34] Erickson HK, Lambert JM. ADME of antibody-maytansinoid conjugates. AAPS J 2012;14:799–805.

[35] Sun X, Widdison W, Mayo M, Wilhelm S, Leece B, Chari R, Singh R, Erickson H. Design of antibody-maytansinoid conjugates allows for efficient detoxification via liver metabolism. Bioconjug Chem 2011;22:728–735.

[36] Senter PD. Potent antibody drug conjugates for cancer therapy. Curr Opin Chem Biol 2009;13:235–244.

[37] Burke PJ, Hamilton JZ, Hunter JH, Jeffrey SC, Doronina SO, Okeley NM, Meyer DW, Senter PD, Lyon RP. Development and pharmacological properties of PEGylated glucuronide-auristatin linkers. Cancer Res 2014;74(19 Suppl):Abstract nr 1786.

[38] Doronina SO, Toki BE, Torgov MY, Mendelsohn BA, Cerveny CG, Chace DF, Deblanc RL, Gearing RP, Bovee TD, Siegall CB, Francisco JA, Wahl AF, Meyer DL, Senter PD. Development of potent monoclonal antibody auristatin conjugates for cancer therapy. Nat Biotechnol 2003;21:778–784.

[39] Lorusso PM, Weiss D, Guardino E, Girish S, Sliwkowski MX. Trastuzumab emtansine: a unique antibody-drug conjugate in development for human epidermal growth factor receptor 2-positive cancer. Clin Cancer Res 2011;17:6437–6447.

[40] Erickson HK, Park PU, Widdison WC, Kovtun YV, Garrett LM, Hoffman K, Lutz RJ, Goldmacher VS, Blattler WA. Antibody-maytansinoid conjugates are activated in targeted cancer cells by lysosomal degradation and linker-dependent intracellular processing. Cancer Res 2006;66:4426–4433.

[41] Erickson HK, Lewis Phillips GD, Leipold DD, Provenzano CA, Mai E, Johnson HA, Gunter B, Audette CA, Gupta M, Pinkas J, Tibbitts J. The effect of different linkers on target cell catabolism and pharmacokinetics/pharmacodynamics of trastuzumab maytansinoid conjugates. Mol Cancer Ther 2012;11:1133–1142.

[42] Shen BQ, Xu K, Liu L, Raab H, Bhakta S, Kenrick M, Parsons-Reponte KL, Tien J, Yu SF, Mai E, Li D, Tibbitts J, Baudys J, Saad OM, Scales SJ, McDonald PJ, Hass PE, Eigenbrot C, Nguyen T, Solis WA, Fuji RN, Flagella KM, Patel D, Spencer SD, Khawli LA, Ebens A, Wong WL, Vandlen R, Kaur S, Sliwkowski MX, Scheller RH, Polakis P, Junutula JR. Conjugation site modulates the *in vivo* stability and therapeutic activity of antibody-drug conjugates. Nat Biotechnol 2012b;30:184–189.

[43] Lyon RP, Setter JR, Bovee TD, Doronina SO, Anderson ME, Leiske CL, Senter PD. Self-stabilizing ADCs: antibody-drug conjugates prepared with maleimido drug-linkers that catalyze their own thiosuccinimide ring hydrolysis. Cancer Res 2013;73 Abstract nr 4333.

[44] Junutula JR, Flagella KM, Graham RA, Parsons KL, Ha E, Raab H, Bhakta S, Nguyen T, Dugger DL, Li G, Mai E, Lewis Phillips GD, Hiraragi H, Fuji RN, Tibbitts J, Vandlen R, Spencer SD, Scheller RH, Polakis P, Sliwkowski MX. Engineered thio-trastuzumab-DM1 conjugate with an improved therapeutic index to target human epidermal growth factor receptor 2-positive breast cancer. Clin Cancer Res 2010;16:4769–4778.

[45] Junutula JR, Raab H, Clark S, Bhakta S, Leipold DD, Weir S, Chen Y, Simpson M, Tsai SP, Dennis MS, Lu Y, Meng YG, Ng C, Yang J, Lee CC, Duenas E, Gorrell J, Katta V, Kim A, McDorman K, Flagella K, Venook R, Ross S, Spencer SD, Lee Wong W, Lowman HB, Vandlen R, Sliwkowski MX, Scheller RH, Polakis P, Mallet W. Site-specific conjugation of a cytotoxic drug to an antibody improves the therapeutic index. Nat Biotechnol 2008;26:925–932.

[46] Lu D, Joshi A, Wang B, Olsen S, Yi JH, Krop IE, Burris HA, Girish S. An integrated multiple-analyte pharmacokinetic model to characterize trastuzumab emtansine (T-DM1) clearance pathways and to evaluate reduced pharmacokinetic sampling in patients with HER2-positive metastatic breast cancer. Clin Pharmacokinet 2013;52:657–672.

[47] Okeley NM, Miyamoto JB, Zhang X, Sanderson RJ, Benjamin DR, Sievers EL, Senter PD, Alley SC. Intracellular activation of SGN-35, a potent anti-CD30 antibody-drug conjugate. Clin Cancer Res 2010;16:888–897.

[48] Girish S, Gupta M, Wang B, Lu D, Krop IE, Vogel CL, Burris Iii HA, Lorusso PM, Yi JH, Saad O, Tong B, Chu YW, Holden S, Joshi A. Clinical pharmacology of trastuzumab emtansine (T-DM1): an antibody-drug conjugate in development for the treatment of HER2-positive cancer. Cancer Chemother Pharmacol 2012;69:1229–1240.

[49] Huang SM, Zhao H, Lee JI, Reynolds K, Zhang L, Temple R, Lesko LJ. Therapeutic protein-drug interactions and implications for drug development. Clin Pharmacol Ther 2010;87:497–503.

[50] Mahmood I, Green MD. Drug interaction studies of therapeutic proteins or monoclonal antibodies. J Clin Pharmacol 2007;47:1540–1554.

[51] Seitz K, Zhou H. Pharmacokinetic drug-drug interaction potentials for therapeutic monoclonal antibodies: reality check. J Clin Pharmacol 2007;47:1104–1118.

[52] Han TH, Grove LE, Lynch CM. Effect of renal impairment on the pharmacokinetics of brentuximab vedotin, an antibody-drug conjugate, in patients with CD30-positive hematologic malignancie. Clin Pharmacol Ther 2013;93:S17.

[53] Sanderson RJ, Hering MA, James SF, Sun MM, Doronina SO, Siadak AW, Senter PD, Wahl AF. *In vivo* drug-linker stability of an anti-CD30 dipeptide-linked auristatin immunoconjugate. Clin Cancer Res 2005;11:843–852.

[54] Chudasama VL, Schaedeli Stark F, Harrold JM, Tibbitts J, Girish SR, Gupta M, Frey N, Mager DE. Semi-mechanistic population pharmacokinetic model of multivalent trastuzumab emtansine in patients with metastatic breast cancer. Clin Pharmacol Ther 2012;92:520–527.

[55] Bender B, Leipold DD, Xu K, Shen BQ, Tibbitts J, Friberg LE. A mechanistic pharmacokinetic model elucidating the disposition of trastuzumab emtansine (T-DM1), an antibody-drug conjugate (ADC) for treatment of metastatic breast cancer. AAPS J 2014;16(5):994–1008.

[56] Lu D, Ng C, Agarwal P, Girish S, Li D, Prabhu S, Nazzal D, Saad O, Dere R, Koppada N, Jin J. Semi-mechanistic multiple-analyte population model of antibody-drug-conjugate pharmacokinetics [abstract]. American Conference on Pharmacometrics (ACoP) 2013; 2013; abstract nr W-006.

[57] Gibiansky L, Gibiansky E. Target-mediated drug disposition model and its approximations for antibody-drug conjugates. J Pharmacokinet Pharmacodyn 2014;41:35–47.

[58] Shah DK, Haddish-Berhane N, Betts A. Bench to bedside translation of antibody drug conjugates using a multiscale mechanistic PK/PD model: a case study with brentuximab-vedotin. J Pharmacokinet Pharmacodyn 2012; 39:643–659.

[59] Shah DK, King LE, Han X, Wentland JA, Zhang Y, Lucas J, Haddish-Berhane N, Betts A, Leal M. A priori prediction of tumor payload concentrations: preclinical case study with an auristatin-based anti-5T4 antibody-drug conjugate. AAPS J 2014;16:452–463.

6

ROLE OF LYMPHATIC SYSTEM IN SUBCUTANEOUS ABSORPTION OF THERAPEUTIC PROTEINS

JIUNN H. LIN[1] AND WEIRONG WANG[2]

[1]3D BioOptimal Co Ltd., Suzhou, China

[2]Janssen Research and Development, LLC, Spring House, PA, USA

6.1 INTRODUCTION

Although oral administration is generally preferred in drug therapy, protein drugs cannot be delivered orally to patients due to rapid degradation in the acidic and protease-rich environment in the gastrointestinal (GI) tract. Therefore, therapeutic proteins can only be injected by parenteral routes, including intravenous (IV) and subcutaneous (SC) injection. Currently, IV administration is the most common route for delivering protein drugs to patients. However, it requires trained and skilled nursing personnel to perform the IV injection in a clinical setting. In addition to long infusion time (30–90 min), IV administration is often associated with the risks of infection, clot formation in blood circulation (thromboembolic events), and other related side effects. In view of the disadvantages associated with the IV injection, it is highly desirable for therapeutic proteins to be self-administered SC by patient themselves providing convenience for patients and improving the quality of life and treatment adherence. As a result, more protein drugs have been approved for SC administration in recent years offering an alternative to current practice of IV administration [1].

While SC administration represents an attractive route of drug delivery for protein drugs, our knowledge of the underlying mechanisms of SC absorption as well as the factors that govern the rate and extent of SC absorption is still limited. Following SC administration, drugs can be transported from the injection site to the blood circulation directly via blood capillaries and indirectly via the lymphatic capillaries (Fig. 6.1). Depending on the molecular weight (size) and biophysical properties of proteins, the relative contribution of blood capillary-mediated transport versus lymphatic capillary-mediated transport to the overall SC absorption may vary significantly. It is generally believed that the lymphatic system plays a quantitatively more important role in the absorption of protein drugs with a molecular weight of greater than 20–30 kDa after SC administration [2–4]. However, this general belief has recently been challenged by some scientists, who believe that direct uptake of protein drugs from the interstitial space into blood capillaries plays a quantitatively more important role than lymphatic capillaries in SC absorption [2–4].

Bioavailability of protein drugs after SC administration is quite variable ranging from 20% to 95% [4, 5]. Some protein drugs have an almost complete bioavailability, while others have poor bioavailability. One possible source of the variation in the bioavailability of protein drugs is the difference in presystemic catabolism of proteins in the interstitial space and/or during the transport in lymphatic vessels after SC administration. However, our knowledge of protein catabolism is also limited. This chapter provides an overview of current knowledge of the processes of protein absorption after SC administration. In addition, the underlying mechanisms that govern the SC absorption as well as the factors that influence the SC absorption will also be discussed.

ADME and Translational Pharmacokinetics/Pharmacodynamics of Therapeutic Proteins: Applications in Drug Discovery and Development, First Edition. Edited by Honghui Zhou and Frank-Peter Theil.

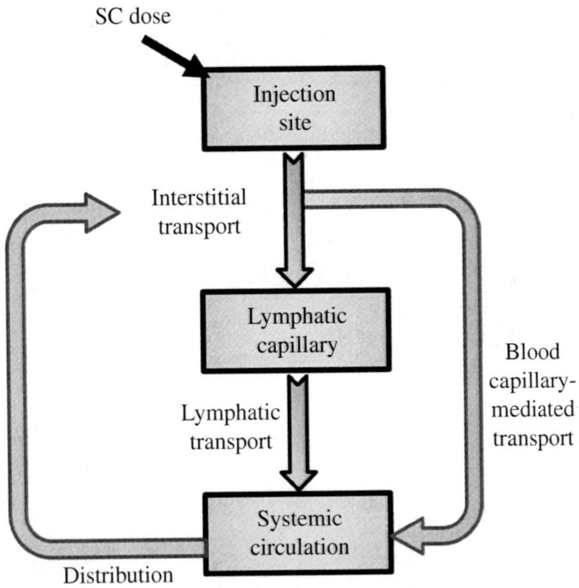

FIGURE 6.1 Scheme of SC absorption of protein drugs via blood capillary-mediated and lymphatic capillary-mediated transport into the systemic circulation.

6.2 PHYSIOLOGY OF SUBCUTANEOUS TISSUE

Subcutaneous tissue, also known as hypodermis, is located beneath the skin. The hypodermis consists primarily of loose connective tissue and lobules of fat, and is highly vascular. Types of cells that are found in the SC tissue are fibroblasts, adipose cells, and macrophages. Blood vessels are extensively distributed throughout the SC tissue forming an intensive network [6]. In addition, lymphatic capillaries are also extensively distributed in the SC tissue in close proximity to blood capillaries and form a lymphatic network that mirrors the blood network. The structural characteristics of blood capillaries that connect arterioles and venules are similar in all SC tissues. The walls of blood capillaries are composed of one-cell thick layer without smooth muscle. The unique structure of blood capillaries enables the exchange of plasma constituents between the blood and the tissues surrounding them.

At the end of blood arterioles, water and many constituents of plasma, including many proteins, are forced out the wall of blood capillaries by hydrostatic pressure through the endothelial pores to interstitium forming interstitial fluid. Reversely, the osmotic pressure drives water and interstitial proteins back into the venous and lymphatic capillaries. In adult humans, every day approximately 20 L of plasma is filtered through blood capillaries. Approximately 17 L of the filtered plasma water get reabsorbed directly into the blood vessels, while the remaining 3 L of interstitial fluid including proteins is reabsorbed into lymphatic capillaries before returning into the blood circulation [6]. Therefore, the main driving forces for the interstitial transport of protein drugs are the hydrostatic and osmotic pressures that occur among blood capillaries, interstitium, and lymphatic capillaries.

The lymphatic system, composed of lymphatic capillaries, collecting vessels, lymph nodes, lymph trunks, and lymph ducts, is a circulation network throughout the body [6]. Lymph forms when interstitial fluid (plasma water and proteins) moves into lymphatic capillaries. The lymph then drains from the capillaries into collecting vessels passing through one or more lymph nodes, where the lymph then drains into large lymph trucks, which in turn leads into the lymph ducts. Eventually, the ducts return the lymph back into the blood circulation at subclavian vein, completing the circuit of lymph transport (the cycle of plasma-tissue fluid-lymph circulation). In humans, there are two drainage areas that make up the entire lymphatic system. Lymph from the head and neck, right arm, and upper right trunk enters the circulation via the right lymph duct at the right subclavian vein. Lymph from all other regions of the body drains into the blood at the joining of the left subclavian vein and left jugular vein, below the collar bone via the left lymph duct, also known as thoracic lymph duct [6].

Unlike the blood circulation, the lymphatic system is not a closed system. The anatomic structures of the lymphatic capillaries are markedly different from those of blood capillaries. The walls of blood capillaries are composed of a sealed endothelium with tight junctions that allow small molecules but limit large molecules to freely diffuse. On the other hand, lymphatic capillaries have discontinuous basement membranes and lack tight junctions between their endothelial cells. The estimated distance of intercellular junction of lymphatic capillaries varies from about 50 nm to several micrometers [6–8]. Because of the differences in the vascular architecture, molecules that are smaller than 5 nm are preferentially absorbed into the blood capillaries, while larger molecules that are greater than 10 nm are favorably absorbed from the interstitium into the more porous lymphatics after SC injection [2].

In summary, the SC tissue is highly vascular and filled with the networks of blood and lymphatic circulation. Because of the distinct differences in the structure of walls between blood and lymphatic capillaries, it is generally believed that most of the plasma water that is filtered out from the blood is reabsorbed from the interstitial space into blood circulation via blood capillaries, while the plasma proteins that are forced out the blood circulation are reabsorbed through the lymphatic capillaries.

6.3 INTERSTITIAL TRANSPORT FROM SC INJECTION SITE

A SC injection is given as a bolus into the interstitial area of hypodermis, which is filled with the networks of blood and lymphatic circulations. After SC administration, therapeutic proteins must be transported from the injection site through interstitial space of the SC tissue to the blood and/or lymphatic capillaries before being absorbed into the systemic circulation. The interstitium can be divided into two compartments: the interstitial fluid and the solid extracellular matrix, which consists of a fibrous collagen network embedded in a gel of proteoglycans that are proteins heavily glycosylated with glycosaminoglycan. Thus, the interstitium is a mixture of fluid space and solid barrier that buffers diffusive and convective forces. As aforementioned, the main driving forces for the interstitial transport of therapeutic proteins are the hydrostatic and osmotic pressures. In addition to the main driving forces, the molecular size (weight) and biophysical properties of therapeutic proteins also play an important role in the interstitial transport.

Molecular size of therapeutic proteins plays an important role in the rate of the interstitial transport. In an *in vitro* study, the diffusion coefficient of macromolecules with variable molecular size was measured in a cell–collagen matrix model by using the method of fluorescence correlation spectroscopy. The results suggested that the diffusion coefficients of biomolecules were inversely correlated with molecular weight and size [9]. The mean diffusion coefficient of Alexa488-dextran (10 kDa), FITC-dextran (40 kDa), and Alexa488-IgG (immunoglobulin G) (150 kDa) was 81.7, 44.7, and 29.6 $\mu m^2/s$, respectively, while the corresponding mean value of Stokes radius (hydrodynamic radius) was 3.56, 6.52, and 9.84 nm. These results are consistent with the notion that the collagen network and proteoglycans in the interstitium serve as physiological barriers to protein transport after SC administration.

While molecular size is an important factor that governs the rate of interstitial transport of proteins, electric charge of therapeutic proteins can also influence their interstitial transport from the SC injection site to blood and lymphatic capillaries. Collagen is positively charged at physiological pH, while proteoglycans are highly negatively charged. Overall, there is a net negative charge in the interstitium [6]. Using an *in vivo* mouse model that can directly measure the interstitial transport rate of macromolecules, the interstitial velocity of anionic dextran (3k-An-Dx; 3 kDa) was estimated to be 42 μm/s, while the interstitial velocity of neutral dextran (3k-Dx; 3 kDa) was 34 μm/s [10]. The anionic dextran moves at a higher velocity through the interstitium than the neutral molecule.

One of the disadvantages of SC administration is the limitation of drug injection volume. Because of the solid structure of extracellular matrix in hypodermis, the maximum allowable drug volume for SC injection is 2 mL. When the drug volume of SC administration is greater than 2 mL, the general practice is that the dose must be divided into two or three syringes to administer at different sites. To overcome the volume limitation, a fully human recombinant DNA-derived hyaluronidase enzyme (rHuPH20) has been developed to temporarily modify the structure of hypodermis by hydrolyzing hyaluronan [11]. Because of rapid turnover rate, the loss of hyaluronan can be restored rapidly by resynthesis. Within 24 h of the injection of rHuPH20, the interstitial matrix structures are restored without histologic alterations or signs of inflammation [12].

In order to deliver a larger amount of dose to patients, SC formulations coformulated with rHuPH20 are currently under investigation for rituximab and trastuzumab [13, 14], and their clinical use have recently been approved in EU. In a clinical phase 1b study, 375–800 mg/m^2 of rituximab was delivered SC in a volume of 4.4–15 mL, when coformulated with rHuPH20 [14]. The increased injection volume is most likely due to an increased interstitial fluid volume at the injection site resulting from the rHuPH20-induced transient disruption of solid extracellular matrix. Therefore, rHuPH20 can be used as a special agent to increase injection volumes of therapeutic proteins for SC injection. Because the treatment of rHuPH20 can transiently widen the channels between collagen fibers and thereby increase the hydraulic conductivity of interstitium, it is hypothesized that treatment with rHuPH20 may increase the rate and extent of SC absorption of therapeutic proteins. However, at the current time, there are still limited clinical data to support this notion, especially on the increase of bioavailability.

In summary, there are many factors that can influence the SC absorption of therapeutic proteins. Molecular size and electric charge are known to influence the interstitial transport and thereby affect the absorption of protein drugs after SC administration. One drawback of SC administration is the limitation of drug injection volume. Hyaluronidase, such as rHuPH20, can be used as a special agent to transiently disrupt the solid extracellular matrix at the injection site and increase the injection drug volume.

6.4 RELATIVE ROLE OF BLOOD AND LYMPHATIC SYSTEMS IN SC ABSORPTION

Unlike the blood system that is a closed system, with the heart as its central motor, the lymphatic vessels do not form a closed circulatory system. There is no central pump and lymphatic vessels produce their own propulsion force via smooth musculatures. Because of the lack of central pump, the flow rate of lymph is approximately 100–500 times slower than the blood flow rate [6]. For example, the hepatic blood flow is in the range of 300–400 mL/min in dogs, while the thoracic lymphatic flow is about 1.5 mL/min in the dog under normal condition [15]. The slow lymphatic flow is one of the causes responsible for the prolonged time (T_{max}) to reach the peak concentration following SC administration of therapeutic proteins.

Lymphatic capillaries generally possess a more irregular and wider lumen than blood capillaries. The most unique structural feature of lymphatic capillaries is the discontinuity of the capillary wall and the ends of endothelial cells are overlapping. The lymphatic capillaries are generally observed in a partially or fully collapsed state. An increase in interstitial pressure causes these junctions to open, thereby permitting the passage of interstitial water and proteins into the lymphatic capillaries. In other words, osmotic pressure is the main driving force causing lymphatic filling. The openings of lymphatic capillary walls have been estimated to be in the range of 15–20 nm to several micrometers [16]. In contrast, the walls of blood capillaries consist of continuous one-cell layer of endothelium with loose cell–cell junctions, which permit the exchanges of plasma constituents between the blood and the interstitial fluid. The physiological upper limit of pore size in the walls of nonfenestrated blood capillaries is approximately 5 nm [17].

Following SC injection, drug molecules can be absorbed either by the blood capillaries or by lymphatic capillaries, or by a combination thereof. In view of the structural differences in the capillary walls between blood and lymphatic capillaries, it is expected that the lymphatic capillaries may play a quantitatively more important role in the absorption of large molecular protein drugs as compared to blood capillaries after SC administration, while small molecular compounds, particularly for lipophilic compounds, will be predominately absorbed by the blood capillaries. Therefore, the relative contribution of blood or lymphatic capillaries to the overall SC absorption of protein drugs may vary significantly depending on their molecular size as well as biophysical properties.

The role of lymphatics in SC absorption of therapeutic proteins has been the subject of intensive investigation. Using the lymphatic-cannulated sheep model, the role of lymphatics in SC absorption has been evaluated by measuring the cumulative recovery of therapeutic proteins in popliteal lymph [18]. The combined results from a number of studies have demonstrated a positive relationship between the molecular weight and cumulative recovery in lymph after SC administration (Fig. 6.2). In this sheep model, small and large molecule compounds were injected SC into the interdigital space of the hind limb, and the peripheral lymph from the popliteal lymph vessel in close proximity to the injection site was collected for measuring the recovery of administered dose [16, 18]. As shown in Figure 6.2, the cumulative recovery of the administered dose in lymph increased with increasing molecular size. Proteins with a molecular weight exceeding 20–30 kDa are preferentially absorbed by lymphatic system, while the SC absorption is almost completely via the lymphatic transport for proteins with a molecular weight of greater than 50–60 kDa. These results strongly suggest that the fraction of dose absorbed by lymphatic capillaries after SC administration increases as the molecular weight of proteins increases.

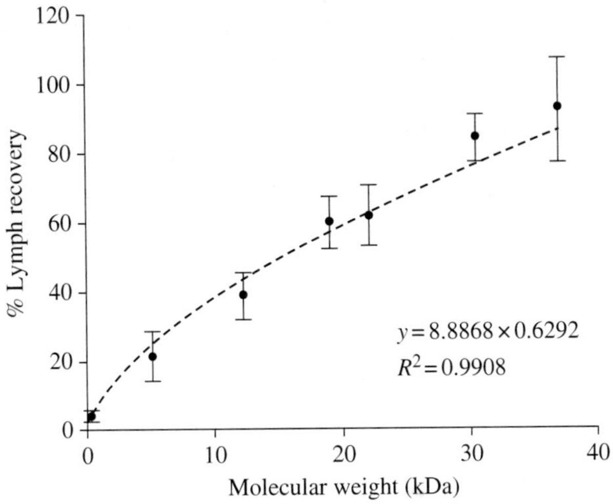

FIGURE 6.2 Relationship between the fraction of the dose absorbed into peripheral lymph and the molecular weight of proteins in sheep. (Modified from Porter and Charman [16] and McLennan et al. [18].)

Although it is generally believed that lymphatic transport plays a quantitatively more important role in the absorption of therapeutic proteins following SC administration, the general belief has been challenged by Kagan et al. [19]. Based on the results of a rat study, the investigators claimed that lymphatic capillaries played only a minor role in SC absorption of proteins. In this study, bovine insulin (5.6 kDa), recombinant human erythropoietin (rhEPO) alfa (30.4 kDa), and bovine albumin (66 kDa) were used as model compounds. The cumulative amount of insulin, erythropoietin, and albumin in lymph collected from the thoracic duct-cannulated rats was very low after SC administration, ranging from 0.07% to 2.0% dose. Because the cumulative lymph recovery of insulin, erythropoietin, and albumin after SC administration was so low, the investigators concluded that the proteins were absorbed directly by the blood capillaries while the lymphatic capillaries did not significantly contribute to the SC absorption of proteins [19]. Similarly, low recovery of therapeutic proteins has also been reported by other investigators. Kojima et al. reported that the recovery of recombinant human tumor necrosis factor (17 kDa) from the thoracic duct-cannulated rats was only 0.03% [20]. Apparently, the role of lymphatic system in SC absorption of protein drugs appears to be much less important in rats than sheep.

However, as opposed to the Kagan's report, a reasonably good recovery of proteins from the thoracic duct-cannulated rats has been reported by Wang et al. [21]. Approximately 30% of the dosed PEG30-EPO (~60 kDa) was recovered in the lymph collected from thoracic duct-cannulated rats. On the other hand, the bioavailability of PEG30-EPO (~60 kDa) was 38% in noncannulated rats following SC administration, while 10% in the thoracic duct-cannulated rats. It is of interest to note that the recovery of dose (30%) collected from thoracic duct lymph is in good agreement with the difference (28%) in the bioavailability of PEG30-EPO between the noncannulated and thoracic duct-cannulated rats. These results suggest that lymphatic capillary-mediated absorption accounted for approximately 80% of the bioavailability of PEG30-EPO after SC administration. Similarly, a reasonably good lymph recovery from the thoracic duct has also been reported for PEGylated polymers in rats [22]. The partial 30-h lymph recovery of the PEGylated polymer (Lys_{16} $[PEG_{2000}]_{32}$) was 29% after SC administration in rats. The PEGylated polymer has a molecular weight of 68 kDa and a radius of 6.7 nm. Moreover, in a recent study, the partial 30-h lymph recovery of trastuzumab (150 kDa) from the thoracic duct in rats was about 27% after SC administration [23]. The results of the above rat studies clearly suggest that the lymphatic uptake contributes significantly to the SC absorption of therapeutic proteins in rats.

The apparent discrepancy can be attributed to the differences in SC injection sites used in these studies. For small animals such as rats, it is technically difficult to collect lymph from peripheral lymphatic vessels near the region where protein drugs are SC injected, and the thoracic duct just above cisterna chyli has been the only site used for lymph collection. Depending on the site of SC injection, the recovery of proteins from the thoracic duct may vary markedly due to the complicated network of lymphatic circulation in rats. The routes of lymphatic drainage in rats have been mapped in detail by Tilney [24]. As shown in Figure 6.3, lymph in the upper region of rat left hind leg is drained into the lymphatic vessels that pass through inguinal and axillary lymph nodes and eventually enters blood circulation via the subclavian duct, bypassing the thoracic duct. On the other hand, lymph in the lower region of rat left hind leg is mainly drained into the lymphatic vessels that pass through popliteal and lilac lymph nodes and enters blood circulation via the thoracic duct, while a small fraction of lymph from the lower region of rat hind leg still travels down the inguinal node-subclavian duct pathway.

Not surprisingly, in the Kagan's and Kojima's studies, the proteins were SC injected in the upper region of rat hind leg [19, 20]. In contrast, the proteins were injected in the lower region of rat left hind leg for those rat studies with reasonably good lymph recovery from the thoracic duct [21–23]. Clearly, the discrepancy in the lymph recovery of proteins in rats can be explained by the difference in the SC injection site. Therefore, the hypothesis that the lymphatic uptake does not significantly contribute to the SC absorption of therapeutic proteins in rats is not valid.

It is of interest to note that even in the thoracic duct-cannulated rats, the bioavailability of PEG30-EPO was about 10% after SC administration [21]. It is not entirely clear whether the 10% bioavailability of PEG30-EPO in the thoracic duct-cannulated rats was absorbed into the systemic circulation directly via blood capillaries, or indirectly via lymphatic capillaries. As shown in Figure 6.3, a small fraction of proteins may be absorbed by lymphatic capillaries and enter the systemic circulation via the inguinal lymph node-subclavian duct pathway that bypasses the thoracic duct when proteins are given SC at the lower region of rat left hind leg. Since PEG30-EPO was given SC at the lower region of rat left hind leg, the 10% SC bioavailability of PEG30-EPO in the thoracic duct-cannulated rats is likely

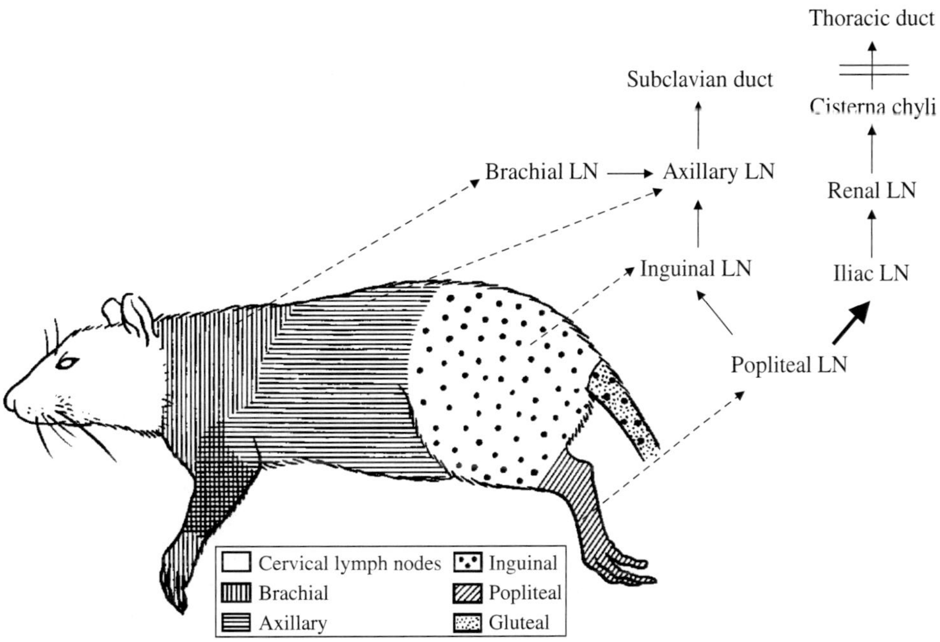

FIGURE 6.3 Lymph drainage paths of the subcutaneous areas in rats. (Modified from Tilney [24].) LN, lymph node.

due to a small fraction that bypassed the thoracic duct after SC administration, rather than direct uptake by blood capillaries. Given the fact that the physiological upper limit of pore size in the capillary walls of blood capillaries in SC tissues is approximately 5 nm [17], and the molecular weight of PEG30-EPO (~60 kDa, with 30 kDa PEG), it is very difficult, if not impossible, that PEG30-EPO can pass through the blood capillary wall and enter into blood circulation.

In addition to the site of SC injection, the amount of protein recovered in lymph can be dramatically different depending on the site of lymph collection. In a study, the pharmacokinetic (PK) of human growth hormone (hGH; 22 kDa) was investigated in sheep following IV and SC administration [25]. The bioavailability of hGH following SC administration in sheep was calculated to be 58%. The fraction of the administered hGH collected in peripheral (popliteal duct) lymph was 61.7%, whereas only 8.6% was collected in central (thoracic duct) lymph. The dramatic difference in the fraction of recovery between the peripheral and central lymph suggests a loss of hGH during lymphatic transport. Although the investigators speculated that the loss of hGH was likely due to a process of clearance (protein catabolism) during the lymphatic transport, it requires further study to explore the mechanism underlying the difference in the lymph recovery.

Recently, the role of neonatal Fc receptor (FcRn) binding in the SC absorption of IgG monoclonal antibody (mAb) has been investigated. In a preclinical study, the PK of 7E3, a model IgG mAb, was studied following IV and SC administration to wild-type and FcRn-deficient mice [3]. The SC bioavailability of 7E3 was reduced significantly from 82.5% in wild-type mice to 28.3% in FcRn-deficient mice. These results strongly suggest that FcRn plays an important role in determining the SC bioavailability of IgG mAbs in mice. However, it is not entirely clear whether the effect of FcRn on the SC bioavailability of 7E3 is due to FcRn-mediated protection from catabolism, or due to FcRn-mediated transport from the injection site to blood circulation.

In another preclinical study, the PK and lymph recovery of trastuzumab from the thoracic duct-cannulated rats over 30 h were characterized after IV and SC administration [23]. A kinetic model containing peripheral and central lymph compartments as well as a central blood compartment with a saturable process of FcRn binding was used to fit all the plasma and lymph concentrations of trastuzumab after IV and SC administration in rats. The model-estimated bioavailability of trastuzumab after SC administration in rats was 85%, which is in agreement with the observed bioavailability (78%) in rats after IV and SC administration. Based on the proposed kinetic model, approximately 53% of the trastuzumab dose was absorbed by lymphatic capillaries into peripheral lymph compartment, while about 32% of the dose was directly absorbed by blood capillaries via the FcRn-binding process into the central compartment of blood circulation. However, the underlying mechanism of FcRn binding for the absorption of mAbs from the SC injection site into the blood circulation remains unknown.

Similarly, a PK model that included the FcRn binding as part of the SC absorption was also developed for the SC absorption of rituximab in rats [26]. The SC bioavailability of rituximab in rats decreased with increasing dose. The SC bioavailability of rituximab was approximately 70% at 1 mg/kg, while 18% at 40 mg/kg. The investigators speculated that the dose-dependent SC absorption of rituximab may be due mainly to a saturable FcRn binding. Based on the proposed PK model, the transport rate constant of the FcRn–rituximab complex directly to the blood circulation system was estimated to be 0.29/day, while the lymph transport rate constant of free rituximab was estimated to be 0.209/day after SC administration at the middle abdomen. Based on the model simulations, the investigators suggest that the FcRn binding plays an important role in the SC absorption of rituximab in rats. However, the validity of the underlying mechanism of a kinetic model cannot just rely on the model predictions without external validation.

In summary, it is generally believed that the lymphatic capillaries, rather than the blood capillaries, play a quantitatively more important role in the SC absorption of protein drugs with a molecular weight of greater than 20–30 kDa. The general belief is consistent with the fact that the physiological upper limit of pore size in the walls of blood capillaries is approximately 5 nm. Recently, it has been proposed that FcRn binding may play an important role in the SC absorption of IgG mAbs. However, the underlying mechanism of the effect of FcRn binding on the SC absorption of IgG mAbs remains unknown. Further studies are required to explore the role of FcRn on SC absorption of mAbs.

6.5 PRESYSTEMIC CATABOLISM IN SC ABSORPTION OF PROTEINS

Although the underlying mechanisms for the SC absorption of therapeutic proteins and oral absorption of conventional small molecule drugs are quite different, the PK principles are equally applicable to the absorption of large and small molecules. When given orally, small drug molecules are absorbed from the small intestine to the systemic circulation via the liver. During the first passage through the intestine and liver, a fraction of drug molecules may be subject to hepatic and intestinal metabolism. The phenomenon of hepatic and intestinal metabolism during the process of absorption is referred to as hepatic and intestinal first-pass metabolism, also known as presystemic metabolism. Thus, presystemic metabolism has to be taken into consideration when estimating the oral bioavailability, which is defined as the fraction of an administered dose of small molecule drug that reaches the systemic circulation.

Like small molecule drugs, the SC bioavailability of a protein drug is also defined as the fraction of an administered dose of the protein drug that reaches the systemic blood circulation. During absorption, a fraction of the administered therapeutic proteins may also be subject to presystemic metabolism (catabolism) in the interstitium before it reaches the lymphatic capillaries. In addition, a fraction of protein drugs may also be degraded in lymphatic vessels during protein transport before it enters blood circulation. If a protein drug is catabolically stable and not degraded, the administered dose of the protein drug should be completely absorbed from the injection site because the dose is directly injected into the SC tissue and there is no physiological limitation on the duration that it can be absorbed. Therefore, in theory, the bioavailability of a catabolically stable protein drug should be 100% after SC administration.

Following SC administration, protein drugs exhibit a wide range of bioavailability in animals and humans, ranging from 20% to 100% [4, 5, 16]. The poor bioavailability of some therapeutic proteins strongly suggests that a significant fraction of these proteins may be degraded before or during lymphatic transport. Lymphatic transport of proteins travels a long way through a successive number of lymph nodes before it reaches the blood circulation. Although the walls of lymphatic vessels are open without cell–cell junctions, lymphatic fluid can only flow into but not out of the lymphatic vessels due to their unique structures. In an *in vitro* study, the leakage of hydrophilic FITC-dextran 12,000 (12 kDa) from the intraluminal space of isolated lymph vessels was found to be only minimal, while FITC-dextran 71,000 (71 kDa) did not leak out of the lymphatic wall [27]. Similarly, no leakage of macromolecules has been demonstrated *in vivo*. When ^{131}I-albumin was infused directly into a lymphatic vessel of dog leg, the albumin was returned virtually without loss to the blood circulation via thoracic duct [28]. Together, these results suggest that the low bioavailability of some protein drugs is not due to the leakage during lymphatic transport.

The low bioavailability of some therapeutic proteins is likely due to protein degradation in the interstitium in the region of SC injection site as well as in lymphatic vessels during the protein transport. The potential of protein degradation during lymphatic transport was first proposed by Charman and coworkers [25]. In a study in sheep, the fraction of hGH collected in peripheral lymph (popliteal duct) draining from the injection site was 61.7%, while 8.6% in central lymph (thoracic duct) after SC administration at the interdigital space of sheep. The authors speculated that the dramatic difference in the fraction of recovery between the popliteal duct and thoracic duct was most likely due to the "clearance" (protein degradation) within the lymphatic vessels.

Recently, Wang et al. have conducted an *in vitro* study to explore the potential of protein catabolism (degradation) in SC tissue and lymphatic nodes using PEG30-EPO and PEG40-EPO as model proteins [21]. After 24-h incubation with SC tissue homogenates, distinct degradation products of PEG30-EPO and PEG40-EPO were observed by using SDS-polyacrylamide gel electrophoresis. Furthermore, profound catabolic activity was also observed in cells derived from lymph nodes. Degradation products of PEG30-EPO and PEG40-EPO were also observed after incubation with lymph node cell suspension. The original concentrations of PEG30-EPO and PEG40-EPO were markedly reduced after 24-h incubation with 2×10^6 lymph node cells, while none left after 24-h incubation with 17×10^6 cells. In another preclinical study, the SC absorption of insulin was investigated in pigs [29]. It was found that a significant amount (21%) of SC administered insulin was degraded at the injection site. Together, these results support the notion that therapeutic proteins may be subject to presystemic catabolism in the SC tissue and the lymph nodes after SC administration.

It is well known that FcRn plays an important role in rescuing IgG molecules from intracellular lysosomal degradation resulting in the persistence of IgG *in vivo* [30]. IgG binds strongly to FcRn at acidic pH (<6.5) but not at neutral pH (7.0–7.5). Upon pinocytosis, IgG molecules enter acidic endosomes and bind to FcRn. A small fraction of IgG molecules that are not bound to FcRn within endosomes undergoes proteolytic degradation in lysosomes, while a majority of the FcRn–IgG complex recycles to the vascular endothelial cell surface, where the IgG molecules are released at the physiological pH in blood circulation. The protective mechanism of FcRn results in the persistence of IgG in plasma leading to a long half-life. The role of FcRn in the regulation of IgG half-life has been unequivocally demonstrated in FcRn-deficient mice. The plasma half-life of IgG was about 218 h in wild type mice, while 2.8 h in FcRn-deficient mice [31]. In addition, FcRn has also been shown to mediate transcytosis of IgG in *in vitro* systems [30]. Therefore, it is possible that FcRn can also mediate direct blood absorption of IgGs from the SC space via tissue-to-blood transcytosis.

FcRn has been shown to be highly expressed in the skin and muscle with lesser amount in liver and adipose tissue in mice [32]. Immunohistochemical analyses of muscle and liver with anti-FcRn F(ab)$_2$ indicated that FcRn is expressed in the endothelium of small arterioles and capillaries, but not in larger blood vessels. In addition to the capillary vascular endothelium of organs/tissues, it has been shown that FcRn is also highly expressed in bone-marrow-derived cells such as monocytes, macrophages, and dendritic cells in various tissues including lymph nodes and SC tissue [33]. Because of high expression levels of FcRn, the skin, muscle, and adipose tissues may play an important role in protecting the proteolytic degradation of IgG. Similarly, the bone-marrow-derived cells in lymph nodes and SC tissue may also play a significant role in salvaging the IgG degradation.

The role of FcRn on the SC absorption of IgG mAbs is best exemplified by the SC bioavailability of 7E3, a model IgG mAb, in wild-type and FcRn-deficient mice [3]. The SC bioavailability of 7E3 was reduced significantly from 82.5% in wild-type mice to 28.3% in FcRn-deficient mice. These results clearly show that FcRn can impact the SC absorption of 7E3 in mice. The influence of FcRn on the SC absorption of IgG mAbs has also been demonstrated by using IgG Fc variants with different binding affinities to mouse FcRn [34]. IgG2a Fc variants I253A/H435A, which did not bind to mouse FcRn at either pH 6.0 or 7.4, had the lowest SC bioavailability (41.8%) in mice, whereas variant N434H, which had increased FcRn-binding affinity at pH6.0, but had no FcRn binding at pH 7.4, had the highest SC bioavailability (94.7%). In addition, the SC bioavailability of the wild-type IgG2a in mice was 76.3%. The improved SC bioavailability could be due to the FcRn-mediated protection from presystemic catabolism or the FcRn-mediated transcytosis directly into blood.

In summary, the catabolism of proteins in SC tissue and lymph node has been demonstrated *in vitro* using tissue homogenates and cell suspension prepared from lymph nodes. Therefore, the poor SC bioavailability of some therapeutic proteins may be, at least partly, due to the presystemic catabolism in the SC tissue of injection site and within the lymphatics. For IgG mAbs, FcRn also plays an important role in SC absorption. However, the exact mechanism of how FcRn impacts SC absorption of mAbs remains unknown.

6.6 EFFECT OF INJECTION SITE ON SC ABSORPTION

It has been suggested that the site of SC injection may have a significant impact on the rate and extent of SC absorption of therapeutic proteins. In a clinical study, PK of hGH (22 kDa) was determined in healthy subjects on two separate occasions following SC injection at abdomen or thigh by Beshyah et al. [35]. The peak concentration (103 μU/mL) and area under the concentration–time curve (AUC) (528 μU·h/mL) of hGH after SC injection into the abdomen were significantly higher than those (41 μU/mL and 239 μU·h/mL) after injection into the thigh. These results suggest that SC injected hGH is better absorbed from the abdominal site than from the thigh. Since hGH is a stress hormone that can raise the concentration of glucose and free fatty acids, it is unexpected that there is no significant difference in plasma glucose and free fatty acids between the two injection sites in this study. Therefore, it is evident that there is a discrepancy between the PK changes and pharmacodynamic responses in this study.

As opposed to Beshyah's report, the SC absorption of rhEPO (30 kDa) from the thigh was better than that from the abdomen [36]. Both the C_{max} and AUC of rhEPO were significantly higher after SC injection into the thigh than into the abdomen. Moreover, in another clinical study, there were no significant differences in the C_{max} and AUC of rhEPO between the abdomen and thigh injection sites (abdomen and thigh) [37]. Similarly, in a clinical study, no significant differences in the C_{max} and AUC of golimumab, an antitumor necrosis factor alpha human IgG1 mAb (150 kDa), was found after SC administration at three different sites (upper arm, abdomen, and thigh) [38]. Similar to humans, there was no significant difference in the bioavailability of darbepoetin alfa (DA; 37 kDa) in sheep after SC administration at interdigital space, abdomen, or shoulder [39]. The bioavailability of DA was essentially complete for each of the three injection sites.

Although the cause for the conflicting reports is not entirely clear, it is possible that the discrepancy of the effect of injection site is due mainly to the interindividual variability in the SC absorption. The SC absorption of therapeutic proteins is known to be quite variable between individual patients. For example, in a crossover clinical study, the PK of rhEPO (30 kDa) has been evaluated in healthy subjects after IV and SC administration [40]. The bioavailability of rhEPO ranged from 18% to 80% following SC administration. Taking the above data together, it is reasonable to conclude that the effect of injection site on the extent of SC absorption of therapeutic proteins may not be clinically significant.

In summary, although it has been suggested that the injection site can affect the SC absorption of therapeutic proteins, there are conflicting reports with respect to the effect. The conflicting reports may be due mainly to the large interindividual variability of the SC absorption. Taking the interindividual variation, it is reasonable to conclude that the effect of injection site on the extent of SC absorption of therapeutic proteins may not be clinically significant.

6.7 CONCLUSIONS

After SC administration, proteins must transport from the injection site through interstitial space. Molecular size appears to be one of the important factors that govern the rate of SC absorption. Proteins with a smaller molecular weight move faster through the interstitial space than those with larger molecular weight. Based on the work conducted in sheep model, it is generally believed that therapeutic proteins with a molecular weight exceeding 20–30 kDa are predominantly taken up by lymphatic capillaries after SC administration. This general belief is consistent with the unique structure of lymphatic capillaries. Recently, it has been suggested that FcRn plays an important role in improving SC absorption of IgG monoclonal antibodies. Although the role of FcRn in SC absorption of IgG mAbs has been demonstrated, it is not entirely clear whether the effect of FcRn on SC absorption is related to the FcRn-mediated protection from proteolytic degradation or related to the FcRn-mediated transport from interstitial space into the blood circulation. Further studies are required to explore the underlying mechanism of FcRn-mediated effect on the SC absorption of IgG mAbs.

It is well known that small molecule drugs are subject to first-pass metabolism during oral absorption. Like small molecule drugs, protein drugs are also subject to presystemic catabolism during SC absorption. Although the potential of protein catabolism (proteolytic degradation) in SC tissue and lymph nodes has been demonstrated *in vitro*, the processes involved in the first-pass catabolism remain unknown. It is highly desirable to identify the major enzymes that are involved in the protein catabolism and to establish an *in vitro* model that can be used to quantitatively predict the magnitude of presystemic catabolism and SC bioavailability of protein drugs.

In spite of considerable efforts, our understanding of the processes of protein absorption after SC administration is still highly limited. Among the available data regarding the SC absorption, many of them are obtained from animal species. Given the species differences in the physiology of hypodermis and lymphatic system as well as FcRn binding, there is still a large gap in our knowledge of how to extrapolate the experimental data from animals to humans [41, 42]. Therefore, continuous efforts are needed to explore the underlying mechanisms of SC absorption of proteins (IgG mAbs and non-IgG proteins). In addition, efforts should be made to understand the similarity and difference in the protein absorption between animals and humans in order to make meaningful extrapolation.

REFERENCES

[1] McDonald TA, Zepeda ML, Tomlinson MJ, Bee WH, Ivens IA. Subcutaneous administration of biotherapeutics: current experience in animal models. Curr Opin Mol Ther 2010;12:461–470.

[2] Lin JH. Pharmacokinetics of biotech drugs: peptides, proteins and monoclonal antibodies. Curr Drug Metab 2009;10:661–691.

[3] Wang W, Wang EQ, Balthasar JP. Monoclonal antibody pharmacokinetics and pharmacodynamics. Clin Pharmcol Ther 2008;84:548–558.

[4] Richter WF, Bhansali SG, Morris ME. Mechanistic determinants of biotherapeutics absorption following SC administration. AAPS J 2012;14:559–570.

[5] Vugmeyster Y, Xu X, Theil FP, Khawli LA, Leach MW. Pharmacokinetics and toxicology of therapeutic proteins: advances and challenges. World J Biol Chem 2012;3:73–92.

[6] Swartz MA. The physiology of lymphatic system. Adv Drug Deliv Rev 2001;50:3–20.

[7] Yang VV, O'Morchoe PJ, O'Morchoe CC. Transport of protein across lymphatic endothelium in the rat kidney. Microvasc Res 1981;21:75–91.

[8] Casley-Smith JR. The fine structure and functioning of tissue channels and lymphatics. Lymphology 1980;13:177–183.

[9] Kihara T, Ito J, Miyake J. Measurement of biomolecular diffusion in extracellular matrix condensed by fibroblast using fluorescence correlation spectroscopy. PLoS One 2013;8:1–8.

[10] Reddy ST, Berk DA, Jain RK, Swartz MA. A sensitive *in vivo* model for quantifying interstitial connective transport of injected macromolecules and nanoparticles. J Appl Physiol 2006;101:1162–1169.

[11] Frost GI. Recombinant human hyaluronidase (rHuPH20): an enabling platform for subcutaneous drug and fluid administration. Expert Opin Drug Deliv 2007;4:427–440.

[12] Bookbinder LH, Hofer A, Haller MF, Zepeda ML, Keller GA, Lim JE, Edgington TS, Shepard HM, Patton JS, Frost GI. A recombinant human enzyme for enhanced interstitial transport of therapeutics. J Control Release 2006;114:230–241.

[13] Salar A, Avivi I, Bittner B, Bouabdallah R, Brewster M, Catalani O, Follows G, Haynes A, Hourcade-Potelleret F, Janikova A, Larouche JF, McIntyre C, Pederson M, Pereira J, Sayyed P, Shpilberg O, Tumyan G. Comparison of subcutaneous versus intravenous administration of rituximab as maintenance treatment for follicular lymphoma: results from a two-stage, phase IB study. J Clin Oncol 2014;32:1782–1791.

[14] Shpilberg O, Jackisch C. Subcutaneous administration of rituximab (Mabthera) and trastuzumab (Herceptin) using hyaluronidase. Br J Cancer 2013;109:1556–1561.

[15] Knott EM, Tune JD, Stoll ST, Downey HF. Increased lymphatic flow in the thoracic duct during manipulative intervention. J Am Osteopathol Assoc 2005;105:447–456.

[16] Porter CJH, Charman SA. Lymphatic transport of proteins after subcutaneous administration. J Pharm Sci 2000; 89:297–310.

[17] Sarin H. Physiologic upper limits of pore size of different blood capillary types and another perspective on the dual pore theory of microvascular permeability. J Angiogenes Res 2010;2:14.

[18] McLennan DN, Porter CJ, Edwards GA, Martin SW, Heatherington AC, Charman SA. Lymphatic absorption is the primary contributor to the systemic availability of epoetin Alfa following subcutaneous administration to sheep. J Pharmacol Exp Ther 2005;313:345–351.

[19] Kagan L, Gershkovich P, Mendelman A, Amsili S, Ezov N, Hoffman A. The role of the lymphatic system in subcutaneous absorption of macromolecules in the rat model. Eur J Pharm Biopharm 2007;67:759–765.

[20] Kojima K, Takahashi T, Nakanishi Y. Lymphatic transport of recombinant human tumor necrosis factor in rats. J Pharmcobiodyn 1988;11:700–706.

[21] Wang W, Chen N, Shen X, Cunningham P, Fauty S, Michel K, Wang B, Hong X, Adreani C, Nunes CN, Johnson CV, Yin K, Groff M, Zou Y, Liu L, Hamuro L, Prueksaritanont T. Lymphatic transport and catabolism of therapeutic proteins after subcutaneous administration to rats and dogs. Drug Metab Dispos 2012;40:952–962.

[22] Kaminskas LM, Kota J, McLeod VM, Kelly BD, Karellas P, Porter CJ. PEGylation of polylysine dendrimers improves absorption and lymphatic targeting SC administration in rats. J Control Release 2009;140:108–116.

[23] Dahlberg AM, Kaminska LM, Smith A, Nicolazzo JA, Porter CJH, Bulitta JB, McIntosh MP. The lymphatic system plays a major role in the intravenous and subcutaneous pharmacokinetics of trastuzumab in rats. Mol Pharm 2014;11:496–504.

[24] Tilney NL. Patterns of lymphatic drainage in the adult laboratory rats. J Anat 1971;109:369–383.

[25] Charman SA, Segrave AM, Edwards GA, Porter CJ. Systemic availability and lymphatic transport of human growth hormone administered by subcutaneously injection. J Pharm Sci 2000;89:168–177.

[26] Kagan L, Turner MR, Balu-Lyer SV, Mager DE. Subcutaneous absorption of monoclonal antibodies: role of dose, site of injection and injection volume on rituximab pharmacokinetics in rats. Pharm Res 2012;29:409–499.

[27] Ono N, Mizuno R, Ohhashi T. Effective permeability of hydrophilic substances through walls of lymph vessels: role of endothelial barrier. Am J Physiol Heart Circ Physiol 2005;289:H1676–H1682.

[28] Patterson RM, Ballard CL, Wasserman K, Mayerson HS. Capillary permeability to macromolecules. Am J Physiol 1955;42:432–433.

[29] Berger M, Halban PA, Girardier L, Seydoux J, Offord RE, Renold AE. Absorption kinetics of subcutaneously injected insulin. Evidence for degradation at the injection site. Diabetologia 1979;17:97–99.

[30] Roopenian DC, Akilesh S. FcRn: the neonatal Fc receptor comes age. Nat Rev Immunol 2007;7:715–725.

[31] Montoyo HP, Vaccaro C, Hafner M, Ober RJ, Mueller W, Ward ES. Conditional deletion of MHC class I-related FcRn reveals the sites of IgG homeostasis in mice. Proc Natl Acad Sci USA 2009;106:2788–2793.

[32] Borvak J, Richardson J, Medesan C, Antohe F, Radu C, Simionescu M, Ghetie V, Ward ES. Functional expression of the MHC class I-related receptor, FcRn, in endothelial cells of mice. Int Immunol 1998;10:1289–1298.

[33] Akilesh S, Christianson GJ, Roopenian DC, Shaw AS. Neonatal FcRn expression in bone marrow-derived cells functions to protect serum IgG from catabolism. J Immunol 2007;179:4580–4588.

[34] Deng R, Meng YG, Hyote K, Lutman J, Lu Y, Lyer S, DeForge LE, Theil FP, Fielder PJ, Prabhu S. Subcutaneous bioavailability of therapeutic antibodies as a function of FcRn binding affinity in mice. MAbs 2012;4:101–109.

[35] Beshyah SA, Anyaoku V, Niththyananthan R, Sharp P, Johnston DG. The effect of subcutaneous injection site on absorption of human growth hormone: abdomen versus thigh. Clin Endocrinol (Oxf) 1991;35:409–412.

[36] Macdougall IC, Jones JM, Robinson MI, Miles JB, Coles GA, Williams JD. Subcutaneous erythropoietin therapy: comparison of three different sites of injections. Contrib Nephrol 1991;88:152–156.

[37] Jensen JD, Jensen LW, Madsen JK. The pharmacokinetics of recombinant human erythropoietin after subcutaneous injection at different sites. Eur J Clin Pharmacol 1994;46:333–337.

[38] Xu Z, Wang Q, Zhuang Y, Frederick B, Yan H, Bouman-Thio E, Marini JC, Keen M, Snead D, Davis HM, Zhou H. Subcutaneous bioavailability of golimumab at 3 different injection sites in healthy volunteers. J Clin Parmacol 2010;50:276–284.

[39] Kota J, Machavaram K, McLennan DN, Edwards GA, Porter CJ, Charman SA. Lymphatic absorption of subcutaneously administered proteins: influence of different injection sites on the absorption of darbepoetin alfa using a sheep model. Drug Metab Dispos 2007;35:2211–2217.

[40] Salmonson T, Danilson BG, Wikstrom B. The pharmacokinetics of recombinant human erythropoietin after intravenous and subcutaneous administration to healthy subjects. Br J Clin Pharm 1990;29:709–713.

[41] Ober RJ, Radu CG, Ghetie V, Ward ES. Differences in promiscuity for antibody-FcRn interactions for therapeutic antibodies. Int Immunolo 2001;13:1551–1559.

[42] Deng R, Iyer S, Theil FP, Mortensen DL, Fielder PJ, Prabhu S. Projecting human pharmacokinetics of therapeutic antibodies from preclinical data: what have we learned? MAbs 2011;3:61–66.

7

BIODISTRIBUTION OF THERAPEUTIC BIOLOGICS: METHODS AND APPLICATIONS IN INFORMING TARGET BIOLOGY, PHARMACOKINETICS, AND DOSING STRATEGIES

SEAN B. JOSEPH, SAILETA PRABHU AND C. ANDREW BOSWELL
Genentech Research and Early Development, South San Francisco, CA, USA

7.1 INTRODUCTION

Monoclonal antibodies have provided many new drug candidates for various diseases encompassing several therapeutic indications including neurology, ophthalmology, immunology, and oncology. The mechanism of action for these biologic molecules typically involves binding a soluble ligand or cell-surface protein in order to block or alter a molecular pathway, induce a desired cellular response, or deplete a target cell type. Many target antigens reside within the interstitium, the fluid-filled compartment between the outer endothelial vessel wall and the plasma membranes of cells. Consequently, compartmental drug concentrations within the interstitial spaces (i.e., biophases) of tissues are often more relevant to predicting drug effect than whole-tissue concentrations.

The overall objective of this chapter is to provide an overview of the biodistribution of antibodies including biological determinants and available technologies for measurement. Emphasis will be placed on how biodistribution data can be measured and interpreted in a manner that is relevant to translational pharmacology. To achieve this, we will discuss the following topics: (i) the molecular (e.g., antigen binding, Fc receptor binding, and physicochemical attributes) and physiological (e.g., tissue-specific blood flow, vascular volume, and interstitial volume) properties that govern antibody biodistribution; (ii) technical considerations for *in vivo* studies (e.g., labeling techniques, time point selection, organ harvest strategies, dosing, and euthanasia techniques) and tissue analysis (e.g., gamma counting, scintillation counting, and imaging techniques); and (iii) helpful tips for data interpretation including calculations and units, blood correction, derivation of interstitial concentrations, confirmation of receptor occupancy, explaining unexpectedly rapid clearance, and assisting in clinical dose selection.

7.2 DETERMINANTS OF ANTIBODY BIODISTRIBUTION

Antibody biodistribution is governed by various molecular and physiological factors, many of which also affect systemic pharmacokinetic (PK) exposure. In contrast to most small molecule drugs, antibodies typically exhibit very slow clearance, resulting in half-lives up to 2–3 weeks in human [1–3]. Nonlinear pharmacokinetics is common for monoclonal antibodies, with dependence on dose/concentration, antigen expression, and species cross-reactivity. Unlike orally administered small molecule drugs, antibodies are usually parenterally administered (i.e., intravenous or subcutaneous) and generally exhibit high bioavailabilities. Antibody metabolism occurs by both specific (e.g., complementarity-determining region (CDR)/receptor-mediated) and nonspecific (e.g., pinocytosis, catabolism by proteolytic enzymes) clearance mechanisms and does not

ADME and Translational Pharmacokinetics/Pharmacodynamics of Therapeutic Proteins: Applications in Drug Discovery and Development,
First Edition. Edited by Honghui Zhou and Frank-Peter Theil.

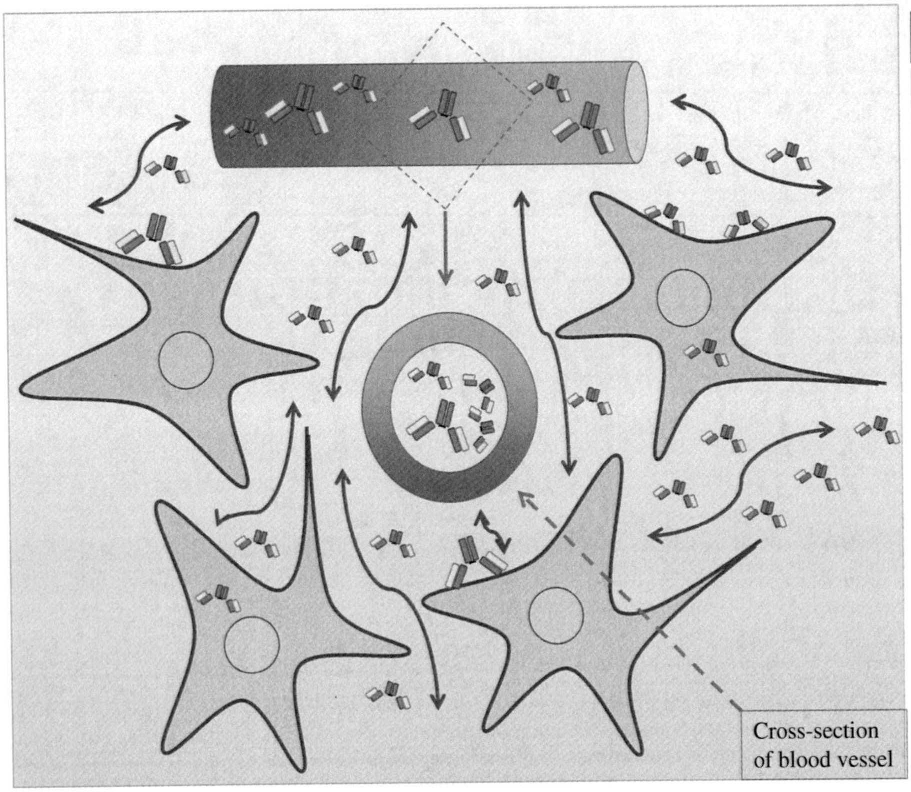

Factors affecting antibody distribution	
Molecular	**Physiological**
FcRn binding affinity	Blood flow
Charge	Vascular volume
Hydrophobicity	Interstitial volume
Antigen density	Lymph flow
Affinity	Vascular permeability

FIGURE 7.1 Factors affecting antibody distribution within tissues. Numerous factors affect the ability of an antibody to extravasate, traverse the interstitium, and interact with cellular FcRn (following pinocytosis) and/or cell-surface antigens in tissues. These factors may be classified into two basic groups: molecular (properties of the antibody itself) and physiological (properties of the tissue).

involve CYP450 or other metabolizing enzymes typical for small molecule metabolism. Intact antibodies are cleared primarily by the reticuloendothelial system (e.g., macrophages) as well as by endothelial cells. The rate of clearance can be affected by saturable target-mediated disposition, Fc receptor interactions, nonspecific physicochemical interactions, and immunogenicity (i.e., antitherapeutic antibody response). The volume of distribution for antibodies is generally much smaller than for small molecule drugs due to limited permeability and half-life extension via Fc receptor-mediated recycling. Overall, the factors affecting antibody biodistribution may be classified into (i) molecular and (ii) physiological properties (Fig. 7.1).

7.2.1 Molecular Properties

*7.2.1.1 **Fc Receptor Interactions: FcRn*** The neonatal Fc receptor (FcRn) plays an important and well-known role in antibody recycling in endothelial and hematopoietic cells and thus influences the systemic pharmacokinetics (PK) of immunoglobulin G (IgG). The long circulation half-life (relative to small molecules of antibodies) are largely governed by the pH-dependent FcRn-mediated protection from lysosomal degradation [4–9]. Specific amino acids within the antibody Fc region that are involved in the Fc–FcRn interaction have been identified [10]. In the absence of target-mediated clearance or at high doses, most antibody clearance typically occurs by nonantigen-mediated (often termed "nonspecific") pathways governed by the reticuloendothelial system [11]. Increasing the binding affinity to FcRn has allowed modest extension of antibody half-life in preclinical studies [8, 12, 13]. Also known as the major histocompatibility complex class I-related receptor, FcRn possesses additional roles for protecting serum albumin from degradation [14] and, to an extent that varies by species, in the passive delivery of IgG from mother to young (i.e., IgG in milk absorbed through gut) [15]. Accordingly, FcRn is expressed widely in endothelial, epithelial, and hematopoietic cells including organs such as skin, muscle, kidney, liver, and placenta [4, 16–18]. Preclinical evaluation of an IgG and its high affinity and low affinity FcRn-binding variants demonstrated that the liver and spleen are the most concentrated sites of IgG catabolism in the absence of FcRn protection, with additional important contributions from skin and gut [19].

Several studies have shown a correlation between the serum half-life and binding affinity of IgGs for FcRn [8, 20–22]. However, this correlation is not always observed, as different human IgG subclasses exhibit different *in vivo* half-lives but appear to bind human FcRn with K_D values that are within the same order of magnitude [23, 24]. A possible explanation of these discrepancies is that the ratio of IgG protected/destroyed depends not only on the affinity of FcRn but also on the competition between the binding of IgG to FcRn and its intracellular degradation before its interaction with FcRn [5]. Several IgG Fc variants with stronger binding to FcRn at pH>7 have shorter *in vivo* half-lives compared with the corresponding wild-type (WT) IgG because they cannot dissociate from FcRn [25, 26]. An ideal IgG Fc variant should possess increased FcRn affinity at acidic pH compared with a WT antibody, thus retaining binding while in the endosome, and having weaker or equivalent affinity at pH>7 to facilitate release back into systemic circulation [8, 25, 26].

7.2.1.2 Fc Receptor Interactions: FcγR In IgG, the Fc domain regulates antibody-based therapy by binding to Fc gamma receptors (FcγRs) and eliciting immune effector functions [1]. Structurally determined differences in FcγR-binding affinity among IgG subclasses result in variations in antibody-dependent cell-mediated cytotoxicity (ADCC) and complement-dependent cytotoxicity [27]. Although these variations in secondary immune function may provide additional pharmacological benefit to antibodies, increasing FcγR interactions could lead to off-target immune-mediated toxicity in some cases. Interaction with FcγR, particularly within liver, represents a major mode of elimination of immune complexes. The importance of Fc effector function on antibody disposition was demonstrated by mutation of a single amino acid critical for FcγR that resulted in a greatly reduced liver uptake of immune complexes [28].

7.2.1.3 Antigen Interactions Target-mediated disposition of antibodies can be greatly affected by the amount of antigen expression *in vivo*. Some antibodies have faster clearance due to target-mediated disposition caused by shed antigen or expression in nontarget tissues. Binding of an antibody to its antigen may prompt internalization (via receptor-mediated endocytosis) and down-modulation. Some antibodies need only bind to a noninternalizing cell-surface receptor and evoke cell death by signaling mechanisms. The tissue distribution profiles of some antibodies are influenced by antigen shedding (i.e., soluble antigen). In general, antibodies having higher affinities for their targets exhibit lower systemic exposure and higher uptake in target-expressing tissues at nontarget-saturating concentrations. Overall, both the systemic and tissue exposures to antibodies may be affected by target-binding affinity, receptor expression, turnover rate, and dose.

7.2.1.4 Physicochemical Interactions Other antibody characteristics that may influence the disposition of antibodies include electrostatic properties and hydrophobicity [29]. Both the overall isoelectric point of an antibody and local charge patches have been associated with clearance and overall antibody disposition [30]. In general, increases in positive charge of antibodies are believed to increase clearance and tissue distribution due to electrostatic attraction to glycosaminoglycans and other negatively charged components of the extracellular matrices (ECMs) within tissues [31]. In the case of antibody–drug conjugates (ADCs), as most small molecule cytotoxins are relatively hydrophobic, it is important to control drug load so that the overall hydrophobicity of the ADC will not lead to shifts in disposition, particularly to the liver [32].

7.2.2 Physiological (Tissue) Properties

The absence of many physiological processes *in vitro* and interspecies differences *in vivo* can confound direct comparisons of *in vitro*, preclinical, and clinical data [31, 33]. A vast array of physiological data for humans and laboratory species is available in the literature [34–37]; however, it should be utilized with an understanding of its limitations. Measurement techniques vary widely, and the use of assumed nominal values is common [31]. Furthermore, the physiologies of disease tissues such as xenograft models are highly variable and largely unknown. Significant physiological variability across species, age, breed, disease status, drug treatment, and time of day [38] motivates direct measurement of relevant physiological properties or processes whenever possible [39]. In terms of antibody distribution within tissues, some of the most influential parameters are blood flow, vascular volume, and interstitial volume (Fig. 7.2) [40].

7.2.2.1 Blood Flow Tissues that are well perfused tend to exhibit higher amounts of antibody distribution. The regional blood flow rates (Q) in various organs and tissues may be measured by sacrificing mice exactly 90 seconds following intravenous bolus injection of ^{86}RbCl [38, 41–44] (Fig. 7.2). Blood flow (Q) may be calculated as follows [38, 39]:

$$Q = \frac{\left(\frac{\text{CPM}_{\text{Rb-86}}}{\text{g tissue}}\right) \times \text{CO}_{\text{total}}}{\left(\frac{\text{CPM}_{\text{Rb-86}}}{\text{total injected dose}}\right)} \tag{7.1}$$

where total cardiac output (CO_{total}) = 8 and 74 mL/min for mice and rats, respectively [37].

7.2.2.2 Vascular Volume Tissues that are rich in vasculature and blood content (e.g., liver, spleen, and kidneys) tend to have higher uptake of antibodies than blood-poor tissues (e.g., muscle, fat, and intestines). The intravascular spaces of rodent tissues may be measured using a previously reported indirect red blood cell (RBC) labeling method [31]. This measurement is based on a clinically utilized blood pool

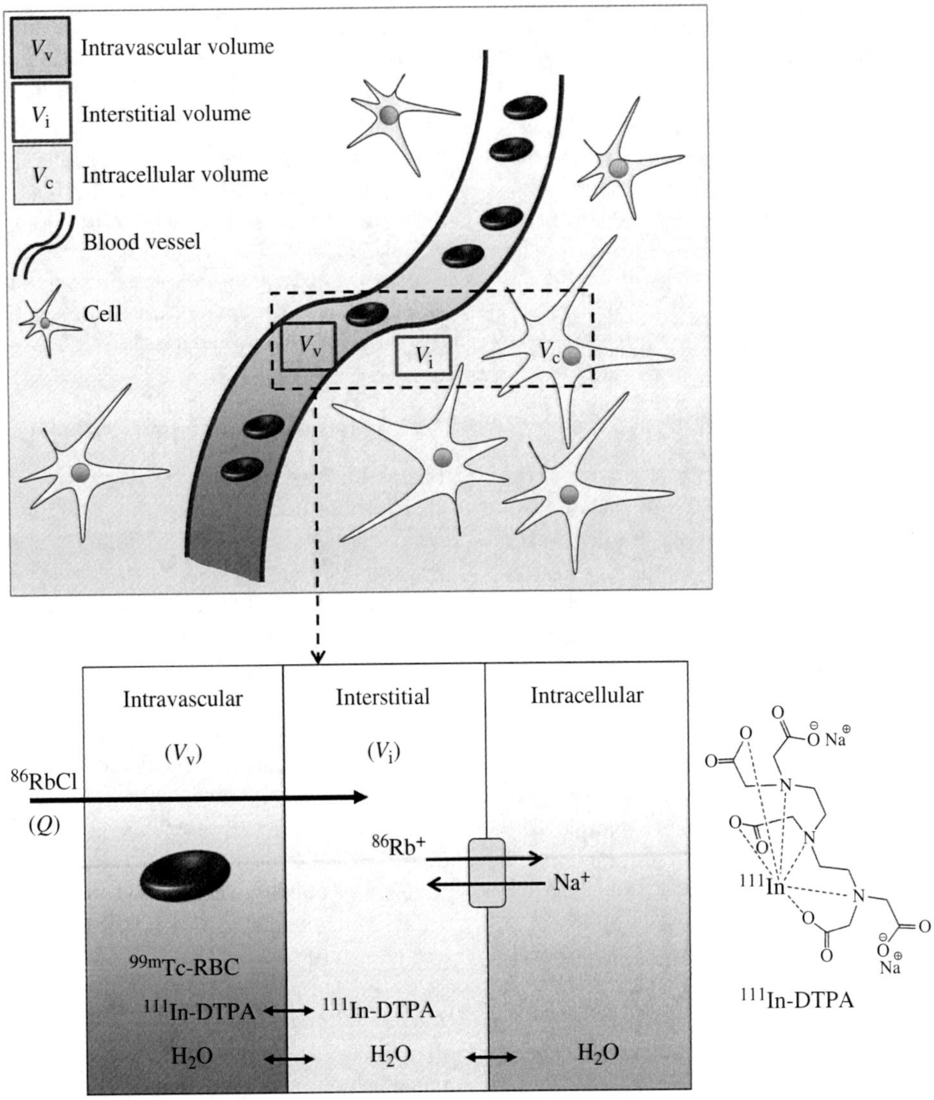

FIGURE 7.2 Conceptual illustration of techniques used to measure physiological parameters relevant to drug uptake in tissues. The tissue is divided into intravascular, interstitial, and intracellular compartments (depicted in left, center, and right, respectively). Vascular volume (V_v) is measured using ^{99m}Tc-labeled red blood cells (RBCs), while the extracellular (i.e., V_v + interstitial (V_i)) space is measured by steady-state infusion of ^{111}In-DTPA. Water molecules freely diffuse between all three compartments. The rate of blood flow (Q) to the tissue is measured as the proportion of a bolus dose of ^{86}Rb$^+$ entering the tissue (possibly entering cells via Na$^+$/K$^+$ ion channels) in a brief time interval. (Adapted from Boswell et al. [40].)

nuclear-imaging protocol and relies on radiolabeling of RBCs with ^{99m}Tc (Fig. 7.2) and measuring the amount of radioactivity in tissues and blood using a gamma counter, yielding vascular volume (V_v) in units of microliters per gram of tissue [38, 39]:

$$V_v = \frac{\left(\dfrac{\text{CPM}_{\text{Tc-99m}}}{\text{g tissue}}\right)}{\left(\dfrac{\text{CPM}_{\text{Tc-99m}}}{\mu\text{L blood}}\right)} \tag{7.2}$$

The indirect method involved transfusion of radiolabeled blood from donor rodents into study (i.e., recipient) rodents, where donor rodents had been subjected to ^{99m}Tc labeling of RBCs *in vivo* following the administration of stannous (Sn^{2+}) pyrophosphate [38, 45]. Use of the clinical Technescan™ PYP™ kit is conceptually based on the original method of Sands et al. for *in situ* (i.e., *in vivo*) RBC labeling with ^{99m}Tc [46, 47]. The previous administration of stannous pyrophosphate, a component of the reconstituted Technescan kit, reduces ^{99m}Tc-pertechnetate intracellularly so that it may bind to the beta chain of hemoglobin [48].

7.2.2.3 Interstitial Volume Numerous drugs are targeted toward receptors that reside within the interstitial space [49]. The interstitium is also referred to as the biophase due to its central role in the biological mechanism of action for many drugs including a number of cancer therapeutic agents [50]. Because most drug targets are located in the interstitial space, interstitial concentrations are often more predictive of drug effect than total tissue concentrations.

Both body fluid balance and maintenance of interstitial hydration are regulated by the distribution volume of macromolecules (e.g., albumin and IgG) [51]. In turn, the distribution volumes of macromolecules are influenced by anionic glycosaminoglycans in the ECM [52]. Following extravasation, antibodies experience various electrostatic forces within the interstitial fluid space due to the presence of negatively charged heparin sulfate and other anionic proteoglycans on the surface of cells within the interstitium [53–55]. Proteoglycans are produced by most mammalian cells as components of the ECM. Diversity in the number of chains, chain lengths, and sulfation patterns allows these macromolecules to make significant contributions to ECM structure and function [54, 55].

The extracellular spaces of rodent tissues may be measured by continuous infusion of the extracellular marker ^{111}In-DTPA (diethylenetriamine pentaacetic acid (DTPA)) [56, 57] (Fig. 7.2). Subtracting the vascular volume (derived from ^{99m}Tc) from the extracellular volume (^{111}In) allows derivation of the pharmacologically relevant quantity, the interstitial volume (V_i) in units of microliters per gram of tissue [39, 56].

$$V_i = \frac{\left(\frac{\text{CPM}_{\text{In-111}}}{\text{g tissue}}\right) - \left(\frac{\text{CPM}_{\text{In-111}}}{\mu\text{L blood}}\right) \times V_v}{\left(\frac{\text{CPM}_{\text{In-111}}}{\mu\text{L plasma}}\right)} \quad (7.3)$$

A similar radiometal–polyaminopolycarboxylate complex, chromium-51-ethylenediaminetetraacetic acid (^{51}Cr-EDTA), has been previously used by others in a similar context [49]. As V_i is calculated from both ^{99m}Tc- and ^{111}In-derived data, the percentages of injected doses (%ID) of ^{111}In-DTPA may also be calculated. Furthermore, because V_v values are generally smaller compared with V_i, the subtractive (blood correction) term in Equation 7.3 often does not drastically affect the calculation. In physiological terms, this suggests that the extracellular volumes (V_e) are approximately equal to interstitial volumes in tissues having lower blood content.

Extracellular volume may be calculated from V_v and V_i if the volume fraction of RBCs (e.g., hematocrit, f) is known:

$$V_e = V_v \times (1 - f) + V_i \quad (7.4)$$

Alternatively, V_e may be calculated by simply omitting the subtractive term in Equation 7.3:

$$V_e = \frac{\left(\frac{\text{CPM}_{\text{In-111}}}{\text{g tissue}}\right)}{\left(\frac{\text{CPM}_{\text{In-111}}}{\mu\text{L plasma}}\right)} \quad (7.5)$$

7.2.2.4 Other Physiological Parameters In addition to blood flow, interstitial volume, and vascular volume, there are many other important physiological parameters that affect antibody uptake within tissues. Vascular permeability is an extremely important parameter, such that physiologically based pharmacokinetic models account for the larger pore sizes in the leakier vasculature of the liver, spleen, and most tumors. Lymphatic flow rate is technically quite challenging to measure due to the limited number of imaging techniques that possess the resolution and depth of detection to quantify this process by which extravasated antibodies are returned to systemic circulation. Interstitial pressure is another important parameter, especially in the context of tumors where it is believed to limit the diffusion of molecules throughout the interstitial space.

7.3 METHODS OF MEASURING ANTIBODY BIODISTRIBUTION

7.3.1 *In Vivo* Study Design Considerations

Both the characteristics of the test molecule and the desired study outcomes must be considered in designing an antibody biodistribution study. Among these considerations are selecting appropriate labeling technique, time points, tissue collection strategy, as well as dosing and euthanasia techniques (Fig. 7.3).

7.3.1.1 Labeling Technique The most obvious consideration in selecting an appropriate radiolabeling method is the physical decay half-life of the radionuclide [58]. For instance, the half-life of iodine-125 is roughly 2 months, while that of indium-111 is less than 3 days. Longer lived metallic radionuclides of interest include gadolinium-153 and lutetium-177 and are governed by similar complexation chemistry as radioindium.

In addition to half-life, the distinction between residualizing (metallic) and nonresidualizing (halogen) radionuclides must be considered. Radiocatabolites of antibodies labeled with radioiodine are rapidly effluxed from cells following receptor-mediated endocytosis and proteolytic catabolism (Fig. 7.4); this is often followed by reentry to systemic circulation, rapid renal filtration, and urinary excretion. In contrast, radiocatabolites of antibodies labeled

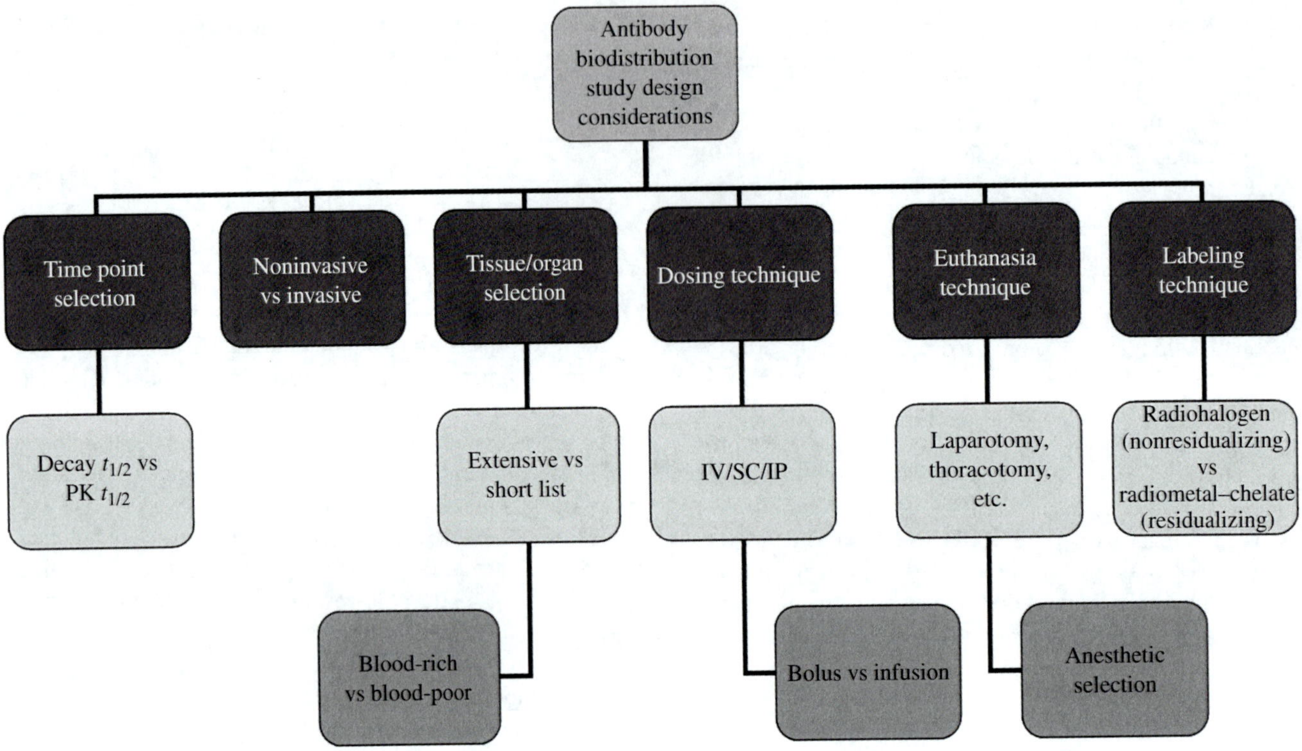

FIGURE 7.3 Diagram of study design considerations for biodistribution studies of antibodies.

with metal radionuclides via 1,4,7,10-tetraazacyclododecane-1,4,7,10-tetraacetic acid (DOTA) or other polyamino-polycarboxylate chelators tend to become trapped inside cells and accumulate in antigen-expressing tissues following receptor-mediated endocytosis [59] due to the residualizing properties of these charged, highly polar probes (Fig. 7.5) [32, 60]. Importantly, while similar pharmacokinetic data in blood and antigen-negative tissues are typically obtained using either radioiodine or radiometal probes, a much different scenario exists in tissues that overexpress the antigen, especially if internalization occurs [61]. Specifically, the true amount of antibody present in tissues that express internalizing antigen is often overestimated due to residualization or trapping of radio-catabolites derived from the cellular metabolism of antibodies labeled with radiometal–chelate complexes. As such, for internalizing antigens, radiometal probes give cumulative uptake in target tissues, whereas radiohalogen probes more closely approximate the "real-time" concentration of antibodies within tissues (i.e., kinetics in the tissue). Given the known normal metabolism of immunoglobulins into their constituent amino acids [62, 63], the primary radiolabeled catabolite of indium-DOTA-labeled antibodies following lysosomal proteolytic degradation is indium-111-DOTA-lysine [64]. See Chapter 19 for a more in-depth discussion of labeling techniques.

7.3.1.2 Time Point Selection One of the most fundamental considerations for preclinical biodistribution studies is time point selection. For noninvasive imaging studies, tissue concentrations may be measured at several different time points, but invasive "cut-and-count" biodistribution studies require animal killing and tissue harvest. Studies often include both early (e.g., 4–8 h postinjection) and late (e.g., 1–3 days postinjection) time points in order to capture tissue levels during both the distribution and elimination phases. As the elimination half-lives of antibodies are often on the order of weeks, having even later time points would be useful; however, technical limitations such as radioactive decay half-life and chemical stability of the label must also be considered.

7.3.1.3 Tissue/Organ Selection Tissue collection strategies for biodistribution studies can vary widely depending on the questions being addressed. If the focus is on a particular disease model such as a tumor xenograft, then collecting only that tissue (along with blood or plasma for exposure) is one possible option. However, collecting additional tissues can provide additional information such as unexpected target expression in healthy tissues, antigen shedding, molecular stability, and clearance mechanisms. In addition to blood and tumor, collecting a short list of tissues including liver, spleen, kidneys, lung, small intestine, and muscle provides insight into how the molecule is distributed into a variety of tissues. Including tissues that are blood-rich (e.g., liver, spleen, kidneys, and lung) as well as blood-poor (e.g., tumor, adipose, and muscle) provides a better opportunity to distinguish between blood pool and true tissue disposition.

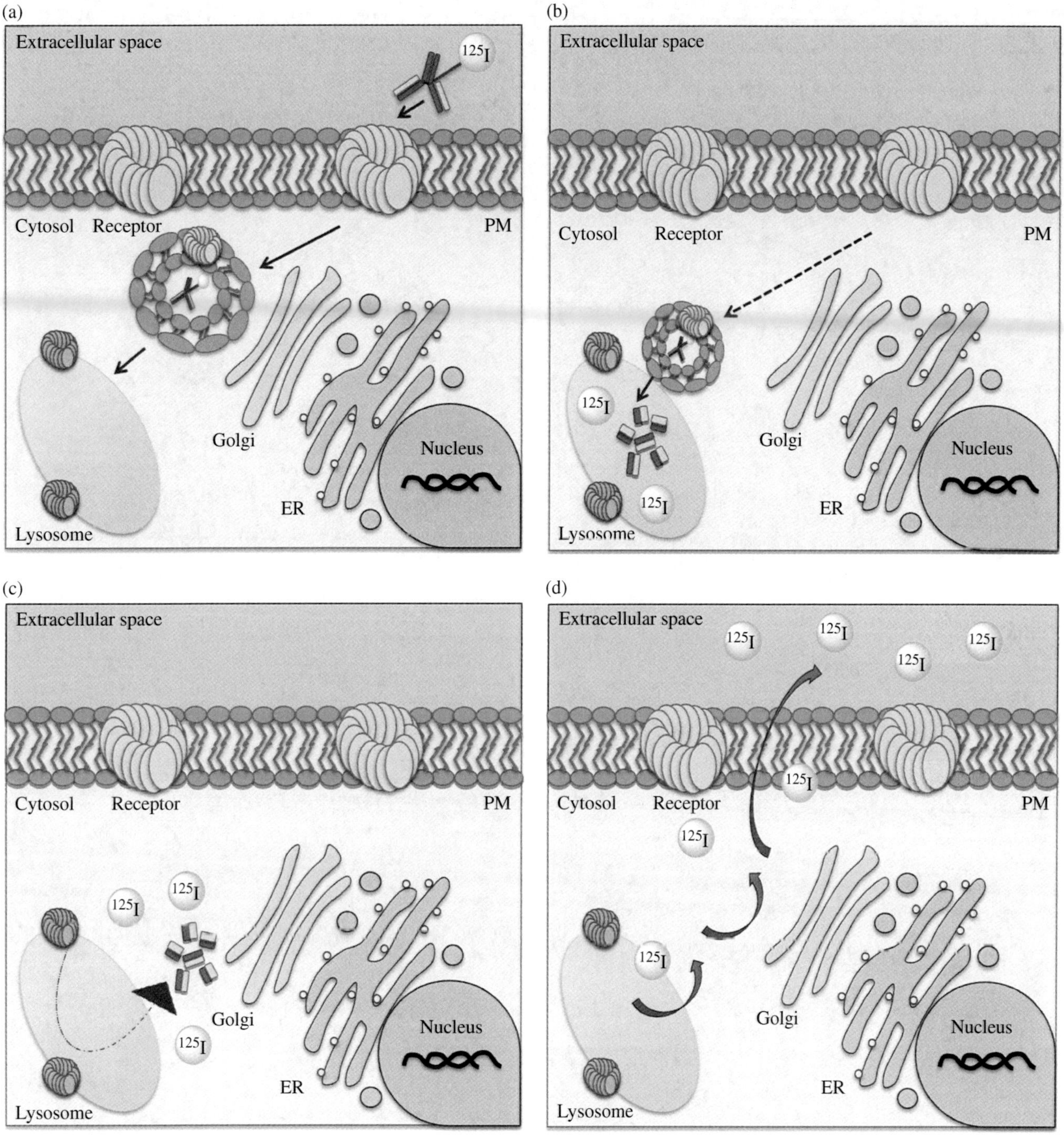

FIGURE 7.4 Cellular metabolism of ^{125}I-labeled (nonresidualizing) antibodies. Processes include receptor-mediated endocytosis (a), lysosomal degradation (b), and efflux of radioiodine (or associated catabolites) from the cell (c, d).

For smaller organs, collecting whole tissues in rodents is feasible. Whole murine livers can easily fit inside test tubes that fit in most gamma counter racks, while rat tissues are usually better accommodated by scintillation vials. However, it is not convenient to collect larger tissues such as muscle and intestine in their entirety, so taking a representative sample is usually the best approach. To maintain consistency, specific regions such as the gastrocnemius (calf) muscle or a specific region of the intestines (e.g., ileum) may be collected. In certain situations, a mass balance is desired, so the remaining "carcass" is included in the analysis as well as excreta (urine and feces) collected using metabolic cages.

Most protocols include excision of tissues with a prompt saline rinse and tissue blot to remove external (but not

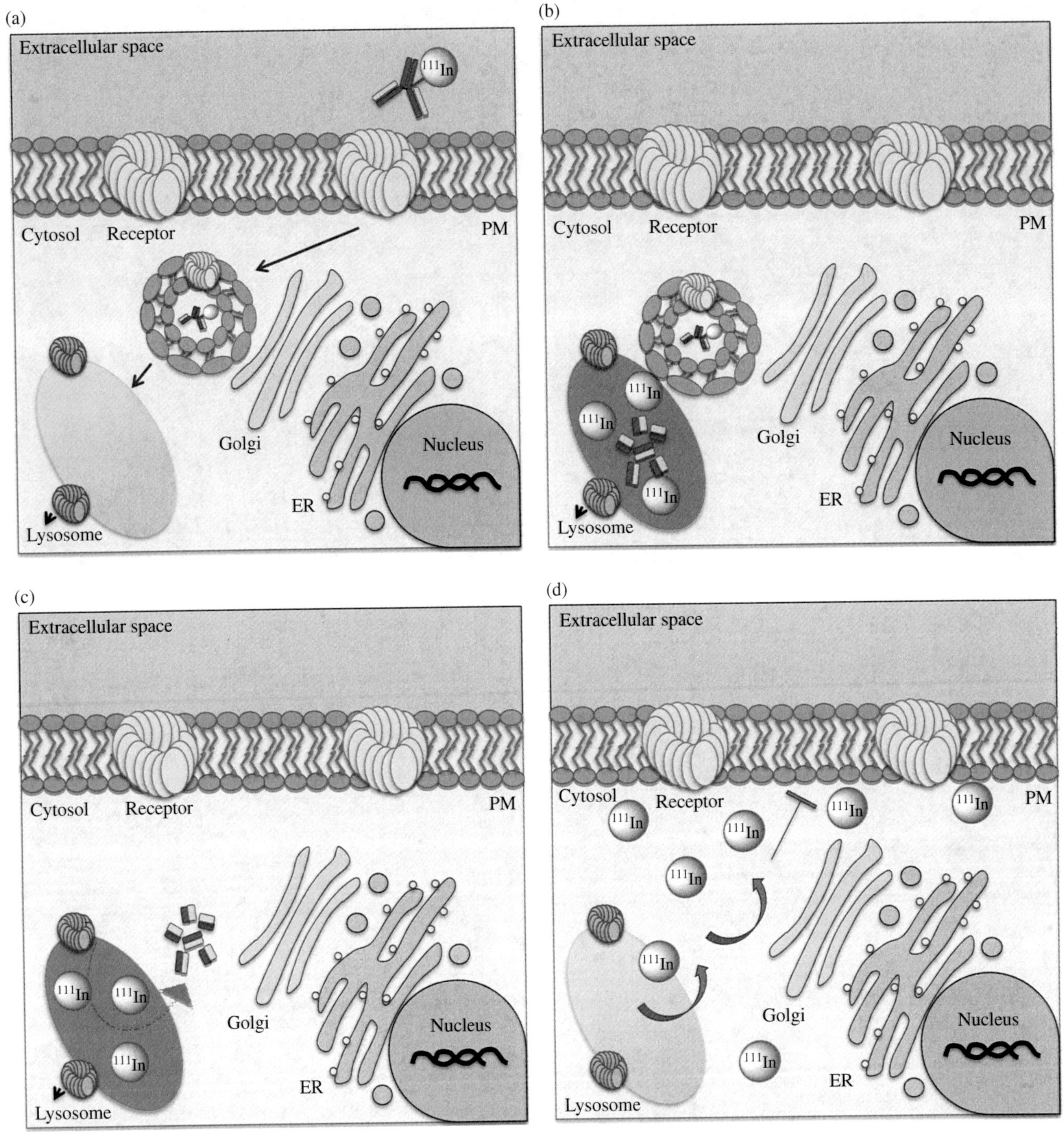

FIGURE 7.5 Cellular metabolism of ^{111}In-labeled (residualizing) antibodies. Processes include receptor-mediated endocytosis (a), lysosomal degradation (b), and retention of radiometal-associated catabolites within the cell (c, d).

internal) blood. While saline perfusion of tissues to remove internal blood is possible, this technique can lead to more variability and experimental error due to inconsistency in perfusion efficiency. Furthermore, it is relatively easy to mathematically correct for the blood contents of tissues (see Section 7.4). Special treatment of certain tissues should also be specified in tissue collection protocols. For instance, it is common to specify removal of food contents from stomach and intestines before analysis.

7.3.1.4 ***Dosing Techniques*** The most common route of administration for biodistribution of antibodies in mice is intravenous (tail vein) bolus injection. The situation is similar for rats, but dosing via cannula is also common due to readily available precannulated animals. However, other routes (e.g., subcutaneous and intraperitoneal) and methods (e.g., continuous infusion and repeated administration) are certainly feasible and probably understudied. In rodents, the typically high bioavailability of most antibodies would likely lead to similar biodistribution profiles at later time points regardless of administration technique. Tail vein dosing in rodents requires nonsedated animals, and is more difficult in mice with pigmentation or in certain transgenic models (e.g., many neurodegenerative models present with tremors and poorly structured veins).

7.3.1.5 Euthanasia Techniques The choice of euthanasia technique is often dictated by local or institutional animal welfare regulations. Killing the animal while under anesthesia is strongly preferred, if not required. Either laparotomy (with incision of the diaphragm) or thoracotomy may be performed to gain access to internal organs. A sample of blood at the time of killing is often desired; this may be collected by cardiac puncture, but care should be taken not to remove blood in excess or with too much suction force in order to avoid partial exsanguination of tissues and/or hemorrhaging or pooling of blood within the lungs. Samples of blood at earlier time points before killing may be collected via retroorbital bleed or, if present, via cannula.

7.3.2 Tissue Analysis

The method of quantitative analysis of antibody uptake in tissues depends largely on the method that was utilized for radiolabeling the molecule (see Chapter 19 for a more in-depth discussion of labeling techniques). Most questions can be answered by labeling with gamma-emitting radionuclides (e.g., ^{125}I, ^{131}I, and ^{111}In) and subsequent analysis in an automatic gamma counter. However, for specialized applications, it may be desirable to use a beta-emitting radionuclide (e.g., ^{3}H and ^{14}C). For instance, if the intent is to follow the metabolic fate of a cytotoxic small molecule drug conjugated to an antibody (i.e., ADC), then beta-emitting radionuclides are necessary to maintain the exact chemical structure of the drug. Moreover, a variety of radionuclides are compatible with various imaging modalities, including beta and gamma emitters for autoradiography, gamma emitters for single-photon emission computed tomography (SPECT) and positron emitters for positron emission tomography (PET). Advantages and disadvantages of each of these techniques are summarized below.

7.3.2.1 Gamma Counting Advantages of radiometric tissue analysis by gamma counting include speed, simplicity, and the ability to provide robust and highly quantitative data. One of the best aspects is that there are virtually no sample processing requirements; whole tissues may be simply inserted whole into test tubes or scintillation vials and counted for radioactivity with no homogenization necessary. The simple nature of this process has coined the term "cut and count." This is due in large part to the higher emission energy of γ photons, relative to β particles, making detection from whole tissues feasible. Raw data are obtained in counts per minute (CPM) and/or disintegrations per minute (DPM).

7.3.2.2 Scintillation Counting Many of the same principles of gamma counting are also applicable to liquid scintillation counting. However, because of the very low energy of β particles, it is necessary to homogenize tissues in a liquid scintillation cocktail. Instead of measuring radioactivity directly, scintillation counters measure the light that is produced when β particles interact with molecules in the cocktail fluid. However, the pigments and debris in homogenized tissues can interfere with this light reaching the detectors, thus it is often necessary to oxidize or bleach the tissue homogenates. It is also common practice to implement quench correction into the data analysis to account for potential underestimation of the amount of radioactivity actually present within pigmented samples.

7.3.2.3 Imaging In addition to the radioanalytical methods above, antibody biodistribution may also be determined by molecular imaging. Whole-body autoradiographic imaging relies on freezing, whole-body sectioning (via a cryotome), and exposure to film to yield two-dimensional high resolution maps of radioactivity within tissue sections (i.e., autoradiographs). These can be coregistered with traditional photographs of the cryosectioned animal in order to assign levels of radioactivity to a particular tissue. The resolution of autoradiography can be far superior to that of noninvasive imaging modalities, but with the disadvantage that an animal must be killed to provide data at a particular time point. The method is also quite labor intensive compared to other biodistribution techniques.

Noninvasive imaging modalities such as SPECT and PET are also available and can be combined with anatomical imaging (e.g., X-ray CT or magnetic resonance imaging) to provide three-dimensional maps of radioactivity throughout a living organism. These techniques offer the ability to perform longitudinal studies (i.e., multiple images over time in a single animal). For more in-depth coverage of molecular imaging, see Chapter 19.

7.4 INTERPRETATION OF BIODISTRIBUTION DATA

Several key concepts should be recognized to aid in interpreting biodistribution data. One critical consideration is the concept of receptor occupancy, which dictates that at higher doses of nonradiolabeled drug there will be competitive inhibition of uptake in receptor-positive tissues. As such, performing biodistribution studies across multiple doses spanning several orders of magnitude can be quite helpful in assessing the levels of target expression in tissues [65–68].

Another consideration is the blood content of tissues. In blood-rich tissues such as liver and kidney, the concentration–time profiles of radiolabeled antibodies often resemble that of blood or plasma (i.e., biphasic clearance) with the maximal concentration occurring immediately at the time of dosing. In contrast, tissues having less blood content (e.g., muscle and fat) often exhibit concentration–time profiles in which the drug concentration does not reach its maximum value until 1–2 days, followed by a slow, gradual clearance.

7.4.1 Calculations and Units

Gamma or scintillation counting yields raw data in CPM and/or DPM, with these two quantities related as follows:

$$\text{Efficiency}\,(\varepsilon) = \frac{\text{CPM}}{\text{DPM}} \tag{7.6}$$

However, it is imperative to remember that ε is gamma counterdependent and can vary from one instrument to another. It is often desirable to derive traditional units of radioactivity from instrument-derived CPM values. The following is always true:

$$1\,\mu\text{Ci} = 2.22e^{6}\text{DPM} \tag{7.7}$$

The following is true for ^{125}I on a particular gamma counter only, but can be derived for other radionuclides using the definition of detection efficiency (ε) in Equation 7.6:

$$1\,\mu\text{Ci} = 1.8e^{6}\text{CPM} \tag{7.8}$$

Combining the knowledge in Equations 7.7 and 7.8, one can deduce that the ε for ^{125}I on this particular gamma counter is equal to 1.8 divided by 2.22 or approximately 80%, meaning that 8 out of 10 gamma photons emitted from ^{125}I-labeled molecules within a tissue are actually detected.

Combining raw counts with measured weights of harvested tissues, it is possible to derive dose-normalized concentrations of radiolabeled drug in tissues. This calculation requires radiometric analysis of a small aliquot of dosing solution alongside the tissues of interest, often referred to as a dosing standard. Dose-normalized concentrations may then be calculated as follows:

$$\%\frac{\text{ID}}{\text{g}} = \left(\frac{\text{CPM}_{\text{tissue}}}{\text{g}_{\text{tissue}} \times \text{CPM}_{\text{ID}}}\right) \times 100 \tag{7.9}$$

For purposes of mass balance, it may also be desirable to express tissue uptake without accounting for the mass of tissues:

$$\%\,\text{ID} = \left(\frac{\text{CPM}_{\text{tissue}}}{\text{CPM}_{\text{ID}}}\right) \times 100 \tag{7.10}$$

Tissue concentrations in %ID/g may easily be converted into more pharmacologically relevant drug concentrations using the following relationship:

$$C_{\text{tissue}} = \%\frac{\text{ID}}{\text{g}} \times \frac{1}{100} \times \text{mass}_{\text{ID}} \tag{7.11}$$

In this case, mass_{ID} is the mass of drug in the injected dose, which may be determined via knowledge of the specific activity of the radiotracer. One convenient method for radiolabeled antibodies is to measure the protein concentrations spectrophotometrically before formulating dosing solutions. The resulting drug concentrations are often reported in units of equivalent mass (e.g., microgram-equivalents per milliliter) as the calculation assumes that the radiolabel is still associated with the antibody.

7.4.2 Compartmental Tissue Concentrations

Tissues may be considered to be comprised of three separate physiological compartments: the intravascular, interstitial, and intracellular spaces (Fig. 7.2). If drug concentrations are measured in terms of total, whole-tissue uptake, then a physiologically based correction is necessary to derive individual compartmental concentrations. Such corrections require knowledge of the relative tissue spaces that are occupied by blood and interstitial fluid. The vascular and interstitial volumes may be utilized to calculate the amounts of antibody in the appropriate compartments.

7.4.3 Blood Correction

Saline perfusion of tissues is often undesirable due to the extra effort and lack of inter-animal consistency. Tissue perfusion not only increases the experimental error in biodistribution studies, but it can also potentially remove "real uptake" in tissues having low affinity interactions and/or leaky vessels (e.g., highly permeable tumors or leaky cerebral capillaries in neurodegenerative models). As such, preclinical methods for measurement of vascular volume (V_v) have been developed to allow mathematical correction for the amount of drug within the blood of tissues [31]. The amount of antibody within the blood of tissues is subtracted from the total tissue uptake:

$$\%\text{ID/g, blood corrected} = \%\text{ID/g} - \%\,\text{ID/mL} \times V_v \tag{7.12}$$

In this case, %ID/mL is the dose-normalized concentration of drug in whole blood that was determined using the same procedure as for %ID/g values for tissues. This correction is more important at early time points (when blood concentrations are higher) and in blood-rich tissues having higher V_v values.

7.4.4 Derivation of Interstitial Concentrations

Methods have been established for the measurement of extracellular volume (V_e) following intravenous infusion of an extracellular marker [33]. Subtracting the vascular volume from the extracellular volume allows derivation of the pharmacologically relevant quantity, the interstitial volume (V_i), which is required for calculation of drug concentrations within the interstitial space. The interstitial concentration is calculated using the fractional interstitial volume (Φ):

$$C_i = \frac{C_{\text{tissue, blood corrected}}}{\Phi} \tag{7.13}$$

The fractional interstitial volume (Φ) may be easily derived from interstitial volume (V_i) data in traditional units of volume per gram of tissue. For example, a tissue comprised of 100 μL of interstitial fluid per gram of tissue has a fractional interstitial volume of 0.100 since each gram of tissue occupies 1000 μL of space (assuming a tissue density of 1 g/mL). The same relationship exists between the fractional vascular volume (γ) and vascular volume (V_v). Once the interstitial concentration of antibody is calculated, its units may be converted from milligrams per milliliter to nanomolar using the molecular weight of approximately 150,000 g/mol for most antibodies.

7.4.5 Confirmation of Receptor Occupancy

Biodistribution studies of antibodies and other target-specific molecules are powerful tools for studying receptor interactions in tissues. One approach is to compare the radioactive uptake of a target-specific antibody to that of a nonbinding, isotype control antibody [69]. Another approach is to use an excess of the nonradiolabeled antibody as a blocking agent [66]. Both radiolabeled and nonradiolabeled antibody molecules can bind antigen, so varying the ratio of these two components will change the amount of bound radioactivity due to competitive inhibition [39]. At high antibody doses when the number of nonradiolabeled antibody molecules is in vast excess of radiolabeled molecules, the amount of bound radioactive antibody within a receptor-expressing tissue will be minimal. It is possible to perform dose escalation studies in which a fixed amount of radiolabeled antibody is coadministered with various levels of nonradiolabeled antibody [65, 68]. Importantly, the dose at which the bound radioactive antibody concentration reaches a minimum corresponds to the lowest dose at which maximal receptor occupancy is achieved.

7.4.6 Explaining Unexpectedly Rapid Clearance

When antibodies exhibit atypical pharmacokinetics, biodistribution studies are useful in determining the factor(s) that drive fast clearance. For molecules with suspected target-mediated clearance, biodistribution studies can identify the tissue(s) to which the radioactive antibody is distributing [65, 67]. This information can inform toxicological concerns and/or confirm clearance mechanisms. If rapid clearance is not mitigated even at saturating doses, then nonspecific, off-target clearance mechanisms (e.g., hydrophobicity and charge) may be influencing pharmacokinetics [70]. In such cases, biodistribution studies may be less informative, particularly if the distribution is occurring broadly across multiple tissues. However, biodistribution studies are capable of identifying nonspecific uptake to a particular organ, such as elevated liver uptake of an ADC relative to its parent unconjugated antibody [32].

7.4.7 Assisting in Clinical Dose Selection

Preclinical biodistribution data in appropriate disease models can guide clinical development by assisting in dose selection. For instance, biodistribution data of an antibody in xenograft-bearing mice was translated into a clinically relevant dose by comparing trough plasma concentrations to phase I clinical data [65, 71].

If the affinity of an antibody is known, then the tumor interstitial concentration necessary for 95% receptor saturation may be calculated by simply multiplying the K_d by 19 (because $0.95 = 19/20$) based on the following mathematical relationship:

$$\%\,\text{Receptor occupancy} = \frac{C_i}{K_d + C_i} \times 100 \tag{7.14}$$

Tumor interstitial concentrations from biodistribution studies at various doses may be directly compared to these values to determine the optimal dose, which may be subsequently allometrically scaled to humans. Alternatively, the tumor-to-plasma ratios may be used to identify the dose that produces the minimum trough concentration to maintain sufficient target saturation.

7.5 CONCLUDING REMARKS

A thorough understanding of the pharmacokinetics and biodistribution of antibody therapeutics can inform their clinical development. Fast clearance or nonlinear plasma kinetic profiles of antibody therapeutics can often be attributed to target-mediated clearance [72] and, in many cases, biodistribution data can identify the tissue in which the target is located [1, 34]. Tissue distribution studies can also reveal nonantigen-dependent, off-target binding, which may also influence plasma clearance [65, 73]. Targets having widespread tissue expression may affect target uptake of a target therapeutic and vice versa. In addition, attributes of the antibody such as size, molecular weight, and target affinity [74, 75], as well as tissue physiology, including the degree of vascularity, antigen density, and necrosis [73, 75–77], can influence antibody distribution to and within target tissues, so it is understandable that some target antigens may be inaccessible to the antibody. The implications of these limitations on therapeutic efficacy warrant a detailed understanding of the relationship between antibody pharmacokinetics, biodistribution, and tissue uptake.

ACKNOWLEDGMENTS

The authors thank Zachary Kawagoe, Jaime Anguiano, Sheila Ulufatu, Michelle Schweiger, Daniela Bumbaca, Victor Yip, Danielle Mandikian, and Simon Williams for technical guidance and/or engaging scientific discussions.

REFERENCES

[1] Lobo ED, Hansen RJ, Balthasar JP. Antibody pharmacokinetics and pharmacodynamics. J Pharm Sci 2004;93:2645–2668.

[2] Reichert JM. Monoclonal antibodies in the clinic. Nat Biotechnol 2001;19:819–822.

[3] Wang W, Wang EQ, Balthasar JP. Monoclonal antibody pharmacokinetics and pharmacodynamics. Clin Pharmacol Ther 2008;84:548–558.

[4] Roopenian DC, Akilesh S. FcRn: the neonatal Fc receptor comes of age. Nat Rev Immunol 2007;7:715–725.

[5] Ghetie V, Ward ES. Transcytosis and catabolism of antibody. Immunol Res 2002;25:97–113.

[6] Junghans RP. Finally! The Brambell receptor (FcRB). Mediator of transmission of immunity and protection from catabolism for IgG. Immunol Res 1997;16:29–57.

[7] Raghavan M, Bonagura VR, Morrison SL, Bjorkman PJ. Analysis of the pH dependence of the neonatal Fc receptor/immunoglobulin G interaction using antibody and receptor variants. Biochemistry 1995;34:14649–14657.

[8] Yeung YA, Leabman MK, Marvin JS, Qiu J, Adams CW, Lien S, Starovasnik MA, Lowman HB. Engineering human IgG1 affinity to human neonatal Fc receptor: impact of affinity improvement on pharmacokinetics in primates. J Immunol 2009;182:7663–7671.

[9] Roopenian DC, Christianson GJ, Sproule TJ, Brown AC, Akilesh S, Jung N, Petkova S, Avanessian L, Choi EY, Shaffer DJ, Eden PA, Anderson CL. The MHC class I-like IgG receptor controls perinatal IgG transport, IgG homeostasis, and fate of IgG-Fc-coupled drugs. J Immunol 2003;170:3528–3533.

[10] Hinton PR, Xiong JM, Johlfs MG, Tang MT, Keller S, Tsurushita N. An engineered human IgG1 antibody with longer serum half-life. J Immunol 2006;176:346–356.

[11] Tabrizi MA, Tseng CM, Roskos LK. Elimination mechanisms of therapeutic monoclonal antibodies. Drug Discov Today 2006;11:81–88.

[12] Deng R, Loyet KM, Lien S, Iyer S, DeForge LE, Theil FP, Lowman HB, Fielder PJ, Prabhu S. Pharmacokinetics of humanized monoclonal anti-tumor necrosis factor-{alpha} antibody and its neonatal Fc receptor variants in mice and cynomolgus monkeys. Drug Metab Dispos 2010;38:600–605.

[13] Zalevsky J, Chamberlain AK, Horton HM, Karki S, Leung IW, Sproule TJ, Lazar GA, Roopenian DC, Desjarlais JR. Enhanced antibody half-life improves *in vivo* activity. Nat Biotechnol 2010;28:157–159.

[14] Anderson CL, Chaudhury C, Kim J, Bronson CL, Wani MA, Mohanty S. Perspective—FcRn transports albumin: relevance to immunology and medicine. Trends Immunol 2006;27:343–348.

[15] Ghetie V, Ward ES. Multiple roles for the major histocompatibility complex class I-related receptor FcRn. Annu Rev Immunol 2000;18:739–766.

[16] Montoyo HP, Vaccaro C, Hafner M, Ober RJ, Mueller W, Ward ES. Conditional deletion of the MHC class I-related receptor FcRn reveals the sites of IgG homeostasis in mice. Proc Natl Acad Sci U S A 2009;106:2788–2793.

[17] Borvak J, Richardson J, Medesan C, Antohe F, Radu C, Simionescu M, Ghetie V, Ward ES. Functional expression of the MHC class I-related receptor, FcRn, in endothelial cells of mice. Int Immunol 1998;10:1289–1298.

[18] Telleman P, Junghans RP. The role of the Brambell receptor (FcRB) in liver: protection of endocytosed immunoglobulin G (IgG) from catabolism in hepatocytes rather than transport of IgG to bile. Immunology 2000;100:245–251.

[19] Yip V, Palma E, Tesar DB, Mundo EE, Bumbaca D, Torres EK, Reyes NA, Shen BQ, Fielder PJ, Prabhu S, Khawli LA, Boswell CA. Quantitative cumulative biodistribution of antibodies in mice: effect of modulating binding affinity to the neonatal Fc receptor. MAbs 2014;6:689–696.

[20] Datta-Mannan A, Witcher DR, Tang Y, Watkins J, Wroblewski VJ. Monoclonal antibody clearance. Impact of modulating the interaction of IgG with the neonatal Fc receptor. J Biol Chem 2007;282:1709–1717.

[21] Ghetie V, Popov S, Borvak J, Radu C, Matesoi D, Medesan C, Ober RJ, Ward ES. Increasing the serum persistence of an IgG fragment by random mutagenesis. Nat Biotechnol 1997;15:637–640.

[22] Suzuki T, Ishii-Watabe A, Tada M, Kobayashi T, Kanayasu-Toyoda T, Kawanishi T, Yamaguchi T. Importance of neonatal FcR in regulating the serum half-life of therapeutic proteins containing the Fc domain of human IgG1: a comparative study of the affinity of monoclonal antibodies and Fc-fusion proteins to human neonatal FcR. J Immunol 2010;184:1968–1976.

[23] Gurbaxani B, Dela Cruz LL, Chintalacharuvu K, Morrison SL. Analysis of a family of antibodies with different half-lives in mice fails to find a correlation between affinity for FcRn and serum half-life. Mol Immunol 2006;43:1462–1473.

[24] West AP Jr, Bjorkman PJ. Crystal structure and immunoglobulin G binding properties of the human major histocompatibility complex-related Fc receptor(,). Biochemistry 2000;39:9698–9708.

[25] Dall'Acqua WF, Woods RM, Ward ES, Palaszynski SR, Patel NK, Brewah YA, Wu H, Kiener PA, Langermann S. Increasing the affinity of a human IgG1 for the neonatal Fc receptor: biological consequences. J Immunol 2002;169:5171–5180.

[26] Gan Z, Ram S, Vaccaro C, Ober RJ, Ward ES. Analyses of the recycling receptor, FcRn, in live cells reveal novel pathways for lysosomal delivery. Traffic 2009;10:600–614.

[27] Jiang FN, Jiang S, Liu D, Richter A, Levy JG. Development of technology for linking photosensitizers to a model monoclonal antibody. J Immunol Methods 1990;134:139–149.

[28] Mortensen DL, Prabhu S, Stefanich EG, Kadkhodayan-Fischer S, Gelzleichter TR, Baker D, Jiang J, Wallace K, Iyer S, Fielder PJ, Putnam WS. Effect of antigen binding affinity and effector function on the pharmacokinetics and pharmacodynamics of anti-IgE monoclonal antibodies. MAbs 2012;4:724–731.

[29] Bumbaca D, Boswell CA, Fielder PJ, Khawli LA. Physiochemical and biochemical factors influencing the pharmacokinetics of antibody therapeutics. AAPS J 2012; 14:554–558.

[30] Boswell CA, Tesar DB, Mukhyala K, Theil FP, Fielder PJ, Khawli LA. Effects of charge on antibody tissue distribution and pharmacokinetics. Bioconjug Chem 2010; 21:2153–2163.

[31] Boswell CA, Ferl GZ, Mundo EE, Schweiger MG, Marik J, Reich MP, Theil FP, Fielder PJ, Khawli LA. Development and evaluation of a novel method for preclinical measurement of tissue vascular volume. Mol Pharm 2010;7:1848–1857.

[32] Boswell CA, Mundo EE, Zhang C, Bumbaca D, Valle NR, Kozak KR, Fourie A, Chuh J, Koppada N, Saad O, Gill H, Shen BQ, Rubinfeld B, Tibbitts J, Kaur S, Theil FP, Fielder PJ, Khawli LA, Lin K. Impact of drug conjugation on pharmacokinetics and tissue distribution of anti-STEAP1 antibody–drug conjugates in rats. Bioconjug Chem 2011;22:1994–2004.

[33] Boswell CA, Ferl GZ, Mundo EE, Bumbaca D, Schweiger MG, Theil FP, Fielder PJ, Khawli LA. Effects of anti-VEGF on predicted antibody biodistribution: roles of vascular volume, interstitial volume, and blood flow. PLoS One 2011; 6:e17874.

[34] Baxter LT, Zhu H, Mackensen DG, Butler WF, Jain RK. Biodistribution of monoclonal antibodies: scale-up from mouse to human using a physiologically based pharmacokinetic model. Cancer Res 1995;55:4611–4622.

[35] Baxter LT, Zhu H, Mackensen DG, Jain RK. Physiologically based pharmacokinetic model for specific and nonspecific monoclonal antibodies and fragments in normal tissues and human tumor xenografts in nude mice. Cancer Res 1994; 54:1517–1528.

[36] Brown RP, Delp MD, Lindstedt SL, Rhomberg LR, Beliles RP. Physiological parameter values for physiologically based pharmacokinetic models. Toxicol Ind Health 1997; 13:407–484.

[37] Davies B, Morris T. Physiological parameters in laboratory animals and humans. Pharm Res 1993;10:1093–1095.

[38] Blumenthal RD, Osorio L, Ochakovskaya R, Ying Z, Goldenberg DM. Regulation of tumour drug delivery by blood flow chronobiology. Eur J Cancer 2000;36:1876–1884.

[39] Boswell CA, Bumbaca D, Fielder PJ, Khawli LA. Compartmental tissue distribution of antibody therapeutics: experimental approaches and interpretations. AAPS J 2012; 14:612–618.

[40] Boswell CA, Mundo EE, Ulufatu S, Bumbaca D, Cahaya HS, Majidy N, Van Hoy M, Schweiger MG, Fielder PJ, Prabhu S, Khawli LA. Comparative physiology of mice and rats: radiometric measurement of vascular parameters in rodent tissues. Mol. Pharmaceutics 2014;11(5):1591–1598. DOI: 10.1021/mp400748t.

[41] Zanelli GD, Fowler JF. The measurement of blood perfusion in experimental tumors by uptake of ^{86}Rb. Cancer Res 1974;34:1451–1456.

[42] Gullino PM, Grantham FH. Studies on the exchange of fluids between host and tumor. III. Regulation of blood flow in hepatomas and other rat tumors. J Natl Cancer Inst 1962;28:211–229.

[43] Cherry SR, Carnochan P, Babich JW, Serafini F, Rowell NP, Watson IA. Quantitative *in vivo* measurements of tumor perfusion using rubidium-81 and positron emission tomography. J Nucl Med 1990;31:1307–1315.

[44] Hammersley PA, McCready VR, Babich JW, Coghlan G. ^{99m}Tc-HMPAO as a tumour blood flow agent. Eur J Nucl Med 1987;13:90–94.

[45] Pavel DG, Zimmer M, Patterson VN. *In vivo* labeling of red blood cells with ^{99m}Tc: a new approach to blood pool visualization. J Nucl Med 1977;18:305–308.

[46] Sands H, Jones PL, Shah SA, Palme D, Vessella RL, Gallagher BM. Correlation of vascular permeability and blood flow with monoclonal antibody uptake by human Clouser and renal cell xenografts. Cancer Res 1988;48:188–193.

[47] Sands H, Shah SA, Gallagher BM. Vascular volume and permeability of human and murine tumors grown in athymic mice. Cancer Lett 1985;27:15–21.

[48] Rehani MM, Sharma SK. Site of Tc-99m binding to the red blood cell: concise communication. J Nucl Med 1980;21:676–678.

[49] Levitt DG. The pharmacokinetics of the interstitial space in humans. BMC Clin Pharmacol 2003;3:3.

[50] Jain RK. Transport of molecules across tumor vasculature. Cancer Metastasis Rev 1987;6:559–593.

[51] Wiig H, Gyenge CC, Tenstad O. The interstitial distribution of macromolecules in rat tumours is influenced by the negatively charged matrix components. J Physiol 2005;567:557–567.

[52] Wiig H, Tenstad O, Bert JL. Effect of hydration on interstitial distribution of charged albumin in rat dermis *in vitro*. J Physiol 2005;569:631–641.

[53] Belting M. Heparan sulfate proteoglycan as a plasma membrane carrier. Trends Biochem Sci 2003;28:145–151.

[54] Hardingham TE, Fosang AJ. Proteoglycans: many forms and many functions. FASEB J 1992;6:861–870.

[55] Kreuger J, Spillmann D, Li JP, Lindahl U. Interactions between heparan sulfate and proteins: the concept of specificity. J Cell Biol 2006;174:323–327.

[56] Shockley TR, Lin K, Sung C, Nagy JA, Tompkins RG, Dedrick RL, Dvorak HF, Yarmush ML. A quantitative analysis of tumor specific monoclonal antibody uptake by human melanoma xenografts: effects of antibody immunological properties and tumor antigen expression levels. Cancer Res 1992;52:357–366.

[57] Sung C, Youle RJ, Dedrick RL. Pharmacokinetic analysis of immunotoxin uptake in solid tumors: role of plasma kinetics, capillary permeability, and binding. Cancer Res 1990;50:7382–7392.

[58] Boswell CA, Brechbiel MW. Development of radioimmunotherapeutic and diagnostic antibodies: an inside-out view. Nucl Med Biol 2007;34:757–778. (Epub 2007).

[59] Wall DA, Maack T. Endocytic uptake, transport, and catabolism of proteins by epithelial cells. Am J Physiol 1985;248:C12–C20.

[60] Shih LB, Thorpe SR, Griffiths GL, Diril H, Ong GL, Hansen HJ, Goldenberg DM, Mattes MJ. The processing and fate of antibodies and their radiolabels bound to the surface of tumor cells *in vitro*: a comparison of nine radiolabels. J Nucl Med 1994;35:899–908.

[61] Perera RM, Zoncu R, Johns TG, Pypaert M, Lee FT, Mellman I, Old LJ, Toomre DK, Scott AM. Internalization, intracellular trafficking, and biodistribution of monoclonal antibody 806: a novel anti-epidermal growth factor receptor antibody. Neoplasia 2007;9:1099–1110.

[62] Morell A, Terry WD, Waldmann TA. Metabolic properties of IgG subclasses in man. J Clin Invest 1970;49:673–680.

[63] Spiegelberg HL, Fishkin BG, Grey HM. Catabolism of human gammaG-immunoglobulins of different heavy chain subclasses. I. Catabolism of gammaG-myeloma proteins in man. J Clin Invest 1968;47:2323–2330.

[64] Rogers BE, Franano FN, Duncan JR, Edwards WB, Anderson CJ, Connett JM, Welch MJ. Identification of metabolites of ^{111}In-diethylenetriaminepentaacetic acid-monoclonal antibodies and antibody fragments *in vivo*. Cancer Res 1995;55:5714s–5720s.

[65] Bumbaca D, Xiang H, Boswell CA, Port RE, Stainton SL, Mundo EE, Ulufatu S, Bagri A, Theil FP, Fielder PJ, Khawli LA, Shen BQ. Maximizing tumour exposure to anti-neuropilin-1 antibody requires saturation of non-tumour tissue antigenic sinks in mice. Br J Pharmacol 2012;166(1):368–377. DOI: 10.1111/j.1476-5381.2011.01777.x.

[66] Pastuskovas CV, Mundo EE, Williams SP, Nayak TK, Ho J, Ulufatu S, Clark S, Ross S, Cheng E, Parsons-Reponte K, Cain G, Van Hoy M, Majidy N, Bheddah S, dela Cruz Chuh J, Kozak KR, Lewin-Koh N, Nauka P, Bumbaca D, Sliwkowski M, Tibbitts J, Theil FP, Fielder PJ, Khawli LA, Boswell CA. Effects of anti-VEGF on pharmacokinetics, biodistribution and tumor penetration of trastuzumab in a preclinical breast cancer model. Mol Cancer Ther 2012;11:752–762.

[67] Boswell CA, Mundo EE, Firestein R, Zhang C, Mao W, Gill H, Young C, Ljumanovic N, Stainton S, Ulufatu S, Fourie A, Kozak KR, Fuji R, Polakis P, Khawli LA, Lin K. An integrated approach to identify normal tissue expression of targets for antibody–drug conjugates: case study of TENB2. Br J Pharmacol 2013;168:445–457.

[68] Boswell CA, Mundo EE, Zhang C, Stainton SL, Yu SF, Lacap JA, Mao W, Kozak KR, Fourie A, Polakis P, Khawli LA, Lin K. Differential effects of predosing on tumor and tissue uptake of an 111In-labeled anti-TENB2 antibody–drug conjugate. J Nucl Med 2012;53:1454–1461.

[69] Bien-Ly N, Yu YJ, Bumbaca D, Elstrott J, Boswell CA, Zhang Y, Luk W, Lu Y, Dennis MS, Weimer RM, Chung I, Watts RJ. Transferrin receptor (TfR) trafficking determines brain uptake of TfR antibody affinity variants. J Exp Med 2014;211:233–244.

[70] Hotzel I, Theil FP, Bernstein LJ, Prabhu S, Deng R, Quintana L, Lutman J, Sibia R, Chan P, Bumbaca D, Fielder P, Carter PJ, Kelley RF. A strategy for risk mitigation of antibodies with fast clearance. MAbs 2012;4:753–760.

[71] Xin Y, Bai S, Damico-Beyer LA, Jin D, Liang WC, Wu Y, Theil FP, Joshi A, Lu Y, Lowe J, Maia M, Brachmann RK, Xiang H. Anti-neuropilin-1 (MNRP1685A): unexpected pharmacokinetic differences across species, from preclinical models to humans. Pharm Res 2012;29:2512–2521.

[72] Lammerts van Bueren JJ, Bleeker WK, Bogh HO, Houtkamp M, Schuurman J, van de Winkel JG, Parren PW. Effect of target dynamics on pharmacokinetics of a novel therapeutic antibody against the epidermal growth factor receptor: implications for the mechanisms of action. Cancer Res 2006;66:7630–7638.

[73] Jain RK. Transport of molecules, particles, and cells in solid tumors. Annu Rev Biomed Eng 1999;1:241–263.

[74] Adams GP, Schier R, McCall AM, Simmons HH, Horak EM, Alpaugh RK, Marks JD, Weiner LM. High affinity restricts the localization and tumor penetration of single-chain fv antibody molecules. Cancer Res 2001;61:4750–4755.

[75] Minchinton AI, Tannock IF. Drug penetration in solid tumours. Nat Rev Cancer 2006;6:583–592.

[76] Thurber GM, Schmidt MM, Wittrup KD. Factors determining antibody distribution in tumors. Trends Pharmacol Sci 2008;29:57–61. . Epub 2008.

[77] Tabrizi M, Funelas C, Suria H. Application of quantitative pharmacology in development of therapeutic monoclonal antibodies. AAPS J 2010;12:592–601. (Epub 2010 Jul 20-24).

8

PREDICTION OF HUMAN PHARMACOKINETICS FOR PROTEIN-BASED BIOLOGIC THERAPEUTICS

CHAO HAN AND CHRISTINA LOURDES MAYER

Janssen Research & Development, LLC, Spring House, PA, USA

8.1 INTRODUCTION

Biologic therapeutics, or biologics, encompass a wide range of products such as vaccines, stem cells, recombinant RNA, proteins, and monoclonal antibodies (mAbs). When the pharmacologic mechanism of a therapeutic, including a biologic, suggests a potential concentration–response relationship, the pharmacokinetics (PK) of the therapeutic agent becomes key in clinical development. Accurate prediction of the PK for a new molecular entity (NME) in humans based on data from *in vitro* and *in vivo* preclinical experiments and preexisting knowledge is critical for a successful first-in-human (FIH) trial and early clinical development [1]. The accuracy of prediction of human PK is directly pertinent to the safety of the FIH trial. A safety margin is often assessed based on the drug exposure in preclinical safety assessment studies and predicted human PK at proposed dose levels before the FIH trial. Accurately predicted human PK is essential for clinical investigators to anticipate pharmacological response for an NME in proof of concept and dose-ranging clinical trials, which play a vital role in late-phase clinical development and registration trials. The projection of the dose level and PK in early clinical development heavily relies on the PK and pharmacokinetic/pharmacodynamic (PK/PD) relationship developed in preclinical studies. Theoretical considerations and common approaches for the practice of human PK prediction for protein-based biologic therapeutics will be discussed in this chapter.

Scientists and researchers have put forth great effort over many decades into understanding the determinants of PK and trying to define a relationship between preclinical species and humans in order to perform reasonably accurate predictions. It has been a major challenge to actually reach this goal. In the 1980s, Prentis *et al* reviewed the performance of drug development by seven UK-owned pharmaceutical companies [2]. One of the major causes of attrition in clinical development was inappropriate PK in human, accounting for 39.4% of failures; in comparison, lack of clinical efficacy accounted for 29.3% of failures. In other words, if the prediction of human PK was improved in early clinical development, a large percentage of the attrition could be avoided. Continuous efforts have been made since the Prentis review to understand the mechanisms of drug absorption, distribution, metabolism, and elimination (ADME) including the improvement of *in vitro* preclinical models, *in vivo* animal models, and predictive mathematical modeling techniques. In 2004, a similar investigation showed that the portion of attrition during clinical development attributed to PK was down to approximately 10% due to successful efforts in elucidating and applying ADME principles; in comparison, the portion attributed to issues with efficacy had only slightly decreased to 26% [3]. These investigations were primarily conducted for small molecule therapeutics; a similar analysis has not been conducted solely for protein-based therapeutics, which may differ in major causes of attrition. A 2010 study of drug development attrition showed that approximately 10% of developed small molecule drugs make it to market, while biologic therapeutics (primarily mAbs with soluble targets) have more than twice this success rate [4]. The relatively high success rates for biologics could be due

ADME and Translational Pharmacokinetics/Pharmacodynamics of Therapeutic Proteins: Applications in Drug Discovery and Development, First Edition. Edited by Honghui Zhou and Frank-Peter Theil.

to generally high target specificity and more favorable and predictable PK of mAbs.

The similarities in drug ADME across mammalian species have been evaluated by numerous studies. Some fundamental approaches for prediction of human PK, namely interspecies scaling and physiologically based PK (PBPK) modeling, have been proposed [5]. Conventional interspecies allometric scaling is a classic, relatively simple mathematical approach that requires attaining relevant data from multiple animal species; selection of relevant preclinical species is critical. PBPK modeling fully incorporates anatomical, physiological, and biological factors relevant to ADME. PBPK models involve high computational complexity and require an extensive amount of high quality experimental data [6]. These two commonly used approaches for prediction of human PK will be discussed in this chapter along with the accompanying concepts and rationale.

8.2 GENERAL ALLOMETRIC SCALING AND INTERSPECIES SCALING METHODS

Allometry, a concept that has existed for almost a century, is the study of the relationship between a living organism's body size to its shape, anatomy, physiology, behavior, and other biological functions [7, 8]. An intriguing allometric relationship between some physiological functions and body weight was first observed by Adolph in 1949 [9]. According to Adolph, the relationship of a measured physiological function and body weight usually follows a power law,

$$\text{Function measurement} \propto \left(\text{Body weight}\right)^{\text{Exponent}},$$

where the exponent can be determined by a series of observations or experiments in subjects with different body weights. Once this relationship is defined, the function measurement in a subject of different body weight could be scaled or predicted.

This allometric relationship is also applicable across different species of land mammals with different body sizes due to their similarities in anatomy, biology, and physiology. Application of the power law to different mammalian species created the foundation of interspecies scaling in pharmacological research.

Dedrick adapted the concept of "scale-up" from chemical engineering to model the disposition of drugs in mammals with different body sizes [10]. In this work, a number of drugs were categorized into metabolized and nonmetabolized, and their PK was studied in different mammalian species. It became apparent that although the physiological processes across selected species are straightforward and predictable, the biochemical processes (e.g., drug metabolism) vary greatly and unpredictably across different species. Nonetheless, the totality of physiological and biochemical processes, such as the blood flow carrying the substrate to a metabolizing organ or the composition and contribution of different drug-metabolizing enzymes, was surprisingly similar. The law of allometric scaling applies only on the basis of the accumulation of observations and knowledge of biology and physiology across mammalian species of different body sizes. Currently, these concepts and theories are heavily utilized when predicting human PK or drug metabolism from experimental data obtained in related animal species.

Interspecies scaling is most commonly used for the prediction of human PK in drug discovery and development for small molecule drugs. An example is dabrafenib, a small molecule BRAF inhibitor used for the treatment of melanoma, for which interspecies scaling was used in early development. The PK of dabrafenib was studied in four preclinical species: mouse, rat, monkey, and dog [11]. The total body clearance of dabrafenib was experimentally determined in each species following a single intravenous (IV) injection of dabrafenib at a clinically relevant dose. A linear relationship was observed between log-transformed clearances and log-transformed body weights and was used to describe the interspecies relationship across selected species (Fig. 8.1). Using the exponent and intercept from the linear regression describing the experimentally determined allometric relationship, clearance in humans was predicted to be 14.8 L/h. This prediction was very close to the observed 12.0 L/h clearance in healthy human subjects in the FIH study [11].

Interspecies allometry is often accurate given similar metabolism and disposition mechanisms across species; however, inconsequential interspecies differences may still exist. For example, the drug-metabolizing enzymes involved in metabolic elimination of dabrafenib differ slightly across the selected preclinical animal species and humans. In humans, dabrafenib is primarily metabolized by CYP2C8 (56%) and CYP3A4 (23%), with minor contribution from CYP2C9 (10%) [12]. The animal species utilized do not share the exact same isozymes and substrate profiles. Nonetheless, the interspecies scaling provided reasonable data for human prediction. Thus, it must follow that the total contribution to metabolic elimination of dabrafenib in the animal species tested is very similar to that in humans, even though the contribution of each isoform of drug metabolizing enzyme may greatly vary. Another example is that the clearance rate of dabrafenib in each individual species deviates from the ideal linear allometric relationship to some extent. This deviation could be due to experimental variability or true species differences. Each animal species may have some uniqueness in ADME mechanisms that is demonstrated during experimental evaluation, resulting in a true shift from the interspecies allometry line. However, when regression is performed using several relevant species, the totality of ADME principles is accounted for and results in relatively accurate prediction of human PK.

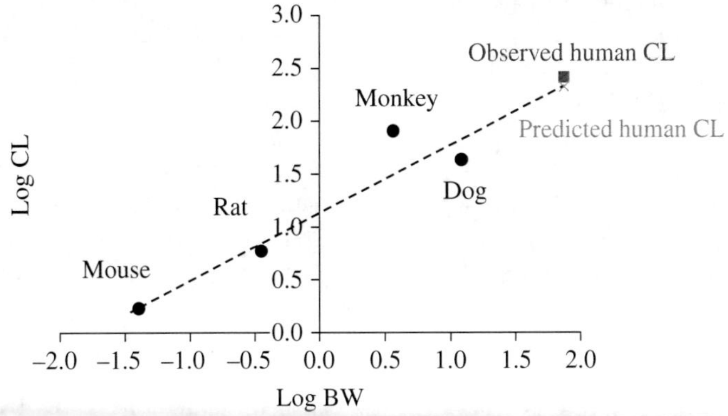

FIGURE 8.1 Allometric scaling of dabrafenib clearance. Dotted line represents the linear relationship between log-transformed body weight (Log BW) and log-transformed clearance (Log CL) in four preclinical species after IV administration of dabrafenib. Observed values for human clearance after IV dosing were consistent with values predicted from preclinical species (Adapted from Denton et al. [11].)

Interspecies allometry assumes biological, physiological, and anatomical similarities underlying similarities in drug metabolism and disposition between animal species and human, rather than differences [5, 10]. Decades of scientific research on physiology and biology of various species has generated a strong understanding of the relationship between biological processes such as drug elimination and body weight. This relationship has proven to be readily applicable to the fate of xenobiotics that are generally thought to be eliminated primarily via the kidneys and liver. However, substantial interspecies differences may violate underlying assumptions for successful allometric predictions, and conventional allometric scaling must be used with caution in these cases.

The above example also demonstrates that the number of species selected for interspecies scaling is important, as it may reveal true differences across species and counteract experimental error. Using multiple species has been encouraged for the prediction of human PK for small molecule therapeutics [13], but it has also been debated whether using multiple species will actually increase the accuracy of predictions [14].

8.3 CONSIDERATIONS FOR INTERSPECIES SCALING OF PROTEIN-BASED BIOLOGIC THERAPEUTICS

Drugs can be classified by molecular size as either small molecule or large molecule drugs; small molecule drugs have a molecular weight less than 1 kDa. Therapeutic drugs can also be subclassified as biologics, indicating production from living organisms or their derivatives; biologics can be composed of carbohydrates, proteins, nucleic acids, cells, or tissue. Examples of therapeutic biologics include vaccines, blood and blood components, somatic cells, gene therapy, recombinant proteins, mAbs, antibody fragments, peptides, fusion proteins, and oligonucleotides [15]. Protein-based biologics can also be categorized by their size and type. For the purpose of the following discussion, small and large proteins are characterized by molecular weights less than or greater than 40 kDa, respectively, and large proteins include both mAbs and non-antibody proteins. Some small protein constructs may be linked and engineered for multiple specificities or half-life extension, resulting in a final molecular size falling into the large protein category. An example would be peginterferon alfa-2a, a pegylated interferon at a total molecular weight of 40 kDa [16]. Another example is the designed ankyrin repeat protein (DARPin) therapeutic platform. Each DARPin unit has a molecular weight of approximately 14–21 kDa, and two or more units may be linked for multiple specificity [17].

While the basic PK concepts dictating small molecule behavior still apply, the factors driving the ADME of protein-based therapeutics are very different [18]. Biodistribution of protein-based biologics is usually limited by polarity, charge, and molecular size. For many mAbs, the distribution volume derived from noncompartmental or compartmental PK modeling methods is close to circulating plasma volume [19]. However, a small apparent volume of distribution does not necessarily reflect true limited tissue distribution. The calculated volume of distribution using a PK model is only a mathematical representation of the molecule's rate of convection to interstitial space versus the rate of lymphatic drainage [20]. Protein-based therapeutics, unlike small molecules, are not typically subject to conventional phase I or II metabolism by metabolizing enzymes to a meaningful extent and are not usually substrates of the typical small molecule drug transporters. Rather, renal excretion and protein catabolism are the major contributors to the elimination of protein-based therapeutics, and both processes are very well conserved across mammalian species.

Renal excretion of proteins occurs through glomerular filtration, tubular proteolytic digestion and reabsorption of the broken down products, and elimination of unabsorbed molecules or fragments in the urine. The efficiency of glomerular filtration for large molecules is directly related to the size and shape of the molecule [21]. Figure 8.2 shows that the glomerular sieving coefficient for a protein decreases with increasing molecular weight [22]. When a protein's molecular weight is greater than 40 kDa, filtration efficiency becomes poor and therefore glomerular filtration may not contribute significantly to elimination. Proteins filtered through the glomeruli are likely subsequently digested in the proximal tubules into small peptides and/or amino acids and reabsorbed back into the circulation by renal uptake transporters. Proteins/peptides that are not susceptible to renal proteases as well as residual proteins will be excreted in the urine.

Protein catabolism through endocytosis is another major pathway for disposition of circulating protein-based therapeutics, particularly for proteins with molecular weight greater than 40 kDa. Endocytosis is an energy dependent process by which larger molecules, such as proteins, are engulfed by cells [23]. The process is utilized in almost all cell types in the body for the convection of large polar molecules that cannot directly cross the lipid bilayer cell membrane by diffusion. Several types of endocytosis processes are common in mammalian cells—pinocytosis, phagocytosis, clathrin-mediated endocytosis, and caveolae [24].

Protein molecules engulfed into the cytosolic space during endocytotic processes are encapsulated by distinct membrane compartments and form vesicles called endosomes, which either internalize large polar molecules at the plasma membrane and recycle them back to the surface (early endosomes) or sort them for degradation (late endosomes and lysosomes). Early endosomes have a tubulovesicular structure and a weakly acidic pH of approximately 6.0–6.5, at which many ligands can dissociate from their receptors, most of the proteins remain stable, and some proteins such as receptors recycle back to the cell surface [25]. Early endosomes also function in transcytotic trafficking of some proteins [26]. Late endosomes, or multivesicular bodies, which exhibit some characteristics of lysosomes, contain many membrane vesicles or membrane lamellae and have a more acidic pH of approximately 5.5. Late endosomes are thought to be the location of final sorting events before lysosomal degradation. Lysosomes have an even lower pH of approximately 5.0 and contain activated proteases that hydrolytically break down proteins into amino acids [27, 28].

Since renal elimination and endocytotic protein catabolism are the primary mechanisms of disposition and elimination for protein-based therapeutics and both functions are highly conserved across mammalian species, the concept of interspecies scaling can be applied. Mordenti and colleagues examined the clearance and volume of distribution of five human therapeutic proteins—recombinant CD4, CD4 immunoglobulin G (IgG), growth hormone, tissue-plasminogen activator, and relaxin—in humans and laboratory animals (mouse, rat, hamster, rabbit, monkey, and dog) [29]. The molecular weights of these protein-based therapeutics range from 6 to 98 kDa. The clearance and volume of distribution data were analyzed as a function of body weight. The analysis revealed that the clearance and

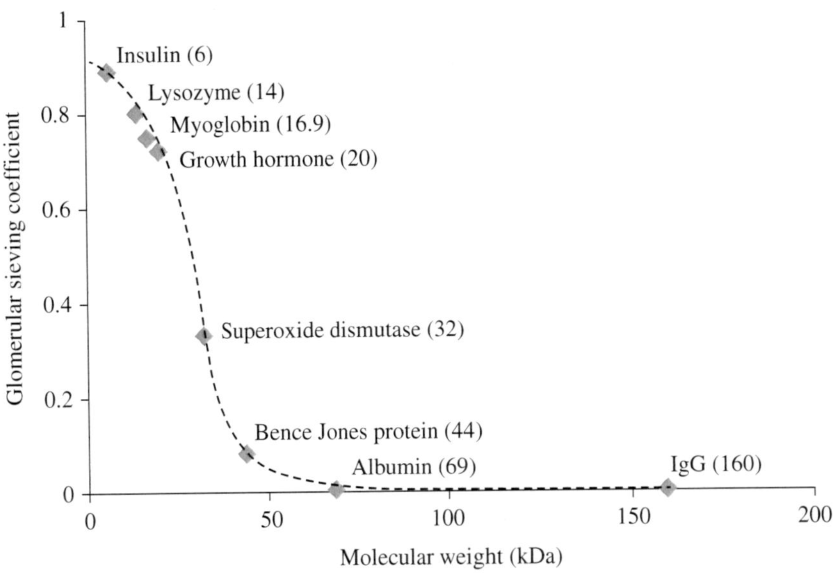

FIGURE 8.2 Molecular size and glomerular filtration for protein therapeutics. Dotted line represents the regression between molecular weight in kilodaltons and glomerular sieving coefficient. Glomerular filtration is most efficient for proteins smaller than 40 kDa; for molecular weights exceeding 40 kDa, the filtration rate falls off sharply. The molecular weight of each test substance is indicated in parentheses. IgG, immunoglobulin G (Data from Braekman [22].)

volume of distribution for each protein were satisfactorily described by the allometric equation [$Y=a\times W^b$], where Y is the PK parameter, W is body weight, and a and b are the allometric scaling factors. A range of values for the allometric exponent b was derived from the analysis: 0.65–0.84 for clearance (mL/min) and 0.83–1.05 for initial volume of distribution (mL). These values are similar to frequently cited exponent values for small molecules and are as expected based on empirical interspecies relationships. The fact that the PK parameters for selected large molecules followed well-defined, size-related physiological relationships further justified rational utilization of allometric scaling methods in early clinical development of protein therapeutics.

Utilization of interspecies scaling based on preclinical data from multiple species for the prediction of human PK was further investigated by Mahmood [30]. In this study, 15 molecules including recombinant human interleukin 2 (rhIL2), factor VIII, digoxin-Fab, PEG interleukin 2 (PEG IL2), erythropoietin-b (EPO-b), factor IX, recombinant erythropoietin (rEPO), vascular endothelial growth factor (VEGF), alteplase, recombinant relexin (rRelexin), BM 06.022 (a nonglycosylated mutant of human tissue-type plasminogen activator), hirudin, atrial natriuretic factor (ANF), lenercept, and saruplase were used for scaling and compared for prediction accuracy. The molecular weight of the molecules ranged from 5 to 7 kDa (rRelexin and Hirudin) to approximately 280 kDa (factor VIII) [31]. Human clearance could be predicted within a onefold difference for most of the molecules, with the exceptions of hirudin, ANF, lenercept, and saruplase.

Relative contribution of renal elimination and protein catabolism (which are well conserved across mammalian species) depends on the properties of the molecule, such as molecular weight. Other factors such as binding to the target or immunogenicity (e.g., lenercept [32]) may also have an impact on the PK and, therefore, on the accuracy of interspecies predictions. Generally, interspecies scaling for prediction of human PK can be applied to protein therapeutics, and previous studies have demonstrated reasonably accurate predictions. Similar to small molecules, the classic multispecies approach is favored for protein therapeutic allometric scaling, as it has demonstrated favorable results [30], but there is also some evidence that single-species scaling from monkeys above saturating concentrations may also be reasonably predictive in some cases and this will be discussed with more mechanistic details in the following sections [33].

8.3.1 Considerations for Interspecies Scaling of mAbs

IgG-based mAbs represent the most successful class of protein-based therapeutics developed thus far. Over 650 biologic therapeutics are in development (excluding vaccines), and more than 50% are mAbs according to the PhRMA report [34]. About 70% of therapeutic mAbs approved or under review in the European Union and United States are of the IgG1 isotype [35]. IgG-based mAbs usually have a molecular weight of approximately 150 kDa. The PK of this class of protein therapeutics exhibits many interesting and unique properties.

Interspecies scaling was used for the prediction of human PK from preclinical data for a small number of mAbs, including ABX-IL-8 [36], bevacizumab [37], and pertuzumab [38], with varied results. When allometric scaling was further examined for a collection of 14 mAbs by Ling et al., the results indicated intriguingly and clearly that using more than one species would not necessarily generate the best prediction [33]. The authors suggested that using cynomolgus monkey in a single-species approach with a fixed exponent would result in the best prediction. Another investigation of simple allometric scaling was conducted by Wang and Prueksaritanont for protein-based therapeutics with an expanded database of 34 therapeutic proteins including 12 mAbs [39]. These findings were in agreement with Ling et al. in concluding that a fixed-exponent approach using data from a single species alone could provide very reasonable predictions of human PK. The general impression from these two studies, then, was that utilizing more than one species, especially small species (e.g., rodents), would not improve predictability. This conclusion, however, seems contrary to prior evidence discussed previously by Mahmood [30]. What, then, makes mAbs different from other protein-based therapeutics? A deeper understanding of mAb elimination could provide the answer.

8.3.1.1 *Factors That Affect the Elimination of mAbs*

The mechanisms for elimination and disposition of therapeutic mAbs have been reviewed [36], and the differences between protein drugs and small molecule drugs were previously mentioned. Phase I and phase II drug-metabolizing pathways catalyzed by cytochrome P450s, other oxidative enzymes, and/or transferases are not meaningful clearance mechanisms for mAbs. In addition, renal elimination contributes very little to the elimination of antibodies due to their large molecular weight (~150 kDa). The primary clearance mechanism for mAbs is via proteolytic catabolism following fluid-phase or receptor-mediated endocytosis [19, 40, 41].

IgG molecules have a roughly Y-shaped construction containing two heavy chains and two light chains as shown in Figure 8.3 [42]. The two arms of the "Y," the Fab fragments (fragment, antigen-binding), contain sites that can bind to specific antigens, and the binding of each arm is typically identical. The Fab region is composed of one constant and one variable domain from each heavy and light chain of the antibody. The base of the "Y," the Fc fragment (fragment, crystallizable), plays several important roles in modulating immune cell activity and maintaining the homeostasis of circulating antibody concentrations. The Fc region is composed of two heavy chains. It is well

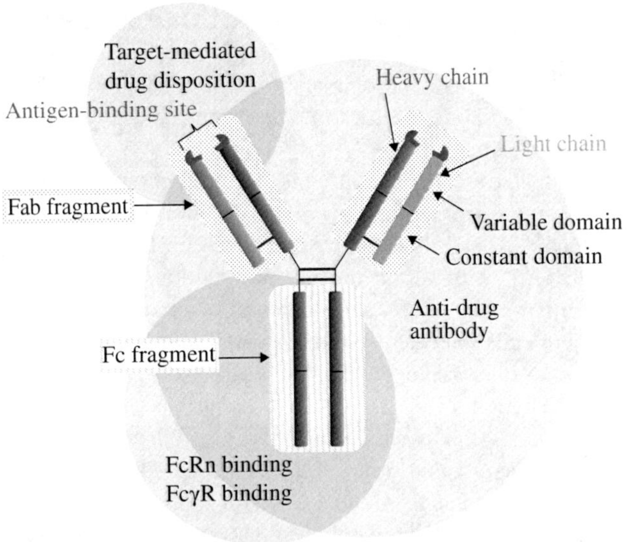

FIGURE 8.3 IgG structure and elements contributing to elimination of monoclonal antibodies. IgG molecules have a Y-shaped construction and contain two heavy chains and two light chains joined by disulfide bonds (solid heavy line). Each chain contains a constant and variable domain (separated by a line). Each arm of the Y is a Fab fragment and is composed of one constant and one variable domain from each heavy and light chain. The two Fab fragments contain antigen-binding sites in the variable domains that are typically identical; binding of these sites to target may result in increased clearance through internalization of the drug–target complex. The base of the Y is the Fc fragment. Binding of the Fc fragment to FcRn results in reduced clearance through antibody recycling; binding to FcγR may result in increased clearance through FcγR-dependent reactions. ADA can form against any part of the molecule and could result in increased clearance. IgG, immunoglobulin G; Fab, antigen-binding fragment; Fc, crystallizable fragment; FcRn, neonatal Fc receptor; FcγR, Fcγ receptor; and ADA, antidrug antibody.

known that the Fc region binds to neonatal Fc receptor (FcRn) expressed on the cell membrane [43].

FcRn, also known as the Brambell receptor, is expressed primarily on the endothelial cells, but also on many other cell types, including monocytes, macrophages, and dendritic cells [44, 45]. FcRn was originally believed to function in the transmission of immunoglobulin from mother to newborns, and investigation of this role led to the discovery that FcRn also effectively protects native IgGs from protein catabolism [46, 47]. In these studies, the administered Fc region of a G globulin had greater than 10 days circulating half-life while the corresponding Fab fragments only had a half-life of approximately 0.2 days. Thus, it was discovered that the Fc region of IgG binds to FcRn during endocytosis. The binding affinity is pH dependent [48], increasing dramatically at lower pH such that the binding complex protects the antibody from proteolytic lysosomal degradation following the endocytotic processes. "Rescued" antibody molecules can then be recycled back to the cell membrane. The Fc–FcRn-binding affinity decreases significantly at physiologic pH of approximately 7.4 at the cell surface, allowing the molecules to be released back into the circulation or into interstitial spaces. These processes are conceptually illustrated in Figure 8.4. This mechanism largely explains the long half-life of IgGs and many IgG-based therapeutic mAbs [54].

Results from many studies support the critical role of FcRn in the disposition of mAbs. The relationship between *in vitro* FcRn-binding affinity at pH approximately 6.0 and the corresponding *in vivo* elimination half-life was investigated for several IgG1-based Fc constructs in cynomolgus monkeys [48]. However, a direct relationship between increased binding affinity to the receptor at this pH and improved PK properties was not observed in this early study. More data from other studies indicate that reengineering of the Fc region for altered FcRn-binding affinity can alter the PK properties of a molecule [55–57]. Another study demonstrated that modification of a few amino acid residues in the Fc-variant region significantly increased the *in vivo* half-life of several mAbs, including bevacizumab, in *hFcRn* transgenic mice and cynomolgus monkeys [58]. With a greater understanding of the FcRn–IgG interaction in the endocytotic process in the past few years, it is now understood that characterizing FcRn–IgG-binding assays must focus on strict pH-dependent nature of the interaction at both acidic (pH 5–6) and neutral pH (pH 7.4) [59].

Other Fc-related functions may also play a role in the disposition of therapeutic mAbs. Fcγ receptors (FcγRs) are among the most important superfamilies of Fc receptors and are responsible for internalization of antibody–antigen complex, for example, during phagocytosis of opsonized microbes [60]. Four classes of FcγRs, FcγR I through FcγR IV, have been reported thus far, each with different binding affinities to different immunoglobulins [61]. In addition to the activation of innate effector pathways and stimulation of antibody-dependent cellular cytotoxicity, FcγRs also play a key role in the regulation of the immune response [62]. The involvement of FcγR in the elimination of antibodies following the interaction with its target was suggested by Reddy in 2000 [63]. Later studies demonstrated that FcγR I- and FcγR III-dependent pathways are responsible for the depletion of anti-CD20 antibody after binding to the target [64].

8.3.1.2 Interspecies Scaling Considerations for mAbs

Interspecies scaling recognizes trends in biological functions across animal species with distinct body sizes and utilizes these trends for predictions. Extra caution must be taken when factors that affect disposition and elimination, and therefore PK, do not obey the laws of allometry. The Fc–FcRn-binding affinities across different species have a very unique pattern [65]. For instance, human IgG binds to mouse FcRn stronger than it binds to human FcRn, while mouse

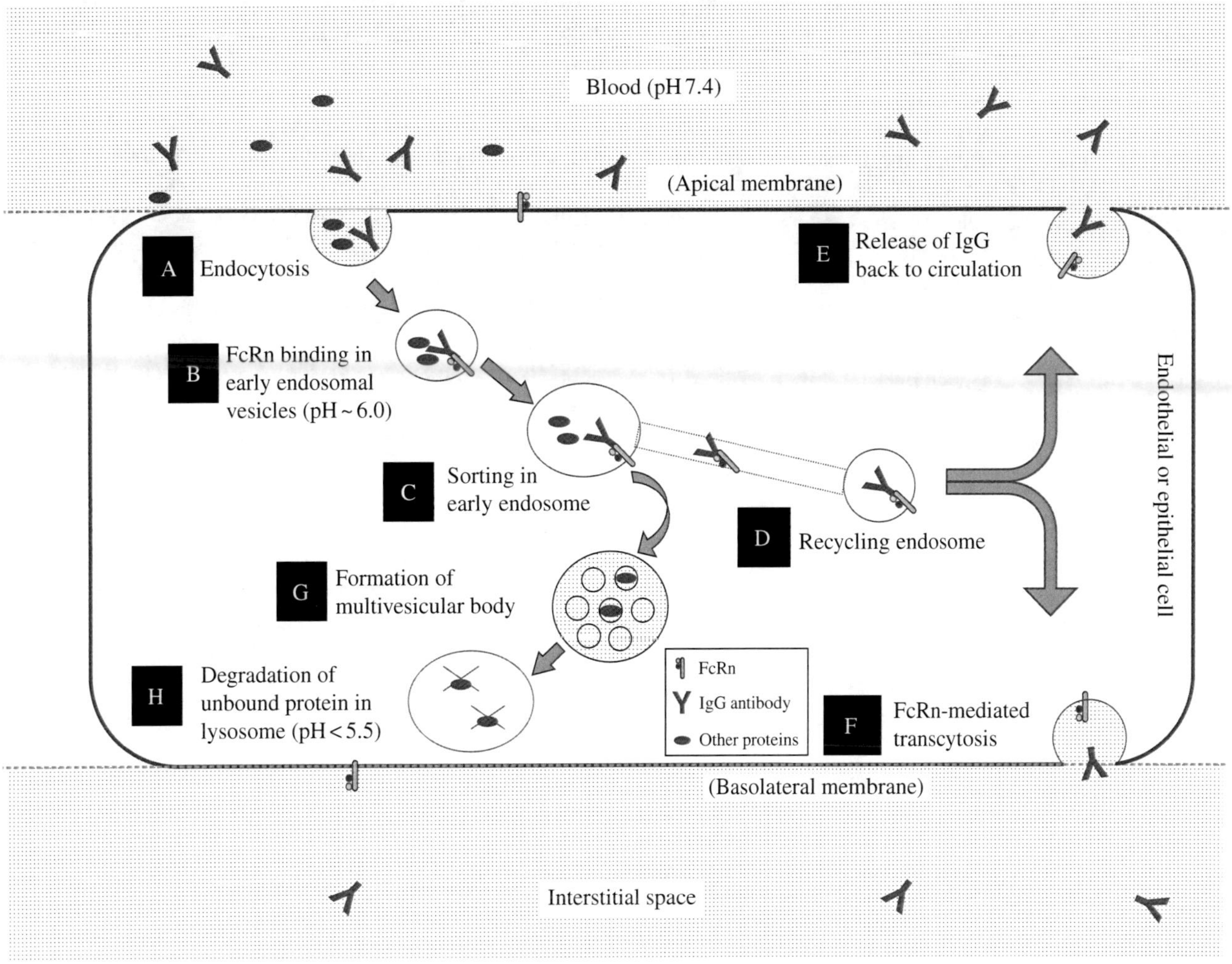

FIGURE 8.4 The fate of IgG upon endocytosis. (A) Extracellular proteins, including IgG, are endocytosed into an endothelial cell. (B) In the early endosomal vesicle, FcRn binds to the Fc region of an endocytosed IgG antibody; this binding occurs at acidic pH (6.0–6.5). (C) In the early sorting endosome, FcRn–IgG complex is segregated from other unbound proteins and is transferred via recycling tubules to (D) a recycling endosome. The complex can then be (E) transported back to the apical plasma membrane where IgG is released upon exposure to a near-neutral pH or (F) be transported to the basolateral membrane, resulting in FcRn-mediated transcytosis. (G) Selective depletion of the early endosome ultimately generates a multivesicular body whose contents, including the remaining unbound proteins, are destined for (E) lysosomal degradation. Similar transcellular pathways for IgG molecules are also found in epithelial and other cell types [49–53].

IgG1 does not bind human FcRn with significant affinity *in vitro* [66]. These species differences do not comply with the concept of allometry and explain why a multispecies approach (or use of the wrong species in this case) may result in inaccurate prediction of human PK. FcγR idiosyncracies in different species are also important. Even though FcγRs are expressed in all studied mammalian species and their ontogeny is similar, differences in the intracellular domain and cellular expression patterns have been observed [62]. Furthermore, human FcγR III binds IgG1 and IgG3, whereas nonhuman primate FcγR III tends to bind human IgG1 and IgG2 [67]. These differences should be fully considered when selecting a species for PK evaluation and interspecies scaling of therapeutic mAbs.

The biggest advantage of therapeutic mAbs is high biologic specificity and slow clearance. Many of the therapeutic mAbs are engineered specifically to bind human target and may not be cross-reactive with animal species. These favored properties paradoxically limit the available preclinical species for interspecies scaling. More published data support single-species allometric scaling using cynomolgus monkeys, since the FcRn affinity is often similar to human and the IgGs exhibit similar target binding. Oitate *et al.* studied the PK data of 24 mAbs in monkeys and humans [68]. The estimated allometric exponents for soluble and membrane-bound targets were consistent with previously published ranges [33, 39]. In another recent study, 13 therapeutic mAbs exhibiting linear PK in the studied dose range were investigated [69]. With this more comprehensive dataset, the authors suggested that using cynomolgus monkey as a single species with a fixed exponent of 0.85 clearly demonstrated reliable prediction [33, 69]. What we learned in these studies was that projection of human clearance using multiple animal species may not be the optimal approach for mAbs because the binding affinity of human IgG1 for murine FcRn is approximately 2.5-fold higher than that for human FcRn.

As a result of the known FcRn-binding properties in wild-type mice, human FcRn-transgenic mice were generated and utilized in PK and PD studies for human antihuman HER2 antibodies [70]. The results demonstrated that the FcRn-humanized mouse is a promising animal model for research and development of human IgG-based therapeutics. Now, a series of humanized transgenic mouse models have been developed: Tg276 and Tg32 homo- and hemi-zygous transgenics. These humanized mouse lines are transgenic for the human gene encoding IgG receptor FcRn large subunit p51 (FCGRT) and are engineered to have a deletion in mouse Fcgrt. They do not express mouse FcRn but do express human FcRn protein [71]. The human FcRn-transgenic mouse has become a valuable tool for assessing the *in vivo* PK of human mAb-based therapeutics, especially when investigating Fc-engineered proteins [72]. The PK of a small number of therapeutic mAbs was investigated in Tg32- and Tg276-transgenic mice, and the results were compared with cynomolgus monkey data and human data from clinical studies [73]. Strong correlations in elimination half-life and clearance were observed between the transgenic mice and monkeys or humans, suggesting great potential for the transgenic mouse model to be used in interspecies scaling. This notion will be further strengthened by accumulation of more preclinical and clinical data.

Interspecies scaling for the prediction of mAbs in humans from data obtained in animal species is practical and applicable. The predictions have been reliable based on limited datasets (compared to small molecule drugs). However, species-specific mechanisms in the disposition of mAb molecules should be fully recognized when applying interspecies allometry [74]. Using wild-type mice in allometric scaling is likely to result in overestimation of the elimination half-life or underestimation of clearance for therapeutic mAbs. Research to date suggests that using cynomolgus monkey as a single species with a fixed exponent will generate the most reliable human predictions for mAbs.

8.3.2 Other Factors that may Affect PK Interspecies Scaling for Protein-Based Therapeutics

Other scientific phenomena may be important to consider as they could affect animal PK in preclinical studies, although their effects may not be translatable or scalable through interspecies scaling.

8.3.2.1 Impact of Antidrug Antibodies on PK Xenogeneic proteins administered to humans and animals are likely to be immunogenic. Antidrug antibodies (ADAs) could develop and lead to an alteration in the clearance of protein-based therapeutics. The immunogenicity potential of therapeutic proteins varies among different species and molecules. The potential for immunogenicity in humans decreases as the nonhuman fraction of the protein is reduced [40]. The potential for immunogenicity of a therapeutic protein in animal species, however, is still difficult to predict. A human protein such as a mAb is xenogeneic when administered to monkeys and likely to be immunogenic. This biologics-specific issue adds another dimension of complexity to species considerations for mAb disposition, although it is generally believed that the occurrence of ADA in animal species does not translate to the occurrence of ADA in humans.

Administered therapeutic mAbs may be recognized by the immune system as foreign and trigger the development of ADA against the therapeutic molecule. The development of ADA can affect PK, PD, and safety profiles [75]. The immunogenicity of mAbs in humans has been studied extensively and monitored closely for effects on efficacy and safety [76, 77]. The development of ADA inevitably impacts the quality of preclinical data for interspecies scaling. For example, the inaccurate prediction of lenercept PK in the Mahmood study may have resulted from its immunogenicity [30, 32]. Nonetheless, few preclinical studies on this topic have been published.

Development of ADA in animals used for preclinical PK studies could cause an apparent increase, decrease, or no change in the clearance of the therapeutic protein (this observation also depends on the nature of the bioassay, i.e., free drug concentration vs total concentration). Rojas and colleagues found that ADA–infliximab complex exhibited much faster clearance than infliximab [78]. The accelerated clearance could be due to more efficient phagocytosis for the complex of larger molecular size. It is also quite reasonable to speculate the involvement of FcγR-mediated processes in the clearance of the complex [63, 64].

When ADA develops in animals following administration of a therapeutic mAb, the titer to the mAb may be detected in serum samples. However, a conclusive result would depend on many factors and is often confounded by residual drug concentration [79]. An aberrant change in the drug concentration–time profile that is variable among individual animals is usually informative of the development of ADA, in our experience. The changes usually occur several days following administration, and the response can be very species dependent and variable among individual animals.

A quantitative relationship between ADA or ADA titers and the extent of its impact on drug concentrations has not been successfully established. Some options may be chosen to deal with situations of positive ADA [74]. One option is to omit the entire profile of the individual animal(s) that tests positive for ADA or demonstrates a distinct pattern of altered elimination. Another option, used when the majority of animals are affected, is to only utilize data up to the time point before ADA development for the calculation of PK parameters. The corresponding approaches are illustrated in Figure 8.5 with simulated data.

8.3.2.2 ***Target-Mediated Disposition*** Target-mediated drug disposition (TMDD) may result in nonlinear PK and often occurs for protein-based therapeutics that bind to cell membrane-bound targets. The drug–target complex may appear to be cleared from the circulation in a rapid but saturable manner. The clearance cascade involves the formation of drug–target complex that is likely internalized and rapidly eliminated through proteolytic processes [80]. The rate of elimination of therapeutic protein by target-mediated pathways is not linear and becomes a function of the drug concentration, target expression density, and turnover rate of the target. Saturation of the process is normally observed at higher doses of the therapeutic protein or following repeated dosing [81].

In general, cross-species activity for therapeutic mAbs is limited to one or a few animal species, and in many cases only monkeys. Even in a cross-reactive species, quantitative measures, such as IC_{50} or binding affinity (K_D), can be quite different from those in humans. In addition, regulation of the target and its expression pattern and intensity *in vivo* are likely different between animal species and humans. Corresponding information in human and animal species is usually unavailable for a quantitative comparison or interspecies scaling, especially for many first-in-class therapeutics in early development. These variables become a hurdle for building the scaling relationship when target-mediated nonlinear PK exists. Mathematical models using a classic linear compartmental model in parallel with a nonlinear clearance process may help attain a better understanding of the clearance processes in clinical settings [82]. Retrospective analysis of clinical and preclinical data for a number of therapeutic mAbs demonstrated great potential for use of the parallel clearance model for the prediction of human PK from preclinical data [83].

Nonlinear PK is not usually apparent for therapeutic proteins binding to soluble ligands. One plausible explanation might be that most soluble targets studied so far have relatively low circulating concentrations, reflecting a low production rate. Thus, their impact on the PK of the therapeutic protein could be negligible at clinically relevant doses. In this context, an assay specific for free drug concentration should be used in PK studies.

Mathematical models describing TMDD have been thoroughly studied and applied to clinical and preclinical PK studies [84, 85]. These models can be used for allometric scaling of the PK for therapeutic proteins that exhibit TMDD. The PK in animal species should be studied across two to three different dose levels to cover a large enough concentration range to decipher nonlinear PK properties. A TMDD model requires more information, such as the measurement of free and/or total target concentrations in addition to drug concentration, which may not be readily available in some cases. Relevant preclinical human data must be available for interspecies scaling. Dong and colleagues utilized a two-compartment model with parallel Michaelis–Menten nonlinear elimination for this purpose [83].

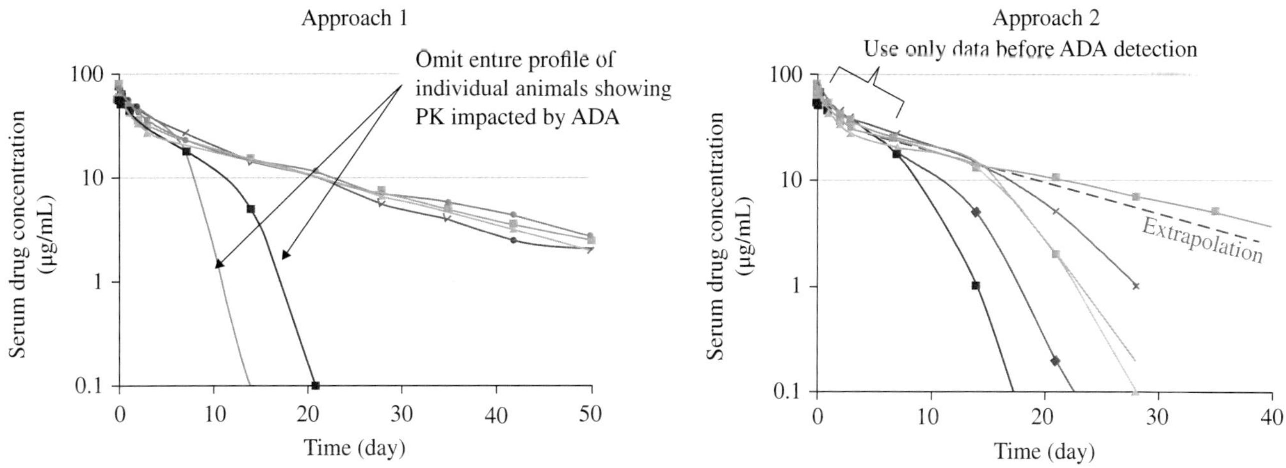

FIGURE 8.5 Simulated PK profiles with varying degrees of ADA impact and two approaches for handling ADA positive data (or data with a distinct pattern of altered elimination). In Approach 1, the entire dataset of the individual animals with aberrant PK attributed to ADA are omitted from noncompartmental analysis. In Approach 2, since the majority of animals showed altered elimination associated with positive ADA, only data up to the time point before ADA development are utilized in the calculation of PK parameters. Approach 2 will result in a greater degree of extrapolation (dashed line) in the terminal phase of the profile.

The model parameters obtained from cynomolgus monkeys were used for interspecies scaling. The results from this retrospective analysis demonstrated that parallel Michaelis–Menten nonlinear elimination model can be used and produces reasonable predictions. Using a Michaelis–Menten nonlinear elimination model to describe the nonlinear PK of TMDD has been theoretically analyzed and proven to be applicable [86]. This model can perform equally or even better than the TMDD model at relatively higher drug concentrations and when rapid dissociation rate constant (k_d) for the drug–target complex disassociation is present. Using these models for the prediction of human PK from data obtained in relevant preclinical species opens another door for allometric scaling. Although only based on a handful of examples for protein-based therapeutics, the modeling approach for allometric scaling of TMDD-related nonlinear PK is schemed for future reference in Figure 8.6.

8.4 PHYSIOLOGICALLY BASED PK MODELING

Mordenti examined the physiologic similarities between mammals and proposed two fundamental approaches for interspecies scaling: PBPK modeling and simple allometric scaling [5]. The PBPK model fully incorporates similar anatomical, physiological, and biochemical details between animal species and humans. Theoretically, PBPK modeling is superior to other types of models in addressing interspecies differences. Such a thorough and complex model, however, requires a large amount of information, and its practicality in the prediction of human PK has been discussed [6]. After decades of accumulation of *in vitro* and *in vivo* human data, the capabilities of PBPK modeling are now extensive and have played a very important role in drug development and regulation. PBPK modeling has been used in predicting the processes of drug clearance, distribution, and absorption; in quantitative prediction of PK-based drug–drug interactions; and in investigating the impact of age, genetics, disease, and formulation for small molecule drugs [87]. Researchers have explored using full PBPK models to predict circulating and tissue concentrations for therapeutic mAbs. Most parameter estimations for mAbs thus far are derived from animal models and validated retrospectively using human data.

As discussed earlier, protein catabolism is a primary mechanism for the elimination of large protein-based therapeutics. This clearance pathway, in general, exists in all tissues. Furthermore, large protein molecules that distribute in tissues and are spared from catabolic clearance and/or target binding will drain back into the systemic circulation through lymphatic circulation. Mammillary models used in conventional PK, by definition, assume that all clearances occur from the central compartment; however, peripheral clearance is known to be significant for large protein molecules such as mAbs [88]. Convection is the predominant mechanism for extravascular distribution of large protein molecules. Baxter et al. used PBPK modeling to reflect the difference in convection coefficients on vascular and lymphatic sides that yield lower concentrations of mAb in the interstitial fluids [89]. Incorporating lymphatic circulation for the drainage of mAb that distributed into tissue space and intratissue FcRn-binding dynamics, Garg and Balthasar utilized a PBPK model to predict IgG tissue concentrations [90]. The concept was validated in wild-type and FcRn-knockout mice.

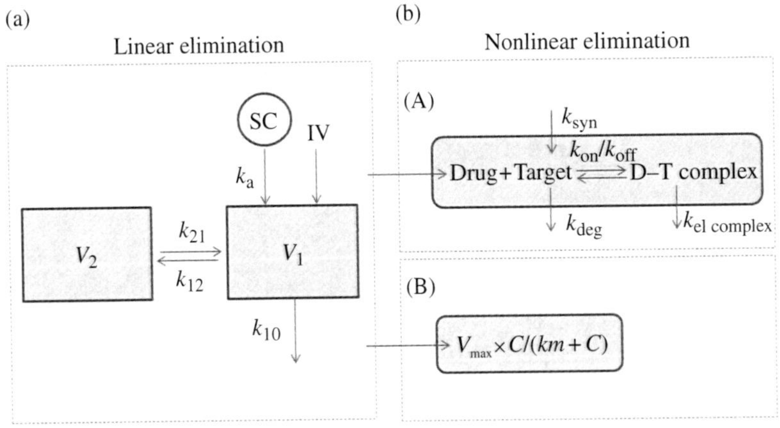

FIGURE 8.6 Mathematical models used to describe the PK of protein therapeutics. (a) A two-compartment PK model is often utilized to describe the absorption, distribution, and linear elimination associated with protein therapeutics. This model can be used to conduct simple allometric scaling. (b) Two PK modeling methods are often utilized to describe the nonlinear elimination of a protein therapeutic. (A) A TMDD model can be utilized when enough information is available and describes drug binding to the target and elimination of both target and drug–target complex. (B) A Michaelis–Menten equation may be used to characterize nonlinear PK more simply, where V_{max} is the maximum rate of nonlinear elimination at saturating concentrations and K_m is the substrate concentration at which the rate of nonlinear elimination is half V_{max}.

The model that incorporated the influence of FcRn on IgG disposition provided more accurate predictions of IgG tissue kinetics in control and FcRn-knockout mice. In Section 8.3.1.2, we discussed that engineering Fc sequences to alter Fc–FcRn binding affinity can affect the rate of clearance of mAbs. Incorporating FcRn-binding affinity data obtained *in vitro*, Chen and Balthasar used a catenary PBPK model to predict extended half-life for a mAb constructed with higher FcRn-binding affinity [91]. The results matched published experimental data very well.

A full PBPK model for mAbs utilizes mass balance to include almost all tissues. The structure of a full PBPK model is usually oriented by relative importance of the tissue/organ in drug metabolism and elimination. For example, the liver and kidneys are always distinctly considered from other tissue/organs for small molecules. However, this structure may not be suitable for all protein-based therapeutics. Building such a model requires the measurements of drug concentrations in numerous organs and tissues. When only serum or plasma concentration data are available, full PBPK models with a diverse array of parameters often require assigning or fixing some of the parameters for accurate fitting or prediction of plasma and tissue concentration–time data, or in some cases, with no fitted parameters [92, 93]. "Lumping" has been a commonly employed approach to reduce the complexity of full PBPK models by grouping tissues that show similar drug-disposition properties together to form fewer compartments [94, 95]. Minimal physiologically based PK (mPBPK) models consider organs and tissues according to the kinetic nature of large protein molecules and lump them into fewer compartments. This approach provides a mathematically and computationally simpler alternative, allowing fitting of only plasma data to generate physiologically relevant parameters while evaluating peripheral clearance [96]. Bringing the value of mPBPK further, Cao et al. established a "second-generation" mPBPK model that incorporates several essential determinants of mAb PK for system-average evaluations of mAb disposition [97]. For more details, please refer to the dedicated chapter on PBPK by Cao and Jusko (Chapter 12).

Utilization of PBPK modeling to provide valuable predictions for drug development and regulation of protein-based large molecule therapeutics is still at its exploratory stage. This modeling approach has already demonstrated its advantage theoretically and practically for some protein-based therapeutics. However, it is unclear whether utilization of PBPK increases the accuracy and precision of PK predictions for mAbs over conventional allometric scaling methods. Studies have demonstrated very reasonable predictions (within twofold of clinical observations) for mAbs using single-species allometric scaling [33, 39, 69]. PBPK modeling may provide additional utility for exploring the ADME properties of protein therapeutics or nontraditional platforms in addition to characterization and prediction of PK/PD relationships. Further accumulation of data, a better understanding of the ADME of large molecules, and development of *in vitro* and preclinical assays to generate relevant data to feed the modeling efforts will greatly advance the quality and predictability of PBPK models.

8.5 PERSPECTIVES BEYOND THE PREDICTION

8.5.1 Prediction of Human PK Serves Different Purposes at Different Stages of Drug Development

The prediction of human PK for protein-based biologic therapeutics serves several purposes during candidate selection and early development. Predicted PK in humans will directly influence the strategy in clinical development, including the selection of dose and therapeutic regimen. In balance with other developability criteria, a molecule with a favorable PK profile (predicted from animal data) will be selected for preclinical and early clinical development [98]. It is essential to understand related therapeutic area landscapes, possible influence of disease status on PK, and desired PK/PD relationship by which a PK profile will be translated into clinical outcomes. Therefore, the application and evaluation of the prediction for candidate selection should always be in conjunction with other developability perspectives—for instance, mechanism of action, target-binding affinity, and tissue distribution, as opposed to only focusing on the half-life and clearance.

Successful preclinical biology research leads to the discovery of potential target-specific protein therapeutics. For biologic therapeutics, off-target adverse effects are usually less problematic or more predictable than for small molecule therapeutics. On the other hand, the favorable target specificity of protein-based therapeutics could have significant impact on the PK as discussed in Section 8.3.2.2. PK must be viewed in conjunction with the production rate of the target and accumulation of drug–target complex concentration in the circulation [99]. Depending on the nature of the target and quality of the therapeutic molecule, predictable PK may not translate into the desired therapeutic outcome [100]. The horizon of preclinical PK studies for protein-based therapeutics in animal species may ultimately be expanded to address many of these therapeutic questions in addition to PK.

With the development of protein-based therapeutics, some new mechanisms of action may require a different view of PK. For example, blinatumomab is a T-cell redirecting oncology therapy that is composed of two different units, one that binds to CD19 on cancer cells while the other binds to CD3 on T-cells to initiate cancer cell killing by the T-cell [101]. According to the mechanism, only a very small ratio of receptor occupancy is required to initiate/activate T-cell-mediated activity. Therefore, clinically efficacious drug concentrations in the circulation are much lower than would be conventionally anticipated [102]. A full understanding of receptor occupancy becomes much more important for such mechanisms. More sophisticated modeling incorporating binding affinities for each target along with PK may be necessary. The conventional approaches for PK prediction may not be able to add much value in these cases. In addition, preclinical PK and safety data must be evaluated for target cross-reactivity and relative binding affinity across selected preclinical species and human cells before the results can be carefully interpreted, to avoid recurrence of the TeGenero[1] case [103].

A reliable or accurate prediction of human PK is always desired. Generally, a predicted value within twofold of the observed value is considered acceptable. By this standard, single-species scaling for mAbs has demonstrated acceptable predictive capabilities. However, as models are built on existing knowledge and preclinical information, there are still unknowns and even unknown unknowns. In most cases, an evaluation of the starting dose and safety margin before the FIH clinical trial is based on predicted PK. The safety consideration is always the first priority in understanding the potential bias in prediction. The importance of human PK prediction from nonclinical data decreases as more relevant clinical PK information becomes available from the FIH study. Extrapolation of PK from healthy subjects to patients or between different populations presents another unique challenge and will not be discussed in this chapter.

8.5.2 Safety Considerations When Predicting Human PK for Protein-Based Therapeutics

It is vital to understand the mechanism of action and potential species differences as already discussed in the previous sections. When information gaps exist between preclinical animal species and human, risk management strategies should be developed. For example, in the case of TMDD, when the relative density of receptor expression is unknown, one should consider using the PK information at saturating dose(s) in preclinical species for the prediction. An overestimation of the exposure in humans is given with such an approach, but it provides the maximal safety measurement of the starting dose for FIH.

The distribution of large protein molecules is usually restricted to the circulation-that is, they distribute into the plasma or serum volume. It has been noted in several publications that the exponent for interspecies scaling of volume of distribution is very close to 1 for protein-based therapeutics and mAbs [19, 33, 69]. This is also consistent with the FDA guidance on recommending the maximum recommended starting dose [104]. The maximum concentration following IV administration of large proteins can be roughly estimated using a very simple formula—dose divided by the circulating serum volume. The result provides a prediction for the highest possible concentration for comparison with the toxicokinetic results and other available safety information to evaluate the safety margin. This result could also serve as an upper limit when assessing modeling and simulation results.

[1]TGN1412 was a CD28-specific mAb superagonist developed by TeGenero Immuno Therapeutics for the treatment of hematologic malignancies. Because of underpredicted biological action in humans, the FIH starting dose derived from toxicology evaluations in monkeys resulted in disastrous cytokine storm in healthy volunteers in 2006.

8.6 CONCLUSIONS

Accurate prediction of human PK plays a critical role in early drug development. Tools such as interspecies scaling and/or PBPK modeling have been developed for this purpose. Although work lies ahead in generating more data and improving the accuracy of predictions for protein-based therapeutics, a thorough understanding of relevant mechanisms and excellent predictive tools already exists. It is essential for us to fully understand the concepts and theories of human PK prediction before employing the available tools. As an example, using cynomolgus monkey as a single species in interspecies scaling counterintuitively provided better predictions of human PK for mAbs. An integrated view of diverse perspectives and considerations when developing a strategy for human PK prediction is always prudent. Thus, it is judicious to consider safety factors, the mechanism of action for the molecule, the clinical development strategy for the disease, and the available interspecies scaling tools when performing prediction of human PK.

REFERENCES

[1] Bearumont K, Smith DA. Does human pharmacokinetic prediction add significant value to compound selection in drug discovery research? Curr Opin Drug Discovery Dev 2009;12(1):67–71.

[2] Prentis RA, Lis Y, Walker SR. Pharmaceutical innovation by the seven UK-owned pharmaceutical companies (1964-1985). Br J Clin Pharmacol 1988;25(3):387–396.

[3] Kola I, Landis J. Can the pharmaceutical industry reduce attrition rates? Nat Rev Drug Discovery 2004;3(8):711–715.

[4] DiMasi JA, Feldman L, Seckler A, Wilson A. Trends in risks associated with new drug development: success rates for investigational drugs. Clin Pharmacol Ther 2010;87(3):272–277.

[5] Mordenti J. Man versus beast: pharmacokinetic scaling in mammals. J Pharm Sci 1986;75(11):1028–1040.

[6] Teitelbaum Z, Lave T, Freijer J, Cohen AF. Risk assessment in extrapolation of pharmacokinetics from preclinical data to humans. Clin Pharmacokinet 2010;49(9):619–632.

[7] Shingleton AW. Allometry: the study of biological scaling. Nat Educ Knowl 2010;3(10):2.

[8] Gayon J. History of the concept of allometry. Am Zool 2000;40:748–758.

[9] Adolph EF. Quantitative relations in the physiological constitutions of mammals. Science 1949;109(2841):579–585.

[10] Dedrick RL. Animal scale-up. J Pharmacokinet Biopharm 1973;93(1):435–461.

[11] Denton DL, Minthorn E, Stanley W, Carson SW, Young GC, Richards-Peterson LE, Botbyl J, Han C, Morrison RA, Blackman SC, Ouellet D. Concomitant oral and intravenous pharmacokinetics of dabrafenib, a BRAF inhibitor, in patients with BRAF V600 mutation-positive solid tumors. J Clin Pharmacol 2013;53(9):955–961.

[12] Lawrence SK, Nguyen D, Bowen C, Richards-Peterson L. The metabolic drug-drug interaction profile of dabrafenib: *in vitro* investigations and quantitative extrapolation of the P450-mediated DDI risk. Drug Metab Dispos 2014;42: 1180–1190. DOI: 10.1124/dmd.114.057778.

[13] Mahmood I, Balian JD. Interspecies scaling: a comparative study for the prediction of clearance and volume using two or more than two species. Life Sci 1996;69:579–585.

[14] Ward KW, Smith BR. A comprehensive quantitative and qualitative evaluation of extrapolation of intravenous pharmacokinetic parameters from rat, dog, and monkey to humans. I. Clearance. Drug Metab Dispos 2004;32(6): 603–611.

[15] US FDA Available at http://www.fda.gov/AboutFDA/CentersOffices/OfficeofMedicalProductsandTobacco/CBER/ucm133077.htm. Accessed 2015 Jan 14.

[16] Hoffmann-La Roche Inc. 2002. Pegasys (peginterferon alfa-2a). Available at http://www.fda.gov/downloads/Drugs/DevelopmentApprovalProcess/HowDrugsareDevelopedandApproved/ApprovalApplications/TherapeuticBiologicApplications/ucm094462.pdf. Accessed 2015 Jun 10.

[17] Stumpp MT, Binz HK, Amstutz P. DARPins: a new generation of protein therapeutics. Drug Discov Today 2008;13(15/16):695–701.

[18] Tang L, Meibohm B. Pharmacokientics of peptides and proteins. In: Meibohm B, editor. *Pharmacokinetics and pharmacodynamics of biotech drugs.* , Chapter 2 Weinheim: Wiley-VCH Verlag GmbH &Co KGaA; 2006. pp. 17–43.

[19] Wang W, Wang EQ, Balthasar JP. Monoclonal antibody pharmacokinetics and pharmacodynamics. Clin Pharmacol Ther 2008;84(5):548–558.

[20] Lobo ED, Hansen RJ, Balthasar JP. Antibody pharmacokinetics and pharmacodynamics. J Pharm Sci 2004;93(11): 2645–2668.

[21] Macck T, Johnson V, Kau ST, Figueiredo J, Sigulem D. Renal filtration, transport, and metabolism of low-molecular-weight proteins: a review. Kidney Int 1979;16(3): 251–270.

[22] Braekman R. In: Reid R, editor. *Pharmacokinetics and Pharmacodynamics of Protein Therapeutics in Peptide and Protein Drug Analysis.* , Chapter 20New York: Marcel Dekker, Inc.; 2000. pp. 633–669.

[23] Marsh M, McMahon HT. The structural era of endocytosis. Science 1999;285(5425):215–220.

[24] Marsh M. *Endocytosis.* Oxford University Press; 2001. ISBN 978-0-19-963851-2.

[25] Mellman I. Endocytosis and molecular sorting. Ann Rev Cell Dev Biol 1996;12:575–625.

[26] Mukherjee S, Ghosh RN, Maxfield FR. Endocytosis. Physiol Rev 1997;77(3):759–803.

[27] Stoorvogel W, Strous GJ, Geuze HJ, Oorschot V, Schwartz AL. Late endosomes derive from early endosomes by maturation. Cell 1991;65(3):417–427.

[28] Gruenberg J, Maxfield FR. Membrane transport in the endocytic pathway. Curr Opin Cell Biol 1995;7(4):552–563.

[29] Mordenti J, Chen SA, Moore JA, Ferraiolo BL, Green JD. Interspecies scaling of clearance and volume of distribution data for five therapeutic proteins. Pharm Res 1991;8(11): 1351–1359.

[30] Mahmood I. Interspecies scaling of protein drugs: prediction of clearance from animals to humans. J Pharm Sci 2004;93(1):177–185.

[31] Hodges WA. Characterization of the recombinant human factor VIII expressed in the milk of transgenic swine. [Thesis for the degree of Master of Science in Chemical Engineering]. Blacksburg (VA): Virginia Polytechnic Institute and State University; 2001.

[32] Richter WF, Gallati H, Schiller CD. Animal pharmacokinetics of the tumor necrosis factor receptor-immunoglobulin fusion protein lenercept and their extrapolation to humans. Drug Metab Dispos 1999;27(1):21–25.

[33] Ling J, Zhou H, Jiao Q, Davis DM. Interspecies scaling of therapeutic monoclonal antibodies: initial look. J Clin Pharmacol 2009;49(12):1382–1402.

[34] Pharmaceutical Research and Manufactures of America. 2013. Medicines in development, biologics – overview. 2013 Report.

[35] The Antibody Society. 2014. Therapeutic monoclonal antibodies approved or in review in the European Union or United States. Available at http://www.antibodysociety.org/news/approved_mabs.php. Accessed 2014 Jul 28.

[36] Tabrizi MA, Tseng CM, Roskos LK. Elimination mechanisms of therapeutic monoclonal antibodies. Drug Discov Today 2006;11(1–2):81–88.

[37] Lin YS, Nguyen C, Mendoza JL, Escandon E, Fei D, Meng YG, Modi NB. Preclinical pharmacokinetics, interspecies scaling, and tissue distribution of a humanized monoclonal antibody against vascular endothelial growth factor. J Pharmacol Exp Ther 1999;288(1):371–378.

[38] Adams CW, Allison DE, Flagella K, Presta L, Clarke J, Dybdal N, McKeever K, Sliwkowski MX. Humanization of a recombinant monoclonal antibody to produce a therapeutic HER dimerization inhibitor, pertuzumab. Cancer Immunol Immunother 2006;55(6):717–727.

[39] Wang W, Prueksaritanont T. Prediction of human clearance of therapeutic proteins: simple of allometric scaling method revisited. Biopharm Drug Dispos 2010;31(4):253–263.

[40] Kuester K, Kloft C. Pharmacokinetics of monoclonal antibodies. In: Meibohm B, editor. *harmacokinetics and Pharmacodynamics of Biotech Drugs.*, Part II, Chapter 3 Weinheim: Wiley-VCH; 2006. pp. 45–91.

[41] Mould D, Breen B. Pharmacokinetics and pharmacodynamics of monoclonal antibodies. Biodrugs 2010;24(1): 23–39.

[42] Janeway C. *Immunobiology*. 5th ed. Garland Publishing; 2001. ISBN 0-8153-3642-X.

[43] Martin WL, West AP Jr, Gan L, Bjorkman PJ. Crystal structure at 2.8 A of an FcRn/heterodimeric Fc complex: mechanism of pH-dependent binding. Mol Cell 2001;7(4): 867–877.

[44] Zhu X, Meng G, Dickinson BL, Li X, Mizoguchi E, Miao L, Wang Y, Robert C, Wu B, Smith PD, Lencer WI, Blumberg RS. MHC class I-related neonatal Fc receptor for IgG is functionally expressed in monocytes, intestinal macrophages and dendritic cells. J Immunol 2001;166(5): 3266–3276.

[45] Brambell FWR, Hemmings WA, Morris IG. Theoretical model of γ-globulin catabolism. Nature 1964;203(4952): 1352–1355.

[46] Brambell FWR. The transmission of immunity from mother to young and the catabolism of immunoglobulins. Lancet 1966;288(7473):1087–1093.

[47] Spiegelberg HL, Weigle WO. The catabolism of homologous and heterologous 7S gamma globulin fragments. J Exp Med 1965;121(3):323–338.

[48] Datta-Mannan A, Witcher DR, Tang Y, Watkins J, Jiang W, Wroblewski VJ. Humanized IgG1 variants with differential binding properties to the neonatal Fc receptor: relationship to pharmacokinetics in mice and primates. Drug Metab Dispos 2007;35(1):86–94.

[49] Ober RJ, Martinez C, Vaccaro C, Zhou J, Ward ES. Visualizing the site and dynamics of IgG salvage by the MHC class I-related receptor, FcRn. J Immunol 2004;172(4):2021–2029.

[50] Ward ES, Martinez C, Vaccaro C, Zhou J, Tang Q, Ober RJ. From sorting endosomes to exocytosis: association of Rab4 and Rab11 GTPases with the Fc receptor, FcRn, during recycling. Mol Biol Cell 2005;16(4):2028–2038.

[51] Delevoye C, Miserey-Lenkei S, Montagnac G, Gilles-Marsens F, Paul-Gilloteaux P, Giordano F, Waharte F, Marks MS, Goud B, Raposo G. Recycling endosome tubule morphogenesis from sorting endosomes requires the kinesin motor KIF13A. Cell Rep 2014;6(3):445–454.

[52] Grant BD, Donaldson JG. Pathways and mechanisms of endocytic recycling. Nat Rev Mol Cell Biol 2009;10 (9):597–608.

[53] Tzaban S, Massol RH, Yen E, Hamman W, Frank SR, Lapierre LA, Hansen SH, Goldenring JR, Blumberg RS, Lencer WI. The recycling and transcytotic pathways for IgG transport by FcRn are distinct and display an inherent polarity. J Cell Biol 2009;185(4):673–684.

[54] Roopenian DC, Akilesh S. FcRn: the neonatal Fc receptor comes of age. Nat Rev 2007;7(9):715–725.

[55] Datta-Mannan A, Witcher DR, Tang Y, Watkins J, Wroblewski VJ. Monoclonal antibody clearance. Impact of modulating the interaction of IgG with the neonatal Fc receptor. J Biol Chem 2007;282(3):1709–1717.

[56] Hinton PR, Xiong JM, Johlfs MG, Tang MT, Keller S, Tsurushita N. An engineered human IgG1 antibody with longer serum half-life. J Immunol 2006;176(1):346–356.

[57] Kenanova VE, Olafsen T, Andersen JT, Sandlie I, Wu AM. Engineering of the Fc region for improved PK (FcRn Interaction). In: Kontermann R, Dübel S, editors. *Antibody Engineering*., Chapter 27New York: Springer; 2010. pp. 411–430.

[58] Zalevsky J, Chamberlain AK, Horton HM, Karki S, Leung IW, Sproule TJ, Lazar GA, Roopenian DC, Desjarlais JR. Enhanced antibody half-life improves *in vivo* activity. Nat Biotechnol 2010;28(2):157–159.

[59] Datta-Mannan A, Wroblewski VJ. Application of FcRn binding assays to guide mAb development. Drug Metab Dispos 2014;42(11):1867–1872.

[60] Fridman W. Fc receptors and immunoglobulin binding factors. FASEB J 1991;5(12):2684–2690.

[61] Nimmerjahn F, Ravetch JV. Fcγ receptors: old friends and new family members. Immunity 2006;24(1):19–28.

[62] Nimmerjahn F, Ravetch JV. Fcγ receptors as regulators of immune response. Nat Rev Immunol 2008;8(1):34–47.

[63] Reddy MP, Kinney CA, Chaikin MA, Payne A, Fishman-Lobell J, Tsui P, Dal Monte PR, Doyle ML, Brigham-Burke MR, Anderson D, Reff M, Newman R, Hanna N, Sweet RW, Truneh A. Elimination of Fc receptor-dependent effector functions of a modified IgG4 monoclonal antibody to human CD4. J Immunol 2000;164(4):1925–1933.

[64] Uchida J, Hamaguchi Y, Oliver JA, Ravetch JV, Poe JC, Haas KM, Tedder TF. The innate mononuclear phagocyte network depletes B lymphocytes through Fc receptor-dependent mechanisms during anti-CD20 antibody immunotherapy. J Exp Med 2004;199(12):1659–1669.

[65] Ober RJ, Radu CG, Ghetie V, Ward ES. Differences in promiscuity for antibody–FcRn interactions across species: implications for therapeutic antibodies. Int Immunol 2001;13 (12):1551–1559.

[66] Andersen JT, Daba MB, Berntzen G, Michaelsen TE, Sandlie I. Cross-species binding analyses of mouse and human neonatal Fc receptor show dramatic differences in immunoglobulin G and albumin binding. J Biol Chem 2010;285 (7):4826–4836.

[67] Rogers KA, Scinicariello F, Attanasio R. IgG Fc receptor III homologues in nonhuman primate species: genetic characterization and ligand interactions. J Immunol 2006;177(6): 3848–3856.

[68] Oitate M, Masubuchi N, Ito T, Yabe Y, Karibe T, Aoki T, Murayama N, Kurihara A, Okudaira N, Izumi T. Prediction of human pharmacokinetics of therapeutic monoclonal antibodies from simple allometry of monkey data. Drug Metab Pharmacokinet 2011;26(4):423–430.

[69] Deng R, Lyer S, Theil F-P, Motensen DL, Fielder PJ, Prabhu S. Projecting human pharmacokinetics of therapeutic antibodies from nonclinical data. What have we learned? MAbs 2011;3(1):61–66.

[70] Petkova SB, Meng Y-JG, Sproule TJ, Christianson GJ, Roopenian DC. Application of humanized FcRn mouse

model for *in vivo* testing of therapeutic antibodies. J Immunol 2007;178 (meeting abstract supplement):S245.

[71] Roopenian DC, Christianson GJ, Sproule TJ. Human FcRn transgenic mice for pharmacokinetic evaluation of therapeutic antibodies. Method Mol Biol 2010;602:93–104.

[72] Proetzel G, Roopenian DC. Humanized FcRn mouse models for evaluating pharmacokinetics of human IgG antibodies. Methods 2014;65:148–153.

[73] Tam SH, McCarthy SG, Brosnan K, Goldberg KM, Scallon BJ. Correlations between pharmacokinetics of IgG antibodies in primates vs. FcRn-transgenic mice reveal a rodent model with predictive capabilities. MAbs 2013;5(3):1–9.

[74] Han C, Zhou H. Monoclonal antibodies: interspecies scaling with minimal preclinical information. Ther Deliv 2010;2(3): 259–268.

[75] Schellekens H. Immunogenicity of therapeutic proteins: clinical implications and future prospects. Clin Ther 2002;24(11):1720–1740.

[76] Shankar G, Pendley C, Stein KE. A risk-based bioanalytical strategy for the assessment of antibody immune responses against biological drugs. Nat Biotechnol 2007;25 (5):555–561.

[77] Shankar G, Arkin S, Cocea L, Devanarayan V, Kirshner S, Kromminga A, Quarmby V, Richards S, Schneider CK, Subramanyam M, Swanson S, Verthelyi D, Yim S, American Association of Pharmaceutical Scientists. Assessment and reporting of the clinical immunogenicity of therapeutic proteins and peptides-harmonized terminology and tactical recommendations. AAPS J 2014;16(4):658–673.

[78] Rojas JR, Taylor RP, Cunningham MR, Rutkoski TJ, Vennarini J, Jang H, Graham MA, Geboes K, Rousselle SD, Wagner CL. Formation, distribution, and elimination of inf liximab and anti-infliximab immune complexes in cynomolgus monkeys. J Pharmacol Exp Ther 2005;313(2): 578–585.

[79] Geng W, Kolaitis G, Kim H-J, Steele TD. 2008. Impact of immunogenicity on pharmacokinetics/toxicokinetics of a human monoclonal antibody in cynomolgus monkeys. Abstract at AAPS/NBC http://abstracts.aaps.org/Published/Browse.aspx?colID=9. Accessed 2015 Jul 1.

[80] Wileman T, Harding C, Stahl P. Receptor-mediated endocytosis. Biochem J 1985;232(1):1–14.

[81] Ng CM, Bai S, Takimoto CH, Tang MT, Tolcher AW. Mechanism-based receptor-binding model to describe the pharmacokinetic and pharmacodynamic of an anti-a5b1 integrin monoclonal antibody (volociximab) in cancer patients. Cancer Chemother Pharmacol 2009;65 (2):207–217.

[82] Bauer RJ, Dedrick RL, White ML, Murray MJ, Garovoy MR. Population pharmacokinetics and pharmacodynamics of the anti-CD11a antibody hu1124 in human subjects with psoriasis. J Pharmacokinet Biopharm 1999;27(4):397–420.

[83] Dong JQ, Salinger DH, Endres CJ, Gibbs JP, Hsu CP, Stouch BJ, Hurh E, Gibbs MA. Quantitative prediction of human pharmacokinetics for monoclonal antibodies: retrospective analysis of monkey as a single species for first-in-human prediction. Clin Pharmacokinet 2011;50(2):131–142.

[84] Mager DE, Krzyzanski W. Quasi-equilibrium pharmacokinetic model for drugs exhibiting. Pharm Res 2005;22(10): 1589–1596.

[85] Gibiansky L, Gibiansky E. Target-mediated drug disposition model: approximations, identifiability of model parameters and applications to the population pharmacokinetic–pharmacodynamic modeling of biologics. Expert Opin Drug Metab Toxicol 2009;5(7):803–812.

[86] Yan X, Mager DE, Krzyzanski W. Selection between Michaelis-Menten and target-mediated drug disposition pharmacokinetic models. J Pharmacokinet Pharmacodyn 2010;37(1):25–47.

[87] Rowland M, Peck C, Tucker G. Physiologically-based pharmacokinetics in drug development and regulatory science. Annu Rev Pharmacol Toxicol 2011;51:45–73.

[88] Lammerts van Bueren JJ, Bleeker WK, Bφgh HO, Houtkamp M, Schuurman J, van de Winkel JG, Parren PW. Effect of target dynamics on pharmacokinetics of a novel therapeutic antibody against the epidermal growth factor receptor: implications for the mechanisms of action. Cancer Res 2006;66(15): 7630–7638.

[89] Baxter LT, Zhu H, Mackensen DG, Jain RK. Physiologically based pharmacokinetic model for specific and nonspecific monoclonal antibodies and fragments in normal tissues and human tumor xenografts in nude mice. Cancer Res 1994;54(6):1517–1528.

[90] Garg A, Balthasar JP. Physiologically-based pharmacokinetic (PBPK) model to predict IgG tissue kinetics in wild-type and FcRn-knockout mice. J Pharmacokinet Pharmacodyn 2007;34(5):687–709.

[91] Chen Y, Balthasar JP. Evaluation of a catenary PBPK model for predicting the *in vivo* disposition of mAbs engineered for high-affinity binding to FcRn. AAPS J 2012;14(4): 850–859.

[92] Shah DK, Betts AM. Towards a platform PBPK model to characterize the plasma and tissue disposition of monoclonal antibodies in preclinical species and human. J Pharmacokinet Pharmacodyn 2012;39(1):67–86.

[93] Abuqayyas L, Balthasar JP. Application of PBPK modeling to predict monoclonal antibody disposition in plasma and tissues in mouse models of human colorectal cancer. J Pharmacokinet Pharmacodyn 2012;39(6):683–710.

[94] Nestorov IA, Aarons LJ, Arundel PA, Rowland M. Lumping of whole-body physiologically based pharmacokinetic models. J Pharmacokinet Biopharm 1998;26(1):21–46.

[95] Pilari S, Huisinga W. Lumping of physiologically-based pharmacokinetic models and a mechanistic derivation of classical compartmental models. J Pharmacokinet Pharmacodyn 2010;37(4):365–405.

[96] Cao Y, Jusko WJ. Applications of minimal physiologically-based pharmacokinetic models. J Pharmacokinet Pharmacodyn 2012;39(6):711–723.

[97] Cao Y, Balthasar JP, Jusko WJ. Second-generation minimal physiologically-based pharmacokinetic model for monoclonal antibodies. J Pharmacokinet Pharmacodyn 2013;40(5): 597–607.

[98] Han C, Wang B. Factors that impact the developability of drug candidates: an overview. In: Wang B, Siahaan T, Soltero RA, editors. *Drug Deliver. Principles and Applications*. Hoboken: Wiley Interscience; 2005. pp. 1–14.

[99] Wang W, Wang X, Doddareddy R, Fink D, McIntosh T, Davis HM, Zhou H. Mechanistic pharmacokinetic/target engagement/pharmacodynamic (PK/TE/PD) modeling in deciphering interplay between a monoclonal antibody and its soluble target in cynomolgus monkeys. AAPS J 2014;16(1):129–139.

[100] Fetterly GJ, Aras U, Meholick PD, Takimoto C, Seetharam S, McIntosh T, de Bono JS, Sandhu SK, Tolcher A, Davis HM, Zhou H, Puchalski TA. Utilizing pharmacokinetics/pharmacodynamics modeling to simultaneously examine free CCL2, total CCL2 and carlumab (CNTO 888) concentration time data. J Clin Pharmacol 2013;53(10):1020–1027.

[101] Zimmerman Z, Maniar T, Nagorsen D. Unleashing the clinical power of T cells: CD19/CD3 bi-specific T cell engager (BiTE®) antibody construct blinatumomab as a potential therapy. Int Immunol 2015;27(1):31–37. ePublish. DOI: 10.1093/intimm/dxu089.

[102] Baeuerle PA, Reinhardt C. Bispecific T-cell engaging antibodies for cancer therapy. Cancer Res 2009;69(12): 4941–4944.

[103] Eastwood D, Findlay L, Poole S, Bird C, Wadhwa M, Moore M, Burns C, Thorpe R, Stebbings R. Monoclonal antibody TGN1412 trial failure explained by species differences in CD28 expression on $CD4^+$ effector memory T-cells. Br J Pharmacol 2010;161(3):512–526.

[104] US FDA. 2005. Guidance for industry. Estimating the maximum safe starting dose in initial clinical trials for therapeutics in adult healthy volunteers.

9

FIXED DOSING VERSUS BODY-SIZE-BASED DOSING FOR THERAPEUTIC BIOLOGICS—A CLINICAL PHARMACOLOGY STRATEGY

DIANE D. WANG, JUSTIN T. HOFFMAN AND KOUROSH PARIVAR
Pfizer Oncology, San Diego, CA, USA

9.1 INTRODUCTION

Over the course of the last two decades, therapeutic biologics, including therapeutic peptides, proteins, and monoclonal antibodies (mAbs), have come to make up a substantial portion of the clinical therapeutic agents within drug development pipelines. Currently, there are more than 100 therapeutic biologics that have been approved for human use in the United States or the European Union, and many more are under development [1].

Compared to conventional small molecule drugs, which are most commonly dosed with fixed doses, therapeutic biologics are often administered based on a metric of the patients' body size, such as body weight (BW) and body surface area (BSA), although this trend appears to be without a clear scientific rationale. Recent reviews of mAbs, therapeutic peptides, and proteins approved for use by the Food and Drug Administration (FDA) found that the majority of these agents are administered in adult patients based on a body-size-dependent dosing approach [2–4]. These observations have been considered artifacts of the extrapolation methods used in choosing the starting doses in first-in-human (FIH) trials based on doses used in preclinical experiments in animals, which generally involves the interspecies allometric scaling of the no observed adverse effect level (NOAEL) or the minimum anticipated biological effect level (MABEL) on a mg/kg basis [2, 4]. Based on the assumption that increased body size would lead to an increase in a patient's drug elimination capacity (clearance) and a therapeutic agent's volume of distribution, it has also been perceived that individualizing dose based on a patient's body size should minimize the interpatient variability in drug exposures, thereby optimizing clinical outcomes [5, 6]. This perception has been furthered by the assumption that since the potential sources of pharmacokinetic (PK) variability in agents such as mAbs are thought to be more limited than that of small molecules (where factors other than body size, such as genetic polymorphisms in metabolic enzymes and transporters, smoking status, concomitant medications, and food habits, can account for significant PK variability) due to the differences in distribution and elimination mechanisms of these drug classes, the relative contribution of body size to the overall population variability would be larger and using a dosing approach that controls for body size would have a more significant clinical impact. The validity of these perceptions and assumptions has been challenged by recent simulation studies investigating the performance of fixed and body-size-based dosing approaches in reducing the intersubject variability in PK and pharmacodynamics (PD) of therapeutic biologics with published population PK and/or population PK/PD models in adult populations in the literature [2–4], and for therapeutic biologics still in the development pipeline [4].

Most therapeutic biologics are target specific with a relatively large therapeutic window and a generally small contribution of body size to PK variability [2, 3], thus the

ADME and Translational Pharmacokinetics/Pharmacodynamics of Therapeutic Proteins: Applications in Drug Discovery and Development, First Edition. Edited by Honghui Zhou and Frank-Peter Theil.

rationale selection of a dosing approach should be made in the context of these unique characteristics. While strong and sometimes proportional relationships between body size and a therapeutic agent's volume of distribution and clearance have been demonstrated across species, that is, from mice or rats to humans, where the body-size range is large, the relationship often seems to lose its clinical significance when the body-size range is restricted to a typical target patient population in adult humans of only two- to threefold [4, 7]. For instance, a recent review of the population PK of mAbs in adult patients has found that the power law exponents fitted to describe the effect of body-size metrics on clearance and volume PK parameters for mAbs are typically in the range of 0.3–0.7, which is smaller than the typical BW exponents used for interspecies allometric scaling (i.e., 0.85 for clearance and 1 for volume of distribution) [4, 8]. In addition, while a statistically significant body-size effect on PK parameters in a population PK analysis may provide supportive evidence for body-size-based dosing, as will be discussed in this chapter, statistical significance alone should not be the basis for selection.

This chapter focuses on the dosing strategies for therapeutic biologics in human adult populations, and it is acknowledged that the strategies for the dosing of pediatric populations may be different. For instance, the relative range of body size in a pediatric population can be considerably wider, depending on the age range, than that in typical adult populations, and body-size-based dosing approaches are generally recommended in these pediatric populations in keeping with available regulatory guidance [4, 9, 10].

9.1.1 Considerations for the Selection of a Dosing Approach

9.1.1.1 Goals of Dosing Approach Selection In general, the aim of any dosing strategy for a therapeutic agent is to optimize the overall clinical performance within an intended indication by maximizing the efficacious effect while minimizing the adverse events within the target patient population. One consideration to the dosing strategy is whether dosing should be based on the patient's body size, or whether the dosing should occur independent of a patient's demographics (fixed dosing). The selection of a dosing approach and regimen is influenced by a multitude of factors including the therapeutic agents' PK and PD properties, its exposure–response relationship for efficacy and safety (including the therapeutic window), the target population demographics, the route of administration, ease of dosing compliance, and pharmacoeconomic considerations.

A fixed-dosing regimen provides the administration of the same amount of a therapeutic agent regardless of a patient's demographics, thus this dosing approach offers advantages in convenience, cost-effectiveness, and compliance. It is more convenient for pharmacists or nurses to prepare and administer a unit dose to patients, rather than individualizing each administration based on body size. This could have potentially important safety implications, as it could avoid or reduce errors in calculating and preparing an individualized dose for each patient. By administering a unit dose, it could also prevent or minimize the wasting of "left-overs" during preparation in contrast to individualized dosing based on body size. This is especially important for therapeutic biologics since their manufacturing is often expensive and time-consuming. Therefore, when there is no significant advantage of one dosing approach over another from a PK and PD perspective, fixed dosing is the approach of choice due to its operational and economic advantages.

A clear scientific rationale for whether a drug should be administered based on a strategy that is independent of patient demographics (fixed dosing) or based on a patient's body size, such as BW and BSA, will mainly depend on the effect of body size on the PK and PD of the therapeutic agent, as well as its therapeutic window. A good dosing strategy should provide reduced interpatient variability in PK and/or PD, and ultimately optimize therapeutic outcomes. For drugs with a wide therapeutic window, the fixed-dosing approach is generally chosen for adult patients, regardless of the influence of body size on PK and PD properties due to its convenience, better compliance, less risk for medical errors, and cost-effectiveness. On the other hand, for drugs with a narrow therapeutic window, the effect of body size, as well as other influential factors, on PK and PD should be carefully evaluated in order to select a dosing strategy that minimizes the variability in drug exposures across the target population and optimizes the risk/benefit ratio of the agent in the target population.

9.1.1.2 Key Parameter in Determining the Magnitude of a Body-Size Effect The determination that a body-size metric, that is, BW or BSA, is a statistically significant covariate on a therapeutic agent's PK parameter(s) does not necessarily justify body-size-based dosing as a better approach than fixed dosing, even purely based on PK and/or PD variability, since simple body-size correction as mg/kg or mg/m^2 could overcorrect the effect of body size on exposure as the relationship is rarely proportional. When the effect of body size on a PK parameter is modeled as a power function, the determinant factor in evaluating the magnitude of a body-size effect on a PK parameter is the value of the exponent (α) on the normalized covariate. For example, the effect of body size on clearance (CL) can be described by a normalized power function as follows:

$$CL = CL_{\text{typical}} \cdot \left(\frac{\text{bodysize}}{\text{bodysize}_{\text{typical}}} \right)^{\alpha} \tag{9.1}$$

Intersubject variability in the area under the concentration–time curve (AUC) is related to the α value, as illustrated in

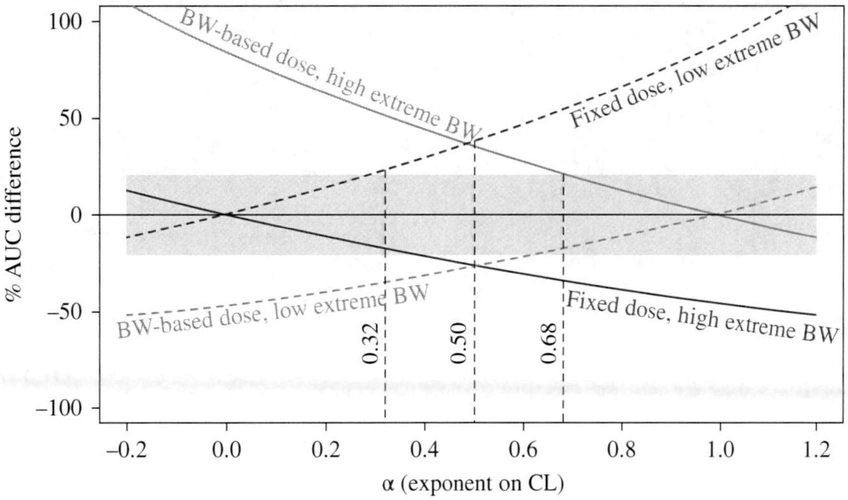

FIGURE 9.1 The percent difference of AUC for patients with extremely low body weight (BW) (40 kg, broken lines) and extremely high BW (140 kg, solid lines) from those for patients with a median BW of 75 kg as a function of the α values following a fixed (red) and a BW-based (green) dose, assuming $CL = CL_{\text{typical}} \cdot \left(\frac{BW}{BW_{\text{typical}}}\right)^{\alpha}$. The shaded area represents AUC values within 100 ± 20% of typical AUC. Wang et al. [2]. Reproduced with permission of *J Clin Pharmacol*. (*See insert for color representation of this figure.*)

Figure 9.1. The percent difference in AUC between typical and extreme BWs has been simulated for low extreme and high extreme BW values across a range of α values for both dosing approaches (BW-based dosing vs a fixed dose set to the typical population BW). The low extreme, typical, and high extreme BW values used in this simulation were set to 40, 75, and 140 kg, respectively, based on simulated patient population as described in Wang et al. [2]. It can be seen that in terms of reducing the intersubject variability, fixed-dosing works the best when *CL* is not affected by BW ($\alpha = 0$), BW-based dosing works the best when α is equal to 1, and the two dosing approaches perform the same when α is equal to 0.5. It is also apparent that fixed dosing provides less variability in exposure when $\alpha < 0.5$, while this is true for BW-based dosing when $\alpha > 0.5$. Depending on the α value, one dosing approach could have an advantage over the other. However, more often than not, the two dosing approaches lead to similar variability in drug exposure across patient population, although they tend to overdose or underdose different body-size patient groups. In general, fixed dosing could lead to overexposure of drug in subjects with lower BW and underexposure of drug in subjects with higher BW relative to typical BW subjects, whereas body-size-based dosing generally does the opposite.

For the purpose of illustrating how this methodology can be used to rationally choose the most appropriate dosing approach in the context of clinical drug development, we will discuss the case of a fictional therapeutic biologic with a well-defined therapeutic window and an available population PK model. The shaded area in Figure 9.1 represents the allowable population variability in drug exposure (AUC) for the fictional biologic that will ensure adequate confidence that it will hit its therapeutic window throughout its target patient population. For this case, the upper and lower bounds of the shaded area cover the ±20% AUC difference between patients with extreme BW and those with typical BW, which is identified as the maximum allowable variability in exposure for this agent. In this context, if the therapeutic agent had a population PK model that used a power model function to evaluate a body-size metric on clearance and reported an α value less than 0.32, then a fixed-dosing approach will keep the patients in the target population with extreme BW within the allowable variability boundary if all other covariates are kept the same. On the other hand, BW-based dosing will provide the same benefit if the agent had a reported α value greater than 0.68. If the α value of the agent fell between 0.32 and 0.68, neither approach can keep the drug exposure (AUC) difference between typical BW and extreme BW patients less than 20%, and consistently hitting the therapeutic window could not be guaranteed.

9.1.1.3 Therapeutic Index Whether a dosing approach is acceptable also depends on the width of the targeted range of exposure as illustrated in Figure 9.1. When a drug has a wide therapeutic window, a drug exposure (AUC) difference of more than ±20% used in our fictional example may still be tolerated without additional safety issues. For example, if a 100% AUC difference is acceptable in terms of efficacy and safety, both dosing approaches should work well for therapeutic biologics with the α range of 0–1.0. Thus, for drugs with a wide therapeutic window, where a larger degree of variability in drug exposure across the target population would be considered acceptable in terms of efficacy and safety, a body-size-based dosing approach would have to outperform the fixed-dosing regimen to an unrealistic margin before its scientific merit could overcome the practical and economic advantages of fixed dosing. On the other hand, for drugs with a narrow therapeutic window, the effect of body size, as well as other influential factors, on PK and PD should be carefully evaluated in order to select a dosing strategy that optimizes the risk/benefit ratio.

9.1.1.4 Contribution of Body Size to Overall PK and/or PD Intersubject Variability Many factors, in addition to body size, may also cause differences in drug exposure among patients. These factors can include intrinsic factors (e.g., age, sex, disease states, organ functions, serum protein level, target antigen expression level, and genetic polymorphisms in targets) as well as environmental factors (e.g., concomitant medications, smoking, and dietary habit). Thus, the relative contribution of body size to the overall intersubject variability in PK parameters (such as CL, V_{max}, Q, V_1, or V_2) also plays a role in rationally determining whether the dose needs to be adjusted based on body-size metrics. When overall intersubject variability is low and unlikely to be clinically relevant or when body size only explains a very small percentage of the overall intersubject variability, then adjusting the dose based on body size would lead to minimal reduction in exposure variability. On the other hand, if body size is a major source for intersubject variability that is clinically relevant, body-size-based dosing may provide significant benefit when supported by other factors, such as a narrow therapeutic window. Depending on the clinical relevance of intersubject variability and the contribution of body-size effects to the overall intersubject variability, body-size-based dosing may or may not be necessary even when it significantly reduces intersubject variability in exposure.

As the ultimate goal of a clinical study is to achieve its efficacy and safety endpoints, efficacy data, safety data, and surrogate PD markers, when available, will supersede PK data when using reductions in population variability to support a selected dosing approach. It is well known that intersubject variability in drug response is generally greater than that in drug exposure measures because of differences in disease state and genetic variations in drug targets [11–13]. Therefore, the clinical benefit, if any, of body-size-based dosing with regard to the reduction of variability in drug exposure measurements could be "diluted" for downstream endpoints of clinical response or PD measurements, due to the additional sources of population variability in drug response. The effect of body size on PD endpoint should be evaluated for each individual therapeutic biologic under development when data are available.

9.1.2 Evaluations of Fixed Dosing versus Body-Size-Based Dosing

To illustrate the methodologies that can be used in investigations to support the rational selection of an appropriate dosing approach, we will review recently published simulation studies that evaluated the performance of fixed dosing versus body-size-based dosing in adult patients receiving mAbs [2, 4] and therapeutic peptides/proteins [3]. Data used in these simulation studies were collected from the population PK/PD studies of therapeutic peptides and proteins published in peer-reviewed journals [2–4] or available to the authors via internal data [4]. The common selection criteria for inclusion of a therapeutic biologic in these simulation studies included the availability of population PK and/or PD models for adult patients or healthy subjects and adequate assessment of the effect of body size on the PK (and/or PD) parameters. As two of the three simulation studies utilized identical methodologies to investigate dosing approach performance in mAbs [2] and therapeutic peptides [3], these two simulation studies will be emphasized here. The third study, by Bai et al. [4], investigating the dosing approach performance in mAbs was essentially a confirmatory analysis of the previous study using a similar methodology and nonpublic proprietary population PK data involving mAbs still in clinical development. Thus, the simulation study by Bai et al. will be referenced when appropriate, but not emphasized.

9.1.2.1 General Methodology For each of the simulation studies, mixed-effect models were used to describe the PK and/or PD of all the selected mAbs, therapeutic peptides, and proteins including the effect of body size on PK and/or PD parameters. Simulation analyses were conducted using NONMEM (version VI, GloboMax, Hanover, Maryland). AUC calculation for each simulated subject was calculated as dose/CL for agents exhibiting linear PK. For agents exhibiting nonlinear PK, the AUC was calculated by integration of the concentration–time curves by trapezoidal method using S-PLUS [3] or by integration of the individual concentration–time profiles using NONMEM [2]. The maximum concentration (C_{max}) for each subject was determined as the maximal concentration from the simulated concentration–time profile of the subject. For PD measurements, C_{max} or the minimum concentration (C_{min}) of the PD marker, whichever reflects the maximal drug effect, was determined from the PD marker-time profile of the subject.

9.1.2.2 Methods of Dosing Approach Performance Evaluation The performance of the two dosing approaches was evaluated by their population performance and individual performance [2–4]. In addition, α value serves as a determinant for dosing approach selection.

9.1.2.2.1 Population Performance Evaluation The population performance of a dosing approach provides an overall picture of PK and/or PD variability across the whole patient population. In other words, how well does a dosing approach accommodate the entire patient population?

Population performance was assessed by comparing the intersubject variability (expressed as CV% (coefficient of variation)) in the exposure (AUC and C_{max}) of 1000 subjects simulated following the two dosing approaches. The dosing approach that produced less intersubject variability provides a better population performance.

Briefly, Monte Carlo simulations were conducted using the final PK and/or PD model reported for each mAb, peptide, or protein to obtain the concentration–time profiles following both fixed-dosing and body-size-based dosing approaches. The dose used for simulation was the dose recommended in the labeling for marketed products or the dose used in the reported clinical trials for biologics still under development. The median value of body size (BW or BSA) was used as the conversion factor for dose determination so that the dose used in fixed-dosing approach is the same as the dose for the subjects with median body size in body-size-based dosing approach. For all simulation studies, 1000 subjects were simulated per dosing approach. The sampling points were chosen based on the PK or PD properties of therapeutic biologics, and the same sampling schedule was used for both dosing approaches.

For each simulation study, with the exception of BW, values of influential covariates were randomly generated using S-PLUS 7.0 (TIBCO, Palo Alto, California) assuming normal or log-normal distribution. The values of parameters used for generating these covariates were selected by trial and error with a goal of reproducing the patient population by matching the median, standard deviation, or the range of covariates to those reported in the corresponding population PK/PD study. BSA was generated assuming normal distribution with a median of 1.82 m^2 and a range of 1.2–2.4 m^2. BW values were generated assuming that transformation of BW by a power function ($Z=BW^{-0.5}$) would follow a normal distribution [2, 14]. The dataset of randomly generated BW (1000 subjects) had a median of 75.7 kg and a range of 38.8–187.2 kg, which covers the range of those reported for the selected mAbs, peptides, and proteins [2, 3].

9.1.2.2.2 Individual Performance Evaluation The individual performance evaluates the effect of body size on exposure measures under extreme conditions. The amount of overdose and underdose to patients with extreme body size is of particular concern both in drug development and for regulatory approval as it may have significant efficacy and safety implications.

Individual performance was evaluated by comparing the percentage difference in the exposure between subjects with extreme body size and typical body size following the two dosing approaches. The dosing approach that resulted in a smaller difference in PK or PD exposure between subjects with extreme body-size subjects and typical body size has better individual performance.

Briefly, to evaluate the individual performance, the PK profiles for the subjects with typical, low extreme, and high extreme body size were simulated. The typical, low extreme, and high extreme BW/BSA used were 75.7 kg/1.8 m^2, 40 kg/1.3 m^2, and 140 kg/2.3 m^2, respectively. For covariates other than BW/BSA, typical values were used. The intersubject variability and residual errors were all set to zero for the simulations conducted for both dosing approaches.

9.1.2.2.3 α Value The simplest way to assess whether fixed dosing or body-size-based dosing may be better in reducing intersubject variability in drug exposure (AUC) is to look at the α value of the body-size effect on *CL* as defined in Equation 9.1 obtained from covariate analysis. However, the covariate models used to characterize the effect of body size on *CL* for the selected molecules are not always modeled in the form of Equation 9.1. Therefore, an effort was made to obtain the α values for all biologics evaluated. For agents where body-size metrics were not found to be a statistically significant covariate on *CL*, a zero value was assigned to α. For those agents whose α values were not reported in the literature, the following steps were used to obtain the α value for each agent: (i) generate a series of *CL* values over the range of body size reported based on the reported original covariate model and (ii) fit the generated *CL* versus body-size data using Equation 9.1 to obtain the α value using WinNonlin 5.2 (Pharsight, Mountain View, California). After obtaining α values for all included therapeutic biologics, the performance of the two dosing approaches in terms of reducing intersubject variability in AUC of each therapeutic biologic is assessed by evaluating the α value based on the following criteria:

- $\alpha < 0.5$—fixed dose performs better
- $\alpha = 0.5$—fixed and body-size-based dosing have similar performance
- $\alpha > 0.5$—body-size-based dosing performs better.

These criteria apply to both population and individual performances.

9.1.2.2.4 Contribution of Body Size to Overall Intersubject Variability Depending on the contribution of the effect of body size to the overall intersubject variability, body-size-based dosing may or may not be necessary even when it significantly reduces intersubject variability in exposure. The α values for the body-size effect on relevant PK parameters often appear to correlate with the relative contribution of body size to the overall variability, although the rank orders of the two are not exactly the same across all tested therapeutic agents [3]. This discrepancy is likely due to the difference in the extent of the contribution of other identified and unidentified factors, such as demographic characteristics and disease conditions, which can differentiate dosing approach selection conclusions based on α values alone.

The contribution of the effect of body size to the overall intersubject variability was assessed by the following simulation approach. Briefly, simulated PK datasets of 100 subjects were generated using the published "final" PK models for the selected therapeutic biologics including all influential covariates following a fixed dose. The relevant covariate values were randomly generated as described above. This simulated dataset was then fitted using both the "final" model and a "reduced" model, which excluded body size as covariate(s) for any PK parameter(s). The percentage change in the intersubject variability of the relevant PK parameters by the "final" model in comparison to the "reduced" model was calculated to determine the contribution of body-size effect to the overall intersubject variability of the relevant PK parameters.

9.1.2.3 Dosing Approach Performance of Monoclonal Antibodies As mentioned previously, the primary scientific rationale for the selection of a dosing approach should be the minimization of population variability in drug exposure (PK) and drug response (PD). The sources for the variability in exposure are likely to be different between mAbs and conventional small molecules due to their different distribution and elimination mechanisms [15]. Unlike small molecules, which are generally distributed into tissues via diffusion and eliminated through metabolism and/or renal excretion, mAbs extravasate mainly by convection and are eliminated by catabolism and/or target-mediated clearance. The FcRn receptor plays an important role in protecting IgG molecules from catabolism [15]. For mAbs undergoing target-mediated clearance, the binding affinity of the mAb to the target and the extent of the target expression could be important determinants of mAb clearance.

Wang et al. [2] investigated the performance of body-size-based and fixed dosing in reducing PK and/or PD variability in adults for 12 mAbs with published population PK and/or PD models (Table 9.1). At the population level, 95th percentile intervals of concentration–time profiles, distribution, and variability of exposure for 1000 simulated subjects after both dosing approaches were examined. As expected, the median concentration–time profiles were superimposed on each other for all the mAbs studied, as the median dose for the two dosing approaches is the same. The focus of this comparison was therefore the 95th percentile intervals of the concentration–time profiles. Overall, the 95th percentile intervals did not differ remarkably between the fixed and BW/BSA-based dosing approaches across all the 12 mAbs studied, although body-size-based dosing resulted in somewhat narrower 95th percentile intervals for some mAbs, such as sibrotuzumab and omalizumab, while fixed-dosing approach did a better job for others such as bevacizumab, golimumab, and alemtuzumab. Analysis of the distribution curves of AUC and C_{max} following simulations of the two dosing approaches showed that the two dosing approaches resulted in different distribution curves of AUC and C_{max} for some mAbs, but these differences were generally small, and neither of the dosing approaches shows a clear advantage in terms of reducing AUC or C_{max} variability. The intersubject variability of exposure following the two dosing approaches is presented in (Fig. 9.2). Body-size-based dosing resulted in less intersubject variability in AUC for 5 mAbs, while the fixed-dosing approach produced less variability for the other 7 mAbs. The average variability in AUC of all the 12 mAbs studied was similar following the two dosing approaches (42.4% by fixed dosing vs 44.2% by BW/BSA-based dosing). Similar results were also obtained for C_{max} (30.1% by fixed dosing and 30.3% by BW/BSA-based dosing).

At the individual level, the difference between the exposure of patients with extreme body sizes from the typical exposure following both approaches was compared (Fig. 9.3). The fixed-dosing approach performed better than the body-size-based dosing with a smaller percent difference in drug exposure (AUC) between subjects with extreme body sizes and typical body size for 7 of 12 mAbs and in C_{max} for 5 of 12 mAbs. In general, the body-size-based dosing tended to overexpose patients with a larger body size, while underexpose patients with a smaller body size. The opposite was true for the fixed-dosing approach, where patients with small body size tended to be overexposed, while patients with larger body sizes were underexposed.

The percent contribution of body-size metrics to the overall intersubject variability in PK parameters was determined for 7 of the mAbs (Table 9.2). As discussed previously, body-size-based dosing resulted in less variability in AUC for 5 of the 12 investigated mAbs, and 3 of those 5 (pertuzumab, efalizumab, and rituximab) are included in Table 9.2. Based on the minimal contribution of body size to the overall intersubject variability in clearance for these agents, the reduction in intersubject variability for these three drugs is limited as reflected by the comparable population performance of the two dosing approaches (Fig. 9.2), and one could argue that the likely benefit of reduced variability in dosing these agents in an individualized manner on a body-size-based approach would be insufficient to overcome the added operational and economic costs associated with this approach unless it was required by a narrow therapeutic window.

For the one mAb for which both population PK and PD models were available in the literature, omalizumab, the performance of each dosing approach was evaluated for a PD endpoint (free IgE levels) as shown in Figure 9.4.

TABLE 9.1 The Population PK/PD Models for the Selected Monoclonal Antibodies

mAb	Type	Dosing	Structure Model	Covariates Median (Range)	Covariate Model	N	α on CL
Rituximab	Chimeric IgG1	mg/m^2	2-CMT, linear	BSA: 1.72 (1.41–2.35)	$CL = CL^* \times (\text{BSA}/1.72)^{1.02}$ $V_1 = V_1^* \times (\text{BSA}/1.72)^{0.73}$	102	1.02
Omalizumab	Humanized IgG1	Based on IgE baseline and BW	TMD	BW: 61.1 (39–150)	$CL/F = CL/F^* \times (\text{BW}/61.1)^{0.911}$ $V = V^* \times (\text{BW}/61.1)^{0.658}$	202	0.911
Efalizumab	Humanized IgG1	mg/kg	1-CMT, linear	BW: 91 (43–192)	$CL/F = CL/F^* \times (\text{BW}/91)^{0.754}$	1088	0.754
Pertuzumab	Humanized IgG1	mg/kg	2-CMT, linear	BW: 69 (45–150.6) BSA: 1.73 (1.40–2.53)	$CL = CL^* \times (\text{BW}/69)^{0.587}$ $V_1 = V_1^* \times (\text{BSA}/1.72)^{1.16}$	153	0.587
Bevacizumab	Humanized IgG1	mg/kg	2-CMT, linear	BW: 74 (49–114)	$CL = CL^* \times (\text{BW}/74)^{0.368}$ $V_1 = V_1^* \times (\text{BW}/74)^{0.411}$	491	0.368
Matuzumab	Humanized IgG1	mg	2-CMT, both linear and nonlinear elimination	BW: 71 (44–125)	$CL_{\text{linear}} = CL_{\text{linear}}^* \times (1 + 0.0087 \times (\text{BW-71}))$ $V_1 = V_1^* \times (1 + 0.0044 \times (\text{BW-71}))$	90	0.622[a]
Sibrotuzumab	Humanized IgG1	mg/m^2 or mg	2-CMT, both linear and nonlinear elimination	BW: 75 (42–134)	$CL_{\text{linear}} = CL_{\text{linear}}^* \times (1 + 0.0182 \times (\text{BW-75}))$ $V_1 = V_1^* \times (1 + 0.0125 \times (\text{BW-75}))$ $V_2 = V_2^* \times (1 + 0.0105 \times (\text{BW-75}))$ $V_{\max} = V_{\max}^* \times (1 + 0.00934 \times (\text{BW-75}))$	60	1.384[a]
Cetuximab	Chimeric IgG1	mg/m^2	2-CMT, Michaelis–Menten elimination	BW: 60 (34–113) IBW 64.2 (43.3–81.3)	$V_{\max} = V_{\max}^* \times (1 + 0.0108 \times (\text{IBW-64}))$ $V_1 = V_1^* \times (1 + 0.0083 \times (\text{BW-60}))$	143	NA
Golimumab[b]	Human IgG1	mg/kg	2-CMT, linear	BW: 72.3 (53–153.8)	$V_1 = V_1^* \times (\text{BW}/74)^{0.86}$	36	0
Infliximab	Chimeric IgG1	mg/kg	2-CMT, linear	BSA: 1.94 (1.37–2.6)	$V_1 = V_1^* \times (\text{BSA}/1.94)^{0.744}$	274	0
Trastuzumab	Humanized IgG1	mg/kg	2-CMT, linear	BW: 70 (42–119)	$V_1 = V_1^* \times (\text{BW}/65)^{0.556}$	476	0
Alemtuzumab	Humanized IgG1	mg	2-CMT Michaelis–Menten elimination	BW: (45–167)	WT is not a covariate	67	0

Wang et al. [2]. Reproduced with permission of *J Clin Pharmacol*.

N: number of patients from which the PK and/or PD data were used for population PK or PD modeling.

[a]The $\boldsymbol{\alpha}$ value was obtained as follows: first a series of CL values over the range of reported body size were generated based on the reported original covariate model; then, the generated CL versus body-size data were fitted using Equation 9.1 to obtain the $\boldsymbol{\alpha}$ value.

[b]Parameter value without body-size correction.

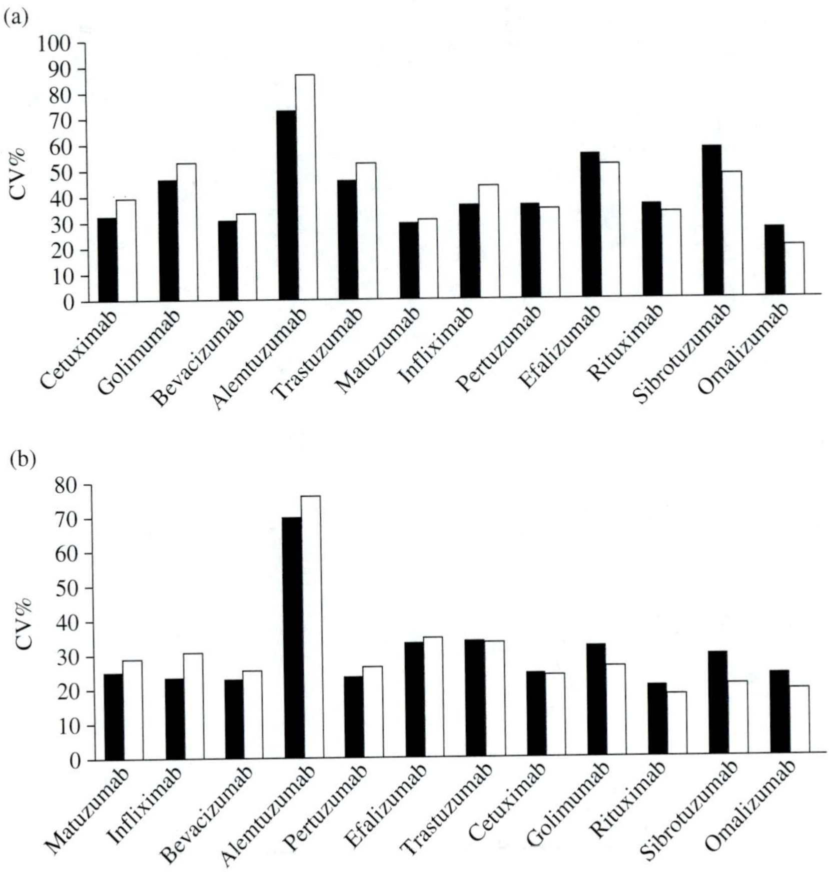

FIGURE 9.2 Comparison of the variability of simulated AUC (a) and C_{max} (b) of 1000 subjects after receiving a single fixed (solid bar) and body weight/body surface area (BW/BSA)-based dose (open bar). (Wang et al. [2]. Reproduced with permission of *J Clin Pharmacol.*)

Comparing the performance of the dosing approaches on omalizumab PK (Figs. 9.2 and 9.3) and PD (Fig. 9.4), it is clear that the body-size-based dosing regimen performs better for both endpoints; however, the differences in performance between the two dosing approaches are smaller for all evaluated measures when modeling to the PD endpoint relative to the PK endpoints. This is consistent with the observation that the increase in sources of variability in PD endpoints relative to PK endpoints often leads to a "dilution" of signal or benefit previously observed for PK endpoints.

The overall results of the simulation study show that the two dosing approaches perform similarly across the mAbs investigated with fixed dosing being better for some mAbs and body-size-based dosing being better for the others. The performance of the dosing approach at the population level and the individual level, and the contribution of body size to the overall intersubject variability were closely associated with the α value on the normalized body-size covariate on clearance of the respective agents, identifying this parameter as a simple early predictor of overall dosing approach performance. At the individual level, the body-size-based dosing approach tended to overexpose patients with a larger body size, while underexposing patients with a smaller body size, while the opposite was true for the fixed-dosing approach. However, the PK variability introduced by either dosing method might be subordinate relative to the variability seen in PD endpoints for safety or efficacy. The simulation study conducted by Bai et al. [4] had similar observations and drew the same conclusions from their unique database of mAbs still in clinical development. Thus, in contrast to the conventional assumptions that seem to be driving dosing approach selections in mAb development, body-size-based dosing strategies did not consistently outperform fixed dosing in reducing intersubject variability in PK and PD endpoints.

9.1.2.4 Dosing Approach Performance of Therapeutic Proteins/Peptides

The clinical benefit of body-size-based dosing for other therapeutic biologics, namely therapeutic peptides and proteins, has been investigated separately from mAbs. These biologics were examined separately due to considerations of the different sources of the variability in exposure between mAbs and therapeutic peptides and proteins since they do not necessarily share the same distribution

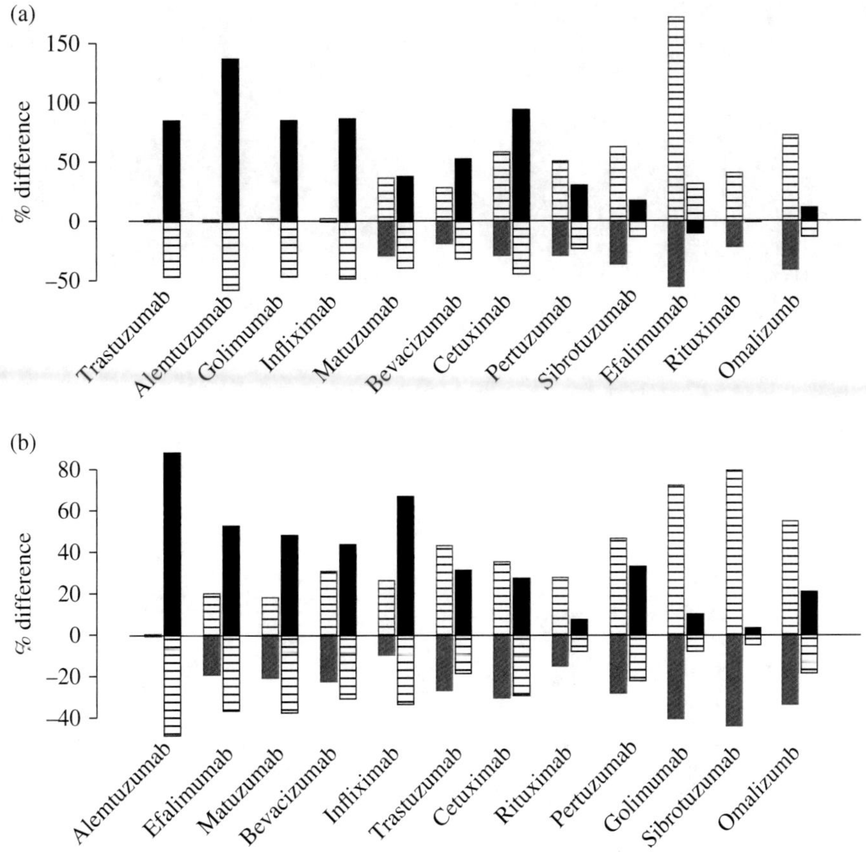

FIGURE 9.3 Comparison of the deviation of AUC (a) and C_{max} (b) for subjects with low (open bar) and high (solid bar) extreme body weight/body surface area (BW/BSA) values from the typical exposure after fixed (red) and BW/BSA-based (black) dosing. (Wang et al. [2]. Reproduced with permission of *J Clin Pharmacol*.) (*See insert for color representation of this figure*.)

TABLE 9.2 Percentage Contribution of Body-Size Measurements to the Overall Intersubject Variability of Pharmacokinetics (PK) Parameters—Monoclonal Antibodies

Monoclonal Antibodies	% Contribution of Body-Size Metric[a] to the Intersubject Variability		α Value	
	CL	V_1	*CL*	V_1
Bevacizumab	2.74	11.1	0.368	0.411
Trastuzumab	NA	4.6	0	0.556
Pertuzumab	11.9	28.3	0.587	1.16
Infliximab	NA	15.1	0	0.744
Golimumab	NA	13.6	0	0.86
Efalizumab	2.9	NA	0.754	0
Rituximab	4.81	22.6	1.02	0.73

Wang et al. [2]. Reproduced with permission of *J Clin Pharmacol*.

[a]Refer to Table 9.1 for specific body-size metric used for each PK parameter.

and elimination mechanisms. In contrast to mAbs, which usually share the same IgG structure with a molecular weight of approximately 150 kDa, therapeutic peptides and proteins comprise a much more diverse group of molecules.

Large therapeutic proteins may share similar distribution and elimination mechanisms to mAbs. They are generally distributed via convection and eliminated via intracellular catabolism following fluid-phase or receptor-mediated endocytosis. However, mAbs differ from most of

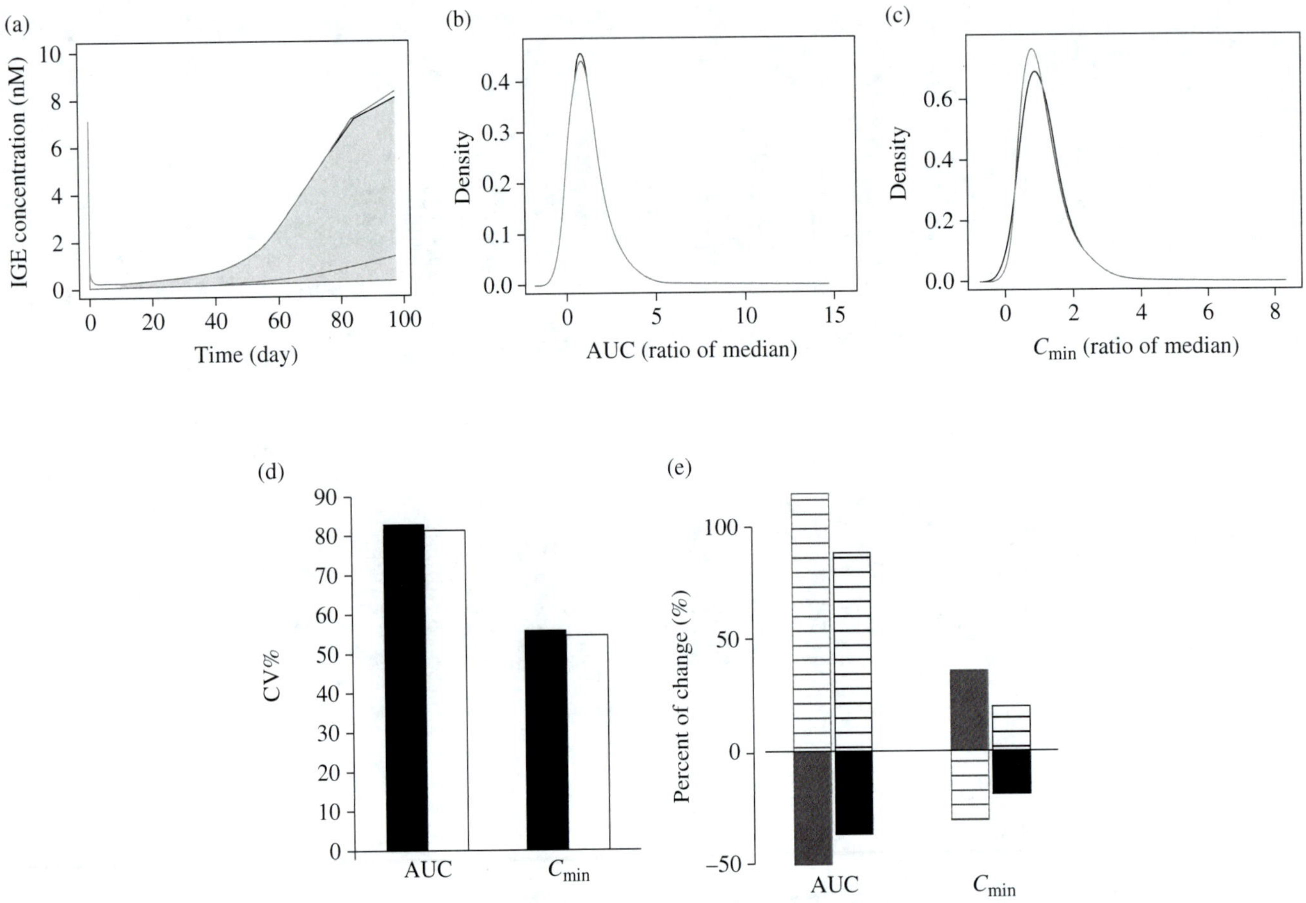

FIGURE 9.4 (a) The median and 95 percentile interval of simulated free IgE concentration–time profile after a fixed (red) and a body weight (BW)-based (blue) dose. (b) The distribution of free IgE AUC (the area between free IgE concentration–time curve and the free IgE baseline) after fixed (red) and BW-based (blue) dosing. (c) The distribution of free IgE C_{min} after fixed (red) and BW-based (blue) dosing. (d) The intersubject variability of AUC and C_{min} across 1000 subjects after fixed (solid bar) and BW-based (open bar) dosing. (e) Deviation of AUC and C_{min} of free IgE for subjects with low (open bar) and high (solid bar) extreme BW from the typical value after a fixed (red) and BW-based (black) dose of omalizumab. (Wang et al. [2]. Reproduced with permission of *J Clin Pharmacol*.) (*See insert for color representation of this figure.*)

therapeutic proteins in that a significant fraction of mAbs is protected from protein catabolism by FcRn-mediated recycling, while most of the therapeutic peptides and proteins do not have such protective mechanisms, with the exception of fusion proteins containing the Fc region of IgGs [16, 17].

For smaller therapeutic peptides and proteins, depending on the molecular size and physiochemical properties (e.g., charge and lipophilicity), renal excretion and diffusion into tissues may play an important role in their overall elimination and distribution mechanisms in addition to catabolism and convection [16].

Zhang et al. [3] compared the performance of body-size-based and fixed dosing in reducing PK and/or PD variability in adults for 18 therapeutic biologics, including 11 therapeutic proteins and 7 peptides, with published population PK and/or PD models (Table 9.3). Among the 18 selected biologics, 12 are administered based on their body size in adult patients. Interestingly, for some products that are administered using body-size-based dosing, such as hematide and onercept, body-size measure had been shown not to be a covariate of any PK parameter (Table 9.3).

As discussed for the work with mAbs by Wang et al. [2], the comparative performance of the two dosing approaches based on AUC can be simply assessed by the α values (the exponent of the power function as defined in Eq. 9.1) reported from the published population PK/PD models, with an α value of 0.5 as the inflection point determining the better performing model. As shown in Table 9.3, 12 of 18 biologics had reported α values less than 0.5, 1 biologic had a reported α value equal to 0.5, and 5 biologics had reported α values greater than 0.5. These observations suggested that fixed dosing would perform better for 12 molecules ($\alpha<0.5$), body-size-based dosing would perform better for 5 molecules ($\alpha>0.5$), and the two dosing approaches would perform similarly for 1 molecule ($\alpha=0.5$).

TABLE 9.3 Population Pharmacokinetic (PK) Models for the Selected Therapeutic Peptides and Proteins

Generic Name	Type	MW (Da)	Dosing	Structure Model	Body Size Covariates, Mean (SD or range)	Body Size Covariate Model	N	α on CL
Abatacept	Fusion Protein	92,300	mg/kg	2-CMT, linear	BW: 78.3 (21.0)	$CL = CL_0 + CL_1 \times (BW/78.3)$	388	0.4[a]
Daptomycin	Peptide	1,620	mg/kg	2-CMT, linear	BW: 75.1 (48.2–152.8)	$Q = 3.46 + 0.0593 \times (BW\text{-}75.1)$ $V_2 = 3.13 + 0.0458 \times (BW\text{-}75.1) \times z^b$	282	0
Darbepoetin alfa	Protein	37,100	μg/kg	2-CMT, linear	BW: 70.8 (36,123)	$CL = CL^* \times (BW/70)^{0.623}$ $V = V^* \times (BW/70)^{0.639}$	140	0.623
Degarelix[c]	Peptide	1,632	mg	3-CMT, linear	No information	Body size is not a covariate	24	0
Emfilermin	Protein	22,007	μg/kg	1-CMT, linear	BW: 62 (48–83)	$V = V^* + 6.7 \times (WT\text{-}62)$	64	0
Enfuvirtide	Peptide	4,492	mg	1-CMT, linear	BW: 72.2 (12.7)	$CL/F = CL_0 + CL_1^* \times (BW/70)$	534	0.5[a]
Erythropoietin alfa	Protein	30,400	Units/kg	1-CMT, linear	BW: 72.2 (18.96)	$CL = CL^* \times (BW/70)^{0.75}$ $V = V^* \times (BW/70)^{1.37}$	48	0.75
Erythropoietin beta	Protein	30,000	μg/kg	1-CMT, linear	BW: 62.0 (51.0–79.0)	$ka = ka^* \times BW^{-1.92}$ $ke = ke^*$ $V = V^* \times (BW)^{0.776}$	48	0.776[d]
Etanercept	Fusion protein	150,000	mg	1-CMT, linear	BW: 73.6 (19.2)	$CL/F = (CL/F)^* \times (BW/70)^{0.75} V/F = (V/F)^* \times (BW/70)$	182	0.75[e]
Hematide[a]	PEGylated peptide	76,000	mg/kg	1-CMT, linear and nonlinear	BW: 76.9 (59–96)	BW is not a covariate	28	0
Lanreotide autogel	Peptide	1,096	mg	3-CMT, linear	BW: 67.1 (12.4)	BW is not a covariate	50	0
Octreotide acetate	Peptide	1,019	μg	1-CMT, linear	BW: 77.1 (51–103)	$V = V^* \times (BW/81)^{0.362}$	59	0
Onercept	Fusion protein	18,000	mg/kg	2-CMT, linear	BW: 73.1 (11.2)	BW is not a covariate	48	0
PEG-interferon alfa-2b	Protein	19,271	μg/kg	1-CMT, linear	BW: 80 (41–149)	$CL = CL^* \times (BW/70)^{0.455}$	817	0.455
Plitidepsin	Peptide	1,110	mg/m²	3-CMT, nonlinear	BSA: 1.78 (1.29–3.32)	BSA is not a covariate	283	0
Recombinant factor VIIa[c]	Protein	50,000	μg/kg	2-CMT, linear	No information	BW is not a covariate	28	0
rhGH[c]	Protein	22,000	mg/kg	1-CMT, nonlinear	BW: 77.9	BW is not a covariate	21	0
u-hFSH	Protein	30,000	IU	1-CMT, linear	BW: 59 (7.4)	$CL/F = (CL/F)^* \times (1 + 0.017 \times (BW\text{-}58.5))$	62	0.99[a]

Zhang et al. [3]. Reproduced with permission of *J Clin Pharmacol*. MW, molecular weight; rhGH, recombinant human growth hormone; and u-hFSH, urinary human follicle-stimulating hormone.

*Parameter value without body size correction.

[a]The α value was obtained as follows: first, a series of CL values over the range of reported body size were generated based on the reported original covariate model; then the generated CL versus body size data were fitted using Equation 9.1 to obtain the α value.

[b]$z = 1.93$ for subjects with a bacterial infection and 1 for noninfected subjects.

[c]Biologics whose population PK models were developed based on data from less than 30 patients.

[d]The α value was derived using equation $ke = CL/V$.

[e]The α value was obtained by trying different values for α during the modeling and found that a value of 0.75 best described the data.

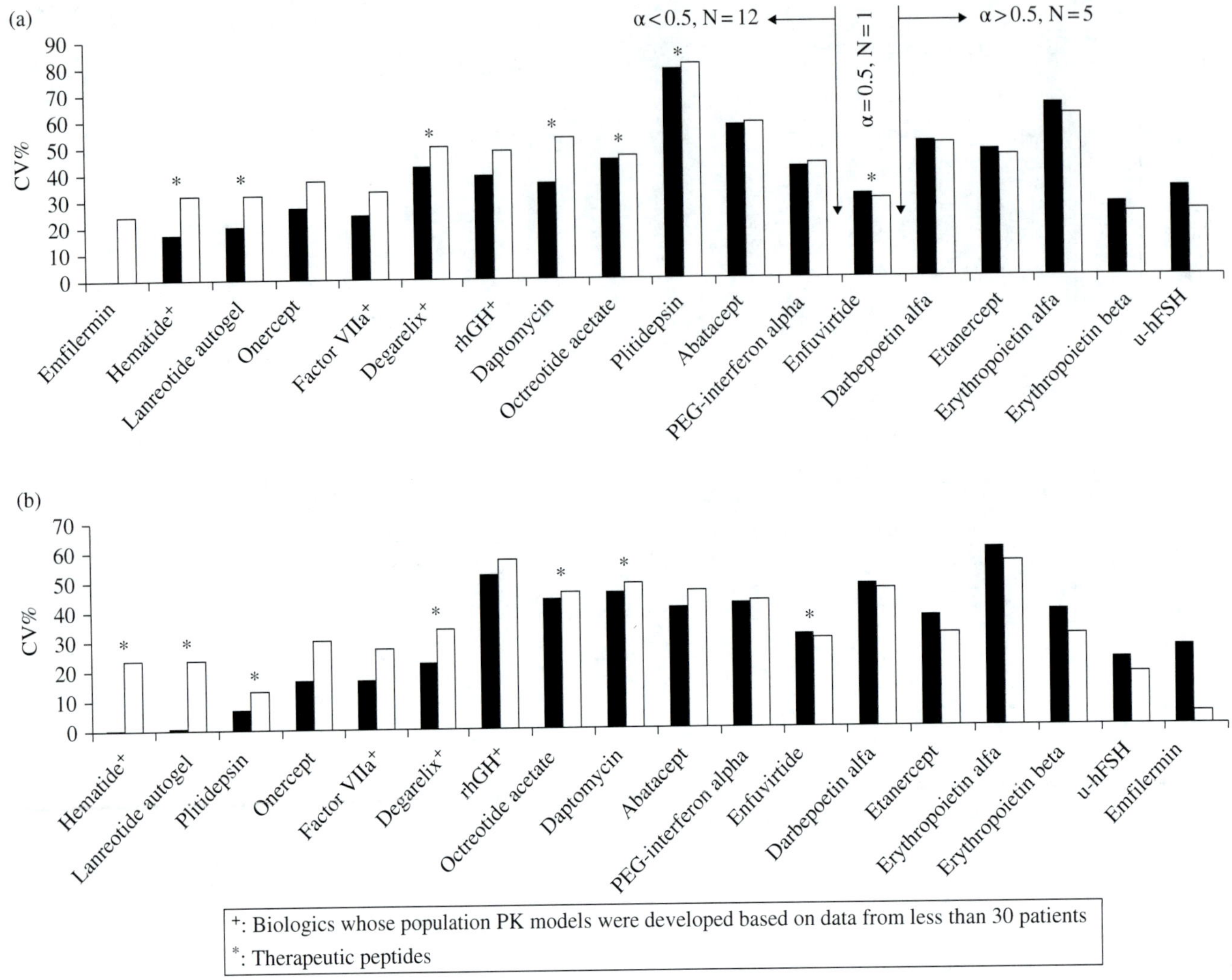

FIGURE 9.5 Comparison of the intersubject variability of simulated (a) AUC and (b) C_{max} of 1000 subjects after receiving a single fixed (solid bar) dose or a body-size (BW/BSA)-based dose (open bar). rhGH, recombinant human growth hormone; u-hFSH, urinary human follicle-stimulating hormone. (Zhang et al. [3]. Reproduced with permission of *J Clin Pharmacol*.)

The results of the simulation studies for comparing the performance of these two dosing approaches at the population level are presented in Figure 9.5. Consistent with the prediction based on the α values, five therapeutic biologics with $\alpha > 0.5$ exhibited less intersubject variability in AUC when body-size-based dosing was adopted, while the other 12 biologics with $\alpha < 0.5$ exhibited less variability in AUC when fixed dosing was used. For the molecule with $\alpha = 0.5$, enfuvirtide, the variability was similar for the two dosing approaches. Similar results were also obtained for individual performance (Fig. 9.6).

The consistent conclusions obtained from the α values and from the simulation studies further reconciled the recommendation of using the α value on clearance to select the optimal dosing approach if AUC is the exposure parameter of the main concern, but what if the PK parameter most closely associated with the determinant safety or efficacy endpoints is C_{max} and not AUC? This approach can also be used for comparative performance evaluation based on C_{max} if body size is only a covariate for the central volume of distribution (V_1) but not for any other parameters, as was the case for emfilermin and octreotide (Table 9.3). However, if a body-size metric is a covariate on more than one PK parameter that could affect C_{max}, simulation needs to be conducted to evaluate the performance of the two dosing approaches in terms of reducing C_{max} variability.

The population and individual performances of the two dosing approaches based on C_{max} were evaluated by simulation studies and shown in Figures 9.5 and 9.6, respectively. At the population level, body-size-based dosing resulted in less intersubject variability in C_{max} for 7 of the 18 selected biologics, while fixed dosing produced less variability in

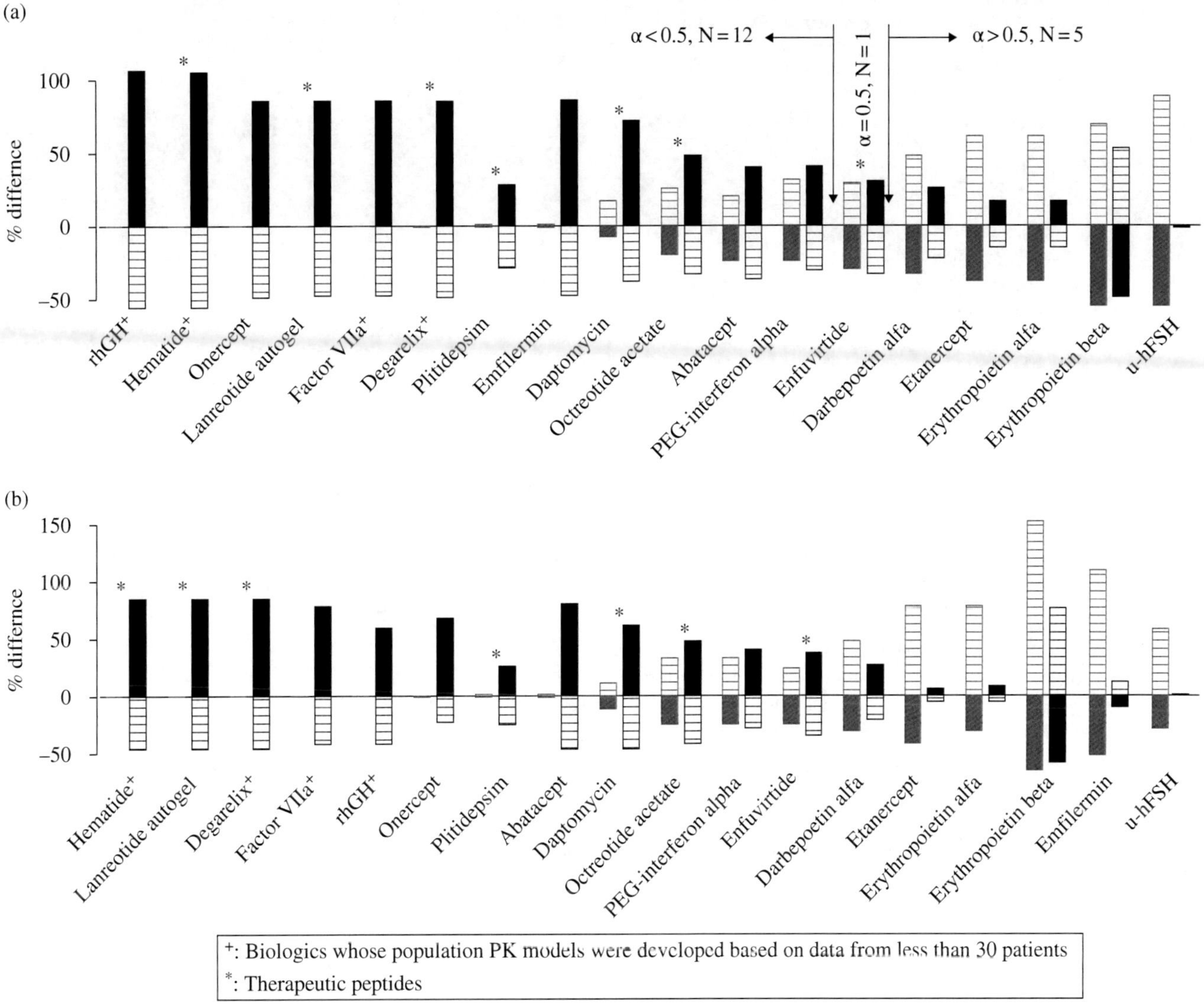

FIGURE 9.6 Comparison of the deviation (% difference) of (a) AUC and (b) C_{max} for subjects with low (open bar) and high (solid bar) extreme body-size (BW/BSA) measurements from the typical values (AUC and C_{max} for subjects with median body size measurements) after a fixed dose (red) or a body-size (BW/BSA)-based dose (black). BSA, body surface area; BW, body weight; rhGH, recombinant human growth hormone; u-hFSH, urinary human follicle-stimulating hormone. (Zhang et al. [3]. Reproduced with permission of *J Clin Pharmacol*.) (*See insert for color representation of this figure.*)

C_{max} for the other 11 biologics (Fig. 9.5). At individual level, body-size-based dosing produced a smaller percentage difference in C_{max} between subjects with extreme and typical body sizes for 6 of 18 biologics, while fixed dosing produced smaller percentage difference for the other 12 biologics (Fig. 9.6). The results from both population- and individual-level evaluations are again very consistent with the only exception of enfuvirtide, whose reported population PK model has no body-size metric captured on the volume of distribution and the reported α value on clearance sits on the discriminating inflection point of 0.5 for dosing approach performance, and the body-size-based dosing was shown to have slightly better individual performance but slightly worse population performance. This discrepancy is likely due to the varying relative contribution of the BW-independent component of the *CL* function ("CL_0") to the overall enfuvirtide *CL* across the BW range (Table 9.3). As was noted by Wang et al. [2] in their work with mAbs, this simulation study with therapeutic peptides and proteins found that, in general, body-size-based dosing tends to over-expose patients with larger body sizes and underexpose patients with smaller body size, while the opposite held true for the fixed-dosing approach.

The percent contribution of body-size metrics to the overall intersubject variability in PK parameters and their corresponding α values are compared for eight of the

TABLE 9.4 Percentage Contribution of Body-Size Measurements to the Overall Intersubject Variability of Pharmacokinetics (PK) Parameters—Therapeutic Proteins and Peptides

Biologics	% Contribution of BW to the Intersubject Variability		α Value	
	CL	V_1	*CL*	V_1
u-hFSH	13.6	NA	0.99	0
Etanercept	18.4	26.9	0.75	1
Erythropoietin alfa	8.48	11	0.75	1.37
Darbepoetin alfa	1.80	0.42	0.623	0.639
Enfuvirtide	3.17	NA	0.5	0
PEG-interferon alfa-2b	2.12	NA	0.455	0
Abatacept	3.46	NA	0.4	0
Emfilermin	NA	4.16	0	1.66

Zhang et al. [3]. Reproduced with permission of *J Clin Pharmacol.*
BW, body weight and u-hFSH, urinary human follicle-stimulating hormone.

therapeutic proteins/peptides in Table 9.4. It was observed that the effect of body size had a small, in some cases moderate, contribution to the overall intersubject variability of major PK parameters, ranging from 1.8% to 18.4% for *CL* and from 0.42% to 26.9% for V_1 (Table 9.4). Interestingly, the α values for the body-size effect on relevant PK parameters appear to correlate with the relative contribution of body size to the overall variability, although the rank orders of the two are not exactly the same. As noted previously, this discrepancy is attributed to the difference in the extent of the contribution of other identified and unidentified factors, such as demographic characteristics and disease conditions, among these biologics.

For the 3 biologics out of the 18 selected for which both population PK and PD models were available in the literature, abatacept, darbepoetin alfa, and etanercept, the performance of each dosing approach was evaluated for a PD endpoint as shown in Figure 9.7. The PD response of abatacept on IL-6 levels was described by an indirect response model, in which the IL-6 degradation rate was stimulated by abatacept according to an E_{max} model [18]. The PD response of darbepoetin on hemoglobin levels was described by a modified indirect response model, where serum darbepoetin levels following weekly administration stimulated the production of hemoglobin through an E_{max} model [19]. A logistic regression model was adopted to describe the exposure–response relationship for etanercept [20]. The cumulative AUC of etanercept was used as the exposure variable, and the American College of Rheumatology response criterion of 20% improvement (ACR20) was used as the binominal clinical outcome. Body-size metrics were not identified as covariates for any PD parameters in any of these three population PD models. As shown in Figures 9.5–9.7, the difference between the performances of the two dosing approaches based on PD is smaller than that based on PK for all these three therapeutic agents. For example, the intersubject variability in drug exposure of etanercept following the two dosing approaches was shown to be 47.5% (fixed dosing) versus 45.7% (body-sized-based dosing) for AUC and 37.4% (fixed dosing) versus 31.4% (body-size-based dosing) for C_{max} at the population level. However, the intersubject variability in its PD measures (ACR20) was 82.2% (fixed dosing) versus 81.9% (body-size-based dosing) for AUC and 70.8% (fixed dosing) versus 70.3% (body-size-based dosing) for C_{max} of the PD effect.

As the ultimate goal of a clinical trial is to achieve its efficacy and safety endpoints, efficacy data, safety data, and data of surrogate PD markers, when available, are more important than PK data alone. The smaller difference in intersubject PD variability between the two dosing approaches suggested that the clinical benefit, if any, of body-size-based dosing could be further "diluted" in terms of drug response or PD measurements as shown in this study for abatacept, darbepoetin alfa, and etanercept.

The overall results of the simulation study were consistent with the prior simulation study by Wang et al. [2] and demonstrated that the two dosing approaches perform similarly across the therapeutic proteins investigated with fixed dosing being better for some and body-size-based dosing being better for the others. However, there was a trend noted among the selected therapeutic peptides, where the fixed-dosing approach performed better for 6 of the 7 included peptides.

9.1.2.5 Relationship between the Type of Biologics and Body-Size Effect on PK Parameters The mAbs investigated by Wang et al. included 3 chimeric, 8 humanized, and 1 fully human mAbs [2]. Seven of the mAbs were known to exhibit linear PK while nonlinear clearance is involved for five of them (Table 9.1). The results of this simulation study demonstrate that there is no intrinsic correlation between the type of mAb and effect of body size on mAb disposition. Therefore, it has been concluded that

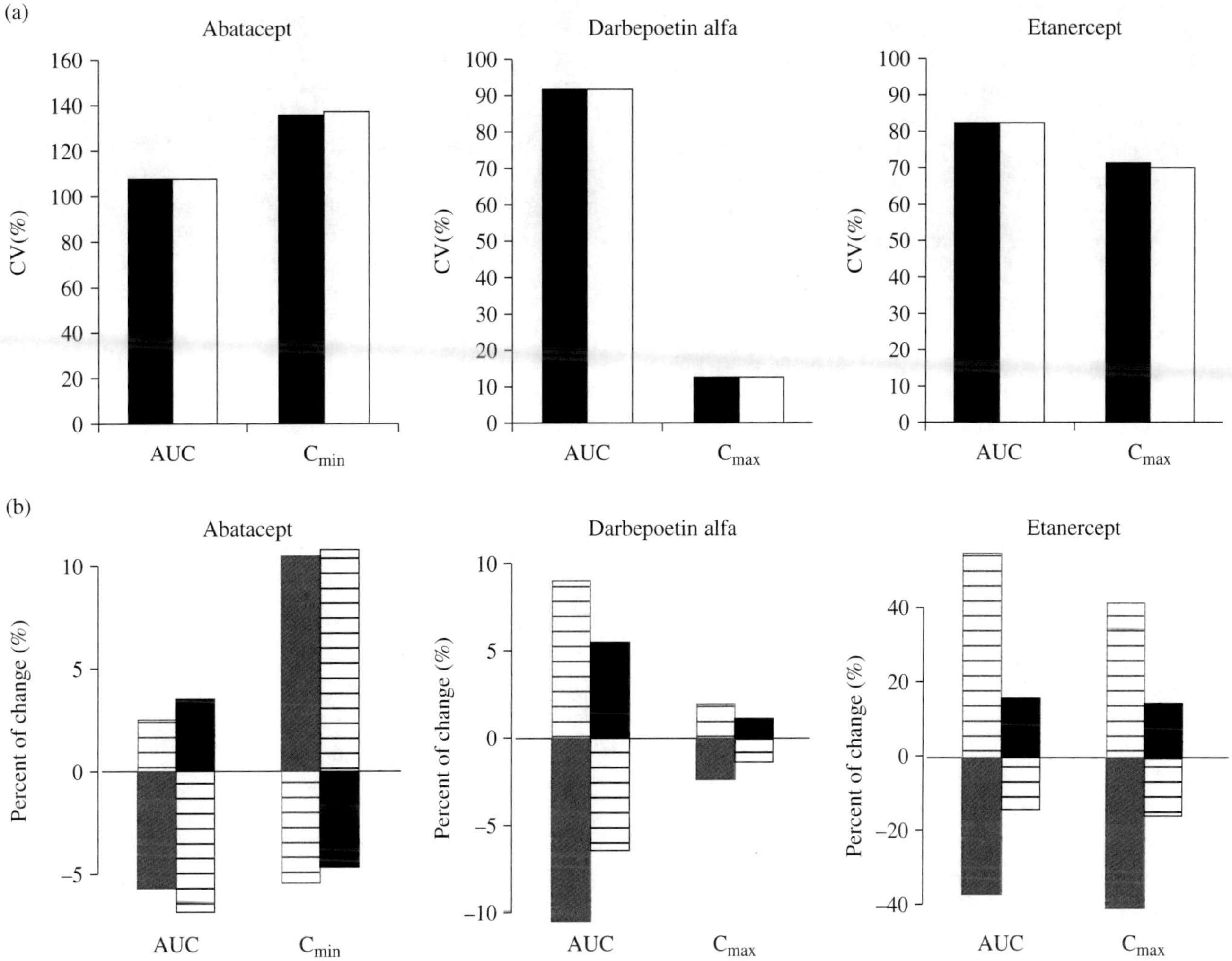

FIGURE 9.7 (a) The intersubject variability of AUC and C_{max} (or C_{min}) of the PD markers for abatacept, darbepoetin alfa, and etanercept across 1000 subjects after fixed (solid bar) and body-size-based (open bar) dosing. (b) Deviation of AUC and C_{max} (or C_{min}) of the PD markers for subjects with low (open bar) and high (solid bar) extreme BW from the typical value after a fixed (grey) and body-size-based (black) dose of abatacept, darbepoetin alfa, and etanercept. The AUC and C_{max} (or C_{min}) of the PD markers are defined as follows: for abatacept, AUC represents the area between the IL-6 baseline and serum IL-6 concentration–time curve and C_{min} represents the minimal concentration of serum IL-6; for darbepoetin, AUC represents the area between the hemoglobin concentration–time curve and hemoglobin baseline, and C_{max} represents the maximal concentration of hemoglobin; for etanercept, AUC represents the area under the probability of achieving American College of Rheumatology response criterion of 20% improvement (ACR20) response–time curve, and C_{max} represents the maximal probability of achieving ACR20 response. (Zhang et al. [3]. Reproduced with permission of *J Clin Pharmacol*.) (*See insert for color representation of this figure.*)

whether a mAb should be dosed based on body size or with a fixed dose cannot be determined by their type and elimination pathways.

The therapeutic biologics investigated by Zhang et al. included 7 therapeutic peptides and 11 therapeutic proteins [3]. Among the 11 therapeutic proteins, including 3 fusion proteins and 1 pegylated protein, no apparent correlation between either the type or the size of therapeutic proteins and the body-size effect on their PK was observed. However, it was noted that the α values of body-size effect on *CL* for the 7 peptides were all less than 0.5 with six having α values equal to 0. As a result, a fixed-dosing approach would perform better for all the peptides evaluated in the simulation study than a body-size-based dosing approach. Whether this can be generalized to other peptides remains a topic for further investigation when population PK data on more therapeutic peptides become available.

9.1.3 Rationale Dosing Approach Selection Strategies Based on Stage of Clinical Development

9.1.3.1 Dosing Approach in Early-Phase Drug Development When a therapeutic protein or mAb is first tested in humans, the effect of body size on PK and/or PD parameters of the agent in humans is unknown. Since no obvious advantage has been identified for one approach over the other in terms of reducing variability in PK/PD measurements at this stage, either dosing approach may be used in FIH and other early-stage trials before the effect of body size on the PK and PD of the agent in adult humans can be evaluated. However, the fixed-dosing approach would be preferred since it offers advantages in ease of preparation, reduced cost, and reduced chance of dosing errors.

9.1.3.2 Dosing Approach in Late-Phase Drug Development The intended dosing approach for the to-be-marketed product is preferably selected before the start of the Phase 3 studies so that the efficacy and safety of the dosing approach can be evaluated in the Phase 3 program to support regulatory approval. Unlike the situation for FIH studies, by the time Phase 1 and/or Phase 2 studies have been completed, data become available to allow for the evaluation of the effect of body size on PK and PD. There is also the potential for data to become available to establish exposure–response relationships for efficacy and safety (including the establishment of a therapeutic window), allowing for the evaluation of dosing approach selection on clinical outcomes.

A full population PK and PD analysis should be conducted including covariate evaluation. If body size is identified as a covariate of PK or PD parameters, population and individual performances of both dosing approaches should be evaluated. If fixed dosing is better or if the two dosing approaches are similar, a fixed-dosing strategy should be used in Phase 3 studies. If body-size-based dosing provides significantly less variability in PK and PD, further evaluation should be conducted using available safety and efficacy data. The selection of the dosing approach in late-phase drug development should largely be driven by the therapeutic window of the agent. The dosing approach that is expected to produce an acceptable safety and efficacy profile may be chosen for the Phase 3 studies.

9.2 CONCLUSIONS

Despite the lack of a clear scientific rationale, the majority of marketed therapeutic biologics are dosed using body-size-based dosing approaches. This chapter has highlighted the methodology that should be used in the rational selection of a dosing approach to therapeutic biologics. The performance of body-size-based dosing and fixed-dosing approaches have been evaluated for mAbs [2, 4] and therapeutic peptides/proteins [3], in terms of their population and individual performances in reducing intersubject PK and/or PD variability in adult patients. The results of these simulation studies have demonstrated that in contrast to conventional assumption, body-size-based dosing did not always result in less intersubject variability in drug exposure and PD measurements when compared to fixed-dosing approaches. Recommendations for the rational selection of dosing approaches for therapeutic biologics in an adult population have been presented based on these results. For adult FIH studies, the fixed-dosing approach is preferred since this approach offers advantages in ease of dosing preparation, reduced cost, and reduced chance of dosing errors. When sufficient data become available, a full assessment of body-size effect in PK and/or PD should be conducted. The final dosing approach for Phase 3 trials in adults should be selected based on the established body-size effect on the PK and PD, the therapeutic window of the therapeutic products, and other factors that may impact the outcome of the study.

REFERENCES

[1] Dimitrov DS. Therapeutic proteins. Methods Mol Biol 2012;899:1–26.

[2] Wang DD, Zhang S, Zhao H, Men AY, Parivar K. Fixed dosing versus body size-based dosing of monoclonal antibodies in adult clinical trials. J Clin Pharmacol 2009;49:1012–1024.

[3] Zhang S, Shi R, Li C, Parivar K, Wang DD. Fixed dosing versus body size-based dosing of therapeutic peptides and proteins in adults. J Clin Pharmacol 2012;52:18–28.

[4] Bai S, Jorga K, Xin Y, Jin D, Zheng Y, Damico-Beyer LA, Gupta M, Tang M, Allison DE, Lu D, Zhang Y, Joshi A, Dresser MJ. A guide to rational dosing of monoclonal antibodies. Clin Pharmacokinet 2012;51(2):119–135.

[5] Gurney H. Dose calculation of anticancer drugs: a review of the current practice and introduction of an alternative. J Clin Oncol 1996;14(9):2590–2611.

[6] Mathijssen RH, de Jong FA, Loos WJ, van der Bol JM, Verwiej J, Sparreboom A. Flat-fixed dosing versus body surface area based dosing of anticancer drugs in adults: does it make a difference? Oncologist 2007;12(8):913–923.

[7] Mahmood I. Pharmacokinetic allometric scaling of antibodies: application to first-in-human dose estimation. J Pharm Sci 2009;98(10):3850–3861.

[8] Deng R, Iyer S, Theil FP. Projecting human pharmacokinetics of therapeutic antibodies from nonclinical data: what have we learned? MAbs 2011;3(1):61–66.

[9] US Food and Drug Administration [FDA]. *Guidance for Industry: E11 Clinical Investigation of Medicinal Products in the Pediatric Population*. Rockville (MD): FDA; 2000. Available at http://www.fda.gov/downloads/Drugs/Guidance ComplianceRegulatoryInformation/Guidances/ UCM073143.pdf. Accessed 2015 Jul 15.

[10] European Medicines Agency. *ICH Topic E11: Clinical Investigation of Medicinal Products in the Paediatric Population*. [document reference CPMP/ICH/2711/99] London: European Medicines Agency; 2001. Available at http://www.ema.europa.eu/docs/en_GB/document_library/Scientific_guideline/2009/09/WC500002926.pdf. Accessed 2015 Jul 15.

[11] Mrozek K, Marcucci G, Paschka P, Whitman SP, Bloomfield CD. Clinical relevance of mutations and gene-expression changes in adult acute myeloid leukemia with normal cytogenetics: are we ready for a prognostically prioritized molecular classification? Blood 2007;109(2):431–448.

[12] Yang CH, Yu CJ, Shih JY. Specific EGFR mutations predict treatment outcome of stage IIIB/IV patients with chemotherapy naïve non-small-cell lung cancer receiving first line gefitinib monotherapy. J Clin Oncol 2008;26(16):2745–2753.

[13] Goncalves A, Esteyries S, Taylor-Smedra B. A polymorphism of EGFR extracellular domain is associated with progression-free survival in metastatic colorectal cancer patients receiving cetuximab-based treatment. BMC Cancer 2008;8:169.

[14] Hermanussen M, Danker-Hopfe H, Weber GW. Body weight and the shape of the natural distribution of weight, in very large samples of German, Austrian, and Norwegian conscripts. Int J Obes Relat Metab Disord 2011;25(10):1550–1553.

[15] Lobo ED, Hansen RJ, Balthasar JP. Antibody pharmacokinetics and pharmacodynamics. J Pharm Sci 2004;93(11):2645–2668.

[16] Lin JH. Pharmacokinetics of biotech drugs: peptides, proteins, and monoclonal antibodies. Curr Drug Metab 2009;10: [661]–691.

[17] Wang W, Wang EQ, Balthasar JP. Monoclonal antibody pharmacokinetics and pharmacodynamics. Clin Pharmacol Ther 2008;84:548–558.

[18] Roy A, Mould DR, Wang XF, Tay L, Raymond R, Pfister M. Modeling and simulation of abatacept exposure and interleukin-6 response in support of recommended doses for rheumatoid arthritis. J Clin Pharmacol 2007;47:1408–1420.

[19] Agoram B, Heatherington AC, Gastonguay MR. Development and evaluation of a population pharmacokinetic-pharmacodynamic model of darbepoetin alfa in patients with nonmyeloidmalignancies undergoing multicycle chemotherapy. AAPS J 2006;8:E552–E563.

[20] Lee H, Kimko HC, Rogge M, Wang D, Nestorov I, Peck CC. Population pharmacokinetic and pharmacodynamic modeling of etanercept using logistic regression analysis. Clin Pharmacol Ther 2003;73:348–365.

10

IMPACT OF DISEASES, COMORBIDITY, AND TARGET PHYSIOLOGY ON ADME, PK, AND PK/PD OF THERAPEUTIC BIOLOGICS

SONGMAO ZHENG, WEIRONG WANG AND HONGHUI ZHOU
Janssen Research and Development, LLC, Spring House, PA, USA

10.1 INTRODUCTION

Therapeutic biologics, primarily represented by monoclonal antibodies (mAbs), have fundamentally changed the pharmacotherapy paradigm in several areas, such as immune-mediated inflammatory diseases and oncology. A mechanistic understanding of the absorption, distribution, metabolism, and excretion (ADME) of a biologic in relevant animal disease models and patients would greatly facilitate moving a candidate biotherapeutic from discovery through preclinical and clinical development. Conceptually, any disease or comorbidity that can regulate or impact the ADME and subsequently the pharmacokinetics/pharmacodynamics (PK/PD) of a biologic would have an impact on the treatment response that may sometimes necessitate a dose alteration in different patient populations, different disease states, patients with varying disease burdens, or in patients with different comorbidities. In particular, target properties (e.g., baseline concentrations, turnover rates, distribution, and fate following drug binding) are important determinants of the treatment response of a therapeutic biologic. The PD of a biologic is not the only aspect driven by target engagement, the PK is also often impacted via target-mediated drug disposition (TMDD) interactions. Other factors, such as immunogenicity, demographics, and concurrent medications, may also alter the ADME, PK, and PK/PD of biologics.

This chapter provides an overview of our current knowledge on how diseases, comorbidity, target physiology, and other treatment interventions (e.g., surgery) impact the ADME, PK, and PK/PD of therapeutic biologics. A diagram of the interplay between these elements is shown in Figure 10.1. The underlying mechanisms associated with these processes will also be discussed.

10.1.1 ADME of Biologics

Like conventional small molecule drugs, the PK of biologics is a collective depiction of ADME processes. Most biologics can only be administered by parenteral routes, that is, intravenous (IV), subcutaneous (SC), or intramuscular (IM). Following SC or IM administration, biologics are absorbed either directly via blood capillaries or indirectly via lymphatic capillaries, depending on size and molecular weight (MW) [1]. Our knowledge on the factors that govern the absorption process of biologics is still limited and requires further scientific efforts to explore even the fundamentals of this important ADME process [2].

Biologics are primarily distributed in the plasma compartment (systemic circulation) and extracellular fluids throughout the body. Given the size and hydrophilicity of biologic molecules, distribution from the systemic circulation to tissues occurs mostly via convective transport through pores on blood capillary walls [3, 4]. The extent to which a biologic distributes to a tissue site largely depends on the size of the molecule, blood/lymph flow, and blood capillary permeability [5]. Importantly, the interaction between a biologic and the attendant target, especially if the target has high abundance or rapid turnover, can significantly change the

ADME and Translational Pharmacokinetics/Pharmacodynamics of Therapeutic Proteins: Applications in Drug Discovery and Development, First Edition. Edited by Honghui Zhou and Frank-Peter Theil.

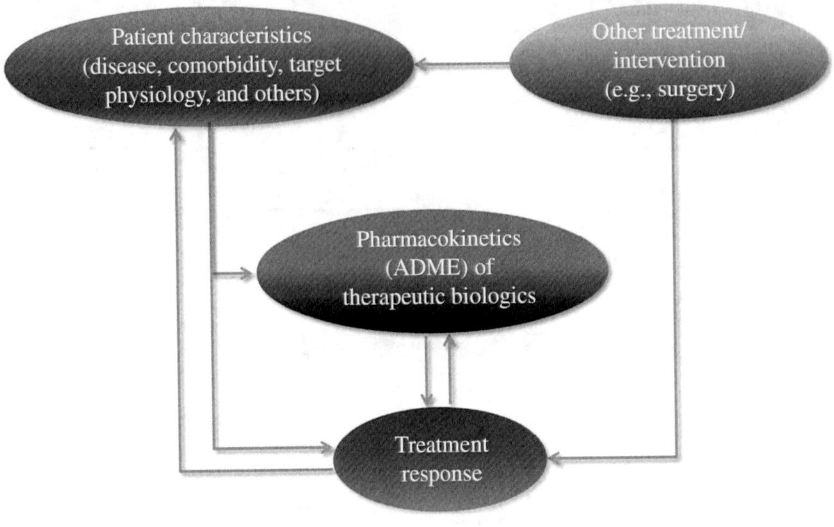

FIGURE 10.1 Interplay of disease, comorbidity, target physiology, and other interventions on the ADME, PK, and PK/PD of therapeutic biologics.

distribution of the therapeutic to the tissue site expressing the target [6]. Except for perhaps some very small peptides, for example, cyclosporine A [7], biologics are not subject to cytochrome P450-mediated metabolism and are also not substrates for small molecule drug transporters. Biologics are generally believed to be catabolized into small peptides and eventually amino acids via proteolysis throughout the body. Renal filtration, however, can be a major elimination mechanism for smaller size biologics [8], but the role of renal clearance for biologics with MW greater than 69 kDa is minimal in healthy subjects [9].

The elimination rates of biologics can vary dramatically for different types of molecules. When compared to other therapeutic proteins, IgG-based drugs (mAbs and Fc-fusion proteins) exhibit sustained persistence in the systemic circulation via a unique "salvage" mechanism involving the neonatal Fc receptor (FcRn or "Brambell receptor") [10]. Theoretically, if the binding of FcRn and mAb is altered, or FcRn expression is affected by some disease state, the disposition of the mAb may be impacted accordingly.

A unique feature of biologics is that they have the potential to be immunogenic. When a nonendogenous biologic drug is injected into an animal or human, an immune response can cause the development of anti-drug antibodies (ADAs) [11]. Consequently, the formation of ADA may be associated with an altered drug clearance, reduced clinical efficacy, or an increased risk for adverse events.

10.1.2 Roles of TMDD for Biologics

Another unique PK characteristic of biologics is the process of TMDD [12–14]. The PK of a conventional small molecule drug is usually independent of the therapeutic target because the fraction of drug bound to the target is usually negligible and TMDD is not a significant clearance mechanism for small molecule drugs, except for a few noted cases such as warfarin [12]. In contrast, TMDD is common for biologics, especially mAbs, due to relatively low nonspecific systemic clearance and high target-binding affinity. Following binding of a therapeutic biologic to a cell-surface receptor, the drug–target complex is usually internalized and subsequently degraded in the lysosomes of the target cells [15]. This target-mediated clearance can be much faster than other nonspecific clearance pathways, especially for mAbs; however, in contrast to other clearance mechanisms, target-mediated clearance has limited capacity. Therefore, the hallmark of TMDD is the observation of nonlinear PK, that is, higher clearance and a larger volume of distribution at lower doses. TMDD is most apparent for mAbs targeting membrane-bound receptors with high abundance or rapid turnover, for example, epidermal growth factor receptor (EGFR), cluster of differentiation 3 (CD3), and cluster of differentiation 20 (CD20) [16]. It is common to see therapeutic target physiology change as a result of disease burden. In such cases, any disease progression or treatment/intervention that affects target status may alter the PK characteristics of drugs subject to TMDD and subsequently impact the PK/PD and treatment response of such drugs.

10.2 IMPACT OF DISEASES AND COMORBIDITY ON ADME AND PK OF THERAPEUTIC BIOLOGICS

10.2.1 Disease and Comorbidity on the Subcutaneous Absorption of Biologics

As discussed earlier, our knowledge of the factors that govern the absorption processes of biologics remains limited. The bioavailability of most macromolecules administered

SC was between 30% and 100% [2], and it is common to see large intersubject variability for SC absorption. Theoretically, the bioavailability of a biologic drug is governed by the time it takes to reach the systemic circulation, and by how susceptible the molecule is to presystemic catabolism. Many diseases and comorbidities are known to produce physiologic changes, such as to blood/lymph flow rates, vascular permeability, and protease activity, but the extent to which these changes impact SC absorption remains poorly understood. Reports on how diseases and comorbidities impact the SC absorption of biologics are sparse, and are primarily based on population PK and covariate analyses.

The most prominent example of such impact is with obesity. Since obesity has been associated with decreased tissue perfusion due to endothelial dysfunction or a decrease in blood capillary density in skeletal muscle [17], it provides a significant example of the impact disease and comorbidity on the SC absorption of a biologic agent. The blood flow and capillary density in adipose tissue at the site of SC administration were shown to be negatively correlated with body mass index (BMI) in two clinical studies [18, 19]. In men administered soluble insulin SC, increasing adiposity prolongs the duration of the early absorption lag phase and reduces the rate of insulin absorption [20]. In addition, body weight was found to be a negative covariate for bioavailability following SC administration of recombinant human erythropoietin (rHuEPO) and polyethylene glycol (PEG)-conjugated interferon alpha (IFNα) in humans [21–23]. In an animal model, obesity in mice produced impaired lymphatic transport [24]. A significant negative correlation between collecting lymphatic vessel function and body weight was reported in mice fed chronically with a high fat diet [25]. It has also been demonstrated in Sprague–Dawley rats administered a PEG-conjugated erythropoietin (EPO) that the "fat" rats, fed with a high fat diet and weighing about twice that of the control animals, had about a fourfold lower bioavailability than that in the control animals [26].

Age is another prominent covariate identified for having an impact on the SC absorption of biologics. In a population PK meta-analysis using data from 16 studies and 427 healthy subjects who received SC doses of rHuEPO, increasing age was associated with a decrease in the first-order absorption rate constant in addition to the decrease in bioavailability accompanying increasing body weight [23]. Another publication also reported an approximate 20–30% reduction in the absorption rate of darbepoetin alfa per decade of age [27]. Age-related reductions in physical activity and lymphatic flow rates may provide a plausible physiological rationale for this relationship. A recent physiologically based SC absorption model also suggested that a change in lymphatic flow rate can significantly impact the SC absorption of mAbs [28].

Besides the physicochemical properties of therapeutic biologics, the extent to which they are susceptible to presystemic catabolism depends on the protease activity at both the SC injection site and during lymphatic transport. The common proteases and peptidases in the SC tissue have been reviewed previously [29]. Under normal conditions, protease activities are not high in the SC space; however, since varying degrees of tissue damage, for example, blood vessel breakage, disorganization of cellular and acellular structures of the SC space, and inflammation, are anticipated following SC drug administration, any resulting damage could potentially change the activity of matrix metalloproteinases (MMPs), clotting cascade proteases (e.g., thrombin), fibrinolytic enzymes (e.g., plasminogen), and collagenases [29, 30]. Disruption of blood vessels could also lead to clot formation, which might entrap and/or destroy the drug administered [31]. Moreover, proteases can be affected by certain disease states and are reported to be upregulated with disease progression [32]. Inflammatory signals are also known to increase local vascular permeability and protease activity. The clinical implications of these structural and physiologic changes on the SC absorption of biologics remain to be explored.

10.2.2 Disease and Comorbidity on the Distribution of Biologics

Following SC administration, the distribution of a biologic drug to tissue sites depends predominantly on blood/lymph flow, blood capillary permeability, and target binding. As in the case of drug absorption, many diseases and comorbidities are known to affect the processes associated with drug distribution. For example, it is well known that inflammation leads to an increase in blood capillary permeability [33], and would be expected to have a broad impact on the distribution of therapeutic biologics distribution, although the magnitude and clinical implications of this impact need further exploration.

The distribution of biologics into solid tumors is of special interest. The development of biologic therapies for cancer is increasing rapidly, primarily due to its high affinity binding to tumor-specific targets. These tumor-specific targets are often associated with relatively high expression levels that can have a profound impact on the distribution of biologics. For example, ^{111}In-labeled daclizumab that targets CD25 was administered to CD25-expressing leukemia patients who subsequently underwent whole-body imaging. In a patient with low tumor burden, daclizumab remained largely in the vascular space, whereas, in a patient with high tumor burden, the antibody was cleared rapidly from the blood compartment and was primarily localized to tumors in the bone marrow and spleen [34]. A similar study showed that ^{131}I-labeled mAb G250 localized specifically to human renal cell carcinoma expressing the cell-surface antigen G250 that is not detected in normal kidney [35]. Similarly, selective localization to tumor tissue was demonstrated in 19 of 20 colorectal carcinoma patients with hepatic metastases who were administered ^{131}I-labeled mAb A33 that binds to the antigen A33 homogeneously expressed by virtually all colon cancers and in the colon mucosa, but not in other epithelial tissues [36].

The unique anatomical and physiological properties of solid tumors, for example, leaky tumor vasculature, lack of lymphatic drainage and elevated interstitial fluid pressure, and tumor heterogeneity, are also key determinants of the distribution of biologics. These factors and their clinical impact have been extensively reviewed in Chapter 16 by Greg Thurber and elsewhere [37], and will not be discussed here.

Another special topic is the distribution of biologics across the blood–brain barrier (BBB), typically characterized by the extremely tight junctions around the brain capillaries. Under normal conditions, the BBB impedes the influx of almost all biologics from blood to brain tissue [38]; however, disruption of the BBB can occur as a result of many brain disorders including tumors and inflammation, [39, 40]. Bevacizumab, a 150-kDa mAb against vascular-endothelial growth factor (VEGF), has demonstrated efficacy in the treatment of patients with recurrent glioblastoma in combination with standard chemotherapy [41]. The efficacy of bevacizumab has been suggested to be related to the compromised integrity of the BBB in glioblastoma patients, thus increasing distribution of bevacizumab to the site of action, despite the hypothesis that the reduction of edema by normalizing and reducing the permeability of the vasculature are the main reasons for the beneficial effects of bevacizumab in glioblastoma patients [42, 43]. Unfortunately, by normalizing the tumor vasculature in glioblastoma patients, bevacizumab may ultimately prevent further access of large molecules such as mAbs to the tumor site. In a preclinical model, preadministration of bevacizumab was shown to decrease tumor uptake of trastuzumab, and this phenomenon was mechanistically attributed to reduced vascular permeability, vascular density, and blood perfusion [44, 45].

10.2.3 Hepatic Impairment

Although the liver has been suggested as a major organ for elimination of many higher MW biologics including mAbs, more typical elimination pathways are usually receptor- or endocytosis-mediated, in contrast to the conventional elimination of small molecule drugs that are metabolizing enzyme- and transporter-mediated [3, 15]. Yang et al. recently performed a systemic analysis to examine whether studies assessing hepatic impairment, routinely conducted for small molecule drugs, are necessary for therapeutic biologics [46]. Among the 91 therapeutic biologics approved by the U.S. FDA (excluding antibody–drug conjugates (ADCs), see below), no dedicated PK trials were conducted in patients with hepatic impairment, and a subgroup ($n=2$) or population PK ($n=5$) analysis was only performed for 7 of them [46]. Based on the outcomes of the subgroup or population PK analyses, hepatic dysfunction showed no major impact on the PK of the therapeutic proteins, and no dose modification was recommended. As a result, dedicated PK trials in patients with hepatic impairment was not considered necessary for therapeutic biologics; however, continued use of subgroup or population PK analyses to further evaluate the potential effect of hepatic impairment is still viewed as important [46]. A similar recommendation was also made by Zhao et al. based on the limited impact of hepatic impairment on the exposure of therapeutic biologics [47].

This rule, however, may not apply to a special class of biologics, that is, ADCs, which contain a highly toxic small molecule drug component that can act as a substrate for metabolizing enzymes and transporters in the liver [47, 48]. For gemtuzumab ozogamicin (GO) (Mylotarg), the first approved ADC product, a total of 11 metabolites were found when it was incubated with human liver microsomes, human liver cytosol, and human leukemia cells [47]. As a result, the product label states that "extra caution should be exercised when administering Mylotarg in patients with hepatic impairment" [49]. The small molecule drug components, monomethyl auristatin E (MMAE) and mertansine (DM1), of two other more recently approved ADCs, brentuximab vedotin and ado-trastuzumab emtansine (T-DM1), respectively, have both been identified as substrates of cytochrome P450 enzymes, and the fecal/biliary route has been shown to be the major elimination pathway for MMAE- and DM1-containing catabolites [48]. For brentuximab vedotin, MMAE exposure has been further shown to increase in patients with hepatic impairment [48]. Dedicated hepatic impairment studies for both brentuximab vedotin and T-DM1 are ongoing, that the data are not yet available [48].

10.2.4 Renal Impairment

The impact of renal impairment on biologics has been effectively reviewed recently [9]. The effect of renal impairment on the PK of biologics is dependent on the ability of the compound to undergo glomerular filtration, which is largely driven by MW. Consequently, the clearance of lower MW biologics would be expected to be impacted more by renal impairment [9]. For example, the PK of anakinra (17.3 kDa) [50], interleukin-10 (IL) (18 kDa) [51], and human growth hormone (22 kDa) [52] showed a high dependency on renal function. But, for the higher MW rHuEPO (34 kDa), renal clearance only accounts for less than 3% of total body clearance, and renal impairment or end-stage renal disease did not affect the PK of this compound [53]. Similarly, pegfilgrastim (39 kDa), a 19-kDa protein with a 20-kDa PEG molecule attached to the N-terminus, does not undergo glomerular filtration, and no discernable correlation was observed between the PK and renal function [54]. These findings are not entirely unexpected. In addition to molecular size, other factors, such as molecular shape, charge, and other biophysical properties, should also be considered.

Renal clearance usually plays a minimal role in the elimination of biologics with MW greater than 69 kDa, for example, antibody-based therapeutics. Indeed, for bevacizumab [55], cetuximab [56], rituximab [57], trastuzumab [58], and ustekinumab [59], renal impairment has been shown to have no clear impact on the PK characteristics of these mAbs [9]; however, there is increasing evidence that some forms of renal disease, such as diabetic nephropathy, may lead to an increase in the renal elimination of IgG [60]. In a recent preclinical study using a streptozotocin (STZ)-induced diabetic nephropathy mouse model, a murine mAb 8C2 exhibited more than a twofold increase in clearance in STZ-treated mice and the 8C2 clearance was highly correlated with the urinary albumin excretion (UAE) rate [61]. These results suggest that severe renal impairment, and especially end-stage renal disease, may still have an impact on the disposition of higher MW biologics, including IgG-based therapeutics.

In humans, proteinuria due to kidney damage is commonly observed in patients with systemic lupus erythematous (SLE) [62]. Indeed, an association between increasing baseline proteinuria and increasing clearance was observed in the population PK analysis of belimumab, a human mAb against soluble B-lymphocyte stimulator (BLyS) for SLE [63]. Since only patients with proteinuria less than 6 g/day at screening were included in the belimumab trials, the estimated effect of proteinuria was relatively small in this report [63]. The effect, however, may become clinically significant in nephropathies such as membranous glomerulonephritis, where patient urine protein levels ≥12 g/day are not uncommon.

10.2.5 Immune-Mediated Inflammatory Diseases

Inflammation can have a significant impact on the elimination of biologics. Proteolytic catabolism within the cells of the reticuloendothelial system (RES) is an important route of elimination for mAbs and disease severity may influence the elimination of mAbs through these RES-mediated mechanisms [64, 65]. It was found that patients with elevated C-reactive protein, a systemic inflammatory biomarker, have accelerated mAb clearance, suggesting that the presence of systemic inflammation may increase the catabolism of mAbs in the RES [66, 67].

Infliximab, a mAb against tumor necrosis factor alpha (TNFα), has been approved to treat a number of inflammatory diseases such as rheumatoid arthritis (RA), inflammatory bowel diseases (IBDs), and ankylosing spondylitis. The elimination half-life of infliximab in RA patients is inversely correlated with the magnitude of the inflammatory burden as measured by the disease activity score, DAS-28 [68]. The clearance of infliximab in IBD patients (in particular, ulcerative colitis patients) was about 50% higher than in RA patients [64]. Since infliximab exhibited linear PK at doses above 1 mg/kg [69], the observed clearance increase is likely not related to TMDD, but instead may be caused by other nonspecific clearance pathways. In severe IBD, the bowel mucosa is diffusely denuded and ulcerated, leading to a massive loss of proteins. This phenomenon is known as "protein-losing enteropathy" (PLE). It was found that the clearance of proteins (including large proteins such as IgG) correlated with the extent, length, and severity of the intestinal lesion in patients with IBD [70, 71]. PLE is a plausible mechanism that contributed to the increased clearance of infliximab in patients with IBD. Accordingly, infliximab was detected in the feces of patients with severe IBD, and the highest concentrations were measured during the first days after the initiation of therapy, that is, the most acute phase of the disease [72].

Similar findings were made for other anti-TNFα therapies as well. For example, Louis and Panes suggested that in cases of a more severe inflammation, there may be more rapid consumption of the therapeutic antibody leading to an insufficient SC bioavailability of adalimumab, another anti-TNFα mAb [73]. Systemic inflammation and/or PLE have also been suggested as possible causes for observed trastuzumab PK exposure differences between metastatic gastric cancer (mGC) and metastatic breast cancer patients [74].

10.2.6 Diabetes

The potential impact of diabetes on mAb PK was reported for ustekinumab [59]. Ustekinumab, an anti-IL-12/IL-23 mAb, has been approved for the treatment of plaque psoriasis and psoriatic arthritis. In both pivotal Phase III studies (PHOENIX 1 and PHOENIX 2) for ustekinumab, 10.6% of the psoriatic patients had diabetic comorbidity. In those patients with diabetic comorbidity, 28.7% had higher apparent clearance of ustekinumab than in patients without diabetic comorbidity [59].

The underlying mechanism for the enhanced clearance in patients with diabetes has not yet been elucidated, although several alternative mechanisms, such as altered renal clearance, decreased glycation of ustekinumab, or altered FcRn binding affinity, have been postulated to explain the enhanced clearance of ustekinumab in diabetics. Altered renal clearance can be largely excluded since the estimated creatinine clearance was not significantly different between patients with and without diabetes [59], although UAE rate was not measured in these studies [61]. Interestingly, serum IgG can be nonenzymatically glycosylated in diabetic patients [75], and this modification may contribute to increased clearance by significantly lowering the antigen affinity of the antibody and increasing the rate of dissociation of the antigen–antibody complex [76]. Although it remains a possible cause of enhanced clearance, to date there is no direct evidence that FcRn binding is altered by glycated IgG. Further investigation will be needed to understand the potential mechanisms of increased ustekinumab clearance in psoriatic patients with diabetic comorbidity.

10.2.7 Immunogenicity

The development of an immune response can have a substantial impact on the safety, efficacy, and PK of therapeutic biologics [3, 11]. Lenercept, a TNF receptor fusion protein, elicits the development of ADA 6–10 days after administration to laboratory animals, and the increase in ADA coincided with the accelerated clearance of lenercept [77]. In the clinical setting, development of ADAs were reported to occur in 19–26% of RA and spondyloarthritis patients treated with infliximab, in 28% of RA patients treated with adalimumab, and in 5% of psoriatic patients treated with ustekinumab [78, 79]. For newer biological drugs in the development for psoriasis, the formation of ADAs occurs in 18–25% of patients treated with certolizumab-pegol, 5.0–9.8% of patients treated with brodalumab, and 5.4% of patients treated with golimumab [79]. Immunogenicity was often associated with lower drug concentrations and lower efficacy. A meta-analysis showed that IBD patients with ADAs to infliximab had a threefold higher risk of loss of clinical response when compared to ADA-negative patients [80].

The immunological impacts of a disease, together with the associated systemic burden, may affect the level of activation or suppression of the host immune system, and thus influence the potential development ADAs [81]. As an example, the presence of inflammatory signals from sites of inflammation can lead to a more potent B-cell production of IgGs against biologic-derived antigens [82]. Patients with systemic malignancy have been shown to have a lower prevalence of ADAs against therapeutic IFNα-2a when compared to patients suffering from viral infections [83].

Administration of concomitant immunomodulators, such as thiopurines or methotrexate (MTX), has also been associated with lower rates of ADAs in a number of studies [84]. Maini et al. observed that the coadministration of low dose (7.5 mg/week) MTX could significantly elevate the systemic levels of infliximab [85]. Concurrently, the incidence of immunogenicity in those patients was also greatly decreased. Similar observations have also occurred with other anti-TNFα biologics, such as adalimumab, golimumab, and certolizumab [81]. It has been hypothesized that the decrease in clearance of anti-TNFα biologics coadministered with MTX might be correlated to a decrease in the incidence of immunogenicity from the immunosuppressive effect of MTX.

Prolonged treatment with biologics can sometimes also induce immune tolerance, that is, diminish the formation of ADAs. For example, regular administration of infliximab appears to promote immune tolerance to this mAb and reduces the development of ADAs in IBD. When compared to an ADA incidence of up to 61% in patients treated with infliximab after five infusions, a lower prevalence (5–10%) of ADAs has been reported following continuous maintenance therapy with infliximab [81]. Prolonged courses of high dose recombinant factor VIII have also been used successfully to induce immune tolerance in patients with hemophilia [86].

10.3 IMPACT OF DISEASE AND TARGET PHYSIOLOGY ON PK AND PK/PD OF THERAPEUTIC BIOLOGICS

Biologics are typically targeted therapies, that is, they modulate a specific target (or targets) to achieve PD effects. The physiology of the target is commonly tempered by the disease burden and treatment/intervention, which in turn could impact the PK and PK/PD of therapeutic biologics. A summary of the therapeutic proteins that fall into this category is presented in Table 10.1. The impact of disease and target physiology on the PK and PK/PD of therapeutic biologics will be discussed in this section.

10.3.1 Biologics against Membrane-Bound Targets

Generally, TMDD is most commonly observed and more conspicuous for biologics against membrane-bound targets. A number of anticancer biologics targeting cell-surface tumor antigens, for example, CD11, CD20, CD22, CD52, HER2, and EGFR, exhibited nonlinear PK. For these mAbs, patients with higher tumor loads tended to show much more rapid and extensive drug distribution and elimination than what was observed in healthy individuals or in patients with lower tumor volume [112].

Target levels are often downregulated following treatment via (i) decreasing synthesis, (ii) increasing the endocytosis rate constant of the drug–target complex compared to that of the target alone, and (iii) decreasing the recycling to degradation ratio, or by a combination of these factors [113]. For example, binding of cetuximab to EGFR has been shown to produce internalization of the antibody–receptor complex, which leads to an overall downregulation of EGFR expression [114]. An interesting recent report showed that treatment with combinations of noncompetitive anti-EGFR mAbs can induce receptor clustering, leading to more effective synergistic receptor downregulation of EGFR when compared to monotherapies [113]. For IgG1 mAbs targeting receptors on the cell membrane, treatment can lead to significant decreases in the target cells via antibody-dependent cellular cytotoxicity (ADCC) or other Fc-mediated effector functions. Rituximab, an anti-CD20 mAb, is one example for this effect. The CD20 antigen is found on the surface of normal and malignant B-lymphocytes. Binding of rituximab produced rapid and sustained depletion of circulating and tissue B-cells, and hence reduction in the CD20 targets on B-cells [115].

Target levels can also be upregulated with treatment. For example, filgrastim is a recombinant human granulocyte colony-stimulating factor (G-CSF) that promotes the

TABLE 10.1 Examples of Therapeutic Biologics That Showed Target-Dependent and/or Treatment-Dependent PK

Target	Drug Name	Indication[a]	Reported Clinical PK	References
α4 integrin	Natalizumab	Relapsing multiple sclerosis	Nonlinear PK over a dose range of 0.03–3 mg/kg, with clearance decreasing from 86.9 to 0.3 mL/h/kg	[87]
CS1	Elotuzumab	Multiple myeloma	Nonlinear PK over the dose range of 0.5–20 mg/kg, with clearance decreasing from 71.4 to 15.7 mL/h	[88]
CD4	Clenoliximab	Rheumatoid arthritis	Nonlinear PK over the dose range of 0.05–15.0 mg/kg; treatment caused an 80% reduction in CD4 density for up to 3 weeks, without depleting T-cells	[89]
CD11a	Efalizumab	Chronic plaque psoriasis	Nonlinear PK over the dose range of 0.1–10 mg/kg, with clearance decreasing from 322 to 6.6 mL/day/kg; treatment caused a rapid reduction in the level of CD11a expression on CD3-positive lymphocytes to about 25% of pretreatment levels	[90]
CD20	Rituximab	Non-Hodgkin's lymphoma	Nonlinear PK, with clearance decreasing from 38.2 mL/h after the first dose to 9.2 mL/h after the fourth dose; may be caused by wipeout of CD20-positive cells after the initial infusion	[91]
	Tositumomab	Non-Hodgkin's lymphoma	Nonlinear PK; a greater tumor burden, splenomegaly, or bone marrow involvement were associated with increased volume of distribution, faster clearance, and shorter half-life	[92]
	Ofatumumab	Chronic lymphocytic leukemia	Nonlinear PK over a dose range of 500–2000 mg, with clearance decreasing from 65 to 215 mL/h after the first dose to 10–28 mL/h after the fourth dose	[93, 94]
CD22	Moxetumomab pasudotox	Hairy cell leukemia	Nonlinear PK over a dose range of 5–50 μg/kg QODx3, with tumor burden being a significant covariate on clearance in HCL patients	[95]
	Inotuzumab ozogamicin	Acute lymphocytic leukemia/ non-Hodgkin's lymphoma	Nonlinear PK over a dose range of 0.4–2.4 mg/m^2, with decreased clearance after multiple dosing, relative to the first dose	[96]
CD33	Gemtuzumab ozogamicin	Acute myelogenous leukemia	Nonlinear PK after administration of the first 9 mg/m^2 dose; increased concentrations were observed after the second dose and may be due to a decrease in clearance by CD33-positive blast cells	[97]
CD52	Alemtuzumab	B-cell chronic lymphocytic leukemia	Nonlinear PK over a dose range of 7.5–240 mg, with maximal rate of elimination demonstrating covariation with white blood cell counts, and CL decreasing as tumor burden decreased	[98]
c-Met	Onartuzumab	Non-small-cell lung cancer/ gastric cancer	Nonlinear PK over a dose range of 1–4 mg/kg, with clearance decreasing from 15.0 to 7.1 mL/day/kg; linear PK over a dose range of 4–30 mg/kg	[99]
EGFR	Cetuximab	Squamous cell carcinoma of the head and neck	Nonlinear PK over a dose range of 50–500 mg/m^2, with clearance decreasing from 83.7 to 20.0 mL/h/m^2	[100]
	Panitumumab	Colorectal cancer	Nonlinear PK over a dose range of 0.75–2 mg/kg, with clearance decreasing from ~75 to ~4 mL/day/kg	[101]
	Necitumumab	Non-small-cell lung cancer	Nonlinear PK over a dose range of 100–1000 mg/week, with clearance decreasing from 53.2 to 13.9 mL/h after the first dose of cycle 1; and from 40.2 to 1.45 mL/h after the last dose of cycle 1	[102]
	Matuzumab	Solid cancer	Nonlinear PK over a dose range of 400–1600 mg/week	[103]
HER2	Trastuzumab	Breast cancer	Nonlinear PK over a dose range of 10–500 mg once weekly, with terminal half-life increasing from 1.1 to 23 days	[104]
	Trastuzumab emtansine (T-DM1)	Breast cancer	Nonlinear PK at doses <1.2 mg/kg/3 weeks, with clearance decreasing from 21.1–27.8 to 7.13–12.7 mL/d/kg at higher doses	[105]
IL-6R	Tocilizumab	Rheumatoid arthritis	Nonlinear PK over a dose range 0.15–28 mg/kg	[106]

(*Continued*)

TABLE 10.1 (Continued)

Target	Drug Name	Indication[a]	Reported Clinical PK	References
PD-L1	MEDI4736	Non-small-cell lung cancer	Nonlinear PK over a dose range of 0.1–10 mg/kg Q2W and approached linearity at ≥3 mg/kg Q2W	[107]
RANKL	Denosumab	Bone metastases	Nonlinear PK under doses of 120 mg/month with a maximal clearance value of 85 mL/h	[108]
5T4	Naptumomab estafenatox	Renal cell carcinoma	Nonlinear PK when comparing the first (11.6–27.4 μg/kg) and second cycles of therapy (8.75–21.0 μg/kg); clearance increased from 0.11 to 6.39 L/h/kg after the second dose	[109]
IgE (soluble)	Omalizumab	Moderate-to-severe persistent allergic asthma	Nonlinear PK at doses lower than 0.5 mg/kg	[110]
Complement protein C5 (soluble)	Eculizumab	Paroxysmal nocturnal hemoglobinuria	Nonlinear PK over a dosing regimen of 600 mg weekly for the first 4 weeks, followed by 900 mg for the fifth dose 1 week later, then 900 mg every 2 weeks thereafter	[111]

5T4, oncofetal trophoblast protein; c-Met, hepatocyte growth factor receptor; CS1, CD2 subset 1; PD-L1, human programmed death-ligand 1; and RANKL, receptor activator of nuclear factor kappa-B ligand.

[a]Not inclusive of all indications.

production of neutrophils by binding to the G-CSF receptor on neutrophils [116]. The increase in filgrastim clearance after multiple dose administrations was attributed to an increased neutrophil count in the bone marrow and blood that yielded a parallel increase in the total G-CSF receptor density, which increased G-CSF receptor-mediated endocytosis [117]. Another example was shown for a "bispecific T-cell engager" (BiTE) antibody fragment, which can connect a T-cell and a cancer cell by simultaneously binding to the T-cell antigen CD3 and a tumor cell target antigen CD19, subsequently triggering T-cell activation to destroy the tumor cell [118]. During T-cell activation, T-cells will proliferate, which will lead to an increase in CD3 levels, which in turn will impact formation of the CD3–BiTE–CD19 complex and the associated tumor cell killing.

Evaluation of both the dynamics of key target cell depletion/repletion and targeting antigen density modulation through the course of treatment is critical for assessing the PK/PD relationship of biologics. Moreover, the alteration of PK due to target modulation can in turn impact target engagement and downstream PD effects, which may produce time-dependent changes in target concentration. Treatment with efalizumab presents a complex case with homeostatic feedback mechanisms. After administration of efalizumab, there is downregulation of the target CD11a [119]. Since there is negative feedback regulation on the production rate of the CD11a precursor, or CD11a, some subjects showed an overshoot (>100% baseline) of the target antigen after discontinuation of treatment with efalizumab [110, 120]. Another study showed that while efalizumab treatment led to a substantial decrease in CD11a and efalizumab-binding sites (EfaBSs), a parallel increase in EfaBS− and EfaBS+ lymphocytes was also observed [121]. The overall effect on target concentration will be the net result of any change to both target-expressing cell number and target number per cell.

Biologics targeting membrane-bound antigens will not necessarily exhibit nonlinear PK over the dose ranges being studied. This finding could be attributed to relatively low target abundance to systemic drug concentrations, or the targets may be located in tissues with very limited blood capillary permeability [14]. More importantly, for therapeutic biologics at clinically efficacious doses, target levels often have a limited impact on systemic drug PK [122] and/or clinical response. For example, onartuzumab exposure was not affected by hepatocyte growth factor receptor (c-Met) diagnostic status, indicating that target-mediated clearance was not a major factor in the linear PK range [123]. For trastuzumab, tumor burden did not influence the PK of this mAb [124], which is not surprising since a biologic that functions as an antagonist often requires greater than 90% receptor occupancy to inhibit downstream signaling and clinical efficacy. As long as there is no dose-limiting toxicity, the systemic therapeutic concentrations of these biologics usually exceed what is necessary for saturation of the available target [125].

It is important to recognize that even when the systemic PK of a biologic appears to be linear, the tissue site target may not necessarily be fully saturated [14]. For example, *ex vivo* analyses of xenograft tumors from zalutumumab-treated nude mice revealed that relatively high antibody plasma concentrations were required for maximum target saturation in high EGFR-expressing human A431 tumors compared to the concentrations needed for target saturation in low EGFR-expressing human xenograft tumors [126]. Saturation of EGFR in normal tissues does not predict saturation of the target in tumor tissue since local antibody concentrations can be more rapidly reduced by antibody internalization in tumors overexpressing EGFR [126]. Consequently, saturation of the receptor in the tumor may not have been achieved, and the therapeutic effect may be impacted [126]. In addition, the coexistence of shed antigens in different physiological compartments of the body (e.g., blood and interstitial spaces) can influence the binding of a drug to the pharmacological (membrane-bound) target and create a complex scenario where receptor occupancy of the membrane-bound target can be affected by levels of the soluble receptor, the drug concentration in plasma and the interstitial spaces, the relative potency of the drug for the soluble and membrane-bound receptor, and differences in the binding kinetics of the soluble and membrane-bound receptors [127].

10.3.2 Biologics against Soluble Targets

Soluble ligands such as cytokines or chemokines are an important class of targets for biologics [128, 129]. For biotherapeutics directed against such soluble ligands, target engagement is reflected by the interplay between drug and the proportion of free to bound target. As a result, therapeutic efficacy is theoretically driven by the magnitude and duration of the reduction in free target *in vivo* [130–132], and can be affected by the "quasi-equilibrium" between drug and bound target [130, 133, 134]. Following dosing with a longer half-life therapeutic biologic, such as a mAbs, an accumulation of bound target is often observed [130–132]. The magnitude of the increase in total target (free + bound) concentrations *in vivo* will depend on target synthesis and clearance rates as well as the elimination rate of the drug–target complex [133].

Therapeutic biologics targeting soluble antigens are more likely to exhibit linear PK [16]. For example, linear clearance was observed for rilonacept (50 μg/kg to 2000 mg [135, 136]), bevacizumab (0.3 and 10 mg/kg [137]), infliximab (1–20 mg/kg [69, 138]), adalimumab (0.5–10 mg/kg [139]), and rilotumumab (0.5–20 mg/kg [140]). One reason for this observed PK is that many of the soluble targets with significant therapeutic interest, for example, TNFα, IFNα, VEGF, and IL-6, have low endogenous levels [16, 128, 129]. Another reason is that therapeutic biologics that bind to a soluble target may exhibit less dramatic differences in elimination compared to those that bind to a membrane-bound target. Of note, not all biologics targeting soluble antigens exhibit linear PK. One notable example is omalizumab, a mAb targeting a relatively abundant soluble target, immunoglobulin E (IgE). Omalizumab exhibits dose-dependent nonlinear elimination related to the formation of large complexes with IgE [133, 141].

Soluble targets (e.g., TNFα, IgE, IL-6, IL-17, and IL-23) are often elevated in various disease conditions and therapeutic intervention can modulate the production of soluble targets. For example, a preclinical mechanism-based PK/PD model for siltuximab, an antihuman IL-6 mAb, suggested that a sustained reduction of IL-6 by siltuximab may lead to an increase in the production of IL-6 [134]. In contrast, prolonged treatment with omalizumab may decrease IgE synthesis [142], presumably mediated through the specific targeting of membrane IgE-bearing B-cells that subsequently induces a state of immune unresponsiveness [143]. Changes in soluble target production usually have limited impact on the PK of therapeutic biologics, but these changes could have a more profound impact on the treatment response.

10.3.3 When Targets Exist as Both Membrane-Bound and Soluble

Many targets can exist as both membrane-bound and soluble forms. For example, c-Met is a membrane-bound receptor, but the extracellular domain (ECD) of the MET receptor can be proteolytically cleaved and shed into circulation in a soluble form (s-Met) [144]. The shed soluble targets are generally not the intended therapeutic targets, but the presence of these ligands can be an indication of membrane-bound targets and therefore potentially useful biomarkers.

The levels of circulating targets can be influenced by disease, and significant attention has been given to the measurement of shed antigen concentrations in blood and the correlation of these concentrations with clinical response [145–147]. Higher plasma levels (2.21 μg/mL) of shed HER2 ectodomain (ECD^{HER2}) have been detected in cancer patients compared to those (<15 ng/mL) in healthy subjects [145, 148]. Similarly, higher levels (up to 15 μmol/L) of circulating CD20 have been reported in patients with chronic lymphocytic leukemia (CLL) and non-Hodgkin's lymphoma (NHL) compared to levels (121–547 nmol/L) in healthy subjects [149, 150]. Interestingly, an ECD^{HER2} plasma level of 500 ng/mL was set as a cutoff value for the stratification and analysis of data from a study showing that the shed ECD^{HER2} level in serum was significantly associated with trastuzumab PK (e.g., patients with any measurable shed ECD^{HER2} serum level had lower mean trough trastuzumab concentrations compared to patients without measurable circulating ECD^{HER2}) and clinical outcome (e.g., patients with disease progression showed a significant increase in shed ECD^{HER2} serum concentrations over time) [146].

Circulating shed targets can also serve as a "target sink" and affect the ability of drugs to bind to the intended targets. Junghans et al. reported that soluble CD25 (soluble T-cell activation antigen (sTAC)) can block daclizumab-binding sites and diminish antibody binding. This study concluded that the *in vivo* activity of daclizumab was inversely correlated with sTAC concentrations [151]. When tumor antigens are shed, high circulating levels of these antigens may affect the distribution of drugs to the tumor site. Moreover, the concentration of shed antigen in the tumor interstitial spaces could be considerably higher than that in the blood, and further compromise the ability of drugs to bind to the intended membrane-bound target [152].

Targets that primarily exist in a soluble form can also have precursors that are membrane-anchored, for example, TNFα and transforming growth factor alpha [153, 154]. While expressed at lower levels, such membrane-anchored precursors may actually contribute to the therapeutic effect, despite having limited impact on drug PK. For example, the role of membrane-anchored TNFα (mTNFα) has been suggested to be distinct in different disease indications, that is, inhibition of both the soluble TNFα and mTNFα was needed for efficacy in a mouse IBD model; however, in RA, inhibition of mTNFα did not appear to be required [155].

10.4 CORRELATION BETWEEN THE PK OF THERAPEUTIC BIOLOGICS AND TREATMENT RESPONSE

As discussed previously, target physiology and many other factors can affect the PK of therapeutic biologics and subsequently impact the treatment response; however, there may not always be reliable methods to fully assess the status of the therapeutic target *in vivo*. Biologics can exhibit highly variable trough levels, especially in cancer patients, even at doses that saturate the target. Following a loading dose of 400 mg/m^2 and 250 mg/m^2 weekly dosing with cetuximab in patients with advanced solid tumors of epithelial origin, although the mean trough value (54,800 ng/mL) was above the 13,000 ng/mL K_m value (the level of cetuximab required for 50% EGFR occupancy) determined by a population PK model, observed trough levels in some patients fell below K_m [100, 156]. This finding indicated that a 250 mg/m^2 weekly dose may be insufficient for receptor saturation in some patients.

The reason(s) for this PK variability is not always apparent. Interestingly, even though no apparent relationship between any PK parameter and target binding (the maximum EGFR membrane-staining intensity in tumor cells) was observed for panitumumab and cetuximab [157, 158], PK characteristics were found to be related to patient response for a number of mAbs targeting membrane-bound tumor antigens. As shown in Figure 10.2, higher trough drug levels were associated with higher clinical responses for at least six mAbs in various disease indications. For cetuximab, a correlation between monotherapy trough levels and antitumor response was reported for patients with advanced epithelial malignancies [100]. In a study of 96 metastatic colorectal cancer patients receiving cetuximab at a loading dose of 400 mg/m^2 followed by weekly infusions of 250 mg/m^2, the median progression-free survival (PFS) of patients who had a residual cetuximab concentration on day 14 below the median value was 3.3 months compared to 7.8 months for the other patients (p=0.004) [168], despite clinical data showing neither response rate nor PFS were correlated with the intensity and extent of tumor EGFR staining. Although the underlying mechanisms for these observations are not entirely clear, they may be related to the challenges of evaluating target expression levels in cancer patients, the difficulty in evaluating tumor burden in clinical practice or that some other target-independent patient characteristics could impact both mAb PK and the treatment response [169, 170]. Regardless of the cause, PK characteristics alone may still provide useful information about the treatment response.

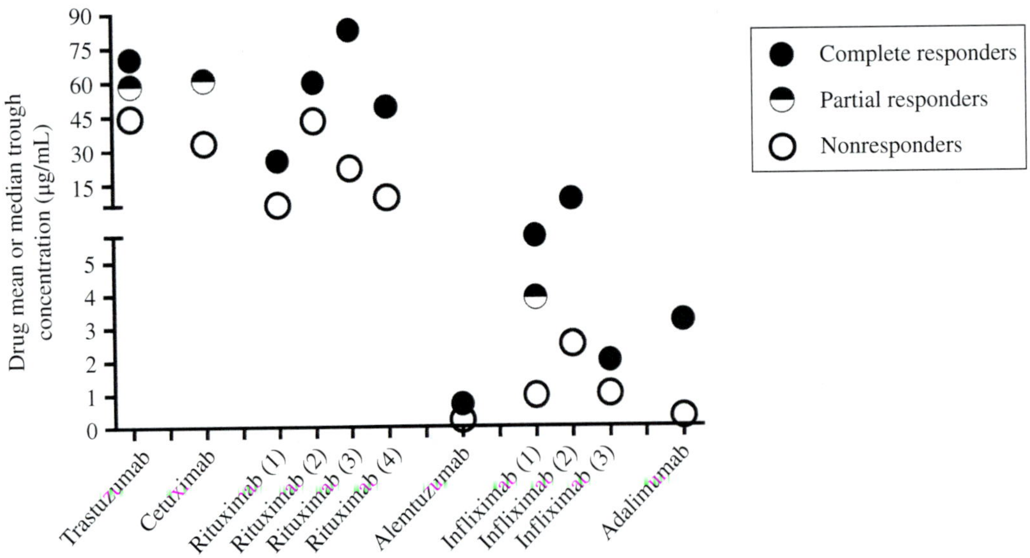

FIGURE 10.2 Mean or median drug trough concentration in relation to patient response for six mAbs: trastuzumab (anti-HER2) [159], cetuximab (anti-EGFR) [99], rituximab (anti-CD20) (1) [89], rituximab (2) [160], rituximab (3) [161], rituximab (4) [162], alemtuzumab (anti-CD52) [163], infliximab (anti-TNFα) (1) [164], infliximab (2) [165], infliximab (3) [166], and adalimumab (anti-TNFα) [167].

A PK assessment could also be highly informative when extrapolating drug effects for different disease indications. Trastuzumab PK exposure in mGC patients was reported to be approximately 30–40% lower than previously stated for patients with metastatic breast cancer receiving the same dosing regimen [74], and mGC patients with lower trastuzumab concentrations were associated with shorter overall survival [171]. Similarly, mGC patients exhibited approximately 50% lower bevacizumab exposure [159] and approximately 37% lower pertuzumab trough concentrations than patients with breast cancer [160]. For trastuzumab, initial studies suggest it is unlikely that the PK differences can be attributed to the target (e.g., similar shed antigen, target expression, and tumor size in mGC and breast cancer patients), but may be more likely due to systemic inflammation and/or gastric protein leakage [161]. The underlying mechanism for faster mAb clearance in mGC patients warrants further research [124, 159].

10.5 OTHER PATIENT CHARACTERISTICS THAT CAN IMPACT THE TREATMENT RESPONSE OF THERAPEUTIC BIOLOGICS

Although the interplay between drug and targets is critical, there are many other important patient characteristics that can impact the treatment response. For example, patient response to a biologic therapy can be significantly impacted by target expression levels, and patient genetic heterogeneity (polymorphism, mutation status, and gene copy number). The assessment of HER2 amplification by fluorescence *in situ* hybridization (FISH) provides a classic example of target expression levels that are used to select patients for trastuzumab treatment. Based on results from 765 patients in three clinical trials, a clinical benefit from trastuzumab therapy appeared to be restricted to patients with HER2 FISH-positive metastatic breast cancer [162]. The underlying biology and pharmacology facilitated the development of a diagnostic test that would identify the 15–20% of patients who would be eligible for trastuzumab therapy to treat their HER2-expressing invasive breast cancers. To our knowledge, it is the first example of the use of patient stratification to enable clinical development in a population with a small responder rate. Other examples are available with the anti-EGFR mAbs. Results from 1121 of the patients participating in the First-Line Erbitux in lung cancer (FLEX) study showed that 31% had high tumor EGFR expression (defined as an immunohistochemistry (IHC) score ≥200 on a scale of 0–300). Compared to patients treated with chemotherapy alone, overall survival was significantly prolonged when cetuximab was added to the chemotherapy regimen in this group (median 12.0 vs 9.6 months; hazard ratio 0.73; $p=0.011$). In contrast, there was no survival benefit when patients with low EGFR expression (IHC score <200) were treated with cetuximab [163]. In line with these clinical findings, cetuximab activity in non-small-cell lung cancer (NSCLC) patient-derived xenograft models ($n=45$) was only demonstrated in tumors that expressed high levels (IHC score of ≥200) of EGFR [164]. In patients treated with panitumumab, responders showed a slightly greater degree of EGFR tumor membrane staining (percentage of cells with positive staining: 94% for the responders and 72% for nonresponders) [165].

Patient genetic heterogeneity is another important factor to consider for treatment response. Cetuximab or panitumumab are only effective in 10–20% of unselected metastatic colorectal cancer patients. Since Kirsten rat sarcoma viral oncogene homolog (KRAS) is a downstream component of the EGFR signaling pathway and cells with mutant KRAS do not respond to anti-EGFR therapies, knowledge of the KRAS mutational status is a key predictor of tumor suitability for anti-EGFR therapy [166, 167]. The presence of a B-Raf proto-oncogene, serine/threonine kinase (BRAF) wild-type is also required for a response to panitumumab or cetuximab treatment in metastatic colorectal cancer [172]. Inevitably, all metastatic colorectal cancer patients eventually develop resistance to anti-EGFR mAbs that occurs through emergence of KRAS mutations in approximately 50% of the cases. The amplification of the MET proto-oncogene is associated with acquired resistance in tumors that do not develop KRAS mutations during anti-EGFR therapy [173]. Gene copy number can be another important predictive factor. With an overall accuracy of 75.9%, a number ≥2.47 EGFR copies/nucleus emerged as the best cutoff value to discriminate responders from nonresponders to panitumumab therapy [174], and these results were generally consistent with an earlier cohort study evaluating EGFR copy number in tumors from 31 patients with metastatic colorectal cancer [175]. Interestingly, in colorectal cancer cell lines, the concentration of cetuximab that completely inhibited proliferation of cells with amplified EGFR copy number did not affect proliferation of cells with unamplified EGFR [175].

The patient response to mAb therapy can be potentially impacted by the Fc gamma receptor (FcγR) genotypes as well. ADCC is an important mechanism for IgG1 mAbs to eliminate antigen-expressing target cells, and is mediated by the Fc domain of IgG that binds to the FcγRs on leukocytes [176]. Certain single-nucleotide polymorphisms (SNPs) in the coding regions of the FcγRIIA and FcγRIIIA genes can impact the binding affinity of FcγR for human IgGs [177]. The SNPs in human FcγRIIA (H131R) and FcγRIIIA (V158F) have been correlated with clinical responses to trastuzumab, rituximab, and cetuximab [178]. In a 2008 report, patients with the high affinity allele for either FcγRIIA (131H/H) or FcγRIIIA (158V/V) had significantly better objective response rates and PFS with trastuzumab therapy than patients with neither genotype (PFS estimates were 30.3 and 12.8 months respectively; $p=0.01$) [179]. Another retrospective, nonrandomized study with 35

HER2-positive breast cancer patients in both neoadjuvant and metastatic settings also demonstrated that the FcγRIIA 131 H/H genotype significantly correlated with clinical response, while the FcγRIIIA158 V/V genotype did not correlate with clinical response [180]; however, the largest analysis in this area did not show any association between FcγRIIIA/IIA genotypes and clinical outcome in trastuzumab-treated HER2-positive breast cancer patients [181].

A beneficial effect of the FcγRIIIA 158-V/V genotype was also reported for rituximab therapy in patients with follicular lymphoma and diffuse large B-cell lymphoma (DLBCL), but larger studies examining the effects of FcγR polymorphisms on the outcomes of follicular lymphoma and CLL patients treated with rituximab and chemotherapy showed no association between FcγR genotype and either response rate or outcome [178]. Similarly for cetuximab, some studies showed that patients with the FcγRIIIA 158-V alleles were more likely to have a response to cetuximab treatment, while other studies reported that the 158-V/V allele was not beneficial [178]. These inconsistent findings are likely related to multiple confounding factors such as inconsistencies in tumor type, the combinations of cytotoxic agents used, the clinical setting, and the clinical benefit parameters measured in different studies [178].

Another example of patient characteristics affecting Fc-mediated effector cell function is related to the observation that some mAbs are susceptible to proteolytic breakdown by tumor-associated (e.g., MMPs) and microbial proteases (e.g., GluV8 of *Staphylococcus aureus* and IdeS of *Streptococcus pyogenes*) [182, 183]. These proteases cleave mAbs at the hinge region between the antigen-binding region and the Fc fragment resulting in a loss of Fc-mediated cell-killing functions without a concomitant loss of antigen-binding capability or circulating antibody half-life [183, 184].

Baseline tumor size, the number of target lesions and the number of lymph node lesions were identified as important covariates in the treatment response to pembrolizumab, a potent antibody against programmed death receptor-1 (PD-1) [185]. Similarly, across multiple cancer types, clinical responses for programmed death-ligand 1 (PD-L1) inhibition by the engineered humanized antibody MPDL3280A were observed in patients with tumors expressing high levels of PD-L1, especially when PD-L1 was expressed by tumor-infiltrating immune cells [186]. Target characteristics may also have an impact on drug exposure and the effect of ADCs. For example, modeling has shown that intracellular gemtuzumab ozogamicin (GO) exposure is linked to a high CD33 antigen production rate and a low ozogamicin efflux through P-glycoprotein. Intracellular exposure was also linked to the initial number of leukemic blast cells, with even a modest reduction in blast burden increasing intracellular drug exposure [187]. A recent study [188] suggested that genetic variations in CD33 could impact the clinical outcome of GO-based therapy in pediatric acute myeloid leukemias (AMLs). If confirmed by more studies, CD33 SNPs could serve as prognostic markers in future prospective trials to assist the selection of patient subsets most suitable for therapies containing GO and possibly other CD33-targeted immunotoxins [188].

Pfreundschuh et al. have shown that treatment with rituximab improves the adverse prognostic effect of maximal tumor diameter (MTD). Due to the linear prognostic effect of MTD on outcome in DLBCL, this study established a cutoff point of 10 cm to delineate those patients with bulky disease who may benefit from rituximab therapy [189]. In addition, a CD20 antigen surface threshold level required for effective rituximab-associated, complement-mediated cytotoxicity has been identified, although a direct correlation between CD20 surface expression and rituximab-associated, complement-mediated cytotoxicity was observed only in rituximab-sensitive cell lines [190].

For biologics directed against soluble targets, the basal level of the target may also affect the clinical response. A high BMI negatively influenced clinical response in a study of RA patients treated with infliximab [191]. Since adipose tissue is a metabolically active source of proinflammatory cytokines (e.g., TNF), obese patients with Crohn's disease would have inherently higher TNF levels than patients with normal weight [64].

10.6 THE INTERPLAY BETWEEN DISEASE, TARGET PHYSIOLOGY, AND PK/PD OF THERAPEUTIC BIOLOGICS: CASE EXAMPLES

A few case examples that demonstrate the interplay between disease, target physiology, and PK/PD of therapeutic biologics are presented with more details in this section.

CD20

Rituximab, a B-cell lymphoma drug that targets CD20, provides one of the best characterized cases reflecting the complex interaction between drug target and clinical response. CD20 is a human B-lymphocyte-restricted differentiation antigen located on pre-B and mature B-lymphocytes, but not on plasma cells [192]. CD20-expressing cells are distributed widely throughout the body [115, 141]. Binding of rituximab results in rapid and sustained depletion of circulating and tissue CD20-expressing B-cells in humans and monkeys, primarily through complement-dependent cytotoxicity and ADCC and, to a lesser degree, via cellular apoptosis by sparing IgG-producing plasma cells [115, 141].

In B-cell lymphoma patients, the mean serum rituximab concentration was inversely correlated with measurements of tumor bulk and the number of circulating B-cells at baseline [91]. A strong inverse correlation between target concentration and drug clearance has also been described in patients treated with rituximab for NHL [91]. In NHL patients, covariate analysis revealed that patients with higher CD19+ B-cell counts or tumor burden at baseline had a higher target-mediated clearance [193]. As CD20-expressing cells were depleted, rituximab elimination decreased following multiple infusions [193]. This correlation was also observed in a murine lymphoma model expressing human CD20. Rituximab concentrations were inversely correlated with tumor burden, and drug exposure influenced mouse response and survival [169].

PK variability is recognized as a major factor affecting the clinical response to rituximab [194]. Higher rituximab trough levels were correlated with longer median PFS [91] and higher response rates [115]. The concept of antigenic mass, which takes into account total tumor load and the expression levels of the CD20 target antigen, offers an explanation for the correlation between rituximab plasma concentrations and treatment responses [194, 195]. In adult patients with relapsed B-cell lymphoma expressing the CD20 antigen, 17 treatment responders had a median rituximab serum level of 82.7 μg/mL before the second infusion versus 17 nonresponders who had a median serum level of 21.9 μg/mL ($p=0.029$) [91].

CD25

Daclizumab is an anti-IL-2Rα (CD25 antigen on the surface of T-lymphocytes) mAb employed to provide immunoprophylaxis of acute rejection after organ transplantation. CD25 is upregulated on activated T-lymphocytes in several autoimmune diseases, and in patients with allograft rejection or graft-versus-host disease (GvHD) [196]. CD25 is also overexpressed in several hematological malignancies. Patients undergoing treatment for GvHD would be expected to have a substantially higher concentration of CD25+ T-cells than either patients undergoing renal transplantation or patients administered daclizumab prophylactically for the prevention of GvHD [197]. Not surprisingly, daclizumab exhibited disease-dependent elimination half-life. The half-lives of daclizumab were 79–94, 165.4, and 480 h for patients receiving treatment for GvHD, prophylaxis for GvHD, and for renal transplantation, respectively [197].

In CD25+ leukemia patients, Koon et al. determined a strong inverse correlation between CD25+ cell expression and apparent daclizumab half-life [34]. One patient with a high WBC count and 100% CD25 expression displayed rapid clearance following the first dose administration of daclizumab, and more than 80% of the dose cleared within 48 h. The PK of daclizumab appeared to follow a more typical two-compartment model and the estimated β half-life stabilized at approximately 480 h after repeated dosing [34].

IgE

IgE is the central mediator driving the inflammatory cascade in patients with allergic (IgE-mediated) asthma [198, 199]. A humanized mAb, omalizumab, interrupts the allergic inflammatory cascade by binding to IgE and preventing it from interacting with the high affinity IgE receptor (FcεRI) on mast cells and basophils [200–202]. In patients with moderate-to-severe persistent allergic asthma, the clinical efficacy of omalizumab can be correlated to the ability of the biologic to suppress free IgE [133, 142, 203, 204]. The value of free IgE assessment for this soluble target was supported by the finding that asthma symptoms reemerged when free IgE returned toward baseline after omalizumab treatment cessation [204].

While free IgE concentrations are suppressed during omalizumab treatment, total IgE (free IgE + omalizumab-bound IgE) levels are increased [133]. Total IgE levels should be sustained at steady state, when a balance between IgE synthesis and degradation rates is maintained. In some cases, however, a decrease in total IgE levels was subsequently observed after the initial increase in total IgE levels, even though overall concentrations of omalizumab were maintained at steady state [142]. Using mechanistic PK/PD modeling, the decrease in total IgE levels was attributed to a decrease in the production rate of IgE [142]. Recently, Chan et al., further suggested that omalizumab may decrease human B-cell synthesis of IgE by specifically targeting membrane IgE-bearing B-cells and inducing a state of immune unresponsiveness [143]. Based on these observations, a total IgE level has been proposed as a biomarker to assess whether IgE production in individual patients has been sufficiently downregulated to consider discontinuation of anti-IgE therapy [142].

CD11b

UK-279,276 is a recombinant glycoprotein and a selective antagonist of CD11b/CD18 integrin on neutrophils. As a result, the biologic blocks the infiltration of activated neutrophils into the site of infarction in preclinical models of acute stroke [205, 206]. UK-279,276 displayed nonlinear PK in doses ranging from 0.06 to 1.5 mg/kg and the duration of CD11b saturation was dose (concentration) dependent [207]. An acute stroke leads to the proliferation of neutrophils and an upregulation of CD11b, altering the PK/PD in patients compared to healthy subjects. A complex mechanistic model characterizing neutrophil maturation and proliferation, CD11b upregulation, and three clearance pathways for UK-279,276, including CD11b-mediated elimination, provided an effective bridge between healthy subjects and patients [205].

EPO

EPO is the primary hormone that stimulates erythroid proliferation and differentiation through interaction with a cell-surface receptor (erythropoietin receptor (EpoR)) on erythroid progenitor cells located primarily in the bone marrow [208–210]. In human and animal studies, the clearance of EPO was associated with dynamic changes in the EpoR pool [208, 211–214]. Previous clinical studies have demonstrated two distinctive PK behaviors of EPO: nonlinear PK and changes in PK following a period of treatment with rHuEPO [214, 215]. The output from PK/PD model that accounted for receptor-mediated endocytosis via EpoR as the primary mechanism for the nonlinear disposition of rHuEPO indicated that only 3.1% of EpoR were occupied at baseline conditions in healthy subjects, and receptor occupancy increased dose-dependently [213]. Following repeated IV bolus tracer doses of biologically active ^{125}I-rHuEPO in sheep, phlebotomy (blood removal from circulation) produced a rapid increase in EPO plasma concentrations, which coincided with a reduction in EPO clearance, presumably due to saturation of EpoR [214]. As plasma EPO returned toward baseline levels, a subsequent increase in EPO clearance was noted. Based on a model that included positive feedback control of the EpoR pool, the initial reduction in EPO plasma clearance was attributed to a transient saturation of EpoR from the phlebotomy-induced stimulation of EPO and expansion of the EpoR pool to compensate for receptor loss and to adjust to a greater need for EpoR progenitor cells to restore hemoglobin concentrations to normal levels [214].

10.7 CONCLUDING REMARKS

Many factors may alter the PK of therapeutic biologics and subsequently affect clinical responses to targeted diseases. The identification and mechanistic understanding of influential covariates in the early stages of biologic drug development, preferably in the preclinical PK/PD stage, would be valuable for characterization of exposure–response relationships and ultimately to inform the selection of dose level/dose regimen for clinical development. Nevertheless, prospective investigations of this sort are not often seen, and in many cases, important covariates (such as disease type, disease burden, target physiology, comorbidity, and immunogenicity) are identified via *post hoc* population PK analyses or via cross-study comparisons. Therapeutically useful prognostic factors to better inform the design of Phase IIb/III studies remain mostly absent for biologics. In recent years, pharmaceutical scientists have come to recognize the need and importance of mechanistic PK/PD investigations that evaluate the sites of action for therapeutic proteins in associated disease models. The efforts and investments in translational PK/PD research and identification of important factors affecting ADME and PK/PD of biologics would no doubt facilitate the effective development of safe and efficacious biologics for patients, and shorten the development cycle from bench to bedside.

ACKNOWLEDGMENTS

The authors thank Dr. Frank-Peter Theil for his insightful scientific discussions. The authors also thank Dr. Kenneth Graham for his excellent editorial assistance for this chapter.

REFERENCES

[1] Porter CJ, Charman SA. Lymphatic transport of proteins after subcutaneous administration. J Pharm Sci 2000;89:297–310.

[2] McDonald TA, Zepeda ML, Tomlinson MJ, Bee WH, Ivens IA. Subcutaneous administration of biotherapeutics: current experience in animal models. Curr Opin Mol Ther 2010;12:461–470.

[3] Wang W, Wang EQ, Balthasar JP. Monoclonal antibody pharmacokinetics and pharmacodynamics. Clin Pharmacol Ther 2008;84:548–558.

[4] Zhou L, Hoofring SA, Wu Y, Vu T, Ma P, Swanson SJ, Chirmule N, Starcevic M. Stratification of antibody-positive subjects by antibody level reveals an impact of immunogenicity on pharmacokinetics. AAPS J 2013;15:30–40.

[5] Garg A, Balthasar JP. Physiologically-based pharmacokinetic (PBPK) model to predict IgG tissue kinetics in wild-type and FcRn-knockout mice. J Pharmacokinet Pharmacodyn 2007;34:687–709.

[6] Abuqayyas L, Balthasar JP. Application of PBPK modeling to predict monoclonal antibody disposition in plasma and tissues in mouse models of human colorectal cancer. J Pharmacokinet Pharmacodyn 2012;39:683–710.

[7] Zheng S, Tasnif Y, Hebert MF, Davis CL, Shitara Y, Calamia JC, Lin YS, Shen DD, Thummel KE. CYP3A5 gene variation influences cyclosporine a metabolite formation and renal cyclosporine disposition. Transplantation 2013;95(6):821–827. Publish Ahead of Print: 10.1097/TP.0b013e31827e6ad9.

[8] Tang L, Persky AM, Hochhaus G, Meibohm B. Pharmacokinetic aspects of biotechnology products. J Pharm Sci 2004;93:2184–2204.

[9] Meibohm B, Zhou H. Characterizing the impact of renal impairment on the clinical pharmacology of biologics. J Clin Pharmacol 2012;52:54S–62S.

[10] Roopenian DC, Akilesh S. FcRn: the neonatal Fc receptor comes of age. Nat Rev Immunol 2007;7:715–725.

[11] Rosenberg AS. Immunogenicity of biological therapeutics: a hierarchy of concerns. Dev Biol (Basel) 2003;112:15–21.

[12] Levy G. Pharmacologic target-mediated drug disposition. Clin Pharmacol Ther 1994;56:248–252.

[13] Mager DE, Jusko WJ. General pharmacokinetic model for drugs exhibiting target-mediated drug disposition. J Pharmacokinet Pharmacodyn 2001;28:507–532.

[14] Cao Y, Jusko WJ. Incorporating target-mediated drug disposition in a minimal physiologically-based pharmacokinetic model for monoclonal antibodies. J Pharmacokinet Pharmacodyn 2014;41:375–387.

[15] Tabrizi MA, Tseng CML, Roskos LK. Elimination mechanisms of therapeutic monoclonal antibodies. Drug Discov Today 2006;11:81–88.

[16] Dostalek M, Gardner I, Gurbaxani BM, Rose RH, Chetty M. Pharmacokinetics, pharmacodynamics and physiologically-based pharmacokinetic modelling of monoclonal antibodies. Clin Pharmacokinet 2013;52:83–124.

[17] Frisbee JC. Hypertension-independent microvascular rarefaction in the obese Zucker rat model of the metabolic syndrome. Microcirculation 2005;12:383–392.

[18] Summers LK, Samra JS, Humphreys SM, Morris RJ, Frayn KN. Subcutaneous abdominal adipose tissue blood flow: variation within and between subjects and relationship to obesity. Clin Sci (Lond) 1996;91:679–683.

[19] Villela NR, Kramer-Aguiar LG, Bottino DA, Wiernsperger N, Bouskela E. Metabolic disturbances linked to obesity: the role of impaired tissue perfusion. Arq Bras Endocrinol Metabol 2009;53:238–245.

[20] Vora JP, Burch A, Peters JR, Owens DR. Relationship between absorption of radiolabeled soluble insulin, subcutaneous blood flow, and anthropometry. Diabetes Care 1992;15:1484–1493.

[21] MacDougall IC, Jones JM, Robinson MI, Miles JB, Coles GA, Williams JD. Subcutaneous erythropoietin therapy: comparison of three different sites of injection. Contrib Nephrol 1991;88:152–156. ; discussion 157–158.

[22] Silva M, Poo J, Wagner F, Jackson M, Cutler D, Grace M, Bordens R, Cullen C, Harvey J, Laughlin M. A randomised trial to compare the pharmacokinetic, pharmacodynamic, and antiviral effects of peginterferon alfa-2b and peginterferon alfa-2a in patients with chronic hepatitis C (COMPARE). J Hepatol 2006;45:204–213.

[23] Olsson-Gisleskog P, Jacqmin P, Perez-Ruixo JJ. Population pharmacokinetics meta-analysis of recombinant human erythropoietin in healthy subjects. Clin Pharmacokinet 2007;46:159–173.

[24] Weitman ES, Aschen SZ, Farias-Eisner G, Albano N, Cuzzone DA, Ghanta S, Zampell JC, Thorek D, Mehrara BJ. Obesity impairs lymphatic fluid transport and dendritic cell migration to lymph nodes. PLoS One 2013;8:e70703.

[25] Blum KS, Karaman S, Proulx ST, Ochsenbein AM, Luciani P, Leroux JC, Wolfrum C, Detmar M. Chronic high-fat diet impairs collecting lymphatic vessel function in mice. PLoS One 2014;9:e94713.

[26] Wang W, Chen N, Shen X, Cunningham P, Fauty S, Michel K, Wang B, Hong X, Adreani C, Nunes CN, Johnson CV, Yin KC, Groff M, Zou Y, Liu L, Hamuro L, Prueksaritanont T. Lymphatic transport and catabolism of therapeutic proteins after subcutaneous administration to rats and dogs. Drug Metab Dispos 2012;40:952–962.

[27] Agoram B, Sutjandra L, Sullivan JT. Population pharmacokinetics of darbepoetin alfa in healthy subjects. Br J Clin Pharmacol 2007;63:41–52.

[28] Zhao L, Ji P, Li Z, Roy P, Sahajwalla CG. The antibody drug absorption following subcutaneous or intramuscular administration and its mathematical description by coupling physiologically based absorption process with the conventional compartment pharmacokinetic model. J Clin Pharmacol 2013;53:314–325.

[29] Mrsny RJ, Daugherty AL. *Proteins and Peptides: Pharmacokinetic, Pharmacodynamic, and Metabolic Outcomes*. New York: Informa Healthcare; 2009. p 80–105.

[30] Riley KN, Herman IM. Collagenase promotes the cellular responses to injury and wound healing *in vivo*. J Burns Wounds 2005;4:e8.

[31] Gurtner GC, Werner S, Barrandon Y, Longaker MT. Wound repair and regeneration. Nature 2008;453:314–321.

[32] Richter WF, Bhansali SG, Morris ME. Mechanistic determinants of biotherapeutics absorption following SC administration. AAPS J 2012;14:559–570.

[33] Abbott NJ. Inflammatory mediators and modulation of blood-brain barrier permeability. Cell Mol Neurobiol 2000;20:131–147.

[34] Koon HB, Severy P, Hagg DS, Butler K, Hill T, Jones AG, Waldmann TA, Junghans RP. Antileukemic effect of daclizumab in CD25 high-expressing leukemias and impact of tumor burden on antibody dosing. Leuk Res 2006;30:190–203.

[35] Oosterwijk E, Bander NH, Divgi CR, Welt S, Wakka JC, Finn RD, Carswell EA, Larson SM, Warnaar SO, Fleuren GJ, Oettgen HF, Old LJ. Antibody localization in human renal cell carcinoma: a phase I study of monoclonal antibody G250. J Clin Oncol 1993;11:738–750.

[36] Welt S, Divgi CR, Real FX, Yeh SD, Garin-Chesa P, Finstad CL, Sakamoto J, Cohen A, Sigurdson ER, Kemeny N, Carswell EA, Oettgen HF, Old LJ. Quantitative analysis of antibody localization in human metastatic colon cancer: a phase I study of monoclonal antibody A33. J Clin Oncol 1990;8:1894–1906.

[37] Wittrup KD, Thurber GM, Schmidt MM, Rhoden JJ. Practical theoretic guidance for the design of tumor-targeting agents. Methods Enzymol 2012;503:255–268.

[38] Cecchelli R, Berezowski V, Lundquist S, Culot M, Renftel M, Dehouck MP, Fenart L. Modelling of the blood-brain barrier in drug discovery and development. Nat Rev Drug Discov 2007;6:650–661.

[39] Lee SW, Kim WJ, Park JA, Choi YK, Kwon YW, Kim KW. Blood-brain barrier interfaces and brain tumors. Arch Pharm Res 2006;29:265–275.

[40] Lossinsky AS, Shivers RR. Structural pathways for macromolecular and cellular transport across the blood-brain barrier during inflammatory conditions. Review. Histol Histopathol 2004;19:535–564.

[41] Vredenburgh JJ, Desjardins A, Herndon JE II, Marcello J, Reardon DA, Quinn JA, Rich JN, Sathornsumetee S, Gururangan S, Sampson J, Wagner M, Bailey L, Bigner DD, Friedman AH, Friedman HS. Bevacizumab plus irinotecan in recurrent glioblastoma multiforme. J Clin Oncol 2007;25:4722–4729.

[42] Desjardins A, Barboriak DP, Herndon JE II, Marcello J, Reardon DA, Quinn JA, Rich JN, Sathornsumetee S, Friedman HS, Vredenburgh JJ. Effect of bevacizumab (BEV) and irinotecan (CPT-11) on dynamic contrast-enhanced magnetic resonance imaging (DCE-MRI) in glioblastoma

(GBM) patients [abstract]. J Clin Oncol (Meeting abstracts) 2008;26(15 Suppl):2026. Available at http://meeting.ascopubs.org/cgi/content/short/26/15_suppl/2026#otherarticles. Accessed 2015 Jul 6.

[43] Jain RK. Normalization of tumor vasculature: an emerging concept in antiangiogenic therapy. Science 2005;307:58–62.

[44] Pastuskovas CV, Mundo EE, Williams SP, Nayak TK, Ho J, Ulufatu S, Clark S, Ross S, Cheng E, Parsons-Reponte K, Cain G, Van Hoy M, Majidy N, Bheddah S, dela Cruz Chuh J, Kozak KR, Lewin-Koh N, Nauka P, Bumbaca D, Sliwkowski M, Tibbitts J, Theil FP, Fielder PJ, Khawli LA, Boswell CA. Effects of anti-VEGF on pharmacokinetics, biodistribution, and tumor penetration of trastuzumab in a preclinical breast cancer model. Mol Cancer Ther 2012;11:752–762.

[45] Abuqayyas L, Balthasar JP. Pharmacokinetic mAb-mAb interaction: anti-VEGF mAb decreases the distribution of anti-CEA mAb into colorectal tumor xenografts. AAPS J 2012;14:445–455.

[46] Yang J, Shord S, Zhao H, Men Y, Rahman A. Are hepatic impairment studies necessary for therapeutic proteins? Clin Ther 2013;35:1444–1451.

[47] Zhao L, Ren TH, Wang DD. Clinical pharmacology considerations in biologics development. Acta Pharmacol Sin 2012;33:1339–1347.

[48] Han TH, Zhao B. Absorption, distribution, metabolism, and excretion considerations for the development of antibody-drug conjugates. Drug Metab Dispos 2014;42:1914–1920.

[49] Mylotarg® (gemtuzumab ozogamicin for Injection) Product's Label. 2010. Available at http://labeling.pfizer.com/show labeling.aspx?id=119. Accessed 2015 Jun 5.

[50] Yang BB, Baughman S, Sullivan JT. Pharmacokinetics of anakinra in subjects with different levels of renal function. Clin Pharmacol Ther 2003;74:85–94.

[51] Andersen SR, Lambrecht LJ, Swan SK, Cutler DL, Radwanski E, Affrime MB, Garaud JJ. Disposition of recombinant human interleukin-10 in subjects with various degrees of renal function. J Clin Pharmacol 1999;39:1015–1020.

[52] Klitgaard T, Nielsen JN, Skettrup MP, Harper A, Lange M. Population pharmacokinetic model for human growth hormone in adult patients in chronic dialysis compared with healthy subjects. Growth Horm IGF Res 2009;19:463–470.

[53] Kindler J, Eckardt KU, Ehmer B, Jandeleit K, Kurtz A, Schreiber A, Scigalla P, Sieberth HG. Single-dose pharmacokinetics of recombinant human erythropoietin in patients with various degrees of renal failure. Nephrol Dial Transplant 1989;4:345–349.

[54] Yang BB, Kido A, Salfi M, Swan S, Sullivan JT. Pharmacokinetics and pharmacodynamics of pegfilgrastim in subjects with various degrees of renal function. J Clin Pharmacol 2008;48:1025–1031.

[55] Garnier-Viougeat N, Rixe O, Paintaud G, Ternant D, Degenne D, Mouawad R, Deray G, Izzedine H. Pharmacokinetics of bevacizumab in haemodialysis. Nephrol Dial Transplant 2007;22:975.

[56] Thariat J, Azzopardi N, Peyrade F, Launay-Vacher V, Santini J, Lecomte T, Etienne-Grimaldi MC, Paintaud G, Milano G. Cetuximab pharmacokinetics in end-stage kidney disease under hemodialysis. J Clin Oncol 2008;26:4223–4225.

[57] Jillella AP, Dainer PM, Kallab AM, Ustun C. Treatment of a patient with end-stage renal disease with Rituximab: pharmacokinetic evaluation suggests Rituximab is not eliminated by hemodialysis. Am J Hematol 2002;71:219–222.

[58] Micallef RA, Barrett-Lee PJ, Donovan K, Ashraf M, Williams L. Trastuzumab in patients on haemodialysis for renal failure. Clin Oncol (R Coll Radiol) 2007;19:559.

[59] Zhu Y, Hu C, Lu M, Liao S, Marini JC, Yohrling J, Yeilding N, Davis HM, Zhou H. Population pharmacokinetic modeling of ustekinumab, a human monoclonal antibody targeting IL-12/23p40, in patients with moderate to severe plaque psoriasis. J Clin Pharmacol 2009;49:162–175.

[60] Bakoush O, Tencer J, Tapia J, Rippe B, Torffvit O. Higher urinary IgM excretion in type 2 diabetic nephropathy compared to type 1 diabetic nephropathy. Kidney Int 2002;61:203–208.

[61] Engler FA, Zheng B, Balthasar JP. Investigation of the influence of nephropathy on monoclonal antibody disposition: a pharmacokinetic study in a mouse model of diabetic nephropathy. Pharm Res 2014;31:1185–1193.

[62] Cameron JS. Lupus nephritis. J Am Soc Nephrol 1999;10: 413–424.

[63] Struemper H, Chen C, Cai W. Population pharmacokinetics of belimumab following intravenous administration in patients with systemic lupus erythematosus. J Clin Pharmacol 2013;53:711–720.

[64] Ordas I, Mould DR, Feagan BG, Sandborn WJ. Anti-TNF monoclonal antibodies in inflammatory bowel disease: pharmacokinetics-based dosing paradigms. Clin Pharmacol Ther 2012;91:635–646.

[65] Zhu X, Meng G, Dickinson BL, Li X, Mizoguchi E, Miao L, Wang Y, Robert C, Wu B, Smith PD, Lencer WI, Blumberg RS. MHC class I-related neonatal Fc receptor for IgG is functionally expressed in monocytes, intestinal macrophages, and dendritic cells. J Immunol 2001;166:3266–3276.

[66] Fasanmade AA, Adedokun OJ, Ford J, Hernandez D, Johanns J, Hu C, Davis HM, Zhou H. Population pharmacokinetic analysis of infliximab in patients with ulcerative colitis. Eur J Clin Pharmacol 2009;65:1211–1228.

[67] Fasanmade AA, Adedokun OJ, Blank M, Zhou H, Davis HM. Pharmacokinetic properties of infliximab in children and adults with Crohn's disease: a retrospective analysis of data from 2 phase III clinical trials. Clin Ther 2011;33:946–964.

[68] Mori S. A relationship between pharmacokinetics (PK) and the efficacy of infliximab for patients with rheumatoid arthritis: characterization of infliximab-resistant cases and PK-based modified therapy. Mod Rheumatol 2007;17:83–91.

[69] REMICADE® (infliximab). 2011. Investigator's brochure, Edition No.12, Release Date: 01 Aug 2011, Janssen Research & Development.

[70] Beeken WL, Busch HJ, Sylwester DL. Intestinal protein loss in Crohn's disease. Gastroenterology 1972;62:207–215.

[71] Kapel N, Meillet D, Favennec L, Magne D, Raichvarg D, Gobert JG. Evaluation of intestinal clearance and faecal excretion of alpha 1-antiproteinase and immunoglobulins during Crohn's disease and ulcerative colitis. Eur J Clin Chem Clin Biochem 1992;30:197–202.

[72] Brandse J, Wildenberg M, de Bruyn J, Wolbink G, Lowenberg M, Ponsioen CY, van den Brink GR, D'Haens GRAM. Fecal loss of infliximab as a cause of lack of response in severe inflammatory bowel disease. Gastroenterology 2013;142(Suppl. 1):S210. DOI: 10.1016/S1873-9946(13)60521-8. Abstract 157.

[73] Louis E, Panes J. Adalimumab in ulcerative colitis: can pharmacodynamics be improved based on pharmacokinetics? Gastroenterology 2012;142:176–178.

[74] European Medicines Agency. 2012. EPAR summary for the public. Available at http://www.ema.europa.eu/docs/en_GB/document_library/EPAR_-_Summary_for_the_public/human/000278/WC500049819.pdf. Accessed 2015 Jun 5.

[75] Kaneshige H. Nonenzymatic glycosylation of serum IgG and its effect on antibody activity in patients with diabetes mellitus. Diabetes 1987;36:822–828.

[76] Kennedy DM, Skillen AW, Self CH. Glycation increases the vascular clearance rate of IgG in mice. Clin Exp Immunol 1993;94:447–451.

[77] Richter WF, Gallati H, Schiller CD. Animal pharmacokinetics of the tumor necrosis factor receptor-immunoglobulin fusion protein lenercept and their extrapolation to humans. Drug Metab Dispos 1999;27:21–25.

[78] Vincent FB, Morand EF, Murphy K, Mackay F, Mariette X, Marcelli C. Antidrug antibodies (ADAb) to tumour necrosis factor (TNF)-specific neutralising agents in chronic inflammatory diseases: a real issue, a clinical perspective. Ann Rheum Dis 2013;72:165–178.

[79] Carrascosa JM, van Doorn MB, Lahfa M, Nestle FO, Jullien D, Prinz JC. Clinical relevance of immunogenicity of biologics in psoriasis: implications for treatment strategies. J Eur Acad Dermatol Venereol 2014;28(11):1424–1430.

[80] Nanda KS, Cheifetz AS, Moss AC. Impact of antibodies to infliximab on clinical outcomes and serum infliximab levels in patients with inflammatory bowel disease (IBD): a meta-analysis. Am J Gastroenterol 2013;108:40–47. ; quiz 48.

[81] Moss AC, Brinks V, Carpenter JF. Review article: immunogenicity of anti-TNF biologics in IBD - the role of patient, product and prescriber factors. Aliment Pharmacol Ther 2013;38:1188–1197.

[82] Batista FD, Harwood NE. The who, how and where of antigen presentation to B cells. Nat Rev Immunol 2009;9:15–27.

[83] Ryff JC. Clinical investigation of the immunogenicity of interferon-alpha 2a. J Interferon Cytokine Res 1997;17(Suppl 1):S29–S33.

[84] Bendtzen K. Anti-TNF-alpha biotherapies: perspectives for evidence-based personalized medicine. Immunotherapy 2012;4:1167–1179.

[85] Maini RN, Breedveld FC, Kalden JR, Smolen JS, Davis D, Macfarlane JD, Antoni C, Leeb B, Elliott MJ, Woody JN, Schaible TF, Feldmann M. Therapeutic efficacy of multiple intravenous infusions of anti-tumor necrosis factor alpha monoclonal antibody combined with low-dose weekly methotrexate in rheumatoid arthritis. Arthritis Rheum 1998;41:1552–1563.

[86] Astermark J, Morado M, Rocino A, van den Berg HM, von Depka M, Gringeri A, Mantovani L, Garrido RP, Schiavoni M, Villar A, Windyga J, EHTSB. Current European practice in immune tolerance induction therapy in patients with haemophilia and inhibitors. Haemophilia 2006;12:363–371.

[87] Sheremata WA, Vollmer TL, Stone LA, Willmer-Hulme AJ, Koller M. A safety and pharmacokinetic study of intravenous natalizumab in patients with MS. Neurology 1999;52:1072–1074.

[88] Zonder JA, Mohrbacher AF, Singhal S, van Rhee F, Bensinger WI, Ding H, Fry J, Afar DE, Singhal AK. A phase 1, multicenter, open-label, dose escalation study of elotuzumab in patients with advanced multiple myeloma. Blood 2012;120:552–559.

[89] Mould DR, Davis CB, Minthorn EA, Kwok DC, Elliott MJ, Luggen ME, Totoritis MC. A population pharmacokinetic-pharmacodynamic analysis of single doses of clenoliximab in patients with rheumatoid arthritis. Clin Pharmacol Ther 1999;66:246–257.

[90] Bauer RJ, Dedrick RL, White ML, Murray MJ, Garovoy MR. Population pharmacokinetics and pharmacodynamics of the anti-CD11a antibody hu1124 in human subjects with psoriasis. J Pharmacokinet Biopharm 1999;27:397–420.

[91] Berinstein NL, Grillo-Lopez AJ, White CA, Bence-Bruckler I, Maloney D, Czuczman M, Green D, Rosenberg J, McLaughlin P, Shen D. Association of serum Rituximab (IDEC-C2B8) concentration and anti-tumor response in the treatment of recurrent low-grade or follicular non-Hodgkin's lymphoma. Ann Oncol 1998;9:995–1001.

[92] BEXXAR (package insert). 2013. Glaxo-Smith-Kline, Research Triangle Park (NC). Available at http://www.gsk.ca/english/docs-pdf/product-monographs/Bexxar.pdf. Accessed 2015 Jun 5.

[93] Coiffier B, Losic N, Ronn BB, Lepretre S, Pedersen LM, Gadeberg O, Frederiksen H, van Oers MH, Wooldridge J, Kloczko J, Holowiecki J, Hellmann A, Walewski J, Robak T, Petersen J. Pharmacokinetics and pharmacokinetic/pharmacodynamic associations of ofatumumab, a human monoclonal CD20 antibody, in patients with relapsed or refractory chronic lymphocytic leukaemia: a phase 1-2 study. Br J Haematol 2010;150:58–71.

[94] Struemper H, Sale M, Patel BR, Ostergaard M, Osterborg A, Wierda WG, Hagenbeek A, Coiffier B, Jewell RC. Population pharmacokinetics of of atumumab in patients with chronic lymphocytic leukemia, follicular lymphoma, and rheumatoid arthritis. J Clin Pharmacol 2014;54:818–827.

[95] Arons E, Stetler-Stevenson M, Wilson WH, FitzGerald DJP, Pastan I. Pharmacokinetic analysis of response in hairy cell leukemia treated by anti-CD22 recombinant immunotoxin moxetumomab pasudotox. Blood 2013;122:2871.

[96] Advani A, Coiffier B, Czuczman MS, Dreyling M, Foran J, Gine E, Gisselbrecht C, Ketterer N, Nasta S, Rohatiner A, Schmidt-Wolf IG, Schuler M, Sierra J, Smith MR, Verhoef G, Winter JN, Boni J, Vandendries E, Shapiro M, Fayad L. Safety, pharmacokinetics, and preliminary clinical activity of inotuzumab ozogamicin, a novel immunoconjugate for the treatment of B-cell non-Hodgkin's lymphoma: results of a phase I study. J Clin Oncol 2010;28:2085–2093.

[97] Dowell JA, Korth-Bradley J, Liu H, King SP, Berger MS. Pharmacokinetics of gemtuzumab ozogamicin, an antibody-targeted chemotherapy agent for the treatment

of patients with acute myeloid leukemia in first relapse. J Clin Pharmacol 2001;41:1206–1214.

[98] Mould DR, Baumann A, Kuhlmann J, Keating MJ, Weitman S, Hillmen P, Brettman LR, Reif S, Bonate PL. Population pharmacokinetics-pharmacodynamics of alemtuzumab (Campath) in patients with chronic lymphocytic leukaemia and its link to treatment response. Br J Clin Pharmacol 2007;64:278–291.

[99] Xin Y, Jin D, Eppler S, Damico-Beyer LA, Joshi A, Davis JD, Kaur S, Nijem I, Bothos J, Peterson A, Patel P, Bai S. Population pharmacokinetic analysis from phase I and phase II studies of the humanized monovalent antibody, onartuzumab (MetMAb), in patients with advanced solid tumors. J. Clin. Pharmacol. 2013;53:1103–1111.

[100] Fracasso PM, Burris H III, Arquette MA, Govindan R, Gao F, Wright LP, Goodner SA, Greco FA, Jones SF, Willcut N, Chodkiewicz C, Pathak A, Springett GM, Simon GR, Sullivan DM, Marcelpoil R, Mayfield SD, Mauro D, Garrett CR. A phase 1 escalating single-dose and weekly fixed-dose study of cetuximab: pharmacokinetic and pharmacodynamic rationale for dosing. Clin Cancer Res 2007;13:986–993.

[101] Yang BB, Lum P, Chen A, Arends R, Roskos L, Smith B, Pérez Ruixo JJ. Pharmacokinetic and pharmacodynamic perspectives on the clinical drug development of panitumumab. Clin Pharmacokinet 2010;49:729–740.

[102] Kuenen B, Witteveen PO, Ruijter R, Giaccone G, Dontabhaktuni A, Fox F, Katz T, Youssoufian H, Zhu J, Rowinsky EK, Voest EE. A phase I pharmacologic study of necitumumab (IMC-11F8), a fully human IgG1 monoclonal antibody directed against EGFR in patients with advanced solid malignancies. Clin Cancer Res 2010;16:1915–1923.

[103] Kuester K, Kovar A, Lupfert C, Brockhaus B, Kloft C. Population pharmacokinetic data analysis of three phase I studies of matuzumab, a humanised anti-EGFR monoclonal antibody in clinical cancer development. Br J Cancer 2008;98:900–906.

[104] Bruno R, Washington CB, Lu JF, Lieberman G, Banken L, Klein P. Population pharmacokinetics of trastuzumab in patients with HER2+ metastatic breast cancer. Cancer Chemother Pharmacol 2005;56:361–369.

[105] Girish S, Gupta M, Wang B, Lu D, Krop IE, Vogel CL, Burris Iii HA, LoRusso PM, Yi JH, Saad O, Tong B, Chu YW, Holden S, Joshi A. Clinical pharmacology of trastuzumab emtansine (T-DM1): an antibody-drug conjugate in development for the treatment of HER2-positive cancer. Cancer Chemother Pharmacol 2012;69:1229–1240.

[106] Frey N, Grange S, Woodworth T. Population pharmacokinetic analysis of tocilizumab in patients with rheumatoid arthritis. J Clin Pharmacol 2010;50:754–766.

[107] Fairman D, Narwal R, Liang M, Robbins PB, Schneider A, Chavez C, Lu H, Pak M, Blake-Haskins A, Vasselli J, Ibrahim RA, Shalabi AM, Roskos L. Pharmacokinetics of MEDI4736, a fully human anti-PDL1 monoclonal antibody, in patients with advanced solid tumors. J Clin Oncol 2014;32(Suppl):5s. Abstract 2602. Available at http://meetinglibrary.asco.org/content/134334-144 Accessed 2015 Jul 6.

[108] Gibiansky L, Sutjandra L, Doshi S, Zheng J, Sohn W, Peterson MC, Jang GR, Chow AT, Pérez-Ruixo JJ. Population pharmacokinetic analysis of denosumab in patients with bone metastases from solid tumours. Clin Pharmacokinet 2012;51:247–260.

[109] Borghaei H, Alpaugh K, Hedlund G, Forsberg G, Langer C, Rogatko A, Hawkins R, Dueland S, Lassen U, Cohen RB. Phase I dose escalation, pharmacokinetic and pharmacodynamic study of naptumomab estafenatox alone in patients with advanced cancer and with docetaxel in patients with advanced non-small-cell lung cancer. J Clin Oncol 2009;27:4116–4123.

[110] Xolair® (Omalizumab). *US Prescribing Information.* South San Francisco (CA): Genentech Inc. & Novartis Pharmaceuticals Corporation: East Hanover (NJ); 2006. . Available at http://www.gene.com/gene/products/information/pdf/xolair-prescribing.pdf. Accessed 2015 Jun 5..

[111] Soliris (eculizumab) Assessment Report (Scientific discussion section) by the European Medicines Agency 2007 Available at http://www.ema.europa.eu/docs/en_GB/document_library/EPAR_-_Scientific_Discussion/human/000791/WC500054212.pdf. Accessed 2015 Jun 5.

[112] Glassman PM, Balthasar JP. Mechanistic considerations for the use of monoclonal antibodies for cancer therapy. Cancer Biol Med 2014;11:20–33.

[113] Spangler JB, Neil JR, Abramovitch S, Yarden Y, White FM, Lauffenburger DA, Wittrup KD. Combination antibody treatment down-regulates epidermal growth factor receptor by inhibiting endosomal recycling. Proc Natl Acad Sci U S A 2010;107:13252–13257.

[114] Harding J, Burtness B. Cetuximab: an epidermal growth factor receptor chemeric human-murine monoclonal antibody. Drugs Today (Barc) 2005;41:107–127.

[115] McLaughlin P, Grillo-Lopez AJ, Link BK, Levy R, Czuczman MS, Williams ME, Heyman MR, Bence-Bruckler I, White CA, Cabanillas F, Jain V, Ho AD, Lister J, Wey K, Shen D, Dallaire BK. Rituximab chimeric anti-CD20 monoclonal antibody therapy for relapsed indolent lymphoma: half of patients respond to a four-dose treatment program. J Clin Oncol 1998;16:2825–2833.

[116] Dale DC, Bonilla MA, Davis MW, Nakanishi AM, Hammond WP, Kurtzberg J, Wang W, Jakubowski A, Winton E, Lalezari P, Robinson W, Glaspy JA, Emerson S, Gabrilove J, Bincent M, Boxer LA. A randomized controlled phase III trial of recombinant human granulocyte colony-stimulating factor (filgrastim) for treatment of severe chronic neutropenia. Blood 1993;81:2496–2502.

[117] Krzyzanski W, Wiczling P, Lowe P, Pigeolet E, Fink M, Berghout A, Balser S. Population modeling of filgrastim PK-PD in healthy adults following intravenous and subcutaneous administrations. J Clin Pharmacol 2010;50:101S–112S.

[118] Baeuerle PA, Reinhardt C. Bispecific T-cell engaging antibodies for cancer therapy. Cancer Res 2009;69:4941–4944.

[119] Ng CM, Joshi A, Dedrick RL, Garovoy MR, Bauer RJ. Pharmacokinetic-pharmacodynamic-efficacy analysis of

efalizumab in patients with moderate to severe psoriasis. Pharm Res 2005;22:1088–1100.

[120] Sharma A, Ebling WF, Jusko WJ. Precursor-dependent indirect pharmacodynamic response model for tolerance and rebound phenomena. J Pharm Sci 1998;87:1577–1584.

[121] Bonnekoh B, Malykh Y, Bockelmann R, Bartsch S, Pommer AJ, Gollnick H. Profiling lymphocyte subpopulations in peripheral blood under efalizumab treatment of psoriasis by multi epitope ligand cartography (MELC) robot microscopy. Eur J Dermatol 2006;16:623–635.

[122] Zheng S, Gaitonde P, Andrew MA, Gibbs MA, Lesko LJ, Schmidt S. Model-based assessment of dosing strategies in children for monoclonal antibodies exhibiting target-mediated drug disposition. CPT Pharmacometrics Syst Pharmacol 2014;3:e138.

[123] Xin Y, Jin D, Eppler S, Damico-Beyer LA, Joshi A, Davis JD, Kaur S, Nijem I, Bothos J, Peterson A, Patel P, Bai S. Population pharmacokinetic analysis from phase I and phase II studies of the humanized monovalent antibody, onartuzumab (MetMAb), in patients with advanced solid tumors. J Clin Pharmacol 2013;53:1103–1111.

[124] Cosson VF, Ng VW, Lehle M, Lum BL. Population pharmacokinetics and exposure-response analyses of trastuzumab in patients with advanced gastric or gastroesophageal junction cancer. Cancer Chemother Pharmacol 2014;73:737–747.

[125] Deng R, Jin F, Prabhu S, Iyer S. Monoclonal antibodies: what are the pharmacokinetic and pharmacodynamic considerations for drug development? Expert Opin Drug Metab Toxicol 2012;8:141–160.

[126] Lammerts van Bueren JJ, Bleeker WK, Bogh HO, Houtkamp M, Schuurman J, van de Winkel JG, Parren PW. Effect of target dynamics on pharmacokinetics of a novel therapeutic antibody against the epidermal growth factor receptor: implications for the mechanisms of action. Cancer Res 2006;66:7630–7638.

[127] Li L, Gardner I, Rose R, Jamei M. Incorporating target shedding into a minimal PBPK-TMDD model for monoclonal antibodies. CPT Pharmacometrics Syst Pharmacol 2014;3:e96.

[128] Chan AC, Carter PJ. Therapeutic antibodies for autoimmunity and inflammation. Nat Rev Immunol 2010;10:301–316.

[129] Scott AM, Wolchok JD, Old LJ. Antibody therapy of cancer. Nat Rev Cancer 2012;12:278–287.

[130] Davda JP, Hansen RJ. Properties of a general PK/PD model of antibody-ligand interactions for therapeutic antibodies that bind to soluble endogenous targets. MAbs 2010;2:576–588.

[131] Tang CY, Prueksaritanont T. Theoretical analysis of interplay of therapeutic protein drug and circulating soluble target: temporal profiles of 'free' and 'total' drug and target. Pharm Res 2011;28:2447–2457.

[132] Zheng S, McIntosh T, Wang W. Utility of free and total target measurements as target engagement and efficacy biomarkers in biotherapeutic development – opportunities and challenges. J Clin Pharmacol 2015;55(Suppl 3):S75–84.

[133] Lowe PJ, Tannenbaum S, Gautier A, Jimenez P. Relationship between omalizumab pharmacokinetics, IgE pharmacodynamics and symptoms in patients with severe persistent allergic (IgE-mediated) asthma. Br J Clin Pharmacol 2009;68:61–76.

[134] Wang W, Wang X, Doddareddy R, Fink D, McIntosh T, Davis HM, Zhou H. Mechanistic pharmacokinetic/target engagement/pharmacodynamic (PK/TE/PD) modeling in deciphering interplay between a monoclonal antibody and its soluble target in cynomolgus monkeys. AAPS J 2014;16:129–139.

[135] Hoffman HM. Rilonacept for the treatment of cryopyrin-associated periodic syndromes (CAPS). Expert Opin Biol Ther 2009;9:519–531.

[136] Radin A, Marbury T, Osgood G, Belomestnov P. Safety and pharmacokinetics of subcutaneously administered rilonacept in patients with well-controlled end-stage renal disease (ESRD). J Clin Pharmacol 2010;50:835–841.

[137] Shih T, Lindley C. Bevacizumab: an angiogenesis inhibitor for the treatment of solid malignancies. Clin Ther 2006; 28:1779–1802.

[138] Siddiqui MA, Scott L. Infliximab. Drugs 2005;65: 2179–2208.

[139] HUMIRA (adalimumab). 2002. Package insert, Abbott Laboratories. Available at http://www.accessdata.fda.gov/drugsatfda_docs/label/2002/adalabb123102LB.htm. Accessed 2015 Jun 5.

[140] Zhu M, Doshi S, Gisleskog PO, Oliner KS, Perez Ruixo JJ, Loh E, Zhang Y. Population pharmacokinetics of rilotumumab, a fully human monoclonal antibody against hepatocyte growth factor, in cancer patients. J Pharm Sci 2014;103:328–336.

[141] Tabrizi M, Bornstein GG, Suria H. Biodistribution mechanisms of therapeutic monoclonal antibodies in health and disease. AAPS J 2010;12:33–43.

[142] Lowe PJ, Renard D. Omalizumab decreases IgE production in patients with allergic (IgE-mediated) asthma; PKPD analysis of a biomarker, total IgE. Br J Clin Pharmacol 2011;72:306–320.

[143] Chan MA, Gigliotti NM, Dotson AL, Rosenwasser LJ. Omalizumab may decrease IgE synthesis by targeting membrane IgE+ human B cells. Clin Transl Allergy 2013;3:29.

[144] Prat M, Crepaldi T, Gandino L, Giordano S, Longati P, Comoglio P. C-terminal truncated forms of Met, the hepatocyte growth factor receptor. Mol Cell Biol 1991;11:5954–5962.

[145] Lennon S, Barton C, Banken L, Gianni L, Marty M, Baselga J, Leyland-Jones B. Utility of serum HER2 extracellular domain assessment in clinical decision making: pooled analysis of four trials of trastuzumab in metastatic breast cancer. J Clin Oncol 2009;27:1685–1693.

[146] Moreno-Aspitia A, Hillman DW, Dyar SH, Tenner KS, Gralow J, Kaufman PA, Davidson NE, Lafky JM, Reinholz MM, Lingle WL, Kutteh LA, Carney WP, Dueck AC, Perez EA. Soluble human epidermal growth factor receptor 2 (HER2) levels in patients with HER2-positive breast cancer receiving chemotherapy with or without trastuzumab: results from North Central Cancer Treatment Group adjuvant trial N9831. Cancer 2013;119:2675–2682.

[147] Pegram MD, Lipton A, Hayes DF, Weber BL, Baselga JM, Tripathy D, Baly D, Baughman SA, Twaddell T, Glaspy JA, Slamon DJ. Phase II study of receptor-enhanced chemosensitivity using recombinant humanized anti-p185HER2/neu

monoclonal antibody plus cisplatin in patients with HER2/neu-overexpressing metastatic breast cancer refractory to chemotherapy treatment. J Clin Oncol 1998;16:2659–2671.

[148] Tokuda Y, Watanabe T, Omuro Y, Ando M, Katsumata N, Okumura A, Ohta M, Fujii H, Sasaki Y, Niwa T, Tajima T. Dose escalation and pharmacokinetic study of a humanized anti-HER2 monoclonal antibody in patients with HER2/neu-overexpressing metastatic breast cancer. Br J Cancer 1999;81:1419–1425.

[149] Manshouri T, Do KA, Wang X, Giles FJ, O'Brien SM, Saffer H, Thomas D, Jilani I, Kantarjian HM, Keating MJ, Albitar M. Circulating CD20 is detectable in the plasma of patients with chronic lymphocytic leukemia and is of prognostic significance. Blood 2003;101:2507–2513.

[150] Giles FJ, Vose JM, Do KA, Johnson MM, Manshouri T, Bociek G, Bierman PJ, O'Brien SM, Keating MJ, Kantarjian HM, Armitage JO, Albitar M. Circulating CD20 and CD52 in patients with non-Hodgkin's lymphoma or Hodgkin's disease. Br J Haematol 2003;123:850–857.

[151] Junghans RP, Carrasquillo JA, Waldmann TA. Impact of antigenemia on the bioactivity of infused anti-Tac antibody: implications for dose selection in antibody immunotherapies. Proc Natl Acad Sci U S A 1998;95:1752–1757.

[152] Zhang Y, Pastan I. High shed antigen levels within tumors: an additional barrier to immunoconjugate therapy. Clin Cancer Res 2008;14:7981–7986.

[153] Pandiella A, Massague J. Cleavage of the membrane precursor for transforming growth factor alpha is a regulated process. Proc Natl Acad Sci U S A 1991;88:1726–1730.

[154] Massague J, Pandiella A. Membrane-anchored growth factors. Annu Rev Biochem 1993;62:515–541.

[155] Horiuchi T, Mitoma H, Harashima S, Tsukamoto H, Shimoda T. Transmembrane TNF-alpha: structure, function and interaction with anti-TNF agents. Rheumatology (Oxford) 2010;49:1215–1228.

[156] Tan AR, Moore DF, Hidalgo M, Doroshow JH, Poplin EA, Goodin S, Mauro D, Rubin EH. Pharmacokinetics of cetuximab after administration of escalating single dosing and weekly fixed dosing in patients with solid tumors. Clin Cancer Res 2006;12:6517–6522.

[157] Ma P, Yang BB, Wang YM, Peterson M, Narayanan A, Sutjandra L, Rodriguez R, Chow A. Population pharmacokinetic analysis of panitumumab in patients with advanced solid tumors. J Clin Pharmacol 2009;49:1142–1156.

[158] Luedke E, Jaime-Ramirez AC, Bhave N, Carson WE III. Monoclonal antibody therapy of pancreatic cancer with cetuximab: potential for immune modulation. J Immunother 2012;35:367–373.

[159] Han K, Jin J, Maia M, Lowe J, Sersch MA, Allison DE. Lower exposure and faster clearance of bevacizumab in gastric cancer and the impact of patient variables: analysis of individual data from AVAGAST phase III trial. AAPS J 2014;16:1056–1063.

[160] Kang Y, Rha SY, Tassone P, Barriuso J, Yu R, Szado T, Garg A, Bang Y. Pertuzumab pharmacokinetics and safety in combination with trastuzumab and chemotherapy in patients with HER2-positive advanced gastric cancer (AGC). Ann Oncol 2013;24 (Suppl 4):iv19. Abstract No.: O-0021.

[161] Li C, Quartino A, Han K, Allison D, Garg A, Mangat R, Lum B. 2014. Pharmacokinetics of Trastuzumab and Bevacizumab in gastric cancer. AAPS NBC 2014.

[162] Mass RD, Press MF, Anderson S, Cobleigh MA, Vogel CL, Dybdal N, Leiberman G, Slamon DJ. Evaluation of clinical outcomes according to HER2 detection by fluorescence *in situ* hybridization in women with metastatic breast cancer treated with trastuzumab. Clin Breast Cancer 2005;6:240–246.

[163] Pirker R, Pereira JR, von Pawel J, Krzakowski M, Ramlau R, Park K, de Marinis F, Eberhardt WE, Paz-Ares L, Störkel S, Schumacher KM, von Heydebreck A, Celik I, O'Byrne KJ. EGFR expression as a predictor of survival for first-line chemotherapy plus cetuximab in patients with advanced non-small-cell lung cancer: analysis of data from the phase 3 FLEX study. Lancet Oncol 2012;13:33–42.

[164] Amendt C, Staub E, Friese-Hamim M, Storkel S, Stroh C. Association of EGFR expression level and cetuximab activity in patient-derived xenograft models of human non-small cell lung cancer. Clin Cancer Res 2014;20(17):4478–4487.

[165] Weiner LM, Belldegrun AS, Crawford J, Tolcher AW, Lockbaum P, Arends RH, Navale L, Amado RG, Schwab G, Figlin RA. Dose and schedule study of panitumumab monotherapy in patients with advanced solid malignancies. Clin Cancer Res 2008;14:502–508.

[166] Amado RG, Wolf M, Peters M, Van Cutsem E, Siena S, Freeman DJ, Juan T, Sikorski R, Suggs S, Radinsky R, Patterson SD, Chang DD. Wild-type KRAS is required for panitumumab efficacy in patients with metastatic colorectal cancer. J Clin Oncol 2008;26:1626–1634.

[167] Karapetis CS, Khambata-Ford S, Jonker DJ, O'Callaghan CJ, Tu D, Tebbutt NC, Simes RJ, Chalchal H, Shapiro JD, Robitaille S, Price TJ, Shepherd L, Au HJ, Langer C, Moore MJ, Zalcberg JR. K-ras mutations and benefit from cetuximab in advanced colorectal cancer. N Engl J Med 2008;359:1757–1765.

[168] Azzopardi N, Lecomte T, Ternant D, Boisdron-Celle M, Piller F, Morel A, Gouilleux-Gruart V, Vignault-Desvignes C, Watier H, Gamelin E, Paintaud G. Cetuximab pharmacokinetics influences progression-free survival of metastatic colorectal cancer patients. Clin Cancer Res 2011;17:6329–6337.

[169] Dayde D, Ternant D, Ohresser M, Lerondel S, Pesnel S, Watier H, Le Pape A, Bardos P, Paintaud G, Cartron G. Tumor burden influences exposure and response to rituximab: pharmacokinetic-pharmacodynamic modeling using a syngeneic bioluminescent murine model expressing human CD20. Blood 2009;113:3765–3772.

[170] Jakobsen JN, Santoni-Rugiu E, Sorensen JB. Changes in epidermal growth factor receptor expression during chemotherapy in non-small cell lung cancer. Cancer Chemother Pharmacol 2014;73:131–137.

[171] Yang J, Zhao H, Garnett C, Rahman A, Gobburu JV, Pierce W, Schechter G, Summers J, Keegan P, Booth B, Wang Y. The combination of exposure-response and case-control analyses in regulatory decision making. J Clin Pharmacol 2013;53:160–166.

[172] Di Nicolantonio F, Martini M, Molinari F, Sartore-Bianchi A, Arena S, Saletti P, De Dosso S, Mazzucchelli L, Frattini M, Siena S, Bardelli A. Wild-type BRAF is required for response to panitumumab or cetuximab in metastatic colorectal cancer. J Clin Oncol 2008;26:5705–5712.

[173] Bardelli A, Corso S, Bertotti A, Hobor S, Valtorta E, Siravegna G, Sartore-Bianchi A, Scala E, Cassingena A, Zecchin D, Apicella M, Migliardi G, Galimi F, Lauricella C, Zanon C, Perera T, Veronese S, Corti G, Amatu A, Gambacorta M, Diaz LA Jr, Sausen M, Velculescu VE, Comoglio P, Trusolino L, Di Nicolantonio F, Giordano S, Siena S. Amplification of the MET receptor drives resistance to anti-EGFR therapies in colorectal cancer. Cancer Discov 2013;3:658–673.

[174] Sartore-Bianchi A, Moroni M, Veronese S, Carnaghi C, Bajetta E, Luppi G, Sobrero A, Barone C, Cascinu S, Colucci G, Cortesi E, Nichelatti M, Gambacorta M, Siena S. Epidermal growth factor receptor gene copy number and clinical outcome of metastatic colorectal cancer treated with panitumumab. J Clin Oncol 2007;25:3238–3245.

[175] Moroni M, Veronese S, Benvenuti S, Marrapese G, Sartore-Bianchi A, Di Nicolantonio F, Gambacorta M, Siena S, Bardelli A. Gene copy number for epidermal growth factor receptor (EGFR) and clinical response to antiEGFR treatment in colorectal cancer: a cohort study. Lancet Oncol 2005;6:279–286.

[176] Ravetch JV, Bolland S. IgG Fc receptors. Annu Rev Immunol 2001;19:275–290.

[177] Warmerdam PA, van de Winkel JG, Vlug A, Westerdaal NA, Capel PJ. A single amino acid in the second Ig-like domain of the human Fc gamma receptor II is critical for human IgG2 binding. J Immunol 1991;147:1338–1343.

[178] Mellor JD, Brown MP, Irving HR, Zalcberg JR, Dobrovic A. A critical review of the role of Fc gamma receptor polymorphisms in the response to monoclonal antibodies in cancer. J Hematol Oncol 2013;6:1.

[179] Musolino A, Naldi N, Bortesi B, Pezzuolo D, Capelletti M, Missale G, Laccabue D, Zerbini A, Camisa R, Bisagni G, Neri TM, Ardizzoni A. Immunoglobulin G fragment C receptor polymorphisms and clinical efficacy of trastuzumab-based therapy in patients with HER-2/neu-positive metastatic breast cancer. J Clin Oncol 2008;26:1789–1796.

[180] Tamura K, Shimizu C, Hojo T, Akashi-Tanaka S, Kinoshita T, Yonemori K, Kouno T, Katsumata N, Ando M, Aogi K, Koizumi F, Nishio K, Fujiwara Y. FcgammaR2A and 3A polymorphisms predict clinical outcome of trastuzumab in both neoadjuvant and metastatic settings in patients with HER2-positive breast cancer. Ann Oncol 2011;22:1302–1307.

[181] Hurvitz SA, Betting DJ, Stern HM, Quinaux E, Stinson J, Seshagiri S, Zhao Y, Buyse M, Mackey J, Driga A, Damaraju S, Sliwkowski MX, Robert NJ, Valero V, Crown J, Falkson C, Brufsky A, Pienkowski T, Eiermann W, Martin M, Bee V, Marathe O, Slamon DJ, Timmerman JM. Analysis of Fcgamma receptor IIIa and IIa polymorphisms: lack of correlation with outcome in trastuzumab-treated breast cancer patients. Clin Cancer Res 2012;18:3478–3486.

[182] Brezski RJ, Jordan RE. Cleavage of IgGs by proteases associated with invasive diseases: an evasion tactic against host immunity? MAbs 2010;2:212–220.

[183] Kinder M, Greenplate AR, Grugan KD, Soring KL, Heeringa KA, McCarthy SG, Bannish G, Perpetua M, Lynch F, Jordan RE, Strohl WR, Brezski RJ. Engineered protease-resistant antibodies with selectable cell-killing functions. J Biol Chem 2013;288:30843–30854.

[184] Brezski RJ, Vafa O, Petrone D, Tam SH, Powers G, Ryan MH, Luongo JL, Oberholtzer A, Knight DM, Jordan RE. Tumor-associated and microbial proteases compromise host IgG effector functions by a single cleavage proximal to the hinge. Proc Natl Acad Sci U S A 2009;106: 17864–17869.

[185] Elassaiss-Schaap J, Lindauer A, Sostelly A, Ahamadi M, Gergich K, Kang P, de Alwis D, de Greef R. Modeling of tumor size reduction patterns in advanced melanoma under treatment with MK-3475, a potent antibody against PD-1. Abstr Ann Meet Popul Approach Group Eur 2014:23. Abstract 3213. ISSN 1871-6032. Available at www.page-meeting.org/?abstract=3213. Accessed 2015 Jun 5.

[186] Herbst RS, Soria JC, Kowanetz M, Fine GD, Hamid O, Gordon MS, Sosman JA, McDermott DF, Powderly JD, Gettinger SN, Kohrt HE, Horn L, Lawrence DP, Rost S, Leabman M, Xiao Y, Mokatrin A, Koeppen H, Hegde PS, Mellman I, Chen DS, Hodi FS. Predictive correlates of response to the anti-PD-L1 antibody MPDL3280A in cancer patients. Nature 2014;515:563–567.

[187] Jager E, van der Velden VH, te Marvelde JG, Walter RB, Agur Z, Vainstein V. Targeted drug delivery by gemtuzumab ozogamicin: mechanism-based mathematical model for treatment strategy improvement and therapy individualization. PLoS One 2011;6:e24265.

[188] Mortland L, Alonzo TA, Walter RB, Gerbing RB, Mitra AK, Pollard JA, Loken MR, Hirsch B, Raimondi S, Franklin J, Pounds S, Cao X, Rubnitz JE, Ribeiro RC, Gamis A, Meshinchi S, Lamba JK. Clinical significance of CD33 nonsynonymous single-nucleotide polymorphisms in pediatric patients with acute myeloid leukemia treated with gemtuzumab-ozogamicin-containing chemotherapy. Clin Cancer Res 2013;19:1620–1627.

[189] Pfreundschuh M, Ho AD, Cavallin-Stahl E, Wolf M, Pettengell R, Vasova I, Belch A, Walewski J, Zinzani PL, Mingrone W, Kvaloy S, Shpilberg O, Jaeger U, Hansen M, Corrado C, Scheliga A, Loeffler M, Kuhnt E, MabThera International Trial (MInT) Group. Prognostic significance of maximum tumour (bulk) diameter in young patients with good-prognosis diffuse large-B-cell lymphoma treated with CHOP-like chemotherapy with or without rituximab: an exploratory analysis of the MabThera International Trial Group (MInT) study. Lancet Oncol 2008;9:435–444.

[190] Tsai PC, Hernandez-Ilizaliturri FJ, Bangia N, Olejniczak SH, Czuczman MS. Regulation of CD20 in rituximab-resistant cell lines and B-cell non-Hodgkin lymphoma. Clin Cancer Res 2012;18:1039–1050.

[191] Klaasen R, Wijbrandts CA, Gerlag DM, Tak PP. Body mass index and clinical response to infliximab in rheumatoid arthritis. Arthritis Rheum 2011;63:359–364.

[192] Nadler LM, Ritz J, Hardy R, Pesando JM, Schlossman SF, Stashenko P. A unique cell surface antigen identifying lymphoid malignancies of B cell origin. J Clin Invest 1981;67:134–140.

[193] Li J, Levi M, Charoin J-E, Frey N, Kheoh T, Ren S, Woo M, Joshi A, Valente N, Jumbe NS. Rituximab exhibits a long half-life based on a population pharmacokinetic analysis in non-Hodgkin's lymphoma (NHL) patients. Blood 2007;110:700, (ASH Annual Meeting Abstracts) [abstract no. 2371].

[194] Cartron G, Trappe RU, Solal-Celigny P, Hallek M. Interindividual variability of response to rituximab: from biological origins to individualized therapies. Clin Cancer Res 2011;17:19–30.

[195] Tobinai K, Igarashi T, Itoh K, Kobayashi Y, Taniwaki M, Ogura M, Kinoshita T, Hotta T, Aikawa K, Tsushita K, Hiraoka A, Matsuno Y, Nakamura S, Mori S, Ohashi Y, IDEC-C2B8 Japan Study Group. Japanese multicenter phase II and pharmacokinetic study of rituximab in relapsed or refractory patients with aggressive B-cell lymphoma. Ann Oncol 2004;15:821–830.

[196] Waldmann TA, Pastan IH, Gansow OA, Junghans RP. The multichain interleukin-2 receptor: a target for immunotherapy. Ann Intern Med 1992;116:148–160.

[197] Mould DR, Sweeney KR. The pharmacokinetics and pharmacodynamics of monoclonal antibodies–mechanistic modeling applied to drug development. Curr Opin Drug Discov Devel 2007;10:84–96.

[198] Burrows B, Martinez FD, Halonen M, Barbee RA, Cline MG. Association of asthma with serum IgE levels and skin-test reactivity to allergens. N Engl J Med 1989;320:271–277.

[199] Holgate S, Casale T, Wenzel S, Bousquet J, Deniz Y, Reisner C. The anti-inflammatory effects of omalizumab confirm the central role of IgE in allergic inflammation. J Allergy Clin Immunol 2005;115:459–465.

[200] Presta LG, Lahr SJ, Shields RL, Porter JP, Gorman CM, Fendly BM, Jardieu PM. Humanization of an antibody directed against IgE. J Immunol 1993;151:2623–2632.

[201] Busse W, Corren J, Lanier BQ, McAlary M, Fowler-Taylor A, Cioppa GD, van As A, Gupta N. Omalizumab, anti-IgE recombinant humanized monoclonal antibody, for the treatment of severe allergic asthma. J Allergy Clin Immunol 2001;108:184–190.

[202] Presta L, Shields R, O'Connell L, Lahr S, Porter J, Gorman C, Jardieu P. The binding site on human immunoglobulin E for its high affinity receptor. J Biol Chem 1994;269:26368–26373.

[203] Soler M, Matz J, Townley R, Buhl R, O'Brien J, Fox H, Thirlwell J, Gupta N, Della Cioppa G. The anti-IgE antibody omalizumab reduces exacerbations and steroid requirement in allergic asthmatics. Eur Respir J 2001;18:254–261.

[204] Slavin RG, Ferioli C, Tannenbaum SJ, Martin C, Blogg M, Lowe PJ. Asthma symptom re-emergence after omalizumab withdrawal correlates well with increasing IgE and decreasing pharmacokinetic concentrations. J Allergy Clin Immunol 2009;123:107–113.e3.

[205] Jonsson EN, Macintyre F, James I, Krams M, Marshall S. Bridging the pharmacokinetics and pharmacodynamics of UK-279,276 across healthy volunteers and stroke patients using a mechanistically based model for target-mediated disposition. Pharm Res 2005;22:1236–1246.

[206] Krams M, Lees KR, Hacke W, Grieve AP, Orgogozo JM, Ford GA, ASTIN Study Investigators. Acute Stroke Therapy by Inhibition of Neutrophils (ASTIN): an adaptive dose-response study of UK-279,276 in acute ischemic stroke. Stroke 2003;34:2543–2548.

[207] Lees KR, Diener HC, Asplund K, Krams M, UK-279,276-301 Study Investigators. UK-279,276, a neutrophil inhibitory glycoprotein, in acute stroke: tolerability and pharmacokinetics. Stroke 2003;34:1704–1709.

[208] Nalbant D, Saleh M, Goldman FD, Widness JA, Veng-Pedersen P. Evidence of receptor-mediated elimination of erythropoietin by analysis of erythropoietin receptor mRNA expression in bone marrow and erythropoietin clearance during anemia. J Pharmacol Exp Ther 2010;333:528–532.

[209] Jacobs K, Shoemaker C, Rudersdorf R, Neill SD, Kaufman RJ, Mufson A, Seehra J, Jones SS, Hewick R, Fritsch EF, Kawakita M, Shimizu T, Miyake T. Isolation and characterization of genomic and cDNA clones of human erythropoietin. Nature 1985;313:806–810.

[210] Sawada K, Krantz SB, Dai CH, Koury ST, Horn ST, Glick AD, Civin CI. Purification of human blood burst-forming units-erythroid and demonstration of the evolution of erythropoietin receptors. J Cell Physiol 1990;142:219–230.

[211] Luo XY, Yang MH, Peng P, Wu LJ, Liu QS, Chen L, Tang Z, Liu NT, Zeng XF, Liu Y, Yuan GH. Anti-erythropoietin receptor antibodies in systemic lupus erythematosus patients with anemia. Lupus 2013;22:121–127.

[212] Woo S, Krzyzanski W, Jusko WJ. Target-mediated pharmacokinetic and pharmacodynamic model of recombinant human erythropoietin (rHuEPO). J Pharmacokinet Pharmacodyn 2007;34:849–868.

[213] Krzyzanski W, Wyska E. Pharmacokinetics and pharmacodynamics of erythropoietin receptor in healthy volunteers. Naunyn Schmiedebergs Arch Pharmacol 2008;377:637–645.

[214] Chapel SH, Veng-Pedersen P, Schmidt RL, Widness JA. Receptor-based model accounts for phlebotomy-induced changes in erythropoietin pharmacokinetics. Exp Hematol 2001;29:425–431.

[215] Widness JA, Veng-Pedersen P, Peters C, Pereira LM, Schmidt RL, Lowe LS. Erythropoietin pharmacokinetics in premature infants: developmental, nonlinearity, and treatment effects. J Appl Physiol (1985) 1996;80:140–148.

11

IMMUNOGENICITY: ITS IMPACT ON ADME OF THERAPEUTIC BIOLOGICS

HARALD KROPSHOFER AND WOLFGANG F. RICHTER
Roche Pharmaceutical Research and Early Development, Basel, Switzerland

11.1 INTRODUCTION

The formation and/or presence of anti-drug antibodies (ADAs) directed against a drug substance is commonly denoted as "immunogenicity" in pharmaceutical sciences. The immunogenicity of therapeutic biologics is an intensively studied feature shared by almost all biologics that have been developed, as yet, although ADA incidence and clinical relevance of ADA vary widely between different classes of biologics or indications.

This chapter addresses essentially three aspects: (i) which factors are key in initiating and orchestrating an ADA response; (ii) the impact of ADA (both neutralizing and non-neutralizing) on the absorption, distribution, and clearance of biologics, such as monoclonal antibodies (mAbs) and other protein drugs; and (iii) how to deal with ADA responses in nonclinical and clinical studies in the establishment of a pharmacokinetic/pharmacodynamic (PK/PD) relationship. The impact of ADA on absorption, distribution, and clearance will be discussed mainly based on nonclinical data.

11.2 IMMUNOGENICITY OF THERAPEUTIC BIOLOGICS

11.2.1 The Underlying Cellular Immunology

Although therapeutic biologics are highly specific for a defined target, immune cells represent an often underestimated site of cellular uptake for biologics: when they enter the human body via the parenteral route, they encounter the sentinels of the immune system in the blood, the skin, or in the lymph ducts long before they engage with their target. The contact between a therapeutic biologic and cells of the immune system is of course even more intense in case the target is expressed on immune cells, such as B-cells, T-cells, or dendritic cells (DC). The most critical immune cells in this context are DC [1], for example, dermal DC in the skin, plasmacytoid DC in the blood, Kupffer cells in the liver, or microglial cells in the brain. They are able to integrate signals derived from therapeutic biologics and their surrounding microenvironment and trigger downstream events leading to induction of immunogenicity or tolerance. Likewise, DC can bind and take up biologics and move to secondary lymphoid organs, such as local lymph nodes or the spleen, where they activate T-lymphocytes (T-cells) or B-lymphocytes (B-cells). It is activated and differentiated B-cells that finally trigger secretion of drug-specific ADA—the measurable hallmark of immunogenicity.

11.2.1.1 Dendritic Cells at Parenteral Entry Sites DC are located at almost all border zones of the human organism to the environment. Biologics, which are administered via the intravenous (i.v.) route, encounter blood-borne DC. Although the number of DC in the blood circulation is low as compared to mucosal compartments or the skin, blood DC are able to capture protein drugs or drug–ADA complexes, denoted as "immune complexes (ICs)," and traffic to the splenic marginal zone (MZ) to prime B-cells for ADA formation.

ADME and Translational Pharmacokinetics/Pharmacodynamics of Therapeutic Proteins: Applications in Drug Discovery and Development, First Edition. Edited by Honghui Zhou and Frank-Peter Theil.

From vaccine research, we know that the density of DC in certain layers of the skin is very high: while subcutaneous (s.c.) fat and muscle tissue contain relatively few DC, the dermis and the epidermis are densely populated by different subsets of DC [2]; while epidermal Langerhans cells are primarily activating cytotoxic T-cells, it is CD14+ dermal DC that stimulate naive CD4+ T-cells relevant for ADA generation.

11.2.1.2 Dendritic Cell Receptors for Drug Uptake In their immature state, DC display high endocytosis activity. They are equipped with two types of uptake pathways for therapeutic biologics and both of them may have profound effects on the PK and PD of biologics: (i) the degradative pathway that channels biologics into acidic endosomal/lysosomal compartments in which they are degraded into peptides that may be loaded onto major histocompatibility complex (MHC) class II molecules for subsequent cell surface presentation of respective MHC–peptide complexes required for priming CD4+ helper T-cells and (ii) the nondegradative pathway that internalizes, recycles, and releases biologics, thereby prolonging their biological half-lives (Table 11.1).

Fueling of the degradative pathway may occur via several types of surface receptor families and receptor-independent macropinocytosis. The best-studied receptors are the following [3]: (i) the pattern-recognition receptors, such as heat-shock protein receptors or toll-like receptors (TLRs) (e.g., the endotoxin-binding TLR-4); (ii) the C-type lectin family (DEC-205 and DC-SIGN); (iii) the Fcγ receptors; (iv) the MHC class II protein family (e.g., HLA-DR or HLA-DQ in humans); (v) the CD1 protein family (e.g., CD1a or CD1d); and (vi) the complement receptors. These receptors enable immature DC to capture and internalize a large variety of therapeutic biologics, for example, therapeutic antibodies, peptides, lipids, and polysaccharides or glycoproteins. Immature DC can also take up therapeutic biologics passively via fluid-phase endocytosis, denoted as macropinocytosis, reaching an uptake rate of up to 10 DC cell volumes per day [4].

The nondegradative pathway of DC is facilitated by essentially three receptors: (i) mannose receptor, (ii) FcγRIIB, and (iii) the neonatal Fc receptor (FcRn). The mannose receptor of DC binds mannosylated proteins on the DC surface and facilitates their endocytosis and sorting to recycling endosomes [5], thereby preventing their premature catabolism in lysosomes. Likewise, the FcγRIIB may capture ICs composed of a protein drug and an ADA, and recycle these complexes to the cell surface. This process preserves IC during transit of DC from peripheral tissues to the spleen. Here, DC carrying IC on their cell surface bound to FcγRIIB can efficiently prime MZ B-cells. This type of B-cell activation is direct and independent of T-cell [6]. In contrast, FcγRIIB on Kupffer cells have been described to bind ICs and reduce their half-lives by directing them into lysosomes where they are degraded. FcRn is known to be the receptor responsible for the long half-life of IgG (21 days in humans) [7]: FcRn binds IgG in early endosomes and protects them from degradation by recycling them back to the cell surface (cf. below). FcRn also recycles small ICs, containing IgG bound to a monomeric antigen, while big multivalent ICs are directed to lysosomes and are degraded there [8].

11.2.1.3 Role of T-Cells and B-Cells Formation of ADA is the culmination of a series of events that lead to priming of naive B-cells, their differentiation, and subsequently the secretion of ADA by highly specialized B-cells, denoted as "effector B-cells," "plasma B-cells," or "plasma cells" [9]. Priming of naive B-cells may be under the control of CD4+ helper T-cells or independent of T-cells. In more than 90% of cases, ADA formation is CD4+ T-cell-dependent.

TABLE 11.1 Human Dendritic Cell Receptors for Binding and Uptake of Biologics

Receptor Family	Receptors	Cognate Pathogen-Derived Ligands	Potential Therapeutic Protein Ligands
Pattern-recognition receptors	TLRs 1–9	Structural proteins, glycans, or DNA/RNA	Heat shock proteins, glycoproteins, protein aggregates
Fc receptor family	FcγR FcεR	IgG IgE	Therapeutic antibodies, Ig conjugates, and Fc fusion proteins
C-type lectin receptors	DEC-205 DC-SIGN DCIR	Mannose-containing glycans Viral and self-glycoproteins Mannose-/fucose-containing glycans	Mannosylated glycoproteins Glycoproteins Mannosylated or fucosylated glycoproteins
HLA class II	HLA-DR, HLA-DQ, HLA-DP	Peptides (12–30 amino acids)	Therapeutic peptides
CD1 family	CD1a, CD1b, CD1c, CD1d, CD1e	Lipids, glycolipids	Alkylated peptides and proteins

A scenario for T-cell-independent ADA generation involves splenic MZ B-cells primed by direct contact with antigen-charged DC. T-cell-independent B-cell activation may be driven by highly ordered structural pattern, such as repeats of protein domains or certain polysaccharides, which are able to cross-link the B-cell receptor (BcR). The linear or three-dimensional structure recognized by the BcR is denoted as B-cell epitope. Furthermore, ADAs formed in the absence of T-cells generally share the IgM isotype and low antigen avidity. Furthermore, these ADA responses lack memory. It is currently hypothesized that aggregates of biologics may activate B-cells T-independently, although it is unknown whether protein aggregates do contain B-cell epitope repeats that are appropriately spaced for priming of BcRs on MZ B-cells [10].

In contrast, the development of long-lived and high avidity ADA of the IgG or IgE subtype associated with immunological memory following the readministration of a therapeutic biologic indicates the progression of a CD4+ T-cell-dependent immune response, relying on "T-cell epitopes" [3]: in contrast to B-cell epitopes (cf. above), T cell epitopes are short (12–30 amino acids) linear peptide fragments, generated by limited proteolysis of a biologic within DC or B-cells. They are bound by MHC class II molecules. The resulting MHC II–peptide complexes are presented on the surface of mature DC and B-cells for recognition by specific receptors on CD4+ T-cells. When CD4+ T-cells recognize these MHC II–peptide complexes on mature DC in the context of costimulatory molecules (see below), they start to proliferate and secrete cytokines. When such activated CD4+ T-cells recognize the same MHC II–peptide complexes on the surface of B-cells, they provide help to those B-cells: activated B-cells finally start to differentiate and undergo class-switching toward plasma cells that secrete ADA of the IgG and IgE isotype.

11.2.1.4 Danger versus Tolerance The development of vaccines revealed that the presence of T- and B-cell epitopes, the uptake by DC followed by T- and B-cell engagement may not suffice to make a vaccine work, for example, trigger the formation of neutralizing antibodies (nAbs) against a target pathogen. The missing component is often an adjuvant: it provides a costimulatory signal that upregulates costimulatory molecules on the surface of DC or B-cells, for example, CD40 and CD86, or on activated T-cells, for example, CD40 ligand. Similar adjuvant-like effects are shared by pathogen-specific molecules, such as endotoxins, heat-shock proteins, bacterial peptides, lipids or polysaccharides, or viral RNA/DNA sequences. In the context of biologics, it may be impurities, excipients, for example, detergents or metal ions, or the modalities of the administration, for example, tissue irritation by a needle, that provide costimulatory signals. According to the so-called danger hypothesis, all factors that lead to the generation of costimulation are denoted as danger signals and are often perceived by pattern-recognition receptors, such as the TLRs [11] (cf. above).

In case administration of a biologic is not associated to a danger signal, immunological tolerance rather than immunogenicity may be the consequence. There are several ways of how biologics may induce peripheral tolerance: (i) Activation of natural CD4+ CD25+ FoxP3+ regulatory T-cells (nTregs): they may suppress bystander effector T-cells directed against unrelated T-cell epitopes of a biologic, thereby rendering the biologic nonimmunogenic. (ii) Inclusion of regulatory T-cell epitopes (Tregitopes): they are derived from the Fc and Fab domains of IgG molecules and are able to induce tolerance to coadministered proteins [12]. Tregitopes may belong to the group of T-cell epitopes that stimulate nTregs. (iii) Induction of suppressor DC: DC that mature in the presence of TH2-type cytokines, such as IL-10, differentiate into suppressor DC that have the license for the generation of inducible regulatory T-cells (iTregs). (iv) Induction of plasmacytoid DC that trigger Tregs and inhibit effector T-cell responses. (v) Oromucosal administration: there is preliminary evidence from studies in mice that therapeutic proteins, such as IFNs (interferons), which are immunogenic in man and mice, become nonimmunogenic upon administration via a mucosal surface. This is most likely accomplished via activation of suppressor DC and iTregs in the intestinal mucosa [13]. (vi) Induction of high dose tolerance: repeated administration of high doses of therapeutic antibodies, such as infliximab or abciximab, reduces the immunogenicity of these biologics [14]. This may be due to the presence of Tregitopes in these biologics and the fact that high copy numbers of HLA–Tregitope complexes on DC may be required for activation of nTregs.

11.2.2 Aspects Facilitating Immune Responses against Biologics

11.2.2.1 Drug Substance-Related Aspects Key factors intrinsic to the drug molecule itself are T- and B-cell epitopes: for example, protein sequences that are normally not effectively presented to T-cells or B-cells because they are absent in human proteins or not abundant enough to overcome the thresholds for T- and B-cell activation. Consequently, nonhuman sequences in chimeric Abs or fusion proteins may bear an increased immunogenicity risk. With regard to T-cell epitopes, short peptide drugs that carry nonhuman sequences may bear an elevated risk compared to protein drugs that carry the same nonhuman sequences as peptide drugs can be bound by MHC class II molecules already on the DC surface without the necessity of prior internalization or proteolytic cleavage. Impurities, such as host cell proteins or side products of chemically synthesized peptide drugs, may give rise to neo-epitopes or have adjuvant-like activity and provide costimulation. With regard to B-cell epitopes, aggregates of biologics may cross-link the BcR, thereby overcoming the need for costimulation or T-cell help [15].

11.2.2.2 Drug Product-Related Aspects In this category, the purity, the formulation, and the dose come into consideration. Particular components of the formulation, such as excipients, for example, detergents, their degradation products or preservatives, may function as danger signals that transmit costimulatory signals to human immune cells (cf. above). Finally, the dose and the dosing regimen play a very important role: at very low doses the number of T-cell epitopes generated may be too low to overcome the thresholds for activation of T-cells, while at very high doses the induction of high dose drug tolerance may suppress immune responses. Likewise, daily or highly frequent and chronic dosing may lead to activation of Tregs, thereby lowering the immunogenicity risk, while dosing periods that are interspersed by drug-free intervals may lead to absence or disappearance of Tregs, thereby increasing the risk for immunogenicity.

11.2.2.3 Target-Related Aspects Biologics with immunomodulatory activity, which are designed to inhibit receptor molecules on immune cells or trigger their depletion, such as anti-CD52 or anti-CD20 mAbs, may increase the threshold for immunogenicity—at least under chronic dosing regimens. Along the same lines of reasoning, nondepleting mAbs targeting surface receptors on DC and other antigen-presenting cells, for example, CD40 or CD25, may also bear an increased potential to become immunogenic.

11.2.2.4 Patient-Related Aspects The genetic background of the patient appears to be a very strong determinant of the immunogenicity of biologics. First and foremost, the human leukocyte antigen (HLA) class II genotype determines whether a biologic carries an appropriate T-cell epitope or not [16]. Likewise, the genes encoding the T-cell receptor (TcR) and BcR repertoires impact the availability of TcRs or BcR that can recognize any T-cell epitope or B-cell epitope, respectively. Lastly, polymorphic cytokine loci codetermine the thresholds for activation of T-cells and B-cells. Further risk factors that are often patient-specific are medicines taken prior or in parallel to the biologic of interest. In contrast, comedication of immunosuppressive drugs, such as methotrexate or cortisone, is known to lower the immunogenicity risk of biologics.

11.2.2.5 Disease-Related Aspects Diseases that may modulate the overall immune status, such as tumors or autoimmune diseases, can have a pronounced impact on the immunogenicity of a biologic—sometimes during the whole period of the disease, in other cases only at certain stages of the disease. A well-studied drug that appears to display significant differences in immunogenicity, depending on the underlying disease, is the anti-CD20 chimeric antibody rituximab: rituximab has a very low immunogenic potential (<2%) in immunocompromised lymphoma and leukemia patients [17], while it may reach up to 10-fold higher immunogenicity rates in the chronic treatment of rheumatoid arthritis patients [18].

11.2.2.6 Interplay of Risk Factors Experience with approximately 150 biologics that have entered the marketplace tells us that the immunogenicity of therapeutic biologics is based on an individual and often unique interplay of several risk factors rather than a single dominant factor. Specific combinations of risk factors may correlate with specific types of immunogenicity-related adverse events (AEs) and impacts on the PK profile of the respective therapeutic biologic.

11.3 IMPACT OF ADA ON ADME

11.3.1 Impact of ADA on Bioanalytical Results

Once ADAs have been formed, the therapeutic biologic can exist in circulation as "free" molecule or in a complex with ADA as an IC. Under these circumstances, a key rule for bioanalytics of therapeutic biologics is of utmost importance: "You have to know what you measure." Therefore, in PK assessments, it is mandatory to understand whether the respective bioanalytical assay determines "free" (i.e., active drug not complexed with ADA or other binding entities such as soluble ligands) or total drug (i.e., sum of "free" and complexed drug) [18, 19]. In the interpretation of bioanalytical results, it is also important to know whether ADA may impact the bioanalytical results in that they block epitopes required for binding or detection in a binding assay. Loss of bioanalytical response in the presence of ADA may be either due to ADA-mediated loss/clearance of therapeutic biologic from circulation, or just due to interference of ADA with the bioanalytical assay. In the latter case, the biologic can still be in circulation in ADA-complexed form, but not detected by the assay. This needs to be considered in PK/PD correlation or in toxicokinetic assessments, as the therapeutic biologic may still be pharmacologically active despite being bound to an ADA as long as the ADA does not neutralize the biological activity. Bioanalytical approaches to address the quantification of therapeutic biologics in the presence of ADA have been discussed in recent reviews [18, 19].

11.3.2 Formation of Immune Complexes

ADAs bind to the immunogenic drug protein to form ICs. ADAs can be neutralizing or nonneutralizing. Neutralizing antibodies (nAbs) bind to an epitope of the therapeutic biologic that is essential for its biological activity (e.g., in the complementarity-determining region (CDR) of a mAb). As a consequence, the biological activity of the therapeutic biologic is "neutralized." By contrast to nAbs, nonneutralizing antibodies bind to epitopes that are nonessential for the biological activity of the drug.

Binding to both nAbs and nonneutralizing antibodies may lead to changes in the clearance of the therapeutic biologic. The impact of ADA (both nAbs and nonneutralizing antibodies) on clearance depends on the amount of ADA in circulation. Low levels of ADA may have no impact on overall clearance, while high levels of ADA may lead to a changed PK profile [19]. Binding of the free therapeutic biologic to ADA and the subsequent clearance of the drug–ADA ICs represent an additional clearance pathway that may alter the PK properties of the drug (Fig. 11.1). For the appropriate interpretation of PK data, the characteristics of the used bioanalytical assay need to be considered. If the bioanalytical assay detects only the free therapeutic biologic, the additional clearance pathway is due to loss of free therapeutic biologic through the formation of ICs. If the bioanalytical assay also detects therapeutic biologic within ICs, the observed clearance may reflect the clearance of both drug and IC. Parallel to the appearance of the additional clearance pathway via IC formation, the role of the therapeutic protein's genuine clearance pathways (e.g., renal elimination) is diminished in proportion to the lowered amount of free drug.

The size of an IC—and along with its ADME properties—depends on the number of antigenic epitopes as well as the stoichiometry of therapeutic biologic and ADA [20–22]. Therapeutic biologics with only one B-cell epitope, that is, one binding site for antibodies, form only small ICs with ADAs, but cannot form larger or cross-linked complexes

FIGURE 11.1 Scheme on the impact of ADA on therapeutic protein (TP) clearance: the concentration of the ADA is influenced by its formation and catabolism rate ($k_{formation}$ and $k_{catabolism}$) as well as by consumption of ADA in the formation of therapeutic protein-antidrug antibodies immune complexes (TP-ADA ICs). TP is cleared via a direct clearance pathway (CL_{TP}) as well as via formation of TP-ADA IC and its subsequent clearance (CL_{IC}).

(Fig. 11.2a). Therapeutic biologics that carry several B-cell epitopes may form larger and cross-linked ICs. It is of note that in mAbs all antigenic epitopes appear twice, as both heavy and light chains are present twice. In case of the presence of several epitopes within a single drug protein molecule, the size of the ICs depends on the drug–ADA ratio: if the ratio is about 1:1, they may form larger ICs (Fig. 11.2c,d). Johansson and coworkers studied *in vitro* the formation of ICs following mixing of a mAb with an anti-idiotype antibody in a 1:1 molar ratio [23]. Ring dimers of one mAb and one anti-idiotype antibody were not found. Rather ring tetramers were observed most frequently (Fig. 11.2c), followed by ring hexamers (Fig. 11.2d) and ring octomers. Rings with 16–20 members were also observed. Linear, open-chain aggregates were observed less frequently than ring aggregates (30.3% vs 67.8% at a concentration of 10 µg/mL for both mAb and anti-idiotype). The most common linear, open-chain aggregate was the trimer (Fig. 11.2b), followed by pentamers and heptamers or even longer aggregates. It is of note that Johansson's *in vitro* study used a very well-defined pair of mAb and anti-ideotype with both molecules carrying two epitopes each. Formation of ADAs to a therapeutic biologic *in vivo* often represents a polyclonal immune response employing a higher number of epitopes due to the potential presence of multiple epitopes on the therapeutic biologic. The higher number of epitopes may lead to the formation of larger, cross-linked ICs [20].

The evaluation of the drug–ADA ratio revealed that a 1:1 ratio of therapeutic biologic and ADA leads to large ICs (see also above). If either therapeutic biologic or ADA is present in excess, only small ICs are formed [20]. When rabbit anti-human serum albumin (HSA) antibodies (ABs) were mixed *in vitro* with multivalent HSA at a 5- to 20-fold excess of HSA, ICs with a molar composition of HSA_1AB_1 and HSA_2AB_2 as well as larger ICs were observed [24]. In addition, some HSA_2AB_1 complexes were found. At an HAS-excess >20-fold, ICs consisted predominantly of HSA_1AB_1 and HSA_2AB_2 complexes. IC size is also dependent on the association constant. When the association constant is low, small ICs tend to be formed [20].

As described above, the size and composition of ICs depend on the molar ratio of both therapeutic biologic and ADA, and the association constant of the ADA. In the course of an immune response in animals or humans, both the molar ratio and the association constant of the ADA may change. Therefore, remodeling of ICs in circulation may occur. Studies on the kinetics of IC remodeling, however, are missing, as yet.

11.3.3 Clearance of Immune Complexes

Clearance of ICs depends on their size. Both being taken up into clearing cells as well as being processed in these cells via the neonatal Fc receptor (FcRn) depends on the IC size.

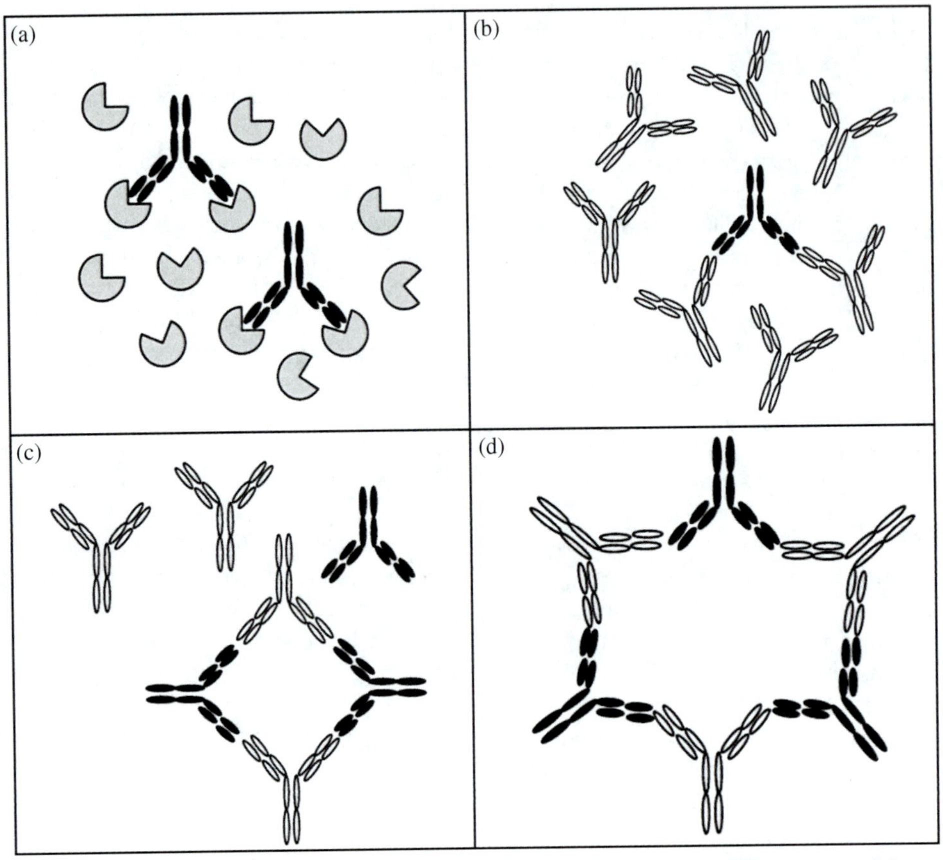

FIGURE 11.2 Schematic presentation of various immune complexes (IC): (a) small IC comprised of an ADA (filled symbol) and two therapeutic biologics with one ADA-binding epitope (open symbol) and (b–d) representative IC formed from a mAb (open symbol) and anti-idiotypic antibodies (closed symbol) as described by Johansson et al. [23]: linear trimer (b), tetramer (c), and hexamer (d).

Small ICs containing only one or two IgG usually persist in circulation [21]. They are not subject to the normal immune clearance mechanisms facilitating uptake into clearing cells, that is, they neither activate complement nor can cross-link Fcγ receptors for cellular uptake. Larger ICs capable of complement activation are taken up primarily by liver and spleen phagocytes, but in the absence of complement also by hematopoietic cells such as DC, macrophages, and monocytes by nonspecific pinocytosis or more efficiently by receptor-mediated uptake [21, 22, 25]. The mechanism of large IC clearance is species dependent. In humans and other primates, IgG1- and IgG3-containing ICs activate the classical complement pathway via binding of the complement component C1q [22]. Following C1q binding, the complement component C3b binds to the Fc parts of immunoglobulins in ICs. ICs carrying C3b bind to the complement receptor 1 (CR1) on erythrocytes. After binding to erythrocytes, the ICs are transported to the liver or spleen, where the ICs are taken up by phagocytes, such as Kupffer cells in liver or macrophages in spleen, via FcγRI, FcγRIIA, FcγRIIIB, CR1, CR2, CR4, or the mannose receptor. While FcγRI has a high affinity to monomeric IgG ($K_d \sim 10^{-9}$ M), FcγRIIA and FcγRIIIB have a low affinity to monomeric IgG ($K_d > 10^{-7}$ and $>10^{-6}$ M, respectively). Due to the low affinity of FcγRIIA and FcγRIIIB, relevant binding and subsequent uptake of ICs require the presence of multimeric ICs with concurrent binding of at least two Fcγ receptors to at least two IgG molecules present in an IC [21]. Thus, the FcγRIIA- and FcγRIIIB-mediated uptake can be expected only for larger ICs formed from therapeutic biologics with multiple epitopes. However, also for FcγRI-mediated uptake, cross-linked ICs facilitate cellular uptake and subsequent lysosomal degradation [26]. The pathway involving erythrocyte binding is unique to humans and monkeys.

In other species, including rodents, complement factor H (CFH) on platelets serves for IC transport in an analogous manner to CR1 on primate erythrocytes [27]. ICs bound to CFH on platelets are transported to liver and spleen phagocytes for elimination. Lovdal and coworkers demonstrated Fcγ receptor-mediated uptake of ICs in Kupffer and endothelial cells of rat liver [28].

Following uptake into phagocytes/hematopoietic cells, ICs reside in the early endosome. At the slightly acidic pH (pH<6.5), the binding the IC to FcγR is lost, while the Fc domain of immunoglobulins present in the IC (either ADAs and, in the case of mAbs, also the therapeutic biologic) can bind to the FcRn [25]. IC handling by FcRn depends on the size of the IC. Monomeric ICs (for scheme, see Fig. 11.2a) undergo FcRn-mediated salvage similar to noncomplexed immunoglobulins. Large, multimeric ICs are transferred to degradation in lysosomes in an FcRn-dependent manner. Clearance of large, multimeric ICs was found to be more rapid in mice with FcRn-containing hematopoietic cells as compared to mice lacking FcRn in hematopoietic cells [25]. *In vitro* studies revealed the mechanism of size-dependent FcRn-mediated sorting of ICs [8]. Endosomal contents destined for salvage, such as monomeric immunoglobulins bound to FcRn, enter narrow-diameter tubules that bud from the early sorting endosome. Multimeric ICs, however, are not found in salvage tubules and, thus, are destined to lysosomal breakdown. Potential reasons for the lack of uptake in salvage tubules include the large size of multimeric ICs or cross-linking of multiple FcRn by multimeric ICs.

11.3.4 Sustaining and Clearing ADAs

The comparative clearance of the therapeutic protein itself and the ICs formed in the presence of ADA determine whether ADAs are sustaining (i.e., reduce clearance) or clearing (i.e., accelerate clearance).

For smaller therapeutic proteins undergoing renal elimination (molecular weight <70 kDa), ADAs are often sustaining. Binding to ADA reduces the "free" fraction of the therapeutic protein undergoing relatively rapid clearance. Binding of an ADA to a therapeutic biologic increases the molecular weight by about 150 kDa so that the resulting ICs are no longer renally eliminated. In addition, monomeric ICs undergo FcRn-mediated recycling. Junghans and Anderson described this sustaining effect of antibodies on smaller proteins in their seminal paper, in which they identified FcRn as the "Brambell receptor" responsible for the long residence time of immunoglobulins, as well as in a companion paper [7, 29]. Free interleukin 2 receptor α (IL-2Rα, T activation antigen (TAC)) undergoes pronounced renal elimination (90% of dose). Binding to an anti-TAC antibody prolongs the half-life of TAC in mice in a relevant manner (38 vs 4.8 h) [29]. The half-life of the TAC/anti-TAC antibody complex was shorter than that of the free anti-TAC antibody in wild-type mice (38 vs 82 h). In FcRn-knockout mice, however, the disposition of both free anti-TAC antibody and TAC/anti-TAC antibody complex was virtually identical, indicating a major role of FcRn in the protection of both the anti-TAC antibody and the complex. The shorter half-life of the complex versus the free anti-TAC antibody suggested that the FcRn-mediated salvage process is less efficient for the complex, either due to lower FcRn binding or due to dissociation of the complex during endosomal recycling. The authors estimated that the TAC/anti-TAC antibody complex undergoes three rounds of endosomal recycling before lysosomal degradation as compared to eight rounds of endosomal recycling for the free anti-TAC antibody. Similar findings were reported for recombinant interleukin 10 (IL-10) when complexed with an anti-IL-10 antibody [30]. The clearance of IL-10, a renally excreted homodimer with an apparent molecular weight of 35 kDa, is markedly reduced in mice after complexation with an anti-IL-10 antibody (5656 and 243 mL/day/kg, respectively; associated half-lives 0.04 and 1.16 days). Similar to the previous example, clearance of the complex was markedly faster than clearance of the free antibody (243 vs 8.5 mL/day/kg). The more rapid clearance of the complex was explained, at least in part, by an 11-fold lower affinity of the complex to mouse FcRn as compared to that of the free anti-IL-10 antibody.

While ADAs may be sustaining for small therapeutic biologics, ADAs against larger therapeutic biologics, such as mAbs, are usually making mAbs cleared more rapidly. ADAs against mAbs are mostly clearing, as they do not reduce clearance via IC formation, as it is the case for the renal clearance of small therapeutic proteins. An exception could be neutralizing ADAs (nAb), which block target binding and, thus, reduce target-mediated drug disposition (TMDD). Nevertheless, ICs formed by a mAb and ADAs are cleared more rapidly than a free mAb. If small ICs are formed (e.g., due to excess of one IC component), the FcRn salvage process is hampered, as described above, but still working [7]. Thus, the IC is cleared somewhat more rapidly as compared to the mAb. If large ICs are formed, the FcRn salvage process is no longer working, as described above. The lack of FcRn salvage and the additional Fcγ receptor-mediated clearance of ICs result in a very rapid clearance of the ICs as compared to the clearance of free mAbs. The overall extent of ADA impact on mAb clearance, however, depends also on the titer and/or formation rate of ADAs [31]. The accelerated clearance is usually associated with a loss in drug efficacy, even if the ADAs are not neutralizing. The formation of ADAs and the associated loss of efficacy have been closely studied for the anti-tumor necrosis factor alpha (TNF-α) mAbs adalimumab and infliximab [32]. Loss of efficacy can be due to accelerated clearance of the mAbs due to IC formation, but can also be due to a direct neutralization of the mAbs by blocking of the binding site by a nAb [33]. It is of note that the formation of ADAs may be also transient so that after a certain period ADA formation diminishes and mAb clearance returns to normal. In a study with adalimumab, in about one-third of ADA-positive patients, the ADA response was found to be transient [34]. Below we provide case examples on the impact of ADAs on mAb clearance both in animal models and in humans.

Case Study I: Accelerated Clearance of mAbs in Animals: When 1.74 mg/kg infliximab were infused into monkeys, followed 30 min later by 0.5 mg/kg radiolabeled anti-infliximab antibodies, ICs were formed within 5 min after the second administration [35]. Large ICs (>670 kDa) were rapidly removed from circulation and were no longer detectable by 24 h. Small ICs were more persistent, with a terminal half-life of 37.5 h (compared to 105 h for infliximab and 86.5 h for control antibody in the absence of ICs). The clearance of infliximab was accelerated by coadministered anti-infliximab antibodies as compared to infliximab in the absence of ADAs.

Case Study II: Formation of ADAs *in vivo* may occur as early as a few days after administration of a test mAb. Formation of ADAs is often evident from an accelerated clearance of the test mAb, which can be observed as early as about 7–10 days after dosing [31, 36]. Following i.v. injection of TNF receptor–Fc fusion protein lenercept to rabbits, circulating lenercept levels declined rapidly with a half-life of about 4 h starting about 7 days after dosing, while ADAs became detectable [36] (Fig. 11.3).

Case Study III: Ng and coworkers studied the impact of ADAs on the PK of adalimumab in cynomolgus monkeys [31]. In addition, they developed a PK model to describe both adalimumab and ADA concentration–time profiles, which allows estimates of ADA formation rate and clearance under the conditions of the study. Thus, the model provided additional insights into ADA formation kinetics and ADA impact on PK, as experimental quantification and assessment of ADA formation rates are hardly possible. The PK model indicated an onset of ADA formation of 8.3 days after dosing; the ADA synthesis rate and elimination half-life were estimated 0.443 log titer/day and 6.2 days, respectively. The ADA-mediated clearance of adalimumab was dependent on ADA levels and could be described by an E_{max} model (E_{max} and EC_{50} 1146 pmol/day/kg/log ADA titer and 7.02 pmol/mL, respectively).

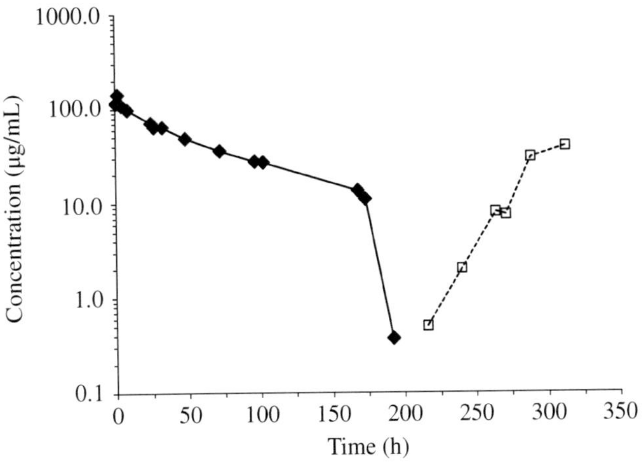

FIGURE 11.3 Rapid clearance of the TNF receptor–Fc fusion protein lenercept followed the onset of an immune response: individual plasma concentrations of lenercept (closed symbols) and anti-lenercept antibodies (open symbols) following a single i.v. dose to a rabbit (5 mg/kg). (Data from [36].)

Case Study IV: Examples for Accelerated Clearance of mAbs in Humans: Also in humans ADA formation can lead to accelerated clearance of mAbs and, thus, to a loss in efficacy. A study with infliximab demonstrated this relationship [37]. ^{99m}Tc-infliximab was administered by i.v. infusion over 2 h to two responder and two nonresponder patients. At start of dosing the responder patients showed nondetectable or low levels of anti-infliximab antibodies, whereas both nonresponder patients showed high levels of anti-infliximab antibodies. During the infliximab infusion, levels of anti-infliximab antibodies declined, and relevant infliximab levels were detectable only after disappearance of anti-infliximab antibodies from circulation. In the responder patient with low antibody levels, only low levels of small ICs were detected during the first hour after dosing. Higher amounts of small ICs were found in one nonresponder, probably 1:1 complexes of infliximab and anti-infliximab antibodies. This patient showed also a more pronounced uptake of radioactivity into liver and spleen by 2 h after the end of infusion as compared to the responder patient, which is consistent with clearance of ICs in these tissues. In the other nonresponder patient, however, large ICs were detectable with various sizes up to a molecular weight of greater than 1000 kDa, associated with a trend to more rapid infliximab clearance. This patient showed also a serious infusion reaction. The formation of large ICs in this patient was consistent with the similar amounts of anti-infliximab antibodies and dosed infliximab. In this patient, the estimated anti-infliximab antibody level was about 24 µg/mL, which was in a similar range as the infliximab level reached in the responder patient in the absence of anti-infliximab antibodies (45 µg/mL).

Case Study V: The impact on ADAs on mAb clearance can be studied under more controlled conditions, when both mAb and ADA are dosed (cf. case study I). Davies et al. studied the clearance of an ^{131}I-mouse antitumor mAb elicited by administration of ^{125}I-human anti-mouse IgG [38]. On Day 0, the patients received 10 mg of ^{131}I-mouse antitumor mAb intraperitoneally. On Days 1 and 2, 18 mg of ^{125}I-human antimouse IgG was administered i.v. to accelerate clearance of the therapeutic mAb, thereby reducing the radiation dose to bone marrow. Following administration of the clearing IgG, large ICs were formed within 5 min and cleared rapidly with a half-life of 11 min in the liver.

The above examples show a marked acceleration of mAb clearance in the presence of ADAs. It is of note, however, that in these examples the acceleration was associated with high levels of ADAs fostering the formation of large, rapidly cleared ICs. If ADAs are formed during therapeutic use of mAb, the levels of ADAs can be lower so that the impact of ADAs on mAb clearance is lower or even missing.

11.3.5 Impact of ADAs on Distribution

Formation of ICs containing a therapeutic biologic and ADAs may change not only the clearance of the therapeutic biologic but also its distribution. The impact of IC formation on distribution is influenced by the difference in size of the ICs compared to the therapeutic biologic. For small therapeutic biologics, formation of ICs may reduce tissue penetration due to the larger size of the ICs as compared to the therapeutic biologic. The change in clearance pathways from predominantly renal clearance to clearance via IC formation affects distribution as well and leads to a higher uptake into liver [30, 38]. The higher uptake into liver is a consequence of the liver being the primary clearance organ for ICs. While for hIL-10 in mice, the kidney is the major organ for clearance and catabolism, and kidney concentrations exceed those in liver by far [30]. Unlike the free hIL-10, antibody-bound hIL-10 was mainly targeted to the liver with minimal uptake into kidneys. Also for large therapeutic biologics, such as mAbs, formation of ICs may lead to an increased uptake into liver and spleen, associated with the more rapid clearance of the ICs as compared to the parent molecule in both organs [35, 39].

ICs may also undergo deposition in tissues. This deposition has been discussed in several reviews [40, 41]. In toxicity studies with therapeutic biologics in laboratory animals, tissue deposition of ICs has been observed in glomerulus, blood vessels, synovium, lung, liver, skin, eye, choroid plexus, or other tissues [22]. In general, small ICs show little risk for tissue deposition, while large ICs may deposit in tissues. This may be particularly the case if large ICs are no longer rapidly cleared due to saturation of the IC clearance pathways. The deposition of ICs in the glomerulus received particular attention in the literature. After administration of a mixture of small and large ICs to mice, deposition of ICs in the glomerulus was observed only as long as large ICs (>mAb_2ADA_2) were present in circulation [40]. In another study, injection of a large dose of small ICs did not lead to the formation of immune deposits. It is of note that these IC deposits in tissue are not static entities, but may undergo remodeling. Already formed deposits may be mobilized again by administration of a large excess of the antigen [42]. On the other hand, tissue deposits may undergo condensation to even larger ICs [40]. Charge–charge interaction between positive charges in the ICs and the negatively charged cell surface appear to promote condensation to larger ICs.

Tissue deposition of ICs may lead to tissue damage such as glomerulonephritis. The safety aspects of tissue deposition of ICs are beyond the scope of this chapter; they have been recently reviewed elsewhere [22].

11.3.6 Impact of ADAs on Absorption

ADAs may impact also the absorption of therapeutic biologics following, for instance, s.c. or intramuscular (i.m.) administration. To the best of our knowledge, studies on the effects of ADAs on absorption are missing. Experimental access to such data, however, appears to be difficult, as the ADA impact on the absorption step needs to be separately assessed from the impact on clearance. Nevertheless, it is expected that ADAs present at the interstitial space of the administration and/or the draining lymphatics form ICs with the administered therapeutic biologic. The impact of ADAs on s.c. absorption may depend on the relative amounts of therapeutic biologic and ADAs at the administration site in the s.c. tissue and the draining lymphatics. Further research is needed to elucidate the impact of ADAs on s.c. and i.m. absorption processes.

11.4 HOW TO DEAL WITH ADME CONSEQUENCES OF IMMUNE RESPONSES?

To establish a causal relationship between unexpected PK/PD profiles and unwanted immune responses, PK/PD sampling needs to be accompanied by sampling for immunogenicity/ADA testing and vice versa—in both nonclinical and clinical studies.

11.4.1 PK Assessment in the Presence of ADAs

Development of ADAs in test animals as well as in human subjects may impact the accuracy of PK assessment regardless of the bioanalytical methods being used. Even though methods to measure both bound and unbound drug are often implemented, the presence of ADAs may hamper the measurable concentrations and alter the PK substantially.

As human subjects in clinical trials are less likely to generate ADAs as compared to animals, in particular in the context of human or humanized biologic drugs, a common strategy employed by pharmacokineticists is to exclude ADA-positive humans from PK assessment—regardless of ADA magnitude. This approach may not be feasible in animals, as a larger fraction of test animals may develop ADAs. In addition, this approach relies on the availability of comprehensive PK data derived from large populations, so they are mainly applied in trials during clinical development rather than nonclinical studies.

In animals with ADA formation, a PK assessment unbiased from ADA formation may be possible, if the onset of ADA impact can be defined, and the observation time before ADA formation is sufficient for PK assessment and drug concentration–time data after the onset of the immune response are disregarded in the PK evaluation [36]. Finally, modeling and simulation approaches to evaluate PK profiles of therapeutic proteins in the presence of ADAs have been described [29, 43, 44].

11.4.2 In-Study Options to Overcome ADA Formation

In the past decade, the field focusing on the immunogenicity of biologics has evolved from merely trying to detect ADA responses into the development of strategies to manipulate the immune system and overcome immune responses. Such immunomodulatory regimens have been pioneered for and are particularly beneficial to therapeutic protein classes that share a high immunogenicity risk, such as replacement proteins or chronically applied therapeutics that lack alternatives. A few typical examples are described in the following sections.

11.4.2.1 Dosing-through and Induction of High Dose Tolerance Interferon beta (IFN-β) is an established chronic treatment for patients with multiple sclerosis. s.c. or i.m. administration of IFN-β is frequently associated with the formation of nAbs that cross-react with other IFN-β drugs and abrogate the treatment effect. A pilot study revealed that infusion of high doses of IFN-β can lead to lowering of nAb titers [45]. These results can be explained in two ways: (i) on short term, infusion of IFN-β is likely to result in the saturation of circulating nAb and binding of excess IFN-β to its receptor, thereby restoring its efficacy, and (ii) on long term, infusion of IFN-β is likely to induce high dose tolerance. Likewise, a high loading dose of an anti-IL-6 receptor mAb overcame the lethal immune response observed before in mice in reprotoxicity studies [46].

11.4.2.2 Immunomodulation to Manage ADAs against Factor VIII About 25% of hemophilia A patients treated with recombinant factor VIII develop nAbs, denoted as "FVIII inhibitors." Intravenous tolerance induction shortly after the detection of nAbs turned out to be successful: provided that the FVIII inhibitor titer is not too high, it is the repeated infusion of moderate to high doses of recombinant or plasma-derived FVIII that leads to FVIII inhibitor disappearance and FVIII recovery [47]. Recently, an alternative treatment proved to rescue patients resistant to the standard tolerance induction: mAb rituximab combined with FVIII was applied to achieve negative inhibitor titers [48]. Although rituximab—that depletes B-lymphocytes—did not show a sustained recovery of FVIII, it converted a high titer inhibitor to a low titer and low responder inhibitor.

11.4.2.3 Tolerance Induction to Enzyme Replacement Therapeutics Infantile Pompe disease (PD), resulting from a deficiency of lysosomal acid α-glucosidase (GAA), requires enzyme replacement therapy with recombinant human α-glucosidase (rhGAA). A subgroup of children suffering from PD develops high titer and sustained nAbs against rhGAA and has very poor prognosis. These patients can be tolerized against rhGAA by therapeutic treatment with a combination of mAb rituximab and methotrexate with or without gammaglobulins. Tolerance persisted even after cessation of immunosuppression and B-cell recovery [49].

11.4.2.4 Immunosuppressives Lowering ADA Titers against Chimeric Antibodies The chimeric mAb infliximab has become a common treatment for refractory Crohn's disease. The development of ADAs against infliximab causes reduced bioavailability and an increased risk of infusion-related reactions (IRRs). Immunosuppressive treatment with azathioprine or mercaptopurine or methotrexate diminished ADA formation, reduced the incidence of IRR, and increased the duration of the therapeutic response [50].

11.4.2.5 Future Approaches to Exploit Immune Tolerance A number of novel approaches to achieve tolerance to therapeutic proteins are currently explored in preclinical models. Some of these approaches involve the design of molecules with reduced immunogenicity. This can be accomplished through either the mutation of immunodominant residues within T- or B-cell epitopes, or replacement of immunogenic human domains by their counterparts derived from other species that possess less immunogenic potential [51]. Alternative approaches foresee tolerogenic administration modes, for example, delivery to nasal and/or gastrointestinal mucosal epithelia or immature DC, which may activate tolerogenic regulatory T-cell populations [52].

11.5 SUMMARY AND CONCLUSIONS

An immune response against a therapeutic biologic can markedly alter its ADME properties. The formation of ICs from therapeutic biologic and ADAs represent an additional clearance pathway for the therapeutic biologic. The resulting ICs may be cleared faster or slower than the therapeutic biologic itself; the associated ADAs are termed clearing and sustained ADAs, respectively. The clearance pathways and organs for the ICs may differ from those of the therapeutic biologic, which may also lead to changes in tissue distribution. The properties of bioanalytical assay(s) must be well understood to allow for a sound PK assessment after the onset of an immune response and formation of ADAs. Strategies are discussed how to deal with PK assessment if ADA formation is observed in a study. In addition, case examples how to mitigate immune responses are presented.

REFERENCES

[1] Steinman RM. Dendritic cells *in vivo*: key target for a new vaccine. Immunity 2008;29:319–324.

[2] Romani N, Thurnher M, Idoyaga J, Steinman RM, Flacher V. Targeting of antigens to skin dendritic cells: possibilities to enhance vaccine efficacy. Immunol Cell Biol 2010;88: 424–430.

[3] Vogt AB, Ploix C, Kropshofer H. Antigen processing for MHC class II. In: Kropshofer H, Vogt AB, editors. *Antigen Presenting Cells*. Weinheim: Wiley-VCH; 2005. p 89–237.

[4] Sallusto F, Cella M, Danieli C, Lanzavecchia A. Dendritic cells use macropinocytosis and the mannose receptor to concentrate macromolecules in the major histocompatibility complex class II compartment: down-regulation by cytokines and bacterial products. J Exp Med 1995;182:389–400.

[5] East L, Isacke CM. The mannose receptor family. Biochim Biophys Acta 2002;1572:364–386.

[6] Bergtold A, Desai DD, Gavhane A, Clynes R. Cell surface recycling of internalized antigen permits dendritic cell priming of B cells. Immunity 2005;23:503–514.

[7] Junghans RP, Anderson CL. The protection receptor for IgG catabolism is the beta2-micro-globulin-containing neonatal intestinal transport receptor. Proc Natl Acad Sci USA 1996;93:5512–5516.

[8] Weflen AW, Baier N, Tang QJ, Van den Hof M, Blumberg RS, Lencer WI, Massol RH. Multivalent immune complexes divert FcRn to lysosomes by exclusion from recycling sorting tubules. Mol Biol Cell 2013;24:2398–2405.

[9] Kropshofer H, Singer T. Overview of cell-based tools for pre-clinical assessment of immunogenicity of biotherapeutics. J Immunotoxicol 2006;3:131–136.

[10] Fehr T, Bachmann MF, Bucher E, Kalinke U, Di Padova FE, Lang AB, Hengartner H, Zinkernagel RM. Role of repetitive antigen patterns for induction of antibodies against antibodies. J Exp Med 1997;185:1785–1792.

[11] Matzinger P. The danger model: a renewed sense of self. Science 2002;296:301–305.

[12] De Groot AS, Moise L, McMurry JA, Wambre E, Van Overtvelt L, Moingeon P, Scott DW, Martin W. Activation of natural regulatory T cells by IgG Fc-derived peptide "Tregitopes". Blood 2008;112:3303–3311.

[13] Meritet JF, Maury C, Tovey MG. Induction of tolerance to recombinant therapeutic proteins. J Interferon Cytokine Res 2001;21:1031–1038.

[14] Wagner CL, Schantz A, Barnathan E, Olson A, Mascelli MA, Ford J, Damaraju L, Schaible T, Maini RN, Tcheng JE. Consequences of immunogenicity to the therapeutic monoclonal antibodies ReoPro and Remicade. Dev Biol 2003; 112:37–53.

[15] Filipe V, Jiskoot W, Basmeleh AH, Halim A, Schellekens H, Brinks V. Immunogenicity of different stressed IgG monoclonal antibody formulations in immune tolerant transgenic mice. MAbs 2012;4:740–752.

[16] Barbosa MD, Vielmetter J, Chu S, Smith DD, Jacinto J. Clinical link between MHC class II haplotype and interferon-beta (IFN-beta) immunogenicity. Clin Immunol 2006;118:42–50.

[17] Piro LD, White CA, Grillo-López AJ, Janakiraman N, Saven A, Beck TM, Varns C, Shuey S, Czuczman M, Lynch JW, Kolitz JE, Jain V. Extended Rituximab (anti-CD20 monoclonal antibody) therapy for relapsed or refractory low-grade or follicular non-Hodgkin's lymphoma. Ann Oncol 1999;10:655–661.

[18] Kelley M, Ahene AB, Gorovits B, Kamerud J, King LE, McIntosh T, Yang J. Theoretical considerations and practical approaches to address the effect of anti-drug antibody (ADA) on quantification of biotherapeutics in circulation. AAPS J 2013;15:646–658.

[19] Sailstad JM, Amaravadi L, Clements-Egan A, Gorovits B, Myler HA, Pillutla RC, Pursuhothama S, Putman M, Rose MK, Sonehara K, Tang L, Wustner JT. A white paper-consensus and recommendations of a global harmonization team on assessing the impact of immunogenicity on pharmacokinetic measurements. AAPS J 2014;16:488–498.

[20] Mannik M. Physicochemical and functional relationships of immune complexes. J Invest Dermatol 1980;74:333–338.

[21] Rehlaender BN, Cho MJ. Antibodies as carrier proteins. Pharm Res 1998;15:1652–1656.

[22] Rojko JL, Evans MG, Price SA, Han B, Waine G, DeWitte M, Haynes J, Freimark B, Martin P, Raymond JT, Evering W, Rebelatto MC, Schenck E, Horvath C. Formation, clearance, deposition, pathogenicity, and identification of biopharmaceutical-related immune complexes: review and case studies. Toxicol Pathol 2014;42:725–764.

[23] Johansson A, Erlandsson A, Eriksson D, Ullén A, Holm P, Sundström BE, Roux KH, Stigbrand T. Idiotypic-anti-idiotypic complexes and their *in vivo* metabolism. Cancer 2002;94:1306–1313.

[24] Arend WP, Mannik M. Determination of soluble immune complex molar composition and antibody association constants by ammonium sulfate precipitation. J Immunol 1974;112:451–461.

[25] Qiao SW, Kobayashi K, Johansen FE, Sollid LM, Andersen JT, Milford E, Roopenian DC, Lencer WI, Blumberg RS. Dependence of antibody-mediated presentation of antigen on FcRn. Proc Natl Acad Sci USA 2008;105:9337–9342.

[26] Harrison PT, Davis W, Norman JC, Hockaday AR, Allen JM. Binding of monomeric immunoglobulin G triggers FcγRI-mediated endocytosis. J Biol Chem 1994;269:24396–24402.

[27] Alexander JJ, Quigg RJ. The simple design of complement factor H: looks can be deceiving. Mol Immunol 2007;44: 123–132.

[28] Løvdal T, Andersen E, Brech A, Berg T. Fc receptor mediated endocytosis of small soluble immunoglobulin G immune complexes in Kupffer and endothelial cells from rat liver. J Cell Sci 2000;113:3255–3266.

[29] Junghans RP, Waldmann TA. Metabolism of Tac (IL2Ralpha): physiology of cell surface shedding and renal catabolism, and suppression of catabolism by antibody binding. J Exp Med 1996;183:1587–1602.

[30] Alvarez HM, So OY, Hsieh S, Shinsky-Bjorde N, Ma H, Song Y, Pang Y, Marian M, Escandón E. Effects of

PEGylation and immune complex formation on the pharmacokinetics and biodistribution of recombinant interleukin 10 in mice. Drug Metab Dispos 2012;40:360–373.

[31] Ng CM, Loyet KM, Iyer S, Fielder PJ, Deng R. Modeling approach to investigate the effect of neonatal Fc receptor binding affinity and anti-therapeutic antibody on the pharmacokinetic of humanized monoclonal anti-tumor necrosis factor-α IgG antibody in cynomolgus monkey. Eur J Pharm Sci 2014;51:51–58.

[32] van Schouwenburg PA, Rispens T, Wolbink GJ. Immunogenicity of anti-TNF biologic therapies for rheumatoid arthritis. Nat Rev Rheumatol 2013;9:164–172.

[33] van Schouwenburg PA, van de Stadt LA, de Jong RN, van Buren EE, Kruithof S, de Groot E, Hart M, van Ham SM, Rispens T, Aarden L, Wolbink GJ, Wouters D. Adalimumab elicits a restricted anti-idiotypic antibody response in autoimmune patients resulting in functional neutralisation. Ann Rheum Dis 2013;72:104–109.

[34] van Schouwenburg PA, Krieckaert CL, Rispens T, Aarden L, Wolbink GJ, Wouters D. Long-term measurement of anti-adalimumab using pH-shift-anti-idiotype antigen binding test shows predictive value and transient antibody formation. Ann Rheum Dis 2013;72:1680–1686.

[35] Rojas JR, Taylor RP, Cunningham MR, Rutkoski TJ, Vennarini J, Jang H, Graham MA, Geboes K, Rousselle SD, Wagner CL. Formation, distribution, and elimination of infliximab and anti-infliximab immune complexes in cynomolgus monkeys. J Pharmacol Exp Ther 2005;313:578–585.

[36] Richter WF, Gallati H, Schiller CD. Animal pharmacokinetics of the tumor necrosis factor receptor-immunoglobulin fusion protein lenercept and their extrapolation to humans. Drug Metab Dispos 1999;27:21–25.

[37] van der Laken CJ, Voskuyl AE, Roos JC, Stigter van Walsum M, de Groot ER, Wolbink G, Dijkmans BA, Aarden LA. Imaging and serum analysis of immune complex formation of radiolabelled infliximab and anti-infliximab in responders and non-responders to therapy for rheumatoid arthritis. Ann Rheum Dis 2007;66:253–256.

[38] Davies KA, Hird V, Stewart S, Sivolapenko GB, Jose P, Epenetos AA, Walport MJ. A study of *in vivo* immune complex formation and clearance in man. J Immunol 1990;144: 4613–4620.

[39] Vugmeyster Y, DeFranco D, Szklut P, Wang Q, Xu X. Biodistribution of [125I]-labeled therapeutic proteins: application in protein drug development beyond oncology. J Pharm Sci 2010;99:1028–1045.

[40] Mannik M. Mechanisms of tissue deposition of immune complexes. J Rheumatol Suppl 1987;13:35–42.

[41] Wener M. Immune complexes in systemic lupus erythematosus. In: Tsokos G, Gordon C, Smolen J, editors. *Systemic Lupus Erythematosus: A Companion to Rheumatology*. Philadelphia (PA): Mosby; 2007. p 214–224.

[42] Mannik M, Striker GE. Removal of glomerular deposits of immune complexes in mice by administration of excess antigen. Lab Invest 1980;42:483–489.

[43] Perez-Ruixo JJ, Ma P, Chow AT. The utility of modeling and simulation approaches to evaluate immunogenicity effect on the therapeutic protein pharmacokinetics. AAPS J 2012;15: 172–182.

[44] Gómez-Mantilla JD, Trocóniz IF, Parra-Guillén Z, Garrido MJ. Review on modeling anti-antibody responses to monoclonal antibodies. J Pharmacokinet Pharmacodyn 2014;41: 523–536.

[45] Millonig A, Rudzki D, Hölzl M, Ehling R, Gneiss C, Künz B, Berger T, Reindl M, Deisenhammer F. High-dose intravenous interferon beta in patients with neutralizing antibodies (HINABS): a pilot study. Mult Scler 2009;15:977–983.

[46] Sakurai T, Takai R, Bürgin H, Shioda A, Sakamoto Y, Amano J, Grimm HP, Richter WF, Higuchi Y, Chiba S, Kawamura A, Suzuki M, Müller L. The effects of interleukin-6 signal blockade on immune system, reproductive and skeletal development in juvenile mice. Birth Defects Res B Dev Reprod Toxicol 2013;98:170–182.

[47] Lillicrap D. The role of immunomodulation in the management of factor VIII inhibitors. Hematology 2006;206: 421–425.

[48] Collins PW, Mathias M, Hanley J, Keeling D, Keenan R, Laffan M, Perry D, Liesner R, UK Haemophilia Centre Doctors' Organisation. Rituximab and immune tolerance in severe hemophilia A: a consecutive national cohort. J Thromb Haemost 2009;7:787–794.

[49] Messinger YH, Mendelsohn NJ, Rhead W, Dimmock D, Hershkovitz E, Champion M, Jones SA, Olson R, White A, Wells C, Bali D, Case LE, Young SP, Rosenberg AS, Kishnani PS. Successful immune tolerance induction to enzyme replacement therapy in CRIM-negative infantile Pompe disease. Genet Med 2012;14:135–142.

[50] Baert F, Noman M, Vermeire S, Van Assche G, D' Haens G, Carbonez A, Rutgeerts P. Influence of immunogenicity on the long-term efficacy of infliximab in Crohn's disease. N Engl J Med 2003;348:601–608.

[51] Barrow RT, Healey JF, Gailani D, Scandella D, Lollar P. Reduction of the antigenicity of factor VIII toward complex inhibitory antibody plasmas using multiply-substituted hybrid human/porcine factor VIII molecules. Blood 2000;95: 564–568.

[52] Ragni MV, Wu W, Liang X, Hsieh CC, Cortese-Hassett A, Lu L. Factor VIII-pulsed dendritic cells reduce anti-factor VIII antibody formation in the hemophilia A mouse model. Exp Hematol 2009;37:744–754.

12

MECHANISTIC PHYSIOLOGICALLY BASED PHARMACOKINETIC MODELS IN DEVELOPMENT OF THERAPEUTIC MONOCLONAL ANTIBODIES

YANGUANG CAO[1,2] AND WILLIAM J. JUSKO[1]

[1] *State University of New York at Buffalo, Buffalo, NY, USA*

[2] *The University of North Carolina at Chapel Hill, Chapel Hill, NC, USA*

12.1 BACKGROUND

Physiologically based pharmacokinetic (PBPK) models, with integrating the structural, *in silico*, and *in vitro* physicochemical data of drugs and the physiological and anatomical features of the body, provide a realistic characterization of the systemic disposition of drugs. PBPK models usually assemble all individual tissues involved in drug disposition and consider the physical volume and blood/lymph flow of each tissue. These physiological constructions allow joint assessments of system- and drug-relevant factors, facilitate translation from *in vitro* to *in vivo*, from animals to human, and permit simulations of various pathophysiological conditions [1, 2]. Furthermore, PBPK models offer a physiologic and mechanistic framework allowing scientists to address a variety of questions about pharmacokinetics (PK), such as drug–drug interactions, disease–drug interactions, and animal to human extrapolation [2, 3]. Compared with classical mammillary models, PBPK models provide considerable details and insights on drug absorption, distribution, biotransformation, and elimination.

There has been a gradual expansion in uses of PBPK models in drug discovery, development, and regulatory science [4]. The numbers of published articles that include "physiologically-based pharmacokinetic model" in the title in MEDLINE are as follows: 10 (1977–1986), 51 (1987–1996), 85 (1997–2006), and 123 (2007–2015) [5]. Recent applications of PBPK models have included lead compound optimization [6], *in vitro*/*in vivo* extrapolation [7], dose selection [8], drug–drug interaction assessment [9, 10], drug–disease interaction evaluation [11], and simulations in special populations [12].

12.2 HISTORY

The concept of PBPK models was first introduced by Teorell in 1937 [13]. Since then, PBPK modeling has been gradually evolving and now has become an important tool in drug discovery and development [1]. Therapeutic monoclonal antibodies (mAbs), as the fastest growing class of new therapeutic molecules, hold great promise for the treatment of a variety of diseases [14, 15]. The application of PBPK models to mAb PK analysis has only a 30-year history. Compared with the wide implementation of PBPK models for small molecules, the development and applications of PBPK models for mAbs is lagging. There are only limited numbers of published PBPK modeling examples for mAbs. The number of published articles for antibodies that include "physiologically-based pharmacokinetics" in the title in MEDLINE is only 15 [5].

It is known that mAbs have several unique PK properties compared with small molecules [16, 17]. There include much less renal and hepatic clearance, neonatal Fc receptor (FcRn) salvation, restricted vascular and cellular permeability, and often nonlinearity associated with target saturation. Thus, PBPK models that are developed for mAbs must manifest these

ADME and Translational Pharmacokinetics/Pharmacodynamics of Therapeutic Proteins: Applications in Drug Discovery and Development, First Edition. Edited by Honghui Zhou and Frank-Peter Theil.

unique PK properties, which result in different model structures and physiological components compared with the typical generic PBPK models for small molecules [18, 19]. Such PBPK models usually have components for the function of FcRn in mAb protection and recycling [18], target-saturable binding [20], endosome sorting and trafficking [21], and extravasation through convection rather than solely diffusion [22, 23]. Because some of these components that are relevant to antibody PK have not been well understood, the currently developed PBPK models for mAbs only reflect the best available knowledge of these physiological components, and such models need improvements as new knowledge becomes available. A recent PBPK model for mAbs is shown in Figure 12.1 [18].

The first PBPK model for immunoglobulin G (IgG) and its relevant fragments ($F(ab')_2$, Fab′) was constructed by Covell et al. in 1986 [22]. In this model, six major organs (liver, muscle, gut, spleen, kidney, and lung) were included and several basic antibody dispositional attributes were incorporated. Extravasation was assumed including both diffusion (*PS*) and convection ($L\cdot(1-\sigma)$), where σ is the reflection coefficient and $(1-\sigma)$ represents the efficiency of convective transport of IgG relative to the lymph flow (*L*). Vascular σ_v was fixed to be 0.95 for all tissues and no lymphatic reflection coefficients were assigned ($\sigma_L = 0$). The latter was modified to a low value ($\sigma_L = 0.2$) in recently developed PBPK models [18, 19]. Experimentally measured lymph flow was not available for individual tissues and it was assumed to be 2% of blood flow for visceral and 4% for nonvisceral tissues in Covell's model. This fraction for lymph flow was borrowed by some of the later developed PBPK models [18, 19].

Baxter et al. developed a more tissue-extensive PBPK model for both specific and nonspecific mAbs and antibody fragments in tumor-bearing mice [23]. This model included a solid tumor compartment where reversible and saturable binding to a specific antigen was defined. The "Two-pore" formalism was employed in this model for IgG extravasation. This theory would account for antibody extravasation at rates higher than convection solely driven by lymph flow. The "Two-pore" theory was proposed by Rippe and Haraldsson in 1994 based on systematic reevaluation of previous published data about capillary transport of macromolecules using heteroporous transport concepts and equations [24]. The equation for the "Two-pore" formalism extravasation model is

$$\text{Extravasation} = \underbrace{PS_L \times \left(1-\frac{C_i}{C_p}\right) \times \frac{Pe_L}{e^{Pe_L}-1}}_{\text{diffusion}} + \underbrace{J_L \times (1-\sigma_L)}_{\text{convection}} + PS_S \times \left(1-\frac{C_i}{C_p}\right) \times \frac{Pe_S}{e^{Pe_S}-1} + J_S \times (1-\sigma_S) \tag{12.1}$$

and

$$J_L = J_{iso} + \alpha_L \times L \tag{12.2a}$$

$$J_S = J_{iso} + \alpha_S \times L \tag{12.2b}$$

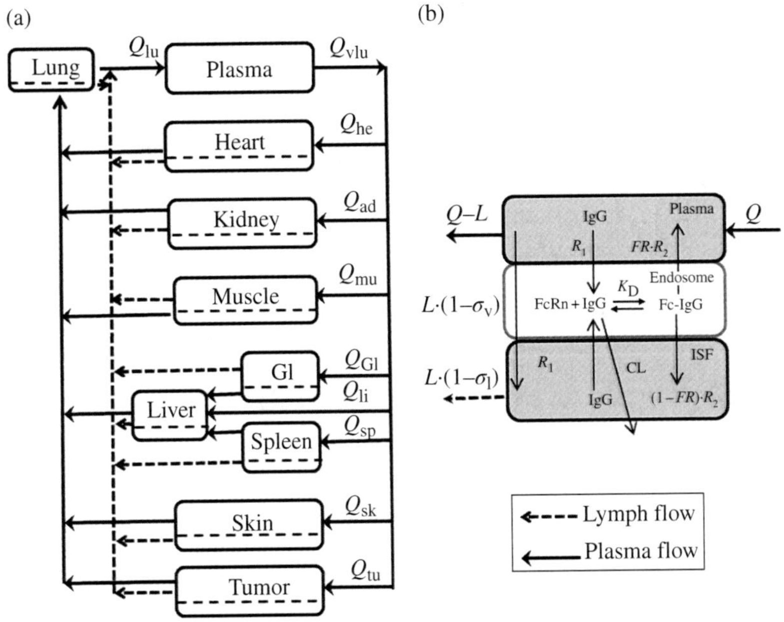

FIGURE 12.1 Structure of physiologically based pharmacokinetic model for IgG disposition (left). Organs are represented by rectangular compartments and connected by blood flow (solid lines) and lymph flow (dashed lines). Each tissue in the model is divided into three sub-compartments (right): vascular, endosomal, and interstitial space (*ISF*). The *Q* and *L* are the plasma and lymph flow, R_1 is the endocytosis rate and R_2 is the recycling rate, σ_V and σ_L are the vascular and lymphatic reflection coefficients, K_D is the dissociation coefficients for IgG binding with FcRn, *CL* is clearance from endosomal compartment, and F_R is the FcRn Recycling fraction. (Adapted from Garg and Balthasar [18].)

where PS is the permeability–surface area product, J_L and J_S are the fluid flow rates across capillary walls in each organ through large and small pores, L is the lymph flow for each organ, and J_{iso} is the fluid recirculation flow rate for each organ. The σ_L and σ_S are osmotic reflection coefficients for large and small pores, and Pe is the Peclet number, representing the ratio of convection to diffusion across large and small pores ($=J\cdot(1-\sigma)/PS$), where σ is the osmotic reflection coefficient. The diffusive component will approach zero when Pe is large at high lymph flow rates. In practice, extravasation will be dominated by convection when $Pe>3$. The α_L and α_S are the fractions of the hydraulic conductivity attributable to the large and small pores. In Baxter's model, two parameters J_{iso} and L were estimated by fitting plasma and a variety of tissue concentration–time data [23]. One major conclusion drawn in Baxter's model was that convection through large pores dominated IgG extravasation and diffusion played a minor role. Baxter et al. used this PBPK model that was developed in mice to further predict human plasma and tissue distribution profiles [25].

Ferl et al. were the first to incorporate FcRn functioning into PBPK models [26]. FcRn, also known as the Brambell receptor, has been recognized to play critical roles in salvage of IgG through its occurrence in pathways of endocytosis in endothelial cells. The function of FcRn was integrated in Ferl's model using a two-compartment submodel to depict the free and FcRn-bound IgG in endosomal space. Free IgG was subject to degradation and bound IgG was recycled back to plasma. Muscle and skin were the only two tissues that enacted FcRn functioning in Ferl's model. However, there were no experimental data for these tissues and "carcass" data from previous studies were "borrowed," which might involve problems for accurate evaluation of FcRn contributions. Endosomal space in endothelial cells was treated as an extension of the organ vasculature and the total IgG concentrations in endosomes was assumed equal to that in organ microvasculature. Davda et al. applied a similar approach incorporating FcRn in endosomes where endosomes was handled as an extension of organ vasculature [27]. Ferl's model also employed the "Two-pore" extravasation process. The reflection coefficient was fixed as 0.98 in all tissues for IgG. Lymph flow and J_{iso} were estimated by fitting plasma and tissue concentration–time data. As in Baxter's model, this model also incorporated saturable antigen binding in subcutaneous xenografts to predict the slow accumulation of specific IgG in solid tumors.

The importance of FcRn for IgG systemic persistence has been long recognized, but the mode of interaction between IgG and FcRn has never been well characterized and there is still no unified model that would adequately explain all available data [28, 29]. Balthasar and coworkers have made long-term efforts to integrate FcRn function into PBPK models to predict mAb systemic disposition [18–20]. The PBPK model in Figure 12.1 is one of their early models. This model defined an endosomal compartment in every tissue, where IgG interacts with FcRn leading to either IgG recycling or transcytosis. The function of FcRn in this model included IgG protection from lysosomal degradation, recycling IgG back to the circulation, and transcytosis into interstitial space. Once inside the endosomal space, the interaction between IgG and FcRn was assumed to achieve an equilibrium state characterized with an equilibrium dissociation constant (K_D). Free IgG in endosomal space would be subject to lysosomal degradation and most of the FcRn-bound IgG would be recycled to the circulation (recycling fraction (F_R) was estimated as 0.715). The "One-pore" formalism was applied in this model for IgG extravasation. Of note, the model was developed based on a broad dataset obtained by stressing the system with either high doses of intravenous IgG (IVIG) or the use of FcRn-knockout mice. This model was further used to predict human plasma profiles by scaling the flow type and the clearance type of parameters, and also used to predict tumor disposition with inclusion of specific antigen binding [20]. Shah and Betts further applied a similar PBPK modeling framework to simultaneously characterize the plasma and tissue disposition of mAbs in both animals and humans [30]. The work indicated that some parameters in this PBPK model may be conserved across species. If this is confirmed, it may greatly facilitate physiologically based interspecies extrapolation of mAb PK.

Chabot et al. proposed a PBPK model with special emphasis on the interfacial exchanges among vascular space, endosomal space, and interstitium [31]. The "One-pore" formalism was employed in this model, but a difference was that this model assumed convection occurred at both capillary and venous pores. About 90% of fluid filtered via capillary pores was recycled back to blood via venous pores in this model. The antibody–FcRn binding was characterized with an equilibrium constant (K_D). The residual radioisotope concentrations were also taken into account in this model by assuming that the degraded fragments continued giving signals when calculating total signals. This fragment pool was subject to first-order decay on a 24-h time scale. In addition, glomerular filtration was specified in kidneys with an estimation of the glomerular sieving coefficient.

Improving FcRn-binding affinity has been explored as a strategy to enhance mAb systemic persistence [32]. However, mixed results have been reported [33–35]. A catenary PBPK model was developed to provide an explanation for the poor correlation between FcRn-binding affinity and mAb systemic persistence [21]. This model applied a series of transit compartments to mimic the brief trafficking of endosomes (~7.5 min). Once in endosomal space, the interaction between mAb and FcRn was characterized with association (k_{on}) and dissociation rate constants (k_{off}), which is different from the previous binding equilibrium assumption. The brief transit for endosomes would not likely allow mAb and FcRn to reach an equilibrium binding before endosomal sorting. The nonequilibrium binding in this PBPK model predicts modest effects of enhanced FcRn affinity to improve IgG systemic persistence relative to previous equilibrium models, and the model predictions showed more consistency with experimental data. This model also provided simulations illustrating different impacts of adjusting k_{on} or k_{off} for the improvement of IgG systemic persistence. This catenary model differed from several previous models for physiological parameters, such as tissue specific FcRn expression, lymph flow, endosomal volume, and uptake rate of IgG into endosomes. A systematic comparison of these PBPK models is provided in Table 12.1.

12.3 PRINCIPLES AND METHODS

The overall principles of PBPK modeling for mAbs are essentially the same as that for small molecules. The PBPK model equations should conform to mass balance. The sum of the antibody amounts in all tissues at a given time plus the eliminated amount up to that time should equal the dose administered. In practice, the PBPK model should account for all distributional spaces as much as possible. Having more tissues in PBPK models allows them to consider a wider range of targets and sites of actions for mAbs for treatment or diagnostic purposes. Checking the experimental and computed mass balance is an important diagnostic approach in PBPK model development. For instance, if the mass balance at an early time after intravenous administration (when an insignificant amount of antibody has been eliminated from the body) is much less than the dose, this may indicate that an important tissue-sequestrating antibody is missed in the PBPK model [36]. If the summed mass is higher than the dose, then a significant bias in the analytical method is implied.

The total amount of antibodies in tissues includes antibodies in tissue vasculature, endosomal space, and interstitium [18, 19]. It has been suggested in previous PBPK models that, for certain high perfusion tissues such as liver and spleen, more than half of antibodies in tissue homogenate come from the tissue vasculature (trapped blood). Thus, it is important to evaluate tissue mAb concentrations by including either trapped blood or after proper tissue perfusion before tissue analysis.

A number of factors should be particularly considered for antibodies in developing PBPK models:

1. **Distribution Space**: Antibodies are known to have limited vascular and cellular permeability [16, 17]. The cellular compartment is generally not included for tissue distribution unless there is a membrane receptor-mediated active pathway, which sometimes applies to tumor cells. In most PBPK models, there are two (vascular and interstitial space) [22, 23] or three (vascular, interstitial, and endosomal space) [18, 19, 26] subcompartments for tissue distribution. The available space in the vascular compartment is normally assumed equal to plasma volume. All interstitial space was thought to be available for antibody distribution in most previous PBPK models, but this may not be true. The distributional fraction for antibodies in interstitial space was suggested to be around 0.4–0.8, which is due to steric, electrostatic, and size exclusions partly related to the presence of collagen and glycosaminoglycan [37, 38]. This ratio may be different among tissues and across species [39]. Therefore, specifically defining the exclusion fraction in each tissue may improve current PBPK models.
2. **Extravasation**: Multiple mechanisms are involved in antibody extravasation, such as diffusion, convection, and FcRn-mediated pinocytosis. Enacting these mechanisms in PBPK models is important to produce reliable predictions. Convection is believed to be the major extravasation mechanism for most tissues in recent PBPK models. Both "Two-pore" [23] and "One-pore" [18] formalism models have been used to account for antibody extravasation.
3. **Lymphatic Distribution**: It is generally believed that convection plays a dominant role in antibody extravasation and lymphatic circulation. Most of the antibodies filtered from the vasculature are returned to blood via lymph. The slow rate of lymphatic flow and the presence of lymph nodes are significant factors for antibodies. The lymphatics should be included in PBPK models. Balthasar's models were the first to incorporate a lymph node compartment in antibody disposition [19–21]. They employed an additional parameter (*Tau*) to account for the transit of antibodies from lymph nodes to the systemic circulation. *Tau* was estimated as 127 [19, 20] and 544 min [21]. The parameter *Tau* was not estimated in Shah's model, but the volume of total lymph nodes was defined to describe lymph recycling [30]. Lymph nodes only contain part of total lymph fluid as the large fraction stays in lymph ducts. Thus, a more physiological approach should use the volume of total lymph fluid, not just the volume of lymph nodes, to account for lymphatic distribution space.
4. **Specific Target Binding**: Antibodies often have high antigen specificity and binding affinity. Interactions with specific antigens can considerably influence antibody distribution and elimination. Site-specific and target-mediated tissue uptake and elimination are often necessary in PBPK models to justify specific tissue accumulation and/or nonlinear clearance [40, 41].

PBPK models are developed mainly for two complementary purposes: PK fitting/parameter estimation and simulation [36]. Most previously developed models for mAbs have been used for both purposes. The general workflow is to

TABLE 12.1 Summary of Previous Physiologically Based Pharmacokinetic Models for Antibodies and Their Fragments

Model [References]	Antibody	Forms	Assay	Tissue Compartment	Extravasation: Convection	Extravasation: Diffusion	Extravasation: Others	FcRn	Notes
Covell et al. [22]	Anti-LYB8.2; nonspecific anti-MOPC-21	IgG $F(ab')_2$ Fab′	^{131}I-	Vascular Interstitium Cellular	$0.5 \cdot L \cdot (1-\sigma) \cdot (C_v + C_i)$ $\sigma = 0.95$ $L = 2\%$ and 4% of plasma flow for visceral and nonvisceral organs	$PS \cdot (C_v - C_i)$ PS was estimated for each organ	NA	NA	A cellular compartment for degradation
Baxter et al. [23]	Anti-CEA, ZCE025; nonspecific anti-MOPC-21	IgG	^{111}In-	Vascular	"Two pore"	$PS_L \cdot (C_v - C_i/R) \cdot F(Pe) + PS_S \cdot (C_v - C_i/R) \cdot F'(Pe)$	NA	NA	T380 colon carcinoma xenograft tumor-bearing mice
		$F(ab')_2$ Fab′		Interstitial	$J_L \cdot (1-\sigma_L) \cdot C_v + J_S \cdot (1-\sigma_S) \cdot C_v$ $\sigma_L = 0.98$; $\sigma_S = 0.26$ L, fitted, $= 0.0008\%$ for skin; $= 0.02\%$ for liver and kidney; 0.07% for tumor (% plasma flow)	$PS_L = 2.66 \times 10^{-6}$ $PS_S = 7.80 \times 10^{-6}$ Based on albumin data, mL/min/g tissue			
Ferl et al. [26]	Anti-CEA, cT84.66	IgG	^{123}I-	Vascular	"Two pore"	$PS_L \cdot (C_v - C_i/R) \cdot F(Pe) + PS_S \cdot (C_v - C/R) \cdot F'(Pe)$	NA	IgG–FcRn: k_{on}, k_{off}	Colorectal LS174T xenograft mice
		$F(ab')_2$ Fab′ scFv Diabody Minibody	^{125}I- ^{131}I-	Endosome (V_{endo} not used) Interstitium	$J_L \cdot (1-\sigma_L) \cdot C_v + J_S \cdot (1-\sigma_S) \cdot C_v$ $\sigma_L = 0.98$; $\sigma_S = 0.26$ L, fixed to Baxter's values	PS_L and PS_S, fixed to Baxter's values			
Balthasar et al. [18]	Anti-human glycoprotein IIB/IIIA, 7E3 (anti-CEA, cT84.66, 8C2)	IgG	^{125}I-	Vascular	"One pore"	NA	Fluid-phase pinocytosis	IgG–FcRn: K_D	Database: wild type
				Endosome ($V_{endo} = 0.5\%$BW)	$L \cdot (1-\sigma_v)$ $\sigma_v = 0.95$				FcRn(−/−) IVIG (tumor-bearing)
				Interstitium	$L = 2\%$ and 4% of plasma flow for visceral and nonvisceral organs				Lymph node was considered

(Continued)

TABLE 12.1 (Continued)

Model [References]	Antibody	Forms	Assay	Tissue Compartment	Extravasation: Convection	Extravasation: Diffusion	Extravasation: Others	FcRn	Notes
Chabot et al. [31]	Anti-LYB8.2; nonspecific anti-MOPC-21, anti-Tac IgG	IgG	^{131}I-	Vascular	"One pore"	$PS_C \cdot (C_v - C_i/R) \cdot F(Pe) + PS_V \cdot (C_v - C_i/R) \cdot F'(Pe)$	Fluid-phase pinocytosis	IgG–FcRn: K_D	Defined capillary and venous convection and diffusion (similar but different with "Two-pore" formalism)
		F(ab′)$_2$		Endosome	$J_C \cdot (1-\sigma_C) \cdot C_v - J_V \cdot (1-\sigma_S) \cdot C_v$	$PS_L = 1.4 \times 10^{-4}$			
		Fab′		(V_{endo}, not reported)	$\sigma_C = 0.897$; $\sigma_v = 0.202$	$PS_S = 4.7 \times 10^{-3}$			
				Interstitial	L=0.024% for liver and GI, and 0.081% for others (of plasma flow)	mL/min/g tissue			
Davda et al. [27]	Anti-TAG-72 IgG CC49	IgG	^{131}I-	Vascular	"Two pore"	$PS_L \cdot (C_v - C_i/R) \cdot F(Pe) + PS_S \cdot (C_v - C_i/R) \cdot F'(Pe)$	NA	Not defined	Endosome compartment was defined as part of vascular compartment
		F(ab′)$_2$		Endosome	$J_L \cdot (1-\sigma_L) \cdot C_v + J_S \cdot (1-\sigma_S) \cdot C_v$	PS_L and PS_S, fixed to literature's values			
		Fab′		(V_{endo} not used) Interstitium	$\sigma_L = 0.98$; $\sigma_S = 0.26$ J_{iso} and L, estimated for each tissue				
Balthasar et al. [21]	Anti-human glycoprotein IIB/IIIA, 7E3	IgG	^{125}I-	Vascular	"One pore"	NA	Fluid phase pinocytosis	IgG–FcRn	Brief endosome transit largely influence FcRn-recycling efficiency
				Endosome	$L \cdot (1-\sigma_v)$			pH-dependent k_{on}, k_{off}	Lymph node was considered
				(V_{endo} ~0.009% tissue)	σ=0.95				
				Interstitium	L=0.2% of plasma flow for each tissue				

develop a PBPK model by fitting plasma and a number of tissue concentration–time profiles to estimate a few numerically identifiable parameters with most other parameters fixed to either hypothetical or literature values. Then, the developed PBPK model is further used to simulate PK beyond the experimental observations for either another species, or pathophysiological conditions, or connection to pharmacodynamic responses. Usually, the available data only support precise estimates of a few parameters in PBPK models; most other parameters are either structurally or numerically unidentifiable. This makes PBPK model parameter estimation itself difficult, particularly for the parameters being insensitive and the objective function surface being flat with change of these parameters, causing the optimizing algorithms to be trapped in local minima [36]. This is quite common for PBPK model estimation because the model is often overparameterized and only certain parameters can be resolved appropriately on the basis of the available data and knowledge. Thus, in modeling practice, most model parameters have to be fixed to either literature reports or hypothetical values leaving a few parameters to estimate. Sensitivity analysis is quite important to obtain more information about parameter reliability and certainty [18, 36].

12.4 CHALLENGES

12.4.1 Physiological Parameters

One advantage of PBPK models is related to its separation of system-dependent and drug-specific parameters, which allow independent evaluations of the influence of system or drug variables on drug disposition. However, this advantage is offset by problems of model overparameterization with certain parameters having high variability and uncertainty. The problem becomes even worse in models for mAbs because many physiological components have not been well characterized and are not readily measurable. For instance, for lymph flow, a proportional value to blood flow was often assumed for each tissue in previous PBPK models, but the values used were often different. The fractions employed or estimated for lymph flow included 0.02% [23], 2% [18, 22], 0.2% [21], and 4% [18, 22] of blood flow in previous models. The volume for endosomal space was another parameter with high uncertainty and it has been assumed as 0.5% [18] and approximately 0.009% [21] of tissue volume in previous PBPK models. The applied values for the rates of endocytosis (uptake) and exocytosis (recycling) were, respectively, about 1.96 and 20.4 day^{-1} in Balthasar's model [18], but were about 13.8 and 5.04 day^{-1} in Ferl's model [26]. These parameters were mostly borrowed from various literature sources or calculated based on allometric relationships. The bias in some parameters would be compensated by other parameters with no notable deviation in fitting experimental data. However, this hidden problem may considerably dampen model performance when it goes beyond the experimental observations. To characterize the potential influence of existing parameter variability on the model predictions, sensitivity analysis should be conducted in model development. Parameters that have great impact on the predictions should be given particular attention and more effort should be made to obtain reliable values of such parameters.

12.4.2 Extravasation Mechanisms

Molecules leave blood vessels (extravasation) mainly by two mechanisms: diffusion and convection [42, 43]. Diffusion depends on the concentration gradient, while convection depends on the interstitial fluid movement caused by pressure gradients. For antibodies, convection is often thought to be the dominant mechanism for extravasation [23, 24]. There are two convection models that have been applied in previous PBPK models: "One-pore" and "Two-pore" formalisms. Sometimes passive diffusion or pinocytosis mediated by FcRn was also included in PBPK models [18, 19]. The mixed extravasation mechanisms are not readily distinguishable when fitting only tissue data, particularly when their relative contributions are heavily dependent on the assumed values for lymph flow, vascular reflection coefficients, and pinocytosis rates into endosomal space. More elaborate experimental designs that separately evaluate these extravasations will be highly valuable for the development of next-generation PBPK models for mAbs.

12.4.3 FcRn Function

FcRn is the most recognized factor affecting IgG clearance [44]. There remains a large knowledge gap in quantitatively translating IgG–FcRn binding affinity to mAb systemic persistence [33–35]. Engineered mAbs with improved FcRn binding often did not show much improvement in their systemic persistence. This poor correlation has been hypothesized to be associated with the brief transit of endosome sorting [21, 45]. Antibody–FcRn binding could not achieve equilibrium during its stay in endosomal space. A catenary PBPK model accommodating the short transit time of endosomes seems to predict a modest effect of binding affinity on systemic clearance [21]. This model may permit broad applications in accounting for pH-dependent binding to simulate the benefit of many engineered mAbs, such as sweeping and recycling antibodies. In order to more accurately account for FcRn function, measuring absolute concentrations of effective FcRn, rate of endocytosis, and concentrations of IgG inside endosomes will be critical for reliable predictions. It remains unclear whether FcRn is involved in IgG tissue uptake, distribution, or subcutaneous absorption. Several recent studies using FcRn-knockout mice model suggested that FcRn involvement in tissue distribution may be bidirectional and tissue specific. Once this is confirmed, the latest PBPK models may accommodate this observation to provide more mechanistic description of FcRn function in IgG systemic disposition.

12.5 SIMPLIFIED PBPK MODELS FOR mAbs

Full PBPK models often accommodate the most currently known factors and mechanisms into the model structure. The complexities in PBPK models sometimes confront overparameterization and result in estimated parameters with high parameter uncertainty. An alternative physiologically based approach for PK analysis is to consider only the essential physiological elements in model construction [46]. A series of "minimal PBPK" modeling approaches has been proposed by us, which takes a reductionist concept to inherit and lump major physiologic attributes from whole-body PBPK models [46, 47]. This minimal modeling approach allows modelers to analyze only plasma concentration–time data and parsimoniously generate physiologically relevant PK parameters. Compared with classical mammillary models, minimal PBPK models provide parameters with low dimensionality and ease of interpretation and, importantly, better PK insights.

12.5.1 Minimal PBPK Models

With considerations of the essential PK attributes for mAbs, a second-generation minimal PBPK model was developed (Fig. 12.2) [47]. This model accommodates antibody-specific PK properties, such as convection as the primary mechanism for extravasation, interstitial fluid as the major extravascular distribution space, and the lymphatics. Two groups of tissues are included in this model according to their vascular endothelial structure: tight (continuous endothelium) and leaky (discontinuous or fenestrated). The tissues with tight endothelial structures are the muscle, skin, adipose, and the brain; the leaky tissues are the liver, spleen, kidney, and the remaining others.

Of note, the vascular endothelial endosome is frequently included in full PBPK models to enact the function of FcRn. It is known that an adult has about 1.0 kg of endothelial cells, which comprise about 1.6×10^{13} cells [48]. Given that endosomes are only a small part of cell space (0.6–2%) [49], the overall volume of vascular endothelial endosome may be about 0.4% of plasma volume and less than 0.1% of interstitial volume. Therefore, from a mass balance perspective, the amount of antibody that resides in the endosomal compartment only accounts for a small fraction of antibody in the system at any moment of time. This analysis is consistent with the conclusions drawn by theoretical analysis of published PBPK modeling parameters [50, 51]. In addition, the transcytosis between plasma and endothelial cells seems quite efficient, which produces fast uptake of mAbs into endothelial endosomes and yields rapid "equilibrium" between endothelial endosomes and plasma. Therefore, the antibody degradation in endothelial lysosomes would operate in a kinetically similar way as antibody systemic clearance (CL_p). Thus, the interaction between mAb and FcRn within endosomal space is assumed primarily influencing mAb systemic elimination and having a minor contribution to antibody interstitial distribution. Herein, higher FcRn affinity would more efficiently reduce mAb lysosome degradation and anticipate a lower value of CL_p in this modeling context. FcRn-binding affinity could be handled as a critical covariate when a population approach is performed using this model.

Nonspecific linear clearance for mAbs would be either from plasma or interstitial fluid. Model A assumed clearance from plasma (CL_p) and Model B assumed clearance from *ISF* (CL_i). The differential equations for Model A are

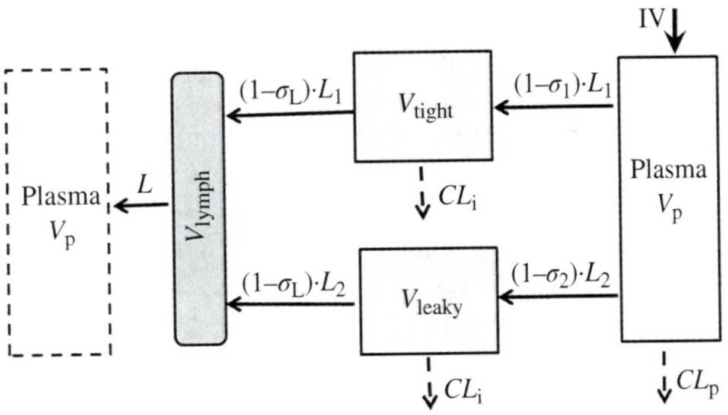

FIGURE 12.2 Second-generation minimal-PBPK model for monoclonal antibody pharmacokinetics. Symbols and physiological restrictions are defined in the text with Equations 12.3–12.8. Clearance is applied either to plasma (CL_p, Model A) or to interstitial fluid (CL_i, Model B). The plasma compartment in the left box represents the venous plasma as in full PBPK models, but is not applied in the present model [47].

$$\frac{dCp}{dt}=\frac{\text{Input}}{V_p}+[C_{lymph}\cdot L-C_p\cdot L_1\cdot(1-\sigma_1) -C_p\cdot L_2\cdot(1-\sigma_2)-C_p\cdot CL_p]/V_p \quad (12.3)$$

$$\frac{dC_{tight}}{dt}=\left[L_1\cdot(1-\sigma_1)\cdot C_p-L_1\cdot(1-\sigma_L)\cdot C_{tight}\right]/V_{tight} \quad (12.4)$$

$$\frac{dC_{leaky}}{dt}=\left[L_2\cdot(1-\sigma_2)\cdot C_p-L_2\cdot(1-\sigma_L)\cdot C_{leaky}\right]/V_{leaky} \quad (12.5)$$

$$\frac{dC_{lymph}}{dt}=[L_1\cdot(1-\sigma_L)\cdot C_{tight}+L_2\cdot(1-\sigma_L)\cdot C_{leaky} -C_{lymph}\cdot L]/V_{lymph} \quad (12.6)$$

The physiological restrictions are V_p is the plasma volume and V_{lymph} is the total lymph volume, and

$$\sigma_1 < 1 \quad (12.7a)$$

and

$$\sigma_2 < 1 \quad (12.7b)$$

$$V_{tight} = 0.65 \cdot ISF \cdot K_p \quad (12.8a)$$

and

$$V_{leaky} = 0.35 \cdot ISF \cdot K_p \quad (12.8b)$$

where C_p and C_{lymph} are antibody concentrations in plasma and lymph, and C_{tight} and C_{leaky} are antibody concentrations in tissue interstitial spaces with continuous endothelium (V_{tight}) and with fenestrated or discontinuous endothelium (V_{leaky}). The L is the total lymph flow and the sum of L_1 and L_2, where $L_1 = 0.33L$ and $L_2 = 0.67L$. The interstitial volume and lymph flow fractions for each group of tissues were obtained from previous values in PBPK models [18, 19]. Similar to Balthasar's model, the "One-pore" formalism was assumed for antibody extravasation. The σ_1 and σ_2 are vascular reflection coefficients for V_{tight} and V_{leaky}. The σ_L is the lymphatic capillary reflection coefficient, which is fixed to a commonly used value of 0.2 in previous PBPK models [18, 19]. The ISF is the total system interstitial fluid and K_p is the available fraction of ISF for antibody distribution. The physiologic parameters for a 70-kg body weight (BW) person are $L = 2.9$ L/day, $ISF = 15.6$ L, $V_{lymph} = 5.2$ L, and $V_{plasma} = 2.6$ L. Also, $K_p = 0.8$ for native IgG_1 and 0.4 for native IgG_4 [36, 37]. For most currently assessed therapeutic mAbs, given that they have similar isoelectric point (*p*I) values (in the range of 8–9) with native IgG_1 [52], K_p is usually set to 0.8 for data analysis.

Only three parameters need to be estimated in this model: σ_1, σ_2, and CL_p (or CL_i). The two clearances are not estimated together, but the model can assess which one works better. Important to note is that this model has one less parameter than the classical 2CM mammillary model that is usually used for antibody PK analysis. One example [53] with minimal PBPK model fitting is shown in Figure 12.3.

The transcapillary escape rate (TER) is the sum of two extravasation routes,

$$TER = L_1 \cdot (1 - \sigma_1) + L_2 \cdot (1 - \sigma_2) \quad (12.9)$$

The concentration ratios at equilibrium between ISF and plasma can be calculated in Model A as

$$(1 - \sigma_1)/(1 - \sigma_L) \text{ for } V_{tight} \quad (12.10a)$$

and

$$(1 - \sigma_2)/(1 - \sigma_L) \text{ for } V_{leaky} \quad (12.10b)$$

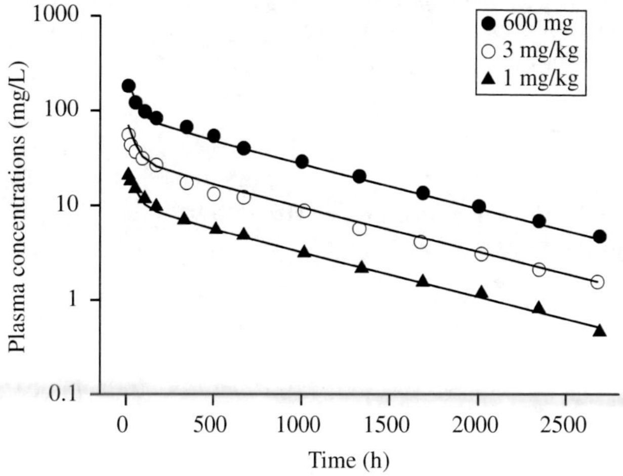

FIGURE 12.3 Pharmacokinetic profiles of canakinumab in human subjects [53]. Symbols are observations and curves are minimal PBPK model fitting (CL_p, Model A). The estimated parameter (mean (CV%)) are: σ_1 0.917 (3.3%), σ_2 0.716 (5.1%), CL_p 7.3 (1.6%) mL/h/70 kg.

In Model B, where clearance from ISF (CL_i) is assumed, the ratios are

$$\frac{C_{tight}}{C_p} = \frac{L_1 \cdot (1 - \sigma_1)}{L_1 \cdot (1 - \sigma_1) + CL_i} \quad (12.11a)$$

and

$$\frac{C_{tight}}{C_p} = \frac{L_2 \cdot (1 - \sigma_2)}{L_2 \cdot (1 - \sigma_2) + CL_i} \quad (12.11b)$$

As shown in Equations 12.9–12.11, the vascular reflection coefficients (σ_1 and σ_2) are parameters that not only reflect transcapillary rates (i.e., distribution rates) but also reveal the extent of distribution. The lower reflection produces a more rapid extravasation, resulting in earlier tissue profile peaking and relatively higher concentrations of mAb in the lumped interstitial compartment. This theoretical prediction is consistent with experimental observations. The "rate-determined extent" is one significant distribution feature for antibodies, which is the fundamental reason why this model has one less systemic parameter than the conventional mammillary model and anticipates more realistic predictions of antibody concentrations in interstitial space.

A key feature of this minimal PBPK model is predicting antibody interstitial concentrations in two groups of lumped tissues by analyzing only plasma data. It is experimentally challenging to obtain samples and measure antibody concentrations in interstitial fluid. Given that many therapeutic mAbs have targets in interstitial space, predicting the interstitial concentrations would help in better understanding interstitial target occupancy and the following pharmacodynamics. In order to assess whether our minimal model gives reasonable predictions of interstitial concentrations, two mAbs (7E3 and 8C2) with measured tissue concentrations [18, 54] were analyzed. By only fitting the plasma profiles, this model predicted tissue concentrations close to experimental measurements.

12.5.2 Survey of mAb PK in Humans with the Minimal PBPK Model

As an intermediate modeling approach between classical mammillary models and whole-body PBPK models, this model offers wider clinical application as it only requires plasma or serum concentrations that are the most common and convenient data to collect in humans. Implementation of this model is relatively easy, unlike full PBPK models that require investment of considerable experience, effort, time, and resources. This model approach, with only plasma data, provides more realistic parameters and predictions than mammillary models, and has become an intermediate alternative before utilizing full PBPK models for mAb PK analysis.

The second-generation minimal PBPK model has been applied to extensively survey mAb PK in human in order to seek general perspectives on mAb distributional and elimination features [55]. Profiles for 72 antibodies were successfully analyzed. The equilibrium concentration ratio (*ISF*/*plasma*) was predicted based on the estimated σ (Eq. 12.10) and clearance values (Eq. 12.11). As shown in Figure 12.4, the concentrations of therapeutic mAbs in tissues with continuous endothelium (V_{tight}) is about 6.3% of that in plasma and this ratio is about 36.0% for tissues with discontinuous or fenestrated endothelium (V_{leaky}).

Aside from predicting interstitial concentrations, the two clearance mechanisms CL_i and CL_p were assessed and compared in terms of their model performance and parameter estimates. For 93% of surveyed mAbs, the model assuming clearance from plasma (CL_p) produced better or at least equivalent model performance than the model with CL_i and yielded most consistent values of vascular reflection coefficients (σ_1 and σ_2) among all antibodies. This may further suggest that CL_p reflects the most common nonspecific clearance for most therapeutic mAbs, meaning that the major site for antibody degradation is likely in blood or tissues where antibodies could reach rapid equilibrium. However, it should be clarified that a model preference for CL_i or CL_p does not necessarily preclude the existence of the other. Both CL_i and CL_p processes could more or less happen. Model preference only suggests the possible location of the primary clearance process. The use of model fitting for such discrimination should be followed by further experimental validation, if possible, when this is an important issue for antibody assessment.

Plasma clearance in this modeling context is closely associated with FcRn salvage and lysosomal degradation. Most of the assessed mAbs have the preferred CL_p assumption. This observation is consistent with lysosomal degradation in endothelial cells playing a primary role in antibody nonspecific clearance. Although more clarification is warranted for the remaining 7% antibodies with CL_i preference, engineering antibodies with higher FcRn affinity does not offer as much promise as antibodies with CL_p preference in terms of improvement of systemic persistence.

Many factors are associated with mAb tissue distribution and systemic clearance [57], including size, hydrophobicity, and charge. Higher *TER* produces higher interstitial concentrations in this modeling context. Of 72 successfully analyzed mAbs, a positive correlation between tissue distribution and systemic clearance was detected (Fig. 12.5). This positive correlation was also indicated in previous studies. An increase in net positive charge increased both tissue retention and systemic clearance [52]. The effect of molecular size was also investigated in this regard showing that a larger molecule would generally result in lower tissue penetration with less systemic clearance [58]. Such positive correlation between distribution and clearance would challenge strategies for improving tissue distribution to enhance target-site exposure as the systemic clearance would probably consequently increase and somewhat offset the improved distribution. An optimum is possible when quantitatively considering all these factors to obtain a balance between distribution and clearance. Whatever factors are involved, this positive correlation appears not applicable to mAbs with high clearances (>35 mL/h/70 kg) (solid symbols in Fig. 12.5), implying that the factors responsible for high clearance may not necessarily increase tissue distribution, in contrast with most other therapeutic mAbs and macromolecules.

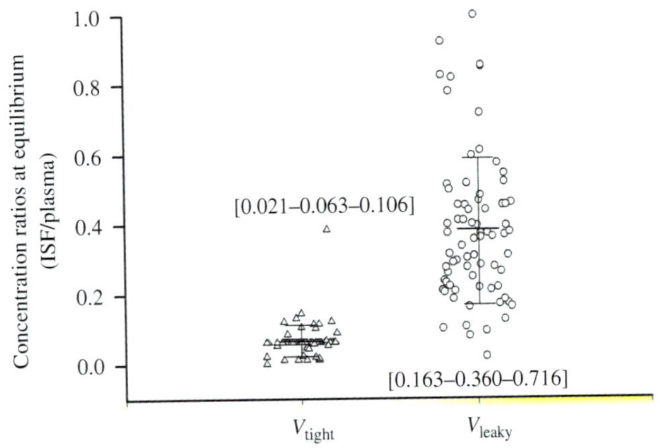

FIGURE 12.4 Predicted mAb concentration ratios between interstitial fluid (*ISF*) and plasma in two groups of lumped tissues: plasma at equilibrium using the model with CL_p. Bars represents mean and standard deviation. Numbers in brackets are [10%–50%–90%] percentiles. (Produced by combining data from Cao and Jusko [55, 56].)

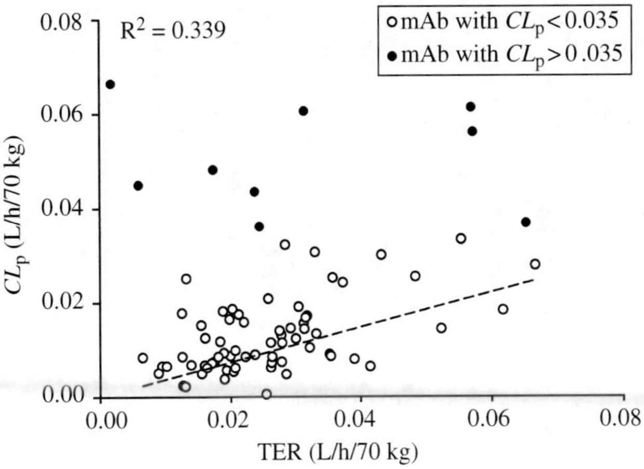

FIGURE 12.5 Correlation between plasma clearance (CL_p) and transcapillary escape rate (*TER*) predicted by the minimal PBPK model. The linear regression line (forced through (0,0); slope=0.34) and correlation coefficient are shown for mAbs with $CL_p < 0.035$ L/h/70 kg. Solid circles: clearance > 0.035 L/h/70 kg; open circles: clearance < 0.035 L/h/70 kg. (Produced by combining data from Cao and Jusko [55, 56].)

12.5.3 Minimal PBPK Model with Target-Mediated Drug Disposition

The term "target-mediated drug disposition" (TMDD) was coined by Levy [59] and the mathematical modeling framework was later established by Mager and Jusko [60]. Drugs exhibiting TMDD often bind with high affinity and to a significant extent (relative to dose) to their target or certain enzymes. Then, the drug–target interaction is reflected in their plasma PK profiles. TMDD is frequently reported for mAbs with strong target binding and high target abundance. Applying TMDD models for mAbs can reveal more insights about their targets and target-binding dynamics, which are essentially associated with pharmacodynamics. It was anticipated that the more realistic predictions of interstitial concentrations by the minimal PBPK model (vs hypothetical predictions with a 2CM mammillary model) would allow this modeling approach to provide more reliable assessment of interstitial target and antibody–target binding dynamics.

The proposed minimal PBPK models provide multiple options to incorporate TMDD [56]. This chapter highlights two situations where the minimal PBPK model enacts TMDD in either plasma (Model C, cTMDD) or interstitial space (Model D, pTMDD). As shown in Figure 12.6, the minimal PBPK model has the same structure and symbol designations as Model A (Fig. 12.2). Plasma clearance (CL_p) is the nonspecific clearance in this model given CL_p was shown to reflect the most common nonspecific clearance according to our assessment of 72 mAbs [54]. The differential equations for these models are not included in this chapter but they are available in our report [56].

The signature profiles for the conventional TMDD model (2CM with TMDD in central compartment) reveal an early quick decline phase at low doses (dose-dependent α-phase) and a prolonged terminal phase. The initial fast decline in conventional TMDD model is fundamentally associated with instant antigen binding at low drug concentrations. The cTMDD model retains almost the same signature profile as the conventional TMDD model. However, for the pTMDD model (when target binding occurs in peripheral tissues), the feature of early fast decline is not present, no matter at how low the dose or concentrations. The α-phases for different doses in the pTMDD model are always parallel. This is because extravasation restricts the initial decline regardless of interstitial target-mediated elimination. The extravasation is a dose-independent process, which yields an initial parallel phase. This feature can help to differentiate the two model conditions and sometimes helps imply target locations. For example, efalizumab [61], an antibody that binds to the α-subunit of LFA-1 (lymphocyte function-associated antigen-1), exhibits a dose-dependent α-phase after intravenous administration. This feature was captured with the cTMDD model, but could not be captured with the pTMDD model. This is consistent with the fact that the target LFA-1 is mainly in blood. However, for zalutumumab [62], an antibody that targets the epidermal growth factor receptor (EGFR), its nonlinear PK profiles exhibit a parallel α-phase. This is most likely associated with the extensive expression of EGFR in extravascular space.

In addition, when targets are in interstitial space, the traditional relationship between plasma concentrations and receptor occupancy (*RO*) is expected to shift from that when targets are in blood. Their relationship would be additionally affected by extravasation rate. In order to simplify their relationship, we assumed quasi-steady-state target binding and equilibrium extravascular distribution ($dA_{ISF\_total}/dt=0$). Then, total antibody in interstitial space (R_{total}) is within the equation:

$$L\cdot(1-\sigma_v)\cdot C_p - L\cdot(1-\sigma_L)\cdot C_{ISF} - \frac{R_{total}\cdot k_{int}\cdot C_{ISF}\cdot V_{ISF}}{K_{ss}+C_{ISF}} = 0 \quad (12.12)$$

This removes the factor "*t*" (time) to produce an explicit equation displaying the intrinsic relationships of other factors. The equation $RO = \frac{C_{ISF}}{K_{ss}+C_{ISF}}$ yields $C_{ISF} = \frac{RO\cdot K_{ss}}{1-RO}$, and substituting C_{ISF} in Equation 12.12, after rearrangement, generates

$$RO = \frac{\left[(1-\sigma_v)\cdot C_p+\omega+\upsilon\right] - \sqrt{\left[(1-\sigma_v)\cdot C_p+\omega+\upsilon\right]^2 - 4\upsilon\cdot(1-\sigma_v)\cdot Cp}}{2\upsilon} \quad (12.13)$$

where $\upsilon = R_{total}\cdot k_{int}\cdot \frac{V_{ISF}}{L}$ and $\omega = K_{ss}\cdot(1-\sigma_L)$.

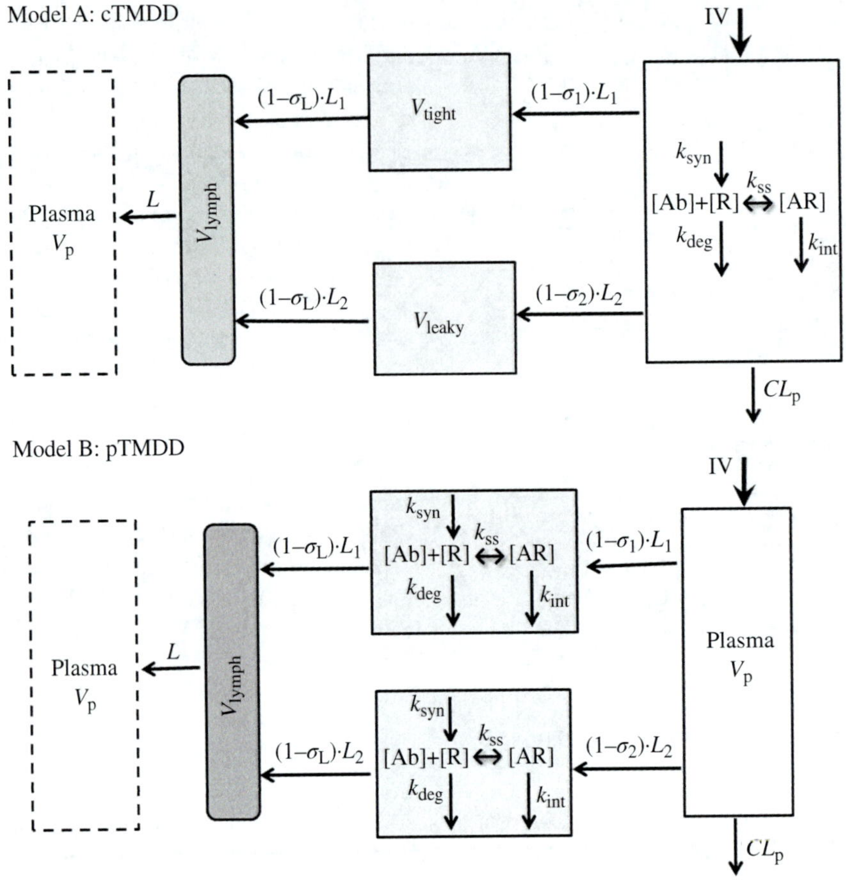

FIGURE 12.6 Model structures of second-generation minimal PBPK models with target-mediated drug disposition in either plasma (cTMDD, Model C) or interstitial fluid (pTMDD, Model D). Rate constants are k_{syn} for target biosynthesis, k_{deg} for target degradation, and k_{int} for antibody-target complex internalization. The K_{ss} is a steady-state constant for antibody-target binding [56].

The correlation between RO and σ_v was then simulated assuming constant plasma concentrations (C_p). The simulations indicated that vascular reflection (σ_v) significantly impacts peripheral RO, particularly for tissues with continuous endothelium (high σ_v). To improve peripheral RO, decreasing σ_v (such as endothelial tight junction openers [63]) appeared to be more efficient for these tissues with high σ_v (i.e., tissues with continuous vascular endothelium). In contrast, for tissues with low σ_v, increasing plasma concentration (or dose) seemed to be an efficient strategy. This analysis would help guide decision making in mAb development.

For the pTMDD model, the apparent target-mediated nonlinear clearance (CL_{TM}) can be derived in a similar approach as the well-stirred hepatic clearance model, where apparent plasma clearance is a function of blood flow and hepatic intrinsic clearance [64]. The well-stirred distribution of antibodies in interstitial space is supported by evidence that antibodies have a much higher diffusivity in interstitial fluid than for perivascular extravasation [65, 66]. Then, the apparent CL_{TM} is

$$CL_{TM} = \frac{L\cdot(1-\sigma_V)\cdot\frac{R_{total}\cdot k_{int}\cdot V_{ISF}}{K_{ss}+C_{ISF}}}{L\cdot(1-\sigma_L)+\frac{R_{total}\cdot k_{int}\cdot V_{ISF}}{K_{ss}+C_{ISF}}} \quad (12.14)$$

This derivation was based on a constant R_{total}, which intrinsically entails an assumption of $k_{deg}=k_{int}$. This assumption supports a simple derivation of CL_{TM}, which otherwise would not be easily solved explicitly. The CL_{TM} reaches its maximum value (CL_{TM_max}) when $C_{ISF}\rightarrow 0$, then

$$CL_{TM\_max} = \frac{L\cdot(1-\sigma_V)\cdot\frac{R_{total}\cdot k_{int}\cdot V_{ISF}}{K_{ss}}}{L\cdot(1-\sigma_L)+\frac{R_{total}\cdot k_{int}\cdot V_{ISF}}{K_{ss}}} \quad (12.15)$$

For a mAb with nonlinear clearance, the clearance usually increases with a decrease in plasma concentration and a maximum clearance is expected when concentrations approach zero. Then, at a given concentration, the clearance saturation ($1 - CL_{TM}/CL_{TM\_max}$) reflects the remaining fraction of nonlinear clearance. When targets are present in plasma, clearance saturation is equivalent to target saturation (*RO*) given that antibody concentrations are usually much higher than target concentrations. If targets exist in interstitial space, the two are no longer equal. Combining Equations 12.14 and 12.15, their relationship is

$$1 - \frac{CL_{TM}}{CL_{TM\_max}} = \frac{\omega \cdot RO}{\omega + \upsilon \cdot (1 - RO)} \quad (12.16)$$

where the parameters pertaining to υ and ω are the same as in Equation 12.13. The relationship between clearance saturation and target saturation was simulated and displayed in Figure 12.7. When TMDD occurs in blood (solid line), the saturation of nonlinear clearance is equal to saturation of targets if considering much higher mAb concentrations than their targets. However, their relationship is dramatically shifted by their target and target-binding parameters when targets are in extravascular space. Of note, misplacing these targets in plasma for modeling analysis would lead to a biased inference of *RO*. As indicated in Figure 12.7, any parameters pertaining to υ and ω (Eq. 12.16) contribute to the shifting. The higher values of R_{total} and k_{int} and higher binding affinity (K_{ss}, K_D) would produce a larger shift. Although σ_v substantially affects interstitial receptor saturation, it does not seem to be associated with such shifting.

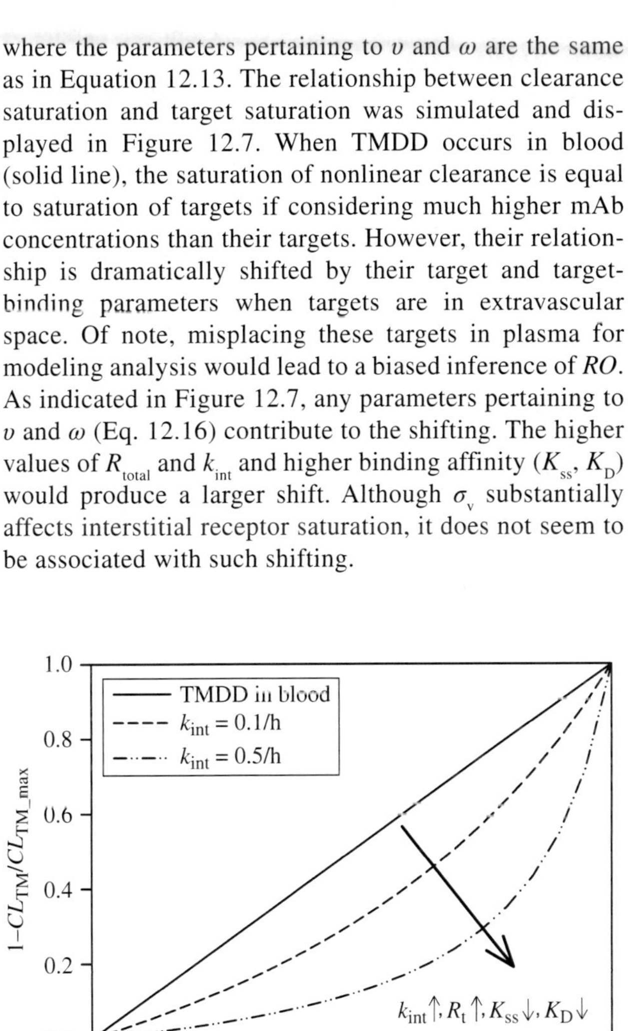

FIGURE 12.7 Simulated relationships between saturation of plasma nonlinear clearance ($1 - CL_{TM}/CL_{TM\_max}$) and saturation of targets (*RO*) in the minimal PBPK model. When the target is in *ISF*, any parameters that are related to target and target binding contribute to the deviation of their relationship from that with target in blood. The k_{int} is receptor-antibody complex internalization rate, R_t is total receptor density, K_D is dissociate rate constant of receptor-target binding, and K_{ss} is steady-state constant for receptor-target binding. The simulation is based on Equation 12.16 [56].

The first convention for TMDD models is to assign target binding in the central compartment of the 2CM mammillary model, but this sometimes yields inconsistent results for antibodies as their targets are often in peripheral tissues where the limited extravasation would not support rapid equilibrium [67]. One important reason for developing physiologically based TMDD models is to obtain more realistic evaluation of target and target binding kinetics. This is particularly important for antibodies with targets in extravascular space. Whole-body PBPK models have incorporated "tissue TMDD" before, mostly for a tumor TMDD model [19]. The minimal PBPK has shown feasibility to extend TMDD models and provide a simple framework for analyzing only plasma data but revealing peripheral target kinetics, which is otherwise not identifiable for peripheral target binding in mammillary models.

12.6 PERSPECTIVES

Despite their considerable potential, PBPK models have not been broadly implemented in therapeutic mAb discovery and development. The major reasons are as follows. (i) The understanding of factors and physiological components affecting antibody PK is incomplete. Many potential factors are still under investigation. Along with the progress of these studies, PBPK models are expected to be further improved by incorporating more mechanistic details and with more certainty in structure and parameters. (ii) The previously developed PBPK models well predicted total tissue concentrations of antibodies, but more effort is needed to use PBPK models to address critical issues in antibody development, such as subcompartment (*ISF*) concentrations, tissue target engagement, first-in-human dose, pH-dependent FcRn or target binding, immunogenicity, and disease influence on antibody disposition. Many of these issues could be addressed with PBPK modeling platforms by integrating more biological and immunological insights. (iii) Similar to PBPK models for small molecules, the structural, methodological, and computational complexity has largely limited its applications.

Minimal PBPK models take a reductionist concept to consider essential physiological components and provide an intermediate modeling approach between noncompartmental and full PBPK models. Such models offer a simple approach to provide physiologically relevant PK analysis based on only plasma data. As often happens in science, "The microscopic complexity seems to have conspired to produce something beautifully simple at the macroscopic level" [68]. This modeling approach has shown potential in addressing several critical issues (e.g., tissue target engagement and clearance mechanisms and predictions of human PK from animals [69]) in antibody development and is valuable for mAb PK analysis in place of or before a full PBPK model is established.

ACKNOWLEDGMENTS

This work was supported by NIH Grants GM57980 and GM24211, FDA Grant U01FD005206, and by the University at Buffalo Center for Protein Therapeutics.

REFERENCES

[1] Peters SA. *Physiologically-Based Pharmacokinetic (PBPK) Modeling and Simulations: Principles, Methods, and Applications in the Pharmaceutical Industry*. John Wiley & Sons, Inc; 2012.

[2] Rostami-Hodjegan A. Physiologically based pharmacokinetics joined with *in vitro-in vivo* extrapolation of ADME: a marriage under the arch of systems pharmacology. Clin Pharmacol Ther 2012;92:50–61.

[3] Smith BJ. An industrial perspective on contemporary applications of PBPK models in drug discovery and development. Biopharm Drug Dispos 2012;33:53–54.

[4] Huang SM, Abernethy DR, Wang Y, Zhao P, Zineh I. The utility of modeling and simulation in drug development and regulatory review. J Pharm Sci 2013;102:2912–2923.

[5] Search: physiologically-based pharmacokinetic model [ti] and yyyy[PDAT]: yyyy[PDAT]. Or physiologically-based pharmacokinetics [ti] and antibody[ab]. 1946. Bethesda (MD): National Library of Medicine. Available at http://pubmed.gov/. Accessed 2015 Aug 18.

[6] Benjamin B, Barman TK, Chaira T, Paliwal JK. Integration of physicochemical and pharmacokinetic parameters in lead optimization: a physiological pharmacokinetic model based approach. Curr Drug Discov Technol 2010;7:143–153.

[7] Malmborg J, Ploeger BA. Predicting human exposure of active drug after oral prodrug administration, using a joined *in vitro*/in silico-*in vivo* extrapolation and physiologically-based pharmacokinetic modeling approach. J Pharmacol Toxicol Methods 2013;67:203–213.

[8] Jones HM, Mayawala K, Poulin P. Dose selection based on physiologically based pharmacokinetic (PBPK) approaches. AAPS J 2013;15:377–387.

[9] Zhao Y, Hu ZY. Physiologically based pharmacokinetic modelling and *in vivo* [I]/K(i) accurately predict P-glycoprotein-mediated drug–drug interactions with dabigatran etexilate. Br J Pharmacol 2014;171:1043–1053.

[10] Brantley SJ, Gufford BT, Dua R, Fediuk DJ, Graf TN, Scarlett YV, Frederick KS, Fisher MB, Oberlies NH, Paine MF. Physiologically based pharmacokinetic modeling framework for quantitative prediction of an herb-drug interaction. CPT Pharmacometrics Syst Pharmacol 2014;3:e107.

[11] Sayama H, Takubo H, Komura H, Kogayu M, Iwaki M. Application of a physiologically based pharmacokinetic model informed by a top-down approach for the prediction of pharmacokinetics in chronic kidney disease patients. AAPS J 2014;16:1018–1028.

[12] Barrett JS, Della Casa Alberighi O, Laer S, Meibohm B. Physiologically based pharmacokinetic (PBPK) modeling in children. Clin Pharmacol Ther 2012;92:40–49.

[13] Teorell T. Kinetics of distribution of substances administered to the body. I. The extravascular modes of administration. Arch Int Pharmacodyn Ther 1937;57:205–225.

[14] Buss NA, Henderson SJ, McFarlane M, Shenton JM, de Haan L. Monoclonal antibody therapeutics: history and future. Curr Opin Pharmacol 2012;12:615–622.

[15] Sliwkowski MX, Mellman I. Antibody therapeutics in cancer. Science 2013;341:1192–1198.

[16] Lobo ED, Hansen RJ, Balthasar JP. Antibody pharmacokinetics and pharmacodynamics. J Pharm Sci 2004;93: 2645–2668.

[17] Mould DR, Green B. Pharmacokinetics and pharmacodynamics of monoclonal antibodies: concepts and lessons for drug development. BioDrugs 2010;24:23–39.

[18] Garg A, Balthasar JP. Physiologically-based pharmacokinetic (PBPK) model to predict IgG tissue kinetics in wild-type and FcRn-knockout mice. J Pharmacokinet Pharmacodyn 2007;34:687–709.

[19] Abuqayyas L, Balthasar JP. Application of PBPK modeling to predict monoclonal antibody disposition in plasma and tissues in mouse models of human colorectal cancer. J Pharmacokinet Pharmacodyn 2012;39:683–710.

[20] Urva SR, Yang VC, Balthasar JP. Physiologically based pharmacokinetic model for T84.66: a monoclonal anti-CEA antibody. J Pharm Sci 2010;99:1582–1600.

[21] Chen Y, Balthasar JP. Evaluation of a catenary PBPK model for predicting the *in vivo* disposition of mAbs engineered for high-affinity binding to FcRn. AAPS J 2012;14:850–859.

[22] Covell DG, Barbet J, Holton OD, Black CD, Parker RJ, Weinstein JN. Pharmacokinetics of monoclonal immunoglobulin G1, $F(ab')_2$, and Fab′ in mice. Cancer Res 1986;46:3969–3978.

[23] Baxter LT, Zhu H, Mackensen DG, Jain RK. Physiologically based pharmacokinetic model for specific and nonspecific monoclonal antibodies and fragments in normal tissues and human tumor xenografts in nude mice. Cancer Res 1994;54:1517–1528.

[24] Rippe B, Haraldsson B. Transport of macromolecules across microvascular walls: the two-pore theory. Physiol Rev 1994;74:163–219.

[25] Baxter LT, Zhu H, Mackensen DG, Butler WF, Jain RK. Biodistribution of monoclonal antibodies: scale-up from mouse to human using a physiologically based pharmacokinetic model. Cancer Res 1995;55:4611–4622.

[26] Ferl GZ, Wu AM, DiStefano JJ III. A predictive model of therapeutic monoclonal antibody dynamics and regulation by the neonatal Fc receptor (FcRn). Ann Biomed Eng 2005;33:1640–1652.

[27] Davda JP, Jain M, Batra SK, Gwilt PR, Robinson DH. A physiologically based pharmacokinetic (PBPK) model to characterize and predict the disposition of monoclonal antibody CC49 and its single chain Fv constructs. Int Immunopharmacol 2008;8:401–413.

[28] Gurbaxani BM, Morrison SL. Development of new models for the analysis of Fc–FcRn interactions. Mol Immunol 2006;43:1379–1389.

[29] Gurbaxani B. Mathematical modeling as accounting: predicting the fate of serum proteins and therapeutic monoclonal antibodies. Clin Immunol 2007;122:121–124.

[30] Shah DK, Betts AM. Towards a platform PBPK model to characterize the plasma and tissue disposition of monoclonal antibodies in preclinical species and human. J Pharmacokinet Pharmacodyn 2012;39:67–86.

[31] Chabot JR, Dettling DE, Jasper PJ, Gomes BC. Comprehensive mechanism-based antibody pharmacokinetic modeling. Conf Proc IEEE Eng Med Biol Soc 2011; 2011:4318–4323.

[32] Kuo TT, Baker K, Yoshida M, Qiao SW, Aveson VG, Lencer WI, Blumberg RS. Neonatal Fc receptor: from immunity to therapeutics. J Clin Immunol 2010;30:777–789.

[33] Yeung YA, Wu X, Reyes AE II, Vernes JM, Lien S, Lowe J, Maia M, Forrest WF, Meng YG, Damico LA, Ferrara N, Lowman HB. A therapeutic anti-VEGF antibody with increased potency independent of pharmacokinetic half-life. Cancer Res 2010;70:3269–3277.

[34] Deng R, Loyet KM, Lien S, Iyer S, DeForge LE, Theil FP, Lowman HB, Fielder PJ, Prabhu S. Pharmacokinetics of humanized monoclonal anti-tumor necrosis factor-{alpha} antibody and its neonatal Fc receptor variants in mice and cynomolgus monkeys. Drug Metab Dispos 2010;38: 600–605.

[35] Datta-Mannan A, Witcher DR, Tang Y, Watkins J, Jiang W, Wroblewski VJ. Humanized IgG1 variants with differential binding properties to the neonatal Fc receptor: relationship to pharmacokinetics in mice and primates. Drug Metab Dispos 2007;35:86–94.

[36] Nestorov I. Whole body pharmacokinetic models. Clin Pharmacokinet 2003;42:883–908.

[37] Wiig H, Tenstad O. Interstitial exclusion of positively and negatively charged IgG in rat skin and muscle. Am J Physiol Heart Circ Physiol 2001;280:H1505–H1512.

[38] Wiig H, Swartz MA. Interstitial fluid and lymph formation and transport: physiological regulation and roles in inflammation and cancer. Physiol Rev 2012;92:1005–1060.

[39] Wiig H, Kaysen GA, al-Bander HA, De Carlo M, Sibley L, Renkin EM. Interstitial exclusion of IgG in rat tissues estimated by continuous infusion. Am J Physiol 1994;266: H212–H219.

[40] Urva SR, Balthasar JP. Target mediated disposition of T84.66, a monoclonal anti-CEA antibody: application in the detection of colorectal cancer xenografts. MAbs 2010;2:67–72.

[41] Xin Y, Bai S, Damico-Beyer LA, Jin D, Liang WC, Wu Y, Theil FP, Joshi A, Lu Y, Lowe J, Maia M, Brachmann RK, Xiang H. Anti-neuropilin-1 (MNRP1685A): unexpected pharmacokinetic differences across species, from preclinical models to humans. Pharm Res 2012;29:2512–2521.

[42] Takakura Y, Mahato RI, Hashida M. Extravasation of macromolecules. Adv Drug Deliv Rev 1998;34:93–108.

[43] Tabrizi M, Bornstein GG, Suria H. Biodistribution mechanisms of therapeutic monoclonal antibodies in health and disease. AAPS J 2010;12:33–43.

[44] Datta-Mannan A, Witcher DR, Tang Y, Watkins J, Wroblewski VJ. Monoclonal antibody clearance. Impact of modulating the interaction of IgG with the neonatal Fc receptor. J Biol Chem 2007;282:1709–1717.

[45] Gurbaxani B, Dostalek M, Gardner I. Are endosomal trafficking parameters better targets for improving mAb pharmacokinetics than FcRn binding affinity? Mol Immunol 2013;56:660–674.

[46] Cao Y, Jusko WJ. Applications of minimal physiologically-based pharmacokinetic models. J Pharmacokinet Pharmacodyn 2012;39:711–723.

[47] Cao Y, Balthasar JP, Jusko WJ. Second-generation minimal physiologically-based pharmacokinetic model for monoclonal antibodies. J Pharmacokinet Pharmacodyn 2013; 40:597–607.

[48] Augustin HG, Kozian DH, Johnson RC. Differentiation of endothelial cells: analysis of the constitutive and activated endothelial cell phenotypes. Bioessays 1994;16:901–906.

[49] Griffiths G, Back R, Marsh M. A quantitative analysis of the endocytic pathway in baby hamster kidney cells. J Cell Biol 1989;109:2703–2720.

[50] Elmeliegy M, Lowe P, Krzyzanski W. Simplification of complex physiologically based pharmacokinetic models of monoclonal antibodies. AAPS J 2014;16:810–842.

[51] Fronton L, Pilari S, Huisinga W. Monoclonal antibody disposition: a simplified PBPK model and its implications for the derivation and interpretation of classical compartment models. J Pharmacokinet Pharmacodyn 2014;41:87–107.

[52] Boswell CA, Tesar DB, Mukhyala K, Theil FP, Fielder PJ, Khawli LA. Effects of charge on antibody tissue distribution and pharmacokinetics. Bioconjug Chem 2010;21:2153–2163.

[53] Chakraborty A, Tannenbaum S, Rordorf C, Lowe PJ, Floch D, Gram H, Roy S. Pharmacokinetic and pharmacodynamic properties of canakinumab, a human anti-interleukin-1beta monoclonal antibody. Clin Pharmacokinet 2012;51:e1–e18.

[54] Abuqayyas L, Balthasar JP. Application of knockout mouse models to investigate the influence of FcgammaR on the tissue distribution and elimination of 8C2, a murine IgG1 monoclonal antibody. Int J Pharm 2012;439:8–16.

[55] Cao Y, Jusko WJ. Survey of monoclonal antibody disposition in man utilizing a minimal physiologically-based pharmacokinetic model. J Pharmacokinet Pharmacodyn 2014;41 (6):571–580.

[56] Cao Y, Jusko WJ. Incorporating target-mediated drug disposition in a minimal physiologically-based pharmacokinetic model for monoclonal antibodies. J Pharmacokinet Pharmacodyn 2014;41:375–387.

[57] Bumbaca D, Boswell CA, Fielder PJ, Khawli LA. Physiochemical and biochemical factors influencing the pharmacokinetics of antibody therapeutics. AAPS J 2012;14:554–558.

[58] Schmidt MM, Wittrup KD. A modeling analysis of the effects of molecular size and binding affinity on tumor targeting. Mol Cancer Ther 2009;8:2861–2871.

[59] Levy G. Pharmacologic target-mediated drug disposition. Clin Pharmacol Ther 1994;56:248–252.

[60] Mager DE, Jusko WJ. General pharmacokinetic model for drugs exhibiting target-mediated drug disposition. J Pharmacokinet Pharmacodyn 2001;28:507–532.

[61] Joshi A, Bauer R, Kuebler P, White M, Leddy C, Compton P, Garovoy M, Kwon P, Walicke P, Dedrick R. An overview of the pharmacokinetics and pharmacodynamics of efalizumab: a monoclonal antibody approved for use in psoriasis. J Clin Pharmacol 2006;46:10–20.

[62] Lammerts van Bueren JJ, Bleeker WK, Bogh HO, Houtkamp M, Schuurman J, van de Winkel JG, Parren PW. Effect of target dynamics on pharmacokinetics of a novel therapeutic antibody against the epidermal growth factor receptor: implications for the mechanisms of action. Cancer Res 2006;66:7630–7638.

[63] Beyer I, van Rensburg R, Strauss R, Li Z, Wang H, Persson J, Yumul R, Feng Q, Song H, Bartek J, Fender P, Lieber A. Epithelial junction opener JO-1 improves monoclonal antibody therapy of cancer. Cancer Res 2011;71:7080–7090.

[64] Wilkinson GR, Shand DG. Commentary: a physiological approach to hepatic drug clearance. Clin Pharmacol Ther 1975;18:377–390.

[65] Nugent LJ, Jain RK. Extravascular diffusion in normal and neoplastic tissues. Cancer Res 1984;44:238–244.

[66] Gerlowski LE, Jain RK. Microvascular permeability of normal and neoplastic tissues. Microvasc Res 1986;31: 288–305.

[67] Fetterly GJ, Aras U, Meholick PD, Takimoto C, Seetharam S, McIntosh T, de Bono JS, Sandhu SK, Tolcher A, Davis HM, Zhou H, Puchalski TA. Utilizing pharmacokinetics/pharmacodynamics modeling to simultaneously examine free CCL2, total CCL2 and carlumab (CNTO 888) concentration time data. J Clin Pharmacol 2013;53:1020–1027.

[68] Gunawardena J. Models in biology: 'accurate descriptions of our pathetic thinking'. BMC Biol 2014;12:29.

[69] Zhao J, Cao Y, Jusko WJ. Across-Species Scaling of Monoclonal Antibody Pharmacokinetics Using a Minimal PBPK Model. Pharm Res 2015 (in press).

13

INTEGRATED QUANTITATION OF BIOTHERAPEUTIC DRUG–TARGET BINDING, BIOMARKERS, AND CLINICAL RESPONSE TO SUPPORT RATIONAL DOSE REGIMEN SELECTION

PHILIP J. LOWE[1], ANNE KÜMMEL[1], CHRISTINA VASALOU[2], SOICHIRO MATSUSHIMA[3] AND ANDREJ SKERJANEC[4]

[1] *Novartis Pharmaceuticals AG, Basel, Switzerland*

[2] *Novartis Institutes for Biomedical Research, Cambridge, MA, USA*

[3] *Novartis Pharmaceuticals K.K., Tokyo, Japan*

[4] *Sandoz Biosimilars, Novartis, Basel, Switzerland*

13.1 INTRODUCTION

Posology, "the study of the dosage of medicines" can be regarded as a scientific field in its own right. Indeed, *en français, la posologie* is on the packaging to describe the dose and regimen to be used. In the development of new medicines, the posology can be critical to get the intended clinical benefit whilst minimizing dose-related unintended or adverse effects. How, then, to get the dose and regimen, *la posologie*, right for all patients?

One technique, pharmacokinetic/pharmacodynamic (PK/PD) model-based analysis, aims to describe, quantitatively, the link between the concentrations of drug in the blood (which perfuses all tissues) with downstream biomarker and clinical responses in said tissues. Once a model has been successfully fitted to PK and PD data, even if the data were not collected at steady state, dose–responses can be projected, through simulation, for the steady-state situation. However, although closer to a physiological truth than simpler dose–response models, they still do not capture intermediate pharmacological mechanisms such as target engagement or receptor occupancy. One technique, developed over the last decade, is to directly quantitate the formation of drug–target complexes, relating this to biomarker changes and clinical outcomes.

This analysis technique has been used to either determine, or aid in the determination of, the dose–response and dose–time–response of several products at Novartis. The posology for canakinumab (Ilaris™) for patients with cryopyrin-associated periodic syndromes was determined by model-based PK/PD analysis and simulations. The approach used two models in parallel. In the first, statistically more robust model, the serum PK drove a probability of inflammatory flares, the primary clinical endpoint. In the second, more mechanistic, narrative generating model, canakinumab captured interleukin-1β reducing levels of the free cytokine, thereby causing reductions in inflammatory biomarkers and the probability of flare [1]. A similar mechanistic IL-1β binding model was developed and used to describe the canakinumab (American College of Rheumatologist's score)

ADME and Translational Pharmacokinetics/Pharmacodynamics of Therapeutic Proteins: Applications in Drug Discovery and Development, First Edition. Edited by Honghui Zhou and Frank-Peter Theil.

ACR dose–response in rheumatoid arthritis [2]. With omalizumab (Xolair™) in asthma and chronic spontaneous urticaria, the binding model has been used to support the dose description [3], an asthma dosing table extension [4, 5], a revision of the asthma dosing table [6], and the posology for chronic spontaneous urticaria [7]. The model is now being used for the next-generation anti-IgE IgG-type monoclonal antibody, QGE031 (ligelizumab).

Naturally, Novartis is not alone in the application of ligand-binding PK/PD models. There are a number of examples where drug–target-binding PK/PD models have been used on preclinical experimental data. Lammerts van Bueren et al. [8] used such a model to quantitate the binding of a monoclonal antibody to the epidermal growth factor receptor, pointing out challenges with regard to quantitating receptor saturation in specific tissues. Similarly, Vugmeyster et al. [9] described a target-binding model for IL-13, concluding that the increase in total target in the serum can aid the understanding of the extent of neutralization occurring in different conditions. On the topic of scaling between species, Betts et al. [10] used a binding model for a monoclonal antibody to Dickkopf-1 in rats and nonhuman primates, predicting doses required for clinical studies in osteoporotic patients. Similarly, Lowe et al. [11] scaled the anti-CD40 monoclonal lucatumumab between nonhuman primates and oncology patients.

The preclinical use of drug–target-binding model continued with human clinical studies. Although Ng et al. in 2005 [12] very nicely described the nonlinear PK of efalizumab linked to CD11a occupancy and turnover and improvement in psoriasis symptoms, this was not a direct target-binding model. The following year, however, they used a full binding model for the anti-CD4 monoclonal TRX1 [13]. In subsequent years, Stefanini et al. [14] were quantitating the capture of vascular-endothelial growth factor with bevacizumab, as were Thai et al. [15] with the fusion protein aflibercept. Sutjandra et al. [16] followed by Gibiansky et al. [17] developed population models for denosumab binding to the receptor activator of nuclear factor κ-B ligand, initially for osteopenia/osteoporosis, then for oncology patients with bone metastases. More recently, Wang et al. [18] described mavrilimumab binding the granulocyte-macrophage colony stimulating factor receptor in rheumatoid arthritis patients, with Fetterly et al. [19] investigating carlumab capturing the chemokine CCL2.

Clearly, the use of direct drug–target-binding models has increased over the years, forming, in effect, a new class to be added to the original effect site and indirect response models. For this chapter, the focus will be on the capture of soluble targets. However, as in the examples above, it is also possible to quantitate the nonlinear target-mediated disposition in the PK to provide an estimate of whole-body target engagement. If related to downstream biomarkers or, preferably, to clinical outcomes, this is just as valuable to the posologist as soluble target capture.

Examples will be given to show, in detail, how to quantitate drug–target binding, integrate this with biomarker measurements, relate this to patient outcomes, and hence enable drug development teams to determine suitable posologies for patients. The examples are as follows: (i) Xolair binding IgE to reduce the itch and hives symptoms of chronic spontaneous urticaria and (ii) a higher affinity anti-IgE, QGE031 (ligelizumab), relating the capture of IgE to downstream basophil and allergen skin prick test responses, in order to decide on the dose range for Phase 2.

13.2 METHODS

Models describing the binding between a drug (warfarin) and targets (albumin and vitamin K 2,3-epoxide reductase) leading to *target-mediated drug disposition* were first described by Levy in 1994 [20]. A general drug–target binding PK/PD model was described by Mager and Jusko in 2001 [21], but there was still little advancement in the use of binding models until quasi-equilibrium solutions enabled more reliable integration and faster computation. Mager and Krzyzanski [22] first reported one of these in 2005, exemplifying its use using ADAPT II for naive pooled data or single individuals. Independently (and in parallel), Hayashi et al. [23] developed a mathematically identical quasi-equilibrium solution for population analyses for use with NONMEM.

13.2.1 Omalizumab, IgE, Itch, and Hives

13.2.1.1 The Omalizumab–IgE Binding Model The population PK/PD model of omalizumab–IgE binding and turnover has evolved over many years [4, 23–25] supporting license applications in various countries. The model specified the 1:1 binding of omalizumab with IgE according to the reversible reaction:

$$\text{omalizumab} + \text{IgE} \leftrightarrows \text{omalizumab} - \text{IgE complexes}$$

Inputs and elimination of omalizumab, IgE, and the complexes were added to this reaction to form a PK–IgE binding and turnover model (Fig. 13.1). In order to quantitate the *in vivo* binding of a drug with a soluble ligand, data on free and/or total drug, plus either total and/or free soluble ligand are required. Having both free and total ligand is nice, but not essential as, through the binding reaction, missing components can be calculated [11]. Indeed, given the difficulties and potential biases associated with free ligand assays, it could be preferable to use solely total drug and total ligand data [26].

The above binding and turnover scheme was translated to a system of three coupled differential equations, one for the

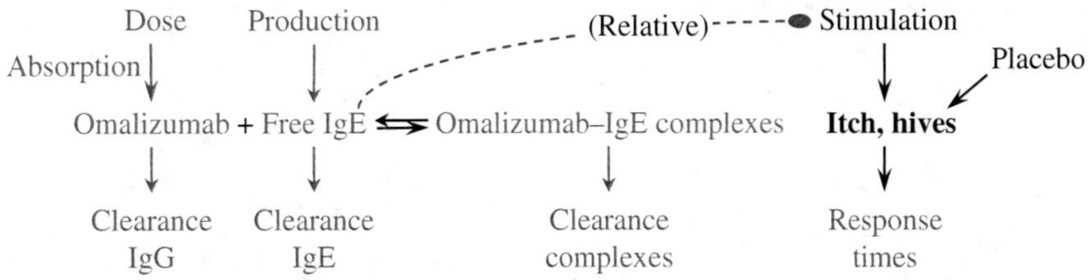

FIGURE 13.1 Drug–target binding model with indirect link to itch and hives symptoms.

subcutaneous administration site, one for total omalizumab (free plus drug–IgE complexes), and another for total IgE (free plus drug–IgE complexes). The coupled differential equations, in terms of molar masses of omalizumab, IgE, and the complexes, were

$$\begin{aligned} S' &= S*(-KA) \\ XT' &= S*KA - XF*CL_X/V_X - CC*CL_C/V_C \\ ET' &= R_E - EF*CL_E/V_E - CC*CL_C/V_C \end{aligned} \quad (13.1)$$

where

$$\begin{aligned} XF &= XT - CC \\ EF &= ET - CC \end{aligned}$$

Here, S is the amount of omalizumab at the subcutaneous site; XT and ET are molar masses of total omalizumab and IgE; XF and EF are free omalizumab and IgE; KA is the absorption rate constant; R_E is the rate of production (i.e., expression) of IgE; and CL_X, CL_E, and CL_C are the clearances of free omalizumab, free IgE, and the complexes, respectively. Although the omalizumab model used a single compartment, there were three volumes for the three components, V_X, V_E, and V_C. Exploratory analyses established that separate V_X and V_E estimates could not be made from the available data; they were assumed the same. The molar mass of the omalizumab–IgE complexes, CC was defined as the solution to the quadratic for the quasi-equilibrium binding:

$$\begin{aligned} CC = \big(&K_D*V_X*V_E/V_C + XT + ET \\ &-((K_D*V_X*V_E/V_C + XT + ET) \\ &**2 - 4*XT*ET)**0.5\big)/2 \end{aligned} \quad (13.2)$$

where

$$K_D = K_{D0}*(XT/ET)**\text{alpha}$$

Here, K_D is the *in vivo*, apparent equilibrium binding dissociation constant and alpha the change in the affinity of binding between omalizumab and IgE as a function of the molar ratio of total omalizumab to total IgE, to take account of the changing stoichiometry of omalizumab to IgE binding from two omalizumab molecules to one IgE when omalizumab concentrations are greater than IgE, to the reverse when IgE exceeds omalizumab as the drug finally clears from the system [27]. Weight-based doses were converted to molar amounts in the program code, then weight-based concentrations were calculated from the molar masses by dividing by the respective volumes, correcting for the molecular mass for each component (free and complex). The free and complex concentrations were then summed to attain total omalizumab and IgE concentrations.

13.2.1.2 Linking IgE with Urticaria Itch and Hives

Extending the model to the effect of omalizumab treatment on urticaria itch and hives, free IgE levels relative to baseline (*rfIgE*) controlled the appearance of symptoms in a traditional indirect response scheme (Fig. 13.1). With omalizumab treatment, free IgE is reduced thereby alleviating symptoms. In the model code, two latent variables, *EFF*, one each for itch and hives, were added as differential equations:

$$EFF' = (STIM - EFF)/XRT \quad (13.3)$$

where

$$\begin{aligned} STIM &= rfIgE**\text{Hill}/(EC_{50}**\text{Hill} + rfIgE**\text{Hill}) \\ rfIgE &= XF/\text{IgE}_0 \\ EFF_{(t=0)} &= 1 \end{aligned}$$

The turnover was parameterized by a drug mean response time, *XRT*, common to itch and hives. The stimulation was by a sigmoidal function with separate EC_{50} values for itch and hives but a common Hill coefficient. The *EFF* drug effect variables could range from 0 to 1. These were multiplied by the maximal possible drug effect (E_{max}) and, since the scores decrease under omalizumab, subtracted from a "placebo or disease natural history" part of the model within a logit-transform.

$$I_{\text{logit}} = PLB - E_{\max} * (1 - EFF) \tag{13.4}$$

where

$$PLB = (I_0 - I_{SS}) * \exp(-(\text{time - } lag) / PRT) + I_{SS}$$

The placebo model was an exponential decrease of itch or hives from an initial state, I_0, to a steady state, I_{SS}, over time, with a time *lag* and placebo response time, *PRT*. After the addition of residual error, the I_{logit} was back-transformed to the original scale.

13.2.2 QGE031 and Omalizumab, IgE, Basophil FcεR1 and Surface IgE, and Allergen Skin Prick Test Response

The one-compartment model described above for omalizumab can be extended to represent the central–peripheral distribution of monoclonal antibodies, as observed in their typically biexponential PK profiles, and the binding and capture of both serum and tissue ligands. A two-compartment binding model has been described for canakinumab and IL-1β [1, 28]. In this, the drug–ligand quasi-equilibrium binding reaction, as described earlier, was replicated to both compartments. This enabled the whole-body production of IL-1β to be estimated based on only total canakinumab and total IL-1β in the serum. However, in this model, it was assumed that the volume of the peripheral compartment was that estimated from the canakinumab PK. Unfortunately, the estimated peripheral volume, 1.7–2.9 L, was far less than the actual 15 L interstitial fluid volume (17.5 L extracellular water minus 2.7 L plasma [29]). This together with the fact that IgG concentrations are known to be lower in interstitial spaces (exemplified by the draining lymphatics [30]) is such that the binding calculations and hence the reported apparent binding parameter are unlikely to be accurate.

To ensure a slightly more realistic representation of interstitial fluid volumes and hence binding calculations, a "physiologic concept" structure was implemented for QGE031 and omalizumab capturing IgE. In this, the clearance was set to occur from both compartments, resulting in the estimation of a larger and more realistic peripheral volume. To account for the fact that the apparent volume for QGE031 or omalizumab–IgE complexes could be smaller than that for the free monoclonals or free IgE, a restriction of the entry of the complexes to the peripheral space was introduced as a reflection coefficient.

The coupled differential and algebraic equations (13.5) then followed the model structure in Figure 13.2. Note that the coupled PK and IgE binding equations were written in terms of mass, whilst basophil FcεR1 and surface IgE (sIgE) represent concentrations.

$$
\begin{aligned}
S' &= S * (-KA) \\
XT' &= S * KA - XF * CL_X / V_C - C_C * CL_C / V_C \\
&\quad - Q * (XF / V_C - XF_P / V_P) \\
&\quad - Q * ((1 - SR) * C_C / V_C - C_P / V_P) \\
ET' &= R_E - EF * CL_E / V_C - C_C * CL_C / V_C \\
&\quad - Q * (EF / V_C - EF_P / V_P) \\
&\quad - Q * ((1 - SR) * C_C / V_C - C_P / V_P) \\
XP' &= -XF_P * CL_X / V_P - C_P * CL_C / V_P \\
&\quad + Q * (XF / V_C - XF_P / V_P) \\
&\quad + Q * ((1 - SR) * C_C / V_C - C_P / V_P) \\
EP' &= R_P - EF_P * CL_E / V_P - C_P * CL_C / V_P \\
&\quad + Q * (EF / V_C - EF_P / V_P) \\
&\quad + Q * ((1 - SR) * C_C / V_C - C_P / V_P) \\
FCER' &= (BFCER * STIM - FCER) \\
&\quad / KBASO
\end{aligned}
\tag{13.5}
$$

where

$$
\begin{aligned}
STIM &= (1 + CEF_P / (CEF_P + EC_{50R})) ** STIMC \\
SIGE &= BSIgE * (FCER / FCER_0) ** SIGES \\
SPTW &= STIMW * (FCER / FCER_0) ** ExpW \\
SPT1' &= (SPTW - SPT1) / (KSPT / 4) \\
SPT2' &= (SPT1 - SPT2) / (KSPT / 4) \\
SPT3' &= (SPT2 - SPT3) / (KSPT / 4) \\
SPT4' &= (SPT3 - SPT4) / (KSPT / 4) \\
WHEA &= SPT4 \\
FLAR &= STIMF * (WHEA / WHE0) ** ExpF
\end{aligned}
$$

As with the previous omalizumab model, the drug transfer from a depot injection site (i.e., subcutaneous, *S*) to the central compartment was a first-order process controlled by *KA*. The total drug in the central compartment, *XT*, gained mass directly from the intravenous infusion, or from the subcutaneous depot site. Free drug, *XF*, was lost at rate CL_X/V_C, whilst that portion that was complexed, C_C, was lost at rate CL_C/V_C. There was also exchange of total drug between the central and peripheral compartments controlled by the intercompartmental flow *Q*. Similarly, the total IgE in the central compartment, *ET*, exchanged with the peripheral compartment. Free IgE was eliminated at rate CL_E/V_C and the complex at rate CL_C/V_C.

From the overall IgE production (E_{PR}), IgE was released by the body into the central compartment at rate

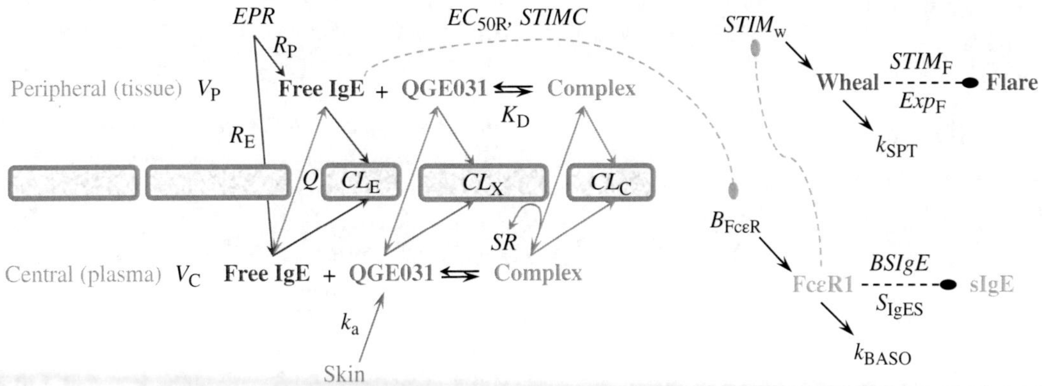

FIGURE 13.2 QGE031 and omalizumab two-compartment binding model structure. The entities modeled (emboldened) were free IgE, the drug (QGE031 or omalizumab), the drug–IgE complexes plus FcεR1 and IgE on the surface of basophils, and the skin prick test wheal and flare. Parameters are in black italicized font, transfer processes as solid arrows and influences or control mechanisms as dotted lines.

R_E in proportion to the volume of this compartment relative to that for the whole body; it was also converted from a mass to a molar amount by accounting for the molecular weight, 190 µg/nmol. Similarly, IgE entered the peripheral compartment at rate R_P.

$$R_E = E_{PR}{}^{*}V_C / V_{SS} / 190,$$
$$R_P = E_{PR}{}^{*}V_P / V_{SS} / 190 \text{ where } V_{SS} = V_C + V_P \qquad (13.6)$$

The peripheral total drug compartment exchanged with the central compartment at rate Q, with free drug and complex being eliminated at rates CL_X and CL_C, respectively. Meanwhile, total IgE was released into and eliminated from the peripheral compartment and exchanged with the central one. The exchange of the drug–IgE complexes was modified by the reflection coefficient *SR* in order to create an effective volume for said complexes smaller than that occupied by the drug and IgE.

The sixth differential equation stated that basophil FcεR1 was introduced at a zero-order rate B_{FCER}/k_{BASO}, that being the baseline divided by the mean residence time of the cell/receptor in that compartment. It should be noted that the mean residence time of measured FcεR1 (and surface IgE) is a composite of the presence of the proteins on the cell surface together with the lifetime of the carrier cell. Without perturbations in cell numbers (or measurements thereof), cell lifetime cannot be divorced from the lifetime of the protein on the surface of said cell. The production of FcεR1 was controlled by the peripheral compartment concentration of free IgE, C_{EFP}, according to a relative (i.e., 0–1) sigmoid relationship, *STIM*, parameterized with a half-max concentration, EC_{50R}, and a shape term *STIMC*. Since surface IgE demonstrated a time profile very similar to that for FcεR1, it was represented algebraically by a power relationship from baseline normalized FcεR1 with a scale parameter S_{IGES} and its own baseline, *BSIgE*. The overall effect is such that, as free IgE is removed from the system, there is downregulation of the number of receptors on the cell surface and, due to the equilibrium between free IgE and Fcε receptors, surface bound IgE decreases.

For the skin prick test equations, the factor driving the wheal response was the baseline normalized basophil FcεR1, $FCER/FCER_0$, parameterized with a scale (baseline) parameter *STIMW* and exponent *ExpW*. *SPT*1 through 4 were differential equations for up to four delay compartments with *KSPT* the mean response time for said response. Flare (*FLAR*) was algebraically represented as a power relationship from relative wheal with a scale parameter *STIMF* and exponent *ExpF*.

The steady-state (initial and final) nanomolar amounts of IgE in the central and peripheral compartment, for FcεR1 and for the SPT, were given by the following equations:

$$\alpha = CL_E / V_C$$
$$\beta = CL_E / V_P$$
$$\chi = Q / V_C$$
$$\delta = Q / V_P$$
$$E_{P0} = \left(R_P + \chi / (\alpha + \chi) * R_E\right) / \left(\beta + \delta - \chi * \delta / (\alpha + \chi)\right)$$
$$E_{T0} = \left(R_E + \delta^{*} E_{P0}\right) / (\alpha + \chi)$$
$$IGE_I = E_{P0} / V_P$$
$$STIM_0 = \left(1 + IGE_I / \left(IGE_I + EC_{50R}\right)\right) ** STIMC$$
$$FCER_0 = B_{FCER} * STIM_0$$
$$SPT1 \ldots 4_0 = STIMW \qquad (13.7)$$

The individual predicted output equations for drug, IgE, basophil FcεR1 and surface IgE, and skin prick test wheal and flare were then as follows:

$$\begin{aligned} C_{TX} &= 0.15 * X_T / V_C / ST \\ C_{TE} &= 190 * E_T / V_C \\ C_{FE} &= 190 * \left(\left(E_T - C_C \right) / V_C \right) \\ C_{FEP} &= 190 * \left(\left(E_{TP} - C_P \right) / V_P \right) \end{aligned} \quad (13.8)$$

In calculating the concentration of total drug in central plasma, C_{TX}, in µg/mL, 0.15 was the molecular mass, 150,000, divided by 10^6 to give the required unit change from nanomolar; X_T was the total drug, C_C the solution to the quadratic expression for the complex, and *ST* the binding stoichiometry. Preliminary modeling indicated that the best fit stoichiometry was two for QGE031 and one for omalizumab. The latter was consistent with the stable cyclic hexamers formed with three molecules each of omalizumab and IgE [27], the former with QGE031 utilizing both of its Fab sites to bind IgE. The concentration of total ligand IgE in central plasma, C_{TE}, in ng/mL, had 190 for the molecular mass of and unit conversion for IgE; E_T is total IgE. The concentration of free IgE in the central plasma was C_{FE}, in ng/mL. The drug and IgE equations in (13.8) were log-transformed to match the data. The basophil and skin prick test equations in (13.5), *FCER* and *SIGE*, *WHEA* and *FLAR* required no further treatment other than log (basophil) or square-root transforms (wheal and flare).

13.2.3 Common Components

13.2.3.1 The Covariate Model The original omalizumab model [23] included only body weight and baseline IgE as covariates; later versions investigated and included others. The generic population parameter, *P*, relationship for the *i*th individual is exemplified with this relationship including body weight and baseline IgE:

$$\begin{aligned} Pi = THETA(1) * (WT/70) ** THETA(2) * (\mathrm{IGE}_0 / 365) \\ ** THETA(2) * EXP(ETA) \end{aligned} \quad (13.9)$$

where the *THETA* represent population mean or covariate relationship parameters to be estimated and *ETA* a Gaussian distribution with mean zero. Although both body weight (*WT*) and baseline IgE (IgE_0) are shown in this illustrative equation, they were not applied to all structural parameters. Body weight centered on 70 kg was specified in an allometric (power) relationship on all clearances and volumes, and on IgE production rate (R_E). Baseline IgE was a function of rate of production and clearance. For QGE031, the absolute bioavailability (BAV) and reflection coefficient parameters were logit-transformed to ensure that the values were between 0 and 1. Parameters that were already powers had normal rather than log-normal distributions.

The observed distribution of baseline IgE in the trials was not normal, so could not be represented with a Gaussian distribution. Baseline IgE was therefore used as a covariate to predict IgE clearance and production rate for any individual patient. Beyond body weight and baseline IgE, other population covariates explored for omalizumab included age, body mass index (BMI), and race. BMI was hypothesized as potentially important given the different fluid volume properties of lean and adipose tissues. As both adult and pediatric patients were studied, age less than 12 was a natural covariate to investigate, although it was hypothesized that much of the age differences should be accounted for through body weight and known differences in baseline IgE levels. The categorical variables age less than 12, sex, and race (black, oriental, and "other" vs the reference Caucasian) were investigated by estimating ratio parameters from the reference 70 kg, 365 ng/mL baseline IgE, male Caucasian adult patient.

Although in theory the binding constant, K_D, between omalizumab and IgE should be identical for all patients as it is a fixed biochemical reaction, the situation *in vivo* is different due to the potential for there to be competition for IgE binding between omalizumab and other binding agents such as IgE receptors. If these other binding agents are not included explicitly in the PK/PD model, the competitive binding will act to make it appear that omalizumab is not able to bind IgE as efficiently; that is, K_D will shift to a higher value. Therefore, interindividual variation in K_D was allowed to take account of the possibility that differing levels of FcεR expressing cells or (unknown) soluble proteins may compete with omalizumab for binding with IgE. Baseline IgE was explored and included as a covariate on K_D as it predicted some of this variation; perhaps indirectly it was informing on the expression level of IgE receptors in the body.

Any remaining variances in the parameters were considered unexplained interpatient variance and specified as log-normal distributions. Correlation between parameters was investigated in preliminary models but, finally, for omalizumab only the covariance between clearance of free omalizumab and its volume, and between the clearance of the complex and IgE production rate, was estimated. For QGE031, an extensive omega correlation matrix was investigated.

13.2.3.2 The Residual Error Model Natural logarithm-transformed data for drug and IgE concentrations, basophil FcεR, and surface IgE were used with corresponding logarithm transforms of the output–concentration functions. Urticaria itch and hives data (range 0–21) were analyzed on the original scale, but with a residual error model constrained by logit transformation. This had a margin of 1 on both sides to avoid numerical issues with observed values of 0 and 21. Skin prick test data were square-root-transformed with a small marginal value of 1 added to avoid numerical issues with zero observed flare and wheal. Given these transformations, all residual error models were therefore additive. The few concentrations outside the quantification range were excluded.

13.2.3.3 Model Evaluation Diagnostic plots were examined to assess model adequacy, possible lack of fit, or violation of assumptions. Plots of observed (OBS) versus population predicted values (PRED) and OBS versus individual predicted values (IPRED) were evaluated for randomness around the line of unity. Plots of weighted residuals (WRES) versus time and WRES versus PRED and time were evaluated for randomness around the zero line.

In addition, statistical shrinkage in maximum *a priori post hoc* Bayesian (empirical) estimates (EBEs) of model parameters used for diagnostic purposes was evaluated [31]. If all individuals' parameters were to be the same as the population estimate, that is, 100% shrunk to the mean value, it would therefore be of no value for determining a relationship with an individual's response. Moreover, to assure appropriate structural model selection, shrinkage in residual variability ("*epsilon*-shrinkage") was also evaluated.

13.2.3.4 Assessment of Model Predictive Performance As described in FDA's 1999 [32] the objective of model validation is to examine whether the model provides a good description of the data in terms of its behavior for the application proposed. The normalized prediction distribution errors (NPDE) [33] were used as a method for assessing the simulation-predictive performance of the final model as it allowed all covariates in the model to be taken into account together with all patients' individualized posologies. The NPDE were plotted to check for no trend to misfit over time and/or per covariate classification.

Although perhaps visually more appealing, visual predictive checks (VPCs) of the shape of the PK or PD profiles, even the covariate- and dose-corrected "prediction-corrected" VPC, are only useful when the regimen and treatment durations are the same. For omalizumab, the individualization via the dosing table allows patients to be four- or two-weekly regimens, and furthermore, the study durations were different for the different studies.

13.2.3.5 Clinical Implications for Posologies Simulations were performed to assess the clinical impact of the knowledge contained within the PK/PD model. These used the same NONMEM control files as for estimation, save for replacing the estimation with a simulation command. The simulations are presented and discussed according to the points arising.

13.2.3.6 Software Used and Validation The analysis was performed using the NONMEM software system, NONMEM VII version 2 extended/super extended (ICON Development Solutions, Ellicott City, MD, USA), the NM-TRAN subroutines version III level 1.1, and the PREDPP model library version V level 1.0 utilizing the Novartis MODESIM high performance computing environment.

For the omalizumab model, the ADVAN6 subroutine utilizing the Runge–Kutta integrator was used, for which a tolerance of 5 was specified. All omalizumab–IgE modeling building used the first-order (FO) method, consistent with the prior use of this model for analyzing omalizumab and IgE data. For the combined QGE031 and omalizumab model, integration used the Livermore solver for ordinary differential equations with automated step size adjustment (LSODA) by specifying ADVAN13, with a tolerance of 6. Estimation was performed using stochastic approximation expectation maximization with interaction (SAEM). Since the model was relatively complex, 10,000 burn-in iterations ensured convergence (assessed visually), followed by 2000 averaging iterations to obtain maximum likelihood parameters. Standard errors were obtained using the $COV procedure.

13.3 RESULTS AND DISCUSSION

13.3.1 Omalizumab Capture of IgE Reducing Itch and Hives

13.3.1.1 Description of Data and Quality of Fit The three studies used for modeling the relationship between free IgE suppression and urticaria itch and hives scores have been published [34–36]. In these, either three or six administrations of placebo, 75, 150, and 300 mg every 4 weeks were investigated. The population model for analyzing omalizumab binding to IgE was unchanged from that previously published. There was no pooled analysis for asthma versus urticaria patient populations, but there were no clinically significant differences in drug and IgE parameters. The omalizumab and IgE quality-of-fit diagnostics and tables of parameters are, therefore, not reproduced here. The major scientific extension was in the quantitative relationship between IgE capture and improvement in the clinical endpoints for urticaria, itch, and hives. The relationships between observed omalizumab concentrations, reduction of free IgE levels relative to baseline, and UAS7 improvement are shown in Figure 13.3. As expected given the binding reaction, the suppression of free IgE was related to omalizumab concentrations. The clinical response related to both drug and to free IgE with a classical sigmoid shape in the local regression smoothing curves. An exploratory fit of a Hill function (which ignored the possibility of time delay hysteresis) estimated a half-maximum EC_{50} of 13% for relative

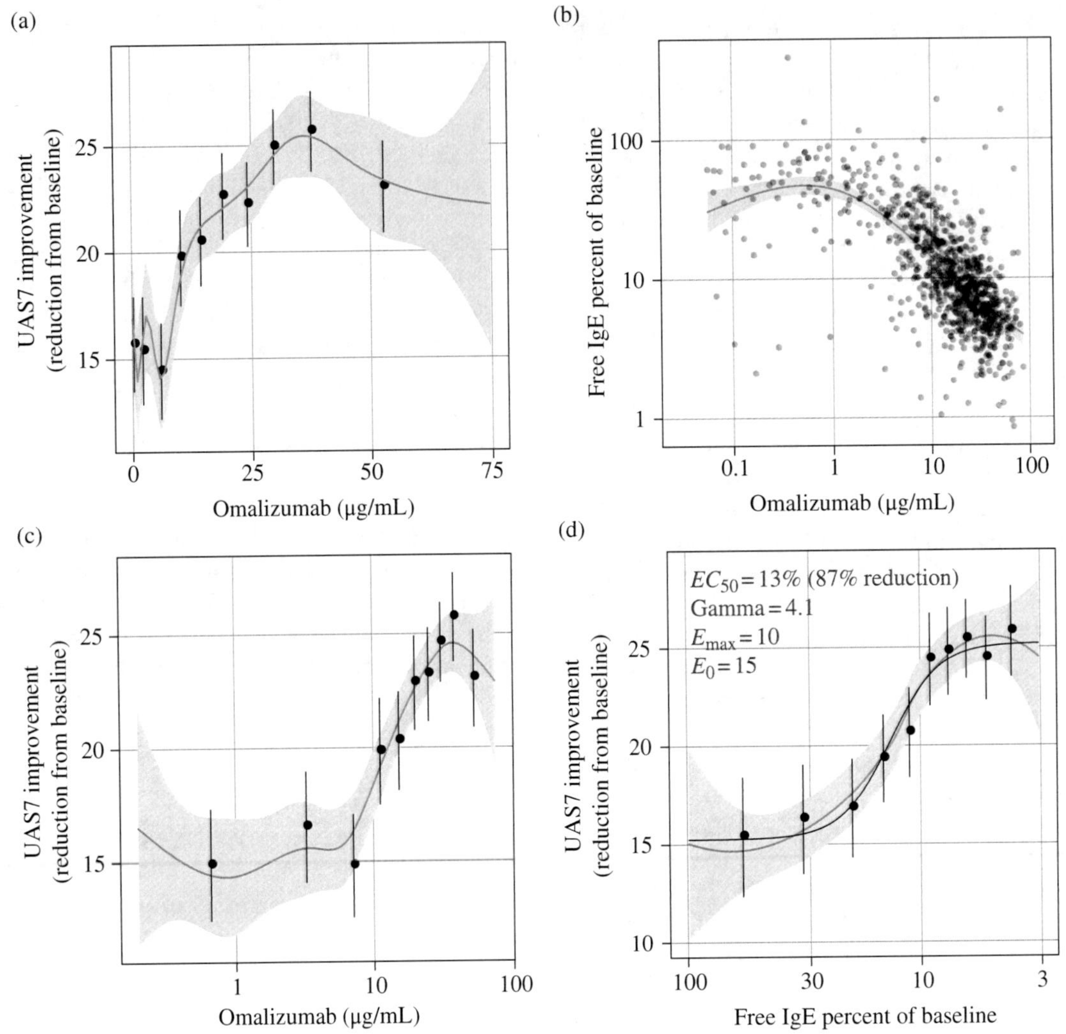

FIGURE 13.3 UAS7 improvement versus omalizumab concentrations and free IgE relative to baseline. The panels display UAS7 improvement versus observed omalizumab concentrations (a, c) and observed individual patient free IgE levels expressed as percent of IgE at baseline (b, d) for trough concentrations at 12 weeks and, for two of the studies, 24 weeks, then at the end of the 20 week followup period. The dots and error bars show the mean and standard deviations of decile groups for omalizumab concentration or free IgE reduction values, respectively. The light gray lines and bands represent *R* loess smoothing splines with 95% confidence intervals based on the nonbinned data. For free IgE reduction, a sigmoidal model was fitted to the nonbinned data. Parameter estimates for the fitted model are indicated in (d) in black. The same dataset as for the population-based PK–IgE analysis was used.

free IgE driving the urticaria symptoms. Estimating this relationship in an indirect response mixed-effect model provided the parameters listed in Table 13.1. There was hysteresis, with a mean response time of 5.7 days following free IgE suppression (which occurs within the first day post dose). The EC_{50} values of 20% (0.2) indicated that half maximal efficacy is expected when free IgE levels are reduced by 80%. In combination with the steep Hill coefficient, the transition from 10% to 90% of maximal efficacy would occur over a very narrow range of 75–84% relative free IgE reduction. The model parameters were precisely estimated, with little shrinkage except in those interindividual variances that were not estimated. It was notable that the unexplained random deviations from the typical values for EC_{50} and for E_{max} were highly correlated between itch and hives; if a patient had a high EC_{50} on itch, the same was true for hives.

The model fitting diagnostics plots in Figure 13.4 showed that individual predictions reproduced the observations, with no substantial bias or trend to misfit over time in the NPDE, save for a slight overprediction of itch scores in the first month. However, not all parameter random distributions were normally distributed. As a consequence, the distribution of NDPE tended to deviate beyond ±2 standard deviations in QQ (quantile-quantile) plots. Since good individual fits were achieved but there was some concern in assuming normally distributed interindividual variation, simulations

TABLE 13.1 Parameter Estimates for Placebo and Relative IgE Itch and Hives Models

	Fixed Effects		Random Effects	
Parameter	Estimate	95% Confidence	Estimate (ω)	Shrinkage (%)
Placebo				
$Itch_0$ (0–21 scale)	12.6	(11.9–13.1)	0.44	8
$Itch_{SS}$ (0–21 scale)	3.8	(2.8–5.0)	3.19	7
Response time, itch (day)	7.1	(4.1–12.2)	5.45	18
$Hives_0$ (0–21 scale)	16.6	(15.8–17.4)	1.66	4
$Hives_{SS}$ (0–21 scale)	4.7	(3.2–6.5)	5.60	4
Response time, hives (day)	10.1	(6.3–16.2)	4.16	16
Lag time (day)	14.3	(8.9–23.2)	6.13	48
IgE link				
E_{max} itch (logit)	1.79	(1.63–1.97)	0.56	2
E_{max} hives (logit)	2.49	(2.25–2.70)	0.49	2
EC_{50} itch (proportion)	0.20	(0.17–0.23)	1.15	25
EC_{50} hives (proportion)	0.20	(0.18–0.23)	1.19	24
Hill coefficient (power)	9.88	(9.67–10.10)	0.1 (fixed)	95
Response time (day)	5.68	(5.24–6.15)	0.1 (fixed)	60
	Placebo (alone)		IgE link	
	Estimate (σ^2)	95% confidence	Estimate (σ^2)	95% confidence
Residual error, hives	0.25	(0.24–0.26)	0.22	(0.21–0.23)
Itch	0.39	(0.37–0.41)	0.29	(0.28–0.29)

The efficacy parameters are given for a patient without angioedema, with a baseline IgE of 80 IU/mL, weighing 80 kg, a BMI of 30 kg/m², and 18 years or older.

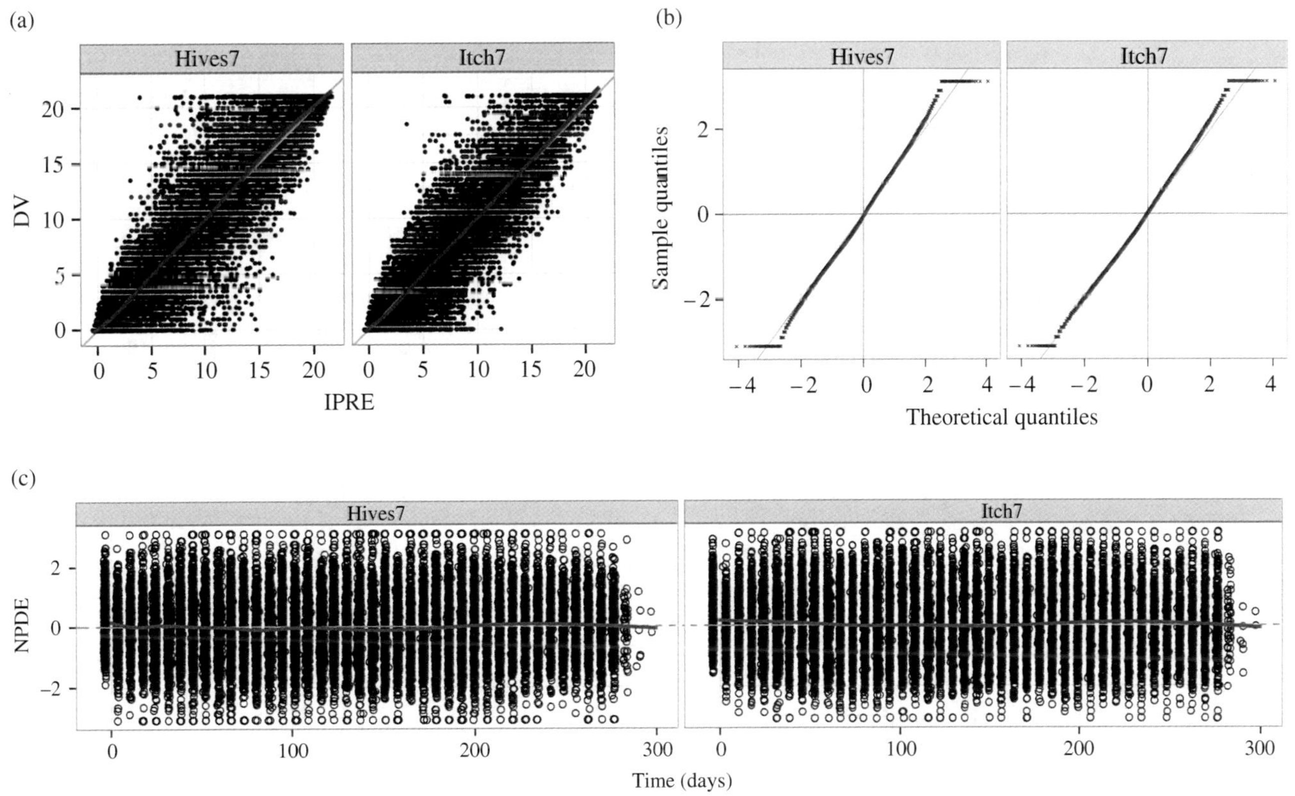

FIGURE 13.4 Quality-of-fit diagnostics. (a) Individual itch and hives predictions versus observations. (b) NPDE over time. (c) QQ-plot of NDPE. The dark gray lines are *R* loess local regressions.

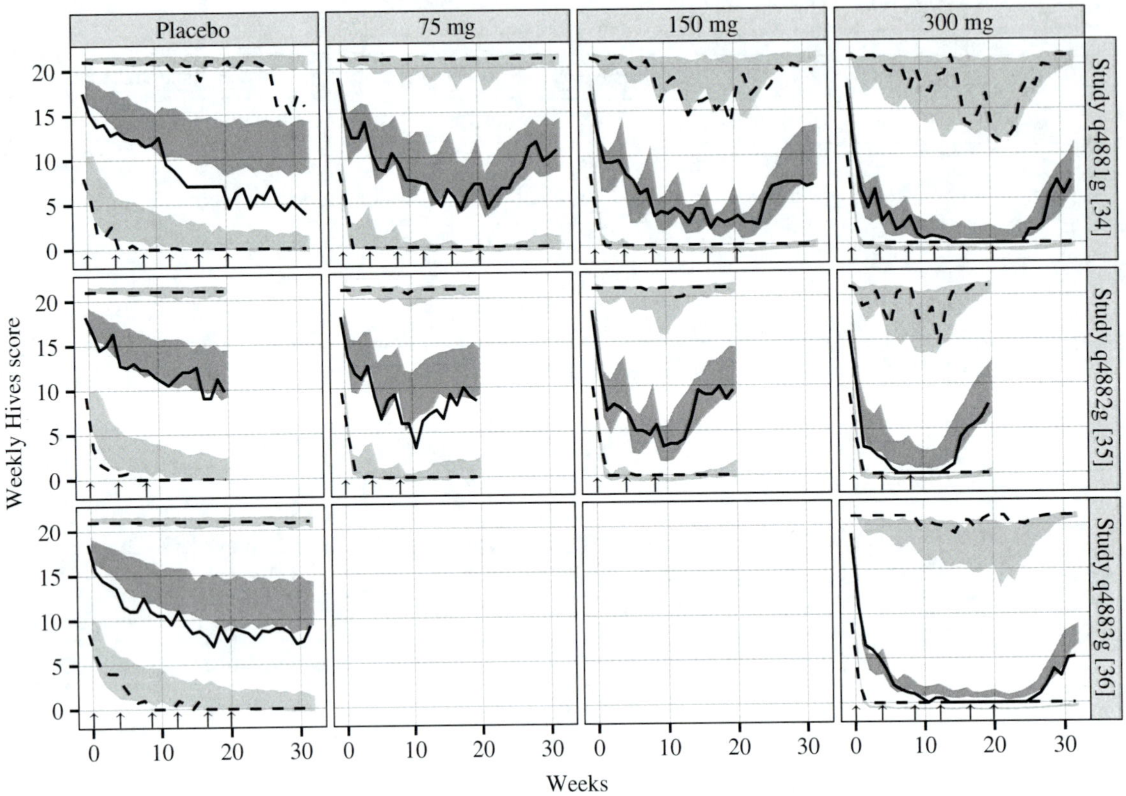

FIGURE 13.5 VPCs for weekly hives score. The observed data median is shown as a solid black line, the 5th and 95th percentiles of the observations as dashed black lines. Ninety-five percent intervals for 200 simulations of each group are indicated as dark gray (median) and light gray (5th and 95th percentile) bands. The arrows indicate omalizumab or placebo administrations.

were run by sampling from the 632 treated patients' individual parameters. VPCs are shown in Figure 13.5 for the weekly hives score; weekly itch was similar, not surprising given the strong correlation between the two variables. Figure 13.6 shows the well-controlled responder rate, a count variable derived from itch and hives where a responder had a sum of weekly itch and hives scores less than or equal to 6. The predictions reproduced the data for the 150 and 300 mg every 4 weeks dosing regimens well, albeit with an underprediction for the placebo effect and, to a lesser extent, for 75 mg.

13.3.1.2 Covariate Effects Some of the patient factors investigated as covariates were declared statistically significant based on their 95% confidence intervals (Table 13.2). However, only body weight, BMI, and baseline IgE influenced the model parameters to any great extent when moving from the 5th to 95th percentiles of these patient factors.

Both BMI and body weight had a pronounced effect on all drug effect parameters. However, they were always counteracting, that is, the sign for the BMI covariate was the opposite of that for the body weight covariate. Considering that these two patient characteristics were to a great extent positively correlated (though not 100%; alternative models with either body weight or BMI were tested), this acted to cancel their effects on the dose required for symptom control. With respect to the two age groups, a significant effect was detected only for the elderly; the elderly responded faster, with a 39% shorter drug response time.

The largest shift was +55% to −30% in EC_{50} for itch and hives, when moving from the lower 5th to upper 95th percentile of baseline IgE. In other words, there was an inverse relationship: a patient with a low baseline IgE level would have a higher EC_{50} value and hence need less omalizumab-induced percent reduction from their baseline value to experience a beneficial effect. It is worth remembering that these EC_{50} were not for omalizumab, but for IgE affecting itch and hives. This, together with the greater dose-for-dose percent suppression of IgE that occurred when IgE molar concentrations approached the point of optimum efficiency, the K_D, started to explain why no IgE-dependent dose adjustment is required. To fully appreciate the implications of multiple parameter covariate influences, however, simulations are required; these are detailed in the next section.

13.3.1.3 Implications of PK/PD for Omalizumab Posology for Urticaria Fitting a kinetic–dynamic population model to drug, biomarker, and clinical data in itself, although useful

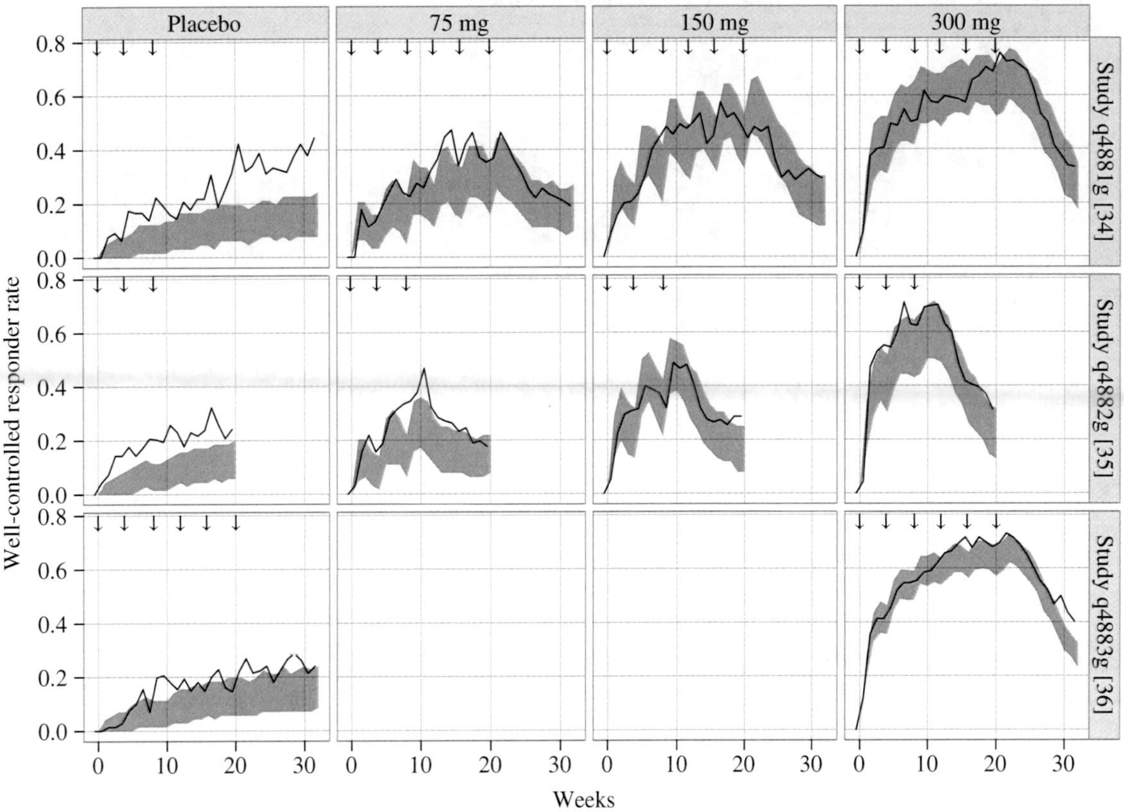

FIGURE 13.6 VPCs for UAS7 less than or equal to 6. The observed responder rate is shown as a solid black line. Ninety-five percent intervals for 200 simulations of each group are indicated as dark gray bands. The arrows indicate omalizumab or placebo administrations.

for documenting properties of a compound, does not make full use of the knowledge contained therein. The clinical implications on the time course and patient responsiveness to the drug are most easily visualized by scenario simulation. Since the results showed that the omalizumab–IgE–urticaria symptoms model was able to describe and simulate the data, model simulations were used to assess (i) the clinical effect of body weight, BMI, and baseline IgE, all of which were found to be influencing the E_{max} and EC_{50} values, and (ii) the tailoring of the posology to maximize the chance for an individual patient to achieve maximal responses.

BMI and body weight are positively correlated, although not perfectly. The extent to which the covariate effects translated into clinical differences for heavy or light patients was not obvious from the equations. The simulations in Figure 13.7 show that no differences were to be expected in well-controlled UAS7 response rates for patients weighing more than 80 kg with a BMI greater than 30. Similarly, also in Figure 13.7, the counteracting effects of the inverse covariate effect of baseline IgE on the IgE EC_{50} for itch and hives was such that there was no difference in well-controlled responder rates for high versus low body burden of the drug target. The effect of baseline IgE on the response time was visible, in that patients with higher levels would be expected to respond sooner.

Also apparent from Figure 13.7 is the observation that, although approximately 50% of patients could be well controlled with 150 mg, there was a higher likelihood of a good response with 300 mg every 4 weeks, especially in the first 2 months. With 150 mg, the well-controlled responder rate increased and decreased in time with omalizumab concentrations and inversely with suppression of free IgE. With 300 mg, there was less "peak-trough" variation within each dosing interval, indicating that this posology was approaching saturation of the response and therefore close to being maximally effective.

13.3.2 QGE031 and Omalizumab Capture of IgE, Reducing Basophil FcεR1, Surface IgE, and Allergen Skin Reactivity

13.3.2.1 Description of Data and Quality of Fit The studies and data used for this PK/PD model analysis have been published [37, 38]. An overview of the PK and biomarker data for the Caucasian subcutaneous multiple dose study is given in Figure 13.8. The Caucasian intravenous and Japanese subcutaneous single-dose studies gave similar PK, IgE, and basophil profiles; albeit with no skin prick testing. Following single or multiple doses of QGE031 or omalizumab, the concentrations of total IgE increased, indicating

TABLE 13.2 Covariate Estimates for Omalizumab IgE Itch and Hives Model

Parameter		Estimate	95% Confidence Interval	Significant?	Shift from 5th to 95th Percentile of Patient Factor or State Change
Baseline angioedema on	E_{max} itch	−0.03	(−0.16 to 0.11)		
	E_{max} hives	−0.07	(−0.20 to 0.055)		
	EC_{50} itch	0.02	(−0.18 to 0.22)		
	EC_{50} hives	0.03	(−0.17 to 0.23)		
	XRT	−0.13	(−0.24 to −0.016)	Yes	−12%
Baseline IgE level on	E_{max} itch	0.05	(−0.0020 to 0.11)		
	E_{max} hives	0.05	(0.00043–0.10)	Yes	−12% to +11%
	EC_{50} itch	−0.17	(−0.24 to −0.10)	Yes	+55% to −30%
	EC_{50} hives	−0.17	(−0.24 to −0.10)	Yes	+55% to −30%
	XRT	−0.13	(−0.17 to −0.082)	Yes	+40% to −24%
Body weight on	E_{max} itch	0.81	(0.21–1.41)	Yes	−27% to +41%
	E_{max} hives	0.75	(0.19–1.32)	Yes	−25% to +37%
	EC_{50} itch	−0.94	(−1.72 to −0.17)	Yes	+45% to −33%
	EC_{50} hives	−0.97	(−1.75 to −0.18)	Yes	+46% to −34%
	XRT	−0.69	(−1.15 to −0.23)	Yes	+31% to −25%
BMI on	E_{max} itch	−0.68	(−1.33 to −0.038)	Yes	+31% to −25%
	E_{max} hives	−0.61	(−1.22 to 0.0011)		
	EC_{50} itch	1.50	(0.65–2.36)	Yes	−41% to +66%
	EC_{50} hives	1.52	(0.64–2.39)	Yes	−42% to +67%
	XRT	0.55	(0.044–1.05)	Yes	−18% to +20%
Age < 18 years on (adolescents)	E_{max} itch	−0.01	(−0.41 to 0.38)		
	E_{max} hives	0.06	(−0.32 to 0.44)		
	EC_{50} itch	0.00	(−0.59 to 0.60)		
	EC_{50} hives	−0.08	(−0.68 to 0.53)		
	XRT	0.03	(−0.26 to 0.33)		
Age ≥ 65 years on (elderly)	E_{max} itch	0.01	(−0.28 to 0.30)		
	E_{max} hives	−0.02	(−0.30 to 0.26)		
	EC_{50} itch	−0.15	(−0.57 to 0.26)		
	EC_{50} hives	−0.16	(−0.58 to 0.26)		
	XRT	−0.49	(−0.80 to −0.18)	Yes	−39%

The covariates were defined as logarithms, hence no difference, log(0), was a ratio of 1. Statistical significance of the covariates was concluded if the 95% confidence intervals of 30,000 samples from the Markov chain did not include zero. The extent of the continuous covariate differences, if significant, was given on the original scale for the 5th and 95th percentiles of the covariate for the population studied. These were 6–646 IU/mL baseline IgE, 54–122 kg body weight, and 21–43 kg/m² BMI.

that IgE had been captured to form drug–IgE complexes. The increase was dose and time dependent, to the extent that IgE was available to be bound. Once the quantity of drug in the body exceeded that of IgE, the total IgE reached a plateau that was in proportion to the baseline level of IgE. Higher drug doses only served to extend the duration of capture. As IgE was captured, the expression of FcεR1 and number of molecules of IgE on the surface of basophils decreased. The return of the basophil biomarkers toward baseline occurred just as total IgE started to decrease, indicating that IgE capture was no longer complete. The timing of the return of the basophil biomarkers to baseline was both drug dose and baseline IgE dependent; in subjects with higher levels of baseline IgE, that is, more IgE in the body, the return to baseline occurred earlier. QGE031 exhibited greater ability than omalizumab to maximally lower FcεR1 and surface IgE, but, once the return started, it was more sudden. This phenomenon was anticipated from calculations based on a binding and turnover model [11, 39] and observations with a prior high affinity anti-IgE [40]. It occurs when a high affinity drug binds to a relatively high expression target, which is turning over relatively rapidly. Effectively, it indicates a disparity between binding potency (as in the K_D) and the concentration of the target, in much the same way that it is difficult to control hydrogen ion concentrations when the pK_a of a buffer is not matched to the pH. The binding system, and hence equations, are the same. Ideally, for maximum control or buffering, the K_D of a drug should match the concentration of its target.

Both QGE031 and omalizumab induced tolerance to allergen in the skin prick test; QGE031 was notably more potent, inducing zero skin reactivity in some subjects. The time course of skin reactivity was notably delayed from the drug and biomarker changes in the central circulation,

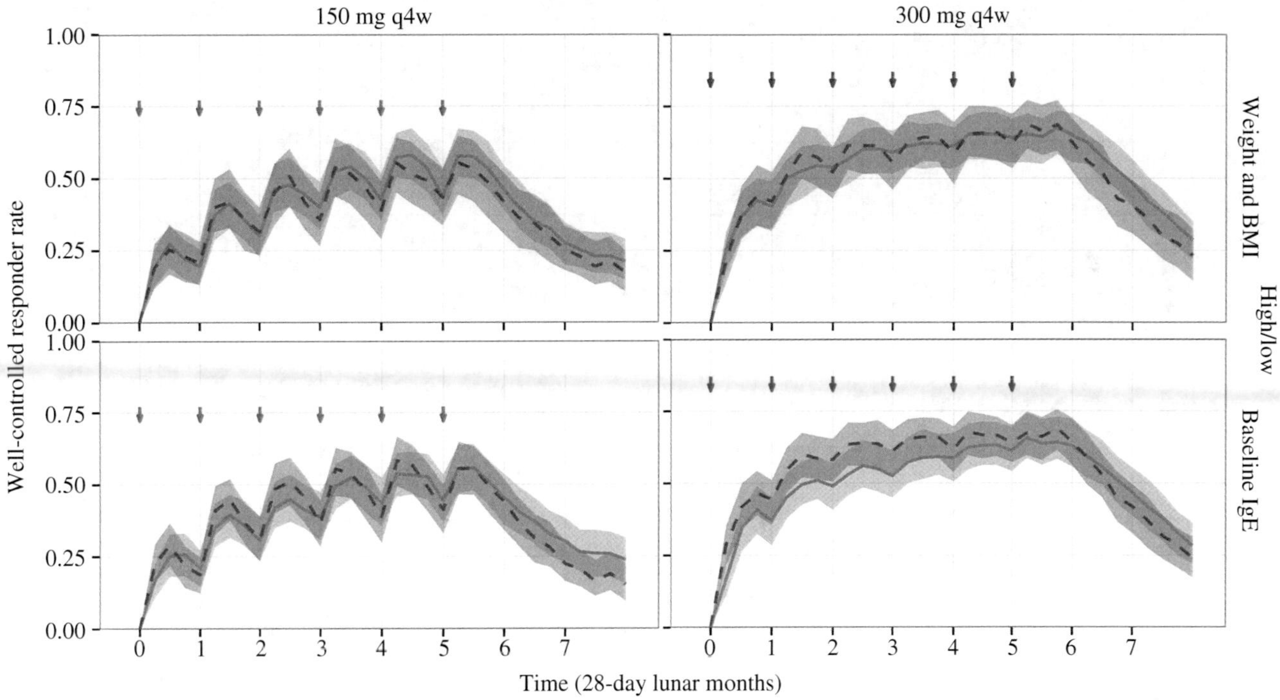

FIGURE 13.7 Well-controlled responder rates for patients with different body weights/BMIs or baseline IgEs. The arrows indicate injection of 150 and 300 mg omalizumab, respectively. Responder rates are subsetted by, top row, high (>80 kg and >30 kg/m²; dashed dark gray) and low (≤80 kg and ≤30 kg/m²; light gray) body weight and BMI, or bottom row, high (>80 IU/mL) or low (≤80 IU/mL) baseline IgE. The lines and bands are medians and 95% intervals of 200 randomly sampled groups of 200 patients drawn from the 632 sets of parameters. Well-controlled responses are defined as UAS7 less than or equal to 6.

indicating an indirect mode of action, either time for the drug to permeate the tissues and/or signal transduction processes. The PK/PD model outlined in Figure 13.2 fitted the data well, as evidenced by the individual predictions following the individual subject data (Fig. 13.8) and the concordance of observations with individual and population predictions over the low-to-high dynamic range of the signals, with time (assuring that the compartmental structure was sufficient) and across the random effects distributions (Fig. 13.9). Since the expected population predictions and NPDE diagnostics utilized Monte Carlo simulation, this gave confidence that the model was suitable for its intended purpose, simulation.

13.3.2.2 Covariates At this early Phase 1 and 2A stage of development in atopic but otherwise healthy subjects (not patients), the exploration of covariates was intentionally restricted to those known from omalizumab to be clinically important: body weight and baseline IgE. The body weight covariate was not estimated but fixed to 0.9 and 1.0 for the clearances and volumes, as per prior reports on omalizumab. Baseline IgE was statistically significant for IgE production, as expected, and also for the non-target-mediated clearance of the drug, CL_X (Table 13.3). In the basophil part of the model, baseline IgE was a significant factor in predicting the ability of IgE to control FcεR1 turnover (EC_{50R}) and on the proportionality factor between FcεR1 and surface IgE (S_{IgES}). For the skin prick testing, there was much unexplained random interindividual variation in both the baseline wheal and flare signals ($STIM_{Wheal,Flare}$) and in the exponential relationships with basophil FcεR1 ($Exp_{Wheal,Flare}$). The only covariates discovered that would explain interindividual variation in these parameters were the baseline wheal and flare measurements.

Regarding QGE031 versus omalizumab, the difference between the two compounds was estimated to be 8.9-fold in the ability to bind and capture IgE (K_D^{app}). This means that, at very low concentrations of IgE, only one-ninth the concentration of QGE031 would be required to neutralize the target. With higher levels of IgE, mass balance limitations would lead to less of a differential in concentration and hence required dose.

Japanese ethnicity was investigated also in atopic but otherwise healthy subjects, prior to the initiation of clinical efficacy studies in Japan. All PK, IgE binding, and basophil PD parameters were evaluated in one step, using SAEM followed by Bayes estimation methods. Many parameters showed statistically significant differences from the Caucasian study populations (Table 13.3). Those that were most marked were drug clearance and ability to bind IgE. Drug clearance was 1.8-fold faster, with slightly

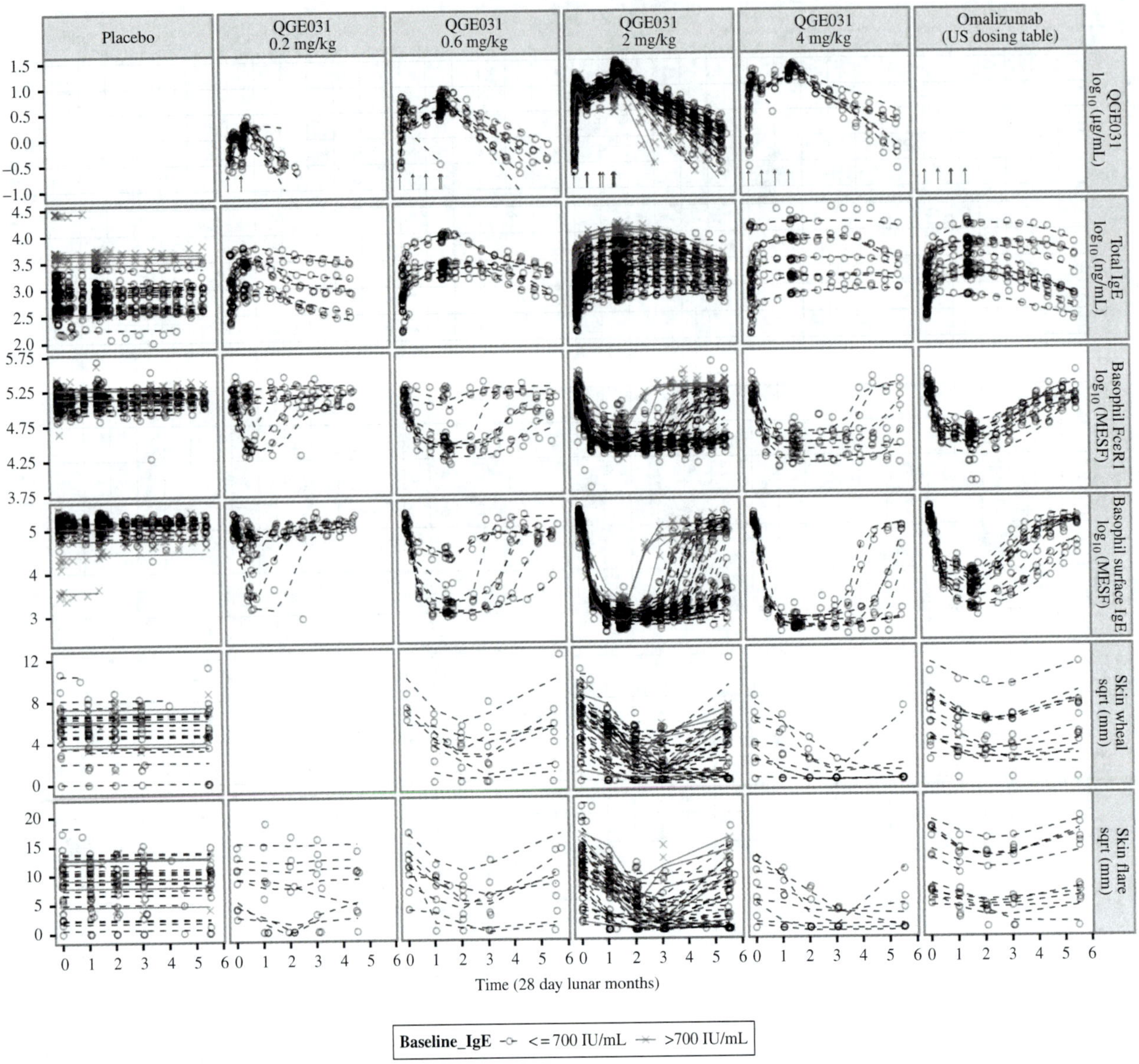

FIGURE 13.8 PK and biomarker data for multiple subcutaneous doses in Caucasian subjects. The colors, symbols and line types refer to the baseline IgE level, dark gray (originally red) solid lines with X symbols being greater than or equal to 700 IU/mL. The units for basophil FcεR1 and surface IgE were molecules of equivalent soluble fluorophore (MESF). Data values for QGE031, total IgE, FcεR1, and surface IgE were log-transformed, with square roots for wheal and flare. The dosing events are shown as arrows in the first row and the curves are the individual predictions.

lower BAV (logit −2.6, Japanese BAV, 97%). In an alternative, slightly worse fitting model, Japanese differences in BAV were estimated without the covariate on clearance and volume—this estimated a Japanese BAV of 47–63% compared with 98–100% for the Caucasians. Whether the effect was on clearance and volume, or on BAV, the fundamental conclusion was that the Japanese experienced lower drug concentrations, dose-for-dose, compared with Caucasians.

In opposition to the lower drug concentrations, the K_D^{app} covariate, with a ratio of 0.24, indicated a 4-fold greater ability of the Japanese to bind and capture IgE. Also notable was the faster turnover of IgE, with both a greater production (E_{PR} 1.9-fold faster) and clearance (2.5-fold faster). In the basophil response system, there were some differences in the baseline levels of FcεR1 and surface IgE; however, these could have been due to either different batches of fluorescent FACS (fluorescence activated cell sorter) assay reagents or to an ethnic difference. Japanese basophils appeared to be more sensitive to IgE with the EC_{50R} covariate ratio of 0.54. How these parameter differences combine to alter the dose–response between Caucasians and Japanese is discussed in the next section.

13.3.2.3 Implications of PK/PD for QGE031 Clinical Development As with the omalizumab urticaria model,

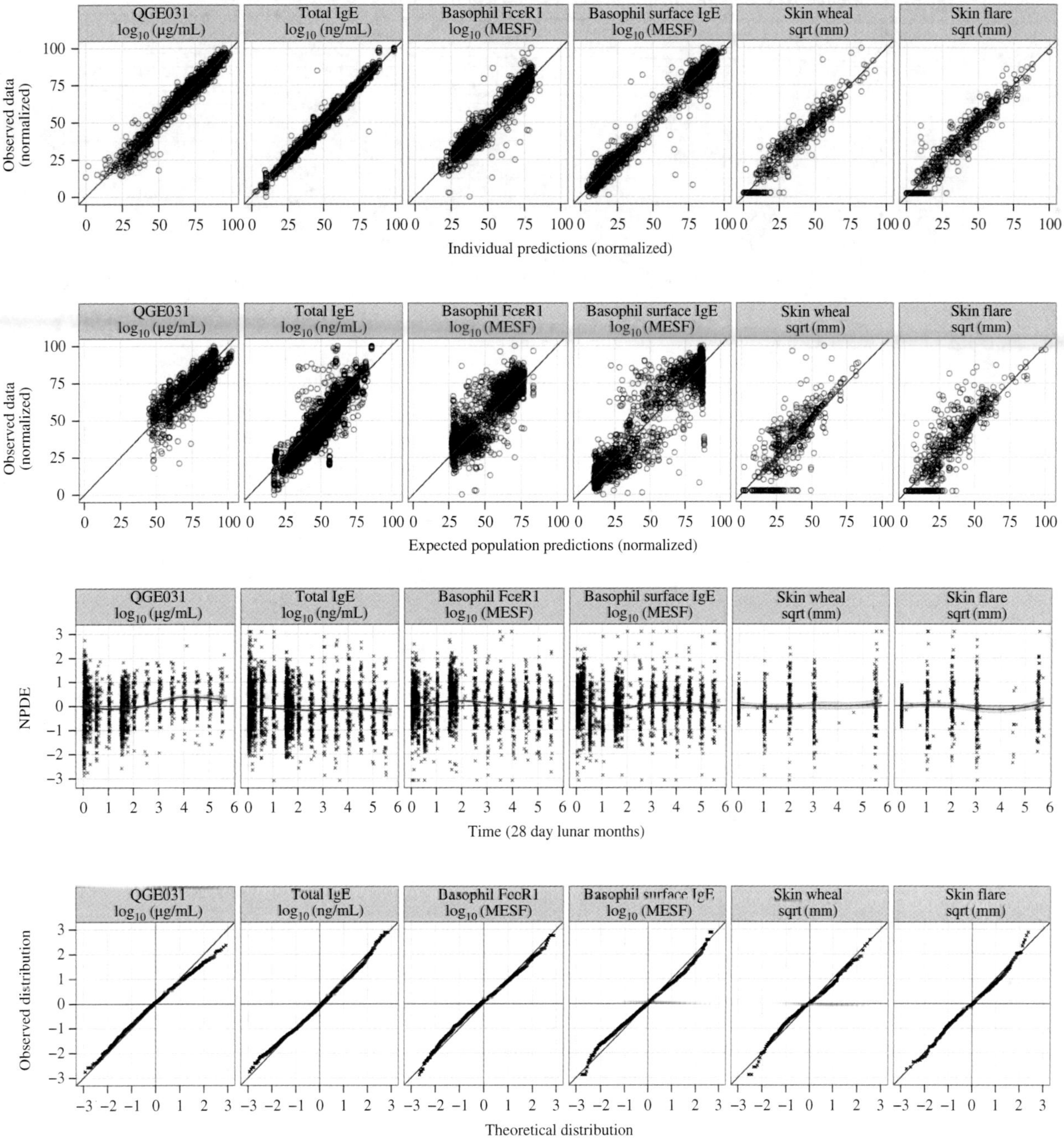

FIGURE 13.9 Quality-of-fit diagnostics. In addition to the individual observed versus predicted plots, the expected population predictions and normalized prediction distribution errors (NPDE) were produced by NONMEM from a sample of 1000 Monte Carlo simulations. The concordance of the random effects distributions was assessed from QQ plots of NPDE (fourth row).

the implications of the knowledge gained on QGE031/omalizumab, IgE, basophil FcεR1, and surface IgE and skin allergen tolerance were explored through Monte Carlo simulations. In these simulations, different posologies were simulated over time to steady state. An example is given in Figure 13.10 for four of the many dose levels simulated. The steady-state trough values for the responses at the last time point were collated to project dose–response for the given dosing frequency. Figure 13.11 shows a selection of the projected dose–responses for basophil FcεR1 and surface IgE and the inhibition of the skin prick allergen-induced wheal. The overlay of the two compounds highlights the expected increase in potency for QGE031 compared with omalizumab. For a lower IgE cohort with less than 250 IU/mL at baseline, a dose of 120 mg every 4 weeks was predicted

TABLE 13.3 Parameter Estimates for QGE031, Omalizumab, IgE, Basophil, and Skin Prick Test Model

	Fixed Effects			Japanese Ratio or Difference[a]			Random Effects	
Parameter	Estimate	95% CI[b]		Estimate	95% CI[b]		Estimate (ω)	Shrinkage (%)
Binding model								
CL_X (L day)	0.33	0.31	0.36	1.8	1.4	2.2	0.41	18
CL_E (L day)	0.89	0.8	1.0	2.5	1.9	3.1	0.47	15
V_C (L)	3.5	3.0	4.0	1.7	1.2	2.4	0.73	15
V_P (L)	17	15	19	1.3	0.95	1.7	0.47	27
Q (L day)	0.86	0.75	1.0	2.1	1.6	3.0	0.62	20
k_a (day)	0.16	0.14	0.20	1.3	0.86	1.9	0.77	16
EPR (μg day)	1800	1600	2100	1.9	1.4	2.4	0.51	14
K_D^{app} (nM)	0.32	0.19	0.45	0.24	0.14	0.42	0.1 fix	—
FcεR1 and surface IgE								
k_{BASO} (days)	3.9	3.5	4.3	1.2	0.97	1.5	0.40	27
BFcεR1 (MESF × 10^{-3})	22	19	25	2.0[c]	1.6	2.4	0.39	17
STIMC	3.5	3.1	4.1	0.15[a]	−0.30	0.59	0.77	22
BSIgE (MESF × 10^{-3})	120	110	130	1.4[c]	1.1	1.7	0.56	3.6
S_{IgES}	3.0	2.8	3.4	−0.31[a]	−0.66	0.022	0.61	25
EC_{50R} (nM)	3.0	2.1	5.7	0.54	0.31	0.88	0.89	17
SR (%)	84	77	91	−1.6[a]	−2.5	−1.1	0.1 fix	—
BAV (%)	100	97	100	−2.6[a]	−6.2	−0.11	0.1 fix	—
Skin test								
k_{SPT} (days)	23	21	26				0.56	13
$STIM_{Wheal}$ (mm)	45	42	48				0.30	10
$STIM_{Flare}$ (mm)	14	13	16				0.48	17
Exp_{Wheal}	1.7	1.5	1.9				1.1	10
Exp_{Flare}	1.17	1.10	1.24				0.31	13
Covariates								
IgE_0 on EPR	0.68	0.62	0.73					
IgE_0 on CL_X	0.084	0.015	0.15					
IgE_0 on K_D	0.17	−0.0056	0.34					
IgE_0 on EC_{50R}	0.35	0.23	0.48					
IgE_0 on S_{IgES}	−0.26	−0.37	−0.16	Residual error			Estimation (σ)	Shrinkage (%)
IgE_0 BFcεR and BSIgE	0.035	−0.043	0.11	QGE031 PK			0.22	8.1
Xolair-QGE K_D ratio	8.9	6.1	14	Total IgE			0.12	7.4
$Wheal_0$ on *ExpW*	0.36	−0.12	0.84	Basophil FcεR1			0.21	6.1
$Flare_0$ on *ExpF*	0.070	−0.014	0.15	surface IgE			0.33	6.2
$Wheal_0$ on *STIMW*	1.86	1.81	1.90	Wheal			0.98	5.8
$Flare_0$ on *STIMF*	1.2	1.0	1.4	Flare			1.6	6.1

For those parameter with covariates, estimates were for 70 kg bodyweight and a baseline IgE of 365 ng/mL (150 IU/mL).

[a] Difference in power or logit; if 95% confidence interval includes zero, then no difference.

[b] Confidence interval.

[c] Please note that since different basophil FACS assay reagents were used for the Japanese study, differences in baselines and ratios may be due to differences in reagent fluorescence intensity. The skin prick test parameters were estimated separately from those for the PK, IgE, basophil, and Japanese part of the model; the prior parameters were fixed in this run.

to be the minimum to fully suppress the skin wheal response. For patients with a higher body burden of IgE (>250 IU/mL cohort), the QGE031 and omalizumab dose–response curves were steeper and overlapping to a greater extent, reflecting increasing molar mass limitations acting in addition to the difference in binding affinity.

From the simulations in Figures 13.10 and 13.11, development decisions could be made on dose ranges to be studied in Phase 2B clinical symptom studies. To properly characterize the concentration– and hence dose–response, one should study an adequate range of exposures, from minimally to maximally effective, including a level above the projected maximum to provide data demonstrating that

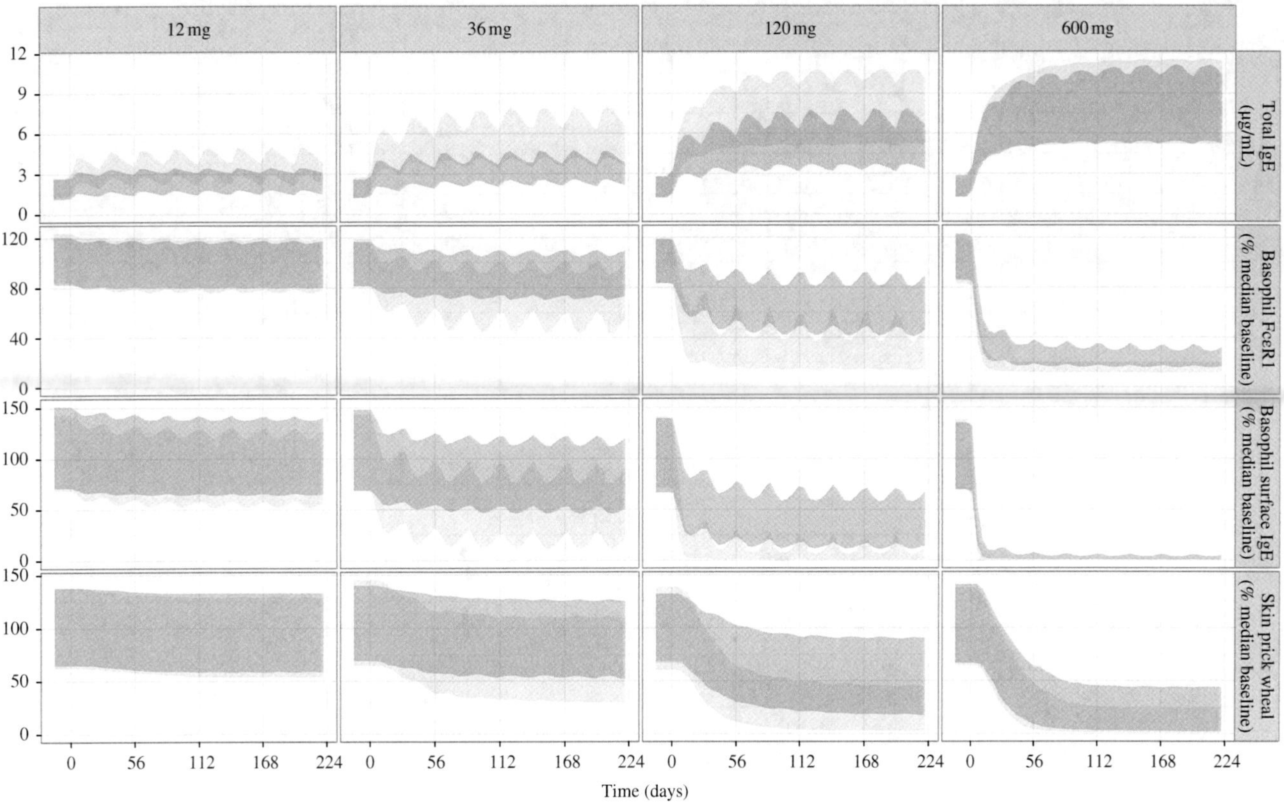

FIGURE 13.10 PK/PD projections for QGE031 and omalizumab. Simulations were for QGE031 given every 4 weeks to atopic subjects with >250 IU/mL baseline IgE levels for drug, total IgE, basophil FcεR1, and surface IgE, and skin prick wheal and flare (drug PK and flare not shown for clarity). Ten replicates of the 110 individuals from the subcutaneous study were simulated, specifying a range of subcutaneous doses administered to atopic subjects every 4 weeks. The shaded band is the interquartile range representing variation between subjects. Within each panel, darker gray represents omalizumab and lighter gray QGE031. Another example (not shown) was run for the lower baseline IgE stratum.

the maximum has been achieved. For the first asthma dose-ranging study, two-weekly administration of 12, 36, 120, 180, and 240 mg were selected, covering a 20-fold range. For two further studies, in asthma and in urticaria, based on these and related simulations, the range was narrowed to 24, 72, and 240 mg q4w. Analyzed together, the studies should provide the necessary data to characterize QGE031 concentration–symptom relationships and enable suitable posologies to be selected for Phase 3.

The analysis of Japanese ethnicity subjects provided a clue on key patient factors to be explored. The differences in the absorption, distribution, and elimination, together with differences in IgE turnover and binding to QGE031 were unexpected. Dose–response simulations displayed the combined effects of the differences in parameters (Fig. 13.12). Although the predicted trough concentrations were significantly lower in the Japanese, total IgE still accumulated in a dose-dependent manner to a level indicative of maximum IgE capture. Basophil FcεR1 and surface IgE decreased with subtly different dose–response shapes, the Japanese being steeper. However, the same posology, 120 mg q2w, should be the minimum required to maximally suppress these downstream biomarkers and hence give the best chance to see clinical responses. These results helped Japanese centers join the global dose-selection studies.

13.4 CONCLUSIONS

In this monograph, a pharmacokinetic/pharmacodynamic process for quantitating drug–target binding and linking with downstream biomarkers and clinical responses has been outlined and exemplified. In this process, one must generate suitable data over a full time course of treatment, including both onset and return toward baseline. Ideally, though not essentially, this should be over multiple dose levels, but as a minimum it should generate sufficient drug concentrations to maximize pharmacodynamic responses. One then fits one or more mathematical models to all data over the full time course, evaluating whether simulations from said models

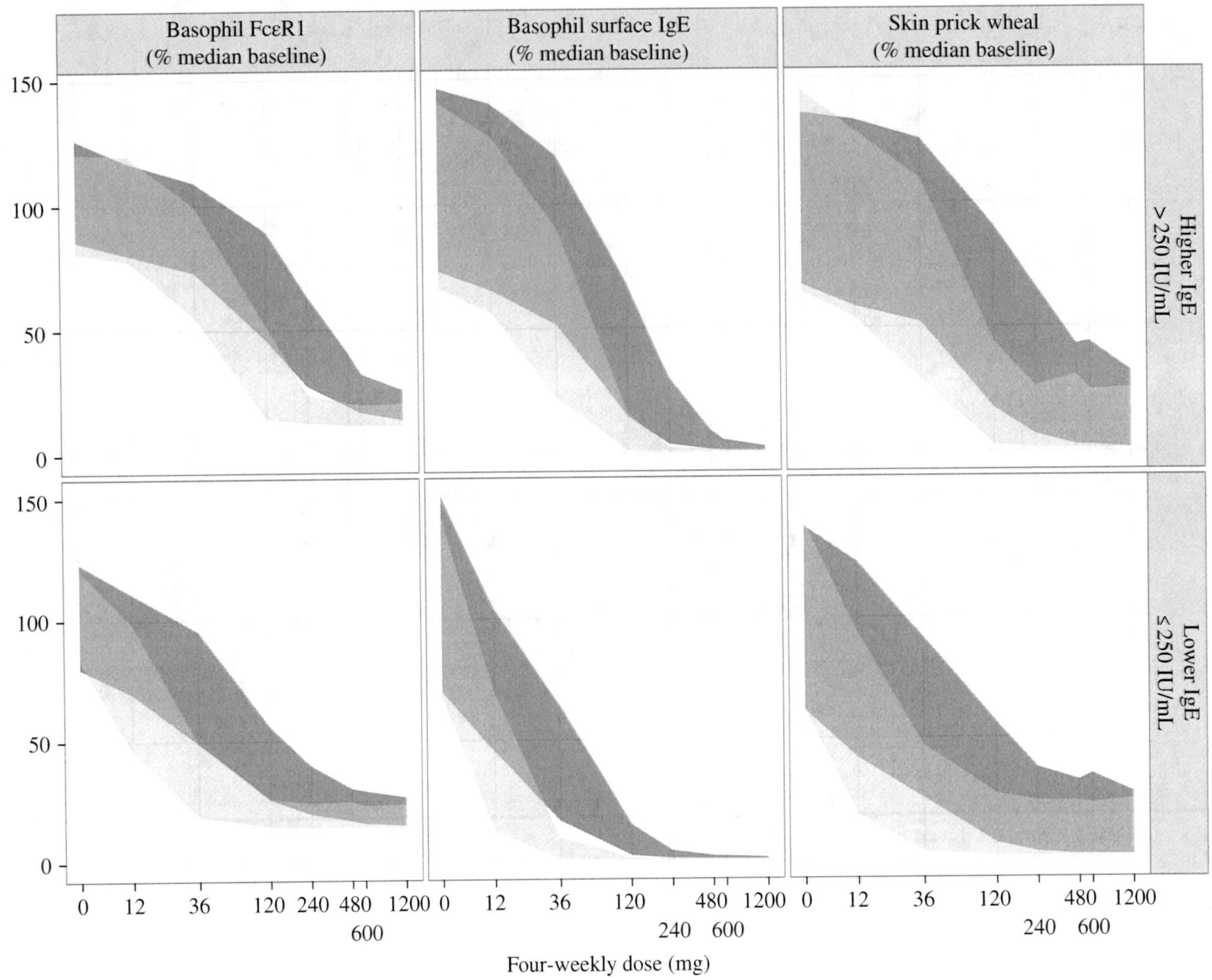

FIGURE 13.11 Projected steady-state QGE031 and omalizumab dose–responses. The biomarker responses at 224 days following four-weekly dosing were taken to plot dose–response curves. The bands are interquartile ranges representing variation between subjects. Dark gray (originally red) represents omalizumab and light gray (originally blue) QGE031.

match the observed data. Only then does one predict new scenarios such as time–response or dose–response.

From the combined QGE031/omalizumab example, steady-state dose–responses relationships were extrapolated from non-steady-state pharmacokinetic and pharmacodynamic study designs and data. The initial pharmacodynamics investigated were circulating biomarkers, FcεR1 and IgE on the surface of basophils, with the total IgE contributing by quantitating the formation of drug–IgE complexes and hence, by calculation, the hidden variable for suppression of free IgE. Although there were ethnic differences between Japanese and Caucasians in pharmacokinetics and IgE binding, when combined, these differences counteracted with an end result that both populations could be studied over the same range of doses.

Importantly, from a clinical (disease) perspective, it was also possible to include a tissue pharmacodynamic response in the model, the allergen skin prick tolerance. Interestingly, this demonstrated a distinct time delay from the circulating biomarkers, taking some weeks to equilibrate. Not only did this generate interesting questions regarding indirect mechanisms of action in tissues, but it also had implications for development planning; if one were to directly measure dose–response earlier than 5–6 months, one would not necessarily propose "the right posology" for chronic treatment.

The model-based quantitation of target binding was directed not solely toward biomarkers and allergen challenge tests, but also to urticaria clinical endpoints of itch (primary), hives, and the combination of the two, the urticaria activity score, UAS7 (key secondaries). A PK/PD model, qualified for simulation, enabled alternative posologies to be explored and proposed to authorities and physicians. The final

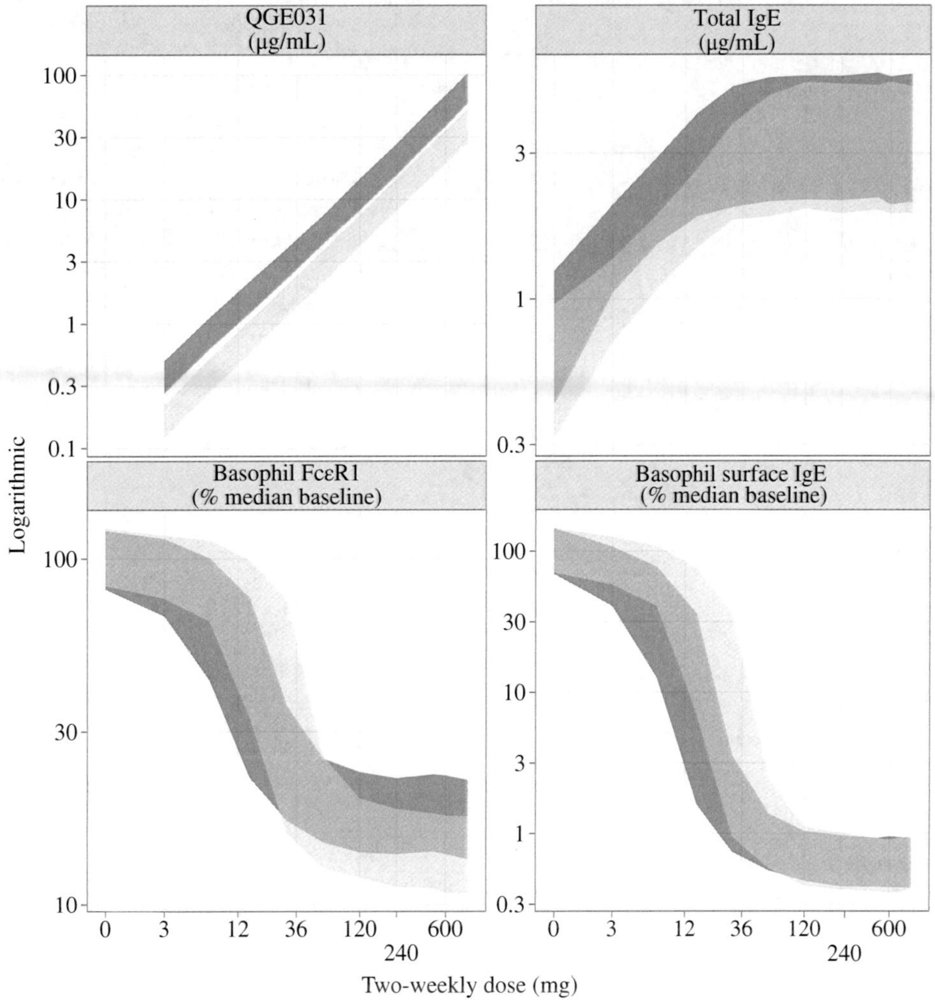

FIGURE 13.12 Projected Japanese versus Caucasian steady state QGE031 dose–responses. The biomarker responses at 168 days following 2 weekly dosing were taken to plot dose–response curves. The bands are interquartile ranges representing variation between subjects. Dark gray (originally red) represents Caucasian and light gray (originally blue) Japanese.

marketed posologies for urticaria varied by region. In the European Union, 300 mg every 4 weeks is authorized; in the United States and Switzerland, both 150 and 300 mg every 4 weeks are available, enabling patients and their attending physicians to adjust the dose as they deem appropriate. Overall, the PK/PD process enabled an integrated quantitation of biotherapeutic drug–target binding, biomarkers, and clinical response to support rational dose regimen selection.

ACKNOWLEDGMENTS

The authors acknowledge the expert help and assistance of their programming colleagues Aurélie Gautier, Claire Petry, and Gregory Pinault for the datasets used for the analyses.

REFERENCES

[1] Lachmann HJ, Lowe P, Felix SD, Rordorf C, Leslie K, Madhoo S, Wittkowski H, Bek S, Hartmann N, Bosset S, Hawkins PN, Jung T. *In vivo* regulation of interleukin 1β in patients with cryopyrin-associated periodic syndromes. J Exp Med 2009;206 (5):1029–1036.

[2] Ait-Oudhia S, Lowe PJ, Mager DE. Bridging clinical outcomes of canakinumab treatment in patients with rheumatoid arthritis with a population model of IL-1β kinetics. CPT Pharmacometrics Syst Pharmacol 2012;1:e4. DOI: 10.1038/psp.2012.6.

[3] Slavin R, Ferioli C, Tannenbaum S, Martin C, Blogg M, Lowe P. Asthma symptom re-emergence after omalizumab withdrawal correlates well with increasing immunoglobulin-E and decreasing pharmacokinetic concentrations. J Allergy Clin Immunol 2009;123:107–113.

[4] Lowe P, Tannenbaum S, Gautier A, Jimenez P. Relationship between omalizumab pharmacokinetics, IgE pharmacodynamics and symptoms in patients with severe persistent allergic (IgE-mediated) asthma. Br J Clin Pharmacol 2009;68 (1):61–76.

[5] EMA. 2010. Procedural steps taken and scientific information after the authorisation. Application II/0019, p. 12 (EMA Xolair Application).

[6] EMA. 2012. Procedural steps taken and scientific information after the authorisation. Application II/0037, p. 5.

[7] EMA. 2014. Application II/0048 Assessment report, Sections 2.3.4–2.3.6. and comments in Section 2.4.3.

[8] Lammerts van Bueren JJ, Bleeker WK, Bøgh HO, Houtkamp M, Schuurman J, van de Winkel JG,, Parren PWHI. Effect of target dynamics on pharmacokinetics of a novel therapeutic antibody against the epidermal growth factor receptor: implications for the mechanisms of action. Cancer Res 2006;66:7630–7638.

[9] Vugmeyster Y, Tian X, Szklu P, Kasaian M, Xu X. Pharmacokinetic and pharmacodynamic modeling of a humanized anti-IL-13 antibody in naïve and ascaria-challenged cynomolgus monkeys. Pharm Res 2008;26:306–315.

[10] Betts AM, Clark TH, Yang J, Treadway JL, Li M, Giovanelli MA, Abdiche Y, Stone DM, Paralkar VM. The application of target Information and preclinical pharmacokinetic/pharmacodynamic modeling in predicting clinical doses of a Dickkopf-1 antibody for osteoporosis. J Pharmacol Exp Ther 2010;333:2–13.

[11] Lowe PJ, Tannenbaum SJ, Wu K, Lloyd P, Sims J. On setting the first dose in man: quantitating biotherapeutic drug–target binding through pharmacokinetic and pharmacodynamic models. Basic Clin Pharmacol Toxicol 2009;106: 195–209.

[12] Ng CM, Joshi A, Dedrick RL, Garovoy MR, Bauer RJ. Pharmacokinetic–pharmacodynamic-efficacy analysis of efalizumab in patients with moderate to severe psoriasis. Pharm Res 2005;22:1088–1100.

[13] Ng CM, Stefanich E, Anand BS, Fielder PJ, Vaickus L. Pharmacokinetics/pharmacodynamics of nondepleting anti-CD4 monoclonal antibody (TRX1) in healthy human volunteers. Pharm Res 2006;23:95–103.

[14] Stefanini MO, Wu FTH, Gabhann FM, Popel AS. Increase of plasma VEGF after intravenous administration of bevacizumab is predicted by a pharmacokinetic model. Cancer Res 2010;70:9886–9894.

[15] Thai HT, Veyrat-Follet C, Vivier N, Dubruc C, Sanderlink G, Mentré F, Comets E. A mechanism-based model for the population pharmacokinetics of free and bound aflibercept in healthy subjects. Br J Clin Pharmacol 2011;72:402–414.

[16] Sutjandra L, Rodriguez RD, Doshi S, Ma M, Peterson MC, Jang GR, Chow AT, Pérez-Ruixo JJ. Population pharmacokinetic meta-analysis of denosumab in healthy subjects and postmenopausal women with osteopenia or osteoporosis. Clin Pharmacokinet 2011;50:793–807.

[17] Gibiansky L, Sutjandra L, Doshi S, Zheng J, Peterson MC, Jang GR, Chow AT, Pérez-Ruixo JJ. Population pharmacokinetic analysis of denosumab in patients with bone metastases from solid tumours. Clin Pharmacokinet 2012;51: 247–260.

[18] Wang B, Lau YY, Liang M, Vainshtein I, Zusmanovich M, Lu H, Magrini F, Sleeman M, Roskos L. Mechanistic modelling of antigen sink effect for mavrilimumab following intravenous administration in patients with rheumatoid arthritis. J Clin Pharmacol 2012;52:1150–1161.

[19] Fetterly GJ, Aras U, Meholick PD, Takimoto C, Seetharam S, McIntosh T, de Bono JS, Sandhu SK, Tolcher A, Davis HM, Zhou H, Puchalski TA. Utilizing pharmacokinetics/pharmacodynamics modeling to simultaneously examine free CCL2, total CCL2 and carlumab (CNTO 888) concentration time data. J Clin Pharmacol 2013;53:1020–1027.

[20] Levy G. Pharmacologic target-mediated drug disposition. Clin Pharmacol Ther 1994;56:248–252.

[21] Mager DE, Jusko WJ. General pharmacokinetic model for drugs exhibiting target-mediated drug disposition. J Pharmacokinet Pharmacodyn 2001;28:507–532.

[22] Mager DE, Krzyzanski W. Quasi-equilibrium pharmacokinetic model for drugs exhibiting target-mediated drug disposition. Pharm Res 2005;22:1589–1596.

[23] Hayashi N, Tsukamoto Y, Sallas WM, Lowe PJ. A mechanism-based binding model for the population pharmacokinetics and pharmacodynamics of omalizumab. Br J Clin Pharmacol 2007;63 (5):548–561.

[24] Meno-Tetang GML,, Lowe PJ. On the prediction of the human response: a recycled mechanistic PK/PD approach. Basic Clin Pharmacol Toxicol 2005;96 (3):182–192.

[25] Lowe P, Renard D. Omalizumab decreases IgE production in patients with allergic (IgE-mediated) asthma; PK/PD analysis of a biomarker, total IgE. Br J Clin Pharmacol 2011;72 (2):306–320.

[26] Staack RF, Jordan G, Heinrich J. Mathematical simulations for bioanalytical assay development: the (un-)necessity and (im-)possibility of free drug quantification. Bioanalysis 2012;4:381–395.

[27] Liu J, Lester P, Builder S, Shire SJ. Characterization of complex formation by humanized anti-IgE monoclonal antibody and monoclonal human IgE. Biochemistry 1995;34: 10474–10482.

[28] Chakraborty A, Tannenbaum S, Rordorf C, Lowe PJ, Floch D, Gram H, Roy S. Pharmacokinetic and pharmacodynamic properties of canakinumab, a human anti-interleukin-1β monoclonal antibody. Clin Pharmacokinet 2012;51 (6):e1–e18.

[29] Levitt DG. The pharmacokinetics of the interstitial space in humans. BMC Clin Pharmacol 2003;3:3.

[30] Olszewski WL, Pasdur J, Kubasiewicz E, Kaleska M, Cooke CJ, Miller NE. Lymph draining from foot joints in rheumatoid arthritis provides insight into local cytokine and chemokine production and transport to lymph nodes. Arthritis Rheum 2001;44:541–549.

[31] Savic RM, Karlsson MO. Importance of shrinkage in empirical bayes estimates for diagnostics: problems and solutions. AAPS J 2009;11 (3):558–569.

[32] FDA. 1999. Guidance for industry: population pharmacokinetics. Available at http://www.fda.gov/downloads/Drugs/Guidances/UCM072137.pdf. Accessed 2014 Sep 25.

[33] Comets E, Brendel K, Mentre F. Model evaluation in nonlinear mixed effect models, with applications to pharmacokinetics. J Soc Fr Stat 2010;151:1.

[34] Saini SS, Bindslev-Jensen C, Maurer M, Grob JJ, Baskan EB, Bradley MS, Canvin J, Rahmaoui A, Georgiou P, Alpan O, Spector S, Rosén K. Efficacy and safety of omalizumab in patients with chronic idiopathic/spontaneous urticaria who remain symptomatic on H1 antihistamines: a randomized, placebo-controlled study. J Invest Dermatol 2015;135:67–75. DOI: 10.1038/jid.2014.306.

[35] Maurer M, Rosén K, Hsieh HJ, Saini S, Grattan C, Gimenéz-Arnau A, Agarwal S, Doyle R, Canvin J, Kaplan A, Casale T. Omalizumab for the treatment of chronic idiopathic or spontaneous urticaria. N Engl J Med 2013;368:924–935.

[36] Kaplan A, Ledford D, Ashby M, Canvin J, Zazzali JL, Conner E, Veith J, Kamath N, Staubach P, Jakob T, Stirling RG, Kuna P, Berger W, Maurer M, Rosén K. Omalizumab in patients with symptomatic chronic idiopathic/spontaneous urticaria despite standard combination therapy. J Allergy Clin Immunol 2013;132:101–109.

[37] Arm JP, Bottoli I, Skerjanec A, Floch D, Groenewegen A, Maahs S, Owen C, Jones I, Lowe PJ. Pharmacokinetics, pharmacodynamics and safety of QGE031 (ligelizumab), a novel high affinity anti-IgE antibody, in atopic subjects. Clin Exp Allergy 2014;44:1371–1385. DOI: 10.1111/cea.12400. Available at 10.1111/cea.12400. Accessed 2015 Jun 25.

[38] Arm JP, Skerjanec A, Groenewegen A, Lowe PJ, Maahs S, Matsushima S, Bottoli I. Safety, tolerability, pharmacokinetics and pharmacodynamics of QGE031, a high affinity anti-IgE antibody in Caucasian and Japanese atopic subjects. Am J Respir Crit Care Med 2014;189:A1325. Available at http://www.atsjournals.org/doi/pdf/10.1164/ajrccm-conference.2014.189.1_MeetingAbstracts.A1325. Accessed 2015 Jun 25.

[39] Roskos L, Klakamp S, Liang M, Arends R, Green L. Molecular engineering II: Antibody affinity. In: Dübel S, editor. *Handbook of Therapeutic Antibodies*. Chapter 7 Weinheim: Wiley-VCH Verlag GmbH & Co. 2007. p 145–169.

[40] Putnam WS, Li J, Haggstrom J, Ng C, Kadkhodayan-Fischer S, Cheu M, Deniz Y, Lowman H, Fielder P, Visich J, Joshi A, Jumbe N. Use of quantitative pharmacology in the development of HAE1, a high-affinity anti-IgE monoclonal antibody. AAPS J 2008;10:425–430.

14

TARGET-DRIVEN PHARMACOKINETICS OF BIOTHERAPEUTICS

WILHELM HUISINGA[1], SASKIA FUHRMANN[2,3], LUDIVINE FRONTON[4] AND BEN-FILLIPPO KRIPPENDORFF[4]

[1] *Universität Potsdam, Potsdam, Germany*
[2] *Universität Potsdam Institute of Biochemistry and Biology, Potsdam, Germany*
[3] *Freie Universitat Berlin, Berlin, Germany*
[4] *Roche Pharmaceutical Research and Early Development, Basel, Switzerland*

14.1 INTRODUCTION

Target-driven pharmacokinetic (PK) is a peculiar characteristic of therapeutic monoclonal antibodies (mAbs).[1] MAbs typically bind with high affinity to specific antigens, such as soluble targets or cell-surface receptors. Unspecific binding is typically considered negligible—in contrast to most small molecule drugs—and target-mediated drug disposition (TMDD) often substantially impacts PK. TMDD includes the processes of binding of the mAb to its antigen and elimination of the mAb–antigen complex. Often, target-driven PK is nonlinear, and this nonlinearity is exploited to infer properties of the targeted systems.

In this chapter, we review different modeling approaches that can be used to analyze, simulate, and infer target-driven PK of mAbs: the classical (whole-body) TMDD modeling approach, including its various approximations (such as the Michaelis–Menten approximation), the cell-level TMDD modeling approach, and the simplified physiologically based pharmacokinetic (PBPK) modeling approach. We discuss the classical whole-body TMDD model and the characteristic features of TMDD profiles. Motivated by the problem of parameter identifiability for this model, we review various reduced TMDD models and their underlying assumptions. Cell-level TMDD and simplified PBPK models are reviewed as two approaches that allow to integrate prior information on the targeted system and the species physiology into the modeling process. The cell-level TMDD approach is particularly suited to translate data from *in vitro* to *in vivo* and to establish a link to the wealth of available systems biology models of pharmacologically relevant targeted systems. The simplified PBPK model for mAbs is particularly suited to make use of the recent promising development of antibody biodistribution coefficients (ABCs) that quantify the target-independent tissue distribution of mAbs. The simplified PBPK model also allows to integrate species-specific data on organ volumes and lymph flows, which supports covariate modeling and extrapolation across species.

For the underlying biological and physiological mechanisms, we refer to the other chapters of this book.

[1] While we focus on therapeutic mAbs in this chapter, analogous results can be expected to hold for other therapeutic proteins.

14.2 SOLUBLE AND MEMBRANE-BOUND TARGETS

For modeling purposes in general, it is important to distinguish the following three categories of targets (cf. [1]):

- circulating soluble targets;
- membrane-bound (cell-surface) targets with minimal or no shedding of the target protein;

ADME and Translational Pharmacokinetics/Pharmacodynamics of Therapeutic Proteins: Applications in Drug Discovery and Development, First Edition. Edited by Honghui Zhou and Frank-Peter Theil.

- membrane-bound targets that shed their extracellular domain physiologically or have a soluble form in the systemic circulation.

Target-driven PK may be observable for mAbs targeting any of the three target categories and the presented modeling approaches are suitable to model TMDD in any of the three cases. For a given target, however, its category (see above) impacts model development and data analysis in two important aspects: (i) the location(s) of the target, in particular, whether the distribution of the target via the plasma and lymph flows has to be taken into account (as for soluble target and shed antigens) and (ii) the way model predictions are compared to the experimental data. In the case of membrane-bound target with no shedding, mAb levels in plasma correspond to the unbound form of the mAb. In the case of soluble targets or membrane-bound target with shedding, free and bound forms of the mAb coexist in plasma, with implications for the bioanalytics [2].

Target-driven PK typically exhibits characteristic features, in particular a nonlinear dependence on dose. The ability to detect this nonlinearity is dependent on the availability of concentration–time data for a wide range of doses. Concentration–time data from clinical studies where doses may be high enough to saturate the target often make it difficult to detect nonlinearity [3]. In addition, soluble targets are often present in low concentrations [1], such that moderate doses might already fully saturate any target-dependent processes. This could be the reason that mostly linear PK models have been reported for these cases [4].

The ability to observe nonlinear target-driven PK experimentally also depends on the mAb form measured in the biological fluid. For example, the concentration–time profile of the free form might show clear nonlinear characteristics, while these might be less visible in the profile of the total concentration, that is, free mAb and mAb bound to the (soluble) target.

14.3 WHOLE-BODY TARGET-MEDIATED DRUG DISPOSITION MODELS AND THEIR APPROXIMATIONS

In this section, we discuss target-driven PK in the context of the well-established class of TMDD models. TMDD models have been designed for drugs that bind with high affinity and to a significant extent (relative to dose) to a pharmacologic target, and as a consequence may exhibit nonlinear PK behavior [5]. Although not specifically designed for antibodies, TMDD models (and various approximations—see below) are now widely used to study the PK of mAbs. The characteristics of target-driven PK are typically interpreted in terms of parameter estimates obtained from experimental data, with a special focus on target-related parameters, such as target saturation and receptor occupancy (in the case of membrane-bound receptors).

A generic TMDD model has been proposed by Mager and Jusko [5]; it is based on a classical two-compartment model extended by two target compartments accounting for the free target and the drug–target complex. It is used to model both, soluble and membrane-bound targets in the absence of target shedding. Since the two target compartments represent the concentration of the target in the entire body, we use the term "whole-body TMDD model" to distinguish it from "cell-level TMDD models" to be presented in Section 14.4. In the sequel, we state all parameters values with specific units for the sake of clarity (choosing (L) for volumes, (nmol) for amounts, (nM) = (nmol/L) for concentrations, and (min) for time).

14.3.1 Generic Whole-Body TMDD Model

A schematic illustration of the generic whole-body TMDD model is shown in Figure 14.1. Drug in the central compartment with apparent volume V_{cen} (L) distributes to and from a peripheral compartment with rate constants k_{cp} and k_{pc} (1/min), respectively. Drug in the central compartment is eliminated unspecifically (i.e., independent of the target) with first-order rate constant k_{degC} (1/min) and binds the free target with association rate constant k_{on} (nM/min) to form a drug–target complex. This complex dissociates with rate constant k_{off} (1/min) or is potentially eliminated with rate constant k_{degRC} (1/min). The target turnover is either (i) explicitly accounted for, more precisely, free target is synthesized with rate k_{syn} (nM/min) and eliminated with rate constant k_{degR} (1/min) or (ii) considered to be absent, that is, the total concentration of target (free and mAb-bound) is assumed to be constant, in which case both k_{syn} and k_{degR} are

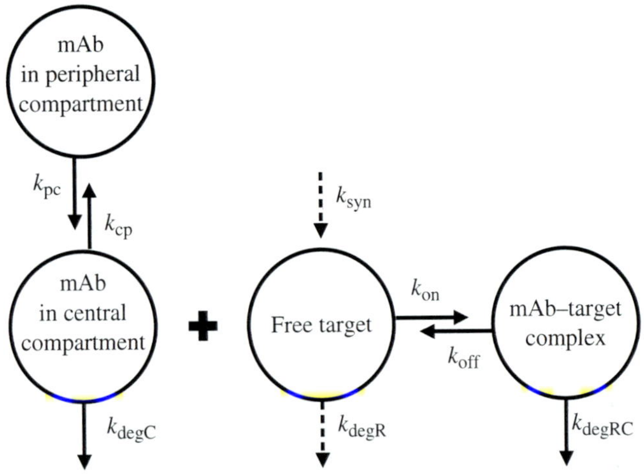

FIGURE 14.1 Generic whole-body TMDD model comprising drug distribution and drug–target interaction processes. See Section 14.3.1 for definitions and description.

set to zero. Denoting the *concentrations* of the free mAb, free target, and the mAb–target complex in the central compartment by C, R, and RC (nM), respectively, and the amount of mAb in the peripheral compartment by A_{per} (nmol), the following system of ordinary differential equations (ODEs) describes the rate of change of the TMDD variables[2]:

$$\frac{\mathrm{d}}{\mathrm{d}t}C = -k_{\text{cp}}C + k_{\text{pc}}\frac{A_{\text{per}}}{V_{\text{cen}}} - k_{\text{degC}}C - k_{\text{on}}R \cdot C + k_{\text{off}}RC + k_{\text{in}}(t)$$

$$\frac{\mathrm{d}}{\mathrm{d}t}A_{\text{per}} = k_{\text{cp}}V_{\text{cen}} \cdot C - k_{\text{pc}}A_{\text{per}}$$

$$\frac{\mathrm{d}}{\mathrm{d}t}R = k_{\text{syn}} - k_{\text{degR}}R - k_{\text{on}}R \cdot C + k_{\text{off}}RC$$

$$\frac{\mathrm{d}}{\mathrm{d}t}RC = k_{\text{on}}R \cdot C - \left(k_{\text{off}} + k_{\text{degRC}}\right)RC$$

with $k_{\text{in}}(t) = r_{\text{inf}}(t) / V_{\text{cen}}$ (nM/min) and infusion rate $r_{\text{inf}}(t)$ (nmol/min).

14.3.2 Characteristics of Target-Driven PK Profiles

Despite the simple model structure, the TMDD model exhibits a surprisingly rich behavior. Illustrative TMDD concentration–time profiles of the free drug in the central (plasma) compartment for a high, medium, and low intravenous (IV) bolus dose are shown in Figure 14.2. A nice and thorough analysis of the characteristic features of TMDD profiles in the absence of a peripheral compartment (i.e., with $k_{\text{cp}} = k_{\text{pc}} = 0$) is presented in Peletier and Gabrielsson [6]. Taking also the distribution to the peripheral compartment into account, concentration–time profiles of the free drug concentration in plasma may exhibit some or all of the following characteristic features (depending on dose relative to target abundance), see Figure 14.2:

- The initial phase is dominated by a rapid binding of free drug to its target (due to the high affinity). While this decay might be hardly visible for large doses,[3] it can have a strong impact for low doses.[4] In both cases, the concentration of free target drops and the concentration of drug–target complex increases correspondingly.

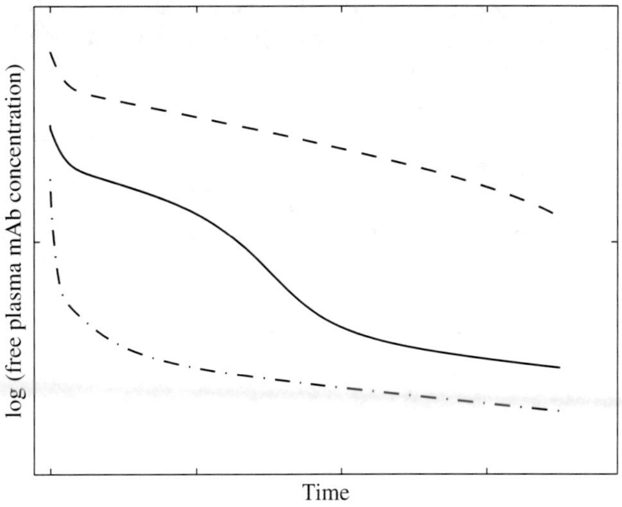

FIGURE 14.2 Illustrative free plasma concentration–time profiles of the generic whole-body TMDD model (as depicted in Fig. 14.1) for a high (dashed line), medium (solid line), and low (dashed-dotted line) IV bolus dose. For a description of the different profiles, see Section 14.3.2.

- Typically, a second phase follows, where drug disposition is effectively linear, representing distribution into the peripheral compartment and drug elimination. For larger doses, the target route is saturated, while for lower doses, it behaves linear. All in all, the TMDD model behaves approximately like a two-compartment model with central and peripheral compartment, and linear elimination from the central compartment. This phase shifts upwards as the dose increases.
- In the mixed phase, the drug concentration is so low that the target route is only partly saturated. This phase shifts to the right as the dose increases.
- In the terminal phase, the target system behaves again approximately linear (including unspecific as well as target-mediated elimination). The slope of this terminal phase is an aggregated parameter, in general depending on all TMDD parameters. Importantly, the terminal slope can be slower, equal, or faster than the approximately linear decay following the initial phase.

For more details on the mathematical aspects for a TMDD model without a peripheral compartment, see [6].

We have seen that target-driven PK can be nonlinear. For small molecule drugs, nonlinear PK is often considered an undesirable property. For mAbs and other therapeutic proteins, however, the characteristic features of the nonlinear target-driven PK are often exploited to infer properties of the targeted system from PK data. A common PD quantity for membrane-bound receptors is the receptor occupancy, that is, the fraction of drug–receptor complexes to the total receptors: $RC / (R + RC)$. In this context, it is worth noting

[2] The initial conditions are $C(0) = \text{dose} / V_{\text{cen}}$ (IV bolus, dose in (nmol)) or $C(0) = 0$ (IV infusion), $A_{\text{per}}(0) = 0$, $R(0) = k_{\text{syn}} / k_{\text{degR}}$, and $RC(0) = 0$.

[3] Zoom into the initial phase of the medium dose in Figure 14.2 for a clear view.

[4] The initial strong decay for the low dose in Figure 14.2 is due to binding of free mAb to the target.

that the generic whole-body TMDD model does not include any endogenous molecules that potentially competitively bind to the target, such as natural ligands for membrane-bound receptors. For mAbs that competitively bind membrane-bound receptors, it would be more natural to consider the relative depletion of natural ligand-bound target as a measure to quantify the drug effect. We have shown in [7] that competitive binding can impact target dynamics, in particular, if the ligand accumulates in the microenvironment of the target.

Moreover, competitive binding also needs to be considered in the case of shed antigens.

14.3.3 Location of the Target: Central versus Peripheral Compartment

In the generic whole-body TMDD model, the drug–target interaction is considered in the central compartment. Whether, for a given target, this is a physiologically plausible assumption depends on the interpretation of the central and peripheral compartment. In many studies, it is noted that estimates of the volume of the central compartment V_{cen} was close to the species' plasma volume, while the peripheral volume V_{per} was a fraction of the interstitial volume. This is often taken as a motivation for a common interpretation that the central compartment is identical to plasma and the peripheral compartment is identical to the interstitial space (or some part of it). This would also be in line with the extravasation being a rate-limiting step of the tissue distribution of therapeutic proteins (cf. Section 14.5). We have shown in [8], however, that this interpretation is *not* consistent with tissue distribution data and insight from a more mechanistic PBPK modeling approach: Although extravasation is the rate-limiting step for tissue distribution, the extent of elimination varies across tissues. In many cases, the tissue concentration–time profiles are just scaled versions of the plasma concentration–time profile, while for tissues, such as the muscle and the skin, the concentration–time profiles often look significantly different (see, e.g., [9, 10]). Consequently, a more consistent interpretation of the whole-body TMDD model is the following [8]:

- The central compartment corresponds to plasma and tissues such as the heart, kidney, liver, lung, and the spleen, which behave kinetically similar to plasma.
- The peripheral compartment corresponds to the most extravasation rate-limited tissues. According to experimental data, these are the tissues such as adipose, bone, muscle, and the skin.

Given the structure of the simplified PBPK model for mAbs, presented in Section 14.5, this interpretation is also in line with our expectations from lumping of small molecule PBPK models.

The generic whole-body TMDD model assumes the target to be present in the central compartment. For a given targeted system, the target might alternatively be assumed to be in the peripheral compartment, or in both. For example, the epidermal growth factor receptor (EGFR) is known to be expressed on skin cells. Consequently, a TMDD model with the target in the peripheral compartment would be physiologically more consistent (see also Section 14.4). For soluble targets in plasma that also may distribute into the interstitial space, a TMDD model with targets in both central and peripheral compartments would be more consistent with the underlying physiology.

14.3.4 Parameter Identifiability and Model Reduction

It was noted that identifiability of TMDD model parameters from PK data alone is difficult, especially when dealing with relatively sparse clinical data, and/or if only either the free or the total drug concentration is available [11], see also [6]. One reason for this identifiability problem is the large timescale difference between the cellular level (drug–target interaction, often on a minutes or hours timescale) and the PK level (characteristic half-lives of days or weeks). There are at least three ways to resolve this problem:

- Refined measurements quantifying in addition to the free drug in plasma the free target and the drug–target complex. Due to various reasons, this is sometimes, but not often an option.
- Model reduction of the whole-body TMDD model to obtain reduced TMDD models with less variables and parameters; this is currently realized most often (see below).
- Integration of alternative sources of data, like *in vitro* data of the targeted system, resulting in cell-level TMDD models (see Section 14.4).

In the literature, various names are associated with such approximate models: Michaelis–Menten approximation, extended Michaelis–Menten approximation, quasi-steady state (QSS) approximation, quasi-equilibrium approximation, rapid-binding model, and total QSS approximation (see, e.g., [6, 7, 11]). Unfortunately, the terminology is not used consistently across publications. In mathematical terms, all reduced TMDD models are based on a model reduction technique called singular perturbation theory; in the biological literature, this technique is known as the QSS approximation. Its underlying assumption is a timescale separation, that is, the assumption that some variables evolve on much faster timescales than others. In our setting, the target dynamics are often on a much faster timescale than drug distribution. As a consequence of the different timescales, the fast variables are "instantaneous" reaching their steady state, while the slow variables are still slowly evolving. This (fast) steady state typically depends on the slow states. It allows us to approximate the fast state as a function of the slow states. Since only the fast variables are considered in steady state (but not the slow ones), this approximation is called "quasi-steady-state approximation"—a more intuitive name would be "partial steady-state approximation." In our context, for example, the drug–target complex *RC* often evolves on a fast timescale, resulting in the QSS approximation RC_{QSS} that depends on the slow free target and free drug variables *R* and *C* (see Eq. 14.1). Based on this approximation, the reduced model is derived by removing the ODE for the fast variables (e.g., delete ODE for *RC*) and replacing the fast variables by their QSS values (e.g., replace *RC* by RC_{QSS}), wherever they appear in the remaining ODEs for the slow variables. This typically results in reduced models with less differential equations and less parameters.

The QSS approximation is called total QSS approximation, if it is applied to a system where the total concentration of some quantity still evolves slowly overtime, while the corresponding free and bound concentrations evolve on a much faster timescale. Typically, the resulting equations contain a square-root term to determine the free concentration as a function of the total concentration (see Sections 14.3.5 and 14.3.7). Mathematically, the total QSS approximation is also based on singular perturbation theory. The application of the QSS approximation to a specific biological system, where a substrate is metabolized by an enzyme, is called the Michaelis–Menten approximation [12, 13]. In the biological and pharmacological community, the term Michaelis–Menten approximation is now widely used to refer to models containing a term of the form $V_{max} \cdot C/(K_M + C)$. The application of the total QSS approximation is called the extended Michaelis–Menten approximation. Finally, in the special situation, where the QSS approximation is applied to a variable that is only involved in reversible reactions, the corresponding QSS is called a quasi-equilibrium and the resulting approximation is called the "quasi-equilibrium approximation" or "rapid-binding model."

Below, we discuss four different reduced TMDD models with both increasingly stronger assumptions and increasingly simpler structure. The reduced models differ in (i) whether target turnover is explicitly considered or assumed to be in QSS and (ii) whether only the elimination aspect is considered, or whether elimination and target binding is considered. This is equivalent to use the QSS or total QSS assumption on the target dynamics. Table 14.1 gives an overview of the different reduced models and their underlying assumptions. For the sake of clarity, we refer to the original whole-body TMDD model as the full TMDD model.

14.3.5 Extended Michaelis–Menten Approximation with Target Turnover

The reduced TMDD model closest to the full TMDD model is based on *Assumption RC: The drug–receptor complex RC is in quasi-steady-state*. This implies

$$RC_{QSS} = \frac{R_{tot} \cdot C}{K_M + C}. \tag{14.1}$$

based on the approximation $R_{tot} = R + RC_{QSS}$. Using also $C_{tot} = C + RC_{QSS}$ (nM), we can establish the ODEs for R_{tot} and C_{tot} from the full TMDD model. We obtain for the rate of change of the total central drug concentration C_{tot} and free peripheral drug amount A_{per}, and the total receptor concentration R_{tot} the following equations[5]:

$$\frac{d}{dt}C_{tot} = -k_{cp}C + k_{pc}\frac{A_{per}}{V_{cen}} - k_{degC}C - k_{degRC}\frac{R_{tot} \cdot C}{K_M + C} + k_{in}(t)$$

$$\frac{d}{dt}A_{per} = k_{cp}V_{cen} \cdot C - k_{pc}A_{per}$$

$$\frac{d}{dt}R_{tot} = k_{syn} - k_{degR}R_{tot} - \left(k_{degRC} - k_{degR}\right)\frac{R_{tot} \cdot C}{K_M + C}$$

$$C = \frac{1}{2}\left(C_{eff} + \sqrt{C_{eff}^2 + 4K_M C_{tot}}\right)$$

with $C_{eff} = C_{tot} - R_{tot} - K_M$ (nM) and Michaelis constant (nM):

$$K_M = \frac{k_{off} + k_{degRC}}{k_{on}}. \tag{14.2}$$

Note that a square-root term as for *C* is typical for a saturable (binding) process. In the case of $R_{tot} = 0$, the corresponding square-root term gives $C = C_{tot}$. In comparison to the full TMDD model, the number of ODEs and the number of parameters are reduced by one.

The ODE for R_{tot} gives further insight into the dynamics of the targeted system. In [6], three scenarios are distinguished, depending on the relative magnitude of the degradation rate constants of the free and drug-bound target:

- if $k_{degRC} = k_{degR}$, then $R_{tot} = R_0$, that is, the total target concentration stays constant;
- if $k_{degRC} < k_{degR}$, then the total target concentration R_{tot} temporarily increases, maximally to the value

$$R_* = k_{syn} / k_{degRC} \tag{14.3}$$

[5] The initial conditions are $C_{tot}(0) = \text{dose} / V_{cen}$ (IV bolus, dose in (nmol)) or $C_{tot}(0) = 0$ (IV infusion), $A_{per}(0) = 0$, and $R_{tot}(0) = k_{syn} / k_{degR}$.

TABLE 14.1 Underlying Assumptions of Different Reduced Models of the Whole-Body TMDD Model

Reduced TMDD Models and Underlying Assumptions		Relevant Extent of Binding to the Target: Yes	Relevant Extent of Binding to the Target: No
Target turnover much faster than PK	Yes	Extended Michaelis–Menten approximation	Michaelis–Menten approximation
	No	Extended Michaelis–Menten approximation with target turnover	Michaelis–Menten approximation with target turnover

Note that the terminology is not uniquely used in the literature (see Section 14.3.4). See Section 14.3.5–14.3.8 for alternatively used model names, such as "quasi-steady state model."

- if $k_{\text{degRC}} > k_{\text{degR}}$, then the total target concentration R_{tot} temporarily decreases.

Thus, mAbs can effectively down- or upregulate the total receptor concentrations. Given the full TMDD model, the validity of *Assumption RC* can easily be verified: The difference

$$\Delta = RC - RC_{\text{QSS}} \tag{14.4}$$

with RC and RC_{QSS} (based on Eq. 14.1) being predicted by the full TMDD model should instantaneously decay. This is exactly the interpretation of the QSS assumption on the drug–target complex: The quantity RC reaches almost instantaneously its partial steady-state RC_{QSS}.

In [14], the extended Michaelis–Menten approximation with target turnover was referred to as "QSS approximation." As mentioned above, this term refers to the approximation technique and is not uniquely characterizing the reduced model. We therefore prefer the term extended Michaelis–Menten approximation with target turnover.

14.3.6 Michaelis–Menten Approximation with Target Turnover

A further reduction is obtained, if we assume in addition to *Assumption RC* (see above) the *Assumption B: The concentration of drug bound to the target is negligible compared to the free drug concentration, that is, $RC \ll C$.* Approximating RC by RC_{QSS}, *Assumptions RC + B* read

$$RC_{\text{QSS}} = \frac{R_{\text{tot}} . C}{K_{\text{M}} + C} \quad \text{and} \quad RC_{\text{QSS}} \ll C. \tag{14.5}$$

A consequence of *Assumption B* is that free and total drug concentration in the central compartment are approximately equal, that is, $C_{\text{tot}} = C + RC_{\text{QSS}} \approx C$. Using this approximation, we obtain for the rate of change of the central concentration C and the peripheral drug amount A_{per} the following equations[6]:

[6] The initial conditions are $C(0) = \text{dose} / V_{\text{cen}}$ (IV bolus, dose in nmol) or $C(0) = 0$ (IV infusion), $A_{\text{per}}(0) = 0$, and $R_{\text{tot}}(0) = k_{\text{syn}} / k_{\text{degR}}$.

$$\frac{\text{d}}{\text{d}t} C = -\left(k_{\text{degC}} + k_{\text{cp}}\right) C + k_{\text{pc}} \frac{A_{\text{per}}}{V_{\text{cen}}} - k_{\text{degRC}} \frac{R_{\text{tot}} \cdot C}{K_{\text{M}} + C} + k_{\text{in}}(t)$$

$$\frac{\text{d}}{\text{d}t} A_{\text{per}} = k_{\text{cp}} V_{\text{cen}} \cdot C - k_{\text{pc}} A_{\text{per}}$$

$$\frac{\text{d}}{\text{d}t} R_{\text{tot}} = k_{\text{syn}} - k_{\text{degR}} R_{\text{tot}} - \left(k_{\text{degRC}} - k_{\text{degR}}\right) \frac{R_{\text{tot}} \cdot C}{K_{\text{M}} + C}$$

with K_{M} defined as above and the maximal elimination capacity (nM/min)

$$V_{\max} = k_{\text{degRC}} \cdot R_{\text{tot}}. \tag{14.6}$$

Note that $V_{\max}$ is generally time dependent, since it depends on the total concentration of receptors, which typically changes over time (see Section 14.3.5). Again, given the full TMDD model, the validity of *Assumption B* can easily be verified. The ratio $\rho = RC_{\text{QSS}} / C$, or equivalently

$$\rho = \frac{R_{\text{tot}}}{K_{\text{M}} + C} \ll 1 \tag{14.7}$$

with R_{tot} and C as predicted by the full TMDD model should be very small.

In [14], the Michaelis–Menten approximation with target turnover was referred to as "Michaelis–Menten Model." The same name is typically used to refer to the model presented in Section 14.3.8. Hence, we prefer to use the addition "with target turnover" to uniquely specify the model.

14.3.7 Extended Michaelis–Menten Approximation

A further reduction is achieved by making the stronger *Assumption R + RC: All receptor species, that is, R and RC are in quasi-steady-state.* As a consequence, also $R_{\text{tot}} = R + RC$ is in QSS with

$$R_{\text{tot,QSS}} = \frac{k_{\text{syn}}\left(K_{\text{M}} + C\right)}{k_{\text{degR}} K_{\text{M}} + k_{\text{degRC}} C} \tag{14.8}$$

Approximating R_{tot} by $R_{tot,QSS}$ in Equation 14.1 results in

$$RC_{QSS} = \frac{k_{syn} / k_{degRC} \cdot C}{k_{degR} K_M / k_{degRC} + C} \tag{14.9}$$

Using $C_{tot} = C + RC_{QSS}$ (nM), we derive an ODE for C_{tot} from the full TMDD model. We obtain for the rate of change of the total central drug concentration C_{tot} and the free peripheral drug amount A_{per} the following equations[7]:

$$\frac{d}{dt} C_{tot} = -k_{cp} C + k_{pc} \frac{A_{per}}{V_{cen}} - k_{degC} C - k_{degRC} \frac{B_{max} \cdot C}{K_M + C} + k_{in}(t)$$

$$\frac{d}{dt} A_{per} = k_{cp} V_{cen} \cdot C - k_{pc} A_{per}$$

$$C = \frac{1}{2}\left(C_{eff} + \sqrt{C_{eff}^2 + 4\left(k_{degR} / k_{degRC}\right) K_M C_{tot}}\right)$$

with $C_{eff} = C_{tot} - B_{max} - (k_{degR} / k_{degRC}) K_M$ (nM) and maximal target capacity B_{max} (nM):

$$B_{max} = \frac{k_{syn}}{k_{degRC}} \tag{14.10}$$

14.3.8 Michaelis–Menten Approximation

The simplest reduced TMDD model is based on *Assumption* $R + RC$ and *Assumption B*, reducing the number of ODEs by two and the number of parameters by three in comparison to the full TMDD model. Hence, it is assumed that all target species are in QSS and that the concentration of drug bound to the target is negligible compared to the free drug concentration. Consequently, we do not distinguish between free and total drug concentration in the central compartment. Using the same approach as before yields for the rate of change of the free = total drug concentration in the central compartment C and the amount A_{per} in the peripheral compartment the following equations[8]:

$$\frac{d}{dt} C = -k_{cp} C + k_{pc} \frac{A_{per}}{V_{cen}} - k_{degC} C + k_{pc} \frac{A_{per}}{V_{cen}} - \frac{V_{max} \cdot C}{K_M + C} + k_{in}(t) \tag{14.11}$$

$$\frac{d}{dt} A_{per} = V_{cen} \cdot k_{cp} C - k_{pc} A_{per}, \tag{14.12}$$

[7] The initial conditions are $C_{tot}(0) = \text{dose} / V_{cen}$ (IV bolus, dose in nmol) or $C_{tot}(0) = 0$ (IV infusion) and $A_{per}(0) = 0$.

[8] The initial conditions are $C(0) = \text{dose} / V_{cen}$ (IV bolus, dose in nmol) or $C(0) = 0$ (IV infusion) and $A_{per}(0) = 0$.

where maximal elimination rate V_{max} is a constant given by

$$V_{max} = k_{syn} \tag{14.13}$$

14.3.9 Model Selection

Typically, reduced TMDD models are used in a data-driven modeling context. Based on the available data and statistical criteria for model comparison, some reduced TMDD model is selected to analyze the data. In this regard, the results presented in [6] are of particular relevance. The authors consider a whole-body TMDD model (in the absence of a peripheral compartment) and analyze the approximation quality of the Michaelis–Menten approximation (also in the absence of a peripheral compartment). This is done by comparing the ability of the Michaelis–Menten approximation to reproduce the characteristic features of typical TMDD profiles (see Section 14.3.2). Noteworthy, the authors conclude that the Michaelis–Menten approximation is in general not able to reproduce the initial decay as well as the slope of the terminal phase of TMDD profiles [6, p. 440]. They further make the crucial observation that by adding a peripheral compartment to the Michaelis–Menten model (while maintaining only a central compartment for the full TMDD model), the Michaelis–Menten model mimics also the slope of the terminal phase to some extent. This remarkable insight has fundamental consequences for the interpretation of the Michaelis-Menten model. For example, in the outlined case, it would not be legitimate to interpret the peripheral compartment of the Michaelis-Menten model in the common terms of drug distribution, since it is rather some artificial compartment compensating for the inability of the reduced model to reproduce certain features of the full model.

Mainly data-driven modeling approaches are more prone to suffer from such problems. In the next sections, we present alternative approaches that aim at reducing this risk by constraining the model based on translating *in vitro* data into the *in vivo* context or integrating prior information on species-dependent data and target-independent distributional processes.

14.4 CELL-LEVEL TARGET-MEDIATED DRUG DISPOSITION MODELS

There is a variety of kinetic models of therapeutically targeted systems in the field of systems biology (e.g., the BioModels database [15]). Systems pharmacology models aim at integrating such kinetic models into the whole-body level. Here, we focus on membrane-bound targets, resulting in cell-level TMDD models, but an analogous approach can be taken for soluble targets (see, e.g., [16]). Cell-level models are particularly suited to translate *in vitro* data into an *in vivo* context and are of increasing relevance for target identification and ranking in drug discovery [17, 18]. A recent prominent example is the use of a kinetic model to identify critical components in ErbB signaling pathways [19].

In this section, we illustrate cell-level TMDD models for antibodies antagonistically inhibiting the EGFR system, a prominent target in cancer therapy. The binding of one of its natural ligands to the EGFR results in the activation of signal transduction pathways that mediate a variety of cellular responses that include cell proliferation, differentiation, survival, and angiogenesis. The cell-level model is based on a generic model of ligand–receptor activation and trafficking [20–22]. In contrast to the whole-body TMDD model, the cell-level model is parameterized using rate constants that have been experimentally determined and validated *in vitro*. The choice of the biological processes considered in the cell-level model is motivated by the expectation to have an impact on the PK of the drug and to provide a link to more detailed systems biology models of downstream signaling pathways (e.g., [23]). For a more detailed and in depth description and analysis of the cell-level TMDD model, see [24].

14.4.1 Cell-Level TMDD Model with a Single-Cell Type

The cellular level is depicted in Figure 14.3a. In the extracellular environment of the targeted system, the natural ligand and the drug competitively bind a membrane-bound target (receptor) with associate rate constants k_{onL} and k_{onC} (1/(nM.min). The membrane-bound ligand–receptor complex dissociates with rate constant k_{offL} (1/min) or is internalized and subsequently degraded with rate constant k_{degRL} (1/min). In contrast, the membrane-bound drug–receptor complex dissociates with rate constant k_{offC} (1/min) or is internalized and subsequently degraded with rate constant k_{degRC} (1/min). The free membrane-bound receptor is synthesized with rate k_{synR} (nmol/min) and internalized with rate constant k_{degR} (1/min). The internalized free receptor is recycled to the membrane with rate constant k_{recyRi} (1/min) or degraded with rate constant k_{degRi} (1/min). We denote the *numbers* of membrane-bound ligand–receptor and drug–receptor complexes by RL and RC (nmol), and the *numbers* of free membrane-bound receptor and free internalized receptor by R and R_{i} (nmol), respectively. Extracellular ligand and drug are assumed to have units in (nM). For the whole-body level, we consider a two-compartment model with central and peripheral volumes V_{cen} and V_{per} (L), intercompartmental clearances q_{pi} and q_{ip} (L/min), and unspecific linear clearance CL_{lin} (L/min). Motivated by [8, 25], we consider the drug–target interaction in the peripheral compartment.

In the cell-level TMDD model, the single-cell level is scaled up to the whole-body level by multiplying the impact of a single cell by the number of accessible cells N_{cell}, see Figure 14.3b for illustration. Hence, the single-cell model describes the behavior of an average cell. Below, we show how this setting can be extended to include different cell types, such as normal and tumor cells. Denoting the total central and free peripheral concentrations of the drug by C_{cen} and C_{per} (nM), respectively, and the receptor species per single cell (nmol) as above, the rate of change of cell-level TMDD model is given by[9]

$$V_{\mathrm{cen}} \frac{\mathrm{d}}{\mathrm{d}t} C_{\mathrm{cen}} = -q_{\mathrm{pi}} C_{\mathrm{cen}} + q_{\mathrm{ip}} C_{\mathrm{per}} - CL_{\mathrm{lin}} \cdot C_{\mathrm{cen}} + k_{\mathrm{in}}(t)$$

$$V_{\mathrm{per}} \frac{\mathrm{d}}{\mathrm{d}t} C_{\mathrm{per}} = +q_{\mathrm{pi}} C_{\mathrm{cen}} - q_{\mathrm{ip}} C_{\mathrm{per}} + N_{\mathrm{cell}} \cdot \underbrace{\left(k_{\mathrm{offC}} \cdot RC - k_{\mathrm{onC}} \cdot R \cdot C_{\mathrm{per}}\right)}_{\text{whole-body single-cell interaction}}$$

$$\frac{\mathrm{d}}{\mathrm{d}t} R = k_{\mathrm{synR}} - k_{\mathrm{onL}} R \cdot L - k_{\mathrm{onC}} R \cdot C_{\mathrm{per}} + k_{\mathrm{offL}} RL + k_{\mathrm{offC}} RC - k_{\mathrm{degR}} \cdot R + k_{\mathrm{recyRi}} \cdot R_{\mathrm{i}}$$

$$\frac{\mathrm{d}}{\mathrm{d}t} R_{\mathrm{i}} = k_{\mathrm{degR}} \cdot R - k_{\mathrm{recyRi}} \cdot R_{\mathrm{i}} - k_{\mathrm{degRi}} \cdot R_{\mathrm{i}}$$

$$\frac{\mathrm{d}}{\mathrm{d}t} RL = k_{\mathrm{onL}} \cdot L \cdot R - k_{\mathrm{offL}} RL - k_{\mathrm{degRL}} RL$$

$$\frac{\mathrm{d}}{\mathrm{d}t} RC = k_{\mathrm{onC}} \cdot C_{\mathrm{per}} \cdot R - k_{\mathrm{offC}} RC - k_{\mathrm{degRC}} \cdot RC$$

with $k_{\mathrm{in}}(t) = r_{\mathrm{inf}}(t)/V_{\mathrm{cen}}$ (nM) and infusion rate $r_{\mathrm{inf}}(t)$ (nmol/min). Since all single-cell parameters are taken from *in vitro* experiments, the only unknown parameters are the number of relevant cells N_{cell} and the target-unrelated parameters: compartment volumes V_{cen} and V_{per}, intercompartmental clearances q_{pi} and q_{pi}, and unspecific clearance CL_{lin}. These parameters are estimated from PK data as before. Since we fixed all parameters of the targeted system, parameter identification is not expected to be a problem in this approach.

We finally remark that alternatively and as in [24], the receptor species might be given in the number of molecules. Then, an additional scaling factor $SF_{\mathrm{unit}} = 10^9 / N_{\mathrm{avog}}$ from the number of molecules (nmol) is needed, where $N_{\mathrm{avog}} = 6.02 \times 10^{23}$ (1/mol) denotes the Avogadro constant. To realize the different units, one simply has to replace N_{cell} by $N_{\mathrm{cell}} \cdot SF_{\mathrm{unit}}$ in the equation above.

14.4.2 Cell-Level TMDD Model with Normal and Tumor Cells

Due to the explicit representation of the cellular level in the model, it is easily possible to include different cell types, see

[9] The initial conditions are $C_{\mathrm{cen}}(0) = \mathrm{dose}/V_{\mathrm{cen}}$ (IV bolus, dose in nmol) or $C_{\mathrm{cen}}(0) = 0$ (IV infusion), and $C_{\mathrm{per}}(0) = 0$, while the initial conditions for the receptor variables are chosen identical to the pretreatment steady-state values.

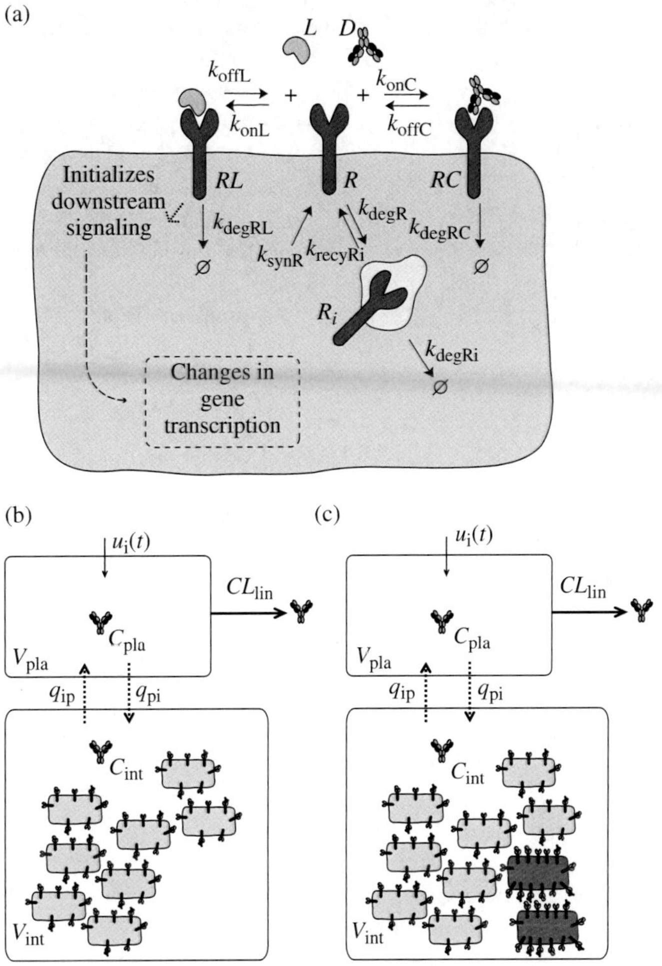

FIGURE 14.3 Schematic illustration of the cell-level TMDD model for analyzing the inhibitory effect on receptor activation of anti-EGFR antibodies. (a) Cell-level kinetic model of receptor activation and inhibition. (b) Cell-level TMDD model including a single-cell type (normal cells). (c) Cell-level TMDD model including normal and tumor cells. See Section 14.4 for definitions and description.

Figure 14.3c for illustration. This allows us to study, for example, the impact of different tumor cell type on drug efficacy. Tumor cells with increased receptor synthesis or decreased receptor internalization rate have been found experimentally [26–28]. In the absence of the drug, both alterations lead to increased receptor levels at the cell surface, often (imprecisely) termed "receptor overexpression." Tumor cells with overexpressed receptors have been observed in certain tumor types, and down-regulation of receptor levels is a therapeutic strategy. Using the subscript N for normal cells and the subscript T for tumor cells, the rate of change of the cell-level TMDD model with normal and tumor cells is given by[10]

$$V_{\text{cen}} \frac{\mathrm{d}}{\mathrm{d}t} C_{\text{cen}} = -q_{\text{pi}} C_{\text{cen}} + q_{\text{ip}} C_{\text{per}} - CL_{\text{lin}} \cdot C_{\text{cen}} + k_{\text{in}}(t)$$

$$\begin{aligned} V_{\text{per}} \frac{\mathrm{d}}{\mathrm{d}t} C_{\text{per}} = &+ q_{\text{pi}} C_{\text{cen}} - q_{\text{ip}} C_{\text{per}} + N_{\text{N}} \cdot \underbrace{\left(k_{\text{offC}} \cdot RC_{\text{N}} - k_{\text{onC}} \cdot R_{\text{N}} \cdot C_{\text{per}}\right)}_{\text{normal cells}} \\ &+ N_{\text{T}} \cdot \underbrace{\left(k_{\text{offC}} \cdot RC_{\text{T}} - k_{\text{onC}} \cdot R_{\text{T}} \cdot C_{\text{per}}\right)}_{\text{tumor cells}} \end{aligned}$$

$$\begin{aligned} \frac{\mathrm{d}}{\mathrm{d}t} R_{\text{N}} = &\, k_{\text{offL}} RL_{\text{N}} - k_{\text{onL}} R_{\text{N}} L + k_{\text{offC}} RC_{\text{N}} - k_{\text{onC}} R_{\text{N}} C_{\text{per}} \\ &+ k_{\text{synR,N}} - k_{\text{degR,N}} R_{\text{N}} + k_{\text{recyRi}} R_{\text{i,N}} \end{aligned}$$

$$\frac{\mathrm{d}}{\mathrm{d}t} R_{\text{i,N}} = k_{\text{degR,N}} \cdot R_{\text{N}} - k_{\text{recyRi}} \cdot R_{\text{i,N}} - k_{\text{degRi}} \cdot R_{\text{i,N}}$$

[10] The initial conditions are $C_{\text{cen}}(0) = \text{dose} / V_{\text{cen}}$ (IV bolus, dose in nmol) or $C_{\text{cen}}(0) = 0$ (IV infusion), and $C_{\text{per}}(0) = 0$, while the initial conditions for the receptor variables (normal and tumor cells) are chosen identical to the pretreatment steady-state values.

$$\frac{\mathrm{d}}{\mathrm{d}t}RL_{\mathrm{N}} = k_{\mathrm{onL}}L \cdot R_{\mathrm{N}} - k_{\mathrm{offL}}RL_{\mathrm{N}} - k_{\mathrm{degRL,N}}RL_{\mathrm{N}}$$

$$\frac{\mathrm{d}}{\mathrm{d}t}RC_{\mathrm{N}} = k_{\mathrm{onC}}C_{\mathrm{per}} \cdot R_{\mathrm{N}} - k_{\mathrm{offC}}RC_{\mathrm{N}} - k_{\mathrm{degRC}} \cdot RC_{\mathrm{N}}$$

$$\frac{\mathrm{d}}{\mathrm{d}t}R_{\mathrm{T}} = k_{\mathrm{offL}}RL_{\mathrm{T}} - k_{\mathrm{onL}}R_{\mathrm{T}}L + k_{\mathrm{offC}}RC_{\mathrm{T}} - k_{\mathrm{onC}}R_{\mathrm{T}}C_{\mathrm{per}} + k_{\mathrm{synR,T}} - k_{\mathrm{degR,T}}R_{\mathrm{T}} + k_{\mathrm{recyRi}}R_{\mathrm{i,T}}$$

$$\frac{\mathrm{d}}{\mathrm{d}t}R_{\mathrm{i,T}} = k_{\mathrm{degR,T}} \cdot R_{\mathrm{T}} - k_{\mathrm{recyRi}} \cdot R_{\mathrm{i,T}} - k_{\mathrm{degRi}} \cdot R_{\mathrm{i,T}}$$

$$\frac{\mathrm{d}}{\mathrm{d}t}RL_{\mathrm{T}} = k_{\mathrm{onL}}L \cdot R_{\mathrm{T}} - k_{\mathrm{offL}}RL_{\mathrm{T}} - k_{\mathrm{degRL,T}}RL_{\mathrm{T}}$$

$$\frac{\mathrm{d}}{\mathrm{d}t}RC_{\mathrm{T}} = k_{\mathrm{onC}}C_{\mathrm{per}} \cdot R_{\mathrm{T}} - k_{\mathrm{offC}}RC_{\mathrm{T}} - k_{\mathrm{degRC}} \cdot RC_{\mathrm{T}}$$

Using the above model, we showed that tumor cells, overexpressing receptor, responded very differently to drugs, depending on the underlying mechanistic cause (increased receptor synthesis or decreased internalization) [24].

In [25], an extended Michaelis–Menten approximation of the whole-body TMDD model with the target in the peripheral compartment was used to analyze an anti-EGFR mAb in cynomolgus monkeys. We have shown in [24] that a cell-level TMDD model parameterized based on, for example, fibroblast cells [21, 22], is also consistent with the experimental PK data. Based on the cell-level predictions of the activated receptor, that is, *RL*, we studied the impact of different drug properties, such as target affinity, internalization rate constant, as well as the dose on the impact of downregulating receptor activity. Including also different tumor cell types, we analyzed the impact of the tumor cell type on different measures of drug efficacy. The cell-level TMDD modeling approach can be seen as an important first step toward integrating more detailed systems biology models of downstream signaling processes relevant to human diseases [19, 29, 30]. Furthermore, we have shown that under certain assumptions, the receptor system equations can be solved analytically, which allows to develop simplified models about its role in signal transduction [31].

14.5 SIMPLIFIED PHYSIOLOGICALLY BASED PHARMACOKINETIC MODEL FOR mAbs

There is a wealth of anatomical and physiological data for various species available (see, e.g., [32, 33]). In addition, during the drug discovery and development process, various drug-specific *in vitro* data are generated. PBPK models aim at integrating such species- and drug-specific data into the modeling process and translate them into the *in vivo* context. This enables *a-priori* predictions, that is, predictions before any *in vivo* experiments [34], and also facilitates covariate modeling [35]. Applying lumping (model reduction) techniques to PBPK models allows to derive simple compartment models that are consistent with the knowledge represented in the PBPK model and, furthermore, gives a possible physiologically based interpretation of classical compartment models [8, 36].

For small molecules, PBPK models have a long history in toxicokinetics [37]. In PK, the use of PBPK models was significantly advanced with the development of *a-priori* methods to predict tissue-to-plasma partition coefficients (see [38] for the seminal work by Poulin and Theil). Today, PBPK models are an integrated part of drug development and regulatory science [39]. With the development of ABCs for mAbs [40]—the analog to tissue-to-plasma partition coefficient of small molecules—a similar advance can be expected for this class of drugs. In the sequel, we review a PBPK modeling approach that allows to leverage this information by directly integrating ABC values. Alternative existing PBPK models are typically much more detailed (requiring to make a number of assumptions on unobservable processes) or are much more coarse-grained (losing almost completely the level of individual organs); see Chapter 12 for details.

In the first part, we describe the PK properties for nontargeting mAbs with the simplified PBPK model. This already enables us to also analyze the impact of target-driven PK by comparing experimental data (with target) to predictions with the simplified PBPK model without a target. In the second part, we describe how to model the drug–target interaction in the PBPK context. Except for a reparameterization in terms of interstitial rather than organ tissue (i.e., interstitial and intracellular) volumes, the outline of the simplified PBPK model closely follows [8].

14.5.1 Target-Independent Pharmacokinetics

The simplified PBPK model comprises the anatomical compartments of total plasma space and the interstitial spaces of the lung, adipose, bone, gut, heart, kidney, liver, muscle, skin, and the spleen, see Figure 14.4. The plasma compartment with volume V_{pla} (L) denotes the total arterial and venous plasma, including the vascular space associated with each organ. For interstitial space of each organ with volume[11] V_{int} (L), an extravasation rate-limited distribution model is assumed. This includes interstitial uptake by convection through large pores and transport

[11] More precisely, we should write $V_{\mathrm{org,int}}$ to denote the interstitial space of the organ "org." For readability, we omit the reference "org" to the organ in the sequel.

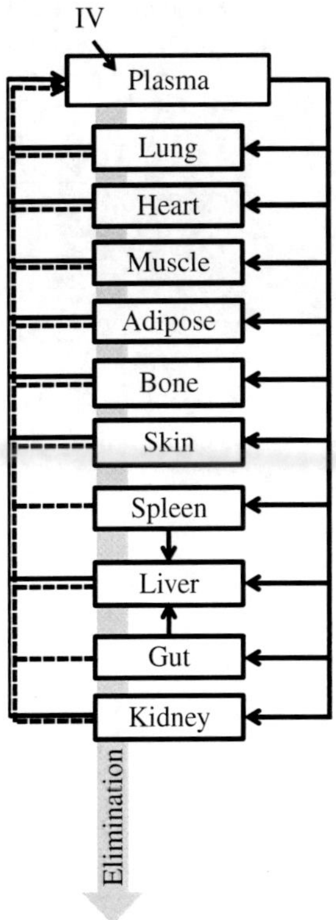

FIGURE 14.4 Structure of the simplified mAb PBPK model. Organs are interconnected by plasma flow (solid lines) and lymph flow (dashed lines). The plasma compartment comprises total arterial and venous plasma and the vascular plasma space associated with the organs. The remaining compartments represent the interstitial space of the organs. Elimination of mAb is accounted for from plasma and/or from organ interstitial space. See Section 14.5.1 for description of the target-independent part and Section 14.5.2 for how to include the drug target interaction

across the endothelial cell layer, and outflow into the plasma via the lymph flow or via cellular trafficking. The extravasation rate-limited distribution model is parameterized in terms of the organ lymph flow L_{org} (L) and some apparent, organ-specific vascular reflection coefficient σ_{vas} (unitless). It accounts for the fact that only the fraction $(1-\sigma_{vas})$ of the plasma concentration is accessible for extravasation. The apparent vascular reflection coefficient is a mixed parameter, depending on drug properties, pore size in the vascular wall, and number of pores of the organ. Analogous to tissue distribution models for small molecule drugs, the outflow via the lymph into the plasma is parameterized in terms of organ-specific partition coefficient K_{int}, which is defined in terms of readily available ABC values (see below). Elimination of therapeutic antibodies is modeled by plasma or intrinsic interstitial organ clearance *CLpla* and/or $CLint_{int}$ (L/min) that subsumes several processes, such as Fcγ receptor-mediated clearance or nonspecific endocytosis, and subsequent degradation within lysosomes. The rate of change of the plasma and interstitial organ concentrations C_{pla} and C_{int} (nM) is given by the following ODEs[12]:

$$V_{pla}\frac{d}{dt}C_{pla} = L_{pla}\cdot\left(C_{in}-\left(1-\sigma_{pla}\right)C_{pla}\right) - CLpla\cdot C_{pla} + k_{in}(t)$$

$$V_{int}\frac{d}{dt}C_{int} = L_{org}\cdot\left(\left(1-\sigma_{vas}\right)C_{pla} - \frac{C_{int}}{K_{int}}\right) - CLint_{int}\cdot C_{int}$$

with $k_{in}(t) = r_{inf}(t)/V_{pla}$ (nM), where $r_{inf}(t)$ (nmol/min) is the infusion rate. The second equation applies to each organ. In the first equation, C_{in} (nM) denotes the inflowing concentration into the plasma compartment that results from the lymph flowing into the plasma:

$$L_{pla}\cdot C_{in} = \sum_{org} L_{org}\cdot\frac{C_{int}}{K_{int}}$$

where the sum is taken over all organs.[13] A main advantage of the simplified PBPK model in comparison to existing, more detailed PBPK models is that it can easily be parameterized in terms of known physiological, anatomical, and mAb-specific parameters. Species-specific data are available from, for example, [32, 33]. As it is commonly assumed, the lymph flow is defined as a percentage of the plasma flow, 2% and 4% of plasma tissue flow for visceral and nonvisceral tissues, respectively (see also [41] and the discussion in [8] for reported differences in lymph flows). Tissue partition coefficients can be directly derived from the readily available $ABC_{tis} = C_{tis}/C_{pla}$ via the relationship $(1-\sigma_{vas})(1-E_{int})K_{int} = ABC_{int}$ with $ABC_{int} = ABC_{tis}/f_{int}$ (cf. [42]) and interstitial volume fractions $f_{int} = V_{int}/V_{tis}$ (see, e.g., [43, Table B-I]). The interstitial extraction ratio E_{int} (unitless) is defined in terms of the intrinsic clearance $CLint_{int}$ in the common way:

$$E_{int} = \frac{CLint_{int}K_{int}}{CLint_{int}K_{int} + L_{tis}} \tag{14.14}$$

Importantly, we have shown in [8] that, based on commonly available tissue data, only the total plasma clearance can be reliably identified. With the advent of richer experimental tissue data, for example, imaging data, the identification of the elimination capacity of different organs, that is, $CLint_{int}$ or E_{int}, becomes

[12] The initial conditions are $C_{pla}(0) = \text{dose}/V_{pla}$ (IV bolus, dose in nmol) or $C_{pla}(0) = 0$ (IV infusion), and $C_{int}(0) = 0$ for all organs.

[13] For the plasma compartment, the total plasma lymph flow L_{pla} and the apparent total reflection coefficient σ_{pla} are defined as $L_{pla} = \sum_{org} L_{org}$ and $L_{pla}\cdot\left(1-\sigma_{pla}\right) = \sum_{org} L_{org}\cdot\left(1-\sigma_{vas}\right)$.

TABLE 14.2 Interstitial-to-Plasma Concentration Ratios $ABC_{\text{int}} = C_{\text{int}} / C_{\text{pla}}$ Derived from the Antibody Biodistribution Coefficients in [40] and Interstitial Volume Fractions from [43] (Table B-I)

Tissue	Adipose	Bone	Gut	Heart	Kidney	Liver	Lung	Muscle	Skin	Spleen
ABC_{int}	0.35	0.70	0.53	0.75	0.61	0.66	0.58	0.32	0.51	0.61

See text for more details.

feasible. By definition, the ABC_{int} values characterize the interstitial-to-plasma concentration ratio of mAbs. As can be inferred from Table 14.2, the concentration in the interstitial space range from 30% to 75% of the plasma concentration. As observed in [40], for a given tissue, the ABC_{tis} values are constant across species. This observation greatly facilitates extrapolation between species and is much different from properties of tissue partition coefficients for small molecule drugs. Vascular reflection coefficients σ_{vas} are chosen based on the tissue properties, such as leakiness and tightness of the vascular wall (see [8]).

As mentioned in the introduction, already a simplified PBPK model without a target can be of use in characterizing the impact of TMDD in experimental data. Based on this model, we have shown in [8] that a common interpretation of classical two-compartment models for mAb disposition—identifying the central compartment with the total plasma volume and the peripheral compartment with the interstitial space (or part of it)—is not consistent with current knowledge. Rather, as stated in Section 14.3.2, the central compartment comprises total plasma and the interstitial spaces of those organs that are kinetically similar to plasma, while the peripheral compartment comprises tissues that are kinetically much slower than the plasma, typically the adipose, bone, muscle, and the skin. In this sense, a PBPK model for mAbs is very similar to that for small molecules (in particular, for drugs with low volume of distribution); only the PK evolves on much slower timescales.

14.5.2 Drug–Target Interaction

The simplified PBPK model can easily be extended to account for the presence of a target. We exemplify the extension for a membrane-bound target by integrating a cell-level model and an extended Michaelis–Menten approximation of the cell-level model (see [24] for details).

For the integration of the cell-level model into the simplified PBPK model, we assume that the target is expressed on the cellular membrane of some organ (e.g., the EGFR in skin tissue). With the notation from Section 14.4.1, we simply have to replace the ODE for a given organ by

$$V_{\text{int}} \frac{\text{d}}{\text{d}t} C_{\text{int}} = L_{\text{org}} \cdot \left((1-\sigma_{\text{vas}}) C_{\text{pla}} - \frac{C_{\text{int}}}{K_{\text{int}}} \right) - CLint_{\text{int}} \cdot C_{\text{int}} + N_{\text{cell}} \cdot (k_{\text{offC}} \cdot RC - k_{\text{onC}} \cdot R \cdot C_{\text{int}})$$

$$\frac{\text{d}}{\text{d}t} R = k_{\text{synR}} - k_{\text{onL}} R \cdot L - k_{\text{onC}} R \cdot C_{\text{int}} + k_{\text{offL}} RL + k_{\text{offC}} RC - k_{\text{degR}} \cdot R + k_{\text{recyRi}} \cdot R_{\text{i}}$$

$$\frac{\text{d}}{\text{d}t} R_{\text{i}} = k_{\text{degR}} \cdot R - k_{\text{recyRi}} \cdot R_{\text{i}} - k_{\text{degRi}} \cdot R_{\text{i}}$$

$$\frac{\text{d}}{\text{d}t} RL = k_{\text{onL}} \cdot L \cdot R - k_{\text{offL}} RL - k_{\text{degRL}} RL$$

$$\frac{\text{d}}{\text{d}t} RC = k_{\text{onC}} \cdot C_{\text{int}} \cdot R - k_{\text{offC}} RC - k_{\text{degRC}} \cdot RC$$

where the receptor species (nmol) and the local ligand concentration (nM) are located on the membrane of the cells of the organs.

Alternatively, we may integrate an extended Michaelis–Menten approximation to the above model, which has proven very good approximation quality to the cell-level model (see [24] for details). This approximation is analogous to the extended Michaelis–Menten approximation in Section 14.3.7 with B_{max} (nmol) (rather than in nM) denoting the maximal target-binding capacity and K_{M} (nM) denoting the corresponding Michaelis constant. For a given organ, we denote the amount of drug associated with the receptor system by A_{RS} (nmol) and the concentration in the interstitial space by C_{int} (nM). For the sake of clarity, we remark that C_{int} denotes the target-independent concentration in the interstitial space. We next define the concentration $C_{\text{int+RS}}$ (nM) as

$$C_{\text{int+RS}} = C_{\text{int}} + A_{\text{RS}} / V_{\text{int}} \tag{14.15}$$

which denotes the total concentration of mAb (interstitial space and receptor system), both with respect to the interstitial volume. Then, the rate of change of the concentration $C_{\text{int+RS}}$ in the given organ is given by the following differential and algebraic equations:

$$V_{\text{int}} \frac{\text{d}}{\text{d}t} C_{\text{int+RS}} = L_{\text{org}} \cdot \left((1-\sigma_{\text{vas}}) C_{\text{pla}} - \frac{C_{\text{int}}}{K_{\text{int}}} \right) - CLint_{\text{int}} \cdot C_{\text{int}} - k_{\text{deg}} \frac{B_{\text{max}} C_{\text{int}}}{K_{\text{M}} + C_{\text{int}}}$$

with interstitial concentration defined by

$$C_{\text{int}} = \frac{1}{2}\left(C_{\text{eff}} + \sqrt{C_{\text{eff}}^2 + 4K_{\text{M}}C_{\text{int+RS}}}\right) \quad (14.16)$$

and $C_{\text{eff}} = C_{\text{int+RS}} - B_{\max} / V_{\text{int}} - K_{\text{M}}$ (nM).

14.6 CONCLUSION: LOOKING AT DATA THROUGH MODELS

Usually, we interpret data in a given context, that is, based on certain assumptions. Mathematical models are a way to express these assumptions in a formal language and in an explicit form. Predictions based on mathematical models are amenable for comparison to experimental data or communication to other researchers [44]. In a way, models are like "glasses" to look at data.

In this chapter, we have presented different modeling approaches capable of simulating, analyzing, and inferring target-driven PK of mAbs. A more detailed, physiologically based representation aims at a more accurate representation of the processes that are expected to be relevant. This enables as well as requires the integration of data from other sources, like *in vitro* data on the target dynamics or anatomical and physiological data on organ volumes and plasma/lymph flows. This can also be seen as a covariate modeling approach and facilitates extrapolation, for example, to other species. At the same time, integration of data from other sources constrains the degrees of freedom of the model by reducing the number of unknown parameters. It can be expected to reduce the likelihood of encountering identifiability problems. Of course, such an approach also increases the number of assumptions made. If the confidence is high in made assumptions and parameterizing data, this is surely the way to take. If confidence is reduced, then one has to balance an increase in detailedness with the increase in the number of assumptions made. An alternative route is to reduce the detailedness of the model, resulting in reduced models with less equations and parameters.

In general, a model cannot be used to verify its own underlying assumptions. For example, the Michaelis–Menten approximation does not allow to verify whether free and drug-bound targets present fast kinetics (QSS assumption) and whether the extent of target binding is negligible. These assumptions can only be verified in the context of more detailed models. As a consequence, rather than explicitly verifying assumption, more often we assume that the model and underlying assumptions are valid and compare the model predictions to experimental data. If the model allows to reproduce the experimental data (with reasonable parameter values/estimates), we consider the underlying assumptions and the represented processes are consistent with the experimental data. Yet, it is important to realize that parameters and considered processes may compensate for unresolved physiological details and therefore introduce some bias, as discussed in the context of the Michaelis–Menten approximation of the TMDD model (see Section 14.3.9).

We have discussed in detail the different characteristics of TMDD profiles that are linked to different phases and concentration ranges. In the context of dose-dependent target-driven PK, it is important that experimental data include both high concentration ranges that saturate the target system and low concentration ranges that allow for linear target dynamics.

In silico modeling is constantly gaining importance to support drug development. For instance, the European Medicines Agency has highlighted the importance of TMDD modeling for biologics in the context of the minimum anticipated biological effect level (MABEL) approach [45]. Here, TMDD models are used to predict receptor occupancy. Since multiple factors affect the relationship between dose and receptor occupancy, TMDD models have been proposed to predict human PK and support safe and effective first-in-human doses within clinical trials [46].

In conclusion, nonlinear PK that is introduced by the target can be well described and explained using mathematical models. Seen through the "glasses" of a model, these nonlinearities allow to obtain precious information about the target if used in the right way.

ACKNOWLEDGMENT

S.F. acknowledges financial support from the Graduate Research Training Program PharMetrX: Pharmacometrics and Computational Disease Modeling, Freie Universität Berlin and Universität Potsdam, Germany (www.pharmetrx.de).

REFERENCES

[1] Kuang B, King L, Wang HF. Therapeutic monoclonal antibody concentration monitoring: free or total? Bioanalysis 2010;2:1125–1140.

[2] Staack RF, Jordan G, Heinrich J. Mathematical simulations for bioanalytical assay development: the (un-)necessity and (im-)possibility of free drug quantification. Bioanalysis 2012;4:381–395.

[3] Dirks N, Meibohm B. Population pharmacokinetics of therapeutic monoclonal antibodies. Clin Pharmacokinet 2010;49:633–659.

[4] Keizer RJ, Huitema AD, Schellens JH, Beijnen JH. Clinical pharmacokinetic of therapeutic monoclonal antibodies. Clin Pharmacokinet 2010;49:493–507.

[5] Mager DE, Jusko WJ. General pharmacokinetic model for drugs exhibiting target-mediated drug disposition. J Pharmacokinet Pharmacodyn 2001;28:507–532.

[6] Peletier LA, Gabrielsson J. Dynamics of target-mediated drug disposition: characteristic profiles and parameter identification. J Pharmacokinet Pharmacodyn 2012;39:429–451.

[7] Krippendorff BF, Küster K, Koft C, Huisinga W. Nonlinear pharmacokinetics of therapeutic proteins resulting from receptor mediated endocytosis. J Pharmacokinet Pharmacodyn 2009;36:239–260.

[8] Fronton L, Pilari S, Huisinga W. Monoclonal antibody disposition: a simplified PBPK model and its implications for the derivation and interpretation of classical compartment models. J Pharmacokinet Pharmacodyn 2014;41:87–107.

[9] Garg A. Investigation of the role of FcRn in the absorption, distribution, and elimination of monoclonal antibodies. Ph.D. thesis. Buffalo (NY): Faculty of the Graduate School of State University of New York; 2007.

[10] Abuqayyas L, Balthasar J. Application of PBPK modeling to predict monoclonal antibody disposition in plasma and tissues in mouse models of human colorectal cancer. J Pharmacokinet Pharmacodyn 2012;39:683–710.

[11] Gibiansky L, Gibiansky E, Kakkar T, Ma P. Approximations of the target-mediated drug disposition model and identifiability of model parameters. J Pharmacokinet Pharmacodyn 2008;35:573–591.

[12] Michaelis L, Menten ML. Die Kinetik der Invertinwirkung. Biochem Z 1913;49:333–369.

[13] Segel LA, Slemrod M. The quasi-steady-state assumption: a case study in perturbation. SIAM Rev 1989;31:446–477.

[14] Gibiansky L, Gibiansky E. Target-mediated drug disposition model: relationships with indirect response models and application to population PK-PD analysis. J Pharmacokinet Pharmacodyn 2009;36:341–351.

[15] European Bioinformatics Institute as part of the European Molecular Biology Laboratory. *BioModels Database*. http://www.ebi.ac.uk/biomodels-main/. Accessed 2015 Jul 7.

[16] Betts A, Clark TH, Yang J, Treadway JL, Li M, Giovanelli MA, Abdiche Y, Stone DM, Paralkar VM. The application of target information and preclinical pharmacokinetic/pharmacodynamic modeling in predicting clinical doses of a dickkopf-1 antibody for osteoporosis. J Pharmacol Exp Ther 2010;[333]:2–13.

[17] Hood L, Perlmutter RM. The impact of systems approaches on biological problems in drug discovery. Nat Biotechnol 2004;22:1215–1217.

[18] Kreeger PK, Lauffenburger DA. Cancer systems biology: a network modeling perspective. Carcinogenesis 2010;31:2–8.

[19] Chen WW, Schoeberl B, Jasper PJ, Niepel M, Nielsen UB, Lauffenburger DA, Sorger PK. Input-output behavior of ErbB signaling pathways as revealed by a mass action model trained against dynamic data. Mol Syst Biol 2009;5:239.

[20] Shankaran H, Wiley HS, Resat H. Cell surface receptors for signal transduction and ligand transport: a design principles study. PLoS Comput Biol 2007;3:e101.

[21] Knauer DJ, Wiley HS, Cunningham DD. Relationship between epidermal growth factor receptor occupancy and mitogenic response. Quantitative analysis using a steady state model system. J Biol Chem 1984;259:5623–5631.

[22] Starbuck C, Lauffenburger DA. Mathematical model for the effects of epidermal growth factor receptor trafficking dynamics on fibroblast proliferation responses. Biotechnol Prog 1992;8:132–143.

[23] Schoeberl B, Eichler-Jonsson C, Gilles ED, Müller G. Computational modeling of the dynamics of the MAP kinase cascade activated by surface and internalized EGF receptors. Nat Biotechnol 2002;20:370–375.

[24] Krippendorff BF, Oyarzun DA, Huisinga W. Predicting the F(ab)-mediated effect of monoclonal antibodies *in vivo* by combining cell-level kinetic and pharmacokinetic modelling. J Pharmacokinet Pharmacodyn 2012;39:125–139.

[25] Lammerts van Bueren JJ, Bleeker WK, Bøgh HO, Houtkamp M, Schuurman J, van de Winkel JG, Parren PW. Effect of target dynamics on pharmacokinetics of a novel therapeutic antibody against the epidermal growth factor receptor: implications for the mechanisms of action. Cancer Res 2006;66:7630–7638.

[26] Merlino G, Xu Y, Ishii S, Clark A, Semba K, Toyoshima K, Yamamoto T, Pastan I. Amplification and enhanced expression of the epidermal growth factor receptor gene in A431 human carcinoma cells. Science 1984;224:417.

[27] Lin C, Chen W, Kruiger W, Stolarsky L, Weber W, Evans RM, Verma IM, Gill GN, Rosenfeld MG. Expression cloning of human EGF receptor complementary DNA: gene amplification and three related messenger RNA products in A431 cells. Science 1984;224:843–848.

[28] Reddy CC, Wells A, Lauffenburger DA. Proliferative response of fibroblasts expressing internalization-deficient epidermal growth factor (EGF) receptors is altered via differential EGF depletion effect. Biotechnol Prog 1994;10:377–384.

[29] Borisov N, Aksamitiene E, Kiyatkin A, Legewie S, Berkhout J, Maiwald T, Kaimachnikov NP, Timmer J, Hoek JB, Kholodenko BN. Systems-level interactions between insulin–EGF networks amplify mitogenic signaling. Mol Syst Biol 2009;5:256.

[30] Kholodenko BN. Cell-signalling dynamics in time and space. Nat Rev Mol Cell Biol 2006;7:165–176.

[31] Oyarzun DA, Bramhall JL, Lopez-Caamal F, Richards FM, Jodrell DI, Krippendorff BF. The EGFR demonstrates linear signal transmission. Integr Biol (Camb) 2014;6:736–742.

[32] Brown RP, Delp MD, Lindstedt SL, Rhomberg LR, Beliles RP. Physiological parameter values for physiologically based pharmacokinetic models. Toxicol Ind Health 1997;13:[407]–484.

[33] International Commission on Radiological Protection (ICRP). 2002. Basic anatomical and physiological data for use in radiological protection: reference values. In: ICRP Publication 89.

[34] Poulin P, Theil FP. Prediction of pharmacokinetics prior to *in vivo* studies. II. Generic physiologically based pharmacokinetic models of drug disposition. J Pharm Sci 2002;91:1358–1370.

[35] Huisinga W, Solms A, Fronton L, Pilari S. Modeling interindividual variability in physiologically-based pharmacokinetics and its link to mechanistic covariate modeling. CPT Pharmacometrics Syst Pharmacol 2012;1:e4.

[36] Pilari S, Huisinga W. Lumping of physiologically based pharmacokinetic models and a mechanistic derivation of classical compartmental models. J Pharmacokinet Pharmacodyn 2010;37:365–405.

[37] Reddy M, Yang RS, Andersen ME, III Clewell HJ. *Physiologically Based Pharmacokinetic Modeling: Science and Applications*. Wiley; 2005.

[38] Poulin P, Theil FP. A priori prediction of tissue:plasma partition coefficients of drugs to facilitate the use of physiologically-based pharmacokinetic models in drug discovery. J Pharm Sci 2000;89:16–35.

[39] Rowland M, Peck C, Tucker G. Physiologically-based pharmacokinetics in drug development and regulatory science. Annu Rev Pharmacol Toxicol 2011;51:45–73.

[40] Shah D, Betts A. Antibody biodistribution coefficients—inferring tissue concentrations of monoclonal antibodies based on the plasma concentrations in several preclinical species and human. MAbs 2013;5:297–305.

[41] Jones H, Mayawala K, Poulin P. Dose selection based on physiologically based pharmacokinetic (PBPK) approaches. AAPS J 2013;15(2):377–387.

[42] Fronton L. Modeling approaches to characterize the disposition of monoclonal antibodies. Ph.D. thesis. Math-Nat Faculty of University of Potsdam/Germany; 2015.

[43] Kawai R, Lemaire M, Steimer J, Bruelisauer A, Niederberger W, Rowland M. Physiologically based pharmacokinetic study on a cyclosporin derivative, SDZ IMM 125. J Pharmacokinet Biopharm 1994;22:327–365.

[44] Rao BM, Lauffenburger DA, Wittrup KD. Integrating cell-level kinetic modeling into the design of engineered protein therapeutics. Nat Biotechnol 2005;23:191–194.

[45] EMEA. 2007. Guideline on the clinical investigation of the pharmacokinetics of therapeutic proteins.

[46] Agoram B. Use of pharmacokinetic/pharmacodynamic modelling for starting dose selection in first-in-human trials of high-risk biologics. Br J Clin Pharmacol 2007;67:53–160.

15

TARGET-DRIVEN PHARMACOKINETICS OF BIOTHERAPEUTICS

GUY M.L. MENO-TETANG
UCB Celltech, Slough, Berkshire, UK

15.1 INTRODUCTION

Biotherapeutics are a class of molecules that are produced by living organisms. They comprise monoclonal antibodies (mAbs), antibody–drug conjugates (ADCs), antibody fragments, fusion proteins, peptides, recombinant cytokines, vaccines, growth factors, oligonucleotides, and deoxyribonucleic acid (DNA) preparations. These modalities exert their biological functions by binding targets either in solution in plasma and interstitial fluid, on circulating immune cells, or on the surface of cells immobilized in tissues. Most antibodies evaluated in oncology, for instance, target various glycoproteins, glycolipids, and carbohydrates that populate the surface of cancerous cells. A few mAbs in oncology target soluble proteins. The distribution and elimination pathways of mAbs along with different mechanisms of action to neutralize their targets are displayed in Figure 15.1. Experience with small molecular weight (MW) compounds has shown that mechanistic understanding of factors that govern drug absorption, distribution, and elimination is critical for developing effective therapies. However, understanding of the pharmacokinetics (PK) of biotherapeutics is still emerging. Due to their large molecular size and specific interactions with their respective targets, the PK of therapeutic biologics is often different from that of small molecules. The absorption, distribution, and elimination of these molecules are influenced by factors such as target-mediated clearance, the FcRn recycling for Fc-containing proteins, immunogenicity, isoform heterogeneity, and metabolic stability in the case of relatively small MW proteins and peptides. These unique features are not shared with small MW compounds. All these processes are capacity limited and may induce nonlinear PK depending on the concentration of the biotherapeutics and its target in relevant disposition compartments. Therefore, the understanding of their PK is critical to give the "right" dose to patients. In drug development, the PK of a new biotherapeutic is first evaluated in animals and its parameters are extrapolated to man. It is now well established that target-mediated clearance and antidrug–antibody-mediated clearance for therapeutic biologics can be species dependent [2]. Whether PK is species dependent is linked to potential differences in target biology (e.g., binding affinity, expression, and target turnover) and host immune response to a biologic. Species differences have also been documented in FcRn/IgG interactions. Therefore, it is paramount to identify the factors that influence the disposition of biotherapeutics to assist in selecting doses for first-in-man studies [3]. Significant variations in PK parameters of mAbs appear imputable to changes in antigen expression [3, 4]. One could then predict that differences in PK behavior of biotherapeutics can be observed between healthy and disease subjects, during the course of the therapy (i.e., as the level of the target subsides), between adults and pediatrics, and also between ethnic groups. Recently, the impact of off-target binding has been documented on the PK of mAbs; although unexpected, this could contribute to nonlinear behavior as well [5, 6].

This chapter reviews published data on the contribution of target biology including expression, turnover, and binding on the PK of fusion proteins, mAbs, ADCs, cytokines, and recombinant proteins. Other factors contributing

ADME and Translational Pharmacokinetics/Pharmacodynamics of Therapeutic Proteins: Applications in Drug Discovery and Development,
First Edition. Edited by Honghui Zhou and Frank-Peter Theil.

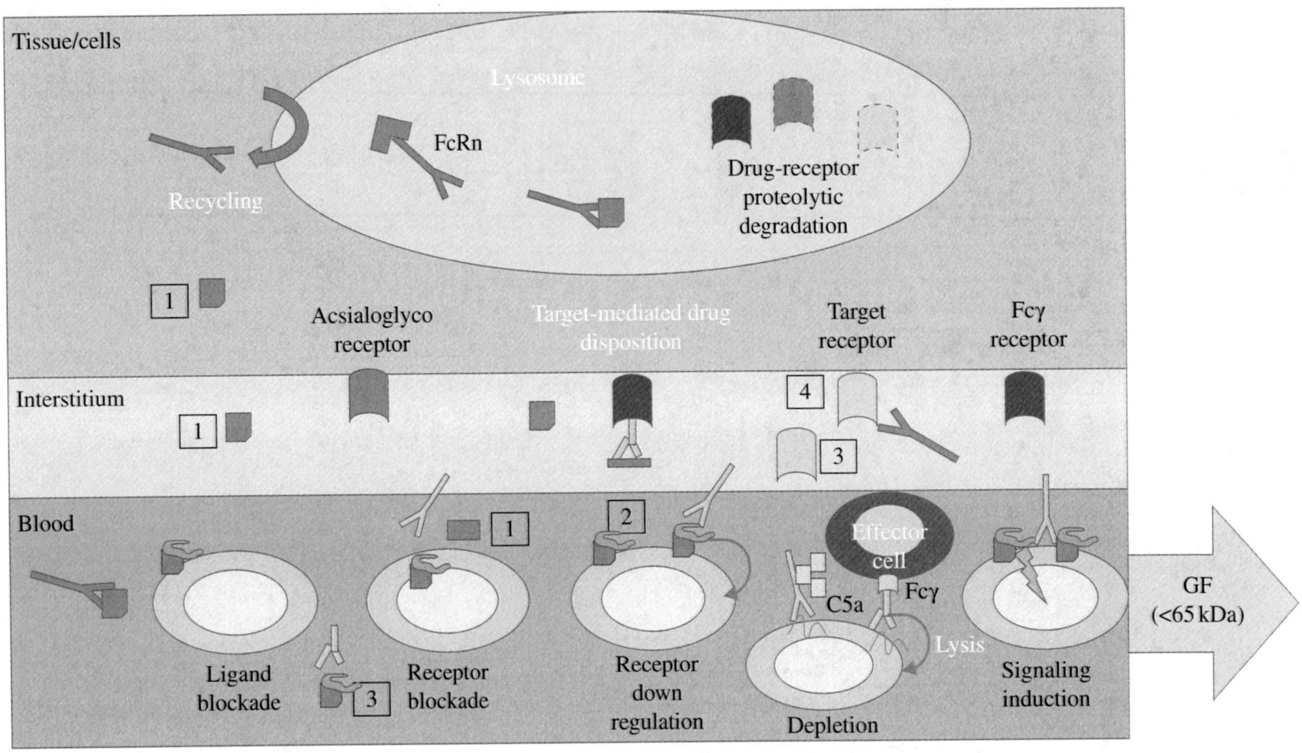

FIGURE 15.1 Distribution and elimination pathways for therapeutic proteins and target localization. 1: soluble target (plasma and interstitium), 2: receptor on circulating cells, 3: soluble receptor, and 4: target on cell membrane. GF: glomerular filtration. (Adapted from Eric Ezan [1].)

to the nonlinear PK of biotherapeutics such as FcRn and Fcγ binding are covered in other chapters of the book.

15.2 PEPTIDE–FC FUSION PROTEINS

Peptide–Fc fusion proteins (or peptibodies) are construct proteins generated by fusing a biologically active peptide with the Fc domain of an immunoglobulin G (IgG). This construct combines the biologic activity of peptides with the stability of mAbs. Wu and Sun [7] have reviewed the PK of peptibodies with a total MW of less than 69 kDa. Romiplostim is composed of a peptide mimetic of thrombopoietin (TPO) fused with the Fc domain of IgG. Romiplostim was approved by the U.S. Food and Drug Administration (FDA) in 2008 and the European Medicines Agency in 2009 for the treatment of chronic immune thrombocytopenia. Romiplostim binds to the TPO receptor with high affinity and promotes generation and maturation of megakaryocytes, the precursors of platelets. It has been shown that romiplostim exhibits nonlinear PK both in preclinical and clinical studies. Systemic clearance and volume of distribution for romiplostim are both dose dependent, due to its binding to the c-Mpl receptor on platelets. Its half-life has been reported to range from 1 to 34 h in humans depending on baseline platelet counts, which influence the clearance of the drug [8, 9].

Trebananib is a peptibody that neutralizes the human Ang-2/Tie2 and human Ang-1/Tie2 interactions on cancer cells with IC_{50} values of 23 and 900 pM, respectively. However, its PK is linear over a wide range of dose levels [10, 11].

CNTO 528 comprises a 20 amino acid erythropoietin mimetic peptide 1 (EMP1) fused to the human IgG1 heavy chain that binds the erythropoietin (EPO) receptor and induces a potent hemoglobin response in animal models and in humans. CNTO 528 exhibits nonlinear PK. Its half-life, which depends on clearance and volume of distribution, varies between 1 and 7 h [12]. The PK of other peptibodies such as CNTO 530, which bind the EPO receptor, dulaglutide, a glucagon-like peptide 1–Fc fusion protein, and ANP–Fc (atrial natriuretic peptide (ANP)) directed against natriuretic peptide receptor-A has linear PK with half-lives ranging a few days [7].

Since fusion proteins contain two domains, one that conveys the biological activity (binding to a ligand or a receptor) and the other improves its stability and half-life (Fc portion), their PK is often complex. Chen et al. [13] have provided a classification of fusion proteins of much larger MW than peptibodies described previously (MW > 60 kDa) into two groups based on the binding properties of the specific functional domain [13]. In the first group, binding to the target does not affect the distribution or elimination of the fusion protein. This group corresponds to most drugs that bind soluble proteins such as the receptor domain in etanercept binding soluble tumor necrosis factor (TNF) or Elspar and Alteplase, which are enzyme drugs, binding asparagine and plasminogen, respectively. Alefacept (MW of 115 kDa) is a dimeric fusion protein produced by Chinese hamster ovary (CHO) cells. It inhibits the interaction between antigen-presenting cells and T-cells by antagonizing the association of their receptors through LFA-3 and CD2. Its PK is also linear [14]. Drug belonging to this group have linear PK since they do not bind to target protein (e.g., intravenous immunoglobulin to treat primary immunodeficiencies) or their target binding does not lead to significant elimination. In the second class, binding to the biological target is responsible for an accelerated loss of its plasma concentration. As these drugs bind to their targets with high affinity and to a significant extent (relative to the dose), high tissue binding and/or elimination leads to an alteration of their plasma concentrations. When the magnitude of the drug target (receptor) levels is similar or larger than the plasma drug levels, drug elimination through receptor-mediated endocytosis (RME) becomes the predominant route of elimination. As a result, their PK is nonlinear with dose-dependent half-lives. This observation has been documented for drugs that bind to cell surface receptors and are internalized and degraded through RME. Examples from this class of fusion proteins include rilonacept, which binds to interleukin 1 (IL-1) receptor on cell surface [15].

15.3 MONOCLONAL ANTIBODIES (mAbs)

Antibodies, also known as immunoglobulins (Igs), are large proteins used by the immune system to identify and neutralize foreign antigens such as bacteria and viruses. They are grouped into five classes according to the structure of their heavy chains: IgA, IgD, IgE, IgG, and IgM. IgG is the predominant class, representing about 80% of Igs in human serum. All of the approved therapeutic antibodies are IgGs. Tables 15.1 and 15.2 represent the binding affinities and PK parameters for several commercialized mAbs and fusion proteins.

15.3.1 Antibodies Absorption

mAbs are predominantly administered subcutaneously (SC) followed by the intravenous (IV) and intramuscular (IM) routes. The bioavailability of mAbs has been reviewed by several authors [4, 17, 18]. These reviews show that the bioavailability of mAbs and other biotherapeutics can vary between 30% and 100%. Although several hypotheses have been proposed to explain the poor bioavailability of some biotherapeutics, its mechanism is not fully understood. The extent of absorption depends on the importance of presystemic antibody degradation by proteolytic enzymes. Using modeling techniques, Kagan et al. [19] have shown that binding (likely to FcRn) is an important determinant of the absorptive kinetics of rituximab in rats. However, proteolytic degradation may also be saturable, leading to a nonlinear bioavailability. Higher degradation is observed with low concentration of mAbs. Following IM administration, the bioavailability of alefacept (human IgG1 fusion protein) in healthy male subjects was approximately 80% [4, 20]. Since targets to some mAbs reside in the interstitial fluid as well as in plasma, it is possible that their expression levels and turnover rates contribute to the nonlinearity observed in their bioavailabilities. However, evidence to support this hypothesis is yet to be provided.

15.3.2 Antibodies Distribution

Distribution of mAbs is a sequential process involving convection, diffusion, extravasation/transcytosis, binding, and catabolism at the binding site. Because of their high MW and poor lipophilicity, the volume of distribution of mAbs is relatively small. In humans, typical values range from 2 to 3 L for the central volume and from 3.5 to 7 L for the volume of distribution at steady state (V_{ss}) [4, 21]. These values indicate that the distribution of mAbs is restricted to extracellular space including the blood stream and the interstitium. A low tissue:blood ratio is observed consistently for most mAbs, typically ranging from 0.1 to 0.5 [22].

Given that plasma volume is 3–5% of total body volume, antibody in plasma may be expected to comprise approximately 20–50% of antibody in the body. This would result in a true V_{ss} for most antibodies about two to five times the plasma volume. When an antibody binds with high affinity to extravascular sites that are associated with high binding capacity, tissue:blood antibody concentration ratios may be much greater than 0.5. For example, it has been shown that the tissue concentrations of a mAb against keratan sulfate were much greater than blood concentrations, with concentration ratios of 5.9, 4.3, 3.5, and 2.6 for lung, esophagus, kidney, and liver, respectively [23]. It has been shown that tissue:blood concentration ratios for some mAbs against endothelial antigens are greater than 1, which has implications that the volume of distribution would be 15-fold higher than the plasma volume [24]. The volume of distribution of some mAbs such as efalizumab exhibits dose dependency due to internalization of the antigen–mAb complex. High affinity mAb–target binding contributes to the apparent volume of mAb distribution, as a high degree of binding leads to a high ratio of the quantity of mAb bound to cellular target proteins, relative to the concentration of mAb in blood. Scheidhauer et al. [25] showed that ^{131}I-labeled rituximab had decreasing ratios of tissue-to-whole-body radioactivity over time in normal organs compared with increasing ratios in tumors, which was attributed to specific mAb binding [25]. The vast majority of PK analyses of antibody drugs have estimated V_{ss} via noncompartmental analysis or via computer fitting with mammillary compartmental models.

TABLE 15.1 Monoclonal Antibodies Fusion Proteins and Their Respective Targets [16]

Antibody	Modality	Type	Target	Target Localization	kD (nM)
Abatacept	Fusion	CTLA-4-IgG1	CD80	Membrane	0.4
Abatacept	Fusion	CTLA-4-IgG1	CD86	Membrane	1.2
Alefacept	Fusion	1-92-LFA-3-IgG1	CD2	Membrane	1
Basiliximab	Chimeric	IgG1	CD25	Membrane	0.1
Catumaxomab	Rat-murine	IgG2	EpCAM-CD3	Membrane	0.2
Daclizumab	Humanized	IgG1	CD25	Membrane	0.2
Gemtuzumab	Humanized	IgG4	CD33	Membrane	0.08
Ibritumomab	Murine	IgG2	CD20	Membrane	14
Alemtuzumab	Humanized	IgG1	CD52	Membrane	10
Efalizumab	Humanized	IgG1	CD11a	Membrane	3
Rituximab	Chimeric	IgG1	CD20	Membrane	8.0
Brentuximab	Chimeric	IgG1	CD30	Membrane	2
Ipilimumab	Human	IgG1	CTLA-4	Membrane	5.25
Muromonab-CD3	Murine	IgG2a	CD3	Membrane	0.83
Ofatumumab	Humanized	IgG1	CD20	Membrane	4.7
Tositumomab	Murine	IgG2	CD20	Membrane	1.4
Trastuzumab	Humanized	IgG1	HER-2	Membrane	5.0
Cetuximab	Chimeric	IgG1	EGFR	Membrane	0.15
Denosumab	Human	IgG2	RANKL	Membrane	0.002
Palivizumab	Humanized	IgG1	RSV	Membrane	0.96
Natalizumab	Humanized	IgG4	α4β1-Integrin	Membrane	0.3
Abciximab	Chimeric	IgG1-Fab	Glycoprotein-IIa–IIIa	Membrane	5
Adalimumab	Human	IgG1	TNF-α	Soluble	0.1
Belimumab	Human	IgG1	BAFF	Soluble	0.2
Bevacizumab	Humanized	IgG1	VEGF	Soluble	1.1
Golimumab	Human	IgG1	TNF-α	Soluble	1.89
Infliximab	Chimeric	IgG1	TNF-α	Soluble	0.044
Infliximab	Chimeric	IgG1	TNF-α	Membrane	1.62
Tocilizumab	Humanized	IgG1	IL-6R	Soluble/Membrane	0.71/2.54
Ustekinumab	Human	IgG1	IL12/IL23	Soluble	NA
Omalizumab	Humanized	IgG1	IgE	Soluble	0.17
Panitumumab	Human	IgG2	EGFR	Membrane	0.05

These analytical approaches are based on the assumption that the site of antibody elimination is in rapid equilibrium with plasma (i.e., it is assumed that all elimination is from the "central" compartment). This assumption may be valid for many antibodies, but incorrect for others (e.g., antibodies that bind to and internalize within cells in tissue sites) [4].

15.3.3 Mechanism of mAb Elimination

There are two distinctive routes for the elimination of therapeutic antibodies. The first is a nonspecific, linear (first-order) clearance pathway, which is mediated by interaction between the Fc region of the antibody and Fc receptors (i.e., FcRn and Fcγ receptors). The second is a clearance pathway that is mediated by the specific interaction between the Fab region of the antibody and its pharmacologic target. In many cases, mAb–target binding precipitates the endocytosis of the mAb–target complex, with subsequent intracellular catabolism and elimination of the antibody. As such, target binding may lead to efficient mAb elimination. With increasing doses of mAb, the target becomes increasingly saturated with antibody, and this saturation leads to decreases in the rate of antibody clearance [26]. This elimination mechanism is nonlinear and often referred to as target-mediated drug disposition (TMDD). The total clearance of an antibody can then be represented by the sum of nonspecific and specific processes: $CL_{\text{TOT}} = CL_{\text{linear}} + CL_{\text{TMDD}}$ (CL = clearance). The specific clearance pathway can be saturable, depending on the dose of mAb, the expression level, and turnover of the target. However, the nonspecific clearance pathway has a large capacity and clearance values associated with this pathway are generally linear in the therapeutic dose range of 1–20 mg/kg for most marketed mAbs. Therefore, suprath-erapeutic doses of mAbs will display linear PK as well. IgGs have a longer half-life of around 21 days in human compared with other isotypes (IgA: 6 days; IgE: 2.5 days; IgM: 5 days; IgD: 3 days) with low clearance values of about 3–5 mL/day/kg in humans in the linear dose range [27]. The low clearance may be due to the protection

TABLE 15.2 Pharmacokinetic Parameters for Monoclonal Antibodies [16]

Antibody	PK	Cmt	$T_{1/2}$ (d)	*CL* (L/d)	V_1 (L)	V_2 (L)	V_{ss} (L)	V_{max} (mg/d)	K_m (mg/L)	CL_{int} (L/d)
Abatacept	Linear	2	16.7	0.346	2.68	2.14	4.82			
Alefacept	Linear		11.1	0.403			6.3			
Basiliximab	Linear	2	8.7	1.33	5.47	1.78	7.25			
Catumaxomab	Linear		2.5							
Daclizumab	Linear	1	17	0.385			10.2			
Gemtuzumab	Linear	1	1.9	2.74	6.58	8.38	14.96			
Ibritumomab	Linear	2	1.17							
Alemtuzumab	Non-LE	2	6		11.3	41.5	52.8	24.48	0.338	72.4
Efalizumab	Non-LE	2	12–44	0.535	5.14	4.18	9.32	2.73	0.0973	28.06
Rituximab	Non-LE	2	8.6	0.257	2.98	3.64	6.62			
Ipilimumab			13.21	0.228			4.4			
Muromonab-CD3			0.75							
Tositumomab			1.17	1.636						
Trastuzumab	Linear	2		0.225	2.95	4.79	7.74			
Cetuximab	Non-LE	2	3–4	0.273	2.83	2.43	5.26	105	74	1.42
Denosumab	Non-LE	2	28	0.066	2.38	1.59	3.97	0.0878	0.164	0.54
Palivizumab			7.5	0.343						
Natalizumab			16.00	0.314						
Abciximab	Linear	1	0.021				9.263			
Adalimumab	Linear	2	14	0.269			6.0			
Belimumab	Linear	2	19.4	0.215	2.56	2.73	5.29			
Bevacizumab	Linear	2	20	0.207	2.66	2.76	5.42			
Canakinumab	Linear	1	26	0.174			6.01			
Certolizumab	Linear	1	14	0.408			6.4			
Eculizumab	Linear	1	11.3	0.528			7.7			
Etanercept	Linear	2	4.26	1.67	5.46	2.01	7.47			
Infliximab	Linear	2	8.0–9.5	0.273	3.06	2.94	6			
Infliximab	Linear	2		0.407	3.29	4.13	7.42			
Tocilizumab	Linear		8–14	0.3	3.5	2.9	6.4	7.5	2.7	
Ustekinumab	Linear	1	15–32	0.465			15.7			
Omalizumab	Non-LE	2	26	0.176			5.9			
Panitumumab	Non-LE	2	7.5	0.273	3.95	2.59	6.54	12.1	0.426	28.40
Golimumab				0.4	3.07	3.68	6.75			

CL_{int}, intrinsic clearance; Cmt, compartment; and non-LE, nonlinear.

provided by the FcRn, a heterodimer comprising a β2-microglobulin (β2 m) light chain and a major histocompatibility complex class I-like heavy chain [28]. FcRn is ubiquitously expressed in cells and tissues including vascular endothelium, professional antigen-presenting cells, adult gut, blood–brain barrier (BBB), kidney, lung, and others [28].

15.3.4 Antibody–Drug Conjugates

ADCs are produced by combining a mAb and a cytotoxic drug (small molecule drug) through a linker. These molecules target antigen that are expressed at higher density on malignant cells as compared to normal tissues. There are currently around 30 ADCs in clinical development for the treatment of blood cancers and solid tumors. Two ADCs, brentuximab vedotin (Adcetris®) and adotrastuzumab emtansine (Kadcyla®), are currently approved by the U.S. FDA [29]. Deslandes [30] has reviewed the PK of ADC constructs. However, given the large number of different targets, differences between ADC constructs, and differences between dosing regimens and patient populations, the comparison of ADC PK is not obvious. Deslandes [30] have analyzed dose/clearance relationships of nine ADCs with complete PK information and found two distinct patterns. In the first group, a decrease in clearance with increasing doses was observed with AVE9633, BT062, MLN2704, and adotrastuzumab emtansine, which suggested target-mediated disposition. Upon increasing dose levels, the curve reaches a plateau when the target-mediated, saturable clearance becomes negligible compared with the target-independent, linear IgG clearance. These profiles were irrespective of the ADC target location (hematologic cancer or solid tumors). A dose-dependent PK profile of ADC constructs may suggest high internalization by the tumor, or the presence of an antigen sink in tissues. This is particularly important for

ADCs that exhibit therapeutic windows that are dependent on the difference in the level and distribution of target between normal tissues and tumors.

The second group of ADC constructs, huC242-DM1, IMGN901, and PSMA-ADC showed no dose-dependent change of clearance [30]. Brentuximab vedotin was administered in a narrow range of three doses; therefore, the effect of dose on clearance could not be evidenced. More detailed analysis should be performed in the context of the characteristics of the targets, such as receptor copy number, heterogeneity, and specificity of expression, internalization rate, and intracellular trafficking; however, these data from patients are usually not available in the literature. The prediction of these effects in the clinic from *in vitro* studies may be biased by other factors such as cytokinetics, cytogenetics, multidrug resistance, and other prognostic factors. The influence of target expression and turnover on ADC PK may also be evaluated by comparing PK parameters between the first cycle of administration and several cycles of therapy. The impact of reducing target expression by the mAb-based therapy gives mechanistic insight on how the target pharmacology drives PK. One can therefore use PK as a biomarker of the effect of mAbs on target biology. This feature is unique to mAbs contrary to small molecules.

15.3.5 Recombinant Proteins

TPO, a 353 amino acid cytokine, is a primary regulator of platelet production. Receptor mediated endocytosis (RME) has been reported to be its major elimination pathway. Jin and Krzyzanski [31] showed, using a mathematical model of human data, that the first-order internalization rate constant of TPO in platelets was 0.1/h. The endogenous binding capacity was 164.0 pM. The second-order binding association constant k_{on} was 0.055/h/pM and the first-order dissociation constant k_{off} was estimated at 2.5/h, which gives an equilibrium dissociation constant K_D as 45.5 pM.

EPO, the primary hormone responsible for the stimulation of erythrocyte production is a heavily glycosylated protein. Veng-Pedersen et al. [32] have shown that EPO is eliminated via a Michaelis–Menten saturable mechanism in sheeps and humans. They speculated that the Michaelis–Menten-type elimination is consistent with the hypothesis that recombinant human erythropoietin (r-HuEPO) is primarily eliminated by irreversible endocytosis of EPO by its receptors on erythroid progenitor cells.

There are three major groups of interferons (INFs), which reflect antigenic and structural differentiation. INF-α (leukocyte INFs) is produced by B lymphocytes, null lymphocytes, and macrophages that can be induced by foreign, virus-infected, tumor, or bacterial cells. INF-β (fibroblast INF) is produced by fibroblasts, epithelial cells, and macrophages that can be induced by viral and other foreign nucleic acids. INF-γ (immune INF) is produced by activated T lymphocytes that are induced by foreign antigens. Mager et al. [33] have developed a pharmacokinetic/pharmacodynamic (PK/PD) model for IFN-1α that depicts receptor binding as a key feature controlling nonlinear elimination, nonstationary kinetics (time-dependent) in monkeys. In their study, repeated daily SC dosing produced modestly elevated IFN-1α and biomarker concentrations as compared to simulated values from single-dose data. The authors were able to describe the data as downregulation of the maximum stimulatory effect of INF-1α on the biomarker neopterin (S_{max}) and a downregulation of its receptors.

15.4 PARAMETERS CONTROLLING TARGET-DRIVEN NONLINEAR PHARMACOKINETICS OF BIOTHERAPEUTICS

15.4.1 Target Localization

mAbs against soluble antigens with low endogenous levels such as TNF-α, IFN-α, VEGF (vascular epidermal growth factor), and IL-5 have usually dose-independent linear clearance, which is primarily driven by nonspecific, target-independent mechanism. As a consequence, exposure to the biotherapeutics as measured by AUC, for instance, increases in proportion with dose. For example, linear PK has been documented for a humanized mAb directed to human IL-5 following IV administration of a 6000-fold dose range (0.05–300 mg/kg) in cynomolgus monkeys [34]. Adalimumab, a mAb against TNF-α, has linear PK in humans. The mean total serum clearance and the estimated mean terminal half-life of adalimumab (0.5–10 mg/kg) range from 0.012 to 0.017 L/h and 10.0 to 13.6 days, respectively, with an overall mean half-life of 12 days [35]. However, mAbs against soluble antigens with high endogenous levels (such as IgE) exhibit nonlinear PK. Target with fast turnover (fast synthesis and fast clearance) might also be responsible for nonlinear PK. Omalizumab has a linear PK only at doses greater than 0.5 mg/kg [36, 37]. It has been shown that soluble antigen in circulation resulting from the cleavage of the antigen from the surface of tumor cells ("shed" antigen), or from secretion of a soluble form of the antigen, can sometimes influence the PK and the efficacy of an ADC. Soluble antigen can bind to the circulating ADC, potentially enhancing clearance of the latter and/or interfering with targeting of the antigen on tumor cells. A correlation between high peripheral blood antigen load and decreased efficacy has been suggested for gemtuzumab ozogamicin in acute myeloid leukemia [38] and for an anti-MUC1 (hCTM01)–calicheamicin conjugate in ovarian cancer [39, 40]. For other antibodies and ADCs, such as trastuzumab and IMGN242, high peripheral blood antigen levels have been correlated with altered PK [41]. The half-life of trastuzumab was reported to be 2 ± 1 days for patients with circulating levels of human epidermal growth factor receptor 2 (HER2) extracellular domain ≥500 ng/ml, as compared with a half-life of 9 ± 5 days for patients with less than 500 ng/ml serum HER2 [41]. Dose-dependent clearance and half-life are more commonly observed in mAbs that interact with cell membrane-expressed antigens. Koon et al. [42] demonstrated a strong inverse correlation between cellular expression of CD25+ and apparent half-life of daclizumab (a mAb specifically binding to CD25) [3, 43]. In patients with advanced tumors overexpressing epidermal growth factor receptors (EGFRs), cetuximab, an anti-EGFR mAb, had an approximately twofold faster clearance at a dose of 100 mg/m^2 compared with the dose of 400 mg/m^2 (0.837 vs 0.374 L/day) [44]. The disposition of murine antihuman CD3 Abs may be determined by the disappearance of their target antigens [3, 45]. In addition, Ng et al. [46] have demonstrated that anti-CD4 mAb had about fivefold faster total clearance at a dose of 1 mg/kg compared with the dose of 10 mg/kg (7.8 ± 0.6 vs 37.4 ± 2.4 mL/day/kg) in healthy volunteers. This suggested that the receptor-mediated clearance was greater at the lower dose since nonspecific clearance is constant across dose levels [3, 47]. When the expression level of a target (i.e., antigen density) is modulated by a mAb, it leads to a time dependency of its PK. Serum concentration levels of rituximab, a chimeric anti-CD20 mAb, were inversely correlated with tumor size and the number of circulating B cells in patients with non-Hodgkin's lymphoma (NHL) [48]. After the first dose of rituximab, target B cells were depleted. As a result, the clearance evaluated in the following treatment cycles was significantly reduced compared with the first cycle (approximately fivefold difference) [3, 49].

15.4.2 Target Affinity

The high affinity binding of mAb to target proteins within solid tumors may act as a barrier to distribution, which is referred to as the "binding-site barrier" hypothesis. Fujimori et al. [50] using modeling analysis have shown that high affinity ($KA > 1.0 \times 10^9$/M) mAbs exhibit heterogeneous tumor distribution, with the majority of the molecules being restricted to the sites proximal to the point of extravasation within the tumor. The results of their simulations suggested that moderate affinity mAb ($KA = 5 \times 10^7$ to 1×10^8/M) would allow optimal distribution. Their predictions have been corroborated by experimental findings of Juweid et al. [48]. Following low doses, mAb intratumoral distribution was limited to areas adjacent to blood vessels, and the extent of tumor distribution was enhanced following high doses of mAb, consistent with the saturation of the binding-site barrier. RME has also been utilized for efficient drug delivery to the target cells with high expression of the receptors. Brain delivery of traditional therapeutic proteins with large MW is almost impossible due to tight junction of the blood-brain-barrier. In order to enhance brain delivery, molecular Trojan horse (MTH) method has been used to deliver therapeutic proteins to brain through receptor-mediated endocytosis and transcytosis. Insulin receptor (IR) and transferrin receptor (TfR) are mostly used with the MTHs because of their high expression levels in the brain endothelial cells. The MTH is generally constructed by fusing a therapeutic protein to each of the heavy chain of a genetically engineered chimeric mAb against the TfR or IR [51].

15.4.3 Target Turnover

HER2 is a member of the human epidermal growth factor receptor family. Amplification or overexpression of this oncogene has been shown to play an important role in the development and progression of certain aggressive types of breast cancer. It is cleaved by proteolysis and can be found in the serum of patients with breast cancer. Some authors have suggested that circulating serum HER2 extracellular domain (HER2 ECD) affect the kinetic profile of trastuzumab given alone [41] or combined with cisplatin [52]. Patients with serum shed ECD above 0.5 mg/L display a shorter half-life for trastuzumab than patients with ECD less than 0.5 mg/L (2.9 ± 3.2 days vs 9.2 ± 5.3 days given alone and (4 ± 2.6 days vs 11 ± 4.4 days) combined with cisplatin). Plasma alemtuzumab concentrations of a patient with low levels of sCD52 were higher than those of a patient with high levels of sCD52 [53].

15.4.4 Target Baseline and Disease Progression

Significant variations in PK parameters of mAbs appear imputable to changes in antigen expression. For instance, gemtuzumab ozogamicin clearance has been shown to decrease by twofold between the two doses. Moreover, a relationship has been shown between the percentage of variation of peripheral blasts cells (the target of gemtuzumab ozogamicin) and the changes in the area under the concentration–time curve (AUC) of the antibody [54]. Therefore, the increase in systemic exposition to gemtuzumab ozogamicin could be related to the disappearance of circulating leukemic targets. Similarly, Berinstein et al. [55] found a significant decrease in rituximab clearance, which in turn resulted in an increase in the terminal half-life between the first and the fourth perfusions. This was attributed to variation of the CD20 antigen expression on peripheral immune cells. An inverse correlation was also reported between the count of blood B cells at baseline and rituximab concentrations determined at various times in patients with NHL. The author showed that the anti-CD20 antibody serum concentrations correlated inversely with the pretreatment tumor mass [55]. These findings show that, for mAb exhibiting TMDD, intra- and interpatient variability in target expression may lead to clinically relevant PK variability.

15.4.5 Off-Target Binding

Despite being specific to a given antigen, mAbs may show off-target binding that may result in faster clearance. Anti-respiratory syncytial virus mAb A4b4 had poor PK in rats and cynomolgus monkeys due to broad nonspecific tissue binding and sequestration [56]. The fast elimination of a humanized antihuman amyloid β peptide (Aβ2) mAb, anti-Aβ Aβ2, in cynomolgus monkeys was linked to off-target binding to cynomolgus monkey fibrinogen [57]. In addition, a humanized anti-fibroblast growth factor receptor 4 (FGFR4) mAb had fast clearance in mice that was attributable to binding to mouse complement component 3 (C3). Hotzel et al. [58] have developed an assay based on ELISA (enzyme-linked immunosorbent assay) detection of nonspecific binding to baculovirus particles that can identify antibodies having an increased risk for fast clearance.

15.5 IMPACT OF TARGET-DRIVEN NONLINEAR PHARMACOKINETICS OF BIOTHERAPEUTICS ON HALOMETRIC SCALING

Interspecies scaling is used to predict human PK for first-in-human clinical trials. In general, mAbs display much higher affinity and specificity for their respective target than small molecules. It has been shown that a therapeutic mAb normally does not cross-react with its target in most animal species, which considerably limits the information available for interspecies scaling. In general, cross-species activity for therapeutic mAbs is limited to one or a few animal species. Even in a cross-reactive species, there may be differences in relevant target interaction parameters such as IC_{50} or K_D. The regulation of the target and its expression level is likely different between animal species and humans. This difference can be quite large. For example, Duconge et al. [59] showed that allometric scaling is not appropriate to predict clearance of a murine anti-EGF/r3mAb in cancer patients. It was hypothesized that this finding is likely due to the low affinity of murine antibodies for human FcRn and the increased target-mediated clearance in cancer patients.

Body weight (BW)-based allometric scaling may be sufficient to scale animal dose to humans or adult doses to children for mAbs with linear kinetics. Such scaling is often accomplished using a simple power model of the form $Y=a*BW^b$, where Y is the parameter of interest (CL or V_{ss}), BW is the body weight, a is the allometric coefficient, and b is the allometric exponent. Literature on adult data suggests that PK parameters for many mAbs change in a less than BW-proportional manner as respective allometric exponents for clearance and volume of distribution were estimated to range from 0.3 to 0.7 [60, 61]. There are only a few examples where strong BW effects were observed as indicated by exponents greater than 0.75 [60]. It should be noted that allometric exponents determined for within species scaling are typically smaller than those observed for between species scaling, which is likely the result of a narrower BW range in adults for a given species (approximately two- to three-fold) [59].

Since BW ranges in children is wider than that in adult, further research is necessary to determine appropriate allometric exponents in children [60]. For example, clearance exponents were estimated to be 0.823 [61] for canakinumab in systemic juvenile idiopathic arthritis (SJIA) patients, whereas a value of 0.75 was found appropriate for palivizumab and infliximab when scaling from adults to children [63, 64].

Using mathematical models and computer simulations, Zheng et al. [65] have investigated the impact of differences in target expression between adults and children on pediatric dosing for mAbs exhibiting nonlinear kinetics. Their results indicate that BW-based dosing is superior to fixed dosing for the same target concentration, whereas the opposite holds true for the same target amount in adults and children.

15.5.1 Ethnic Differences

The evaluation of drugs in Japan often requires bridging PK studies. Chiba et al. [66] have evaluated the PK of several mAbs between Caucasian and Japanese subjects. They found three mAbs that showed differences in clearance between the two ethnic groups: ibritumomab tiuxetan and rituximab, which target cell surface antigens, and ustekinumab targets soluble antigens. The ethnic differences in the PK of ibritumomab tiuxetan and rituximab were explained by differences between Japanese and Caucasians with regard to CD20 levels. Ethnic differences in cell surface expression of the molecular targets of rituximab and ibritumomab tiuxetan could affect their exposure because the PK of mAbs is associated with the immune complex concentration. The authors conclude that the difference in receptor expression level depends on the target disease and, most importantly, dose adjustment might not be necessary given the small difference in clearance between the two ethnic groups.

15.6 CONCLUSIONS AND PERSPECTIVES

Biotherapeutics display both linear and nonlinear kinetics. The extent of nonlinearity often depends on target receptor availability, the efficiency of drug–receptor complex endocytosis, the affinity to the target, and the dose levels administered. Available data show that biotherapeutics that bind to ligands on cell membranes will display more often nonlinear volume of distribution and clearance. In general, biotherapeutics that bind to soluble ligands will exhibit linear PK unless the concentration of the ligand is high as compared to the biotherapeutics concentration. The PK of biotherapeutics that exhibit TMDD are controlled by the level of target

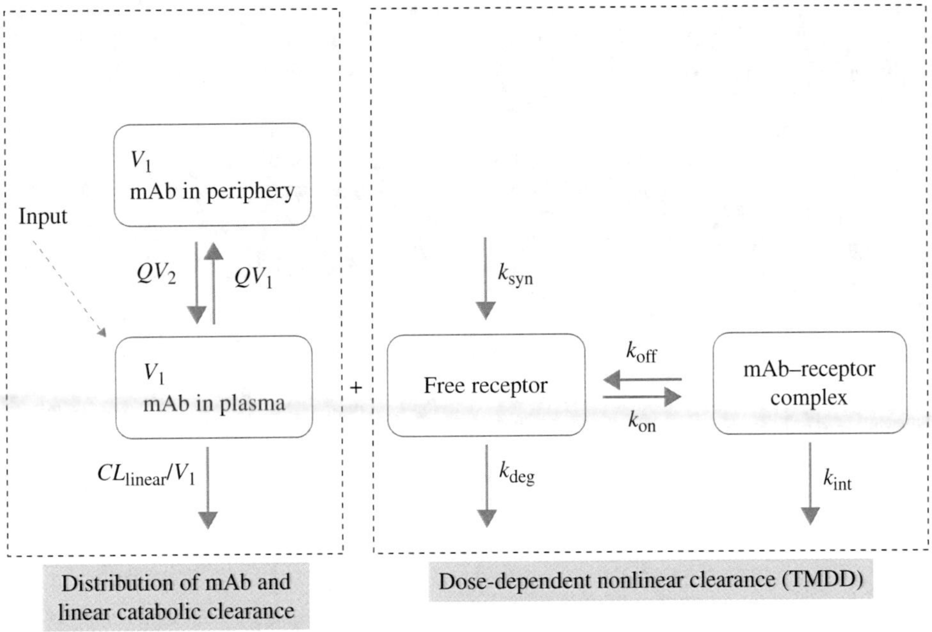

FIGURE 15.2 General pharmacokinetic model of target-mediated drug disposition. V_1, volume of central compartment; V_2, volume of peripheral compartment; Q, intercompartmental clearance; CL_{linear}, linear clearance; k_{syn}, rate constant for receptor synthesis; k_{deg}, rate constant for receptor degradation; k_{on}, rate constant for association of the complex; k_{off}, rate constant for dissociation of the complex; k_{int}, rate of internalization of the complex.

expression, turn over, and affinity. In addition to nonlinear PK, TMDD may be responsible for time-dependent PK, which may necessitate changing dosing regimen as the level of target is reduced as seen with an efficacious treatment. The characterization of the PK of biotherapeutics has to be performed in the context of PK/PD evaluations. This necessitates the use of mathematical models as some parameters such as the rate of production of the target or the rate of inter nalization of the biotherapeutics–ligand complex may not always be experimentally accessible (Fig. 15.2). TMDD is often the source of inter- and intrasubject variability in response to treatments with biotherapeutics. This phenomenon should be considered when selecting the animal species for the evaluation of toxicity, scaling PK parameters between animals and humans, between healthy and disease subjects, between adults and pediatrics, and to a lesser extend between ethnic groups. Improvement in the present understanding of antibody PK/PD will significantly contribute to the development of biotherapeutics.

REFERENCES

[1] Ezan E. Pharmacokinetic studies of protein drugs: past, present and future. Adv Drug Deliv Rev 2013;65:1065–1073.

[2] Andersen JT, Daba MB, Berntzen G, Michaelsen TE, Sandlie I. Cross-species binding analyses of mouse and human neonatal Fc receptor show dramatic differences in immunoglobulin G and albumin binding. Biol Chem 2010;285:4826–4836.

[3] Deng R, Jin F, Prabhu S, Iyer S. Monoclonal antibodies: what are the pharmacokinetic and pharmacodynamic considerations for drug development? Expert Opin Drug Metab Toxicol 2012;8:141–160.

[4] Lobo ED, Hansen RJ, Balthasar JP. Antibody pharmacokinetics and pharmacodynamics. J Pharm Sci 2004;93:2645–2668.

[5] Freeman DJ, McDorman K, Ogbagabriel S, Kozlosky C, Yang BB, Doshi S, Perez-Ruxio JJ, Fanslow W, Starnes C, Radinsky R. Tumor penetration and epidermal growth factor receptor saturation by panitumumab correlate with antitumor activity in a preclinical model of human cancer. Mol Cancer 2012;11:47.

[6] Van CE, Peeters M, Siena S, Humblet Y, Hendlisz A, Neyns B, Canon JL, Van Laethem JL, Maurel J, Richardson G, Wolf M, Amado RG. Open-label phase III trial of panitumumab plus best supportive care compared with best supportive care alone in patients with chemotherapy-refractory metastatic colorectal cancer. J Clin Oncol 2007;25:1658–1664.

[7] Wu B, Sun YN. Pharmacokinetics of peptide-Fc fusion proteins. J Pharm Sci 2014;103:53–64.

[8] Bussel JB, Buchanan GR, Nugent DJ, Gnarra DJ, Bomgaars LR, Blanchette VS, Wang YM, Nie K, Jun S. A randomized, double-blind study of romiplostim to determine its safety and efficacy in children with immune thrombocytopenia. Blood 2011;118:28–36.

[9] Kumagai Y, Fujita T, Ozaki M, Sahashi K, Ohkura M, Ohtsu T, Arai Y, Sonehara Y, Nichol JL. Pharmacodynamics and pharmacokinetics of AMG 531, a thrombopoiesis-stimulating peptibody, in healthy Japanese subjects: a randomized, placebo-controlled study. J Clin Pharmacol 2007;47:1489–1497.

[10] Oliner J, Min H, Leal J, Yu D, Rao S, You E, Tang X, Kim H, Meyer S, Han SJ, Hawkins N, Rosenfeld R, Davy E, Graham K, Jacobsen F, Stevenson S, Ho J, Chen Q, Hartmann T, Michaels M, Kelley M, Li L, Sitney K, Martin F, Sun JR, Zhang N, Lu J, Estrada J, Kumar R, Coxon A, Kaufman S, Pretorius J, Scully S, Cattley R, Payton M, Coats S, Nguyen L, Desilva B, Ndifor A, Hayward I, Radinsky R, Boone T, Kendall R. Suppression of angiogenesis and tumor growth by selective inhibition of angiopoietin-2. Cancer Cell 2004;6:507–516.

[11] Herbst RS, Hong D, Chap L, Kurzrock R, Jackson E, Silverman JM, Rasmussen E, Sun YN, Zhong D, Hwang YC, Evelhoch JL, Oliner JD, Le N, Rosen LS. Safety, pharmacokinetics, and antitumor activity of AMG 386, a selective angiopoietin inhibitor, in adult patients with advanced solid tumors. J Clin Oncol 2009;27:3557–3565.

[12] Bouman-Thio E, Franson K, Miller B, Getsy J, Cohen A, Bai SA, Yohrling J, Frederick B, Marciniak S, Jiao Q, Jang H, Davis H, Burggraaf J. A phase I, single and fractionated, ascending-dose study evaluating the safety, pharmacokinetics, pharmacodynamics, and immunogenicity of an erythropoietin mimetic antibody fusion protein (CNTO 528) in healthy male subjects. J Clin Pharmacol 2008;48:1197–1207.

[13] Chen X, Zaro JL, Shen WC. Pharmacokinetics of recombinant bifunctional fusion proteins. Expert Opin Drug Metab Toxicol 2012;8:581–595.

[14] Biogen. 2003. Clinical pharmacology review of alefacept. Available at http://www.fda.gov/downloads/Drugs/DevelopmentApprovalProcess/HowDrugsareDevelopedandApproved/ApprovalApplications/TherapeuticBiologicApplications/ucm086010.pdf. Accessed 2015 Jun 10.

[15] Hoffman HM, Throne ML, Amar NJ, Sebai M, Kivitz AJ, Kavanaugh A, Weinstein SP, Belomestnov P, Yancopoulos GD, Stahl N, Mellis SJ. Efficacy and safety of rilonacept (interleukin-1 Trap) in patients with cryopyrin-associated periodic syndromes: results from two sequential placebo-controlled studies. Arthritis Rheum 2008;58:2443–2452.

[16] Dostalek M, Gardner I, Gurbaxani BM, Rose RH, Chetty M. Pharmacokinetics, pharmacodynamics and physiologically-based pharmacokinetic modelling of monoclonal antibodies. Clin Pharmacokinet 2013;52:83–124.

[17] Richter WF, Bhansali SG, Morris ME. Mechanistic determinants of biotherapeutics absorption following SC administration. AAPS J 2012;14:559–570.

[18] Richter WF, Jacobsen B. Subcutaneous absorption of biotherapeutics: knowns and unknowns. Drug Metab Dispos 2014;42:1881–1889.

[19] Kagan L, Turner MR, Balu-Iyer SV, Mager DE. Subcutaneous absorption of monoclonal antibodies: role of dose, site of injection, and injection volume on rituximab pharmacokinetics in rats. Pharm Res 2012;29:490–499.

[20] Vaishnaw AK, TenHoor CN. Pharmacokinetics, biologic activity, and tolerability of alefacept by intravenous and intramuscular administration. J Pharmacokinet Pharmacodyn 2002;29:415–426.

[21] Keizer RJ, Huitema AD, Schellens JH, Beijnen JH. Clinical pharmacokinetics of therapeutic monoclonal antibodies. Clin Pharmacokinet 2010;49:493–507.

[22] Baxter LT, Zhu H, Mackensen DG, Jain RK. Physiologically based pharmacokinetic model for specific and nonspecific monoclonal antibodies and fragments in normal tissues and human tumor xenografts in nude mice. Cancer Res 1994;54: 1517–1528.

[23] Kairemo KJ, Lappalainen AK, Kaapa E, Laitinen OM, Hyytinen T, Karonen SL, Grönblad M. *In vivo* detection of intervertebral disk injury using a radiolabeled monoclonal antibody against keratan sulfate. J Nucl Med 2001;42: 476–482.

[24] Danilov SM, Gavrilyuk VD, Franke FE, Pauls K, Harshaw DW, McDonald TD, Miletich DJ, Muzykantov VR. Lung uptake of antibodies to endothelial antigens: key determinants of vascular immunotargeting. Am J Physiol Lung Cell Mol Physiol 2001;280:L1335–L1347.

[25] Scheidhauer K, Wolf I, Baumgartl HJ, Von Schilling C, Schmidt B, Reidel G, Peschel C, Schwaiger M. Biodistribution and kinetics of (131)I-labelled anti-CD20 MAB IDEC-C2B8 (rituximab) in relapsed non-Hodgkin's lymphoma. Eur J Nucl Med Mol Imaging 2002;29: 1276–1282.

[26] Glassman PM, Balthasar JP. Mechanistic considerations for the use of monoclonal antibodies for cancer therapy. Cancer Biol Med 2014;11:20–33.

[27] Morell A, Terry WD, Waldmann TA. Metabolic properties of IgG subclasses in man. J Clin Invest 1970;49:673–680.

[28] Junghans RP. Finally! The Brambell receptor (FcRB). Mediator of transmission of immunity and protection from catabolism for IgG. Immunol Res 1997;16:29–57.

[29] Mullard A. Maturing antibody-drug conjugate pipeline hits 30. Nat Rev Drug Discov 2013;12:329–332.

[30] Deslandes A. Comparative clinical pharmacokinetics of antibody-drug conjugates in first-in-human phase 1 studies. MAbs 2014;6:859–870.

[31] Jin F, Krzyzanski W. Pharmacokinetic model of target-mediated disposition of thrombopoietin. AAPS J 2004;6:86–93.

[32] Veng-Pedersen P, Widness JA, Pereira LM, Schmidt RL, Lowe LS. A comparison of nonlinear pharmacokinetics of erythropoietin in sheep and humans. Biopharm Drug Dispos 1999;20:217–223.

[33] Mager DE, Neuteboom B, Efthymiopoulos C, Munafo A, Jusko WJ. Receptor-mediated pharmacokinetics and pharmacodynamics of interferon-beta1a in monkeys. J Pharmacol Exp Ther 2003;306:262–270.

[34] Zia-Amirhosseini P, Minthorn E, Benincosa LJ, Hart TK, Hottenstein CS, Tobia LA, Davis CB. Pharmacokinetics and pharmacodynamics of SB-240563, a humanized monoclonal antibody directed to human interleukin-5, in monkeys. J Pharmacol Exp Ther 1999;291:1060–1067.

[35] den Broeder A, van de Putte L, Rau R, Schattenkirchner M, Van Riel P, Sander O, Binder C, Fenner H, Bankmann Y,

Velagapudi R, Kempeni J, Kupper H. A single dose, placebo controlled study of the fully human anti-tumor necrosis factor-alpha antibody adalimumab (D2E7) in patients with rheumatoid arthritis. J Rheumatol 2002;29:2288–2298.

[36] Arceci RJ, Sande J, Lange B, Shannon K, Franklin J, Hutchinson R, Vik TA, Flowers D, Aplenc R, Berger MS, Sherman ML, Smith FO, Bernstein I, Sievers EL. Safety and efficacy of gemtuzumab ozogamicin in pediatric patients with advanced CD33+ acute myeloid leukemia. Blood 2005;106: 1183–1188.

[37] Meno-Tetang GM, Lowe PJ. On the prediction of the human response: a recycled mechanistic pharmacokinetic/pharmacodynamic approach. Basic Clin Pharmacol Toxicol 2005;96:182–192.

[38] van der Velden VH, Boeckx N, Jedema I, te Marvelde JG, Hoogeveen PG, Boogaerts M, van Dongen JJ. High CD33-antigen loads in peripheral blood limit the efficacy of gemtuzumab ozogamicin (Mylotarg) treatment in acute myeloid leukemia patients. Leukemia 2004;18:983–988.

[39] Hamann PR, Hinman LM, Beyer CF, Lindh D, Upeslacis J, Shochat D, Mountain A. A calicheamicin conjugate with a fully humanized anti-MUC1 antibody shows potent antitumor effects in breast and ovarian tumor xenografts. Bioconjug Chem 2005;16:354–360.

[40] Davies Q, Perkins AC, Frier M, Watson S, Lalani E, Symonds EM. The effect of circulating antigen on the biodistribution of the engineered human antibody hCTM01 in a nude mice model. Eur J Nucl Med 1997;24:206–209.

[41] Baselga J, Tripathy D, Mendelsohn J, Baughman S, Benz CC, Dantis L, Sklarin NT, Seidman AD, Hudis CA, Moore J, Rosen PP, Twaddell T, Henderson IC, Norton L. Phase II study of weekly intravenous recombinant humanized anti-p185HER2 monoclonal antibody in patients with HER2/neu-overexpressing metastatic breast cancer. J Clin Oncol 1996;14:737–744.

[42] Koon HB, Severy P, Hagg DS, Butler K, Hill T, Jones AG, Waldmann TA, Junghans RP. Antileukemic effect of daclizumab in CD25 high-expressing leukemias and impact of tumor burden on antibody dosing. Leuk Res 2006;30 :190–203.

[43] Sievers EL, Larson RA, Stadtmauer EA, Estey E, Löwenberg B, Dombret H, Karanes C, Theobald M, Bennett JM, Sherman ML, Berger MS, Eten CB, Loken MR, van Dongen JJ, Bernstein ID, Appelbaum FR, Mylotarg Study Group. Efficacy and safety of gemtuzumab ozogamicin in patients with CD33-positive acute myeloid leukemia in first relapse. J Clin Oncol 2001;19:3244–3254.

[44] Krop IE, Beeram M, Modi S, Jones SF, Holden SN, Yu W, Girish S, Tibbitts J, Yi JH, Sliwkowski MX, Jacobson F, Lutzker SG, Burris HA. Phase I study of trastuzumab-DM1, an HER2 antibody-drug conjugate, given every 3 weeks to patients with HER2-positive metastatic breast cancer. J Clin Oncol 2010;28:2698–2704.

[45] Teicher BA. Antibody-drug conjugate targets. Curr Cancer Drug Targets 2009;9:982–1004.

[46] Ng CM, Stefanich E, Anand BS, Fielder PJ, Vaickus L. Pharmacokinetics/pharmacodynamics of nondepleting anti-CD4 monoclonal antibody (TRX1) in healthy human volunteers. Pharm Res 2006;23:95–103.

[47] Jain RK. Normalization of tumor vasculature: an emerging concept in antiangiogenic therapy. Science 2005;307: 58–62.

[48] Juweid M, Neumann R, Paik C, Perez-Bacete MJ, Sato J, van Osdol W, Weinstein JN. Micropharmacology of monoclonal antibodies in solid tumors: direct experimental evidence for a binding site barrier. Cancer Res 1992;52: 5144–5153.

[49] Mukherjee S, Richardson AM, Rodriguez-Canales J, Ylaya K, Erickson HS, Player A, Kawasaki ES, Pinto PA, Choyke PL, Merino MJ, Albert PS, Chuaqui RF, Emmert-Buck MR. Identification of EpCAM as a molecular target of prostate cancer stroma. Am J Pathol 2009;175: 2277–2287.

[50] Fujimori K, Covell DG, Fletcher JE, Weinstein JN. A modeling analysis of monoclonal antibody percolation through tumors: a binding-site barrier. J Nucl Med 1990;31: 1191–1198.

[51] Xiao G, Gan LS. Receptor-mediated endocytosis and brain delivery of therapeutic biologics. Int J Cell Biol 2013;2013: 703545.

[52] Pegram MD, Lipton A, Hayes DF, Weber BL, Baselga JM, Tripathy D, Baly D, Baughman SA, Twaddell T, Glaspy JA, Slamon DJ. Phase II study of receptor-enhanced chemosensitivity using recombinant humanized anti-p185HER2/neu monoclonal antibody plus cisplatin in patients with HER2/neu-overexpressing metastatic breast cancer refractory to chemotherapy treatment. J Clin Oncol 1998;16: 2659–2671.

[53] Albitar M, Do KA, Johnson MM, Giles FJ, Jilani I, O'Brien S, Cortes J, Thomas D, Rassenti LZ, Kipps TJ, Kantarjian HM, Keating M. Free circulating soluble CD52 as a tumor marker in chronic lymphocytic leukemia and its implication in therapy with anti-CD52 antibodies. Cancer 2004;101:999–1008.

[54] Dowell JA, Korth-Bradley J, Liu H, King SP, Berger MS. Pharmacokinetics of gemtuzumab ozogamicin, an antibody-targeted chemotherapy agent for the treatment of patients with acute myeloid leukemia in first relapse. J Clin Pharmacol 2001;41:1206–1214.

[55] Berinstein NL, Grillo-Lopez AJ, White CA, Bence-Bruckler I, Maloney D, Czuczman M, Green D, Rosenberg J, McLaughlin P, Shen D. Association of serum Rituximab (IDEC-C2B8) concentration and anti-tumor response in the treatment of recurrent low-grade or follicular non-Hodgkin's lymphoma. Ann Oncol 1998;9:995–1001.

[56] Wu H, Pfarr DS, Johnson S, Brewah YA, Woods RM, Patel NK, et al. Development of motavizumab, an ultra-potent antibody for the prevention of respiratory syncytial virus infection in the upper and lower respiratory tract. J Mol Biol 2007;368:652–65.

[57] Vugmeyster Y, Szklut P, Wensel D, Ross J, Xu X, Awwad M, et al. Complex pharmacokinetics of a humanized antibody against human amyloid beta peptide, anti-abeta Ab2, in non-clinical species. Pharm Res 2011;28:1696–706.

[58] Hotzel I, Theil FP, Bernstein LJ, Prabhu S, Deng R, Quintana L, Lutman J, Sibia R, Chan P, Bumbaca D, Fielder P, Carter PJ, Kelley RF. A strategy for risk mitigation of antibodies with fast clearance. MAbs 2012;4:753–760.

[59] Duconge J, Fernandez-Sanchez E, Alvarez D. Interspecies scaling of the monoclonal anti-EGF receptor ior EGF/r3 antibody disposition using allometric paradigm: is it really suitable? Biopharm Drug Dispos 2004;25: 177–186.

[60] Bai S, Jorga K, Xin Y, Jin D, Zheng Y, Damico-Beyer LA, Gupta M, Tang M, Allison DE, Lu D, Zhang Y, Joshi A, Dresser MJ. A guide to rational dosing of monoclonal antibodies. Clin Pharmacokinet 2012;51:119–135.

[61] Wang DD, Zhang S, Zhao H, Men AY, Parivar K. Fixed dosing versus body size-based dosing of monoclonal antibodies in adult clinical trials. J Clin Pharmacol 2009;49: 1012–1024.

[62] European Medicines Agency. 2015. Available at http://www.ema.europa.eu/docs/en_GB/document_library/EPAR_-_Assessment_Report_-_Variation/human/001109/WC500152041.pdf.Accessed 2015 Jul 7.

[63] Xu Z, Davis HM, Zhou H. Rational development and utilization of antibody-based therapeutic proteins in pediatrics. Pharmacol Ther 2013;137:225–247.

[64] Robbie GJ, Zhao L, Mondick J, Losonsky G, Roskos LK. Population pharmacokinetics of palivizumab, a humanized anti-respiratory syncytial virus monoclonal antibody, in adults and children. Antimicrob Agents Chemother 2012;56:4927–4936.

[65] Zheng S, Gaitonde P, Andrew MA, Gibbs MA, Lesko LJ, Schmidt S. Model-based assessment of dosing strategies in children for monoclonal antibodies exhibiting target-mediated drug disposition. CPT Pharmacometrics Syst Pharmacol 2014;3:e138.

[66] Chiba K, Yoshitsugu H, Kyosaka Y, Iida S, Yoneyama K, Tanigawa T, Fukushima T, Hiraoka M. A comprehensive review of the pharmacokinetics of approved therapeutic monoclonal antibodies in Japan: are Japanese phase I studies still needed? J Clin Pharmacol 2014;54 (5):483–494.

16

TUMOR EFFECT-SITE PHARMACOKINETICS: MECHANISMS AND IMPACT ON EFFICACY

Greg M. Thurber
University of Michigan, Ann Arbor, MI, USA

16.1 INTRODUCTION

Biologics, particularly monoclonal antibodies, form a rapidly growing field of cancer therapeutics. These molecules have exquisite specificity for their targets, thereby reducing off-target effects. Their large size also provides many opportunities for engineering multiple functions into a single molecule including immune cell recruitment, multiple and independent binding sites for separate targets, and carrier payloads of radiation or small molecule therapeutics. For all their benefits, however, there are inherent challenges in managing the pharmacokinetics of biologics. In this chapter, we will focus on how mechanisms of distribution and efficacy play out at the site of action for cancer biologics: the tumor.

The chapter is organized into two main sections. The first describes the distribution of antibodies and other macromolecules in tumors comprising the effects of tissue physiology, drug properties, and their interaction. A brief review of fluid transport in healthy and tumor tissues is followed by mechanisms that determine antibody distribution. The second section details how this distribution ultimately impacts efficacy. The section starts with an overview of cell-killing mechanisms leading into a discussion on how particular delivery issues impact each of these strategies. The term antibody will be interchanged with macromolecule and biologic throughout the chapter, but most of the concepts apply equally to other binding proteins including alternate scaffolds, antibody fragments, and peptides.

16.2 TUMOR PHARMACOKINETICS

16.2.1 Tissue Physiology, Fluid Balance, and Macromolecular Transport

The physiology of tumors is significantly different than healthy tissue. In healthy tissue, blood is pumped to all the organs in the body through progressively smaller arteries, arterioles, and capillaries, at which point the large surface area of these vessels maximizes exchange of oxygen, nutrients, and waste products with the tissue. The large hydrostatic pressure in the arteries forces blood toward the lower pressure at the venous end of the capillaries. This high pressure in the vessel also forces some fluid from the plasma into the surrounding tissue. The small gaps between the endothelial cells and glycocalyx [1] act as a filter, keeping larger proteins within the vessel lumen. As fluid continues to leak out along the length of the vessel, the concentration of macromolecules increases. Oncotic pressure (osmotic pressure exerted by macromolecules) rises toward the end of the vessel while the hydrostatic pressure drops to the lower venous level. Starling's equation describes how eventually the oncotic pressure can be higher than the hydrostatic pressure, and some of the fluid is resorbed from the interstitial space within the tissue. Any fluid that is not resorbed exits the tissue through the lymph vessels. This keeps the interstitial pressure of most tissues close to zero. This well-ordered system bathes the tissue in interstitial fluid that is constantly exchanged as a small amount of plasma leaks into the interstitial space and leaves as lymph. The efficient vascular system maintains adequate blood flow (often millimeters per second velocity in capillaries [2–4]),

ADME and Translational Pharmacokinetics/Pharmacodynamics of Therapeutic Proteins: Applications in Drug Discovery and Development, First Edition. Edited by Honghui Zhou and Frank-Peter Theil.

and the whole system is able to effectively remove metabolic wastes and deliver oxygen, nutrients, and drugs to the tissue.

There are multiple failures in this delivery system within tumors. First, there are structural abnormalities that affect blood velocity. The progressively smaller vessels ending in capillaries are not found within tumors. Rather, a heterogeneous and stochastic infiltration of vessels provides hypervascularization in some areas and no vessels, resulting in necrosis, in others [5]. Blood flow velocities are often much slower, erratic, and can even reverse direction [6]. Tumors are inflammatory environments [7], and the action of several mediators of angiogenesis, particularly vascular-endothelial growth factor (VEGF), increases the permeability of the vessels to macromolecules including the formation of fenestrae. These immature tumor vessels also lack pericyte coverage and therefore form loose junctions between the endothelial cells [8]. The increased macromolecular permeability disrupts the resorption of fluid. This alone would not be problematic if the lymph system was capable of removing the extra fluid. However, the lymph system is nonfunctional or nonexistent in tumors [2]. While lymph vessels are a major conduit for local metastases, this is prevalent only in the tumor periphery. The solid stress of disorganized and dividing tumor cells can collapse blood and lymph vessels [9], causing the extravasated fluid to build within the tumor. The interstitial pressure increases from close to zero until it approaches the pressure within the blood vessels (elevated interstitial pressure) [10, 11].

Molecules are transported through tissue by two main physical mechanisms—convection and diffusion. Convection results from the bulk transport of fluid from an area of high pressure to low pressure carrying with it all the molecules dissolved in that fluid. If fluid flows from an area of high concentration to tissue of lower concentration, the concentration within that tissue will increase. In other words, the change in concentration over time ($\partial[C]/\partial t$ in Eq. 16.1) will be positive. Diffusion results from random Brownian motion caused by thermal movement in a liquid. On average, more molecules in a highly concentrated region will randomly move to an area of low concentration compared to the number moving in the reverse direction. Therefore, molecules always diffuse from a region of higher concentration to lower concentration according to Fick's law (Eq. 16.2).

$$\frac{\partial[C]}{\partial t} = v\left([C]_{\text{upstream}} - [C]_{\text{downstream}}\right) = -v\frac{\partial[C]}{\partial z} \quad (16.1)$$

$$\frac{\partial[C]}{\partial t} = -D\frac{\partial^2[C]}{\partial z^2} \quad (16.2)$$

where C is the concentration (mol/L) of the antibody, v the fluid velocity (mm/s), z the distance, and D the diffusion coefficient (cm^2/s).

Elevated interstitial pressure within tumors limits convection in the tissue. Note that this does not raise interstitial pressure above the blood vessel pressure, since these vessels are the source of interstitial fluid and the higher pressure [9]. Blood-borne molecules do not have to travel "upstream" against convection. However, this elevated interstitial pressure reduces flow from the vessel into the interstitium. Therefore, they lose the benefit of flowing "downstream." In the absence of significant convection to transport molecules in the tissue, diffusion is the dominant mechanism. The rate of diffusion is dependent on the molecular weight of a compound, so larger molecules such as antibodies are slow to diffuse within tumor tissue. The movement of molecules by convection is proportional to the distance ($1/\partial z$) while, for diffusion, it is proportional to the distance squared ($1/\partial z^2$). Diffusion is therefore inefficient at transporting molecules long distances. However, due to the poor blood flow, incomplete vascularization, shunting, and necrosis within tumors, molecules must diffuse long distances to reach all of the cells. While most cells lie within one or two cell diameters of a capillary in healthy tissue (10–20 μm), diffusion distances of 100 μm or more exist in tumors [12].

As molecules diffuse through the tissue, they encounter a very heterogeneous tumor microenvironment. Depending on the grade and type of tumor, cells may be organized into sheets or consist of a disorganized mixture of cell types. Additionally, endothelial cells, pericytes, tumor-associated macrophages (TAMs), carcinoma-associated fibroblasts (CAFs), and other inflammatory cells reside within the tissue [7]. The consistency can vary from diffuse cells in a pleural effusion to hard desmoplastic tumors. Much of this is driven by the variability in the extracellular matrix of the tumor. This is constantly being reorganized by matrix metalloproteinases (MMPs) that can promote metastasis. The irregularity in oxygen and nutrient supply, diversity of genetic clones, and different cell types result in heterogeneous cell packing throughout the tumor.

Decades of research have been conducted studying the mechanisms, progression, and variability within the tumor microenvironment. In the face of such complexity, a single drug or antibody cannot be designed to independently address all these challenges. However, there are a few dominant rates that determine the majority of antibody behavior in tumors. Tools exist to engineer molecules that overcome the obstacles inherent in these rates, and many molecules have demonstrated promising outcomes in the clinic.

16.2.2 Tumor Transport—An Overview

In its simplest form, tumor pharmacokinetics can be described as a competition between delivery and clearance. Delivery occurs in four major steps: blood flow, extravasation, interstitial transport, and binding (Fig. 16.1). Concurrently, drugs are being cleared from the blood and locally within the tissue. Fast delivery and slow clearance result in a high tumoral concentration, and the reverse is true for slow delivery and

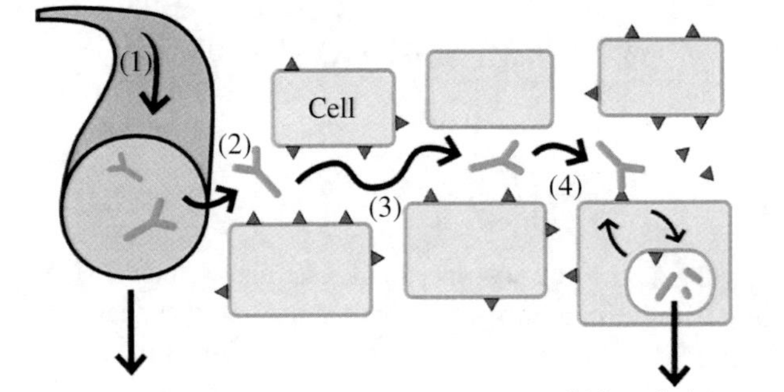

FIGURE 16.1 Steps in systemic targeting—all agents entering a tumor must (1) flow through the local blood vessels, (2) extravasate into the tissue, (3) diffuse through the interstitium, and (4) bind to their target. As these steps are occurring, the molecules are cleared (a) from the systemic circulation and (b) locally within the tissue.

fast clearance. Distinct from the total amount of antibody in the tumor is the distribution, which is a function of diffusion and binding. Since most antibodies bind quickly to their target relative to diffusion, these molecules are immobilized as soon as they exit a blood vessel, trapping them in a perivascular distribution.

Delivery for larger tumors is almost exclusively from tumor blood vessels. (Smaller micrometastases can have significant uptake due to diffusion from their surface [13].) There are two important mechanisms of clearance for antibodies, however. The first is systemic clearance, or removal of the macromolecular drug from the plasma. If the plasma concentration drops below the concentration in the tissue, molecules will stop diffusing from the plasma into the tumor and reverse direction. While the bound antibody will remain immobilized to its target, no more free antibody will exist to diffuse deeper into the tissue and bind more cells. This rapid clearance is often seen for small proteins and peptides that are quickly filtered by the kidneys. Antibodies, however, have some of the longest plasma half lives of any molecule [14]. With multiple doses, a continuous concentration of drug is present in the blood. This is where local clearance becomes dominant. Antibodies typically bind cell surface proteins that are internalized, and the antibody is degraded. If the rate at which antibodies are internalized and degraded equals the rate at which the antibodies exit the blood vessel, no free drug will remain intact to target cells deeper in the tissue.

The specific properties of an antibody determine the exact behavior in the tissue, and a more detailed understanding of the mechanisms provides ways to overcome some of these delivery limitations.

16.2.3 Mechanisms of Tumor Transport

16.2.3.1 Blood Flow versus Extravasation All systemic molecules that target a tumor must undergo four major steps in transport: blood flow to the tissue, extravasation outside the vessel, diffusion through the tissue, and binding at the site of action. It is the relative rates of these steps, all within the tumor, that determine uptake. This is sometimes a point of confusion, since the rates are often compared to healthy tissue instead of with each other. For example, although blood flow in tumors is slow (compared to healthy tissue) and macromolecular permeability is high (compared to healthy tissue), the flow of antibodies in tumor vessels is much faster than the rate at which they extravasate. The ratio of these two rates, defined as the vessel depletion number, can be used to quantify this result:

$$\delta \equiv \frac{PS/V}{Q(1-H)} \qquad (16.3)$$

where P is the vessel permeability, S/V the blood vessel surface area per tumor volume, and their product determines the rate of extravasation. Q is the volumetric blood flow (volume of whole blood per volume of tissue per second) and H the hematocrit (since protein therapeutics are generally excluded from blood cells). A list of typical parameter values can be found in Table 16.1. Depicting the vessels as idealized cylinders within the tissue, this ratio can also be expressed on a microscopic (single vessel) scale:

$$\delta = \frac{2PL}{v(1-H)R_{cap}} \qquad (16.4)$$

where L is the characteristic length of the vessel, v the blood velocity, and R_{cap} the capillary radius.

The vessel depletion number can be used to calculate the amount of drug that exits the blood vessels on a single pass through a tumor. This is also known as the extraction fraction. For antibodies in tumors, this value is close to zero. So little of the antibody leaks out into the tumor that the plasma concentration exiting the tumor is not measurably different than the entering concentration [28, 29]. The equation describing the extraction fraction is [30]

$$E = 1 - \exp(-\delta) \qquad (16.5)$$

Because of the low extraction fraction, the concentration within tumor vessels does not change significantly with changes in blood flow. Therefore, the variable and low blood flow within the tumor does not have a significant effect on total tumor uptake. The surface area of functional vessels does.

16.2.3.2 Extravasation versus Diffusion The permeability of blood vessels has been modeled using several different forms including meshes, one, and two pore models [1]. Here, the analysis is based on an effective permeability from experimental measurements, so the actual mechanism (convection vs diffusion across the capillary endothelium) does not impact the results.

TABLE 16.1 Typical Parameters for Antibodies

Symbol	Parameter	Typical Value	References
P	Permeability	3×10^{-3} μm/s	[15]
S/V	Vessel surface area per tumor volume	$60\,cm^2/cm^3$	[16]
Q	Blood flow per tumor volume	$0.1\,min^{-1}$	[17]
H	Hematocrit	0.45	[18]
R_{cap}	Capillary radius	10 μm	[19]
R	Half intercapillary distance	60 μm	[12]
ε	Void fraction	0.2	[20]
D	Diffusion coefficient	$10\,\mu m^2/s$	[21]
k_{on}	Binding rate	$10^5\,M^{-1}\,s^{-1}$	[22]
k_{off}	Dissociation rate	$1\times10^{-4}\,s^{-1}$	
K_d	Dissociation constant (k_{off}/k_{on})	1 nM	[23]
$[T]$	Target density (receptors per cell multiplied by cells/mL)	800 nM	[24, 25]
$[Ab]_0$	Initial plasma concentration	5 mg/kg or ~500 nM	[26]
A	Fraction of alpha clearance	0.66	[19]
k_α	Alpha phase clearance	$1.6\times10^{-4}\,s^{-1}$ (1.2 h half-life)	
k_β	Beta phase clearance	$1.2\times10^{-6}\,s^{-1}$ (160 h half-life)	
k_e	Internalization rate	$1.3\times10^{-5}\,s^{-1}$ (15 h half-life)	[27]

After molecules extravasate into the interstitial space around the vessel, they diffuse deeper into the tissue. The ratio between these two rates determines the concentration in the tissue relative to the blood. If a molecule has very high permeability relative to diffusion, it will rapidly equilibrate across the blood vessel wall, and the tissue concentration outside the vessel will approach the plasma concentration. If molecules quickly diffuse away from the vessel after slow extravasation, the tissue concentration can be much lower than that in the blood. The ratio of these rates is known as a mass transfer Biot number [31] (named after a French physicist):

$$Bi \equiv \frac{2PR_{cap}}{\varepsilon D} \tag{16.6}$$

where ε is the tumor void fraction (interstitial volume divided by the tissue volume).

The drop in concentration from the plasma to just outside the blood vessel can be estimated using the Biot number. At subsaturating concentrations, the concentration right outside the vessel, $[C]_{tissue}$, relative to the plasma concentration, $[C]_{plasma}$, is approximately

$$\frac{[C]_{tissue}}{[C]_{plasma}} \approx \frac{1}{1+\left(1/Bi\right)} \tag{16.7}$$

If the permeability is very high (e.g., many small molecule drugs), the Biot number is large, and the tissue concentration is similar to the plasma concentration. For Biot numbers much less than 1, the ratio between the two concentrations approaches the value of the Biot number itself.

A typical Biot number is approximately 0.02 for a monoclonal antibody. This means that the average interstitial concentration is approximately 50 times lower than the plasma concentration. This phenomenon is often masked by high binding within the tissue. Although the free concentration in the tissue is low, the bound concentration can be very high. The large concentration difference between the plasma and interstitium may be surprising compared to small molecules that rapidly extravasate, but this high ratio is often used for intravital microscopy experiments where macromolecules delineate the vasculature simply based on slow extravasation into the tissue [32]. The low interstitial concentration relative to the plasma is a major reason why micromolar plasma concentrations are required *in vivo* while only nanomolar concentrations are required to saturate cells *in vitro*.

16.2.3.3 Diffusion versus Binding Antibodies, due to their large size, diffuse through the interstitial spaces between cells. The diffusion rate is slower than the diffusion rate in water for several reasons. The extracellular matrix proteins can impede the free movement of the molecules, the molecules are excluded from cells so they have less area to diffuse, and they must go around the cells rather than in a straight path (tortuosity). All these factors contribute to an effective diffusion coefficient that is several-fold lower than the diffusion coefficient in water. However, experimental measurements have demonstrated that the transport in tissue can be reasonably represented by Fickian diffusion with an effective diffusion coefficient of approximately $10\,\mu m^2/s$ [33–35].

Despite the diffusion of macromolecules being "slow" relative to small molecules, absolute terms can again be misleading. A nonbinding antibody can diffuse 100 μm—far enough to reach most cells in a tumor—within minutes. The heterogeneous delivery seen within most tissues is due to the relative rates of diffusion versus binding. This can slow down the penetration rate by many orders of magnitude and, as discussed below, may prevent penetration indefinitely.

If a molecule is immobilized (e.g., by target binding) at a rate much faster than the rate of diffusion, that molecule will be trapped just outside the vessel resulting in a perivascular distribution. If the molecule diffuses through the tissue at a rate much faster than it is immobilized, the distribution will be much more even throughout the tissue. The ratio between these two rates, binding and diffusion, is called the Damköhler number (named after a German chemist):

$$Da \equiv \frac{k_{on}[T]R^2}{\varepsilon D} \tag{16.8}$$

where k_{on} is the binding/association rate constant, $[T]$ the target concentration in the tissue, and R the diffusion distance (typically half the intercapillary distance). For values greater than 1, the protein does not significantly reach the distance R within the tissue.

The reversible nature of binding results in four modes of diffusion with binding depending on the association and dissociation rates of a molecule (Fig. 16.2a). Most antibodies fall within the "fast-on, slow-off" category due to their high affinity and high binding-site density. High affinity antibodies bind the first antigen they encounter in the tissue and are effectively irreversibly bound. The target antigen will be internalized before the antibody would dissociate from the surface. The only way antibodies penetrate deeper into the tissue is if the first layer of cells is saturated, then more antibodies can diffuse to the next cell layer. The antibodies travel as a saturation front moving through the tissue. Although antigens exist throughout the tumor, the antibodies are immobilized only around vessels (Fig. 16.2b).

Low affinity antibodies are able to dissociate (before internalization) and diffuse deeper into the tissue [37–39]. Slow-on and slow-off molecules are able to diffuse evenly in the tissue before binding, resulting in a more homogeneous delivery. This includes macromolecules [13] and small molecule drugs [36] with slow cellular uptake rates. Finally, molecules that bind slowly and dissociate quickly have very low affinities and are essentially nonbinding molecules.

The heterogeneous distribution of antibodies in tumors is often referred to as a "binding-site barrier [40]." Unfortunately, this term is often misinterpreted to mean a physical barrier. In fact, it is a dynamic and moving "front" in the tissue, and the distance this front travels is dependent on multiple factors including the dose, target density, internalization rate of the target, and time after injection [41]. It is also not unique to macromolecules, since small molecules that show rapid tissue binding relative to diffusion also show this binding-site front in tumors [36].

16.2.3.4 Rate-Limiting Step in Uptake Of the four steps that all molecules undergo to localize in tissue, often one is much slower than the others, and this limits uptake in the tissue. Given that blood flow is faster than extravasation for antibodies (low vessel depletion number), diffusion is faster than extravasation (low Biot number), and binding is faster than diffusion (large Damköhler number), extravasation is the slowest rate in uptake. This limits the amount of antibody reaching the tissue, and it is the ratio of extravasation (uptake) versus clearance that ultimately determines the overall tissue concentration.

16.2.4 Revisiting Tumor Transport Theory

Now that the steps in uptake have been discussed in more detail, the uptake and clearance rates can be quantified to accurately predict the penetration depth of an antibody into tumor tissue. The two limitations discussed in the overview were: (i) systemic clearance faster than uptake and (ii) local clearance faster than uptake.

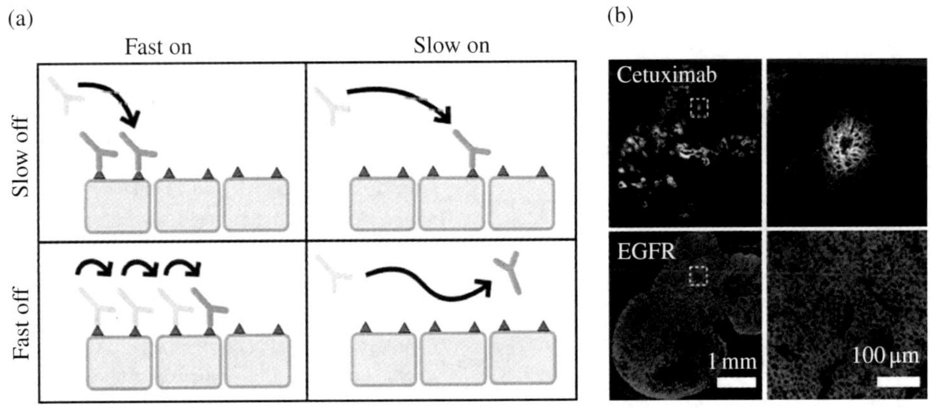

FIGURE 16.2 Tumoral distribution—there are four modes of binding and diffusion in tissue depending on the binding rate and dissociation (off) rate. (a) Due to high target concentrations and association rates relative to diffusion, most antibodies are "fast-on, slow-off" molecules (large Damköhler number). Some molecules, owing to faster diffusion or slower binding, are "slow-on, slow-off" molecules with a small Damköhler number. The reversible nature of binding yields two additional possibilities. If the molecule can completely dissociate quickly, it can diffuse deeper and more homogeneously in the tissue (fast-on, fast-off). Molecules that bind slowly and dissociate rapidly behave as nonbinding agents. (b) Whole tumor (left) and magnification of box (right) showing fluorescent cetuximab in an A431 tumor [36]. *Ex vivo* staining with a noncompetitive anti-EGFR antibody shows that antigen exists throughout the tumor (except in some large areas of necrosis). An isolated capillary (box) shows how the antibody only penetrates a couple of cell layers.

To determine the amount of time it takes to saturate a target within the tissue, the total number of targets (antigen molecules) is divided by the rate at which antibodies are delivered. Considering an infusion (with constant plasma concentration), the saturation time is

$$t_{sat} = \frac{[T]}{[Ab]_0 \, PS/V} \tag{16.9}$$

where $[T]$ is the target concentration and $[Ab]$ the antibody concentration in the plasma.

Because diffusion is much faster than extravasation (i.e., low Biot number), the exact geometry of the vessels does not have a major impact on the results. If we assume an "idealized" Krogh cylinder for vessels (a cylindrical blood vessel surrounded by a layer of tissue) within the tumor:

$$t_{sat} = \frac{R^2[T]}{2PR_{cap}[Ab]_0} \tag{16.10}$$

High affinity antibodies travel as a saturation front moving through the tissue, and if the free molecules are cleared from the plasma faster than the saturation front can reach all cells, this front will stall without any more unbound antibody to bind new target (Fig. 16.3, left). The ratio of uptake to plasma clearance defines the clearance modulus [19]:

$$\Gamma \equiv \frac{R^2[T]}{2PR_{cap}[Ab]_0\left(\dfrac{A}{k_\alpha}+\dfrac{B}{k_\beta}\right)} \tag{16.11}$$

where R is the penetration distance, $[Ab]_0$ the initial antibody concentration in the plasma, A the fraction alpha phase, k_α the alpha-phase clearance rate, B the fraction of beta-phase clearance, and k_β the beta-phase clearance rate. To convert to half-lives:

$$t_{1/2,\alpha} = \frac{0.693}{k_\alpha} \quad \text{and} \quad t_{1/2,\beta} = \frac{0.693}{k_\beta} \tag{16.12}$$

The clearance modulus is the ratio of the characteristic time to saturate the tumor versus the characteristic time of antibody in the plasma. If the value is greater than 1, antibodies will not reach the distance R. The plasma characteristic time is the clearance half-life if the concentration drops according to an exponential decay, or the weighted clearance half-life for the more commonly observed biexponential decay:

$$[C]_{plasma} = [C]_{plasma,0}\left(Ae^{-k_\alpha t} + Be^{-k_\beta t}\right) \quad \text{and} \quad A + B = 1 \tag{16.13}$$

Rapid clearance can be problematic for smaller proteins and protein fragments, such as scFvs, Fab, and $F(ab')_2$ fragments [42]. While the molecular weight dependence is not a sharp cutoff, globular proteins approximately less than 60 kDa in size, linear polymers approximately less than 25 kDa, and nanoparticles approximately less than 5.5 nm are rapidly filtered by the kidneys and removed from the plasma [21, 43].

For most monoclonal antibodies, rapid clearance is not an issue. These molecules are some of the longest circulating proteins in the body due to their stability and large size to avoid renal filtration. Importantly, not only do the molecular properties determine the long half-life in the blood, but the FcRn recycling system reduces nonspecific metabolism in many body tissues [44]. The long half-life combined with multiple doses sustains a high concentration in the blood of treated patients. However, a second mechanism can limit penetration of these molecules.

Local clearance (metabolism within the target tissue) can also limit therapeutic protein distribution within tumors. After antibodies bind their target, they are internalized from the cell surface at variable rates [45, 46]. Although higher order clustering of cell surface receptors can drive rapid internalization [47, 48], bivalent interactions typically do not have this effect. The internalization rate of the bound antibody is often similar to the rate of the antigen itself. This can range from minutes to days depending on the target [27, 49, 50]. Constitutive turnover of the cell membrane results in a moderate internalization half-life of approximately 15 h [27].

Similar to systemic clearance, if the time required to saturate the tissue is longer than the time required for the tissue to degrade the protein, the tissue will never be saturated. The time for degradation is $1/k_e$, where k_e is the net internalization rate for the target. The ratio of saturation time to degradation time is known as the Thiele modulus (named after

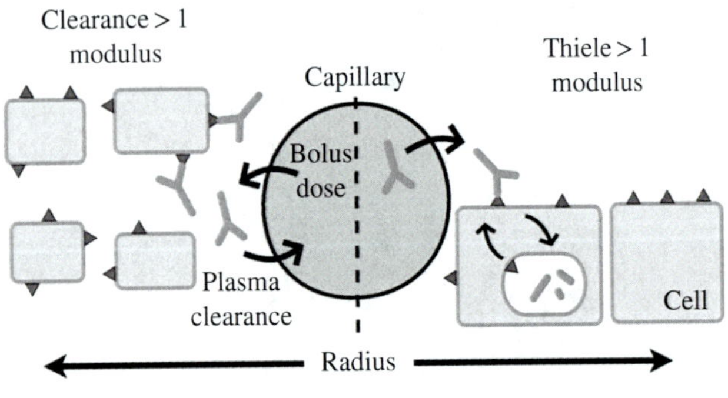

FIGURE 16.3 Competition between uptake and clearance—if an antibody is cleared from the plasma faster than it can penetrate the tissue, no more free antibody remains to target antigen farther from the vessel (left). This can be predicted when the clearance modulus is greater than unity. If antibodies are internalized and degraded in the tissue faster than they are taken up in the tumor, no intact antibody will be able to reach all the cells in the tissue even if dosed continuously (right). This occurs when the Thiele modulus is greater than 1.

E.W. Thiele who described the diffusion of chemicals in a catalyst pellet). The Thiele modulus for a tumor is [19]

$$\phi^2 \equiv \frac{k_e R^2 [T]}{2PR_{cap}[Ab]_0} \tag{16.14}$$

where the $1/k_e$ in the denominator has been moved to the numerator, and the Thiele modulus is written as the square of the value based on historical convention [51]. If the Thiele modulus is greater than 1, antibody will be able to penetrate the tissue to the distance R. If the value is less than 1, the antibody will be consumed by local internalization and degradation before reaching this distance (Fig. 16.3, right). Typical values for these dimensionless numbers are found in Table 16.2.

16.2.5 Impact of Drug Targeting Parameters on Distribution

16.2.5.1 Dose One of the most straightforward approaches to targeting all cells within a tumor is increasing the dose. A larger dose increases the concentration gradient, driving more drugs into the tumor and saturating the target sites. The plasma concentration, $[Ab]_0$, is in the denominator of both the Thiele and clearance moduli, and experimental results with several tumor models validate this prediction [52, 53]. Major tradeoffs include increased cost, dosing frequency, and potentially toxicity.

16.2.5.2 Affinity The ability to readily engineer the affinity of therapeutic proteins raises the question of what the optimal affinity should be. This depends on the particular application, which will be discussed in the next section. Here, the impact on distribution is addressed.

First, for a low affinity antibody to diffuse deeper into the tissue compared to a high affinity antibody, the dissociation rate must be faster than the internalization rate. Otherwise, the antibody will be internalized before it dissociates. Since monoclonal IgG antibodies are bivalent, both arms must dissociate together, which often occurs only for a very weak monovalent affinity. This is one reason why the vast majority of antibodies fall into the "fast-on/slow-off" category (Fig. 16.2). Finally, the dissociation constant, K_d, must be larger than the local free antibody concentration. If not, the high concentration will cause saturation of the receptor, and the distribution will be similar to a high affinity antibody [54].

If all the above conditions are met, the antibody may distribute more evenly throughout the tissue. The diffusion rate and binding kinetics determine how the antibody "jumps" through the tissue for a "fast-on/fast-off" antibody (Fig. 16.2). Conceptually, the diffusion coefficient (D) versus the binding rate ($k_{on}[T]/\varepsilon$) determines the distance of each "jump," and the dissociation rate (k_{off}) determines the number of "jumps" in a given time. Therefore, the diffusion distance for a low affinity molecule depends on the K_d ($K_d = k_{off}/k_{on}$). To diffuse a given distance R before internalization, the following version of the Thiele modulus must be less than 1:

$$\varphi^2_{\text{low affinity}} \equiv \frac{k_e R^2 \left([T]/\varepsilon \right)}{D[K_d]} \tag{16.15}$$

To diffuse this distance before the molecule is cleared from the plasma, an alternate version of the clearance modulus is used:

$$\Gamma_{\text{low affinity}} \equiv \frac{R^2 \left([T]/\varepsilon \right)}{D[K_d]\left(\frac{A}{k_\alpha} + \frac{B}{k_\beta} \right)} \tag{16.16}$$

In practice, the effective K_d often needs to be similar to the target concentration to yield homogeneous distribution in the tumor, and this is seen experimentally [37, 38].

One critical point must be mentioned about the impact of affinity. Although lower affinity antibodies have been shown to distribute more homogeneously throughout tumors, this does not increase the concentration in the tumor. In fact, it is often reduced for lower affinity antibodies. It only changes the *distribution* of the antibodies that do reach the tumor. For example, if a high affinity antibody binds 100% of the targets on 10% of the

TABLE 16.2 Sample Calculations for Dimensionless Numbers

Uptake Ratios			Uptake versus Clearance	
Vessel depletion number Extravasation versus blood flow	Biot number Extravasation versus diffusion	Damköhler number Binding versus diffusion	Clearance modulus Plasma clearance versus uptake	Thiele modulus Local clearance versus uptake
$\delta \equiv \frac{PS/V}{Q(1-H)}$	$Bi \equiv \frac{2PR_{cap}}{\varepsilon D}$	$Da \equiv \frac{k_{on}[T]R^2}{\varepsilon D}$	$\Gamma \equiv \frac{R^2[T]}{2PR_{cap}[Ab]_0\left(\frac{A}{k_\alpha}+\frac{B}{k_\beta}\right)}$	$\varphi^2 \equiv \frac{k_e R^2[T]}{2PR_{cap}[Ab]_0}$
$=0.02$	$=0.03$	$=144$	$=0.3$	$=1.3$

Values from Table 16.1 are used.

cells adjacent to the vessel, an equivalent dose of a low affinity antibody would bind 10% of the target on 100% of the cells.

16.2.5.3 Molecular Weight The size of a targeting protein has multiple effects on the distribution. One of the most important was already mentioned in reference to the molecular weight: larger proteins are generally cleared slower from the blood. Smaller protein scaffolds have been investigated for increased penetration into tissue [55]. They have similar binding kinetics but larger diffusion coefficients, thereby decreasing the Damköhler number. While this does increase the penetration depth, often the penalty of faster clearance outweighs any potential benefit. An insightful analysis of targeting molecules across a wide size range indicated that scFvs have the worst combination of properties [21]. They are small enough to be cleared rapidly by the kidneys but large enough that they are taken up slowly by the tumor. Either larger or smaller protein scaffolds have more efficient uptake.

16.2.6 Experimental Validation and Comparison with Small Molecules

The impact of targeting parameters on distribution has been quantified in tumor spheroids where the environment can be more precisely controlled [39, 56, 57]. Lower affinity molecules distribute more homogeneously but at a lower level of saturation. Slowing down internalization, either through modifying the temperature or selecting a more slowly internalizing target, achieves higher penetration depths.

Many xenograft experiments by several laboratories have been used to validate the theoretical predictions. These include radiolabeled [52], fluorescently labeled [53], and *ex vivo*-stained antibodies [58]. In these experiments, it is important to differentiate the antibody from the label. Fluorophores and radiolabels can be residualizing (trapped within the cells) even after the intact antibody has been degraded. The secondary labeling method (binding intact antibody *ex vivo*) even shows a "receding" antibody binding front [58] whereas residualizing labels get trapped at the farthest distance they penetrate.

Clinical data on the distribution of antibodies are more difficult to obtain, but the heterogeneous distribution has also been confirmed within these tumors. Microscopic models of antibody distribution [59] can be integrated to yield the bulk tissue concentration, and these values also agree with data collected in the clinic [60].

Given the vast experience and literature on small molecule distribution, it is prudent to compare the two classes of drugs. First, the extravasation rate of small molecules is much higher than macromolecules (often 100- to 1000-fold faster). This can sometimes deplete the concentration of small molecules along the length of capillaries (high extraction fraction), reducing uptake in tumors due to poor blood flow. The rapid extravasation equilibrates the concentration across the blood vessel wall (large Biot number). The plasma concentration is therefore often more relevant to the tissue concentration for small molecules than macromolecules. Nonspecific mechanisms of uptake (e.g., protein binding and uptake within lipid membranes) often dominate over specific target interaction. This leads to distribution based on physicochemical properties (such as the $\log P$) more than specific target interactions. This is another reason why target-mediated drug distribution can be more prevalent with macromolecules [61]. The specific target interactions are often equal or larger in magnitude than nonspecific uptake, resulting in a greater impact on overall systemic distribution.

Macromolecules reside within the extracellular space, and this accounts for the parameter epsilon, which is required to convert the overall tissue concentration into an "effective" concentration in the interstitium (Fig. 16.4). This is analogous to R values used to convert tissue concentrations to "effective" free drug concentrations [62]. The reversible nature of these protein and lipid interactions allows the small molecule to eventually reach most of the tissue. Some notable exceptions include molecules with high target densities including DNA and microtubules [63, 64]. These drugs often show strong gradients in the tissue for the same reasons as macromolecules. Finally, metabolism of small molecule drugs often predominates in the liver. If there is no local metabolism (Thiele modulus ~ 0) and the drug is dosed continuously (clearance modulus ~ 0), the small molecule drug can efficiently reach most cells in the tumor. This can be impacted by drug transporters, but these are beyond the scope of this chapter.

16.3 IMPACT OF TUMOR PHARMACOKINETICS ON EFFICACY

The impact of tumor pharmacokinetics on the efficacy of therapeutic proteins is heavily dependent on the mechanism of cell killing. Therefore, this section starts with a brief overview of the major therapeutic mechanisms (both clinical and experimental).

16.3.1 Overview of Cell-Killing Mechanisms

Multiple regions of an antibody can exert therapeutic effects, and often clinical antibodies utilize several mechanisms [65]. Starting with the variable region, binding to a specific receptor can produce several effects. First, this may induce specific signal blockade as in the case of cetuximab and trastuzumab. These agents prevent binding of growth factors or receptor dimerization, thereby shutting down the growth signals of these pathways. Downregulation itself can reduce the number of receptors available for signaling. Finally, binding may induce a signal that sensitizes a cell to other therapies or directly induces apoptosis, a mechanism studied with rituximab [66].

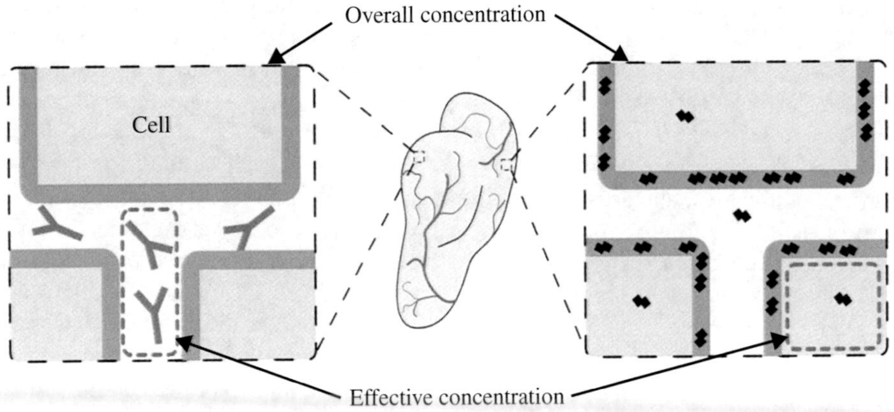

Antibody			Small molecule		
$\varepsilon = 0.1$			$R = 50$		
Overall concentration	Antibody/ tissue volume	1 nM	Overall concentration	Total drug/ tissue volume	100 nM
Effective concentration	Antibody/ interstitial volume	10 nM	Effective concentration	Free drug/ tissue volume	2 nM

FIGURE 16.4 Void fraction—the void fraction for a macromolecule is required in many calculations to compensate for the fact that the molecule transports through the extracellular space. The void fraction adjusts the overall concentration in the tissue to an effective concentration in the interstitial space (left). This is analogous to *R* values used for small molecule drugs. Molecules bound to proteins or in membranes are excluded to adjust the overall tissue concentration to the "effective" free drug concentration able to engage its target (right).

The Fc region and other secondary binding sites (in the case of bispecific antibodies) are responsible for multiple effects. First, they can recruit immune cells bearing Fc gamma receptors that then kill the cell through antibody-dependent cell-mediated cytotoxicity (ADCC) [67, 68]. The Fc region can also recruit complement factors (depending on the particular subclass) resulting in complement-dependent cytotoxicity (CDC) [69, 70]. The multitude of possible bispecific antibodies can work in a similar manner, recruiting additional proteins, cells, or more efficiently downregulating receptors [71, 72].

Antibodies can also carry toxic payloads including antibody–drug conjugates (ADCs), immunotoxins (ITs), and radioimmunotherapy (RIT). ADCs are currently available for solid tumors [73] and lymphomas [74] with many others in the pharmaceutical pipeline. These agents typically must be internalized and degraded for release of the small molecule drug, which then diffuses to its site of action within the cell. Immunotoxins are under active investigation, and these molecules often have intracellular targets that are usually reached after internalization and escape to the cytosol [75].

RIT is only clinically available for leukemias and lymphomas. Several types of radioactive isotopes have been investigated for therapy. Current clinically approved RIT drugs use beta emitters [76] that have fairly long path lengths from the site of decay (~5 mm), while other experimental isotopes have much shorter path lengths (e.g., alpha particles with ~100 μm path length or Auger electron emitters <1 μm) [77]. The path length of the isotope has a direct impact on the relevance of heterogeneity within the tumor.

Finally, some antibodies have nontoxic or indirect mechanisms of action. This includes bevacizumab, which binds the soluble ligand VEGF, preventing it from interacting with its receptor. This can "normalize" the tumor vasculature, affecting the local physiology and delivery of drugs. Other antibodies have been found to cause this same effect indirectly [78].

16.3.2 Pharmacokinetic Impact on Efficacy

16.3.2.1 Dose Increasing the dose will increase the penetration depth of a high affinity antibody until all cells are targeted and there is no heterogeneity due to binding [52, 53]. For signaling blockage of receptors, often a saturating dose is required for efficacy [79]. However, the doses required for saturation are often very high. Clinical concentrations of antibodies in the plasma are often in the micromolar range [80] even though 1–10 nM may be sufficient to saturate the receptors *in vitro*. The need for such a high dose is illustrated by the clearance modulus and Thiele modulus [19, 53]. The slow extravasation of antibodies compared to plasma and/or local clearance requires a large dose to drive a sufficient number of molecules into the tumor for saturation [81, 82]. Doses significantly above saturation are likely detrimental to efficacy, however, with little to no therapeutic benefit and the potential for increased toxicity.

*16.3.2.2 **Affinity*** Lower affinity antibodies can distribute more evenly in tissue at subsaturating doses, albeit at lower fractional coverage of surface receptors [37–39]. Due to the bivalent nature of antibodies, this often requires very low monovalent affinity [83]. The impact on efficacy depends on the mechanism of cell death. In principle, a highly toxic compound that does not require targeting of a million receptors per cell, for example, would be more effective if that same dose were spread more evenly throughout the tissue. A high affinity antibody would deliver a much higher dose than necessary to cells adjacent to blood vessels (cellular overkill) while completely avoiding cells farther away. The tumor would then have to be treated for a longer time (to kill cells layer by layer) or with a higher dose (and higher corresponding toxicities).

In practice, lower affinity improving efficacy is uncommon for several reasons. First, receptor targeting often requires saturating doses as discussed above, in which case there is no binding-site front. For ADCC, CDC, and cell-signaling blockade mechanisms, low affinity reduces the efficacy of these molecules [23, 84] once they reach the site of action, reducing any benefit of more uniform delivery. For ADCs, highly toxic compounds must be used to obtain efficient cell killing [85], so it is unknown whether there is sufficient "overkill" in cells near the vessels. One important point of consideration is off-target toxicity. If significant shed complexes form in the blood and are taken up in the liver, a moderate affinity antibody (with a high avidity for the target cell surface but low monovalent affinity for shed antigen) may strike a balance between avoiding liver toxicity while maintaining tumor targeting [86].

In RIT, isotopes that spread their cytotoxic ionizations over a path length of millimeters would not be strongly impacted by heterogeneity on the hundreds of microns length scale. For alpha emitters and Auger electron emitters, however, a more uniform distribution is required for radiation to reach all cells. However, both cold (unlabeled) and hot (radiolabeled) antibodies contribute to the distribution profile, complicating the picture. A large amount of "cold" dosing will enhance distribution of the radiolabeled molecules by occupying some of the binding sites. In summary, while there is a theoretical possibility of more efficient cell killing with lower affinity targeting, a lower potency often results in a negligible or negative impact at the target site.

*16.3.2.3 **Internalization Rate*** The impact of target internalization on efficacy is highly dependent on the mechanism of cell killing. Slow internalization (or efficient recycling [49]) allows deeper penetration due to a lower Thiele modulus, but many ADCs must be internalized to deliver their payload to the site of action. The same is true for Auger electron emitters in RIT, which must decay next to the nucleus to damage DNA. For cell-signaling blockade, internalization can often drive downregulation of the receptor, enhancing efficacy. However, if the antibody dissociates and the receptor is efficiently recycled, internalization could be detrimental. Ideally, internalization would be just fast enough to achieve cellular toxicity.

Several mechanisms require the antibody to remain on the cell surface to engage other cells or molecules. These include ADCC, CDC, some bispecific applications, and pretargeting approaches [87, 88]. Here, a much slower rate of internalization would allow more time to engage Fc gamma receptors, complement factors, or other bispecific targets. Some ADCs with labile linkers have also been tested that do not require internalization for drug release [89], although the released molecules must then enter cells.

For antibodies with toxic payloads, the fate of the drug or radioisotope is also important. Some molecules are trapped within cells for extensive periods of time (residualizing drugs/probes/isotopes) while others leak out of the cell quickly. For radioisotopes, many metals such as indium are trapped for extended periods of time while iodine rapidly leaks from the cells [90]. In principle, small molecule drugs from ADCs can also diffuse to nearby cells. However, the targets of many of these drugs are present at high concentrations, such as microtubule-binding sites or DNA. The diffusion of these small molecules must also overcome the obstacle of a binding-site front in the tissue.

*16.3.2.4 **Shedding*** Like internalization, shedding can have both positive and negative impacts depending on the particular system being investigated. Many times, shed antigen has a negative impact on efficacy. For disrupting cell signaling, inducing ADCC or CDC, or delivery of intracellular payloads [91], binding to antigen that has detached from the cell surface cannot accomplish the desired impact but raises the necessary dose for saturation.

Underscoring the need to monitor cellular processing of antibodies under development, shed antigen appears to help in some instances [92]. If the internalization rate is sufficiently high to prevent penetration into the tissue (large Thiele modulus), the shed antigen can "rescue" the antibody from degradation. The shed antigen–antibody complex can then penetrate deeper into the tissue and target cells upon dissociation.

*16.3.2.5 **Stability*** Under most circumstances, it is desired that biologics are stable systemically and in the tumor microenvironment to allow adequate targeting. This reduces both plasma clearance and local metabolism in the clearance modulus and Thiele modulus. Stability can be compromised by the inflammatory environment of tumors with TAMs engulfing molecules nonspecifically [13, 36] and a variety of extracellular proteases degrading macromolecules in the interstitium. One of the most well-studied families is the MMPs. These proteins are secreted to degrade extracellular matrix and release growth factors in both wounding healing and tumorigenesis [93].

For ADCs, stability is important up to the point of cellular targeting, when the ADC can be degraded intracellularly to deliver the payload. A problem with labile linkers is the possibility of accumulating significant unconjugated antibody in the blood after deconjugation [94]. If there is a mixture of unconjugated and drug-loaded antibody in the blood and the total antibody dose is saturating, it is expected that the tumor cell surface will be targeted with an equivalent ratio as to what is in the plasma. For example, if the plasma has 30% unconjugated antibody and 70% ADC, then the cell surface would likely have 70% ADC and 30% unconjugated antibody, all other things being equal. If the total dose is subsaturating, the Krogh cylinder model predicts that the same amount of ADC will reach the tumor, but the distribution will be spread over more cells as the unconjugated antibody occupies a fraction of the binding sites.

In several instances, investigators have taken advantage of labile linkers for efficacy. "Prodrug" formulations of antibodies block the binding-site before cleavage by tumor-associated proteases [95, 96]. This is primarily to reduce off-target binding, but in principle it can be used to obtain more homogeneous distribution if the cleavage kinetics are slower than diffusion. ADCs that typically require internalization have also been targeted to extracellular matrix proteins [89]. Following cleavage of the linker, the small molecule drug is able to enter cells and bind its target. It is important to note that the distribution of the small molecule drug is also subject to the same principles of diffusion versus target binding. Small molecule drugs with high target densities (e.g., microtubules or DNA) often show gradients in tumors [63, 64, 97]. Other small molecules have been shown to distribute more homogeneously if they enter cells slowly since they are able to diffuse farther before engaging their target [36].

16.3.2.6 Plasma Clearance Slower plasma clearance can lower the dose required for saturation according to the clearance modulus, and more stable linkers can reduce clearance of ADCs by deconjugation. Two additional factors should be considered when optimizing plasma clearance for target-site effect: increased whole antibody clearance from modification and toxicity to the clearance organs. Particularly for ADCs, an increase in the number of drugs per antibody can alter the physicochemical properties of the molecule (such as surface charge [98]) and increase clearance [99]. This effect is not unique to drugs but occurs with other small molecule modifications and fluorophores.

Excretion is critical for these molecules, particularly ADCs and radiolabeled antibodies, due to toxicity. Overlabeling antibodies can increase clearance by the liver, concentrating the drug in a single organ relative to the more diffuse degradation of unlabeled antibodies. Detoxification is an important step in minimizing off-target activity [100]. RIT is particularly problematic, since much of the dose never reaches the tumor, and decay in the periphery can be very toxic. Smaller, more rapidly cleared agents are filtered by the kidney but taken up in the proximal tubule by several mechanisms including degradation/amino acid transporters [101] and scavengers of peptides and larger proteins [102].

16.3.2.7 Modifying Tumor Physiology The poor drug transport properties in tumors have led investigators to modify the tumor microenvironment to improve drug delivery. One of the most well studied mechanisms arose from antiangiogenic treatments targeted at VEGF [103]. Originally developed to block the formation of new tumor blood vessels, these therapies also significantly modify the tumor vasculature, reducing macromolecular permeability, lowering interstitial pressure, increasing blood flow, and increasing oxygenation [11, 104]. These effects have far-reaching implications; for example, higher oxygen levels can increase drug sensitivity [105]. Many small molecule drugs are limited by blood flow to the tumor (high extraction fraction), and the increased blood flow delivery provides a synergistic effect. It is important to note that it reduces macromolecular permeability, and since this is the rate-limiting step in antibody targeting, it lowers uptake. While the percentage of functional vessels within the tumor can increase due to pruning of nonfunctional capillaries, the total functional surface area is time dependent and also decreases, lowering delivery [106, 107]. Reduced delivery of macromolecules does not automatically result in lower efficacy (since competing effects could include increased sensitivity of the cells due to higher oxygenation). However, this antagonism is important when considering drug combinations.

In contrast to reduced permeability of anti-VEGF therapy, other investigators have increased the vascular permeability [108, 109]. This strategy can improve the delivery of molecules by increasing the rate-limiting step of antibody targeting. An optimal balance must be achieved, however, since large increases in vascular permeability can induce vessel collapse [110]. While this has its own therapeutic benefit, it would antagonize the delivery of therapeutic antibodies.

16.4 CONCLUSIONS

Tumor targeting with protein therapeutics is a complicated multistep process occurring in the presence of large physiological variability within the tissue. By examining the fundamental rates in uptake, however, the distribution within tumors can be quantified, and the effect of modifications to dose, molecular weight, binding affinity, and other parameters can be predicted with reasonable accuracy. These factors must then be combined with the impact on pharmacodynamics for a complete view of efficacy within the tumor microenvironment.

The effect of distribution on efficacy is dependent on the particular system being studied, so no universal answer works in every case. Rather than striving to always increase affinity, improve stability, slow plasma clearance, and use the maximum tolerated dose, a look at the therapeutic mechanism of action and the biology of the target can help elucidate the best targeting strategy. The quantitative parameters outlined in the first section can provide some insight into the tradeoffs of each parameter, helping to design experiments and interpret *in vivo* results. Ultimately, these considerations can help focus design efforts and improve the efficiency of development of these complex molecules.

REFERENCES

[1] Michel CC, Curry FE. Microvascular permeability. Physiol Rev 1999;79 (3):703–761.

[2] Jain RK. Transport of molecules, particles, and cells in solid tumors. Annu Rev Biomed Eng 1999;01:241–263.

[3] Jain RK, Baxter L. Mechanisms of heterogeneous distribution of monoclonal antibodies and other macromolecules in tumors: significance of elevated interstitial pressure. Cancer Res 1988;48:7022–7032.

[4] Eggleton CD, Vadapalli A, Roy TK, Popel AS. Calculations of intracapillary oxygen tension distributions in muscle. Math Biosci 2000;167 (2):123–143.

[5] Ahlstrom H, Christofferson R, Lorelius L. Vascularization of the continuous human colonic cancer cell line LS 174T deposited subcutaneously in nude rats. APMIS 1988;96:701–710.

[6] Pries AR, Cornelissen AJM, Sloot AA, Hinkeldey M, Dreher MR, Hopfner M, Dewhirst MW, Secomb TW. Structural adaptation and heterogeneity of normal and tumor microvascular networks. PLoS Comput Biol 2009;5 (5):11.

[7] Bissell MJ, Radisky D. Putting tumours in context. Nat Rev Cancer 2001;1 (1):46–54.

[8] Winkler F, Kozin SV, Tong RT, Chae SS, Booth MF, Garkavtsev I, Xu L, Hicklin DJ, Fukumura D, di Tomaso E, Munn LL, Jain RK. Kinetics of vascular normalization by VEGFR2 blockade governs brain tumor response to radiation: role of oxygenation, angiopoietin-1, and matrix metalloproteinases. Cancer Cell 2004;6 (6):553–563.

[9] Boucher Y, Jain RK. Microvascular pressure is the principal driving force for interstitial hypertension in solid tumors: implications for vascular collapse. Cancer Res 1992;52: 5110–5114.

[10] Baish JW, Netti PA, Jain RK. Transmural coupling of fluid flow in microcirculatory network and interstitium in tumors. Microvasc Res 1997;53 (2):128–141.

[11] Goel S, Duda DG, Xu L, Munn LL, Boucher Y, Fukumura D, Jain RK. Normalization of the vasculature for treatment of cancer and other diseases. Physiol Rev 2011;91 (3):1071–1121.

[12] Baish JW, Gazit Y, Berk DA, Nozue M, Baxter LT, Jain RK. Role of tumor vascular architecture in nutrient and drug delivery: an invasion percolation-based network model. Microvasc Res 1996;51 (3):327–346.

[13] Thurber G, Figueiredo J, Weissleder R. Multicolor fluorescent intravital live microscopy (FILM) for surgical tumor resection in a mouse xenograft model. PLoS One 2009;4 (11):e8053.

[14] Ghetie V, Popov S, Borvak J, Radu C, Matesoi D, Medesan C, Ober RJ, Ward ES. Increasing the serum persistence of an IgG fragment by random mutagenesis. Nat Biotechnol 1997;15 (7):637–640.

[15] Yuan F, Dellian M, Fukumura D, Leunig M, Berk DA, Torchilin VP, Jain RK. Vascular permeability in a human tumor xenograft: molecular size dependence and cutoff size. Cancer Res 1995;55:3752–3756.

[16] Barth PJ, GhafouriSanati H, Kohler HH, Bittinger A, Riedmiller H. Histological determinants of the vascular surface in prostatic carcinoma. Urol Res 1997;25 (5):303–308.

[17] Vaupel P, Kallinowski F, Okunieff P. Blood-flow, oxygen and nutrient supply, and metabolic microenvironment of human-tumors – a review. Cancer Res 1989;49 (23): 6449–6465.

[18] Green E. *Biology of the Laboratory Mouse*. New York: Dover Publications & The Jackson Laboratory; 1966.

[19] Thurber GM, Zajic SC, Wittrup KD. Theoretic criteria for antibody penetration into solid tumors and micrometastases. J Nucl Med 2007;48 (6):995–999.

[20] Krol A, Nagaraj S, Dewhirst M, Yuan F. Available volume fraction of macromolecules in tumor tissues. FASEB J 2000;14 (4):A167.

[21] Schmidt MM, Wittrup KD. A modeling analysis of the effects of molecular size and binding affinity on tumor targeting. Mol Cancer Ther 2009;8 (10):2861.

[22] Schier R, McCall A, Adams GP, Marshall KW, Merritt H, Yim M, Crawford RS, Weiner LM, Marks C, Marks JD. Isolation of picomolar affinity Anti-c-erbB-2 single-chain Fv by molecular evolution of the complementarity determining regions in the center of the antibody binding site. J Mol Biol 1996;263 (4):551–567.

[23] Tang Y, Lou J, Alpaugh RK, Robinson MK, Marks JD, Weiner LM. Regulation of antibody-dependent cellular cytotoxicity by IgG intrinsic and apparent affinity for target antigen. J Immunol 2007;179 (5):2815–2823.

[24] Lyng H, Haraldseth O, Rofstad EK. Measurement of cell density and necrotic fraction in human melanoma xenografts by diffusion weighted magnetic resonance imaging. Magn Reson Med 2000;43 (6):828–836.

[25] Thurber GM, Weissleder R. Quantitating antibody uptake *in vivo*: conditional dependence on antigen expression levels. Mol Imaging Biol 2011;13 (4):623–632.

[26] Kaliss N, Pressman D. Plasma and blood volumes of mouse organs as determined with radioactive iodoproteins. Proc Soc Exp Biol Med 1950;75:16–20.

[27] Schmidt MM, Thurber GM, Wittrup KD. Kinetics of anti-carcinoembryonic antigen antibody internalization: effects

of affinity, bivalency, and stability. Cancer Immunol Immunother 2008;57:1879–1890.

[28] Heijn M, Roberge S, Jain RK. Cellular membrane permeability of anthracyclines does not correlate with their delivery in a tissue-isolated tumor. Cancer Res 1999;59 (17):4458–4463.

[29] Baxter L, Jain RK. Transport of fluid and macromolecules in tumors: 1. Role of interstitial pressure and convection. Microvasc Res 1989;37:77–104.

[30] Tofts PS. Modeling tracer kinetics in dynamic Gd-DTPA MR imaging. J Magn Reson Imaging 1997;7 (1):91–101.

[31] Deen W. *Analysis of Transport Phenomena*. Oxford University Press; 1998. p 597.

[32] Dreher MR, Liu W, Michelich CR, Dewhirst MW, Yuan F, Chilkoti A. Tumor vascular permeability, accumulation, and penetration of macromolecular drug carriers. J Natl Cancer Inst 2006;98 (5):335–344.

[33] Berk DA, Yuan F, Leunig M, Jain RK. Direct *in vivo* measurement of targeted binding in a human tumor xenograft. Proc Natl Acad Sci U S A 1997;94 (5):1785–1790.

[34] Thurber GM, Schmidt MM, Wittrup KD. Antibody tumor penetration: transport opposed by systemic and antigen-mediated clearance. Adv Drug Deliv Rev 2008;60 (12):1421–1434.

[35] Brown EB, Boucher Y, Nasser S, Jain RK. Measurement of macromolecular diffusion coefficients in human tumors. Microvasc Res 2004;67 (3):231–236.

[36] Bhatnagar S, Deschenes E, Liao J, Cilliers C, Thurber GM. Multichannel imaging to quantify four classes of pharmacokinetic distribution in tumors. J Pharm Sci 2014;103 (10): 3276–3286.

[37] Adams G, Schier R, McCall AM, Simmons HH, Horak EM, Alpaugh RK, Marks JD, Weiner LM. High affinity restricts the localization and tumor penetration of single-chain Fv antibody molecules. Cancer Res 2001;61:4750–4755.

[38] Rudnick SI, Lou J, Shaller CC, Tang Y, Klein-Szanto AJ, Weiner LM, Marks JD, Adams GP. Influence of affinity and antigen internalization on the uptake and penetration of anti-HER2 antibodies in solid tumors. Cancer Res 2011;71 (6):2250–2259.

[39] Thurber GM, Wittrup KD. Quantitative spatiotemporal analysis of antibody fragment diffusion and endocytic consumption in tumor spheroids. Cancer Res 2008;68: 3334–3341.

[40] Fujimori K, Covell DG, Fletcher JE, Weinstein JN. A modeling analysis of monoclonal antibody percolation through tumors: a binding-site barrier. J Nucl Med 1990;31 (7):1191–1198.

[41] Wittrup KD, Thurber GM, Schmidt MM, Rhoden JJ. Practical theoretic guidance for the design of tumor-targeting agents. In: Wittrup KD, Verdine GL, editors. *Methods in Enzymology: Protein Engineering for Therapeutics*., Vol. 203, Pt BVolume 503, 2012. Methods in Enzymology; p 255–268.

[42] Wu AM, Senter PD. Arming antibodies: prospects and challenges for immunoconjugates. Nat Biotechnol 2005;23 (9):1137–1146.

[43] Choi HS, Liu W, Liu F, Nasr K, Misra P, Bawendi MG, Frangioni JV. Design considerations for tumour-targeted nanoparticles. Nat Nanotechnol 2010;5 (1):42–47.

[44] Garg A, Balthasar JP. Physiologically-based pharmacokinetic (PBPK) model to predict IgG tissue kinetics in wild-type and FcRn-knockout mice. J Pharmacokinet Pharmacodyn 2007;34 (5):687–709.

[45] Kyriakos R, Shih LB, Ong GL, Patel K, Goldenberg DM, Mattes MJ. The fate of antibodies bound to the surface of tumor cells *in vitro*. Cancer Res 1992;52:835–842.

[46] Mattes MJ, Griffiths GL, Diril H, Goldenberg DM, Ong GL, Shih LB. Processing of antibody–radioisotope conjugates after binding to the surface of tumor cells. Cancer 1994;73: 787–793.

[47] Mayor S, Rothberg K, Maxfield F. Sequestration of GPI-anchored proteins in caveolae triggered by cross-linking. Science 1994;264 (5167):1948–1951.

[48] Maxfield F, McGraw T. Endocytic recycling. Nat Rev Mol Cell Biol 2004;5:121–132.

[49] Austin CD, De Mazière AM, Pisacane PI, van Dijk SM, Eigenbrot C, Sliwkowski MX, Klumperman J, Scheller RH. Endocytosis and sorting of ErbB2 and the site of action of cancer therapeutics trastuzumab and geldanamycin. Mol Biol Cell 2004;15 (12):5268–5282.

[50] Ackerman ME, Chalouni C, Schmidt MM, Raman VV, Ritter G, Old LJ, Mellman I, Wittrup KD. A33 antigen displays persistent surface expression. Cancer Immunol Immunother 2008;57 (7):1017–1027.

[51] Thiele EW. Relation between catalytic activity and size of particle. Ind Eng Chem 1939;31 (7):916–920.

[52] Blumenthal RD, Fand I, Sharkey RM, Boerman OC, Kashi R, Goldenberg DM. The effect of antibody protein dose on the uniformity of tumor distribution of radioantibodies – an autoradiographic study. Cancer Immunol Immunother 1991;33 (6):351–358.

[53] Rhoden JJ, Wittrup KD. Dose dependence of intratumoral perivascular distribution of monoclonal antibodies. J Pharm Sci 2012;101 (2):860–867.

[54] Thurber G, Schmidt M, Wittrup KD. Factors determining antibody distribution in tumors. Trends Pharmacol Sci 2008;29 (2):57–61.

[55] Yokota T, Milenic D, Whitlow M, Schlom J. Rapid tumor penetration of a single-chain Fv and comparison with other immunoglobulin forms. Cancer Res 1992;52:3402–3408.

[56] Ackerman ME, Pawlowski D, Wittrup KD. Effect of antigen turnover rate and expression level on antibody penetration into tumor spheroids. Mol Cancer Ther 2008;7 (7):2233–2240.

[57] Sutherland R. Cell and environment interactions in tumor microregions: the multicell spheroid model. Science 1988;240:177–184.

[58] Baker J, Lindquist KE, Huxham LA, Kyle AH, Sy JT, Minchinton AI. Direct visualization of heterogeneous extravascular distribution of trastuzumab in human epidermal growth factor receptor type 2 overexpressing xenografts. Clin Cancer Res 2008;14 (7):2171–2179.

[59] Thurber GM, Weissleder R. A systems approach for tumor pharmacokinetics. PLoS One 2011;6 (9):e24696.

[60] Thurber GM, Wittrup KD. A mechanistic compartmental model for total antibody uptake in tumors. J Theor Biol 2012;314:57–68.

[61] Mager DE, Jusko WJ. General pharmacokinetic model for drugs exhibiting target-mediated drug disposition. J Pharmacokinet Pharmacodyn 2001;28 (6):507–532.

[62] Poulin P, Theil FP. A priori prediction of tissue : plasma partition coefficients of drugs to facilitate the use of physiologically-based pharmacokinetic models in drug discovery. J Pharm Sci 2000;89 (1):16–35.

[63] Primeau AJ, Rendon A, Hedley D, Lilge L, Tannock IF. The distribution of the anticancer drug doxorubicin in relation to blood vessels in solid tumors. Clin Cancer Res 2005;11 (24):8782–8788.

[64] Kyle AH, Huxham LA, Yeoman DM, Minchinton AI. Limited tissue penetration of taxanes: a mechanism for resistance in solid tumors. Clin Cancer Res 2007;13 (9):2804–2810.

[65] Chan AC, Carter PJ. Therapeutic antibodies for autoimmunity and inflammation. Nat Rev Immunol 2010;10 (5):301–316.

[66] Bezombes C, Fournie JJ, Laurent G. Direct effect of rituximab in B-cell-derived lymphoid neoplasias: mechanism, regulation, and perspectives. Mol Cancer Res 2011;9 (11): 1435–1442.

[67] Vincenzi B, Zoccoli A, Pantano F, Venditti O, Galluzzo S. CETUXIMAB: from bench to bedside. Curr Cancer Drug Targets 2010;10 (1):80–95.

[68] Spector NL, Blackwell KL. Understanding the mechanisms behind trastuzumab therapy for human epidermal growth factor receptor 2-positive breast cancer. J Nucl Med 2009;27 (34):5838–5847.

[69] Shuptrine CW, Surana R, Weiner LM. Monoclonal antibodies for the treatment of cancer. Semin Cancer Biol 2012;22 (1):3–13.

[70] Dechant M, Weisner W, Berger S, Peipp M, Beyer T, Schneider-Merck T, Lammerts van Bueren JJ, Bleeker WK, Parren PW, van de Winkel JG, Valerius T. Complement-dependent tumor cell lysis triggered by combinations of epidermal growth factor receptor antibodies. Cancer Res 2008;68 (13):4998–5003.

[71] Baeuerle PA, Reinhardt C. Bispecific T-cell engaging antibodies for cancer therapy. Cancer Res 2009;69 (12): 4941–4944.

[72] Lu D, Zhang H, Ludwig D, Persaud A, Jimenez X, Burtrum D, Balderes P, Liu M, Bohlen P, Witte L, Zhu Z. Simultaneous blockade of both the epidermal growth factor receptor and the insulin-like growth factor receptor signaling pathways in cancer cells with a fully human recombinant bispecific antibody. J Biol Chem 2004;279 (4):2856–2865.

[73] Verma S, Miles D, Gianni L, Krop IE, Welslau M, Baselga J, Pegram M, Oh DY, Diéras V, Guardino E, Fang L, Lu MW, Olsen S, Blackwell K, EMILIA Study Group. Trastuzumab emtansine for HER2-positive advanced breast cancer. N Engl J Med 2012;367 (19):1783–1791.

[74] Younes A, Bartlett NL, Leonard JP, Kennedy DA, Lynch CM, Sievers EL, Forero-Torres A. Brentuximab vedotin (SGN-35) for relapsed CD30-positive lymphomas. N Engl J Med 2010;363 (19):1812–1821.

[75] Choudhary S, Mathew M, Verma RS. Therapeutic potential of anticancer immunotoxins. Drug Discov Today 2011;16 (11–12):495–503.

[76] Davies AJ. Radioimmunotherapy for B-cell lymphoma: Y-90 ibritumomab tiuxetan and I-131 tositumomab. Oncogene 2007;26 (25):3614–3628.

[77] Speer TW. *Targeted Radionuclide Therapy*. Philadelphia (PA): Wolters Kluwer Health/Lippincott Williams & Wilkins; 2011., pp xvii, 537 p.

[78] Izumi Y, Xu L, di Tomaso E, Fukumura D, Jain RK. Tumour biology: herceptin acts as an anti-angiogenic cocktail. Nature 2002;416 (6878):279–280.

[79] Schoeberl B, Eichler-Jonsson C, Gilles ED, Muller G. Computational modeling of the dynamics of the MAP kinase cascade activated by surface and internalized EGF receptors. Nat Biotechnol 2002;20 (4):370–375.

[80] Baselga J, Carbonell X, Castañeda-Soto NJ, Clemens M, Green M, Harvey V, Morales S, Barton C, Ghahramani P. Phase II study of efficacy, safety, and pharmacokinetics of trastuzumab monotherapy administered on a 3-weekly schedule. J Clin Oncol 2005;23 (10):2162–2171.

[81] Fenwick J, Philpott G, Connett J. Biodistribution and histological localization of anti-human colon cancer monoclonal antibody (MAb) 1A3: the influence of administered MAb dose on tumor uptake. Int J Cancer 1989;44:1017–1027.

[82] Schroff R, Morgan AC Jr, Woodhouse CS, Abrams PG, Farrell MM, Carpenter BE, Oldham RK, Foon KA. Monoclonal antibody therapy in malignant melanoma: factors effecting *in vivo* localization. J Biol Response Mod 1987;6:457–472.

[83] Rudnick SI, Adams GP. Affinity and avidity in antibody-based tumor targeting. Cancer Biother Radiopharm 2009;24 (2):155–161.

[84] Zhou Y, Goenaga AL, Harms BD, Zou H, Lou J, Conrad F, Adams GP, Schoeberl B, Nielsen UB, Marks JD. Impact of intrinsic affinity on functional binding and biological activity of EGFR antibodies. Mol Cancer Ther 2012;11 (7): 1467–1476.

[85] Ducry L, Stump B. Antibody–drug conjugates: linking cytotoxic payloads to monoclonal antibodies. Bioconjug Chem 2010;21 (1):5–13.

[86] Cao Y, Marks JD, Huang Q, Rudnick SI, Xiong C, Hittelman WN, Wen X, Marks JW, Cheung LH, Boland K, Li C, Adams GP, Rosenblum MG. Single-chain antibody-based immunotoxins targeting Her2/neu: design optimization and impact of affinity on antitumor efficacy and off-target toxicity. Mol Cancer Ther 2012;11 (1):143–153.

[87] Sharkey RM, Karacay H, Cardillo TM, Chang CH, McBride WJ, Rossi EA, Horak ID, Goldenberg DM. Improving the delivery of radionuclides for imaging and therapy of cancer using pretargeting methods. Clin Cancer Res 2005;11 (19):7109S–7121S.

[88] Orcutt KD, Rhoden JJ, Ruiz-Yi B, Frangioni JV, Wittrup KD. Effect of small-molecule-binding affinity on tumor uptake *in vivo*: a systematic study using a pretargeted bispecific antibody. Mol Cancer Ther 2012;11 (6):1365–1372.

[89] Perrino E, Steiner M, Krall N, Bernardes GJ, Pretto F, Casi G, Neri D. Curative properties of noninternalizing antibody–drug conjugates based on maytansinoids. Cancer Res 2014;74 (9):2569–2578.

[90] Press OW, Shan D, Howell-Clark J, Eary J, Appelbaum FR, Matthews D, King DJ, Haines AM, Hamann P, Hinman L, Shochat D, Bernstein ID. Comparative metabolism and retention of iodine-125, yttrium-90, and indium-111 radioimmunoconjugates by cancer cells. Cancer Res 1996;56 (9): 2123–2129.

[91] Zhang Y, Xiang L, Hassan R, Pastan I. Immunotoxin and Taxol synergy results from a decrease in shed mesothelin levels in the extracellular space of tumors. Proc Natl Acad Sci U S A 2007;104 (43):17099–17104.

[92] Pak Y, Zhang YJ, Pastan I, Lee B. Antigen shedding may improve efficiencies for delivery of antibody-based anticancer agents in solid tumors. Cancer Res 2012;72 (13):3143–3152.

[93] Murphy G, Nagase H. Progress in matrix metalloproteinase research. Mol Aspects Med 2008;29 (5):290–308.

[94] Boswell CA, Mundo EE, Zhang C, Bumbaca D, Valle NR, Kozak KR, Fourie A, Chuh J, Koppada N, Saad O, Gill H, Shen BQ, Rubinfeld B, Tibbitts J, Kaur S, Theil FP, Fielder PJ, Khawli LA, Lin K. Impact of drug conjugation on pharmacokinetics and tissue distribution of anti-STEAP1 antibody–drug conjugates in rats. Bioconjug Chem 2011;22 (10):1994–2004.

[95] Erster O, Thomas JM, Hamzah J, Jabaiah AM, Getz JA, Schoep TD, Hall SS, Ruoslahti E, Daugherty PS. Site-specific targeting of antibody activity *in vivo* mediated by disease-associated proteases. J Control Release 2012;161 (3):804–812.

[96] Desnoyers LR, Vasiljeva O, Richardson JH, Yang A, Menendez EE, Liang TW, Wong C, Bessette PH, Kamath K, Moore SJ, Sagert JG, Hostetter DR, Han F, Gee J, Flandez J, Markham K, Nguyen M, Krimm M, Wong KR, Liu S, Daugherty PS, West JW, Lowman HB. Tumor-specific activation of an EGFR-targeting probody enhances therapeutic index. Sci Transl Med 2013;5 (207):207ra144.

[97] Minchinton AI, Tannock IF. Drug penetration in solid tumours. Nat Rev Cancer 2006;6 (8):583–592.

[98] Boswell CA, Tesar DB, Mukhyala K, Theil FP, Fielder PJ, Khawli LA. Effects of charge on antibody tissue distribution and pharmacokinetics. Bioconjug Chem 2010;21 (12): 2153–2163.

[99] Hamblett KJ, Senter PD, Chace DF, Sun MM, Lenox J, Cerveny CG, Kissler KM, Bernhardt SX, Kopcha AK, Zabinski RF, Meyer DL, Francisco JA. Effects of drug loading on the antitumor activity of a monoclonal antibody drug conjugate. Clin Cancer Res 2004;10 (20):7063–7070.

[100] Sun X, Widdison W, Mayo M, Wilhelm S, Leece B, Chari R, Singh R, Erickson H. Design of antibody–maytansinoid conjugates allows for efficient detoxification via liver metabolism. Bioconjug Chem 2011;22 (4):728–735.

[101] Reckelhoff JF, Baylis C. Proximal tubular metalloprotease activity is decreased in the senescent rat-kidney. Life Sci 1992;50 (13):959–963.

[102] Vegt E, Melis M, Eek A, de Visser M, Brom M, Oyen WJ, Gotthardt M, de Jong M, Boerman OC. Renal uptake of different radiolabelled peptides is mediated by megalin: SPECT and biodistribution studies in megalin-deficient mice. Eur J Nucl Med Mol Imaging 2011;38 (4):623–632.

[103] Chung AS, Lee J, Ferrara N. Targeting the tumour vasculature: insights from physiological angiogenesis. Nat Rev Cancer 2010;10 (7):505–514.

[104] Jain RK, Tong RT, Munn LL. Effect of vascular normalization, by antiangiogenic therapy on interstitial hypertension, peritumor edema, and lymphatic metastasis: insights from a mathematical model. Cancer Res 2007;67 (6): 2729–2735.

[105] Turley RS, Fontanella AN, Padussis JC, Toshimitsu H, Tokuhisa Y, Cho EH, Hanna G, Beasley GM, Augustine CK, Dewhirst MW, Tyler DS. Bevacizumab-induced alterations in vascular permeability and drug delivery: a novel approach to augment regional chemotherapy for in-transit melanoma. Clin Cancer Res 2012;18 (12):3328–3339.

[106] Nakahara T, Norberg SM, Shalinsky DR, Hu-Lowe DD, McDonald DM. Effect of inhibition of vascular endothelial growth factor signaling on distribution of extravasated antibodies in tumors. Cancer Res 2006;66 (3):1434–1445.

[107] Stylianopoulos T, Jain RK. Combining two strategies to improve perfusion and drug delivery in solid tumors. Proc Natl Acad Sci U S A 2013;110 (46):18632–18637.

[108] Sugahara KN, Teesalu T, Karmali PP, Kotamraju VR, Agemy L, Greenwald DR, Ruoslahti E. Coadministration of a tumor-penetrating peptide enhances the efficacy of cancer drugs. Science 2010;328 (5981):1031–1035.

[109] Shin TH, Sung ES, Kim YJ, Kim KS, Kim SH, Kim SK, Lee YD, Kim YS. Enhancement of the tumor penetration of monoclonal antibody by fusion of a neuropilin-targeting peptide improves the antitumor efficacy. Mol Cancer Ther 2014;13 (3):651–661.

[110] Zhao LL, Ching LM, Kestell P, Kelland LR, Baguley BC. Mechanisms of tumor vascular shutdown induced by 5,6-dimethylxanthenone-4-acetic acid (DMXAA): increased tumor vascular permeability. Int J Cancer 2005;116 (2): 322–326.

17

BRAIN EFFECT SITE PHARMACOKINETICS: DELIVERY OF BIOLOGICS ACROSS THE BLOOD–BRAIN BARRIER

Gert Fricker and Anne Mahringer
Ruprecht-Karls University Heidelberg, Heidelberg, Germany

In 2007, William Pardridge gave an excellent summary on drug delivery to the central nervous system (CNS) [1]. He stated that out of greater than 7000 drugs in the Comprehensive Medicinal Chemistry database only 5% are used for CNS treatment, and that these drugs are limited to depression, schizophrenia, and insomnia [2]. The fact that so few drugs cross the blood–brain barrier (BBB) appears problematic considering that the number of people with a CNS disorder will grow with an aging population. The number of people older than 65 years will increase by 50% by 2020, and one out of every three individuals will have a CNS condition during their lifetime [3]. A major reason for the very low accessibility of the brain is the so-called blood–brain barrier made up of brain capillaries (Fig. 17.1). In the human brain, these capillaries form a network of approximately 600 km length, which operates as a regulator of ion balance, a mediator of nutrient transport, and a barrier to xenobiotics and potentially harmful molecules. The capillaries are formed by brain microvessel endothelial cells, which are surrounded by pericytes and a basal lamina of 30–40 nm thickness and are covered by astrocyte foot processes being in close neighborhood to neurons. The basal membrane consists of heparin sulfate proteoglycans, fibronectin, collagen type IV, laminin, and other extracellular matrix proteins. The endothelial cells are connected to each other by extremely tight junctions and adherence junctions, preventing almost completely paracellular movement across the barrier. The whole cellular complex is termed "neurovascular unit." However, due to its protective function, the BBB also represents a major obstacle for the development of CNS drugs. Although several carrier proteins and receptors are expressed in the BBB, which serve mainly for nutrient transport across the barrier (Fig. 17.2), most of the small molecules and almost all large molecule drugs, for example, recombinant peptides or antisense agents, are normally excluded from the brain [1, 4, 5]. Beside the tight junctions, the expression of various ABC export pumps recognizing an abundance of small molecules, but also some macromolecules, constitutes a major hurdle for efficient CNS drug delivery. In humans, particularly p-glycoprotein (ABCB1) and breast cancer-resistance protein (ABCG2) are involved in drug export back into the blood circulation (for a review, see [6–8]).

In the past decade, biologics became a new therapeutic option for the treatment of a variety of CNS-related diseases including neurodegenerative diseases, stroke, or brain tumors. In general, the number of biotherapeutic agents entering the pharmaceutical sector is rapidly increasing and, according to EvaluatePharma's "World Preview 2014" report, the annual growth rate of biologics is expected to be 8.5% from 2008 to 2014, 8–10 times greater than the growth rate of small molecules. Distinct from chemically synthesized drugs, biologics are defined as engineered macromolecular medicinal products, such as nucleic acid-based or recombinant therapeutic proteins including antibodies. Despite promising *in vitro* results, the *in vivo* efficacy of such brain-directed drugs remains mostly disappointing after intravenous (IV) administration because they are not able to cross the BBB.

ADME and Translational Pharmacokinetics/Pharmacodynamics of Therapeutic Proteins: Applications in Drug Discovery and Development,
First Edition. Edited by Honghui Zhou and Frank-Peter Theil.

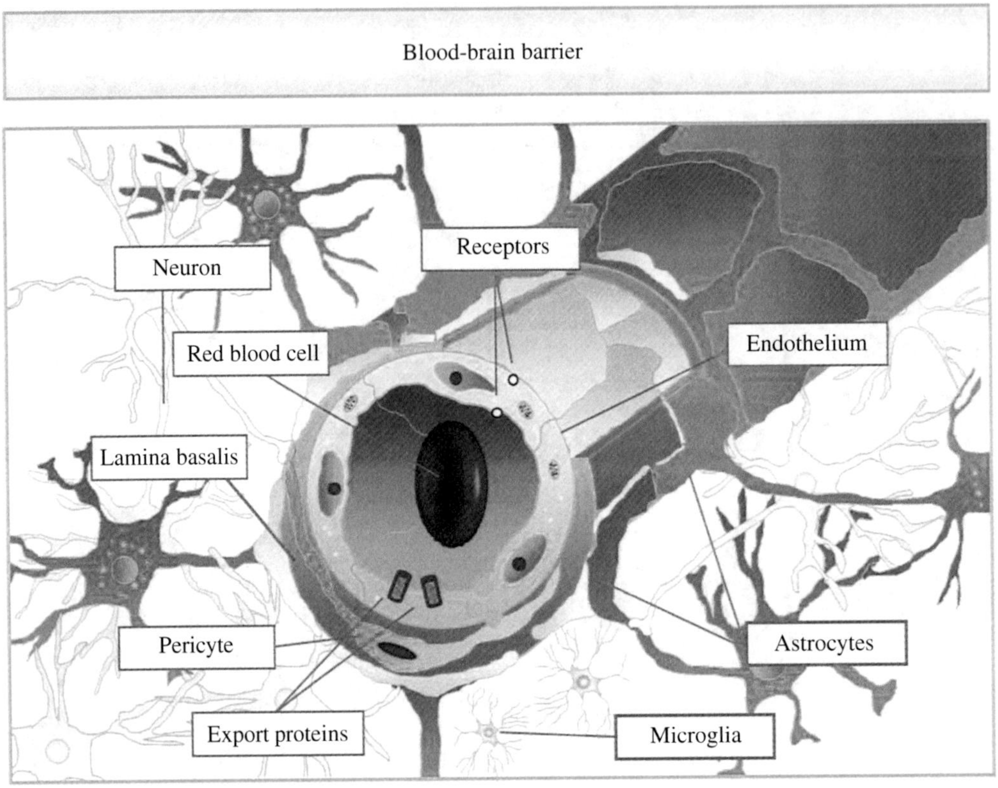

FIGURE 17.1 Cross-section of a brain capillary. Endothelial cells are surrounded by pericytes and a basal lamina covered by astrocyte endfeet. Endothelial cells are connected by very tight junctions and express various export proteins in their luminal membrane preventing uptake of many drugs and drug candidates.

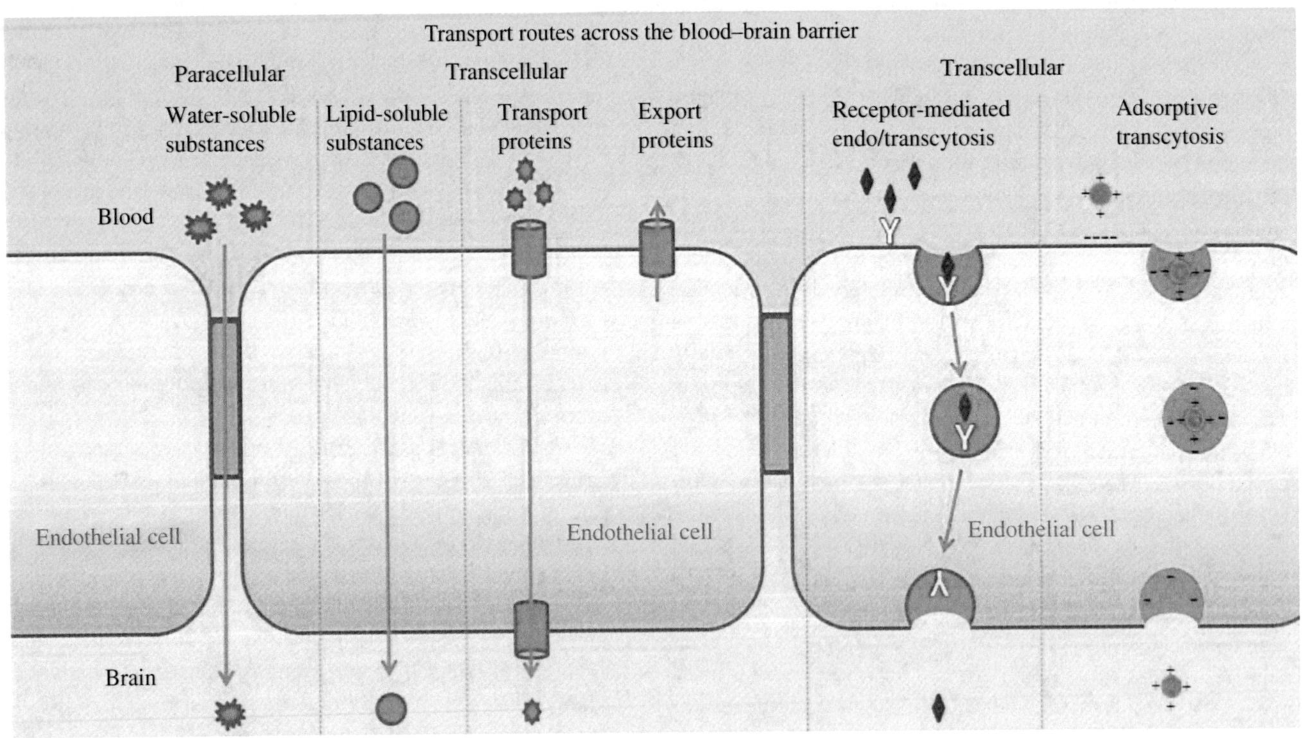

FIGURE 17.2 Transport routes across the blood–brain barrier. Substances may be transported across the endothelial cells of the BBB via paracellular or transcellular pathways. While the paracellular route can be neglected, transcellular passage may occur by passive diffusion, transport protein mediated, or by cytotic processes. In addition, blood to brain transport may be prevented by export proteins including p-glycoprotein, breast cancer-resistance protein, and MRP.

Hence, the understanding of the morphology of the BBB as well as the molecular and cellular mechanisms that determine its function is an inevitable prerequisite for successful delivery of macromolecules to the brain. Here, we review present knowledge about receptors in brain capillary endothelial cells (BCECs) and delivery systems, which can be used to move biologics across the barrier and have an impact on their therapeutic efficacy.

17.1 CYTOTIC PROCESSES AT THE BBB

Beside membrane transport proteins for small molecular nutrients such as glucose or amino acids, the BBB contains a variety of receptors, which might be utilized for the transfer of macromolecules. These receptors act in three different ways: (i) only endocytosis into the brain capillary endothelium without net transport across the barrier, (ii) transcytosis of their ligands from the blood circulation into the brain, and (iii) reverse transcytosis from brain to blood. One of these receptors is the insulin receptor, which transports insulin of peripheral origin into the brain [9, 10]. Further receptors at the BBB are insulin-like growth factor I and II receptors (IGFIR, IGFIIR) [1, 11], low density lipoprotein (LDL) receptor, leptin receptor (OBR [12]), or the receptor of advanced glycation endproducts (RAGE, [13]). The latter is of particular interest in the pathogenesis of Alzheimer's disease as it is involved in the cerebral homeostasis and clearance of amyloid-β [14, 15]. In general, these receptors may provide targets for the brain-directed delivery of drugs, which under normal circumstances do not cross the BBB, including large biopharmaceutics. For example, it had been suggested that the Fc receptor (FcR; [16]) moves IgG molecules from the brain into the blood. However, these findings are questioned by others [17] showing that there was no significant difference in brain distribution of IgG in FcRn-deficient mice and C57BL/6J control mice.

The scavenger receptor (SR) at the BBB does not mediate transcytosis but only endocytosis into the brain microvessel endothelium. It transports acetylated LDL into endothelial cells but not across the barrier [18]. Further receptors at the BBB are the low density lipoprotein receptor-related Proteins 1 (LRP1) and 2 (LRP2, megalin; [19]). Ligands of LRP1 are melanotransferrin (p97) or the synthetic peptide Angiopep-2 [20, 21], but their rate of transcytosis seems to be lower than that of antibodies targeting the transferrin receptor [5]. The transferrin receptor is expressed at luminal as well as at the abluminal membrane of endothelial cells in the BBB and acts in a bidirectional way [16, 22]. It moves apotransferrin in the brain-to-blood direction and holotransferrin from blood to brain.

In addition to these specific receptor-mediated internalization mechanisms, there are two other pathways mediated by caveolae or plasmalemmal vesicles and clathrin-coated pits/vesicles (reviewed in [23]). Caveolae-mediated permeation across endothelial cells has been described as bulk-phase or fluid-phase transcytosis, which is independent of interactions between the transported molecules and the caveolar vesicle membrane. In brain capillaries, the occurrence of caveolae is relatively low, and thus it is not really clear to what extent this mechanism plays a role for transfer across the BBB [24]. In contrast, the expression density of clathrin-coated pits/vesicles at the BBB seems to be much higher [25]. But, because of the negative surface charge of the clathrin-coated pits, only very few of the plasma proteins can be transcytosed randomly within the fluid phase of clathrin-coated vesicles. However, this pathway is of interest for the transport of positively charged molecules including artificially cationized proteins, such as albumin [26]. Electrostatic interactions may occur between the positively charged protein moieties and negatively charged membrane surface regions on the endothelial cells [27]. Thus, all these mechanisms may be used to transport macromolecules in either one or both directions across the BBB.

17.2 RECEPTORS AT THE BBB AS TARGETS FOR BIOLOGICS

17.2.1 Transferrin Receptor

The high expression of the transferrin receptor makes it an ideal candidate as target for the delivery of macromolecules. Beside its natural ligands, it also recognizes other macromolecules as well as colloidal delivery systems, such as liposomes or nanoparticles, coupled to either transferrin or antibodies directed versus the receptor. The receptor exists in two subtypes TfR1 and TfR2 [28, 29]. At the BBB, TfR1 is expressed [30]. But, although the engagement of the TfR has been used in several previous studies as an approach to facilitate the movement of large molecules across the BBB, the understanding of the determinants of effective transport is limited [31–34]. Shortly after the detection of the receptor at the BBB [35], it was shown that it is able to transfer small molecules coupled to antireceptor antibodies such as methotrexate [36] or small antisense oligonucleotides [37]. Recently, Pardridge published a profound overview on the receptor and its transport capacity [38]. He showed that brain uptake of murine anti-rat TfR antibody OX26 is approximately 0.44% injected dose/gram brain, which is about half of the amount of Diazepam, which is almost 100% extracted by brain at single pass.

Several examples have been published showing successful transfer of macromolecules via the transferrin receptor across the BBB, including a conjugate of epidermal growth factor EGF [39]. Radiolabeled human EGF was biotinylated with NHS-PEG3400-biotin, and then conjugated to a complex of a monoclonal antibody (OX 26) versus the rat transferrin receptor, which was coupled to streptavidin. *In vivo* studies demonstrated that this conjugate could be used for imaging brain tumors beyond the BBB.

Yu et al. [32] showed that reducing the affinity of an anti-TfR antibody enhanced receptor-mediated transcytosis across the BBB into mouse brain, thereby reaching therapeutically relevant drug concentrations. Anti-TfR antibodies exhibiting a high affinity to TfR remained associated with the BBB, whereas lower affinity anti-TfR antibody variants showed a broad distribution 24 h after dosing. In a further study, bispecific antibodies have been tailored against TfR and γ-secretase (BACE1: β-amyloid cleaving enzyme-1), which is a prime therapeutic target for Alzheimer's disease. The therapeutic effect of an anti-BACE1 antibody in inhibiting Aβ production had been demonstrated *in vivo* [40]. The antibodies traverse the BBB and reduce β-amyloid levels in the brain [41]. Limits of this approach are as follows: changing the affinity of the antibody construct to the transferrin receptor significantly altered the intracellular trafficking of TfR. High affinity binding to TfR resulted in a dose-dependent reduction of TfR levels in the brain. These antibodies facilitated the trafficking of TfR to lysosomes and thus induced the degradation of TfR. As the authors say, their findings "reveal a fundamental cellular principle with translational implications, namely that TfR cellular trafficking is modulated by TfR antibody affinity…current therapeutic strategies targeting TfR with high-affinity antibodies using chronic dosing paradigms may be severely hindered by the gradual loss of TfR, resulting in limited brain antibody uptake and also impacting physiological iron transport into the brain."

A similar study showed that a difference between monovalent and bivalent antibody influenced the efficacy and that the binding mode to the TfR is absolutely crucial for successful transport of antibodies across the BBB [42]. A TfR–antibody conjugate with an anti-amyloid-β antibody, which used a monovalent binding mode to the TfR increased amyloid-β target engagement in a mouse model of Alzheimer's disease by 55-fold compared to the parent antibody, whereas a bivalent binding mode led to lysosome sorting.

17.2.2 Insulin Receptor

Shortly after the mention of insulin receptors at the BBB [43–45], it was suggested to utilize them for drug delivery across the BBB [45, 46]. Wu et al. [47] aimed to deliver a radiopharmaceutical peptide (Aβ1–40) to the brain by coupling it to a monoclonal antibody versus the insulin receptor. They observed a marked increase in the peptide in rhesus monkey brain within 3 h after IV administration, whereas no uptake of the peptide was seen in the absence of the delivery system. The same group demonstrated nonviral gene transfer (plasmids encoding either luciferase or β-galactosidase) to primate brain after encapsulation into PEGylated immunoliposome, which had been coupled to a monoclonal antibody to the human insulin receptor. The level of luciferase gene expression in the brain was shown to be 50-fold higher in the rhesus monkey as compared to the rat, and the neuronal expression of the β-galactosidase gene in brain was demonstrated by histochemistry and confocal microscopy [48]. This delivery approach was developed further by the production of a genetically engineered fusion protein, where the amino terminus of human brain-derived neurotrophic factor (BDNF) was fused to the carboxyl terminus of the heavy chain of a chimeric human insulin receptor monoclonal antibody (HIRMAb). The pharmacokinetics of the fusion protein was examined in rhesus monkeys where therapeutic levels of BDNF could be produced in brain following IV administration. Consequently, it was postulated that neurotrophins, such as BDNF, can be reformulated to enable these molecules to cross the human BBB and that such fusion proteins may represent a novel class of neurotherapeutics [49]. This strategy was also used to couple α-L-iduronidase, an enzyme that can be utilized for the treatment of the lysosomal storage disease mucopolysaccharidosis (MPS) type I, also known as Hurler's syndrome. The specific activity of the affinity purified fusion protein was comparable to the specific activity of recombinant enzyme. The fusion construct was again rapidly transported into the brain of rhesus monkey following IV administration [50]. Several further studies demonstrated successful application of this strategy for delivery of other biologics including erythropoietin (EPO) [51], the decoy receptor human tumor necrosis factor receptor (TNFR) [52], or Iduronate 2-sulfatase for the treatment of MPS type II [53].

However, a drawback of these studies is given by the fact that most of the experiments used a mouse antibody, which could lead to immunogenic reactions in humans. As one way to address this disadvantage, the antibody was reengineered. A chimeric antibody [54] and a fully humanized form of the antibody against the human insulin receptor have been created [55]. A detailed summarizing description of these modifications is given by Jones and Shusta [56].

17.2.3 Insulin-Like Growth Factor Receptor

The insulin-like growth factor II (IGF-II) receptor is also known as mannose 6-phosphate (M6P) receptor. M6P plays a role in the recognition of lysosomal enzymes and their sorting to lysosomes. Obviously, this receptor is mainly active in newborns as transport of β-glucuronidase across the BBB was only observed in newborn but not in adult mice [11].

17.2.4 LDL Receptor

The presence of an LDL binding receptor (LDLR) was first described in bovine brain capillaries by Méresse et al. [57]. Apparently, the expression of this receptor is modulated by soluble factors released from astrocytes [58]. A recent study gave evidence that LDLR was involved in the uptake of siRNA, when a cholesterol-conjugated 21/23-mer siRNA targeting OAT3 (organic anion transporter 3) mRNA was IV injected into mice after its incorporation into extracted endogenous lipoproteins [59]. Consequently, it was postulated that siRNA may be delivered into BCECs *in vivo* by using endogenous lipoprotein, which could make this strategy useful as a new gene-silencing therapy for diseases involving BCECs. Furthermore, from phage-display biopanning, a series of peptide ligands for LDLR was developed with improved receptor-binding affinity. A single peptide and its analogs were identified, which was demonstrated to efficiently and quickly cross CNS barriers. The binding of this peptide on the extracellular LDLR domain was studied in NMR-oriented structural studies and docking experiments [60]. In an attempt to link a biologic to an LDLR ligand, the lysosomal enzyme α-L-iduronidase (IDUA) was coupled to a receptor-binding peptide from apolipoprotein E (Apo-E) in order to treat MPS type I in a mouse model [31]. Two fusion candidates were generated, showing receptor-mediated binding, endocytosis, and transendothelial transport into nonendothelium perivascular cells, neurons, and astrocytes within 2 days of treatment as well as appropriate lysosomal enzyme trafficking and biological function. Five months after long-term delivery of one of the conjugates, 2–3% of normal brain enzyme activities were obtained in these mice, and the fusion enzyme was detected in neurons and astrocytes throughout the brain.

A further approach to use the LDLR as a route for drug delivery was the development of dual-targeting paclitaxel-loaded nanoparticles for brain tumor treatment, which were decorated with a peptide with special affinity for the receptor to transport the drug across the BBB, and then target brain tumor cells. Cellular uptake mechanism experiments showed that uptake of these nanoparticles by endothelial cells and glioma C6 cells was energy-dependent and caveolae- and clathrin-mediated endocytosis pathways were involved. The nanoparticles significantly increased the transport ratio of Paclitaxel across the BBB and induced apoptosis of C6 glioma cells below the BBB, and these effects were significantly inhibited by excess of free decorating peptide. *In vivo* fluorescence imaging indicated that the nanoparticles labeled with a near-infrared dye permeated across the BBB and accumulated at the glioma site. The median survival time of glioma-bearing mice administered with dual-targeting nanoparticle-bound paclitaxel was significantly prolonged and the treatment induced significantly more cell apoptosis and tumor necrosis than other treatments [61].

17.2.5 Low Density Lipoprotein Receptor-Related Protein 1

Low density lipoprotein receptor-related protein 1 (LRP1/CD91/α2-macroglobulin receptor), which is structurally similar to the LDL receptor, is a multifunctional scavenger transporter and signaling receptor [62, 63]. It appears to transport a variety of ligands across the BBB including amyloid-β [64–66], tissue-type plasminogen activator [67], Apo-E2 and Apo-E3 in free form as well as in complexes with amyloid-β [68], receptor-associated protein [69], and others. In plasma, a soluble form of LRP1 is the major transport protein for peripheral Aβ; in brain endothelium, it is postulated to mediate Aβ efflux across the BBB [70].

17.2.6 Low Density Lipoprotein Receptor-Related Protein 2

LRP2 or megalin is expressed primarily in a subset of epithelial and endothelial cell layers including renal proximal tubules, the ciliary body of the eye, thyroid colloid, epididymis, alveolae, and brain vasculature [71, 72]. Studies by Zlokovic et al. [73], and Shayo et al. [74], gave evidence that megalin mediates cellular uptake and transport of Apo-J and A-β (1–40)/Apo-J complex at the cerebral vascular endothelium.

17.2.7 Leptin Receptor (OBR)

Leptin, a peptide hormone consisting of 167 amino acids, is a major regulator of body weight [12] and controls together with the peptide hormone ghrelin hunger and satiety. Zhang et al. [75] suggested a negative feedback loop between leptin and bodyweight. Fat-induced leptin crosses the BBB by transcytosis and interacts with leptin receptors in the arcuate nucleus to inhibit feeding and increase thermogenesis, which decreases fat mass. Absent or impaired leptin receptors are discussed to be a cause of obesity in humans. In a recent *in vitro* study [76], a 30 amino acid peptide derived from leptin was used as brain-targeting ligand. Dendrigraft poly-L-lysine (DGL) was used as nonviral gene vector and DGL-PEG-Leptin30 was complexed with plasmid DNA to yield nanoparticles. The targeted particles were transported across an *in vitro* BBB model and accumulated in rodent brains after IV administration resulting in relatively high gene transfection efficiency.

17.2.8 Receptor of Advanced Glycation Endproducts

RAGE is a 35-kD transmembrane receptor of the immunoglobulin superfamily, which was first characterized by Neeper et al. [13]. It is a multiligand receptor involved in inflammatory disorders, diabetic complications, tumor growth, and Alzheimer's disease [15]. Similar to LRP1, RAGE appears to interact with amyloid β resulting in the transport of amyloid-β across the BBB [77, 78]. While the majority of findings claim export of amyloid-β through the BBB, there are also findings describing apical-to-basolateral transport of amyloid-β peptides through BBB cells by this receptor, making it interesting as drug target [79]. New findings suggest that amyloid-β reduces p-glycoprotein in the BBB through a RAGE-NF-kB signaling pathway [80].

17.2.9 Scavenger Receptor (SR)

The presence of an SR being responsible for the uptake of polyanions such as succinylated proteins was first shown in cultured bovine brain microvessel endothelial cells [81], when large succinylated proteins were taken up, whereas no significant uptake was observed for native proteins and small succinylated proteins. RT–PCR studies confirmed the expression of SR types I and II in cerebral microvessels [82]. In porcine BCECs, SR class B, type I is expressed and contributes to selective uptake of HDL-associated vitamin E [83]. SR class B, type I may also play a role in the passage of surface modified nanoparticles, which can be used for the delivery of biologics, across the BBB [84, 85].

17.3 "TROJAN HORSE" APPROACHES TO TARGET BBB RECEPTORS

Within the past 2 decades, it became obvious that a "Trojan horse" approach might be useful to deliver macromolecular drugs to the CNS, by coupling the respective drug to an endogenous substrate or an antibody versus one of the abovementioned receptors in the luminal membrane of the BBB. One of the first examples showing the usefulness of this approach was the linkage of monobiotinylated vasoactive intestinal peptide (VIP) to a covalent conjugate of avidin and the OX26 monoclonal antibody to the rat transferrin receptor. Systemic infusion of a low dose chimeric peptide in rats resulted in an *in vivo* CNS pharmacologic effect, namely a 65% increase in cerebral blood flow. Without the brain transport vector, the VIP was ineffective [86]. In another study, a plasmid bearing a fusion gene consisting of transferrin and the enzyme α-L-iduronidase has been constructed, which *in vivo* resulted in the production of high levels of an enzymatically active protein that was transported into the CNS by transferrin receptor-mediated endocytosis in mice [87]. Furthermore, an anti-amyloid-β antibody was made, which uses a monovalent binding mode to the transferrin receptor. This construct increased β-amyloid target engagement in a mouse model of Alzheimer's disease by 55-fold compared to the parent antibody. Interestingly, monovalent binding facilitated transcellular transport, whereas a bivalent binding mode led to lysosome sorting [42].

The most potent Trojan horses known include monoclonal antibodies for the human insulin receptor, the murine OX26 monoclonal antibody to be used in rats, or the rat 8D3 antibody that targets the mouse transferrin receptor. Bacterial β-galactosidase being conjugated to the rat monoclonal antibody versus the transferrin receptor via a streptavidin-biotin linkage was successfully delivered to the brain of mice as determined by the measurement of enzyme activity in the CNS [88]. An elegant study demonstrated the development of a murine monoclonal antibody to the human insulin receptor to be used as shuttle across the BBB. Humanization of the antibody was achieved by complementarity determining region grafting on the FR (framework regions) of the human B43 IgG heavy chain and the human REI κ light chain [55]. The affinity of this humanized antibody to the human insulin receptor was somewhat lower as compared to the murine monoclonal antibody, but IV injection of ^{125}I-labeled humanized antibody to rhesus monkeys resulted in a rapid transport into all parts of the primate brain. The antibody was rapidly taken up into the gray and white matter of the brain and the recovered amount was approximately 1% of the injected dose, which is very high for a large molecule. The list in Table 17.1 shows some examples of fusion conjugates of biologics with such antibodies that have been suggested to be delivered to the brain.

However, some data in primates are conflicting. While the mentioned studies indicated brain penetration of insulin-receptor–antibody fusion proteins, there are also data showing lack of a pharmacological action in primates [101].

An antibody recognizing amyloid plaques at Alzheimer's disease is gantenerumab [102]. A bispecific antibody was reengineered with one domain being a monovalent single-chain transferrin receptor monoclonal antibody [42]. However, brain concentration of the monovalent antibody was very low with a brain uptake of less than 0.1% injected dose/g in the mouse. Another bispecific antibody, where a therapeutic arm is combined with a BBB-transcytosing arm, was recently designed by Farrington et al. [103]. BBB-permeable single-domain antibody FC5 was isolated by phenotypic panning of a naive llama single-domain antibody phage-display library. FC5 was engineered as a mono- and bivalent fusion with the human Fc domain in order to use it as a modular brain delivery platform. The bivalent fusion of FC5 with Fc increased the rate of transcytosis across brain endothelial monolayers by 25% compared with monovalent fusion. In rats, an up to a 30-fold enhanced apparent brain exposure of FC5 compared with control domain antibody–Fc fusions was seen after systemic dosing. The pharmacological potency of this construct was evaluated in a model of inflammatory pain using the BBB impermeable neuropeptides dalargin and neuropeptide Y chemically conjugated with FC5–Fc fusion proteins. Improved serum pharmacokinetics of Fc-fused FC5 contributed to a 60-fold increase in pharmacological potency compared with the single-domain version of FC5; bivalent and monovalent FC5 fusions with Fc exhibited similar potency.

Tumor necrosis factor α (TNFα) is of special interest in ischemic stroke. This proinflammatory cytokine is synthesized in brain within 1 h of an acute experimental ischemic

TABLE 17.1 Examples of fusion conjugates of biologics with antibodies that have been suggested to be delivered to the brain

Ligand	Vector	Receptor	References	Indication
Vasoactive intestinal peptide	OX26 MAb	Rat TfR	[86]	Vasodilatation, ischemic brain damage
α-L-Iduronidase	HIRMAb Murine TfRMAb	Human insulin receptor Murine transferrin receptor	[50] [89, 90]	Mucopolysaccharidosis type I, Hurler's syndrome
Glial-derived neurotrophic factor (GDNF)	MAb	Human insulin receptor	[91, 92]	Parkinson, stroke, motor neuron disease
Brain-derived neurotrophic factor (BDNF)	OX26 MAb	Rat transferrin receptor	[93, 94]	Stroke
Arylsulfatase a	MAb	Human insulin receptor	[95]	Metachromatic leukodystrophy (lysosomal storage disease)
N-Sulfo-glucosamine sulfohydrolase (sulfamidase)	MAb	Human insulin receptor	[96]	Mucopolysaccharidosis type IIIA (Sanfilippo type A syndrome)
TNF-receptor (decoy receptor)	MAb	Human insulin receptor	[52]	Stroke, traumatic brain injury, spinal cord injury, neurodegeneration, depression
Recombinant human soluble CD4	128.1 MAb OX26 MAb	Human TfR Rat TfR	[97]	HIV
Human epidermal growth factor	OX26 MAb	Rat TfR	[39]	Brain tumor diagnosis
Nerve growth factor	OX26 MAb	Rat TfR	[98]	Huntington's disease
PNA antisense to luciferase mRNA	OX26 MAb	Rat TfR	[99]	Diagnostic
PNA antisense to Huntington gene	8D3 MAb	Murine TfR	[100]	Huntington's disease

stroke. Etanercept is a decoy receptor-type TNF inhibitor (TNFI), which is widely used to suppress TNFα action in inflammation in peripheral organs. However, etanercept is not applicable for the treatment of brain stroke because it does not cross the BBB. In order to enable delivery of TNFI, type II human TNFR was fused to the genetically engineered chimeric monoclonal antibody (mAb) against the mouse TfR (cTfRMAb–TNFR fusion protein) [104]. Forty-five minutes after IV administration of 1 mg/kg of this construct, the fusion protein caused a 40–50% reduction in hemispheric, cortical, subcortical stroke volumes, and neural deficit. Treatment with 1 mg/kg etanercept itself had no significant changes in either stroke volume or neural deficit measurements.

In addition, a fusion protein between TNFR and antibody versus the human IR, HIRMAb-TNFR, has been engineered [52]. TNFR–Fc fusion protein did not cross the primate BBB *in vivo*, but the uptake of the HIRMAb–TNFR fusion protein was high and 3% of the injected dose was taken up by the primate brain. The TNFR was selectively targeted to brain, relative to peripheral organs, following fusion to the HIRMAb. In a similar way, a fusion construct HIRMAb–EPO was engineered by fusing human EPO to the carboxyl terminus of the heavy chain of a chimeric monoclonal antibody against the human insulin receptor [51]. EPO is a neurotrophic factor that could be developed as a drug for diverse brain disorders. Studies in rhesus monkeys showed that the fusion protein was selectively transported across the BBB. The permeability–surface area product ratio, as a measure for permeability, between HIRMAb–EPO and the unmodified EPO increased approximately 3–10 times in brain compared to other organs.

A general concern of all fusion proteins is their safety. A recent study demonstrated safety of a TfRMAb–GDNF fusion protein following 12 weeks of twice-weekly IV injection of 2 mg/kg/injection in mice [105]. Organ histology revealed no pathologic changes in brain, pancreas, liver, kidney, spleen, and heart. Body weight did not change, serum parameters remained unaltered, plasma clearance and volume of distribution did not change, and brain uptake of the fusion protein was constant indicating no change in receptor expression. Furthermore, plasma antibody titers against the fusion protein were very low and were not related to the clearance of the protein. Similar observations were made for other fusion proteins [34].

In addition to antibodies, several peptides have been identified to be receptor ligands and to have high permeability across the BBB [20, 21]. For example, therapeutic peptides and proteins have been conjugated to Angiopep-2, an LRP1-ligand for efficient brain delivery. Aprotinin, a pancreatic trypsin inhibitor, which contains the KPI (Kunitz protease inhibitor) sequence, is a ligand of LRP [106]. Transport of aprotinin across the BBB is mediated by LRP [107]. By aligning the amino acid sequence of aprotinin with the Kunitz domain of human proteins, the family of Angiopep peptides was identified [20]. Among these peptides, Angiopep-2 showed the best ability, with three to seven times higher endocytosis in comparison to aprotinin. *In situ*, brain distribution of Angiopep-2 was much higher than that of aprotinin.

Besides the delivery of various small molecules such as doxorubicin or paclitaxel, the Angiopep-2 system has also been used to deliver macromolecules. For example, genetic fusion of the gene encoding human catalase and the gene encoding Angiopep-2 was performed in order to transfer catalase into BCECs for the treatment of oxidative stress caused by excessive production of hydrogen peroxide. The fusion protein retained the same specific enzymatic activity of the native enzyme. About 0.1 μM of the fusion protein entered brain endothelial cells within 15 min, while internalization of the native protein was not observed. Treatment of the cells with 20 units of the fusion protein for 30 min showed protection against H_2O_2 up to 5.0 mM, whereas this protective effect was not observed from treatment with the native protein [108].

Another option for drug delivery to the CNS offer cell-penetrating peptides, which are quite heterogeneous in size (10–27 amino acid residues), but all possess positive charges. Cell-penetrating peptides derived from natural proteins include the transcription-activating factor (Tat), penetration, and the so-called Syn-B vectors as well as engineered short peptides such as the homoarginine vectors, transportan or sequence signal-based peptide (SBP), and fusion sequence-based peptide [23]. The exact mechanisms, by which these peptides are internalized and carry their payload, are still under discussion and may be different for the distinct peptides, but several studies indicate a crucial role of basic residues in the translocating ability of these molecules [109–112].

However, one drawback of "Trojan horses," where the drug of interest is coupled directly to the vector is the limited extent of drug loading per shuttle molecule. But a significant advancement of this technology consists in the linkage of colloidal carriers to a brain-directed vector such as polymeric nanoparticles or liposomes, which might contain thousands of drug molecules. For example, for daunomycin-carrying liposomes coupled to the murine OX26 monoclonal antibody versus the rat transferrin receptor, it has been calculated that one liposome might host ≥10,000 drug molecules [113].

17.4 COLLOIDAL CARRIERS FOR DRUG DELIVERY

Colloidal carriers used for brain-directed drug delivery are mainly polymeric nanoparticles, liposomes, or solid lipid nanoparticles. These carriers are particles in a size range from 1 to 1000 nm, in which therapeutics can be covalently attached, entrapped, or be absorbed. They offer several advantages versus the simple Trojan horse approach. Many thousands of drug molecules can be incorporated, the stability of the drug may be retained, its membrane barrier permeation limiting characteristics may be masked, and thus the system may allow access across the previously impermeable membrane of brain endothelial cells. Once the colloidal carrier reaches brain tissue, drugs may be released by desorption, diffusion across the carrier matrix, or by erosion/degradation of the particle. On the other hand, such carriers need to fulfill a series of requirements to be used a drug shuttles. They must be nontoxic, nonimmunogenic, and noninflammatory, the material should ideally be biodegradable, and the particles need to be nonthrombogenic. In addition, all colloidal carriers have to be stable in blood, which means that they may not be opsonized by blood proteins leading to recognition by the reticuloendothelial system, not activate neutrophil blood cells, or lead to platelet aggregation. Avoidance of interaction with the blood components is normally achieved by PEGylation through coupling of polyethylene glycol (PEG) chains to the surface of the colloidal [114, 115], resulting in a steric stabilization and a prolonged circulation half-life of the colloidal carriers. Liposomes fulfill most of the listed requirements. But, this is not enough to achieve targeting of the BBB. Significant progress has again been made by attaching BBB-directed targeting vector. For example, liposomes have been coupled to thiolated above-mentioned OX26 monoclonal antibody versus the rat transferrin receptor using a bifunctional 2000-Da PEG, which contained a lipid for liposome incorporation at one side and a maleimide for antibody coupling at the other side. These liposomes were able to deliver their load (daunomycin) to the CNS of rats after IV administration [113]. This concept was successfully extended to the CNS delivery of a biologic [116] in an experimental 6-hydroxydopamine model of Parkinson's disease in order to normalize tyrosine hydroxylase activity in the striatum of adult rats. The tyrosine hydroxylase expression plasmid was incorporated into immunoliposomes that were targeted to the rat TfR to undergo both receptor-mediated transcytosis across the BBB and receptor-mediated endocytosis into neurons behind the BBB by accessing the TfR. Thus, the striatal tyrosine hydroxylase activity ipsilateral to the intracerebral injection of the neurotoxin could be normalized.

In a similar way, plasmids encoding either luciferase or β-galactosidase had been incorporated into PEGylated immunoliposomes, which were administered to rhesus monkeys and targeted to the brain with a monoclonal antibody to the human insulin receptor [48]. The rate of transport of this construct across the primate BBB was nearly 10-fold greater than the rate of transport of an antihuman transferrin receptor monoclonal antibody across the primate BBB *in vivo* [117]. The gene expression of luciferase was 50-fold higher in the brain of rhesus monkeys than in brain of rats, where OX26 had been coupled to the liposomes as the targeting vector. Neuronal expression of the β-galactosidase gene in the brain of the monkeys was demonstrated by histochemistry and confocal microscopy. The study also revealed a global delivery of the exogenous gene to the primate brain. Gene expression was higher in gray matter as compared to white matter, which could be explained with an approximate threefold greater vascular density of gray matter relative to white matter [118].

Significant targeting to the brain of mice of double-stranded oligodeoxynucleotide/polyethylenimine complexes by biotinylated PEG-stabilized liposomes, which had been coupled to anti-mouse transferrin receptor antibody 8D3, has been presented by Ko et al. [119]. PEGylated liposomes without the vector showed virtually no brain uptake after correction for organ plasma volume. However, 8D3-targeted liposomes showed increased brain uptake of 0.33%ID/g at 1 h after IV bolus injection. The authors explicitly emphasize the delivery of intact double-stranded native, phosphodiester-based oligodeoxynucleotide because the delivery system protects the integrity in the systemic circulation during the delivery process. 8D3-coupled liposomes were also useful to deliver bacterial β-galactosidase together with a brain-specific promoter taken from the 5′ flanking sequence of the human glial fibrillary acidic protein (GFAP) gene [120]. Confocal microscopy colocalized immunoreactive bacterial β-galactosidase together with immuno GFAP in brain astrocytes.

Beside liposomes, nanoparticles made from biodegradable polymers may be used for brain delivery of biologics. Many factors may influence the BBB transport mechanism of such particles, including the type of polymers, size of the NP, types of surfactants, or the drug molecule. Common polymers include alkylcanoacrylates, polylactide, or lactide/glycolide copolymers degrading within days or weeks dependent on the kind of polymer used. Polymeric nanoparticles are fabricated either by emulsion polymerization, interfacial polymerization, desolvation evaporation, or solvent deposition. A potential disadvantage of such particles may be the requirement of free radicals, radiation, UV, or hydroxyl ions light to trigger the polymerization process, which makes the incorporation of sensitive peptides and proteins difficult. Similar to liposomes, nanoparticles need a brain-directed vector to achieve effective drug delivery. A very elegant method is the incorporation or coating with surfactants such as polysorbates or poloxamers. It has been suggested that apolipoproteins, preferentially Apo-E, adsorb to surfactant-coated nanoparticles during their circulation in blood. Consequently, polysorbate-coated nanoparticles are subject to the same endocytotic process as LDLs undergo at the BBB [121]. Alternatively, Apo-E or fragments thereof as well as Apo-AI or Apo-B100 have been coupled directly to nanoparticles consisting of alkylacrylates or albumin [122]. Electron microscopy as well as laser scanning fluorescence microscopy gave clear evidence for appearance of surfactant treated and/or Apo-E-covered particles beyond the BBB [123, 124].

Interestingly, drugs including biologics appear not necessarily be incorporated into the nanoparticles, but may also just be adsorbed. For example, nerve growth factor A (NGF) has been described to be adsorbed to poly(butylcyanoacrylate) (PBCA) nanoparticles coated with polysorbate and to exert a significant central pharmacological effect after IV administration [125]. NGF is essential for the survival of both peripheral ganglion cells and central cholinergic neurons in the basal forebrain. It is able to prevent the degradation of cholinergic neurons in adult rats having experimental lesions mimicking the cholinergic deficit in Alzheimer's disease [126]. Accelerated loss of central cholinergic neurons during Alzheimer's disease may suggest a possible therapeutic benefit from treatment with NGF. NGF adsorbed to polysorbate-80-coated PBCA nanoparticles was administered in a model of acute scopolamine-induced amnesia in rats as well as in a model of 1-methyl-4-phenyl-1,2,3,6-tetrahydropyridine (MPTP)-induced Parkinsonian syndrome. IV administration of nanoparticle-bound NGF successfully reversed scopolamine-induced amnesia and improved recognition and memory. In addition, a significant reduction in the basic symptoms of Parkinsonism (oligokinesia, rigidity, and tremor) was observed. Efficient transport of NGF across the BBB was confirmed by measurement of NGF concentrations in the murine brain. The highest cerebral concentration of NGF was observed 45 min after administration, but, after 24 h, NGF concentration in brain was still higher in animals treated with particulate NGF compared to animals treated with free NGF (Fig. 17.3).

The abovementioned membrane penetrating carrier peptide Angiopep-2 is also used for the delivery of nanoparticles across the BBB. For example, a PAMAM-PEG-Angiopep/DNA nanoparticulate system used polyamidoamine (PAMAM) coupled to PEG and DNA to deliver DNA to brain glioma [127, 128]. The PAMAM-PEG-Angiopep/DNA particles accumulated especially at the tumor site. Glioma-bearing mice showed a significantly prolonged median survival time when treated with these particles. In a similar approach, an angiopep-conjugated DGL-based gene delivery system was manufactured, in which the therapeutic gene encoding human glial cell line-derived neurotrophic factor (hGDNF) was incapsulated. In a rotenone-induced chronic model of Parkinson's disease, rats showed best improved locomotor activity and apparent recovery of dopaminergic neurons when treated with these particles compared to those in other groups.

17.5 OTHER BRAIN-DIRECTED CARRIERS

Besides apolipoprotein, antibody, and Angiopep-2 modified carriers, several other systems have been engineered to deliver biologics across the BBB. For example, brain-targeting

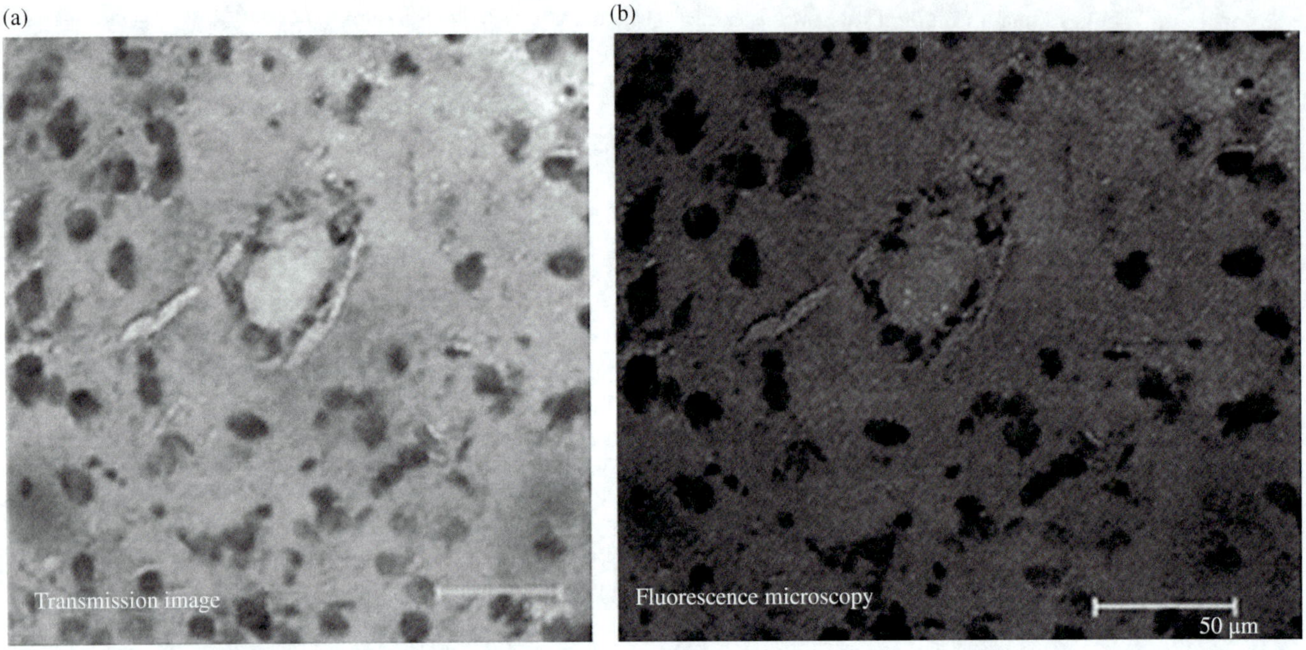

FIGURE 17.3 Passage of surfactant treated nanoparticles across the blood–brain barrier. Sixty minutes after the administration of polysorbate-80-coated fluorescence-labeled nanoparticles to rats by IV injection, particle-associated fluorescence can be assigned to the brain capillary lumen, the endothelial cells, and the perivascular brain tissue. (a) Fluorescence image and (b) transmission image. Reimold et al. [123]. Reproduced with permission of *Eur J Pharm Biopharm.* (*See insert for color representation of this figure.*)

peptide sequences can be directly incorporated into the therapeutic protein. A recently published study demonstrates the successful delivery of the endopeptidase neprilysin, which may be used to reduce the accumulation of Aβ. The authors characterized a recombinant neprilysin that contained a 38 amino acid brain-targeting domain (ASN12) for pharmacology in a mouse as well as efficacy in 2 APPtg mouse models of AD. The recombinant ASN12 transited to the brain with a $t_{1/2}$ of 24 h and accumulated to 1.7% of injected dose at 24 h following IV delivery. Pharmacodynamic evaluation was performed in a tg2576 APPtg mouse with the prion promoter APP695 SWE mutation and in the Line41 mThy1 APP751 mutation mouse. Treatment of either APPtg mouse resulted in reduced Aβ, increased neuronal synapses, and improved learning and memory. The Line41 APPtg mice showed increased levels of C-terminal neuropeptide Y fragments and increased neurogenesis [129].

A 29 amino acid peptide derived from the rabies virus glycoprotein (RVG29) was coupled to PAMAM dendrimers through bifunctional PEG, then complexed with DNA, yielding PAMAM-PEG-RVG29/DNA nanoparticles [130]. These particles were endocytosed by BCECs through a clathrin- and caveolae-mediated energy-depending process, and cellular uptake could be inhibited by free virus glycoprotein and GABA but not by nicotinic acetylcholine receptor agonists/antagonists. PAMAM-PEG-RVG29/DNA NPs showed higher capability to cross the BBB than PAMAM/DNA particles in an *in vitro* BBB model. *In vivo* imaging showed that the NPs were preferably accumulated in brain. Report gene expression of PAMAM-PEG-RVG29/DNA particles was observed in brain and was significantly higher than unmodified NPs.

17.6 STEM CELL-MEDIATED DRUG DELIVERY

The identification of stem cells in the CNS and in association with gliomas has spawned an entire field of research and intense investigation to use them for therapy and as delivery vehicle for biologics. They are ideal candidates for cell-based gene therapy overcoming limitations of viral vectors due to their ability to cross the BBB [131–133]. This makes them particularly interesting for the therapy of glioblastoma multiforme, because of their predicate to home at tumor site, but also for neurodegenerative disorders [134].

Neural stem cells were first used for tumor therapy, but they may also be afflicted with limitations, such as potential risk of tumorigenesis, need for engineering by viral vectors, risk of host insertional mutagenesis, and the lack of being autologous [133]. An alternative are, for example, mesenchymal stem cells [135], which can be obtained from blood cells, bone marrow, or peripheral tissues. Genetically modified stem cells have been used to express suicide genes (cytosine deaminase–uracil phosphoribosyltransferase) to target glioma [136]. The therapeutic stem cells still had tumor tropism when injected to a distant intracranial site and effectively inhibited glioblastoma growth after 5-fluorocytosine therapy.

17.7 FOCUSED ULTRASOUND AND MICROBUBBLES

Among the different strategies to deliver macromolecule therapeutics into the CNS, the focused ultrasound sonication (FUS) together with microbubbles (MBs) gains increasingly interest for the delivery of small molecules and biologics. This approach has been shown to be effective in transiently disrupting the BBB for noninvasive drug delivery [137, 138]. For example, MRI-guided FUS efficiently delivered systemically administered Herceptin [139] to mouse brain or anti-Aβ antibodies to targeted brain regions of the TgCRND8 mouse model of Alzheimer's disease, reducing plaque load within 4 days [140, 141]. A further example is EPO, which could be used as neuroprotective agent against cerebral ischemia/reperfusion-induced brain injury. The MB/focused ultrasound approach significantly increased the cerebral content of EPO by bettering vascular permeability a rodent cerebral infarct model. In acute phase, both significant improvement in neurological score and reduction in infarct volume were found as compared to groups without ultrasound treatment. In chronic phase, long-term behavioral recovery and neuronal loss in brain cortex was significantly improved in the MB/focused ultrasound group [142].

17.8 CONCLUSIONS AND PERSPECTIVES

For a long time, receptor-mediated endocytosis and transcytosis were neglected in the field of CNS drug delivery. However, advances in understanding and defining these processes will provide us with the exceptional possibility to explore new therapeutic avenues. Endocytosis and transcytosis are fundamental processes to take up and transport proteins across endothelial and epithelial cells. This is also the case at the BBB. An increasing number of investigations on targeted drug delivery systems support the concept of the Trojan horse approach using appropriate vectors to carry drugs of interest across the BBB by cytotic mechanisms. With the possibility to engineer new ligands, antibodies, and colloidal carriers, novel and promising therapeutic options have been opened utilizing such processes for delivery of small and large molecules to the brain. Theoretically, almost every recombinant therapeutic protein can be reengineered as bifunctional antibody penetrating the BBB [143]. However, there are still open questions to be answered. For example, in the case of the fusion proteins, cell lines have to be established, which guarantee sufficient production to fulfill large-scale market demands. In addition, there is still not enough information about potential adverse events at chronic use of such delivery systems. As receptor-mediated endocytosis is generally a specific and saturable process, it has to be clarified what happens to the natural ligands of these receptors at long-term drug administration. In addition, for most of the introduced colloidal carrier systems, a detailed toxicological evaluation is still lacking. Furthermore, the available *in vitro* models of the BBB need to be improved to strengthen their predictive value. Systems displaying the complete neurovascular unit may be better suited to reflect the *in vivo* situation than the presently used monocultures of endothelial cells or dual cultures of endothelial cells and astrocytes. However, even if broader use of the above-discussed systems for the administration of biologics will require some time, the continuously ongoing optimization of shuttles and carrier systems by improved surface properties and signal structures will lead to new biocompatible composites, offering novel possibilities of drug delivery to the CNS.

REFERENCES

[1] Pardridge WM. Blood–brain barrier delivery. Drug Discov Today 2007;12:54–61.

[2] Ghose AK, Viswanadhan VN, Wendoloski JJ. A knowledge-based approach in designing combinatorial or medicinal chemistry libraries for drug discovery. 1. A qualitative and quantitative characterization of known drug databases. J Comb Chem 1999;1:55–68.

[3] Regier DA, Boyd JH, Burke JD Jr, Rae DS, Myers JK, Kramer M, Robins LN, George LK, Karno M, Locke BZ. One-month prevalence of mental disorders in the United States. Based on five epidemiologic catchment area sites. Arch Gen Psychiatry 1988;45:977–986.

[4] Pardridge WM. The blood-brain barrier: bottleneck in brain drug development. NeuroRx 2005;2 (1):3–14.

[5] Pardridge WM. Drug transport across the blood-brain barrier. J Cereb Blood Flow Metab 2012;32 (11):1959–1972.

[6] Bauer B, Hartz AM, Fricker G, Miller DS. Modulation of p-glycoprotein transport function at the blood-brain barrier. Exp Biol Med (Maywood) 2005;230:118–127.

[7] Mahringer A, Ott M, Reimold I, Reichel V, Fricker G. The ABC of the blood-brain barrier – regulation of drug efflux pumps. Curr Pharm Des 2011;17:2762–2770.

[8] Neuwelt EA, Bauer B, Fahlke C, Fricker G, Iadecola C, Janigro D, Leybaert L, Molnár Z, O'Donnell ME, Povlishock JT, Saunders NR, Sharp F, Stanimirovic D, Watts RJ, Drewes LR. Engaging neuroscience to advance translational research in brain barrier biology. Nat Rev Neurosci 2011;12:169–182.

[9] Pardridge WM. Strategies for the delivery of drugs through the blood-brain barrier. In: *Annual Reports in Medicinal Chemistry-20*. Academic Press; 1985. p 305–313.

[10] Frank HJ, Pardridge WM, Morris WL, Rosenfeld RG, Choi TB. Binding and internalization of insulin and insulin-like growth factors by isolated brain microvessels. Diabetes 1986;35:654–661.

[11] Urayama A, Grubb JH, Sly WS, Banks WA. Developmentally regulated mannose 6-phosphate receptor-mediated transport of a lysosomal enzyme across the blood-brain barrier. Proc Natl Acad Sci U S A 2004;101:12658–12663.

[12] Banks WA, Lebel CR. Strategies for the delivery of leptin to the CNS. J Drug Target 2002;10:297–308.

[13] Neeper M, Schmidt AM, Brett J, Yan SD, Wang F, Pan YC, Elliston K, Stern D, Shaw A. Cloning and expression of a cell surface receptor for advanced glycosylation end products of proteins. J Biol Chem 1992;267:14998–15004.

[14] Kook SY, Hong HS, Moon M, Ha CM, Chang S, Mook-Jung I. $A\beta_{1-42}$-RAGE interaction disrupts tight junctions of the blood-brain barrier via Ca^{2+}-calcineurin signaling. J Neurosci 2012;32 (26):8845–8854.

[15] Deane RJ. Is RAGE still a therapeutic target for Alzheimer's disease? Future Med Chem 2012;4:915–925.

[16] Zhang Y, Pardridge WM. Mediated efflux of IgG molecules from brain to blood across the blood-brain barrier. J Neuroimmunol 2001;114:168–172.

[17] Garg A, Balthasar JP. Investigation of the influence of FcRn on the distribution of IgG to the brain. AAPS J 2009;11 (3):553–557.

[18] de Vries HE, Kuiper J, de Boer AG, van Berkel TJ, Breimer DD. Characterization of the scavenger receptor on bovine cerebral endothelial cells *in vitro*. J Neurochem 1993;61:1813–1821.

[19] Bu G, Geuze HJ, Strous GJ, Schwartz AL. 39 kDa receptor-associated protein is an ER-resident protein and molecular chaperon for LDL receptor-related protein. EMBO J 1995;14: 2269–2280.

[20] Demeule M, Régina A, Ché C, Poirier J, Nguyen T, Gabathuler R, Castaigne JP, Béliveau R. Identification and design of peptides as a new drug delivery system for the brain. J Pharmacol Exp Ther 2008;324 (3):1064–1072.

[21] Demeule M, Currie JC, Bertrand Y, Ché C, Nguyen T, Régina A, Gabathuler R, Castaigne JP, Béliveau R. Involvement of the low-density lipoprotein receptor-related protein in the transcytosis of the brain delivery vector angiopep-2. J Neurochem 2008;106:1534–1544.

[22] Skarlatos S, Yoshikawa T, Pardridge WM. Transport of [125I]transferrin through the rat blood-brain barrier. Brain Res 1995;683:164–171.

[23] Hervé F, Ghinea N, Scherrmann JM. CNS delivery via adsorptive transcytosis. AAPS J 2008;10:455–472.

[24] Tuma P, Hubbard AL. Transcytosis: crossing cellular barriers. Physiol Rev 2003;83 (3):871–932.

[25] Simionescu M, Ghinea N, Fixman A, Lasser M, Kukes L, Simionescu N, Palade GE. The cerebral microvasculature of the rat: structure and luminal surface properties during early development. J Submicrosc Cytol Pathol 1988;20 (2):243–261.

[26] Bickel U, Yoshikawa T, Pardridge WM. Delivery of peptides and proteins through the blood-brain barrier. Adv Drug Deliv Rev 2001;46 (1–3):247–279.

[27] Kumagai AK, Eisenberg JB, Pardridge WM. Absorptive-mediated endocytosis of cationized albumin and a beta-endorphin-cationized albumin chimeric peptide by isolated brain capillaries. Model system of blood-brain barrier transport. J Biol Chem 1987;262 (31):15214–15219.

[28] Cheng Y, Zak O, Aisen P, Harrison SC, Walz T. Structure of the human transferrin receptor-transferrin complex. Cell 2004;116:565–576.

[29] Kawabata H, Yang R, Hirama T, Vuong PT, Kawano S, Gombart AF, Koeffler HP. Molecular cloning of transferrin receptor 2. A new member of the transferrin receptor-like family. J Biol Chem 1999;274:20826–20832.

[30] Li JY, Boado RJ, Pardridge WM. Blood-brain barrier genomics. J Cereb Blood Flow Metab 2001;21:61–68.

[31] Wang D, El-Amouri SS, Dai M, Kuan CY, Hui DY, Brady RO, Pan D. Engineering a lysosomal enzyme with a derivative of receptor-binding domain of apoE enables delivery across the blood-brain barrier. Proc Natl Acad Sci U S A. 2013;110: 2999–3004.

[32] Yu YJ, Zhang Y, Kenrick M, Hoyte K, Luk W, Lu Y, Atwal J, Elliott JM, Prabhu S, Watts RJ, Dennis MS. Boosting brain uptake of a therapeutic antibody by reducing its affinity for a transcytosis target. Sci Transl Med 2011;3 (84):84ra44.

[33] Staquicini FI, Ozawa MG, Moya CA, Driessen WH, Barbu EM, Nishimori H, Soghomonyan S, Flores LG II, Liang X, Paolillo V, Alauddin MM, Basilion JP, Furnari FB, Bogler O, Lang FF, Aldape KD, Fuller GN, Höök M, Gelovani JG, Sidman RL, Cavenee WK, Pasqualini R, Arap W. Systemic combinatorial peptide selection yields a non-canonical iron-mimicry mechanism for targeting tumors in a mouse model of human glioblastoma. J Clin Invest 2011;121 (1):161–173.

[34] Sumbria RK, Hui EK, Lu JZ, Boado RJ, Pardridge WM. Disaggregation of amyloid plaque in brain of Alzheimer's disease transgenic mice with daily subcutaneous administration of a tetravalent bispecific antibody that targets the transferrin receptor and the Abeta amyloid peptide. Mol Pharm 2013;10: 3507–3513.

[35] Jefferies WA, Brandon MR, Hunt SV, Williams AF, Gatter KC, Mason DY. Transferrin receptor on endothelium of brain capillaries. Nature 1984;312 (5990):162–163.

[36] Friden PM, Walus LR, Musso GF, Taylor MA, Malfroy B, Starzyk RM. Anti-transferrin receptor antibody and antibody-drug conjugates cross the blood-brain barrier. Proc Natl Acad Sci U S A 1991;88 (11):4771–4775.

[37] Boado RJ, Tsukamoto H, Pardridge WM. Drug delivery of antisense molecules to the brain for treatment of Alzheimer's disease and cerebral AIDS. J Pharm Sci 1998;87 (11): 1308–1315.

[38] Pardridge WM. Blood-brain barrier drug delivery of IgG fusion proteins with a transferrin receptor monoclonal antibody. Expert Opin Drug Deliv 2015;12 (2):207–222. [Epub ahead of print].

[39] Kurihara A, Deguchi Y, Pardridge WM. Epidermal growth factor radiopharmaceuticals: 111In chelation, conjugation to a blood-brain barrier delivery vector via a biotin-polyethylene linker, pharmacokinetics, and *in vivo* imaging of experimental brain tumors. Bioconjug Chem 1999;10:502–511.

[40] Atwal JK, Chen Y, Chiu C, Mortensen DL, Meilandt WJ, Liu Y, Heise CE, Hoyte K, Luk W, Lu Y, Peng K, Wu P, Rouge L, Zhang Y, Lazarus RA, Scearce-Levie K, Wang W, Wu Y, Tessier-Lavigne M, Watts RJ. A therapeutic antibody targeting BACE1 inhibits amyloid-ß production *in vivo*. Sci Transl Med 2011;3 (84), Article ID 84ra43.

[41] Bien-Ly N, Yu YJ, Bumbaca D, Elstrott J, Boswell CA, Zhang Y, Luk W, Lu Y, Dennis MS, Weimer RM, Chung I,

Watts RJ. Transferrin receptor (TfR) trafficking determines brain uptake of TfR antibody affinity variants. J Exp Med 2014;211 (2):233–244.

[42] Niewoehner J, Bohrmann B, Collin L, Urich E, Sade H, Maier P, Rueger P, Stracke JO, Lau W, Tissot AC, Loetscher H, Ghosh A, Freskgård PO. Increased brain penetration and potency of a therapeutic antibody using a monovalent molecular shuttle. Neuron 2014;81:49–60.

[43] Van Houten M, Posner BI. Insulin binds to brain blood vessels *in vivo*. Nature 1979;282:623–625.

[44] Pardridge WM, Frank HJL, Cornford EM, Braun LD, Crane D, Oldendorf WH. Neuropeptides and the blood brain barrier. In: Martin JB, Reichlin S, Bick KL, editors. *Neurosecretion and Brain Peptides*. New York: Raven Press; 1981. p 321–328.

[45] Pardridge WM, Eisenberg J, Yang J. Human blood brain barrier insulin receptor. J Neurochem 1985;44:1771–1778.

[46] Ayre SG, Skaletski B, Mosnaim AD. Blood-brain barrier passage of azidothymidine in rats: effect of insulin. Res Commun Chem Pathol Pharmacol 1989;63 (1):45–52.

[47] Wu D, Yang J, Pardridge WM. Drug targeting of a peptide radiopharmaceutical through the primate blood-brain barrier *in vivo* with a monoclonal antibody to the human insulin receptor. J Clin Invest 1997;100:1804–1812.

[48] Zhang Y, Schlachetzki F, Pardridge WM. Global non-viral gene transfer to the primate brain following intravenous administration. Mol Ther 2003;7:11–18.

[49] Boado RJ, Zhang Y, Zhang Y, Pardridge WM. Genetic engineering, expression, and activity of a fusion protein of a human neurotrophin and a molecular Trojan horse for delivery across the human blood-brain barrier. Biotechnol Bioeng 2007;97:1376–1386.

[50] Boado RJ, Zhang Y, Zhang Y, Xia CF, Wang Y, Pardridge WM. Genetic engineering of a lysosomal enzyme fusion protein for targeted delivery across the human blood-brain barrier. Biotechnol Bioeng 2008;99:475–484.

[51] Boado RJ, Hui EK, Lu JZ, Pardridge WM. Drug targeting of erythropoietin across the primate blood-brain barrier with an IgG molecular Trojan horse. J Pharmacol Exp Ther 2010; 333:961–969.

[52] Boado RJ, Hui EK, Lu JZ, Zhou QH, Pardridge WM. Selective targeting of a TNFR decoy receptor pharmaceutical to the primate brain as a receptor-specific IgG fusion protein. J Biotechnol 2010;146:84–91.

[53] Boado RJ, Ka-Wai Hui E, Zhiqiang Lu J, Pardridge WM. Insulin receptor antibody-iduronate 2-sulfatase fusion protein: pharmacokinetics, anti-drug antibody, and safety pharmacology in Rhesus monkeys. Biotechnol Bioeng 2014;111 (11):2317–2325.

[54] Coloma MJ, Lee HJ, Kurihara A, Landaw EM, Boado RJ, Morrison SL, Pardridge WM. Transport across the primate blood-brain barrier of a genetically engineered chimeric monoclonal antibody to the human insulin receptor. Pharm Res 2000;17:266–274.

[55] Boado RJ, Zhang Y, Zhang Y, Pardridge WM. Humanization of anti-human insulin receptor antibody for drug targeting across the human blood-brain barrier. Biotechnol Bioeng 2007;96:381–391.

[56] Jones AR, Shusta EV. Blood-brain barrier transport of therapeutics via receptor-mediation. Pharm Res 2007;24: 1759–1771.

[57] Méresse S, Delbart C, Fruchart JC, Cecchelli R. Low-density lipoprotein receptor on endothelium of brain capillaries. J Neurochem 1989;53:340–345.

[58] Lucarelli M, Borrelli V, Fiori A, Cucina A, Granata F, Potenza RL, Scarpa S, Cavallaro A, Strom R. The expression of native and oxidized LDL receptors in brain microvessels is specifically enhanced by astrocytes-derived soluble factor(s). FEBS Lett 2002;522:19–23.

[59] Kuwahara H, Nishina K, Yoshida K, Nishina T, Yamamoto M, Saito Y, Piao W, Yoshida M, Mizusawa H, Yokota T. Efficient *in vivo* delivery of siRNA into brain capillary endothelial cells along with endogenous lipoprotein. Mol Ther 2011;19 :2213–2221.

[60] Malcor JD, Payrot N, David M, Faucon A, Abouzid K, Jacquot G, Floquet N, Debarbieux F, Rougon G, Martinez J, Khrestchatisky M, Vlieghe P, Lisowski V. Chemical optimization of new ligands of the low-density lipoprotein receptor as potential vectors for central nervous system targeting. J Med Chem 2012;55:2227–2241.

[61] Zhang B, Sun X, Mei H, Wang Y, Liao Z, Chen J, Zhang Q, Hu Y, Pang Z, Jiang X. LDLR-mediated peptide-22-conjugated nanoparticles for dual-targeting therapy of brain glioma. Biomaterials 2013;34:9171–9182.

[62] Dieckmann M, Dietrich MF, Herz J. Lipoprotein receptors--an evolutionarily ancient multifunctional receptor family. Biol Chem 2010;391 (11):1341–1363.

[63] Boucher P, Herz J. Signaling through LRP1: protection from atherosclerosis and beyond. Biochem Pharmacol 2011;81 (1):1–5.

[64] Shibata M, Yamada S, Kumar SR, Calero M, Bading J, Frangione B, Holtzman DM, Miller CA, Strickland DK, Ghiso J, Zlokovic BV. Clearance of Alzheimer's amyloid-ss(1-40) peptide from brain by LDL receptor-related protein-1 at the blood-brain barrier. J Clin Invest 2000;106:1489–1499.

[65] Davis J, Xu F, Deane R, Romanov G, Previti ML, Zeigler K, Zlokovic BV, Van Nostrand WE. Early-onset and robust cerebral microvascular accumulation of amyloid beta-protein in transgenic mice expressing low levels of a vasculotropic Dutch/Iowa mutant form of amyloid beta-protein precursor. J Biol Chem 2004;279:20296–20306.

[66] Bell RD, Sagare AP, Friedman AE, Bedi GS, Holtzman DM, Deane R, Zlokovic BV. Transport pathways for clearance of human Alzheimer's amyloid beta-peptide and apolipoproteins E and J in the mouse central nervous system. J Cereb Blood Flow Metab 2007;27:909–918.

[67] Benchenane K, Berezowski V, Ali C, Fernández-Monreal M, López-Atalaya JP, Brillault J, Chuquet J, Nouvelot A, MacKenzie ET, Bu G, Cecchelli R, Touzani O, Vivien D. Tissue-type plasminogen activator crosses the intact blood-brain barrier by low-density lipoprotein receptor-related protein-mediated transcytosis. Circulation 2005;111:2241–2249.

[68] Deane R, Sagare A, Hamm K, Parisi M, Lane S, Finn MB, Holtzman DM, Zlokovic BV. ApoE isoform-specific disruption of myeloid beta peptide clearance from mouse brain. J Clin Invest 2008;118:4002–4013.

[69] Pan W, Kastin AJ, Zankel TC, van Kerkhof P, Terasaki T, Bu G. Efficient transfer of receptor-associated protein (RAP) across the blood-brain barrier. J Cell Sci 2004;117:5071–5078.

[70] Sagare AP, Deane R, Zlokovic BV. Low-density lipoprotein receptor-related protein 1: a physiological Aβ homeostatic mechanism with multiple therapeutic opportunities. Pharmacol Ther 2012;136:94–105.

[71] Orlando RA, Farquhar MG. Functional domains of the receptor-associated protein (RAP). Proc Natl Acad Sci U S A 1994;91:3161–3165.

[72] Zheng G, Bachinsky DR, Stamenkovic I, Strickland DK, Brown D, Andres G, McCluskey RT. Organ distribution in rats of two members of the low-density lipoprotein receptor gene family, gp330 and LRP/alpha 2MR, and the receptor-associated protein (RAP). J Histochem Cytochem 1994;42:531–542.

[73] Zlokovic BV, Martel CL, Matsubara E, McComb JG, Zheng G, McCluskey RT, Frangione B, Ghiso J. Glycoprotein 330/megalin: probable role in receptor-mediated transport of apolipoprotein J alone and in a complex with Alzheimer disease amyloid β at the blood-brain and blood-cerebrospinal fluid barriers. Proc Natl Acad Sci U S A 1996;93:4229–4234.

[74] Shayo M, McLay RN, Kastin AJ, Banks WA. The putative blood-brain barrier transporter for the β-amyloid binding protein apolipoprotein J is saturated at physiological concentrations. Life Sci 1997;60:PL115–PL118.

[75] Zhang Y, Proenca R, Maffei M, Barone M, Leopold L, Friedman JM. Positional cloning of the mouse obese gene and its human homologue. Nature 1994;372:425–432.

[76] Liu Y, Li J, Shao K, Huang R, Ye L, Lou J, Jiang C. A leptin derived 30-amino-acid peptide modified pegylated poly-L-lysine dendrigraft for brain targeted gene delivery. Biomaterials 2010;31:5246–5257.

[77] Deane R, Du Yan S, Submamaryan RK, LaRue B, Jovanovic S, Hogg E, Welch D, Manness L, Lin C, Yu J, Zhu H, Ghiso J, Frangione B, Stern A, Schmidt AM, Armstrong DL, Arnold B, Liliensiek B, Nawroth P, Hofman F, Kindy M, Stern D, Zlokovic B. RAGE mediates amyloid-beta peptide transport across the blood-brain barrier and accumulation in brain. Nat Med 2003;9:907–913.

[78] Deane R, Bell RD, Sagare A, Zlokovic BV. Clearance of amyloid-beta peptide across the blood-brain barrier: implication for therapies in Alzheimer's disease. CNS Neurol Disord Drug Targets 2009;8:16–30.

[79] Candela P, Gosselet F, Saint-Pol J, Sevin E, Boucau MC, Boulanger E, Cecchelli R, Fenart L. Apical-to-basolateral transport of amyloid-β peptides through blood-brain barrier cells is mediated by the receptor for advanced glycation end-products and is restricted by P-glycoprotein. J Alzheimers Dis 2010;22:849–859.

[80] Park R, Kook SY, Park JC, Mook-Jung I. Aβ1-42 reduces P-glycoprotein in the blood-brain barrier through RAGE-NF-κB signaling. Cell Death Dis 2014;5:e1299.

[81] Tokuda H, Masuda S, Takakura Y, Sezaki H, Hashida M. Specific uptake of succinylated proteins via a scavenger receptor-mediated mechanism in cultured brain microvessel endothelial cells. Biochem Biophys Res Commun 1993;196: 18–24.

[82] Lucarelli M, Gennarelli M, Cardelli P, Novelli G, Scarpa S, Dallapiccola B, Strom R. Expression of receptors for native and chemically modified low-density lipoproteins in brain microvessels. FEBS Lett 1997;401:53–58.

[83] Goti D, Hrzenjak A, Levak-Frank S, Frank S, van der Westhuyzen DR, Malle E, Sattler W. Scavenger receptor class B, type I is expressed in porcine brain capillary endothelial cells and contributes to selective uptake of HDL-associated vitamin E. J Neurochem 2001;76:498–508.

[84] Petri B, Bootz A, Khalansky A, Hekmatara T, Müller R, Uhl R, Kreuter J, Gelperina S. Chemotherapy of brain tumour using doxorubicin bound to surfactant-coated poly(butyl cyanoacrylate) nanoparticles: revisiting the role of surfactants. J Control Release 2007;117:51–58.

[85] Kreuter J. Mechanism of polymeric nanoparticle-based drug transport across the blood-brain barrier (BBB). J Microencapsul 2013;30:49–54.

[86] Bickel U, Yoshikawa T, Landaw EM, Faull KF, Pardridge WM. Pharmacologic effects *in vivo* in brain by vector-mediated peptide drug delivery. Proc Natl Acad Sci U S A 1993;90:2618–2622.

[87] Osborn MJ, McElmurry RT, Peacock B, Tolar J, Blazar BR. Targeting of the CNS in MPS-IH using a nonviral transferrin-alpha-L-iduronidase fusion gene product. Mol Ther 2008;16: 1459–1466.

[88] Zhang Y, Pardridge WM. Delivery of beta-galactosidase to mouse brain via the blood-brain barrier transferrin receptor. J Pharmacol Exp Ther 2005;313:1075–1081.

[89] Boado RJ, Hui EK, Lu JZ, Zhou QH, Pardridge WM. Reversal of lysosomal storage in brain of adult MPS-I mice with intravenous Trojan horse-iduronidase fusion protein. Mol Pharm 2011;8:1342–1350.

[90] Boado RJ, Hui EK, Lu JZ, Pardridge WM. Glycemic control and chronic dosing of rhesus monkeys with a fusion protein of iduronidase and a monoclonal antibody against the human insulin receptor. Drug Metab Dispos 2012;40:2021–2025.

[91] Boado RJ, Zhang Y, Zhang Y, Wang Y, Pardridge WM. GDNF fusion protein for targeted-drug delivery across the human blood-brain barrier. Biotechnol Bioeng 2008;100:387–396.

[92] Pardridge WM, Boado RJ. Pharmacokinetics and safety in rhesus monkeys of a monoclonal antibody-GDNF fusion protein for targeted blood-brain barrier delivery. Pharm Res 2009;26:2227–2236.

[93] Wu D, Pardridge WM. Neuroprotection with noninvasive neurotrophin delivery to the brain. Proc Natl Acad Sci U S A 1999;96:254–259.

[94] Zhang Y, Pardridge WM. Blood-brain barrier targeting of BDNF improves motor function in rats with middle cerebral artery occlusion. Brain Res 2006;1111:227–229.

[95] Boado RJ, Lu JZ, Hui EK, Sumbria RK, Pardridge WM. Pharmacokinetics and brain uptake in the rhesus monkey of a fusion protein of arylsulfatase a and a mono-clonal antibody

against the human insulin receptor. Biotechnol Bioeng 2013;110:1456–1465.

[96] Boado RJ, Lu JZ, Hui EK, Pardridge WM. Insulin receptor antibody-sulfamidase fusion protein penetrates the primate blood-brain barrier and reduces glycosoaminoglycans in Sanfilippo type a cells. Mol Pharm 2014;11:2928–2934.

[97] Walus LR, Pardridge WM, Starzyk RM, Friden PM. Enhanced uptake of rsCD4 across the rodent and primate blood-brain barrier after conjugation to anti-transferrin receptor antibodies. J Pharmacol Exp Ther 1996;277: 1067–1075.

[98] Kordower JH, Charles V, Bayer R, Bartus RT, Putney S, Walus LR, Friden PM. Intravenous administration of a transferrin receptor antibody-nerve growth factor conjugate prevents the degeneration of cholinergic striatal neurons in a model of Huntington disease. Proc Natl Acad Sci U S A 1994;91:9077–9080.

[99] Shi N, Boado RJ, Pardridge WM. Antisense imaging of gene expression in the brain *in vivo*. Proc Natl Acad Sci U S A 2000;97:14709–14714.

[100] Lee HJ, Boado RJ, Braasch DA, Corey DR, Pardridge WM. Imaging gene expression in the brain *in vivo* in a transgenic mouse model of Huntington's disease with an antisense radiopharmaceutical and drug-targeting technology. J Nucl Med 2002;43:948–956.

[101] Ohshima-Hosoyama S, Simmons HA, Goecks N, Joers V, Swanson CR, Bondarenko V, Velotta R, Brunner K, Wood LD, Hruban RH, Emborg ME. A monoclonal antibody-GDNF fusion protein is not neuroprotective and is associated with proliferative pancreatic lesions in parkinsonian monkeys. PLoS One 2012;7 (6):e39036.

[102] Bohrmann B, Baumann K, Benz J, Gerber F, Huber W, Knoflach F, Messer J, Oroszlan K, Rauchenberger R, Richter WF, Rothe C, Urban M, Bardroff M, Winter M, Nordstedt C, Loetscher H. Gantenerumab: a novel human anti-abeta antibody demonstrates sustained cerebral amyloid-beta binding and elicits cellmediated removal of human amyloid-beta. J Alzheimers Dis 2012;28:49–69.

[103] Farrington GK, Caram-Salas N, Haqqani AS, Brunette E, Eldredge J, Pepinsky B, Antognetti G, Baumann E, Ding W, Garber E, Jiang S, Delaney C, Boileau E, Sisk WP, Stanimirovic DB. A novel platform for engineering blood-brain barrier-crossing bispecific biologics. FASEB J. 2014;28 (11):4764–4778. [Epub ahead of print].

[104] Sumbria RK, Boado RJ, Pardridge WM. Brain protection fromstroke with intravenous TNFalpha decoy receptor-Trojan horse fusion protein. J Cereb Blood Flow Metab 2012;32:1933–1938.

[105] Zhou QH, Boado RJ, Hui EK, Lu JZ, Pardridge WM. Chronic dosing of mice with a transferrin receptor monoclonal antibody-glial-derived neurotrophic factor fusion protein. Drug Metab Dispos 2011;39:1149–1154.

[106] Hussain MM, Strickland DK, Bakillah A. The mammalian low-density lipoprotein receptor family. Annu Rev Nutr 1999;19:141–172.

[107] Dehouck MP, Jolliet-Riant P, Brée F, Fruchart JC, Cecchelli R, Tillement JP. Drug transfer across the blood brain barrier: correlation between *in vitro* and *in vivo* models. J Neurochem 1992;58:1790–1797.

[108] Yainoy S, Houbloyfa P, Eiamphungporn W, Isarankura-Na-Ayudhya C, Prachayasittikul V. Engineering of chimeric catalase-Angiopep-2 for intracellular protection of brain endothelial cells against oxidative stress. Int J Biol Macromol 2014;68:60–66.

[109] Vivès E, Brodin P, Lebleu B. A truncated HIV-1 Tat protein basic domain rapidly translocates through the plasma membrane and accumulates in the cell nucleus. J Biol Chem 1997;272:16010–16017.

[110] Wender PA, Mitchell DJ, Pattabiraman K, Pelkey ET, Steinman L, Rothbard JB. The design, synthesis, and evaluation of molecules that enable or enhance cellular uptake: peptoid molecular transporters. Proc Natl Acad Sci U S A 2000;97:13003–13008.

[111] Drin G, Mazel M, Clair P, Mathieu D, Kaczorek M, Temsamani J. Physico-chemical requirements for cellular uptake of pAntp peptide. Role of lipid-binding affinity. Eur J Biochem 2001;268:1304–1314.

[112] Dietz GP, Bähr M. Synthesis of cell-penetrating peptides and their application in neurobiology. Methods Mol Biol 2007;399:181–198.

[113] Huwyler J, Wu D, Pardridge WM. Brain drug delivery of small molecules using immunoliposomes. Proc Natl Acad Sci U S A 1996;93:14164–14169.

[114] Papahadjopoulos D, Allen TM, Gabizon A, Mayhew E, Matthay K, Huang SK, Lee KD, Woodle MC, Lasic DD, Redemann C, Martin FJ. Sterically stabilized liposomes: improvements in pharmacokinetics and antitumor therapeutic efficacy. Proc Natl Acad Sci U S A. 1991;88: 11460–11464.

[115] Woodle MC, Lasic DD. Sterically stabilized liposomes. Biochim Biophys Acta 1992;1113:171–199.

[116] Zhang Y, Calon F, Zhu C, Boado RJ, Pardridge WM. Intravenous nonviral gene therapy causes normalization of striatal tyrosine hydroxylase and reversal of motor impairment in experimental parkinsonism. Hum Gene Ther 2003;14:1–12.

[117] Pardridge WM. *Brain Drug Targeting: The Future of Brain Development*. Cambridge: Cambridge University Press; 2001. p 1–370.

[118] Lierse W, Horstmann E. Quantitative anatomy of the cerebral vascular bed with especial emphasis on homogeneity and inhomogeneity in small parts of the gray and white matter. Acta Neurol 1959;14:15–19.

[119] Ko YT, Bhattacharya R, Bickel U. Liposome encapsulated polyethylenimine/ODN polyplexes for brain targeting. J Control Release 2009;133 (3):230–237.

[120] Shi N, Zhang Y, Zhu C, Boado RJ, Pardridge WM. Brain-specific expression of an exogenous gene after i.v. administration. Proc Natl Acad Sci U S A 2001;98 (22): 12754–12759.

[121] Kreuter J. Nanoparticulate systems for brain delivery of drugs. Adv Drug Deliv Rev 2001;47:65–81.

[122] Kreuter J, Hekmatara T, Dreis S, Vogel T, Gelperina S, Langer K. Covalent attachment of apolipoprotein A-I and

apolipoprotein B-100 to albumin nanoparticles enables drug transport into the brain. J Control Release 2007;118: 54–58.

[123] Reimold I, Domke D, Bender J, Seyfried CA, Radunz HE, Fricker G. Delivery of nanoparticles to the brain detected by fluorescence microscopy. Eur J Pharm Biopharm 2008;70 (2):627–632.

[124] Zensi A, Begley D, Pontikis C, Legros C, Mihoreanu L, Wagner S, Büchel C, von Briesen H, Kreuter J. Albumin nanoparticles targeted with Apo E enter the CNS by transcytosis and are delivered to neurones. J Control Release 2009;137:78–86.

[125] Kurakhmaeva KB, Djindjikhashvili IA, Petrov VE, Balabanyan VU, Voronina TA, Trofimov SS, Kreuter J, Gelperina S, Begley D, Alyautdin RN. Brain targeting of nerve growth factor using poly(butyl cyanoacrylate) nanoparticles. J Drug Target 2009;17:564–574.

[126] Bonner LT, Peskind ER. Pharmacologic treatments of dementia. Med Clin North Am 2002;86:657–674.

[127] Ke W, Shao K, Huang R, Han L, Liu Y, Li J, Kuang Y, Ye L, Lou J, Jiang C. Gene delivery targeted to the brain using an Angiopep-conjugated polyethyleneglycol-modified polyamidoamine dendrimer. Biomaterials 2009;30:6976–6985.

[128] Huang S. Dual targeting effect of Angiopep-2-modified, DNA-loaded nanoparticles for glioma. Biomaterials 2011; 32:6832–6838.

[129] Spencer B, Verma I, Desplats P, Morvinski D, Rockenstein E, Adame A, Masliah E. A neuroprotective brain-penetrating endopeptidase fusion protein ameliorates Alzheimer disease pathology and restores neurogenesis. J Biol Chem 2014;289: [17917]–17931.

[130] Liu Y, Huang R, Han L, Ke W, Shao K, Ye L, Lou J, Jiang C. Brain-targeting gene delivery and cellular internalization mechanisms for modified rabies virus glycoprotein RVG29 nanoparticles. Biomaterials 2009;30:4195–4202.

[131] Binello E, Germano IM. Targeting glioma stem cells: a novel framework for brain tumors. Cancer Sci 2011;102 (11):1958–1966.

[132] Binello E, Germano IM. Stem cells as therapeutic vehicles for the treatment of high-grade gliomas. Neuro Oncol 2012;14 (3):256–265.

[133] Germano IM, Binello E. Stem cells and gliomas: past, present, and future. J Neurooncol 2014;119 (3):547–555.

[134] Aleynik A, Gernavage KM, Mourad YS, Sherman LS, Liu K, Gubenko YA, Rameshwar P. Stem cell delivery of therapies for brain disorders. Clin Transl Med 2014;3:24.

[135] Nakamura K, Ito Y, Kawano Y, Kurozumi K, Kobune M, Tsuda H, Bizen A, Honmou O, Niitsu Y, Hamada H. Antitumor effect of genetically engineered mesenchymal stem cells in a rat glioma model. Gene Ther 2009;11:1155–1164.

[136] Altanerova V, Cihova M, Babic M, Rychly B, Ondicova K, Mravec B, Altaner C. Human adipose tissue-derived mesenchymal stem cells expressing yeast cytosinedeaminase::uracil phosphoribosyltransferase inhibit intracerebral rat glioblastoma. Int J Cancer 2012;130:2455–2463.

[137] Hynynen K, McDannold N, Vykhodtseva N, Jolesz FA. Noninvasive MR imaging-guided focal opening of the blood–brain barrier in rabbits. Radiology 2001;220:640–646.

[138] Yang FY, Wang HE, Lin GL, Lin HH, Wong TT. Evaluation of the increase in permeability of the blood–brain barrier during tumor progression after pulsed focused ultrasound. Int J Nanomedicine 2012;7:723–730.

[139] Kinoshita M, McDannold N, Jolesz FA, Hynynen K. Noninvasive localized delivery of Herceptin to the mouse brain by MRI-guided focused ultrasound-induced blood-brain barrier disruption. Proc Natl Acad Sci U S A 2006;103:11719–11723.

[140] Jordão JF, Ayala-Grosso CA, Markham K, Huang Y, Chopra R, McLaurin J, Hynynen K, Aubert I. Antibodies targeted to the brain with image-guided focused ultrasound reduces amyloid-beta plaque load in the TgCRND8 mouse model of Alzheimer's disease. PLoS One 2010;5:e10549.

[141] Jordão JF, Thévenot E, Markham-Coultes K, Scarcelli T, Weng YQ, Xhima K, O'Reilly M, Huang Y, McLaurin J, Hynynen K, Aubert I. Amyloid-β plaque reduction, endogenous antibody delivery and glial activation by brain-targeted, transcranial focused ultrasound. Exp Neurol 2013;248: 16–29.

[142] Wu SK, Yang MT, Kang KH, Liou HC, Lu DH, Fu WM, Lin WL. Targeted delivery of erythropoietin by transcranial focused ultrasound for neuroprotection against ischemia/reperfusion-induced neuronal injury: a long-term and short-term study. PLoS One 2014;9 (2):e90107.

[143] Pardridge WM, Boado RJ. Reengineering biopharmaceuticals for targeted delivery across the blood-brain barrier. Methods Enzymol 2012;503:269–292.

18

MOLECULAR PATHOLOGY TECHNIQUES IN THE PRECLINICAL DEVELOPMENT OF THERAPEUTIC BIOLOGICS

Thierry Flandre[1], Sarah Taplin[1], Stewart Jones[2] and Peter Lloyd[3]

[1] *Novartis Pharma AG, Basel, Switzerland*

[2] *AstraZeneca, Cambridge, UK*

[3] *KinDyn Consulting Ltd., Horsham, UK*

18.1 INTRODUCTION

The tools and technologies used to describe the absorption, distribution, metabolism, and excretion (ADME) of small molecular weight chemical entities have been well characterized; however, understanding the ADME of large protein- and nucleic-acid-based therapeutics requires a different approach. For example, while traditional radiolabeled studies may provide information regarding the initial distribution of a therapeutic protein, the ability to fully characterize ADME using this approach is quite limited due to recycling of the radiolabel or entrapment of released label in a particular target cell type. Understanding the fate of the molecular probe is therefore critical in this assessment [1] and it is prudent to assume that not all molecules of a certain class will behave in the same way. For example, subtle changes in two monoclonal antibodies (mAbs) with identical FcRn binding domain may dictate that these behave very differently in terms of their pharmacokinetic (PK) characteristics [2]. In addition, while protein therapeutics offer the possibility of exquisite target selectivity, there are examples of off-target binding that affect the PK of the therapeutic protein [3, 4]. The anatomical and physicochemical properties of the environment into which these large molecules may distribute will also play a role in affecting delivery to a particular site, such as glycan chains on cell surfaces and proteoglycans within the extracellular matrix [5–7]. Consequently, a range of molecular pathology-based cellular and biological techniques, which have typically been used to facilitate the interpretation of histopathology data, by providing further insight into the molecular events in the pathogenesis of disease [8], have more recently been applied to fully characterize the ADME properties of protein therapeutics.

In this chapter, we review a range of molecular pathology-based techniques that are currently used in preclinical animal studies to assist in answering specific scientific questions related to target biology (expression) and biotherapeutic distribution (including off-target binding); see overview Table 18.1. We also highlight the importance of using a combination of appropriate complementary technologies rather than a single technique and an integrated approach that combines PK, target expression, concentration, receptor occupancy, pathology, immunogenicity, and other relevant information within the same study to characterize and interpret the ADME profile of a biotherapeutic. This integrated approach can also serve to maximize the information generated in a single study and can limit the number of animals used.

ADME and Translational Pharmacokinetics/Pharmacodynamics of Therapeutic Proteins: Applications in Drug Discovery and Development, First Edition. Edited by Honghui Zhou and Frank-Peter Theil.

TABLE 18.1 Comparison of Methods

Techniques	Product Detected	Pros	Cons	Type of Target and Localization
ELISA	Protein	Established specific method with quantification and high throughput	Sensitivity depending on antibody. Blood compartments[b] only or aggressive protocol for tissue	Ligand level in blood $\gg$ receptor in tissue[a]
Immunoassay-based method (MSD, luminex)	Protein	Specific method with high throughput and quantification	Sensitivity and specificity depending on antibody. Blood compartments only or aggressive protocol for tissue	Ligand level in blood $\gg$ receptor in tissue[a]
FACS	Protein	Established sensitive and specific method with quantification	Low throughput. Blood compartments only or aggressive protocol for tissue	Receptor level in blood cells $\gg$ tissue[a]
Gene microarray	DNA or RNA	Established sensitive method with high throughput	Lower specificity than qRT-PCR with no quantification.	Receptor (and ligand) expression in tissue[a]
Protein microarray	Protein	Established sensitive method with high throughput and quantification	Aggressive protocol for tissue	Receptor and ligand level in blood or tissue[a]
IHC	Protein	Established specific method and high throughput	Low sensitivity and semiquantitative	Receptor (and ligand) expression in cells/tissues
ISH (and *in situ* PCR)	RNA or DNA	Established specific and sensitive method	Semiquantitative	Receptor (and ligand) expression in cells/tissues
LCM	RNA, DNA, or protein	To capture cells in tissues	No direct detection without secondary technique. Complex protocol with low throughput	Dissection of cells from tissues
LSC	Protein	Specific method with quantification	Sensitivity depending on antibody used	Receptor level in cells/tissues
Mass spectrometry (LC–MS)	Protein	Established sensitive and specific method with quantification	Blood compartments only or aggressive protocol for tissue	Ligand level in blood $\gg$ receptor in tissue[a]
Real-time immunoprotein PCR	Protein	Sensitive and specific method. Much more than IHC or ELISA	Complex technique. Difficulty of getting reagents (antibodies)	Ligand and receptor level in cells/tissues
Real-time qRT-PCR	RNA	Sensitive and specific established method with possibility of absolute quantification	Cost	Receptor (and ligand) expression in tissue $\gg$ cells
Northern blot	RNA	Established specific method	Low specificity with no quantification and low throughput	Receptor (and ligand) expression in tissue[a]
Western blot	Protein	Established specific method	Specificity depending on antibody used. Only semiquantitative and low throughput	Receptor and ligand expression in tissue[a]

[a] Individual cells could also be evaluated but need to be combined with LCM.
[b] Blood, serum and plasma.

18.2 TARGET EXPRESSION PROFILING

Target expression profiling determines the location of target production (either soluble or cell surface ligand) and occasionally the target's receptor (when target is a ligand for the receptor) within human and animal tissues and the impact of these data can be viewed across all phases of the discovery process for therapeutic biologics [9, 10]. Preclinically, target and receptor expression data assist in the selection of markers for potential therapeutic intervention, supports the identification of relevant toxicology species, provides further insights into intended pharmacological functions, facilitates the interpretation of any unanticipated *in vivo* toxicity findings, and recognizes potential tissue sinks that may alter PK profiles. Clinically, target expression data can assist in the molecular stratification of disease states, patient selection, and provide confirmation of anticipated pharmacology.

A range of techniques has been used to assess the level of target expression, at both the DNA/RNA and protein level, using whole tissue extracts. DNA/RNA profiling technologies currently include Northern blots, gene microarrays, and real-time quantitative reverse transcriptase polymerase chain reaction (qRT-PCR). Protein detection, which is often considered more physiologically relevant, generally comprises immunoassay-based technologies that depend on the exquisite specificity and sensitivity of antibody reagents raised against a target protein and include Western blot, protein arrays, protein polymerase chain reaction (PCR), enzyme-linked immunosorbent assay (ELISA), and flow cytometry. To facilitate cellular localization to specific cell types within tissues, laser scanning cytometry (LSC) or laser capture microdissection (LCM) has been combined with these methods, or direct application of *in situ* hybridization (ISH) for RNA detection and immunohistochemistry (IHC) for protein localization on tissue sections can be employed. All of these approaches can be coupled to mass spectrometry for accurate protein identification and quantitation.

18.2.1 Detection of *DNA/RNA*-Based Target Expression Using *Whole Tissue Extracts*

18.2.1.1 Northern Blot Northern blot analysis is a traditional and reliable technique for the detection of RNA (primarily mRNA) in cells or tissues. The method involves extraction of RNA from cells or tissues followed by gel electrophoresis (separation of RNA by size), transfer to nylon membranes, and hybridization using labeled (radioactive or fluorescent) probes complementary to either the full or partial sequence of the target gene. The primary advantages of the Northern blot techniques include (i) detection of alternative splice variants through information on size of RNA molecules, (ii) capacity to detect slight changes in the levels of gene expression, and (iii) absolute quantification of RNA when associated with other secondary methods. However, due to the limitations of high sensitivity to degradation by RNAases and low throughput, this method has largely been replaced by faster, more advanced high throughput techniques.

18.2.1.2 Gene Microarray Gene (DNA) microarray/microchip analysis is a screen for target detection and is also an appropriate assay for the identification of all genes and pathways modulated by the drug in both *in vitro* and *in vivo* models. It can be indicative of the mechanism of action, efficacy, and toxicity of drugs and permits the identification of biomarkers. It was originally developed from Southern blotting methods and is a more advanced high throughput technique that simultaneously screens hundreds to thousands of presynthesized "reporter" cDNA (complementary DNA) oligonucleotides (partial gene sequences of 25mers) attached to a solid chemical matrix of glass, silicon (i.e., Affymetrix chip), or microscopic beads (i.e., Illumina). In summary, target mRNA is extracted from cells or tissues, converted to cDNA using reverse transcriptase PCR (RT-PCR), labeled with a fluorescent chromophore and then hybridized to the respective synthetic complementary "reporter" cDNA fragment immobilized on the microarray. The intensity of fluorescence for each "reporter" cDNA is then measured with the signal intensity directly proportional to the quantity of labeled target cDNA bound to "reporter" cDNA. A relative or semiquantitative assessment of fluorescence is performed by normalizing values using constant parameters such as ribosomal gene 18S or comparable parameters such as pretreatment samples or tumor tissue versus normal tissue. The advantages of microarray techniques include (i) combination of low cost with high throughput, (ii) detection of single-nucleotide polymorphisms or genotyping, (iii) signature profiling based on single or multiple genes of diseases, and (iv) identification of biochemical pathways and biomarkers [11]. However, microarrays usually require significant amounts of total RNA (10 μg per assay) and do not provide absolute quantification of RNA levels. Furthermore, if gene expression is confined to a few specific but critical cells this assay would report expression as either very low or even absent. Since no information is provided for cellular localization, additional techniques such as IHC or ISH in combination with laser capture technology would be required to allow for more precise analysis of gene expression at the cellular level.

18.2.1.3 Real-Time Quantitative Reverse Transcriptase Polymerase Chain Reaction Real-time qRT-PCR is a highly sensitive, quantitative, and qualitative technique for assessing the level of specific mRNA in cells and tissues [12–14]. More specifically, total mRNA is extracted from cells or tissues converted to cDNA using reverse transcriptase and then amplified using real-time quantitative polymerase chain reaction (qPCR) methods. Hybridization to target cDNA using a highly specific probe labeled with a fluorescent chromophore is the most frequent tool to detect and measure product. Standard RT-PCR methods generate copies of RNA at an exponential rate until arrest following the consumption of reagents and the accumulation of inhibitors; this can be variable and result in unreliable quantification of the initial level of RNA template. With real-time PCR, the assay is adapted to enable the product to be quantified during the exponential phase only and extrapolated back to the starting quantity of RNA template. Among the various types of real-time qPCR methods, two are more frequently used. The first method uses the SYBRgreen® fluorescent dye, and the second uses a specific TaqMan® probe bearing a fluorescent molecule at the 5′end and a quenching molecule at the 3′end. The SYBRgreen dye binds the double-stranded DNA generated during the amplification steps, which then emits a fluorescent signal. The TaqMan probe hybridizes the targeted sequence during the annealing step, which is cleaved during the amplification step, liberating the fluorescent molecule that is no more quenched and then emits fluorescence. The quantification is based on the cycle threshold (Ct), which is defined as the number of PCR cycles required to reach a certain level of fluorescent signal. The Ct value is directly proportional to the amount of starting template such that higher quantities of mRNA template in the starting material facilitate faster fluorescence signals and lower Ct values. Absolute quantification is then based on serially diluted standards of known concentrations that are used to generate a standard curve, which defines a linear relationship between Ct and initial quantity of mRNA. There are various different protocols for normalizing fluorescence that allow for a more accurate quantification of mRNA [15]. Normalization can be accomplished using controls including (i) a comparable tissue in terms of weight or volume (easy but inaccurate as it is an estimate of cell number in a determined volume), (ii) total RNA (simple but influenced by cellular processes and RNA quality), (iii) reference housekeeping genes such as glyceraldehyde 3-phosphate dehydrogenase (GAPDH) or β-actin (these internal controls are subject to the same conditions but need to be well validated as they may be affected by treatment), and (iv) ribosomal RNA (28S, the most representative of RNA integrity but are transcribed with a polymerase different from mRNA). Note that the optimal approach utilizes multiple housekeeping genes to normalize results from geometric means to mRNA expression levels. For publication of qRT-PCR data, it is recommended to follow the specific guidelines published by Bustin et al. [16]. The advantages of real-time qRT-PCR include (i) high sensitivity, (ii) wide dynamic range, (iii) accurate quantitative measurement, (iv) low coefficient of variation, and (v) high throughput.

18.2.2 Detection of *Protein*-Based Target Expression Using *Whole Tissue Extracts*

18.2.2.1 Western Blot The Western blot (protein immunoblot) is a widely used analytical technique used to detect specific proteins in a sample of tissue homogenate or extract. Gel electrophoresis, which is voltage driven, is applied first to separate native proteins by isoelectric point (pI), molecular weight, electric charge, or a combination of these factors. Separation depends on the treatment of the sample and the nature of the gel such as sodium dodecyl sulfate–polyacrylamide gel electrophoresis (SDS–PAGE) that maintains polypeptides in a denatured state following reduction to remove secondary and tertiary structures, and thus allows separation of proteins by their molecular weight. Sampled proteins become negatively charged and migrate toward the positively charged electrode. Smaller proteins migrate faster through the mesh and the proteins are thus separated according to size (kDa). The concentration of acrylamide determines the resolution of the gel, the greater the acrylamide concentration the better the resolution of lower molecular weight proteins. It is also possible to use two-dimensional polyacrylamide gel electrophoresis for separating complex protein mixtures using two different parameters such as isoelectric point (relative number of positively and negatively charged amino acids) and molecular weight. In principle, this method allows for the separation of all cellular proteins on a single large gel and can often distinguish between different isoforms of a particular protein. The proteins are then transferred via electro-blotting from within the gel to a nitrocellulose or polyvinylidene difluoride membrane where they are attached by hydrophobic and charged interactions in the same position they held within the gel. A blocking step using protein (bovine serum albumin (BSA) or nonfat dry milk) in Tris-buffered saline and detergent (Tween 20 or Triton X-100) is performed to prevent nonspecific binding. The proteins are then detected with antibodies specific to the target protein [17, 18]. Although Western blotting is an example of a method that can be used to detect analytes at the femtogram level, it has not been adopted as a diagnostic tool because it is only qualitative and cannot be adapted into a multiplex format.

18.2.2.2 Protein Microarray The protein microarray (or protein chip) is a rapid and economical, high throughput automated method used to determine protein interactions and function. This highly sensitive assay generally consists of a supporting surface such as a glass slide, nitrocellulose membrane, bead, or microtiter plate that contains an array of bound proteins or other molecules. Labeled probes are then bound to these immobilized proteins that are detected using various published methods. There are three types of protein microarrays: (i) analytical microarrays (capture arrays) comprise a library of antibodies, aptamers, affibodies, peptide–MHC complexes, or lectins attached to a supporting surface that captures and detects specific proteins from a complex protein solution such as a cell lysate. These microarrays provide information about protein expression levels, binding affinities, and specificities and are therefore especially useful in comparing the protein expression in different matrices and in the identification and profiling of diseased tissues. (ii) Functional protein microarrays (target protein arrays) consist of large numbers of purified proteins (functional full-length or protein domains) immobilized on a platform to determine protein–protein, protein–DNA, protein–RNA, protein–phospholipid, and protein–small molecule interactions. These protein chips are currently used to understand drug target identification, protein interactions, biochemical activity, and immune responses [19]. (iii) Reverse-phase protein microarrays (RPPAs) involve attachment of complex cellular/tissue lysates to a solid matrix, which is then probed with labeled antibodies against a target protein of interest. RPPAs therefore allow for the investigation of altered proteins as a result of disease, such as posttranslational modifications, and protein quantification within the sample lysates using reference peptides. Results from these assays can be readily measured with most commercial scanners.

18.2.2.3 Real-Time Immuno Protein PCR Immuno-PCR has been developed for the sensitive detection of protein analytes using a proximity ligation assay that combines the specificity of antibody–protein binding (immunological techniques) with the sensitivity of detecting reporter nucleic acid by real-time PCR (molecular-based technology) [5, 20–24]. Several independent studies have consistently demonstrated the advantages of this technique over traditional ELISA, some achieving more than a 50,000-fold increase in sensitivity over the equivalent immunoassay [25]. Furthermore, immuno-PCR can be used in a multiplex format and offers accurate quantitation. However, Immuno-PCR has not yet replaced ELISA as the assay format of choice in industry due to the technical challenges associated with linking oligonucleotides to antibodies. Several methods have been developed, although they tend to be very complex, unreliable, and often result in cross-linking. To overcome these problems, an intermediate step is sometimes used (such as the streptavidin–biotin system) but the optimum method is one where the antibody is directly linked to the oligonucleotide [26]. This is a complex technique requiring specialist knowledge and equipment and is very time-consuming. Therefore, immuno-PCRs potential as a diagnostic tool has not yet been realized.

18.2.2.4 Enzyme-Linked Immunosorbent Assays The ELISA is a widely applied method for the detection of a protein antigen from within a complex matrix. The sample with an unknown amount of antigen is immobilized on a solid support either nonspecifically (simple adsorption to the surface) or specifically (captured by another antibody specific to the same antigen). The detection antibody, covalently linked to an enzyme, forms a complex with the antigen and in the presence of a substrate produces a visible signal (color change), which is measured by a spectrometer; the signal intensity correlates with the quantity of antigen in the sample. Absolute quantification is based on a standard curve, which is prepared by serial dilution of a standard antigen at known concentrations and is therefore quantitative around the nanogram to picogram range. There are three types of ELISA. (i) Indirect ELISA: an antigen of interest is first attached to a solid matrix through charge interactions. A primary antibody is then used to specifically bind to the test antigen, which in turn is detected by an enzyme-labeled secondary antibody. The main disadvantage of the indirect ELISA is the lack of specificity especially when complex matrices are used as the source of test antigen. (ii) Sandwich ELISA: a specific "capture antibody" is first attached to a solid matrix and any nonspecific binding sites on the surface are then blocked. The antigen of interest within a complex sample then binds to the "capture antibody" and is subsequently detected by an enzyme-labeled secondary antibody forming a "sandwich." (iii) Competitive ELISA: an antigen of interest is first attached to a solid matrix through charge interactions. An unlabeled antibody is then incubated in the presence of the antigen to form antibody/antigen complexes that are added to an antigen-coated surface. Higher levels of antigen in the sample lead to the formation of more antigen/antibody complexes and less unbound antibodies available for antigen binding, hence "competition." As before, detection occurs through the use of an enzyme-labeled secondary antibody. Note that some competitive ELISA assays include enzyme-linked antigen rather than enzyme-linked antibody where the labeled antigen competes with the sample test antigen for primary antibody-binding sites.

18.2.2.5 Flow Cytometry and Fluorescence-Assisted Cell Sorting Flow cytometry is a laser-based, biophysical technology employed in cell counting, cell sorting, biomarker detection, and protein engineering, by suspending cells in a stream of fluid and passing them by an electronic detection system. It allows simultaneous multiparametric analysis of the physical and chemical characteristics of up to thousands of cells per second. A common variation is to physically sort cells based on their properties, so as to purify populations of interest, this specialized type of flow cytometry is called fluorescence-activated cell sorting (FACS). It provides a method for sorting a heterogeneous mixture of biological cells into two or more containers, one cell at a time, based on the specific light-scattering and fluorescent characteristics of each cell. It is a useful scientific instrument as it provides fast, objective, and quantitative recording of fluorescent signals from individual cells as well as physical separation of cells of particular interest [27]. Labels in flow cytometry can include fluorophores, quantum dots, and isotopes. Fluorophores are attached to an antibody that recognizes a target feature on or in the cell and each fluorophore has a characteristic peak excitation and emission wavelength that often overlaps. Consequently, the combination of labels that can be used depends on the wavelength of the lamp(s) or laser(s) used to excite the fluorochromes and on the detectors available [28]. Quantum dots are sometimes used in place of traditional fluorophores because of their narrower emission peaks. Isotope (lanthanide) labeling and mass cytometry is fundamentally different from flow cytometry since cells are introduced ionized, and associated isotopes are quantified via time-of-flight mass spectrometry. Although this method permits the use of a large number of labels, it currently has lower throughput capacity than flow cytometry. It also destroys the analyzed cells, precluding their recovery by sorting [29].

18.2.2.6 Protein Liquid Chromatography Mass Spectrometry Liquid chromatography mass spectrometry (LC–MS) can be used to quantify specific proteins [30–33], including the therapeutic protein of interest in tissues, although one must be careful in interpreting these data when immuno-capture for protein enrichment is used, since only protein with the required binding epitope is detected. In addition, when using protein LC–MS to quantify specific proteins in tissue, the need to either remove or correct for blood contamination should also be considered. Briefly, the tissue is homogenized and proteins solubilized. The proteins are then cleaved enzymatically and preidentified specific signature peptides within the protein of interest are used to quantify the amount of protein present in the tissue. A control protein, ideally a stable isotope labeled version of the protein of interest, is often used to monitor the efficiency of the extraction process and to control for ionization and ion suppression effects in the mass spectrometer. This approach may offer a distinct advantage for modified therapeutic proteins and novel scaffolds where there is some concern regarding the stability of the molecule *in vivo*. For example, cleavage of a terminal peptide could be detected using this approach if both core and terminal peptides are part of the unique LC–MS signature. In contrast, an ELISA-based assay that is reliant on binding epitopes on the terminal peptide for either the capture or detection reagents would detect rapid clearance of the protein (small peptides are cleared more quickly) and this may be interpreted as rapid clearance of the whole protein if this is the only assay available.

18.2.3 Localization of *DNA/RNA* and *Protein*-Based Target Expression at the Cellular Level Using *Tissue Sections*

18.2.3.1 Laser Capture Microdissection and Laser Scanning Cytometry LCM and LSC are rapid and reliable methods for obtaining pure cell populations of interest from complex tissue sections under direct microscopic visualization [34, 35]. These are powerful tools for qualitative and quantitative analysis of target in paraffin-embedded and/or frozen tissue sections and combine the strengths of flow cytometry, immunofluorescent, or chromagen-based detection methods. It incorporates both an inverted light microscope and a near-infrared laser to facilitate the visualization and procurement of cells. Microscopy (fluorescence, bright field, differential interference contrast, and phase contrast microscopy) with a software interface is first used to visualize cells within tissue sections (typically 5–50 μm thick), target cells are then isolated (cut from the tissue) using an ultraviolet pulsed laser (355 nm) and then collected directly onto membranes/film, dropped by gravity into a capture device or propelled under pressure into a collection cap. Recovered cells can be analyzed for DNA, RNA, protein content, automated determination of cell/nuclear counts, cell area, stromal elements, and labeling intensity in target-labeled tissue sections [36] and used to construct cell-specific cDNA libraries [35, 37–40]. However, it is necessary to follow strict protocols pertaining to fixation, preparation, and handling of tissue samples to be microdissected [41].

18.2.3.2 In Situ Hybridization ISH techniques utilize labeled mRNA or DNA probes to localize target gene transcripts within individual cells of tissue sections [42]. Tissue or cell samples are either frozen-fixed or formalin-fixed paraffin embedded to facilitate cross-linking and preserve the target within the tissue and then incubated with a labeled probe that is either complementary DNA (oligonucleotide) or complementary RNA (riboprobe). Hybridization of the probe to the exact target sequence is manipulated through variation in temperature and buffer stringency (salt and detergent concentration) to prevent nonidentical DNA/RNA interactions. The probes are labeled with radio-, fluorescent-, or antigen-labeled bases (e.g., digoxigenin), which provide enhanced spatial resolution for cellular mRNA distribution with lower sensitivity, then localized and quantified in the tissue using (i) autoradiography (more sensitive and informative but has limited spatial resolution due to scatter of the signal), (ii) fluorescence microscopy, or (iii) IHC. More advanced ISH-based technologies now exist such as RNAscope® (Advanced Cell Diagnostic) to visualize single to multiple RNA molecules per cell in fresh frozen- or formalin-fixed paraffin-embedded tissues or cells required for colocalization studies. As before, cells/tissue samples are fixed, then permeabilized using proteases and incubated with multiple oligonucleotide probes (~1 kb) for hybridization to the target RNA. Specificity of oligonucleotide probe binding is provided through the application of preamplifiers and amplifiers (up to 20 of each) that favor hybridization to target probe pairs at specific temperatures. The labeled probe, which is conjugated to either a chromogenic or fluorescent molecule, is hybridized to the amplifier and can then be visualized using standard bright-field or fluorescent microscopy. The advantages of this new technology includes (i) high sensitivity with detection of single RNA molecules per cell, (ii) high specificity, (iii) simultaneous analysis of multiple targets on the same cell/tissue sample, and (iv) no requirement for a RNA free environment. Note that *in situ* PCR is another commonly used method to detect and quantify DNA and RNA in tissue sections or individual cells [43].

18.2.3.3 Immunohistochemistry and Immunocytochemistry IHC and immunocytochemistry (ICC) combine anatomical, immunological, and biochemical techniques to identify discrete antigens in tissues/cells by the interaction of target antigens with specific antibodies tagged with a visible label [44]. IHC makes it possible to visualize the distribution and localization of specific cellular components within cells and in the proper tissue context. Targets can be localized using a single tissue or multiple tissue specimens in high throughput tissue microarrays [45], and double IHC labeling [46] can also be applied to colocalize targets. The various stages of the IHC process are as follows. (i) Sample preparation: human or animal tissue samples are collected and then rapidly preserved to prevent the breakdown of cellular protein and maintain tissue architecture. Fixation methods are critical and must be optimized based on the target antigen. The most commonly used fixative is formalin/formaldehyde, a semireversible, covalent cross-linking reagent; other fixatives are available and selection depends on the antigens being investigated. Fixed tissue samples are then embedded in paraffin to maintain the natural architecture and facilitate tissue sectioning. Formalin-fixed, paraffin-embedded tissues are then sectioned (4–5 μm) with a microtome and mounted onto adhesive-treated glass slides and dried in preparation for deparaffinization (using Xylene), which must be completely removed for the antibodies to access the target antigens. Since formaldehyde fixation cross-links proteins in tissue samples, and may mask antigen presentation, treatment (to unmask the antibody epitopes) either by heat or enzymatic degradation is often required. Note that for many specialist assays (including tissue cross-reactivity (TCR) studies), the tissues must be frozen (snap-frozen in liquid nitrogen) before a gentle fixation step (to preserve morphology) without compromising antigenicity. Frozen/fixed tissues are then sectioned with a precooled cryostat and mounted onto adhesive glass slides. These sections are often dried overnight at room temperature and fixed by immersion in precooled (−20 °C) acetone. (ii) Tissue blocking: high levels of background staining may mask detection of a specific antigen or create false positives. This commonly occurs in staining approaches that use biotin, peroxidases, or phosphatases for the amplification or enzymatic detection of target antigens. Quenching or masking the endogenous forms of these proteins must therefore be physically blocked or chemically inhibited. Background staining may also be observed with antibodies that partially or weakly bind to sites on nonspecific proteins (also called reactive sites) that are similar to the cognate binding sites on the target antigen. To reduce background staining, the samples should be incubated with a buffer that blocks the reactive sites to which the primary or secondary antibodies may otherwise bind. Common blocking buffers include normal serum, nonfat dry milk, BSA or gelatin, and commercial blocking buffers with proprietary formulations. (iii) Sample detection and visualization: antibody-mediated antigen detection approaches are separated into direct and indirect methods. Most indirect methods employ the inherent binding affinity of avidin to biotin to localize the detection antibody to the target antigen and amplify the signal that is detected. IHC target antigens are detected through either chromogenic or fluorescent means, and the type of readout depends on the experimental design. For fluorescent detection, the primary or secondary antibody is conjugated to a fluorophore that is detected by fluorescent microscopy. Chromogenic detection is based on the activities of enzymes, most often horseradish peroxidase or alkaline phosphatase, which form colored, insoluble precipitates upon the addition of substrate, such as diaminobenzidine (DAB) and nitro-blue tetrazolium/5-bromo-4-chloro-3′-indolyphosphate (NBT/BCIP), respectively [47]. Single-step counterstains, added after antibody staining, give contrast result to the primary stain and can be cell structure-specific. Samples are then viewed by light or fluorescent microscopy depending on the antibody detection method applied.

18.3 OFF-TARGET BINDING OF THE THERAPEUTIC BIOLOGIC REAGENT

Another key aspect in the development of therapeutic biologics is determining the profile for off-target binding with the therapeutic product in human and animal tissues to assist in the interpretation of any unanticipated toxicity seen in the *in vivo* toxicology studies and to understand the potential for unanticipated toxicities in human tissues before First Time in Man (FTIM) studies. A range of assays including TCR studies and protein and cellular microarray technologies using the therapeutic candidate have been used to determine off-target binding. In the event of unexpected off-target binding, other more sophisticated technologies such as protein pull-down assays coupled with IHC double labeling may be employed to further define the protein of interest.

18.3.1 Tissue Cross-Reactivity Study

TCR studies are a key component of the preclinical toxicology program for Investigational New Drug (IND) submissions and must be Good Laboratory Practice (GLP) compliant and completed before FTIM clinical trials [48–53]. The primary aim of the TCR study is to identify the potential for unexpected, off-target binding of therapeutic mAbs (and other antibody-like products) to antigens in human tissues.

Although on-target binding data are also generated, this is a secondary endpoint that simply supplements the knowledge of target distribution. In these studies, direct/indirect immunohistochemical methods (as outlined above) are applied using a labeled therapeutic antibody (biotin or fluorescein isothiocyanate (FITC)) and about 40 different normal frozen human tissues from three donors to account for potential genetic polymorphism. Tissues from a pharmacologically relevant animal species can also be included to assist in the interpretation of the *in vivo* toxicology data and to bridge to the human data. A non-GLP validation phase is always conducted before the start of the GLP TCR study to determine appropriate positive/negative control tissues, cells, and reagents and to optimize the immunohistochemical methods. Note that in cases where the clinical candidate is deemed a poor IHC reagent in the validation phase, a GLP TCR study may not be possible. All binding is evaluated and interpreted in the context of the overall preclinical data.

18.3.2 Protein Microarray

The protein microarray (or protein chip), as described above, is a rapid, high throughput screen that can also be used to determine the potential for off-target binding of therapeutic candidates to thousands of targets in a single assay. Often this assay is used very early in development as a component of the characterization process for multiple candidates and is used to rank the best candidates for progression to the next phase of development. However, this assay is highly sensitive and the conditions of the assay often result in candidates artificially binding to a significant number of unexpected targets. Further investigation is required to understand the benefit of this screen in selecting the best candidates.

18.3.3 Cell Microarray Technology (Retrogenix)

Cell microarray technology is an accurate and effective solution for determining more biologically relevant binding of therapeutic antibodies and proteins to both expected (on-target) and unexpected (off-target) proteins in a single assay. The technology utilizes high density arrays of expression vectors encoding up to and in excess of more than 3500 genes (65% of known proteins). Briefly, human cells are grown on the surface of the array that contains the expression vectors [54]. The transfected cells then express the target protein and enable posttranslational modifications before being presented on the cell surface as an overexpressed full-length human membrane protein in its native form. The vectors are specifically mapped on the array to allow identification of the expressed protein and assess the potential for binding of the therapeutic biologic using established commercial software packages.

18.3.4 Protein Pull-Down Assays

Protein fishing is a novel method for capturing proteins through very specific protein–protein interactions such as on-/off-target binding of therapeutic biologics or immune complexes. Many different approaches have been identified all utilizing the concept of selectively capturing "bait" proteins (protein of interest), attached to "prey" proteins (binding proteins) [55]. The basic principle of the method involves incubation of tissue or cell lysate with free or tethered antibodies to capture the protein of interest. These complexes can then be purified either by SDS–PAGE or through washing away unbound protein from the immobilized antibody chambers. The protein complex can be further characterized through incubation with labeled proteins and fluorescent microscopy, immunoblot analysis, or using LC–MS techniques. Note that a new method has recently been described that utilizes a protein containing a site-specifically incorporated 3-azidotyrosine (N3-Y) and high performance magnetic beads for immobilization of the protein via an azido group [56].

18.4 BIODISTRIBUTION OF THERAPEUTIC BIOLOGIC REAGENT

During the development process, it may be necessary to determine the biodistribution of the therapeutic product in animal tissues to identify any potential tissue sinks that may alter the PK profile of a biotherapeutic product. Methods used to determine the biodistribution of a therapeutic protein-based molecule in treated animals include whole-body autoradiography (WBA) and IHC using either chromagen or fluorescence-based detection methods. For RNA- or DNA-based products, qRT-PCR can be used to quantify the levels of DNA in tissues and in the event of persistence in key tissues such as the reproductive organs, then an assessment for insertion into the genome may be required using sophisticated molecular techniques. Note that in all of these assays, it is prudent to perfuse anesthetized animals with sterile saline using a peristaltic pump before tissue sampling to clear the vasculature/systemic circulation of contaminating therapeutic product.

18.4.1 Whole-Body Autoradiography

WBA is a technique used to determine the tissue distribution of radiolabeled test items following dosing in laboratory animals [57–60]. It is generally performed using rodents; however, other species such as the marmoset can be used [61]. Standard protocols typically involve sacrificing animals at appropriate intervals of time after dosing which are then immediately frozen and embedded in a suitable medium such as carboxymethylcellulose. Whole-body sagittal sections are obtained using a cryomicrotome and exposed to phosphor image plates for digital imaging [62]. Since this technique is quantitative, it is possible to relate the signal to the concentration of radioactivity in the organ or tissue [63]. Other purposes for conducting WBA include the identification of organs where radioactivity persists long after dosing, the determination of radioactive isotope concentration in target-expressing organs, and investigation of biodistribution of drug in fetuses following passage across the placental barrier in pregnant animals [64]. In man, imaging techniques such as positron emission tomography and NMR spectrometry allow noninvasive drug tissue distribution studies [65, 66].

18.4.2 Biodistribution: Immunohistochemistry Methods for Protein-Based Therapeutic Products

IHC methods, as outlined above, can be used to understand the systemic distribution of biotherapeutic-based products following *in vivo* administration in animal studies. Note that biodistribution can refer to either endogenous trafficking (blood flow) of the therapeutic product or on/off-target mediated disposition. In brief, a labeled detection antibody, raised against the therapeutic product, is incubated with tissues of interest harvested from treated and control animals, and the presence of staining is directly compared with a negative control antibody (same structure/backbone). This has successfully been applied to investigate the tissue biodistribution in mice treated with trastuzumab and cetuximab [67]. Although the procedure follows the basic principles of IHC, there are several technical challenges related to the selection of appropriate control materials and reagents that should be considered in the study design to facilitate interpretation. First, the labeled detection antibody raised against the therapeutic product, which in the majority of cases is human(ized), should have no cross-reactivity with any endogenous target protein in the selected toxicology species. Secondly, the selection of suitable positive control material is critical in assay validation; cell lines used to engineer the human therapeutic product are the most relevant with the corresponding wild-type cell line as the negative control. Thirdly, tissue perfusion on anesthetized animals should be considered if high levels of therapeutic product are present in the systemic circulation immediately before tissue harvest as this will hinder the interpretation of target-mediated disposition. IHC methods can also be used to determine the biodistribution of DNA/RNA-based therapeutic products (e.g., gene therapy). In this situation, IHC is used to detect translated product (protein) within cells using an antibody raised against the expressed target protein. Note that this method only informs on biodistribution to cells where protein translation has occurred and not on the overall systemic distribution of the DNA/RNA-based therapeutic product.

18.4.3 Biodistribution: Quantitative PCR Methods DNA/RNA-Based Therapeutic Products

Preclinical studies assessing the safety profile of DNA/RNA-based therapeutic products such as gene therapy, cell-based therapies, and DNA vaccines require inclusion of specific endpoints to address both biodistribution and persistence using methods that are both sensitive and quantitative [68] such as the highly sensitive amplification method qPCR. Although no absolute level of sensitivity is specified in the guidelines, the assay must be robust with sufficient validation to demonstrate that the highest level of sensitivity has been attained. To facilitate interpretation of the qPCR data, it is critical to establish a rigorous practice for tissue harvest at the time of necropsy to prevent cross-contamination between the different organs and potential for false positives. First, treated animals should be anesthetized and perfused to remove blood containing the DNA/RNA-based therapeutic product from the systemic circulation, and secondly, tissues should be harvested using a strict protocol that utilizes clean (disposable) instruments for each tissue. The DNA is extracted from the snap-frozen tissues (stored at −80 °C) and analyzed for yield and quality before inclusion in the qPCR assays. Target-specific primers/probes are used to detect vector sequence using established chemistries (e.g., TaqMan) to compare the levels of the DNA/RNA-based therapeutic products across tissues from treated and control animals. The amount of DNA/RNA-based therapeutic product present within a known amount of genomic DNA (normally 1 μg) is determined in tissues from treated animals. A risk assessment for the integration of vector DNA into the host species genome can also be made in combination with a time course for tissue sampling to understand vector persistence. Further controls include carefully assessing the effects of genomic DNA inhibition/interference on assay sensitivity and, in some instances, this may require reducing the amount of genomic DNA included in each PCR reaction (up to a cumulative total of the required amount to be analyzed per tissue sample). Note that multiple samples per tissue may also help control for sampling variability and cross-contamination.

18.5 DISCUSSION

18.5.1 Considerations in the Interpretation of Molecular Pathology-Based Data

Interpretation of data from molecular pathology-based assays should consider limitations in the methods applied and be reviewed in the context of known target biology and all other relevant data generated from complementary assays. More specifically, potential complications in assay interpretation can be addressed in advance by selection of the right method(s), understanding the limitations of the assay(s) and careful attention to study design. For example, if large numbers of samples require screening for target expression, then high throughput assays with high sensitivity, such as PCR-based methods and gene/protein microarray, are generally appropriate. However, these typically use whole tissue extracts and may not necessarily highlight tissues that only express target in a few critical cells; this would therefore need to be combined with other assays such as IHC/ISH that localize expression to specific cell types in tissue sections. New techniques combining sensitivity and specificity are currently being validated and will complement existing assays.

Assay validation is the next critical step in facilitating accurate interpretation of data. Positive and negative control materials (cells or tissues) should be well characterized in terms of target expression at both the mRNA level via Northern blot, ISH, or qRT-PCR, and the protein level, using IHC or Western blot. Although cells overexpressing the target protein often serve as a suitable technical control, artificially elevated levels of target may not sufficiently optimize conditions for test tissues where endogenous levels of target expression is likely to be much lower. Therefore, healthy or diseased tissues, which more closely reflect the *in vivo* situation, are preferable when available. Reagents used in the assay should also be well validated and control steps applied to understand the technical issues that create false-positive and false-negative results. Examples include controlling for RNAase activity and degradation of the target (Northern blot/ISH), cross-linking of target antigens following fixation (IHC) and binding of reagents, such as antibodies, to nonspecific targets (IHC or FACS).

Even when well validated, there may still be discrepancy in the results obtained from complementary assays. For example, it is possible that the levels of target mRNA and protein expression may not be comparable within the same tissue/cell sample since both are regulated independently through different mechanisms. In these cases, it is important to consider the biological plausibility based on known target biology/turnover/recycling within the cell. Note that this has been observed for some interleukin receptors [69].

The translation of well validated *in vitro* data to the *in vivo* situation should be carefully considered. For example, in TCR studies, the cytoplasmic compartment of cells is artificially exposed to the therapeutic antibody (in cut tissue sections) during the immunohistochemical reaction. If staining is seen in the cytoplasm, how relevant is this finding if only the cell membrane is exposed *in vivo*? It is therefore important to use a combination of appropriate complementary technologies rather than a single technique and use an integrated approach that reviews PK, target expression/concentration, receptor occupancy, pathology, immunogenicity, and other relevant information to characterize and interpret the ADME profile of a biotherapeutic.

18.5.2 Examples of Molecular Pathology Methods Used in Preclinical Development

18.5.2.1 Target Expression Profiling for Antibody–Drug Conjugates (ADCs) Antibody–drug conjugates (ADCs) are a new class of highly potent biopharmaceutical drugs designed as a targeted therapy for the treatment of people with life-threatening conditions such as cancer [70]. ADCs are complex molecules composed of an antibody (a whole mAb or an antibody fragment such as a single-chain variable fragment (scFv)) linked, via a stable, chemical, linker with labile bonds, to a biologically active cytotoxic (anticancer) payload or drug. By combining the unique targeting capabilities of mAbs with the cancer-killing ability of cytotoxic drugs, ADCs allow sensitive discrimination between healthy and diseased tissue. This means that, in contrast to traditional chemotherapeutic agents, ADCs target and attack the cancer cell so that healthy cells are less severely affected. However, the targeted protein is often also expressed on normal tissues, and therefore it is essential to understand the quantitative and qualitative expression profile in normal tissues and the ability of the candidate to bind the target in these tissues. This is determined using a combination of techniques including IHC, ISH, qRT-PCR, and the TCR study.

18.5.2.2 Off-Target Binding in TCR Study (CMV) A fully humanized mAb (IgG1) with high binding affinity to human cytomegalovirus (CMV) glycoprotein for the treatment of CMV-infected pregnant women in the prevention of congenital infection and disease was evaluated in a standard GLP TCR study, conducted as part of a preclinical research program. An unexpected off-target staining with the anti-CMV mAb was seen in the keratinized keratinocytes of the stratum corneum and granulosum of the epidermis of human skin (two of three donors) and the keratinized epithelial Hassal's corpuscles of the human thymus (two of three donors) at concentrations of 10 and 30 μg/mL. These findings were not present in rat tissues (used as the *in vivo* toxicology species). This was considered to represent off-target binding, so further investigations were conducted to identify the protein. Fresh human skin samples were first prepared by separating the epidermal tissue from the underlying dermis. The dermal and epidermal protein extracts were then applied to a pull-down assay using the anti-CMV mAb and isotype control antibodies immobilized on beads. Two proteins were extracted and subsequently identified as keratin 5 and 14 using liquid chromatography–tandem mass spectrometry (LC–MS/MS). Cytokeratin 5 was then confirmed using both Western blot methods and colocalization with the anti-CMV mAb using immunohistochemical techniques. In conclusion, off-target binding of the anti-CMV mAb to cytokeratin 5 directly or as a complex involving cytokeratin 5 and cytokeratin 14 was observed. Since the safety risk of binding the epithelial components of the skin and thymus in fetal tissues is unclear, the candidate was replaced by another anti-CMV mAb that showed no off-target binding [71].

18.5.2.3 Biodistribution (Anti-CCL21 mAb) Anti-CCL21 mAb is a fully humanized mAb (IgG1) with high-binding affinity to the human chemokine (C–C motif) ligand 21 (CCL21) thought to play a role in lung inflammation and fibrosis following activation of resident CCR7-expressing fibroblasts and myofibroblasts. The kinetic profile of the anti-CCL21 mAb in nonhuman primates was characterized using an integrated approach combining PK, pharmacodynamic (PD), immunogenicity, IHC, and tissue gene-expressing profiling data from *in vivo* single- and multiple-dose toxicology studies in the cynomolgus monkey. In these studies, the anti-CCL21 mAb demonstrated rapid clearance (faster than the typical kinetics expected for an IgG1) following intravenous administration at doses of 10 and 50 mg/kg; PK/PD modeling showed that this clearance was related to the mAb–ligand complex with evidence of target saturation (plateau of total CCL21 levels) observed after weekly doses of the highest dose (50 mg/kg) only.

Biodistribution of the anti-CCL21 mAb to tissues expressing high levels of target (lymph nodes and spleen) was investigated using immunohistochemical methods. Briefly, SP2/0 cells producing recombinant human monoclonal IgG1 antibody were selected as a positive control for the IHC methods and goat antihuman IgG Fc fragment (Jackson ImmunoResearch Laboratories, Suffolk, UK) used as the detection antibody for the anti-CCL21 mAb; note that there was no apparent cross-reactivity with endogenous cynomolgus monkey IgG. To ensure the specificity of the staining, a negative mouse cell line (expressing mouse IgG1) was applied. A goat antirabbit IgG antibody was also used as a control antibody at the same concentration as the detection antibody. All slides were assessed and graded according to the intensity of the staining, and the staining pattern and distribution of any immunohistochemical stained cells within the tissue were also described. Staining for the anti-CCL21 mAb using IHC showed that drug accumulation occurred in the spleen (red pulp, marginal zone, or germinal center), lymph nodes (parafollicular and medulla), lungs (bronchus associated lymphoid tissue (BALT)), gastrointestinal tract (Peyer's patches or gastrointestinal associated lymphoid tissue (GALT)), liver, and thymus. This staining was considered to be related to the anti-CCL21 mAb bound to (i) circulating CCL21, (ii) cells producing CCL21 such as fibroblastic reticular cells or high endothelial venules [72], and (iii) CCL21 bound to its receptor (CCR7) that is expressed on semimature and mature DCs [73], thymocytes at different developmental stages [74], naive B and T cells [75, 76], Treg cells [77], and a subpopulation of memory T cells known as central memory T cells [76]. In summary, there is a large pool of CCL21 available for binding the anti-CCL21 mAb, either through production or release from storage sites, which is not depleted after repeated dosing with the anti-CCL21 mAb. In human disease states, CCL21 is believed to be expressed at 10-fold higher concentration compared with cynomolgus monkeys. A model describing the PK–PD behavior of the anti-CCL21 mAb and its binding to CCL21 suggests that large doses requiring frequent administration would be required to maintain suppression of CCL21 in the clinical setting.

18.6 CONCLUSION

Complex protein- and nucleic acid-based therapeutics will continue to be a major area of research, and more "hybrid" molecules, which are a mixture of small chemical entities and protein therapeutics, will enter clinical testing. This will require sophisticated analytics to characterize the distribution, metabolism, and elimination of these molecules and their subunits. However, it is clear that neither a unified approach nor a single technology will be sufficient to obtain a thorough understanding of target biology and biotherapeutic biodistribution but rather a complementary range of assays, including molecular pathology-based techniques, to answer a unique set of scientific questions. It is also important that an integrated approach using PK, target expression/concentration, receptor occupancy, pathology, immunogenicity, and other relevant information within the same study is necessary to characterize and interpret the ADME profile of a biotherapeutic.

REFERENCES

[1] Williams SP. Tissue distribution studies of protein therapeutics using molecular probes: molecular imaging. AAPS J 2012;14 (3):389–399.

[2] Wang W, Lu P, Fang Y, Hamuro L, Pittman T, Carr B, Hochman J, Prueksaritanont T. Monoclonal antibodies with identical Fc sequences can bind to FcRn differentially with pharmacokinetic consequences. Drug Metab Dispos 2011;39 (9):1469–1477.

[3] Bumbaca D, Wong A, Drake E, Reyes AE, Lin BC, Stephan JP, Desnoyers L, Shen BQ, Dennis MS. Highly specific off-target binding identified and eliminated during the humanization of an antibody against FGF receptor 4. MAbs 2011;3 (4):376–386.

[4] Vugmeyster Y, Szklut P, Wensel D, Ross J, Xu X, Awwad M, Gill D, Tchistiakov L, Warner G. Complex pharmacokinetics of a humanized antibody against human amyloid beta peptide, anti-abeta Ab2, in nonclinical species. Pharm Res 2011;28 (7):1696–1706.

[5] Boswell CA, Mundo EE, Zhang C, Stainton SL, Yu SF, Lacap JA, Mao W, Kozak KR, Fourie A, Polakis P, Khawli LA, Lin K. Differential effects of predosing on tumor and tissue uptake of an 111In-labeled anti-TENB2 antibody-drug conjugate. J Nucl Med 2012;53 (9):1454–1461.

[6] Bumbaca D, Boswell CA, Fielder PJ, Khawli LA. Physicochemical and biochemical factors influencing the pharmacokinetics of antibody therapeutics. AAPS J 2012;14 (3):554–558.

[7] Tabrizi M, Bornstein GG, Suria H. Biodistribution mechanisms of therapeutic monoclonal antibodies in health and disease. AAPS J 2010;12 (1):33–43.

[8] Gillett NA, Chan CM. Molecular pathology in the preclinical development of biopharmaceuticals. Toxicol Pathol 1999;27 (1):48–52.

[9] Bates S. The role of gene expression profiling in drug discovery. Curr Opin Pharmacol 2011;11 (5):549–556.

[10] Chan JN, Nislow C, Emili A. Recent advances and method development for drug target identification. Trends Pharmacol Sci 2010;31 (2):82–88.

[11] Shoemaker DD, Linsley PS. Recent developments in DNA microarrays. Curr Opin Microbiol 2002;5 (3):334–337.

[12] Heid CA, Stevens J, Livak KJ, Williams PM. Real time quantitative PCR. Genome Res 1996;6 (10):986–994.

[13] Ginzinger DG. Gene quantification using real-time quantitative PCR: an emerging technology hits the mainstream. Exp Hematol 2002;30 (6):503–512.

[14] Giulietti A, Overbergh L, Valckx D, Decallonne B, Bouillon R, Mathieu C. An overview of real-time quantitative PCR: applications to quantify cytokine gene expression. Methods (Duluth) 2001;25 (4):386–401.

[15] Huggett J, Dheda K, Bustin S, Zumla A. Real-time RT-PCR normalisation; strategies and considerations. Genes Immun 2005;6 (4):279–284.

[16] Bustin SA, Benes V, Garson JA, Hellemans J, Huggett J, Kubista M, Mueller R, Nolan T, Pfaffl MW, Shipley GL, Vandesompele J, Wittwer CT. The MIQE guidelines – minimum information for publication of quantitative real-time PCR experiments. Clin Chem 2009;55 (4):611–622.

[17] Towbin H, Staehelin T, Gordon J. Electrophoretic transfer of proteins from polyacrylamide gels to nitrocellulose sheets: procedure and some applications. Proc Natl Acad Sci U S A 1979;76 (9):4350–4354.

[18] Renart J, Reiser J, Stark GR. Transfer of proteins from gels to diazobenzyloxymethyl-paper and detection with antisera: a method for studying antibody specificity and antigen structure. Proc Natl Acad Sci U S A 1979;76 (7):3116–3120.

[19] Tao SC, Chen CS, Zhu H. Applications of protein microarray technology. Comb Chem High Throughput Screen 2007;10 (8):706–718.

[20] Sano T, Smith C, Cantor C. *Immuno-PCR: Very Sensitive Antigen Detection by Means of Specific Antibody-DNA Conjugates*. American Association for the Advancement of Science; 1992.

[21] Niemeyer M, Adler M, Wacker R. Immuno-PCR: high sensitivity detection of proteins by nucleic acid amplification. Trends Biotechnol 2005;23:208–216.

[22] Adler M. Immuno-PCR as a clinical laboratory tool. Adv Clin Chem 2005;39:239–292.

[23] Fredriksson S, Gullberg M, Jarvius J, Olsson C, Pietras K, Gústafsdóttir SM, Ostman A, Landegren U. Protein detection using proximity-dependent DNA ligation assays. Nat Biotechnol 2002;20:473–477.

[24] Gullberg M, Gústafsdóttir SM, Schallmeiner E, Jarvius J, Bjarnegård M, Betsholtz C, Landegren U, Fredriksson S. Cytokine detection by antibody-based proximity ligation. Proc Natl Acad Sci U S A 2004;101:8420–8424.

[25] Saito K, Kobayashi D, Sasaki M, Araake H, Kida T, Yagihashi A, Yajima T, Kameshima H, Watanabe N. Detection of human serum tumor necrosis factor in healthy donors, using a highly sensitive immuno-PCR assay. Clin Chem 1999;45:665–669.

[26] Lind K, Kubista M. Development and evaluation of three real-time immune-PCR assemblages for quantification of PSA. J Immunol Methods 2005;304:107–116.

[27] Herzenberg LA, Julius MH, Masuda T. Demonstration that antigen-binding cells are precursors of antibody-producing cells after purification with a fluorescence-activated cell sorter. Proc Natl Acad Sci U S A 1972;69 (7):1934–1938.

[28] Loken MR. *Immunofluorescence Techniques in Flow Cytometry and Sorting*. 2nd ed. Wiley; 1990. p 341–353.

[29] Ornatsky O, Bandura D, Baranov V, Nitz M, Winnik MA, Tanner S. Highly multiparametric analysis by mass cytometry. J Immunol Methods 2010;361 (1–2):1–20.

[30] Neubert H, Palandra J, Fernandez Ocna M. Quantification of biotherapeutic targets: new opportunities with immunoaffinity LC-MS/MS. Bioanalysis 2014;6 (13):1731–1733.

[31] Woods AG, Sokolowska I, Wetie AG, Wormwood K, Aslebagh R, Patel S, Darie CC. Mass spectrometry for proteomics-based investigation. Adv Exp Med Biol 2014; 806:1–32.

[32] Xie F, Liu T, Qian WJ, Petyuk VA, Smith RD. Liquid chromatography-mass spectrometry-based quantitative proteomics. J Biol Chem 2011;286:25443–25449.

[33] Duan X, Abuqayyas L, Dai L, Balthasar JP, Qu J. High-throughput method development for sensitive, accurate, and reproducible quantification of therapeutic monoclonal antibodies in tissues using orthogonal array optimization and nano liquid chromatography/selected reaction monitoring mass spectrometry. Anal Chem 2012;84:4373–4382.

[34] Espina V, Heiby M, Pierobon M, Liotta LA. Laser capture micro-dissection technology. Expert Rev Mol Diagn 2007;7 (5):647–657.

[35] Emmert-Buck MR, Bonner RF, Smith PD, Chuaqui RF, Zhuang Z, Goldstein SR, Weiss RA, Liotta LA. Laser capture microdissection. Science 1996;274 (5289):998–1001.

[36] Peterson RA, Krull DL, Butler L. Applications of laser scanning cytometry in immunohistochemistry and routine histopathology. Toxicol Pathol 2008;36 (1):117–132.

[37] Krizman DB, Chuaqui RF, Meltzer PS, Trent JM, Duray PH, Linehan WM, Liotta LA, Emmert-Buck MR. Construction of a representative cDNA library from prostatic intraepithelial neoplasia. Cancer Res 1996;56:5380–5383.

[38] Simone NL, Bonner RF, Gillespie JW, Emmert-Buck MR, Liotta LA. Lasercapture microdissection: opening the microscopic frontier to molecular analysis. Trends Genet 1998;14: 272–276.

[39] Simone NL, Remaley AT, Charboneau L, Petricoin EF III, Glickman JW, Emmert-Buck MR, Fleisher TA, Liotta LA. Sensitive immunoassay of tissue cell proteins procured by laser capture microdissection. Am J Pathol 2000;156:445–452.

[40] Banks RE, Dunn MJ, Forbes MA, Stanley A, Pappin D, Naven T, Gough M, Harnden P, Selby PJ. The potential use of laser capture microdissection to selectively obtain distinct populations of cells for proteomic analysis: preliminary findings. Electrophoresis 1999;20:689–700.

[41] Charboneau L, Paweletz CP, Liotta LA. Laser capture microdissection. *Curr Protoc Cell Biol* Chapter 2: Unit 2.5 2001.

[42] Gall JG, Pardue ML. Formation and detection of RNA-DNA hybrid molecules in cytological preparations. Proc Natl Acad Sci U S A 1969;63 (2):378–383.

[43] Bagasra O. Protocols for the *in situ* PCR-amplification and detection of mRNA and DNA sequences Nucleic acid based molecular biology. Nat Protoc 2007;2:2782–2795.

[44] Ramos Vara JA, Miller MA. When tissue antigens and antibodies get along: revisiting the technical aspects of immunohistochemistry--the red, brown, and blue technique. Vet Pathol 2014;51 (1):42–87.

[45] Braunschweig T, Chung JY, Hewitt SM. Tissue microarrays: bridging the gap between research and the clinic. Expert Rev Proteomics 2005;2 (3):325–336.

[46] Chen X, Cho DB, Yang PC. Double staining immunohistochemistry. N Am J Med Sci 2010;2 (5):241–245.

[47] Pilling AM, Harman RM, Jones SA, McCormack NA, Lavender D, Haworth R. The assessment of local tolerance, acute toxicity, and DNA biodistribution following particle-mediated delivery of a DNA vaccine to minipigs. Toxicol Pathol 2002;30 (3):298–305.

[48] ICHS6. 2012. Addendum to preclinical safety evaluation of biotechnology-derived pharmaceuticals: tissue cross reactivity studies (note 1).

[49] EMEA. 2008. Guideline on development, production, characterization and specifications for monoclonal antibodies and related products. December 2008, (EMEA/CHMP/BWP/157653/2007).

[50] US FDA/CBER. 1997. Points to consider in the manufacture and testing of monoclonal antibody products for human use.

[51] Leach M, Halpern WG, Johnson CW, Rojko JL, MacLachlan TK, Chan CM, Galbreath EJ, Ndifor AM, Blanset DL, Polack E, Cavagnaro JA. Use of TCR studies in the development of antibody-based biopharmaceuticals: history, experience, methodology and future directions. Toxicol Pathol 2010;38:1138–1166.

[52] Bussiere JL, Leach MW, Price KD, Mounho BJ, Lightfoot-Dunn R. Survey results of the use of TCR IHC assay. Reg Toxicol Pharmacol 2011;59:493–502.

[53] Geoly FJ. Regulatory forum opinion piece: TCR studies. What constitutes an adequate positive control and how do we report positive staining? Toxicol Pathol 2013;42:1–3.

[54] Turner L, Lavstsen T, Berger SS, Wang CW, Petersen JEV, Avril M, Brazier AJ, Freeth J, Jespersen JS, Nielsen MA, Magistrado P, Lusingu J, Smith JD, Higgins MK, Theander TG. Severe malaria is associated with parasite binding to endothelial protein C receptor. Nature 2013;490:502–505.

[55] Jain A, Liu R, Ramani B, Arauz E, Ishitsuka Y, Ragunathan K, Park J, Chen J, Xiang YK, Ha T. Probing cellular protein complexes using single-molecule pull-down. Nature 2011;473 (7348):484–488.

[56] Ikeda-Boku A, Kondo K, Ohno S, Yoshida E, Yokogawa T, Hayashi N, Nishikawa K. Protein fishing using magnetic nanobeads containing calmodulin site-specifically immobilized via an azido group. J Biochem 2013;154:159–165.

[57] Potchoiba MJ, West M, Nocerini MR. Quantitative comparison of autoradioluminographic and radiometric tissue distribution studies using carbon-14 labeled xenobiotics. Drug Metab Dispos 1998;26:272–277.

[58] Solon EG. Autoradiography: high resolution molecular imaging in pharmaceutical discovery and development. Expert Opin Drug Discov 2007;2 (4):503–514.

[59] Pellegatti M. Preclinical *in vivo* ADME studies in drug development: a critical review. Expert Opin Drug Metab Toxicol 2012;8 (2):161–172.

[60] Erickson HK, Lambert JM. ADME of antibody-maytansinoid conjugates. AAPS J 2012;14 (4):799–805.

[61] Wuis EW, Rijntjes NV, Van der Kleijn E. Whole body autoradiography of 14C-dantrolene in the marmoset monkey. Pharmacol Toxicol 1989;64 (1):156–158.

[62] Solon EG, Balani SK, Lee FW. Whole body autoradiography in drug discovery. Curr Drug Metab 2002;3: 451–462.

[63] Solon EG1, Kraus L. Quantitative whole-body autoradiography in the pharmaceutical industry. Survey results on study design, methods, and regulatory compliance. J Pharmacol Toxicol Methods 2001;46 (2):73–81.

[64] He K, Lago MW, Iyer RA, Shyu WC, Humphreys WG, Christopher LJ. Lacteal secretion, fetal and maternal tissue distribution of dasatinib in rats. Drug Metab Dispos 2008;36:2564–2570.

[65] Lanao JM, Fraile MA. Drug tissue distribution: study methods and therapeutic implications. Curr Pharm Des 2005;11:3829–3845.

[66] Langer O, Muller M. Methods to asses tissue-specific distribution and metabolism of drugs. Curr Drug Metab 2004;5:463–481.

[67] Lee CM, Tannock IF. The distribution of the therapeutic monoclonal antibodies cetuximab and trastuzumab within solid tumors. BMC Cancer 2010;10:255.

[68] FDA/CBER, 2006. Guidance for Industry: Gene Therapy Clinical Trials – Observing Subjects for Delayed Adverse Events

[69] Yu A, Malek TR. The proteasome regulates receptor-mediated endocytosis of interleukin-2. J Biol Chem 2001;276 (1):381–385.

[70] Ducry L, Stump B. Antibody – drug conjugates: linking cytotoxic payloads to monoclonal antibodies. Bioconjug Chem 2010;21 (1):5–13.

[71] Flandre T, Thoree V, Jones S, and Hey A. When a biologics drug targets a viral disease: from TCR validation to off-target staining. Proceeding of 10th European Congress of Toxicologic Pathology; Lake Maggiore. 2012. p. 68.

[72] Carlsen HS, Haraldsen G, Brandtzaeg P, Baekkevold ES. Disparate lymphoid chemokine expression in mice and men: no evidence of CCL21 synthesis by human high endothelial venules. Blood 2005;106 (2):444–446.

[73] Ohl L, Mohaupt M, Czeloth N, Hintzen G, Kiafard Z, Zwirner J, Blankenstein T, Henning G, Förster R. CCR7 governs skin dendritic cell migration under inflammatory and steady-state conditions. Immunity 2004;21 (2):279–288.

[74] Misslitz A, Pabst O, Hintzen G, Ohl L, Kremmer E, Petrie HT, Förster R. Thymic T cell development and progenitor localization depend on CCR7. J Exp Med 2004;200 (4):481–491.

[75] Reif K, Ekland EH, Ohl L, Nakano H, Lipp M, Förster R, Cyster JG. Balanced responsiveness to chemoattractants from adjacent zones determines B-cell position. Nature 2002;416 (6876):94–99.

[76] Sallusto F, Lenig D, Forster R, Lipp M, Lanzavecchia A. Two subsets of memory T lymphocytes with distinct homing potentials and effector functions. Nature 1999;401 (6754):708–712.

[77] Szanya V, Ermann J, Taylor C, Holness C, Fathman CG. The subpopulation of CD4+CD25+ splenocytes that delays adoptive transfer of diabetes expresses L-selectin and high levels of CCR7. J Immunol 2002;169 (5):2461–2465.

19

LABELING AND IMAGING TECHNIQUES FOR QUANTIFICATION OF THERAPEUTIC BIOLOGICS

Julie K. Jang, David Canter, Peisheng Hu, Alan L. Epstein and Leslie A. Khawli

Keck School of Medicine of the University of Southern California, Los Angeles, CA, USA

19.1 INTRODUCTION

The labeling of therapeutic biologics, such as antibodies or proteins, is an invaluable technique that has been extensively used to analyze pharmacokinetics (PK) and metabolism, to measure absolute concentrations in tissues, as well as to facilitate imaging studies. Proteins possess unique sequences of amino acid residues making them perfect vehicles for labeling. Generally, the type of label (e.g., radionuclide or fluorescent) and the protein conjugation strategy used must be carefully considered and tailored for each application. Radionuclides are used extensively, but they have a short shelf-life and require special handling and disposal. Today, the two main methods of radiolabeling proteins are halogenation and the complexation of metallic radioisotopes. Radiohalogens, such as radioiodines, can be directly conjugated to proteins in one step through the formation of a stable covalent bond. An alternative approach to the direct method is the indirect incorporation of radiohalogens into small organic molecules. These molecules contain prosthetic groups allowing for halogenation and conjugation with proteins under mild conditions. Conversely, a few of the radiometallic nuclides (e.g., Tc-99m) can bind to proteins directly through coordination bonds with functional groups on the protein, or indirectly through the use of a bifunctional chelator. However, the stability of such conjugates is usually an issue, since most proteins including monoclonal antibodies (mAbs) do not possess metal coordination sites. One solution to this problem is to attach a bifunctional chelator, which can be covalently conjugated to proteins and will chelate a radiometal, thereby forming a stable [1, 2]. The advancement of available nuclear imaging instrumentation combined with well-characterized radiolabeling chemistry for antibodies and other molecules has enabled sensitive, dynamic, and quantitative measurements of radiolabeled probes with quantitative whole-body autoradiography (QWBA), single-photon emission computed tomography (SPECT), and positron emission tomography (PET) imaging *in vivo*.

In recent years, fluorescence, chemifluorescence, and chemiluminescence emerged as alternative technologies to the traditional radioisotope-based systems. Significant progress has been achieved to attach these labels to antibodies providing improved sensitivity and quantitative analysis for a wide range of research applications (e.g., flow cytometry) and optical imaging (OI; e.g., fluorescence, bioluminescence, near-infrared (NIR), and multispectral) [3]. Furthermore, numerous multimodal contrast agents, such as iron oxide magnetic nanoparticles (MNPs) conjugated to antibodies, are currently in development to generate products that have a variety of *in vitro* and *in vivo* applications detectable by magnetic resonance imaging (MRI) [4, 5]. These newer imaging techniques offer the advantage of using nonionizing radiation, which may have less of a negative impact on health and the environment than the use of radioisotopes. While convenience, speed, and safety are important factors in deciding which imaging technology to use, there are still various areas whereby radioisotopes continue to offer significant advantages over nonradioactive methods. Radioisotopes currently convey greater resolution

ADME and Translational Pharmacokinetics/Pharmacodynamics of Therapeutic Proteins: Applications in Drug Discovery and Development, First Edition. Edited by Honghui Zhou and Frank-Peter Theil.

and sensitivity over nonradioactive methods, and remain the general molecular imaging modality used in the clinic.

In this chapter, considerations on the choice of different labels suitable for various detection and imaging methods are discussed. Labeling methods of protein therapeutics are described in terms of direct versus indirect methods and in terms of radiohalogens (e.g., radioiodine), trivalent cation radiometals (e.g., indium, copper, and zirconium), and fluorophores. The intention of this chapter is to act as a guide for scientists to find the optimal combination of labeling techniques and labels suitable for pharmaceutical application. Widespread use of labels and labeling methods resulted in increased requirements on chemists to select the most appropriate methods for stable attachment of a given label while maintaining the structure of the parental protein, which affects the *in vitro* and *in vivo* behavior, including biodistribution, PK, and imaging results, of the labeled protein. For this reason, the advantages and the limitations of each labeling method are discussed. Emphasis is given to the use of numerous imaging techniques for assessing the PK and biodistribution of biologic proteins, with a focus on PET and SPECT and a brief discussion of the role of OI in this application. The advancement of commercially available molecular imaging instrumentation combined with well-characterized labeling chemistry for proteins has enabled detailed *in vivo*, dynamic, quantitative measurements of radio- or fluorophore-labeled proteins in preclinical models as well as in the clinic. Different approaches to measuring biodistribution and PK of labeled biologic molecules are described in Chapter 7 by Boswell et al. in this book.

19.2 NEW AND CONVENTIONAL METHODS FOR LABELING OF BIOLOGICS

A wide range of mAbs have been produced and shown to express high affinity to antigens with variable specificity. Once labeled, many of them have been suggested as potential candidates for diagnostic imaging or therapy. Today, there are two radioimmunotherapeutics (RITs) targeting the CD20 antigen, ibritumomab tiuxetan (Zevalin®), and tositumomab (Bexxar®) that gained approval for the treatment of indolent B-cell lymphoma. Another radiolabeled mAb product, satumomab pendetide (Oncoscint™), has also been authorized in some European countries for diagnostic use in clinical oncology for colorectal and ovarian cancer [6]. The success of these molecules encouraged the development of a number of radioantibodies, such that there are at least 20 radioantibodies currently in different stages of development for solid tumors and hematological malignancies [7, 8]. Thus, this section presents a general overview with basic information to acquaint the reader with some of the chemical aspects of the labeling of protein therapeutics with radiohalogens, metal chelates, and fluorophores. It elucidates the selection of labels, their diagnostic and therapeutic applications, as well as the prime techniques related to protein modification and the subsequent process of labeling. The comparative assessment of advantages and disadvantages of commonly used labels is also made.

19.2.1 Choice of Labels

The advent of target-specific mAbs stimulated a great interest in the use of radiolabeled antibodies to image biodistribution in the body. However, insufficient uptake ratios of target to nontarget organs (T/N) still affect the performance of radioimmunoimaging with mAbs. Many problems, such as the characterization of antigens and cell receptors, the preparation of optimal mAbs, and the development of the chemical modification and radiolabeling methods for mAbs, remain to be optimized [9]. The radiolabeling of proteins is considered in terms of the choice of radionuclide, the method of conjugation, and effect of conjugation on protein PK and biodistribution. The choice of radionuclide is dependent on its nuclear properties, including its physical half-life, its production factors, available instruments for detection (e.g., counting or imaging), and its effect on the biological half-life of the labeled radiopharmaceutical [10, 11]. Subsequently, the properties of biodistribution and PK play a major role in influencing and determining the efficacy and safety of a drug in various diseases.

When considering the selection of a suitable radionuclide for labeling proteins, there may be one of two possible aims in mind—either to perform a biodistribution radioimaging study or to destroy diseased tissues using radiotherapy. This section deals with the radiohalogen and radiometal nuclides for biodistribution imaging. A number of radionuclides, each with different half-lives and particle emission properties, can be produced in any radioactive element group (halogens and radiometals) in the periodic table. While iodine-125, iodine-131, indium-111, technecium-99m, copper-64, and yitrium-90 have been the radionuclides primarily used in protein labeling, there are a number of different radionuclides that could be used for unique or specific applications. The choice of radionuclide to use for a particular application will be dependent on the nuclear emission properties, physical half-life, decay characteristics, daughter nuclide characteristics, and cost and availability [12]. While many of the radiohalogen and radiometal nuclides are not commercially available, most are obtainable through activation of the appropriate target materials using reactor or cyclotron irradiations. However, it must be emphasized that even though a radionuclide can be produced, issues such as physical and chemical forms of the desired product, ease of separation (i.e., purity) and quantity of radionuclide needed, cost, and environmental impact and safety may preclude its use in an application [13]. Production of these radionuclides is outside the scope of this chapter, but can be found in a number of articles from the literature [12–21].

19.2.1.1 ***Radiohalogens*** The majority of radiolabeling studies used radiohalogens as a radioactive label. These radionuclides have some advantages, such as their commercial availability via routine production in reactors and simple methodology for labeling antibodies. Radiohalogenation of the protein molecule is usually accomplished by utilizing effective oxidizing agents, such as iodogen or chloramine-T [9]. Both procedures yield proteins with high concentrations of radiolabeled product and ideal amounts of radioactivity per protein molecule (i.e., high specific activity). As a group, radiohalogens may be particularly useful for radiolabeling of proteins because (i) their chemistry is well understood, (ii) they form stable covalent bonds, (iii) their steric and electronic nature can be expected to cause minimal alteration to the protein, (iv) highly specific activity radiolabeling can be accomplished, and (v) radionuclides with many different half-lives and photon or particle emissions are obtainable [12].

Radioiodination of proteins has been the most widely used method of radiolabeling for immunological purposes; these coupling reactions are relatively easy to carry out in a routine manner and lead to high protein-specific radioactivity and high radiochemical yields [22–24]. In a broader perspective, the advantages of using photon- and positron-emitting radionuclides rather than weak beta-particle emitters such as tritium or carbon-14 for radiolabeling proteins go further than radiochemical yield. Measuring biological samples labeled with photon or positron (e.g., gamma rays) emitting radionuclides can be done more easily than measuring samples labeled with beta-emitting radionuclides, which would require tissue homogenization and mixing with scintillation cocktails for liquid scintillation counting [12]. Furthermore, many applications of radiolabeled proteins require that highly specific activities be obtained as imaging agents for the visualization and detection at very low levels of protein, which can be achieved with gamma ray-emitting radionuclides [25].

The half-lives and primary emissions of halogen radionuclides that have an application in protein labeling are given in Table 19.1. Radiolabeled proteins can be divided into two groups according to use—those used for *in vitro* assessment and those used *in vivo*. Those that have been used *in vitro* as radiotracers for radioimmunoassays and radio-receptor assays demand critical control of the biological and immunological properties of the tracer [11]. For example, iodine-125 has been the primary radionuclide for *in vitro* studies due to its 60-day half-life and low energy photon (gamma ray: 27–32 keV) emissions. The long half-life permits the preparation and storage of labeled protein for extended periods before usage. The low energy gamma emission makes it particularly attractive for storage, as minimal radiation damage to the labeled protein is expected. Experimentally, highly specific activities are obtained with iodine-125, and its cost and availability are also very reasonable. For these reasons, there are no other halogens radionuclides that can readily be used in the place of iodine-125 for *in vitro* studies [12, 26].

On the other hand, the current use of *in vivo* radiotracers involves the pragmatic exploitation of easily produced and readily available combinations of radionuclide and substrate [11]. For example, the halogen radionuclides that exhibit gamma ray emissions are very important in applications of radiolabeled proteins for biodistribution, PK, and nuclear imaging studies. In addition, the application of radionuclides for the measurement of receptor–ligand binding *in vivo* [27] has become less restricted through the advent of new SPECT and PET imaging techniques. From experiments using I-125 (60-day half-life) and I-131 (8-day half-life) labeled proteins, biodistribution data, PK data, and data on the metabolic fate of the radiolabel can be obtained by sacrificing animals, excising tissue samples, and counting the radioactivity in a gamma counter. However, I-125 and I-131 are not optimal to assess these parameters in patients. I-125, which can be imaged by a thin crystal gamma camera when evaluating small animals such as mice [28], is of little or no value for imaging patients due to the attenuation of the photons through tissue [12]. I-131 is often used for imaging with conventional gamma cameras, but its high energy gamma emissions (364 and 637 keV) are not optimally counted, and its beta-particle emission results in a higher radiation dose to the tissues than is desirable for non-therapeutic purposes. An alternate radionuclide of iodine, I-123, has a reasonable half-life of 13 h and a nearly ideal gamma emission (159 keV) in high abundance for imaging with current gamma cameras; but its high production expense and poor availability significantly limit its current application [9].

Another area that holds a great promise for new applications of radiohalogens is PET imaging and measurements with I-124 (Table 19.1). I-124 (4-day half-life; 2.13 MeV positron energy; 0.6–1.7 MeV gamma rays) offers a chemical bridge between SPECT and PET versions of the same reagent. Because PET works by detecting 511 keV gamma rays that are produced by the annihilation of positrons ($\beta+$) emitted by a radionuclide and nearby electrons, an ideal PET radionuclide should have a low $\beta+$ energy and high $\beta+$ branching ratio. Unfortunately, I-124 has particularly high energy positrons and extraneous gamma rays. The result is significant blurring in small-animal imaging, but can be improved with advanced reconstruction methods [29]. With the long half-life of I-124 of 4 days, slow metabolic processes can be examined as opposed to I-123 (13.2 h half-life), which is used for conventional scintigraphy with SPECT [30]. In addition, quantitative imaging and biodistribution with I-124 appear to be ideal with mAbs since its half-life matches the

TABLE 19.1 Major Properties and Applications of Radiohalogens Commonly Used in Protein Labeling and Molecular Imaging

	I-125	I-123	I-131	I-124	At-211
Half-life	60 d	13 h	8 d	4.2 d	7.2 h
Gamma-ray energy (keV)	35	159	364	511	500–900
Common labeling methods	• Direct labeling • Via prosthetic group	• Direct labeling • Via prosthetic group	• Direct labeling • Via prosthetic group	• Direct labeling • Via prosthetic group	• Direct labeling • Via prosthetic group
Applications	• SPECT imaging • Cut-and-count (animal studies) • Autoradiography • *In vitro* binding studies	• SPECT imaging	• SPECT imaging over multiple days • Therapy • Cut-and-count	• PET imaging over multiple days • Cut-and-count	• X-ray energy offers targeted radiotherapy application • High energy gamma offers SPECT imaging potential

TABLE 19.2 Major Properties and Application of Radiometal Chelates Commonly Used in Protein Labeling and Molecular Imaging

	Tc-99m	In-111	Ga-67	Ga-68	Cu-64	Zr-89
Half-life	6 h	2.8 d	3.3 d	68 min	12.7 h	3.3 d
Gamma-ray energy (keV)	140	173, 247	93, 185, 300	511	511	909
Common labeling methods	Direct labeling	Chelation (DTPA, DOTA)	Chelation (DOTA, NOTA, DFO)	Chelation (DOTA, NOTA, DFO)	Chelation (DTPA, DOTA, TETA)	Chelation (DFO)
Applications	Same day SPECT imaging	SPECT imaging over multiple days	Potential for SPECT imaging over multiple days	PET imaging	PET imaging	PET imaging

biological half-life of antibodies. Given these benefits, PET studies of I-124 radiolabeled antibodies play an important role by providing information on the presence, efficacy, tissue distribution profile, and PK of biologics. Preliminary studies on patients using this approach have been reported elsewhere [31].

Nevertheless, the use of radioiodine has shown pronounced drawbacks for use in radiolabeling proteins [9, 24]. Significant *in vivo* instability caused by deiodination, or release of free iodine from the protein molecule reduces the measured uptake of the labeled protein into target tissues and the T/N ratio. Furthermore, the release of free radioiodine results in high thyroid uptake, thus increasing the localized radiation exposure to this critical organ. For this reason, any administration of radioiodine labeled antibody requires preblocking of the thyroid gland with an excess of cold iodine to prevent subsequent uptake of radiolabel. When I-131 is used as an imaging label, its long half-life of 8 days unnecessarily prolongs *in vivo* radiation exposure, and the emission of beta particles, delivers a high radiation dose to patients. The reader is referred to a review by Wilbur [12] for a thorough overview of radiohalogens for protein labeling.

Another halogen radionuclide that holds potential for some therapy applications is astatine-211 [12, 32] (Table 19.1). The 7.2-h half-life offers many potential advantages for targeted α-particle radiotherapy, which is extremely cytotoxic. As a consequence of its versatile chemistry and intermediate half-life relative to other α-particle emitters, a wide variety of At-211-labeled species have been synthesized and evaluated as targeted radiotherapeutics including peptides, antibodies, and antibody fragments [33]. However, investigations on its therapeutic usefulness have been very limited due to the difficulty in obtaining the radionuclide. At present, at least 16 different At-211-labeled antibodies have been studied [34], reflecting the high level of interest in this α-particle emitter for targeted radiotherapy.

19.2.1.2 Radiometals Radiometals are radioactive isotopes that can be harnessed for applications in diagnostic imaging techniques, such as SPECT and PET, as well as cancer therapy. Some examples of radiometals that can be used for PET imaging are Ga-68, Cu-64, Y-86, and Zr-89. These metals provide sensitive and quantitative images of a variety of molecular processes and targets. SPECT is an older imaging modality than PET, and since its inception in the 1960s, Tc-99m has been the workhorse isotope of SPECT. More recently, the radiometals Ga-67, In-111, and Lu-177 have been increasingly used for SPECT imaging in chelator-based radiopharmaceuticals. Appropriate choice of these radionuclides allows for tailoring the properties of the labeled protein to the application required (see Table 19.2). In recent years, the applications of metallic nuclides in radioimmunolocalization have greatly increased.

Indium-111 has been extensively tried in radioimmunolocalization studies and has excellent nuclear properties for radioimaging (68-h half-life, 173 and 247 keV gamma rays, and no beta emission). The most common method for attaching radioindium to protein is through chelation using a chelator previously conjugated to an antibody or protein before the insertion of the radioprobe. One attractive site-specific labeling strategy makes use of carbonyl chemistry and His-tags that are part of the expressed protein [29, 35], but conventional covalently coupled chelators are also available [36]. Diethylenetriaminepentaacetic acid (DTPA) and its derivatives are the most popular chelates employed for In-111, as well as some other metallic nuclides [36]. Besides In-111, other gamma-emitting metallic nuclides such as Tc-99m, Ga-67, Cu-67, Y-90, Re-186, and Au-199, are also potential candidates for labeling proteins. Tc-99m (6-h half-life; 140 keV gamma energy) is ubiquitous in diagnostic imaging because it can be produced locally in a relatively inexpensive Mo-99/Tc-99m generator.

Driven by the increasing availability of preclinical and clinical PET scanners, the use of "nonstandard PET nuclides" has been growing exponentially in the past decade [37]. Largely complementary to the roles of the four standard PET nuclides (O-15, N-13, C-11, and F-18), nonstandard PET nuclides enable the novel design and synthesis of a wider range of PET tracers to probe a variety of biological events. However, due to their emission of high energy positrons and cascade gamma rays, nonstandard PET nuclides with half-lives ranging from seconds to days must be judiciously chosen for specific applications. This section describes the nonstandard PET nuclides, Cu-64, Ga-68, and Zr-89, that have recently been reported for preclinical or clinical PET studies focusing on the unique features of their productions, radiochemical procedures, and applications. Interested readers are referred to recent excellent review articles for more in-depth discussions of this topic [38–40].

Copper has several radioisotopes including Cu-60, Cu-61, Cu-62, Cu-64, and Cu-67. Commercially available medical cyclotrons have the capability to produce Cu-60 (23.4 min half-life), Cu-61 (3.3-h half-life), and Cu-64 (12.7-h half-life) for PET imaging. Cu-64 is very popular and considerable effort has gone into making Cu-64-producing cyclotrons more widely available in the United States. Cu-64 has been applied to imaging the biodistribution and tissue kinetics of several antibodies in human and nonhuman primates [29]. Despite the excellent image quality at time points up to 24 h, the duration of the Cu-64 studies is limited to 3 days at most due to its half-life. Its optimal application may lie with smaller antibody fragments and peptides, which tend to have shorter biological half-lives than whole antibodies.

Given the sensitivity and quantitation benefits of PET compared to SPECT systems, the use of positron-emitting rather than single-photon isotopes has become more compelling, and protein labeling with PET isotopes has been an active area in drug development [29]. As a possible alternative to Tc-99m, Ga-68 (68-min half-life) is a PET isotope produced in a convenient generator from Ge-68 (9-month half-life), which eliminates the need of an on-site cyclotron and makes Ga-68 an easier and cheaper alternative to F-18. Beyond its convenient availability, Ga-68 combines excellent imaging properties and a simple noncovalent point-of-use labeling strategy applied to protein reagents [41].

Due to its relatively long half-life and low positron emission energy, Zr-89 (3.3-day half-life) is an ideal radionuclide for the labeling of compounds with long blood circulation times, such as antibodies and other proteins. Combining its reasonably long half-life with excellent PET image quality and quantitation, Zr-89 is a very significant and newly available reagent for imaging tissue distribution of protein therapeutics and diagnostics in both preclinical and clinical settings [39, 42–50]. Although Zr-89 emits an extraneous high energy gamma ray (909 keV), the resulting image quality is good and the overall radiation burden is well tolerated [29, 51].

19.2.1.3 Criteria for Choosing Radionuclides in Preclinical Development The availability of radionuclides with potential preclinical applications continues to expand with advances in radiochemistry. Due to the full spectrum of half-lives (minutes to days) and increased accessibility of radionuclides, many of the new radiopharmaceuticals based on these radionuclides are under preclinical and clinical development. With the use of radionuclides for biodistribution or radioimaging of labeled biotherapeutic proteins in mind, the following points are the considerations for their use [9, 20, 52–56]. (i) Radionuclide should have a physical half-life of about 6–200 h. The physical half-life should be sufficiently long to allow for imaging at the time when the T/N ratio reaches a maximum. However, if the half-life is too long, the radionuclide will cause an excessive radiation dose to the patient in clinical settings. (ii) Gamma-energy range should be between 100 and 300 keV. The gamma ray energy should match the image device. In conventional radioimaging, the range of 100–300 keV is most appropriate for external scanning. (iii) High single-energy gamma density is desired for achieving high imaging resolution. (iv) The radionuclide should be produced in a carrier-free form. Since the number of binding sites on a protein or chelator is limited, only a carrier-free radionuclide can yield a labeled antibody of high specific activity. (v) The radionuclide–protein or radionuclide–chelate protein should have satisfactory *in vivo* chemical stability. The ultimate consequence of *in vivo* chemical stability of the label is demonstrated in terms of T/N ratio by means of tissue biodistribution and external imaging. (vi) Radionuclide should be accessible and at reasonable cost.

Beyond the parameters discussed above, there are other variables that may affect the T/N ratio or efficacy of the localization of the radiolabeled proteins, such as chemical impurities of the radiolabel, immunoreactivity influenced by the process of radiolabeling, uptake and catabolism of protein-bound and unbound radionuclides, and plasma and whole-body clearance [57].

19.2.1.4 Other Labels Beyond the development of probes suitable for PET and SPECT imaging alone, recent work has aimed to synthesize fluorescent imaging probes applicable in OI, as this imaging modality provides both preclinical and clinical advantages [58]. The vast selection of fluorophores provides a great deal of flexibility and variation in fluorescence applications. Fluorophores can be divided into three general groups: organic dyes, biological fluorophores, and quantum dots (QDs). (i) Synthetic organic dyes, such as fluorescein, were the first fluorescent compounds used in biological research. Derivatives of these original compounds have been produced to improve their bioconjugation, photostability, and solubility, especially fluorescein isothiocyanate and rhodamine. The small size of these fluorophores is a benefit over biological fluorophores for bioconjugation strategies because they can be cross-linked to proteins without interfering with biological function. (ii) Biological fluorophores are protein derivatives of green fluorescent protein and phycobiliproteins (allophycocyanin, phycocyanin, phycoerythrin, and phycoerythrocyanin) designed for use in biological expression systems. The benefit of these types of fluorophores is that expression plasmids can be introduced into either bacteria, cells, organs, or whole organisms, to drive expression of that fluorophore either alone or fused to a protein of interest in the context of the biological processes studied. (iii) QDs are nanocrystals with unique chemical properties that provide tight control over the spectral characteristics of the fluorophore. QD are nanoscale-sized (2–50 nm) semiconductors that, when excited, emit fluorescence at a wavelength based on the size of the particle. Smaller QDs emit higher energy than large QDs, and therefore the emitted light shifts from blue to red as the size of the nanocrystal increases. QD have been reported to be more photostable than other fluorophores, as one *in vivo* imaging study showed that QDs remained fluorescent for 4 months [59]. While the use of QDs in biological applications is increasing, there are reports of cell toxicity in response to the breakdown of the particles and their use can be cost-prohibitive. For all three groups of probes that can be used in the development of imaging agent, substances emitting light in the NIR and infrared spectrum (700–900 nm) are most useful, as light at these wavelengths exhibits the highest tissue permeability of several millimeters to centimeters *in vivo* [60].

Superparamagnetic iron oxide nanoparticles, on the other hand, are detectable by MRI, enabling a triple-modality imaging with PET/OI and MRI when these nanoparticles are combined with fluorescent dyes and radionuclides [61, 62]. QDs as well as iron oxide nanoparticles have to be coated with biocompatible materials to render them amenable for *in vivo* application. This coating can consist of different materials such as SiO_2 or other inorganic material, dextran, micelles, or polyethylene glycols that enable a chemical modification of the surface of the particles with dyes, radiolabels, and targeting vectors. An alternative approach involves the encapsulation of the fluorophore within the particle coating, which has been shown to result in a much higher fluorescence signal and photostability than a superficial dye conjugation [61].

19.2.2 Labeling Strategies of Biologics

Labels may alter the protein biodistribution through nonspecific changes in bulk, charge, or hydrophobic interactions [63]. This has been a major barrier to the adoption of many otherwise excellent labels. Controlling the labeling sites and limiting the stoichiometry should minimize the risk of immunogenicity and the problems of batch-to-batch heterogeneity, aiding the validation of labels in clinical and other critical settings [64–67]. With optical and radioactive labels alike, incorporating more labels (up to 10) per protein [68] gives rise to a temptingly "brighter" protein with higher specific activity or fluorescent yield, but risks compromised behavior *in vivo* and requires even more cautious validation. One site-specific label per protein is ideal [29].

In general, the labeling procedures should minimally alter the properties of the molecule of interest, and the physical half-life of a radionuclide should match the biological half-life of the molecule to be labeled. However, a label may directly perturb the function of a protein, and the reaction conditions used to introduce the label may inadvertently promote undesirable changes such as oxidation, deamidation, side-chain isomerization, or aggregation [69]. The basic absence of gross changes in PK or molecular weight is not always sufficient as characterization of labeled proteins, and binding or other functional assays are needed to assess the integrity (e.g., immunoreactivity) of a labeled probe [70]. Methods for labeling of proteins, in general, should be rapid and give high yields. The labeled protein that is obtained should be of high specific activity and should be labeled in a manner that results in a stable attachment of the label. Optimization of reaction parameters is almost always a requisite of radiolabeling because radionuclides are generally dilute and contain many minor impurities from processing of the target material and in the chemicals used, which are present in more mass than the radionuclides itself [12]. Any label should remain coupled to its conjugate protein for the duration of the experiment and, ideally, make no difference to its behavior [29]. This is a fundamental criterion in choosing an appropriate label for a particular study. Uncoupled label will remain optically active or radioactive, and be detected in any biodistribution images, but it no longer reveals the presence of the therapeutic protein. The distribution of excreted or catabolized labels can confound the biodistribution and imaging of certain tissues, especially in and around the hepatobiliary system and gut, the kidneys, and urinary bladder. When imaging abdominal sites, for example, this can be a limitation and has been a major driver in the selection among labels [71]. Thus, appropriate choice of the label allows tailoring the properties of the labeled protein to the application required. Today, several labels, including radiohalogens and radiometals, have extended the possibilities of detection and imaging techniques, and in turn, resulted in the need for the development of chemical methods for their conjugation.

19.2.2.1 Radiohalogenation of Proteins Methods of radiohalogenation of proteins have been of interest for a variety of applications for several decades. Although many different radiohalogens have been used to radiolabel proteins, the most studies used radionuclides of iodine, principally I-125, I-123, and I-131. The most common procedure employed in the radiohalogenation of proteins, often referred to as "direct" radiolabeling, has been the reaction of an *in situ*-prepared electrophilic radioiodine species with functional groups on a target protein. Unfortunately, while direct labeling works very well for radioiodine, it is generally of little or no value for radiohalogenation with other elements in the halogen group. For example, direct labeling with bromine, chlorine, and fluorine radionuclides requires harsh oxidizing conditions that cause denaturation of proteins, except when enzymes, such as lactoperoxidase, are used as oxidants. Routine use of enzymes for radiohalogenation is not presently practical due to the cost and availability of the enzymes. Direct labeling of proteins with astatine nuclides can also be accomplished, but the astatine–protein bond produced has been found to be relatively less stable than that of iodine–protein. Of the halogens, iodine is most likely to support a positive charge and thus is the most reactive toward electrophilic addition or substitution reactions, allowing the reactions to occur rapidly under mild conditions [11]. The phenolic ring of tyrosine residues are generally highly reactive toward electrophilic reagents and are readily iodinated by I^+, followed by the imidazole ring of histidine residues, the benzene ring of phenylalanine residues, the indole ring of tryptophan residues, and the sulfhydryl groups of cysteine residues (Fig. 19.1a). Studies on the mechanism of reaction of iodine with tyrosine and other phenols indicate that the phenolic ring is attacked at the ortho positions and can produce both the monoiodo and the diiodo derivatives [12].

An alternative to direct radiohalogen labeling of proteins is conjugation of a small radiohalogenated small molecule to the protein, referred to as the indirect method. In this section, we will direct our attention to a number of different direct and indirect methods for protein radioiodination. In particular, the iodinating agents most widely used to produce electrophilic addition or substitution are iodine (I_2), iodine

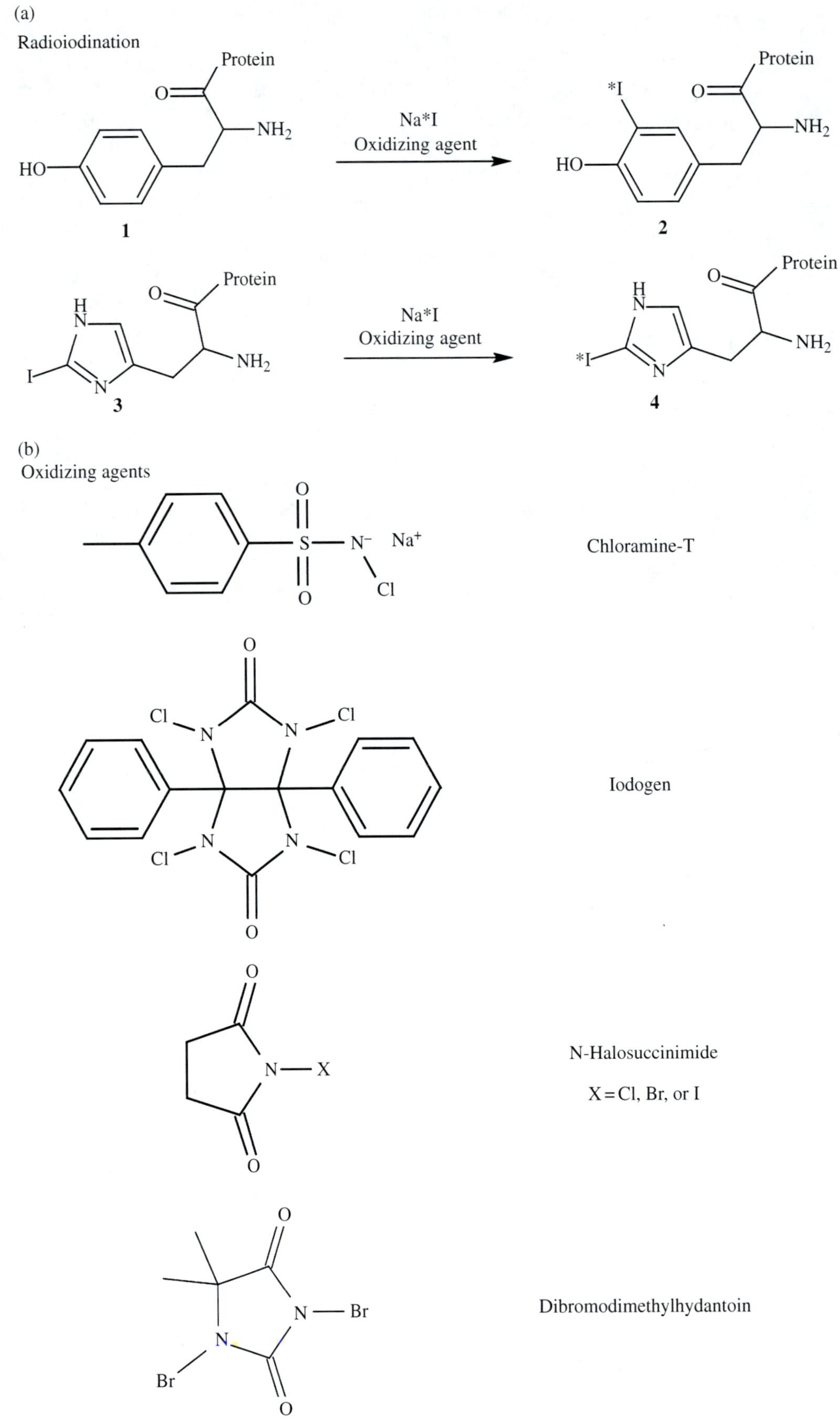

FIGURE 19.1 Radioiodination of tyrosine and histidine residues on proteins. (a) Reaction for the radioiodination on tyrosine and histidine residues. (b) Common *N*-haloamine oxidizing reagents. **1**=tyrosine; **2**=iodinated tyrosine; **3**=histidine; and **4**=iodinated histidine.

monochloride (ICl), oxidizing reagents (e.g., chloramine-T and iodogen), oxidative enzymes, and prelabeled activated small molecules. The reader is referred to a review by Wilbur [12] for a thorough overview of the current radioiodination methods and reagents for conjugate labeling. Importantly, all radioiodination procedures should be performed in a well-ventilated fume hood and all radioactive materials should be shielded by small lead containers. A lead glass should be used to reduce radiation exposure.

19.2.2.1.1 Radioiodination Using Iodine In the early days of radioiodination, molecular I_2 was the most frequently used labeling agent. Today, radiohalogens are generally obtained as their halide salts (e.g., radioactive NaI) and must be oxidized to form an electrophilic halogenating species to label molecules. The oxidants used are ammonium persulfate, nitrous acid, hydrogen peroxide, ferric sulfate, or a mixture of iodide/iodate and chromic acid. However, to avoid protein damage by such strong oxidants, the dihalogen (I_2) formed must be extracted and added to the protein in a buffer. Today, the preferred approach is to generate the electrophilic radioiodination reagent *in situ*. *In situ* oxidation can be accomplished by using I_2, but the disadvantages of this method include lower radiolabeling yield (theoretical maximal yield is 50% because only half of the total I_2 is converted to the positively charged iodinating species), loss of volatile iodine in the oxidation step, and increased radiation exposure to personnel [72].

19.2.2.1.2 Radioiodination Using Iodine Monochloride Several large and small molecules can be labeled with ICl, which is a more powerful iodinating species than I_2. It has been shown that ICl increases the theoretical maximum radiolabeling yield to 100% as compared to I_2 solutions because all iodine atoms can be incorporated. Radioiodine is mixed with ICI before its addition to the target protein [11, 73]. Since carrier ICI is added, the specific activity of the product is lower than that obtained with other methods. The mechanism of this reaction suggests that either H_2OI^+ or ICl might be the electrophile in the rate-determining step [74, 75]. Until recently, other procedures have been extended to the iodination of phenolic compounds via *in situ*-generated ICl using NaI/$FeCl_3$ [76], $NaIO_4$/NaCl/silica sulfuric acid [77], and trichloroisocyanuric acid/I_2/wet SiO_2 [78]. However, iodination using ICl is not a widely used methodology due to being generally cumbersome and costly, its use of a number of hazardous chemicals, and inferior performance in large-scale labeling. The development of quick, inexpensive, widely applicable, and environmentally benign iodinating agents is therefore still an active area of research.

19.2.2.1.3 Radioiodination Using Oxidizing Reagents The most common commercially available oxidizing reagents used are the *N*-haloamine compounds shown in Figure 19.1b. Chloramine-T, the sodium salt of sodium *N*-chlorotoluenesulfonamide, has been used as an oxidant for halides and its chemistry is well known. This reagent is water-soluble and has been utilized for *in situ* oxidation of iodide, bromide, and astatide [12]. In aqueous solutions, it forms HOCl, which is thought to be the actual oxidizing species. This reacts with the radioactive iodide present to form some electrophilic iodine species, which has been suggested to be H_2OI^+ at low pH and ICl at higher pH. In addition to being a powerful oxidizing agent, chloramine-T can also cause a number of undesirable side reactions, including chlorination of aromatic rings, oxidation of thiol groups, and cleavage of tryptophanyl peptide bonds [79]. Nevertheless, use of appropriate molar ratios can produce satisfactory products even with the most sensitive target molecules. Many studies reported that equimolar amounts of chloramine-T added in small aliquots produced radiolabeled proteins that retained their biological activity [9, 80–83]. In order to stop radioiodination, an aliquot of reducing agent, such as sodium metabisulfite, can be added to assure complete cessation of the oxidative process. A version of chloramine-T attached to a solid support, called iodobeads, has become available [84] and allows the oxidizing reagent to be physically removed from solution.

Iodogen (1,3,4,6-tetrachloro-3α,6α-diphenylglycoluril) has also been used extensively in radioiodinations. It is a water-insoluble oxidizing agent that can react with radioiodine to form a highly reactive halogen species [12, 85]. However, the preferred approach is to coat iodogen on the surface of the reaction vessel, thus permitting radiolabeling with very little exposure of the protein to the oxidizing agent. This latter fact has made iodogen attractive because it is readily separated from the radiolabeled protein with less oxidative damage. The major advantage of iodination with chloramine-T or iodogen is their reactivity that allows the rapid preparation of abundant yields of product with high specific activity. As a side note, labeling with iodogen is milder and easier to perform than with chloramine-T.

19.2.2.1.4 Radioiodination Using Enzymes This method employs enzymes called peroxidases that, in the presence of hydrogen peroxide, iodinate tyrosines. Endogenous enzymes from this family iodinate thyroid peptides and proteins in the body. In particular, lactoperoxidase has been put to use for radioiodination and radiobromination, where I^- and Br^- can be enzymatically oxidized to active I^+ and Br^+ [86]. While the use of enzymes in radiohalogenation offers a very mild method of oxidation, routine use of enzymes to radiohalogenate proteins is unlikely as the enzymes can themselves be radiohalogenated, problems exist with purification and separation of the enzyme from the product (unless the enzyme is immobilized on a solid support), and the enzymes are more expensive than the commonly used organic oxidants (e.g., chloramine-T and iodogen) [11, 12].

19.2.2.1.5 Radioiodination Using Prelabeled Small Molecules Some disadvantages of the "direct" radiolabeling methods include exposure of the protein to harsh oxidants and reductants during radiolabeling, the possibility of nonspecific radiolabeling, and denaturation of the protein resulting in low yields of radiolabeled protein. Another significant problem related to the direct radioiodination of proteins is *in vivo* deiodination resulting in significant uptake of radioiodine in the stomach (caused by HCl production) and in the thyroid. These problems can be minimized if a preradiolabeled prosthetic group is used in the indirect radioiodination of the protein. A prosthetic group for radioiodination contains some type of activated or nonactivated aromatic moiety to receive the label and some connecting bridge to covalently attach it to functional groups on proteins. Thus, the "indirect" method for radiolabeling proteins generally follow one of two approaches: (i) radiolabel the prosthetic molecule, then couple it to the protein in two separate reactions, and thereby avoid exposure of the protein to oxidation, or (ii) couple the prosthetic group to the protein, then do the labeling. In the second approach, the protein will be exposed to the oxidating reagent, but the presence of the aromatic group in the prosthesis permits the use of milder radioiodinating conditions. In general, radioiodination of the prosthetic group has different requirements of oxidizing reagents. In preparing radiolabeled small molecules for protein conjugation, many studies have used chloramine-T, tert-butyl hydroperoxide (TBHP) in chloroform/acetic acid solution, and *N*-chlorosuccinimide/acetic acid in methanol for oxidation of radiobromide, radioiodide, and astatide [87–90]. A benefit of *N*-chlorosuccinimide/acetic acid over the other reagents is that it does not chlorinate phenolic compounds, and thus, does not dilute the specific activity of most radioiodinated-labeled compounds [12]. When a small molecule is radiolabeled before conjugation, it is necessary to quench any unreacted halogen since excess halogen could react with the protein during the subsequent conjugation step. If one is concerned about deleterious effects of metabisulfite on the protein, excess tyrosine can be added to trap the excess electrophilic radiohalogen.

Understanding the special sites of reactivity, known as available functional groups, on a biologic protein is the key to choosing the best method for conjugation, whether that be for radiolabeling, cross-linking, or covalent modification. There are many reviews that describe the chemistry of protein radiolabeling through conjugation of small molecules [91, 92]. The principal reactive groups on proteins that are used in conjugation with other small molecules are amines, sulfhydryls (thiols), and oxidized carbohydrate moieties. Indeed, a number of activated and nonactivated aromatic ring small molecules can be radiolabeled and are subdivided on the basis of which protein functional groups are to be conjugated with the radiolabeled small molecule.

19.2.2.1.5.1 Primary Amines as Protein Radiolabeling Sites The most common functional group used for conjugation of radioiodinated small molecules to proteins is primary amines. These occur on the positively charged ε-amino group of lysine residues and the N-terminus of each polypeptide chain. They are abundant, widely distributed, and easily modified because of their reactivity and their location on the surface of protein and mAbs. The need to radiolabel proteins with high specific radioactivity initially led scientists to use prelabeled activated aromatic ring molecules. These small molecules substituted with OH (phenols), NH_2 (anilines), or NHR or NR_2 (*N*-alkylanilines) react rapidly with electrophilic halogens. Such radiolabels need to be appropriately reactive to the target functional group (e.g., N-hydroxysuccinimide esters (NHS esters) or isothiocyanates in the case of primary amines). Examples of labels include the Bolton–Hunter reagent, *N*-succinimidyl 3-(4-hydroxyphenyl) propionate (Fig. 19.2a) [93], and Wood's reagent, methyl 4-hydroxybenzimidate hydrochloride (Fig. 19.2b) [94], which are iodinated and purified from oxidizing and reducing agents before being conjugated to the protein. In this way, the impurities in the iodide solution, the oxidizing agents such as chloramine-T, and the reducing agents such as metabisulfite do not come into contact with the target molecule. The reaction results in the formation of an amide bond with the lysine groups of the target molecule. This allows the iodination of molecules that do not contain available tyrosine residues or with biological activity that might be lowered by alteration at the tyrosine moiety. The yield is usually lower than that obtained with the direct labeling with chloramine-T due to competitive hydrolysis of the active ester [11]. Small molecules that contain amine-reactive functionality such as imidate esters, aldehydes, and isothiocyanate groups have also been used in protein conjugation [95].

While the radioiodination of activated aromatic ring conjugates has been very useful, not all halogens can be introduced by the use of activated aromatic compounds (e.g., Bolton–Hunter reagent). For example, astatine labeling of phenolic compounds is not practical, as the phenolic ring is too activated to produce stable compounds with astatine nuclides [92]. Furthermore, halogenated phenols and anilines are susceptible to dehalogenation by nucleophiles, and radioiodinated tyrosine derivatives are susceptible to *in vivo* enzymatic dehalogenation [96]. Because of their higher stability, nonactivated aromatic ring compounds are good alternatives to activated aromatic rings for halogens [12]. Thus, there have been numerous studies involving radiohalogenations of aromatic compounds via aryl metallic intermediates, for example, aryl tin [97–99]. For example, aryl-tin functional group in *N*-succinimidyl 3-(t-*n*-butylstannyl) benzoate (STB, Fig. 19.2c) has been used extensively for this purpose. Like the Bolton–Hunter reagent, STB can be radioiodinated, and subsequently attached to the target protein by the reaction of the succinimidyl ester present in the labeling reagent with amine groups on the target protein [88, 99–101].

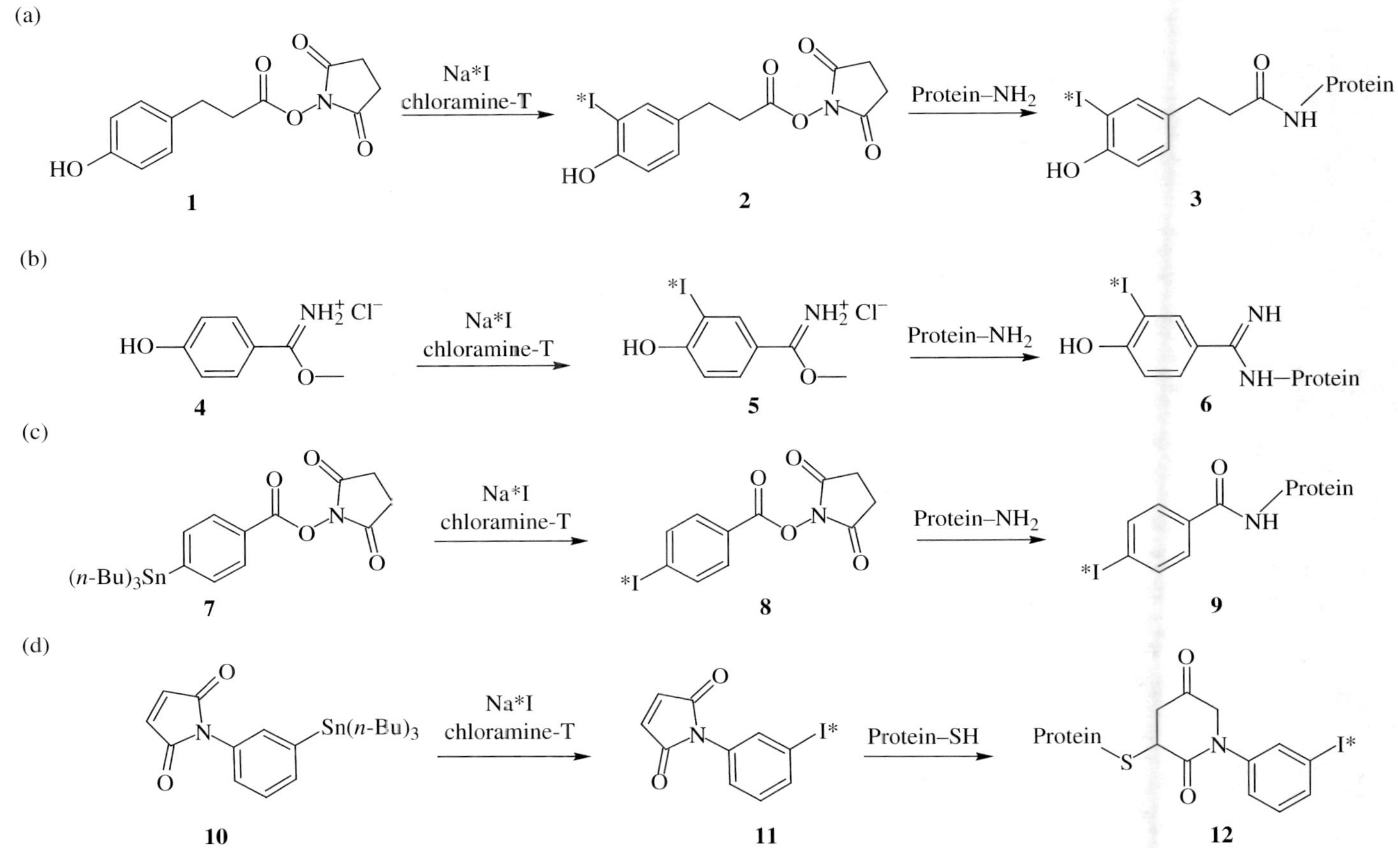

FIGURE 19.2 (a) Radioiodination of proteins via prosthetic groups using Bolton–Hunter reagent. **1**=Bolton–Hunter reagent (*N*-succinimidyl 3-(4-hydroxyphenyl)propionate; **2**=radioiodinated Bolton–Hunter reagent; **3**=radioiodinated conjugate via amino group of lysine residues. (b) Radioiodination of proteins via prosthetic groups using Wood's reagent. **4**=Wood's reagent (methyl 4-hydroxybenximidate hydrochloride); **5**=radioiodinated Wood's reagent; **6**=radioiodinated conjugate via amino residues. (c) Radioiodination of amine-reactive aryl-tin compounds for the radiolabeling of proteins. **7**=STB (*N*-succinimidyl 3-(t-*n*-butylstannyl)benzoate); **8**=[*I]SIB (*N*-succinimidyl 3-iodobenzoate); **9**=radioiodinated conjugate via amino residues. (d) Radioiodinaticn of tin compounds for the radiolabeling of thiol-activated proteins. **10**=TPM (*N*-[*m*-t-(*n*-butyl)stannylphenyl] maleimide); **11**=*m*-[*I]IPM (*N*-(*m*-[*I]iodophenyl)maleimide); **12**=radioiodinated conjugate via thiol residues.

19.2.2.1.5.2 Thiols as Protein Radiolabeling Sites The second useful functional groups for covalently radiolabeling antibodies are thiols. These groups are nucleophiles such as amines; however, they can be reacted selectively over amines on a protein, particularly if the amines are protonated through adjustment of the pH (7 or less). Thiols occur on cysteine residues and exist as disulfide bonds in native proteins, including antibodies, that stabilize the whole-molecule structure. Hinge-region disulfides can be selectively reduced with reagents such as mercaptoethanol or dithiothreitol to produce free thiols available for targeted labeling. Proteins that do not contain native disulfide bonds can be reacted with compounds such as 2-iminothiolane to introduce thiol groups on primary amines [12, 102, 103]. Alternatively, a variety of bifunctional linkers can also be incorporated into proteins to introduce thiols. Discussion of the applications of bifunctional linkers is outside the scope of this chapter, but can be found in numerous reviews [90, 95, 104].

Iodoacetamide and maleimide containing molecules are well known for their thiol specificity [105]. These conjugation reactions are generally conducted at pH ≤ 7 and are complete within a few minutes. For example, two different metallic derivatives, acetyl mercuric and aryl tin, of *N*-phenylmaleimide can be prepared and radioiodinated to give *N*-iodophenylmaleimide [106–108]. Subsequent conjugation of proteins would be accomplished by allowing the maleimide group to react with thiols on the protein (Fig. 19.2d). Thiols can be introduced onto a protein using a commercially available bifunctional coupling reagent, such as *N*-succinimidyl 3-(2-pyridyldithio)propionate (SPDP) at the optimal pH range of 6.0–7.5, or by treatment of intact antibody with dithiothreitol at pH 8 [106].

19.2.2.1.5.3 Oxidized Carbohydrates as Protein Radiolabeling Sites The third functional groups for radiolabeling antibodies are the carbohydrate moieties attached to the hinge region or the Fc portion of some proteins, such as mAbs. The fact that the carbohydrates (e.g., polysaccharides) are located at these distinct areas of the antibody structures, away from the biologically active binding sites (including regions responsible for antigen binding and conveying effector functions), makes these functionalities attractive for site-specific conjugation of prelabeled molecules [109]. Radiolabeling the polysaccharide moieties requires oxidation of the *cis*-diols to create active aldehyde functionalities (–CHO). The aldehyde groups can be coupled directly to primary amines to produce imines, followed by a second step involving reduction of the imines to form a stable conjugate. A number of radiohalogenated amine-containing compounds (e.g., dilactitol iodotyramine [110] and iodohistamine) have been prepared that could potentially be used to radiolabel oxidized carbohydrate moieties. The preparation and conjugation of a radioiodoaniline with carbohydrate-oxidized mAb has been described as well (Fig. 19.3) [101]. In this reaction, radioiodoaniline is prepared from the corresponding aryl-tin derivative. Modification of antibody carbohydrates by oxidation with $NaIO_4$ and conjugation with radio-iodoaniline is then followed by reduction with $NaBH_4$ to form the stable radioimmunoconjugate.

19.2.2.2 Radiometal Labeling of Proteins The attachment of radiohalogens through direct labeling, as described for iodine, is only possible if the nuclides can undergo electrophilic substitutions. Furthermore, these radiohalogens must be capable of being attached under physiological conditions due to the heat-labile nature of proteins. Direct labeling is thought of as advantageous owing to the simplicity of such processes. However, radioactive metal ions are generally more difficult to attach via direct labeling strategies due to their reduced reactivity compared to halogens. Nevertheless, methods for direct approaches for labeling proteins with metallic radionuclides are available. For successful direct radiolabeling to occur, sufficient available

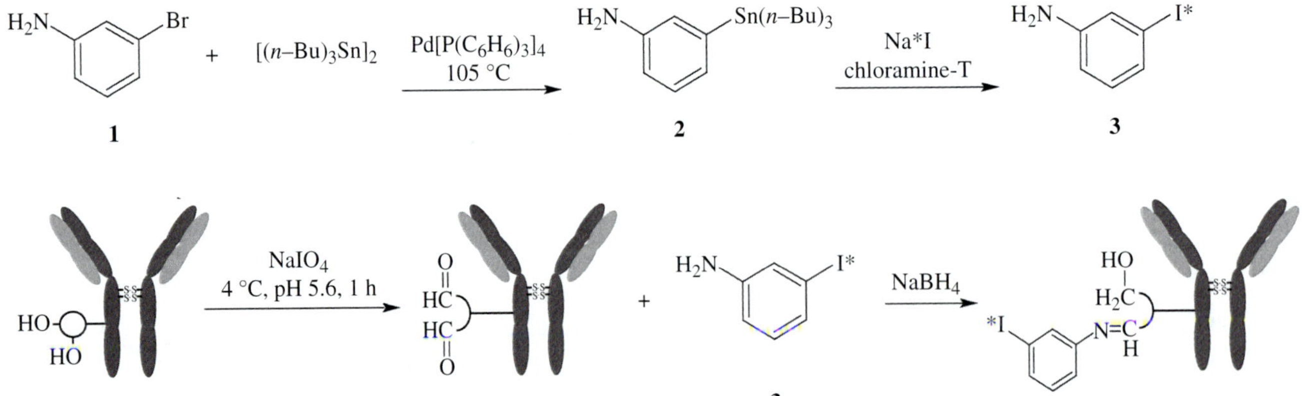

FIGURE 19.3 Preparation and conjugation of radioiodinated iodoaniline with carbohydrate-oxidized antibody. **1** = *m*-bromoaniline; **2** = *m*-aminophenyl tributylstannane intermediate; **3** = *m*-[*I]iodoaniline.

thiol groups are necessary to complex with the generated metal radionuclide. Poor site specificity, and lack of stability when coupling with thiol groups are major shortcomings of direct radiolabeling of antibodies with metal nuclides [111]. Approaches to counter these shortcomings have involved reducing the disulfide linkages to produce thiol groups. However this method, if uncontrolled, is likely to alter the spatial conformation of the protein [112]. These disadvantages have led to the development of indirect radiolabeling methods that link radiometals to proteins using bifunctional chelators. These chelators consists of two functional groups—one group for the attachment to the protein and a chelating unit that carries the radionuclide. As with prosthetic groups in the indirect radioiodination of proteins, chelators have to be stable against hydrolysis under physiological conditions and must not alter the biological properties and specificity of the protein required for target binding. Due to the diversity of the metallic radionuclides, many different variations of bifunctional chelators are used depending on the choice of radionuclide. The size, charge, and electron configuration of the metallic radionuclide will determine the coordination number (varying from 2 to 8) required of a bifunctional chelator in order to accommodate the radiometal [111].

19.2.2.2.1 Acyclic Bifunctional Chelators The first chelating agents developed for coupling radionuclides to biomolecules were the acyclic chelating agents, which lack a ring system in their coordinating form as compared to macrocyclic chelators. The most common acyclic chelators are derivatives of DTPA and EDTA (ethylenediaminetetraacetic acid), which are capable of binding a wide range of metals including the widely used In-111, Y-90, and Tc-99m. Binding of DTPA to a protein, such as an antibody, is possible by the use of the coupling agent isobutyl chloroformate. Attached to the primary amines on proteins, the conjugated acyclic chelating system possesses six or eight donor sites for interaction with the radiometal [113]. A major advantage of using DTPA analogs as bifunctional chelators is mild reaction conditions, which is of particular benefit owing to the labile nature of proteins. Zevalin (Y-90-ibritumomab tiuxetan) is an example of a therapeutic drug utilizing the conjugation of mAb ibritumomab to the DTPA derivative known as tiuxetan, for labeling with 90-yttrium. Zevalin received FDA approval in 2002 for the treatment of non-Hodgkin's lymphoma and is currently still being marketed. Compared to the antibody drug rituximab, it shows greater response rates and highlights the use of anti-CD20 radiotherapy in medicine [114].

Although DTPA has been extensively studied, its low *in vivo* stability diminishes its potential use due to dissociation of the radionuclide from the chelating agent. Another negative aspect of conjugation with DTPA is the synthesis of undesired conjugates, such as double-substituted DTPA analogs (i.e., a single DTPA molecule conjugated to two protein molecules) obtained with the DTPA dianhydride. These negative aspects have led to the synthesis of DTPA derivatives such as tetra-*t*-Bu-DTPA (Fig. 19.4a) that contains only one free carboxyl group, thus eliminating the possibility of forming double-substituted DTPA derivatives. However, the same radionuclide dissociation issues associated with DTPA are equally problematic in the derivatives of DTPA and occur due to the opening of the chelate ring. Alternative methods and modifications to DTPA have been sought in order to provide a more appropriate hindrance to the opening of the acyclic backbone once conjugated. An example is cyclohexane-1,2-diamine *N,N,N′,N′*-tetraacetate (CHX-DTPA) (Fig. 19.4a), which has a decreased likelihood of dissociation. Compared to DTPA, CHX-DTPA and the methyl derivative MX-DTPA (*p*-isothiocyanatobenzyl-diethylenetriamine-pentaacetic acid) (Fig. 19.4a) possess greater *in vivo* stability and are generally preferred in the radiolabeling of mAbs [115]. An antibody that targets human epidermal growth factor receptor 2 (HER2), a member of the epidermal growth factor receptor family, has been conjugated to a variant of the maleimide derivative of CHX-DTPA. This conjugated protein has been used to provide imaging of high contrast in breast cancer, suggesting that CHX-DTPA has a promising role in clinical molecular imaging [116].

19.2.2.2.2 Macrocyclic Bifunctional Chelators Another strategy to overcome the dissociation problems with bifunctional DTPA derivatives is the use of macrocyclic chelators made up of a tetraza- or triazamacrocyclic ring. 1,4,7,10-Tetraazacyclododecane-*N,N′,N″,N‴*-tetraacetic acid (DOTA, Fig. 19.4b) is a valuable substitute for DTPA and is one of the primary workhorse chelators for radiometal chemistry. Bifunctional DOTA and its commercially available derivatives can form thermodynamically stable complexes with divalent and trivalent radiometals, including In-111, Y-86, Y-90, Bi-213, and Ac-225, a quality that makes them very useful in molecular imaging and radiotherapy [117]. However, there are also consequences to this beneficial *in vivo* stability. In comparison to DTPA, incorporation of the radiolabel to DOTA occurs more slowly and results in a lower yield owing to DOTA's rigid structure [118].

The first approach in conjugating DOTA to a protein is to create an amide linkage via the activation of one of the carboxyl groups in DOTA [119]. However, derivatives of DOTA have improved upon this conjugation by using added linker side chains that will bind to proteins without altering their biological activity [120]. The addition of a side chain may allow for better delivery and targeting properties of the radiolabeled antibody. An example of this is the maleimidocysteineamido-DOTA derivatives that inhibit the release of the bifunctional chelate–radiometal complex in a pH-dependent fashion. This is thought to improve tumor

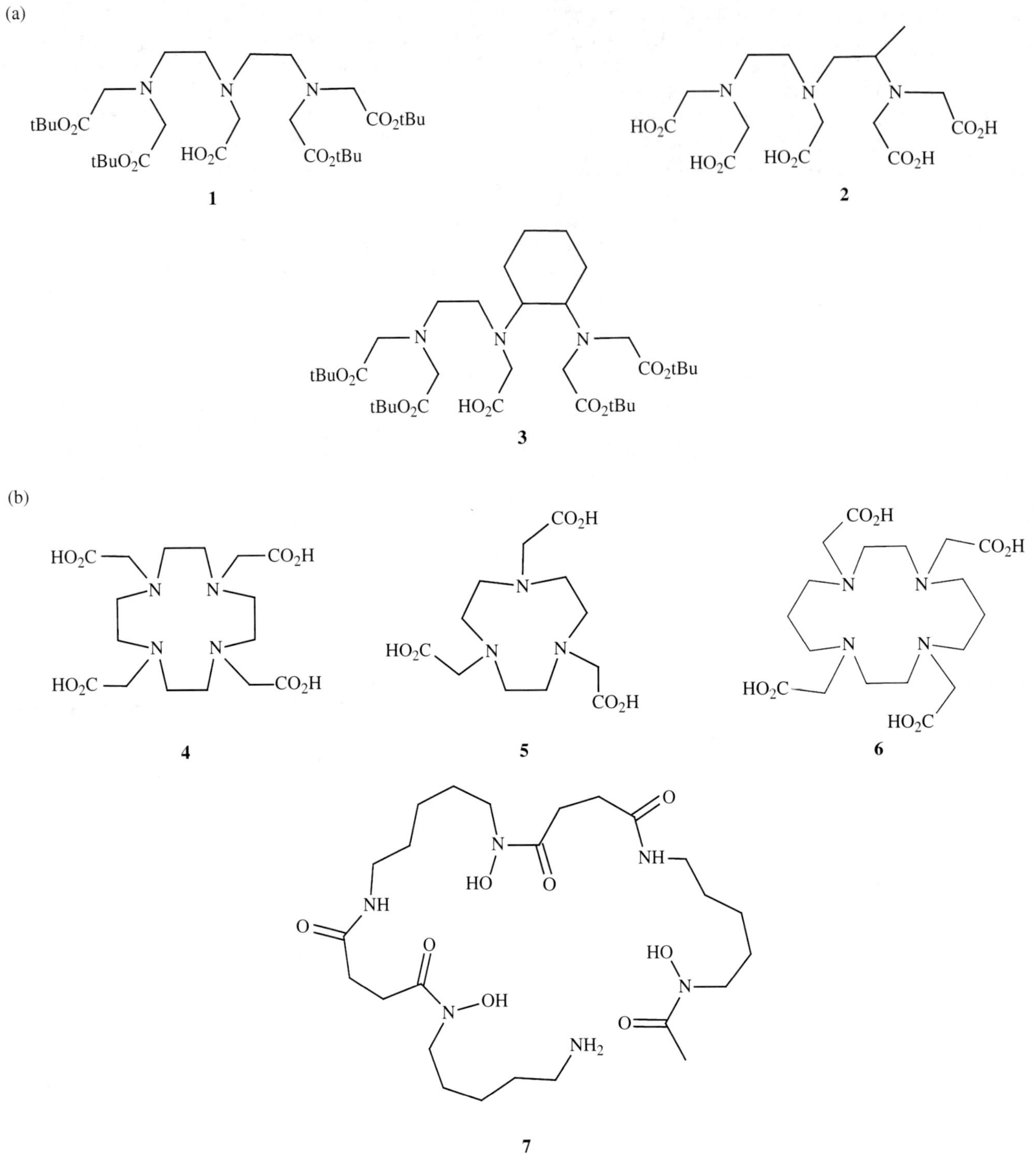

FIGURE 19.4 Bifunctional chelating agents. (a) Structures of DTPA derivatives. **1** = DTPA-tetra (*t*-butyl ester); **2** = MX-DTPA; **3** = CHX-DTPA. (b) Other common macrocyclic bifunctional chelating agents. **4** = DOTA; **5** = NOTA; **6** = TETA; **7** = desferrioxamine B (DFO).

uptake owing to the acidic pH often found in solid tumors, resulting in an accumulation of radiometal at the tumor site [111, 121].

One of the oldest derivative of DOTA is 1,4,7-triazacyclononane-1,4,7-triacetic acid (NOTA, Fig. 19.4b), which is a hexadentate macrocyclic chelator, and is one of the most successful chelators for use with Ga-67, Ga-68, and Cu-64. It has been widely reported that NOTA has a much better stability with Cu-64 than DOTA, with less accumulation of the radionuclide appearing in other organs [122, 123].

Another common chelating agent is 1,4,8,11-tetraazacyclotetradecane-1,4,8,11-tetraacetic acid (TETA, Fig. 19.4b). Compared to the acyclic bifunctional chelators and DOTA, TETA, especially the benzyl TETA derivative, is a more stable chelating agent for copper radionuclides for protein labeling [124, 125]. Use of TETA on peptides with Cu-64 suggests that TETA may have a potential in the molecular imaging of tumors [126].

Another useful chelator for protein labeling is deferoxamine (DFO) or Desferal, a bacterial siderophore that natively binds Fe^{3+} and also has been used extensively with isotopes of gallium and zirconium (Fig. 19.4b). DFO is the only competent Zr-89 chelator available for radiolabeling. Zr^{4+} is a highly charged, hard metal ion with a relatively small ionic radius and is prone to forming insoluble polynuclear hydroxide species in aqueous solution, making it difficult to radiolabel effectively [117]. For many years, the best chelator for Zr-89 had been DTPA, which forms a thermodynamically stable complex with Zr^{4+}, but studies have shown that radiolabeling DTPA with Zr-89 is inefficient (yields < 1%) and *in vivo* stability is exceptionally poor [127]. In contrast, DFO can radiolabel with Zr-89 in quantitative yields (>99%) and is thought to bind Zr-89 with its three hydroxamate groups in a hexadentate fashion [29]. Zr-89-DFO-Zevalin was the first Zr-89 antibody conjugate imaged in humans and was shown to be a suitable PET surrogate for Y-90-Zevalin dosimetry [127]. A number of other studies have been performed using Zr-89-DFO-antibody conjugates with success, such as Zr-89-DFO-U36, Zr-89-DFO-bevacizumab, Zr-89-DFO-J591, Zr-89-DFO-TRC, and Zr-89-DFO-trastuzumab [46, 128–130].

The high 4+ charge makes Zr-89 challenging to incorporate into bifunctional chelators that form stable complexes easily with other common 3+ cationic metal ions (e.g., In^{3+}, Ga^{3+}, Y^{3+} and Lu^{3+}). *In vivo*, Zr-89 that is lost from a bifunctional chelator typically localizes in bone, as demonstrated by the highly unstable Zr-89-DOTA- and Zr-89-DTPA-based antibody conjugates that showed significant Zr-89 accretion in bone 72 h postinjection [131]. Although DFO is an excellent Zr-89 chelator, some decomposition can be observed overtime *in vivo* as Zr-89 slowly accumulates in bone [29]. The design of novel chelators for Zr-89 with improved solubility, *in vivo* stability, and chelation properties would be opportune, considering that DFO is the only current option. The reader is referred to two recent reviews [111, 117] that provide a convenient and accessible overview of the field of radiometal chelating agents.

19.2.2.3 Fluorescent Labeling of Proteins The coupling of fluorescent moieties to proteins to create labeled protein reagents has become a routine and important procedure in the biological sciences and medicine [66]. Molecules that absorb in the NIR region (700–1000 nm) can be efficiently used to visualize and investigate *in vivo* molecular targets because most tissues generate little NIR fluorescence. The most common organic NIR fluorophores are small-molecule organic dyes such as polymethines. Their physical properties, biodistribution, PK, and applications for *in vivo* fluorescence imaging have been summarized in a recent review [59]. Often, a succinimidyl-ester functional group is attached to a fluorophore to allow subsequent conjugation to primary amines on proteins. The presence of multiple primary amines, especially primary amines in the antibody active site, can result in fluorophore conjugation that changes antigen-binding characteristics, and in the extreme, completely inactivates the antibody [132, 133]. Steric hindrance and the absence of additional reactive sites on the fluorophore are presumed to limit the degree of antibody modification by the conjugation reaction. Furthermore, antibodies react with fluorophores at different rates and retain biological activity at different degrees of fluorophore labeling. Thus, conjugation requires a suboptimal fluorophore to protein ratio for the specific coupling reaction of interest. Moreover, the coupling reaction results in a population of antibodies having a heterogeneous distribution in labeling where the number of fluorescence molecules per antibody is variable. Lastly, the presence of multiple fluorophores in close proximity can actually decrease fluorescence through quenching mechanisms (i.e., increased labeling may produce a reagent that is dimmer than one with less labeling) [134, 135]. Chemists continue to develop new NIR fluorochromes with improved fluorescence quantum yield and physical properties, high chemical and photostability, low aggregation tendencies, and low cytotoxicity [60, 136]. Some of these new small-molecule organic dyes have yet to be evaluated for *in vivo* fluorescence imaging.

19.3 MOLECULAR IMAGING FOR THE STUDY OF PK AND BIODISTRIBUTION OF BIOLOGICS

Molecular imaging is a viable method for studying the biodistribution and PK of biologic drug candidates [137–139]. A major task in preclinical drug development is to define the precise relationship between PK and PD (pharmacodynamics): (i) how many reaches the target relative to nontarget tissues and (ii) what are the resulting pharmacological effects. The ability to quantitatively image the biodistribution of therapeutic biologic proteins in a noninvasive fashion can aid in the development of new biologics, dose optimization, and treatment monitoring. Normally, for experimental purposes, biodistribution data must be obtained by dissecting the animal, collecting plasma or tissues, and analysis via radiometric counting measurements ("cut-and-count"). The key advantages of noninvasive imaging are that imaging is less time-consuming and uses less animals and reagents [138].

With regard to clinical uses, there is a critical need to establish an effective noninvasive tool to clearly diagnose diseases as well as provide treatment. If a highly effective noninvasive method existed, a patient could potentially receive effective treatment at an earlier stage with concise, targeted drug delivery. Currently, several noninvasive imaging modalities have been applied in preclinical and clinical drug development, including MRI, X-ray, computed tomography (CT), PET, SPECT, electron microscopy, and ultrasound. Among them, PET, SPECT, and OI are regarded as quantitative or semiquantitative imaging modalities that employ radiotracers or optical tracers to image biodistribution of the labeled proteins, while X-ray, CT, and MRI are normally used for anatomical imaging purposes [140]. The primary modalities utilized for *in vivo* imaging of protein kinetics currently are SPECT and PET [139]. Some representative applications of these imaging modalities in the area of biodistribution and PK are provided in this section along with a review of the basic principles of SPECT, PET, and fluorescence imaging techniques

19.3.1 SPECT Imaging

SPECT is a nuclear medicine tomographic imaging technique using gamma rays. Normally, a marker radioisotope is attached to a protein molecule. After being administered *in vivo*, the gamma emission of the isotope from the radiolabeled protein allows the drug to be seen by a gamma camera. Since the source of SPECT images are gamma ray emissions, three-dimensional images can be acquired with the gamma camera, which permit accurate localization of organs [138]. This information is typically presented as cross-sectional slices through the subject, but can be reformatted or manipulated as required. Therefore, SPECT can be used to provide information about localized function within internal organs, making it particularly useful for tumor or brain imaging. As an example, 3D-acquisition SPECT of the liver was performed to investigate the correlation between tumor accumulation of In-111-bevacizumab and VEGF-A expression in patients with colorectal liver metastases [140]. In another example, several studies applied SPECT imaging to evaluate mesothelin expression in tumors through the use of In-111-labeled antibodies. These findings indicated that the antimesothelin antibody may be developed into a diagnostic agent for imaging mesothelin-expressing cancers [141]. The major advantages of radionuclide-based molecular imaging techniques (SPECT and PET) over optical modalities (e.g., optical and MRI) are that they are very sensitive (down to the picomolar level), quantitative, and there is no tissue penetration limit [138]. However, one disadvantage is that the resolution (typically > 1 mm) of either SPECT or PET is not as high as the other imaging modalities such as MRI.

Radiolabeling of antibodies with radioiodine may also be applied to radioimmunoscintigraphy and SPECT imaging to obtain preclinical information on biodistribution, PK, and tumor detection of several genetically engineered antibodies [142–144]. The imaging signal from radioiodine in target tissues reflects mostly intact protein and partial degradation products. Eventually, iodine catabolites leave the cell, diminishing signal in target tissues and increasing the amount of radioiodine available for thyroid uptake. Images from radioiodinated proteins typically include conspicuous thyroid glands due to the catabolized iodine. The stable attachment of residualizing radiometals (i.e., radiometals that accumulate and get trapped in the cell) to antibodies has been pursued to circumvent the pitfalls of radioiodine, especially for internalizing antibodies. Antibodies labeled with metal radionuclides via DTPA, DOTA, or DFO tend to accumulate in antigen-expressing tissues following receptor-mediated endocytosis. This accumulation is caused by the residualizing properties of the charged, highly polar catabolites (e.g., charged lysine adducts). Signal remains visible along with that of intact or partially degraded protein. Thus, the imaging signal of a residualizing label represents the current or final location of the protein, approximating the total amount accumulated at that site (Fig. 19.5a) [145]. Residualization also occurs to some extent in organs of mAb catabolism such as the liver, the kidney, and the spleen. Such is the case for Zevalin, which utilizes Y-90 chelated by a DTPA analog. The residualization benefits of Y-90 and In-111 over I-131 have been demonstrated by clinically comparing tumor uptake kinetics of Y-90 and I-131 labeled antibodies and by imaging studies with In-111 and I-131 [29].

19.3.2 PET Imaging

In contrast to SPECT tracers that emit gamma rays that are measured directly, PET tracers emit positrons that annihilate to emit pairs of gamma photons. The biodistribution study of therapeutic biologics by PET methodology is quantitatively measured by radiotracer distribution in organs/tissues. Similar to SPECT, PET uses radioactive tracers and a scanner to record data that a computer constructs into two- or three-dimensional images. In other words, the three-dimensional image created from a SPECT or PET scan maps the radionuclide–protein conjugate within the subject (Fig. 19.5b–e) [45, 146–148]. Biodistribution and PK could be recorded by blood sampling at predetermined time points, preferably associated with radiometabolite evaluation with respect to time. Although collecting blood samples has been used as a method in SPECT imaging to estimate blood concentration, the heart has been used in PET to determine blood concentrations after injection of radiolabeled proteins. For example, animals injected intravenously with I-124-labeled antibody and imaged serially up to 7 days postinjection showed that PET blood curves agreed well with direct measurements within 12% at all time points [149].

The distribution of large proteins is determined by the rate of extravasation in tissue, the rate of distribution within

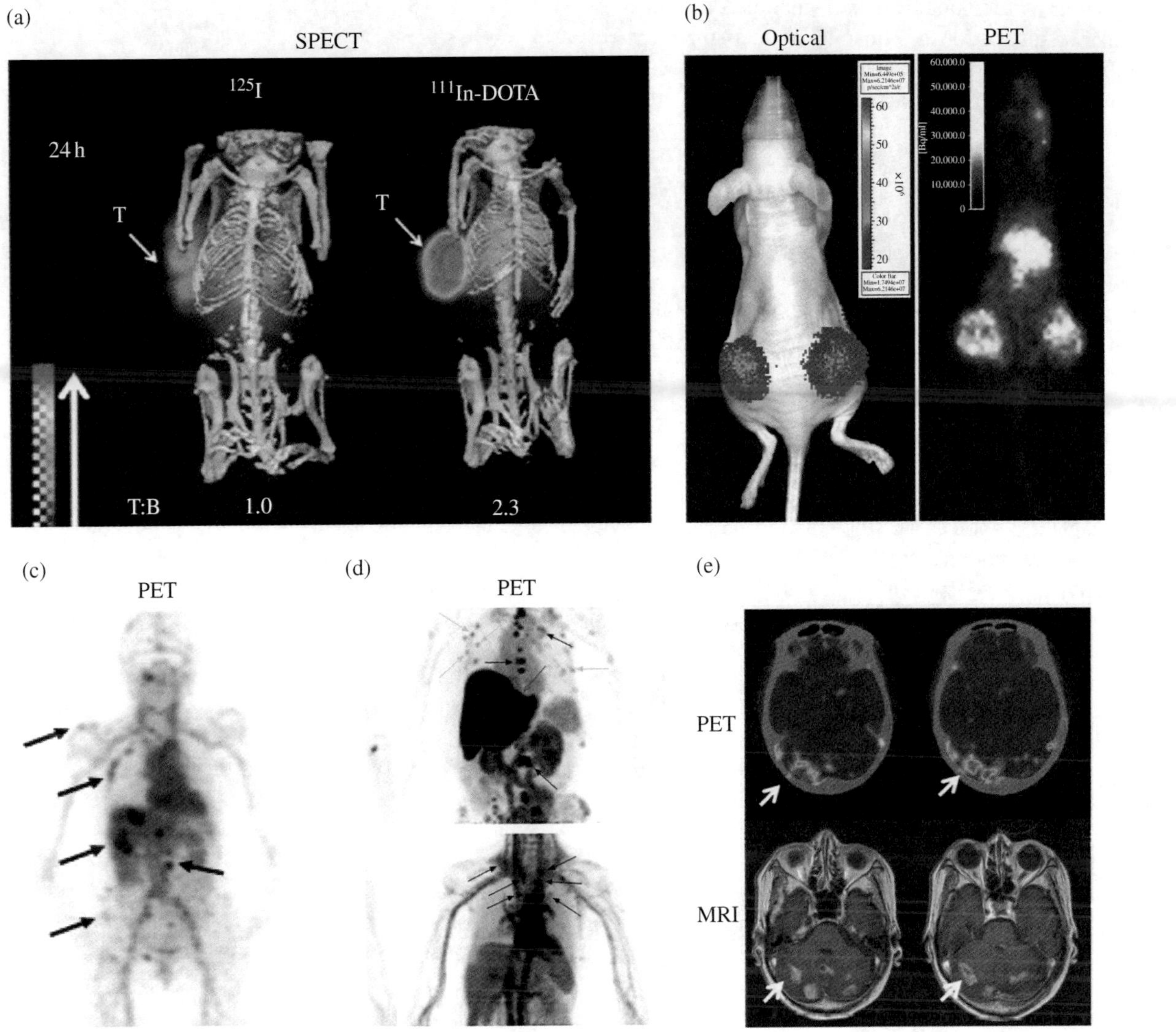

FIGURE 19.5 Examples of radiolabeled antibodies in molecular imaging of tumors. (a) SPECT-CT imaging of ^{125}I-trastuzumab and ^{111}In-DOTA-trastuzumab in KPL-4 xenograft-bearing mice, T:B = tumor-to-blood ratios of radioactive uptake [145]. (b) Dual-mode (optical and PET) imaging of ^{89}Zr-bevacizumab-IRDye800CW in a xenograft-bearing mouse [146]. (c) Five-day PET scan of ^{89}Zr-trastuzumab in a patient with liver and bone metastases, as indicated by the arrows [45]. (d) One-day PET scan of ^{64}Cu-DOTA-trastuzumab in two different patients. Black and blue arrows point to tumor lesions, and red arrow points liver uptake of the radiolabel [147]. (e) ^{64}Cu-DOTA-trastuzumab PET scans (top) of HER2-positive metastatic brain lesions in two patients compared to lesions detected by MRI [148]. (*See insert for color representation of this figure.*)

tissue, binding affinity in tissue, the rates of elimination from tissue, route of injection, and species [138]. Normally, the biodistribution of proteins requires sacrificing a number of rodents and quantitating the tissue concentration by counting the radioactivity emitted from the tissue. Micro-PET imaging provides a noninvasive way of quantifying the concentration of proteins in tissues, a technique that is especially useful for monitoring the kinetics in target organ such as tumor. For instance, a mAb was conjugated to DOTA and labeled with Cu-64. Serial PET imaging revealed that the tumor uptake of the tracer was higher than most organs at late time points, which provided excellent tumor contrast. Biodistribution data as measured by gamma counting were consistent with PET findings [150]. In another example, PET imaging was used to calculate tumor uptake and compared with *ex vivo* biodistribution (i.e., "cut-and-count") after injection of Zr-89-labeled bevacizumab. This study demonstrated that the quantization of radiolabeled bevacizumab in tumor using noninvasive imaging is convenient and validated by *ex vivo* biodistribution studies [128].

New procedures have been established in recent years for assembling a large amount of positron emitters for preclinical and clinical PET use, such as Zr-89, Y-90, and I-124 [150]. In general, I-124 is the radionuclide of choice in combination with noninternalizing antibodies, and Zr-89 is particularly suitable for PET imaging of internalizing mAbs due to its ability to be residualized within cells [2]. Readers are referred to a review on the future directions of these radionuclides and PET imaging [150].

19.3.3 Optical Imaging

OI relies on the detection of photons in the visible, ultraviolet, and NIR portion of the electromagnetic spectrum produced by bioluminescence or fluorescence [139, 151]. Bioluminescence is a chemical process wherein light is emitted during the interaction of a protein produced in engineered cells and an administered substrate, as seen in firefly luciferase and luciferin [152]. Fluorescence imaging uses probes that emit light (fluoresce) after the excitation at a fluorophore-specific wavelength. Depending on the fluorophore, OI of fluorophore-labeled antibodies possibly offers higher sensitivity and temporal resolution than PET in small animals, but unlike PET is limited to a few centimeters of tissue depth [139]. New developments in the chemistry of fluorophores have resulted in a series of fluorochromes with emissions extending from the ultraviolet into the NIR portion of the electromagnetic spectrum [153]. The longer wavelengths of the NIR region are advantageous for imaging because tissue absorption of signal and auto-fluorescence from tissue are minimized at these wavelengths. However, even in the NIR range (700–1000 nm), deep internal organs remain difficult to detect. Furthermore, while three-dimensional image reconstruction of fluorophores has been demonstrated, results with OI are almost exclusively two-dimensional (Fig. 19.5b) [146]. Generally, only organs/areas of interest on or near the surface can be detected reliably with fluorescent labeling [154, 155].

The advantages of OI for assessing antibody PK *in vivo* are the conjugation chemistry and cost, and lack of exposure to ionizing radiation [138]. The fluorophore conjugation is often carried out with a commercially available kit and can be performed in most laboratories without special license, equipment, hazardous precautions, or specific training. The detection systems are generally 30–50% as expensive as those for small-animal SPECT or PET, and planar fluorescence images are generated within seconds or minutes, typically with a photograph overlay. OI experiments can also accommodate four to five animals per scan, allowing an efficient researcher to collect data from 100 animal images per hour, whereas PET and SPECT imaging experiments are often limited to one to two animals per 10- to 30-min scan. Such throughput and cost advantages combined with expected improvements in quantification from advances in instrumentation provide an exciting outlook for this approach [139]. The reader is referred to an excellent review article [156] for more in-depth discussion of fluorescent imaging.

19.4 CONCLUSIONS AND PERSPECTIVES

Biodistribution studies for biologics are routinely carried out using radioactive iodine isotopes. However, iodine-labeled proteins show rapid loss of the radiolabel from the target tissue. The iodotyrosine formed upon proteolysis diffuses out of the cell/tissue, making results difficult to interpret on an appropriate time scale for the evaluation of antibody localization. Hence, standard radioiodinated mAbs do not demonstrate the residualizing capacity of the radiometal complexes of In-111 or Y-90 with DOTA or DTPA [157]. Chemists are currently investigating the use of trifunctional chelating agents that have the residualizing advantages of DOTA but with the capability to be radioiodinated. These chelating agents bind to the protein as usual but have two groups to bind to radiolabels—one to a radiohalogen and the other to chelate a radiometal. The benefit of these trifunctional chelating agents is that they would allow the delivery of two different radionuclides (i.e., a combination of radiohalogen and radiometal). Currently, a SIB-DOTA prosthetic group is under investigation (Fig. 19.6) for the radiolabeling of a halogen and a metal within the same molecule [158]. This chelating agent combines the features of the prototypical, dehalogenation-resistant *N*-succinimidyl 3-iodobenzoate (SIB) with DOTA. More recently, other investigators have reported the synthesis of a novel probe, HIP-DOTA (Fig. 19.6), for labeling antibodies with radioiodine such that residualization occurs in a similar manner as for radiometals [145]. An additional advancement found in this work is that the HIP-DOTA system was designed for the labeling of thiols rather than lysine residues. This allows site-specific radiolabeling of modified mAbs with engineered cysteine residues and reduced disruption of mAb-binding characteristics [153]. Overall, the use of these trifunctional chelators in antibody labeling is a valuable preclinical tool for studying the biodistribution, metabolism, and excretion of antibody therapeutics. Furthermore, the potential of delivering a mixture of radiolabels, each with their own unique properties, may potentially be useful for combining molecular imaging and targeted radioimmunotherapy.

With the current advances in biotherapeutics and antibody-directed drug delivery systems, it has become fundamentally important to be able to screen patients before specific antibody therapies, and to design patient-tailored therapeutic regimens in order to avoid unnecessary toxicities [148]. Noninvasive detection of various molecular markers of diseases can allow for much earlier diagnosis, earlier treatment, and better prognosis that will eventually lead to personalized medicine. Molecular imaging takes advantage of the traditional diagnostic imaging techniques and introduces molecular imaging probes to measure the expression of indicative molecular markers at different stages of diseases. It has been widely reported that various imaging modalities demonstrate

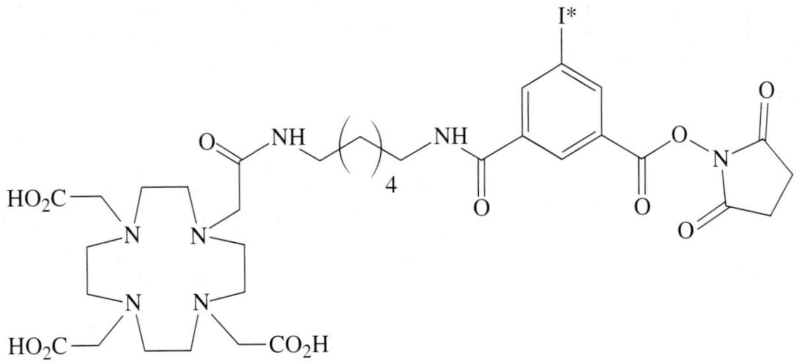

1

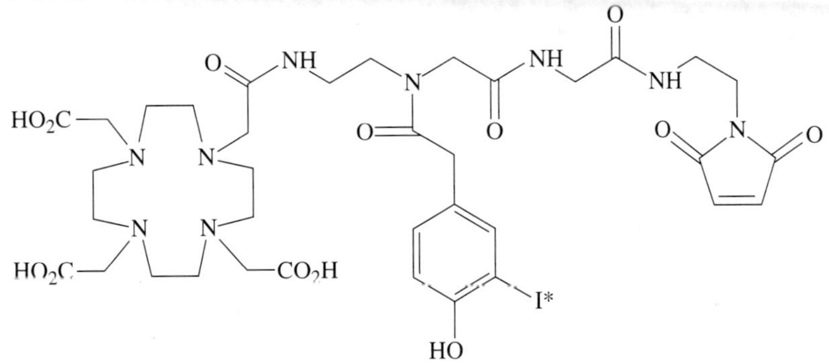

2

FIGURE 19.6 Trifunctional chelating agents. **1** = SIB-DOTA; **2** = HIP-DOTA.

unique advantages in drug development. In this chapter, we presented the distribution and PK studies using PET, SPECT, as well as OI. Among those imaging modalities, PET and SPECT imaging are used clinically, while optical imaging is still in development for use in the clinic with almost all of the studies using this modality still in preclinical standing [138]. In spite of its shortcomings, the field of optical imaging is advancing along with radioimaging, and it will be exciting to see how these imaging modalities will be combined or used individually to benefit personalized medicine.

REFERENCES

[1] Hiltunen JV. Search for new and improved radiolabeling methods for monoclonal antibodies: a review of different methods. Acta Oncol 1993;32:831–839.

[2] Verel I, Visser GW, Boerman OC, van Eerd JE, Finn R, Boellaard R, Vosjan MJ, Stigter-van Walsum M, Snow GB, van Dongen GA. Long-lived positron emitters zirconium-89 and iodine-124 for scouting of therapeutic radioimmunoconjugates with PET. Cancer Biother Radiopharm 2003; 18:655–656.

[3] Thurber GM, Weissleder R. Quantitating antibody uptake *in vivo*: conditional dependence on antigen expression levels. Mol Imaging Biol 2011;13:623–632.

[4] Wankhede M, Bouras A, Kaluzova M, Hadjipanayis CG. Magnetic nanoparticles: an emerging technology for malignant brain tumor imaging and therapy. Expert Rev Clin Pharmacol 2012;5:173–186.

[5] Arruebo M, Valladares M, Gonzalez-Fernandez A. Antibody-conjugated nanoparticles for biomedical applications. J Nanomater 2009;37:1–24.

[6] Bohdiewicz PJ, Scott GC, Juni JE, Fink-Bennett D, Wilner F, Nagle C, Dworkin HJ. Indium-111 OncoScint CR/OV and F 18 FDG in colorectal and ovarian carcinoma recurrences. Early observations. Clin Nucl Med 1995;20:230–236.

[7] Hughes B. Antibody–drug conjugates for cancer: poised to deliver? Nat Rev Drug Discov 2010;9:665–667.

[8] Prabhu S, Boswell CA, Leipold D, Khawli LA, Li D, Lu D, Theil FP, Joshi A, Lum B. Antibody delivery of drugs and radionuclides: factors influencing clinical pharmacology. Ther Deliv 2011;2:769–791.

[9] Yuanfang L, Chuanchu W. Radiolabeling of monoclonal antibodies with metal chelates. Pure Appl Chem 1991;63: 427–463.

[10] Wagner HN, Emmons H. Characteristics of an ideal radiopharmaceutical. In: *Radioactive Pharmaceuticals*. Oak Ridge (TN): Atomic Energy Commission; 1968. p 1–32.

[11] Eckelman WC, Paik CH, Reba RC. Radiolabeling of antibodies. Cancer Res 1980;40:3036–3042.

[12] Wilbur DS. Radiohalogenation of proteins: an overview of radionuclides, labeling methods, and reagents for conjugate labeling. Bioconjug Chem 1992;3:443–470.

[13] Yano Y. Radionuclide generators: current and future applications in nuclear medicine. In: *Radiopharmaceuticals*. New York: Society of Nuclear Medicine; 1975. p 236–245.

[14] Quaim SM, Stocklin G. Production of some medically important short lived neutron-deficient radioisotopes of halogens. Radiochim Acta 1983;34:25–40.

[15] Lambrecht RM. Radionuclide generators. Radiochim Acta 1983;34:9–24.

[16] Ruth TJ, Pate BD, Robertson R, Porter JK. Radionuclide production for the biosciences. Nucl Med Biol 1989;16: 323–336.

[17] Waters SL, Silvester DJ. Inorganic cyclotron radionuclides. Radiochim Acta 1982;30:163–173.

[18] Kirby HW. Production, isolation, and purification of astatine isotopes. In: *Gmelin Handbook of Inorganic Chemistry: At-Astatine*. New York: Springer-Verlag; 1985. p 95–106.

[19] Kilbourn MR. Production of fluorine-18. In: *Fluorine-18 Labeling of Radiopharmaceuticals*. Washington (DC): National Academy Press; 1990. p 4–14.

[20] Moerlein SM, Welch MJ. The chemistry of gallium and indium as related to radiopharmaceutical production. Int J Nucl Med Biol 1981;8:277–287.

[21] Pandey MK, Engelbrecht HP, Byrne JP, Packard AB, DeGrado TR. Production of 89Zr via the 89Y(p,n)89Zr reaction in aqueous solution: effect of solution composition on in-target chemistry. Nucl Med Biol 2014;41:309–316.

[22] Pressman D, Keighley G. The zone of activity of antibodies as determined by the use of radioactive tracers; the zone of activity of nephritoxic antikidney serum. J Immunol 1948;59:141–146.

[23] Bale WF, Spar IL. Studies directed toward the use of antibodies as carriers of radioactivity for therapy. Adv Biol Med Phys 1957;5:285–356.

[24] Williams LE, Wu AM, Kenanova VE, Olafsen T, Yazaki PJ. Numerical comparison of iodine-based and indium-based antibody biodistributions. Cancer Biother Radiopharm 2014;29:91–98.

[25] Izzo JL, Bale WF, Izzo MJ, Roncone A. High specific activity labeling of insulin with I-131. J Biol Chem 1964;239: 3743–3748.

[26] Kakabakos SE, Livaniou E, Evangelatos SA, Evangelatos GP, Ithakissios DS. Isolation of mono- and di-iodine 125 tyramines for conjugation labeling. Eur J Nucl Med Mol Imaging 1991;18:952–954.

[27] Schafer DE. Measurement of receptor-ligand binding: theory and practice. In: *Lecture Notes in Biomathematics: Tracer Kinetics and Physiologic Modeling*. New York: Springer-Verlag; 1983. p 445–507.

[28] Boswell CA, Ferl GZ, Mundo EE, Bumbaca D, Schweiger MG, Theil FP, Fielder PJ, Khawli LA. Effects of anti-VEGF on predicted antibody biodistribution: roles of vascular volume, interstitial volume, and blood flow. PLoS One 2011;6:e17874.

[29] Williams SP. Tissue distribution studies of protein therapeutics using molecular probes: molecular imaging. AAPS J 2012;14:389–399.

[30] Preylowski V, Schlögl S, Schoenahl F, Jörg G, Samnick S, Buck AK, Lassmann M. Is the image quality of I-124-PET impaired by an automatic correction of prompt gammas? PLoS One 2013;8:e71729.

[31] Westera G, Reist HW, Buchegger F, Heusser CH, Hardman N, Pfeiffer A, Sharma HL, von Schulthess GK, Mach JP. Radioimmuno positron emission tomography with monoclonal antibodies: a new approach to quantifying *in vivo* tumour concentration and biodistribution for radioimmunotherapy. Nucl Med Commun 1991;12:429–437.

[32] Epstein AL, Khawli LA. Tumor biology and monoclonal antibodies: overview of basic principles and clinical considerations. Antibody Immunoconjug Radiopharm 1991;4: 373–384.

[33] Zalutsky MR, Pruszynski M. Astatine-211: production and availability. Curr Radiopharm 2011;4:177–185.

[34] Zalutsky MR, Reardon DA, Pozzi OR, Vaidyanathan G, Bigner DD. Targeted α-particle radiotherapy with 211At-labeled monoclonal antibodies. Nucl Med Biol 2007;34: 779–785.

[35] Waibel R, Alberto R, Willuda J, Finnern R, Schibli R, Stichelberger A, Eglil A, Abram U, Mach JP, Plückthun A, Schubiger PA. Stable one-step technetium-99 m labeling of His-tagged recombinant proteins with a novel Tc(I)-carbonyl complex. Nat Biotechnol 1999;17:897–901.

[36] Hnatowidh DJ, Childs RL, Lanteigne D, Najafi A. The preparation of DTPA coupled antibodies radiolabeled with metallic radionuclides: an improved method. J Immunol Methods 1983;65:147–157.

[37] Hao G, Singh AN, Liu W, Sun X. PET with non-standard nuclides. Curr Top Med Chem 2010;10:1096–1112.

[38] Ikotun OF, Lapi SE. The rise of metal radionuclides in medical imaging: copper-64, zirconium-89 and yttrium-86. Future Med Chem 2011;3:599–621.

[39] Holland JP, Williamson MJ, Lewis JS. Unconventional nuclides for radiopharmaceuticals. Mol Imaging 2010; 9:1–20.

[40] Shokeen M, Anderson CJ. Molecular imaging of cancer with copper-64 radiopharmaceuticals and positron emission tomography (PET). Acc Chem Res 2009;42:832–841.

[41] Baum RP, Prasad V, Müller D, Schuchardt C, Orlova A, Wennborg A, Tolmachev V, Feldwisch J. Molecular imaging of HER2-expressing malignant tumors in breast cancer patients using synthetic 111In- or 68Ga-labeled affibody molecules. J Nucl Med 2010;51:892–897.

[42] Xing Y, Zhao J, Shi X, Conti PS. Recent development of radiolabeled nanoparticles for PET imaging. Austin J Nanomed Nanotechnol 2014;2:1–10.

[43] Aerts HJ, Dubois L, Perk L, Vermaelen P, van Dongen GA, Wouters BG, Lambin P. Disparity between *in vivo* EGFR expression and 89Zr-labeled cetuximab uptake assessed with PET. J Nucl Med 2009;50:123–131.

[44] Dijkers EC, Kosterink JG, Rademaker AP, Perk LR, van Dongen GA, Bart J, de Jong JR, de Vries EG, Lub-de Hooge MN. Development and characterization of clinical-grade

89Zr-trastuzumab for HER2/neu immunoPET imaging. J Nucl Med 2009;50:974–981.

[45] Dijkers EC, Oude Munnink TH, Kosterink JG, Brouwers AH, Jager PL, de Jong JR, van Dongen GA, Schröder CP, Lub-de Hooge MN, de Vries EG. Biodistribution of 89Zr-trastuzumab and PET imaging of HER2-positive lesions in patients with metastatic breast cancer. Clin Pharmacol Ther 2010;87:586–592.

[46] Börjesson PK, Jauw YW, Boellaard R, de Bree R, Comans EF, Roos JC, Castelijns JA, Vosjan MJ, Kummer JA, Leemans CR, Lammertsma AA, van Dongen GA. Performance of immuno-positron emission tomography with zirconium-89-labeled chimeric monoclonal antibody U36 in the detection of lymph node metastases in head and neck cancer patients. Clin Cancer Res 2006;12:2133–2140.

[47] Gaykema SB, Brouwers AH, Hovenga S, Lub-de Hooge MN, deVries EG, Schroder CP. Zirconium-89-trastuzumab positron emission tomography as a tool to solve a clinical dilemma in a patient with breast cancer. J Clin Oncol 2012;30: e74–e75.

[48] Oude Munnink TH, Arjaans ME, Timmer-Bosscha H, Schroder CP, Hesselink JW, Vedelaar SR, Walenkamp AM, Reiss M, Gregory RC, Lub-de Hooge MN, de Vries EG. PET with the 89Zr-labeled transforming growth factor-beta antibody fresolimumab in tumor models. J Nucl Med 2011;52:2001–2008.

[49] Heuveling DA, Visser GW, Baclayon M, Roos WH, Wuite GJ, Hoekstra OS, Leemans CR, de Bree R, van Dongen GA. 89Zr-nanocolloidal albumin-based PET/CT lymphoscintigraphy for sentinel node detection in head and neck cancer: preclinical results. J Nucl Med 2011;52:1580–1584.

[50] Nagengast WB, Lub-de Hooge MN, Oosting SF, den Dunnen WF, Warnders FJ, Brouwers AH, de Jong JR, Price PM, Hollema H, Hospers GA, Elsinga PH, Hesselink JW, Gietema JA, de Vries EG. VEGF-PET imaging is a noninvasive biomarker showing differential changes in the tumor during sunitinib treatment. Cancer Res 2011;71:143–153.

[51] Borjesson PK, Jauw YW, de Bree R, Roos JC, Castelijns JA, Leemans CR, van Dongen GA, Boellaard R. Radiation dosimetry of 89Zr-labeled chimeric monoclonal antibody U36 as used for immuno-PET in head and neck cancer patients. J Nucl Med 2009;50:1828–1836.

[52] Hnatowich DJ. Labeled proteins in nuclear medicine-current status. In: *Current Applications in Radiopharmacology: Proceedings of the Fourth International Symposium on Radiopharmacology*. Oxford: Pergamon Press; 1986. p 257.

[53] Gobuty AH, Kim EE, Weiner RE. Radiolabeled monoclonal antibodies: radiochemical pharmacokinetic and clinical challenges. J Nucl Med 1985;26:546–548.

[54] Hnatowich DJ, Griffin TW, Kosciuczyk C, Rusckowski M, Childs RL, Mattis JA, Shealy D, Doherty PW. Pharmacokinetics of an In-111 labeled monoclonal antibody in cancer patients. J Nucl Med 1985;26:849–858.

[55] Fawwaz RA, Wang TST, Estabrook A, Rosen JM, Hardy MA, Alderson PO, Srivastava SC, Richards P, Ferrone S. Immunoreactivity and biodistribution of In-111 labeled monoclonal antibody to a human high molecular weight melanoma associated antigen. J Nucl Med 1985;26:488–492.

[56] Paik CH, Hong JJ, Ebbert MA, Heald SC, Reba RC, Eckelman WC. Relative reactivity of DTPA, immunoreactive antibody-DTPA conjugates, and nonimmunoreactive antibody-DTPA conjugates toward indium-111. J Nucl Med 1985;26: 482–487.

[57] Wessels BW, Rogus RD. Radionuclide selection and model absorbed dose calculations for radiolabeled tumor associated antibodies. Med Phys 1984;11:638–645.

[58] Xu H, Baidoo K, Gunn AJ, Boswell CA, Milenic DE, Choyke PL, Brechbiel MW. Design, synthesis, and characterization of a dual modality positron emission tomography and fluorescence imaging agent for monoclonal antibody tumor-targeted imaging. J Med Chem 2007;50:4759–4765.

[59] Ballou B, Lagerholm BC, Ernst LA, Bruchez MP, Waggoner AS. Noninvasive imaging of quantum dots in mice. Bioconjug Chem 2004;15:79–86.

[60] Shah K, Weissleder R. Molecular optical imaging: applications leading to the development of present day therapeutics. NeuroRx 2005;2:215–225.

[61] Kim JS, Kim YH, Kim JH, Kang KW, Tae EL, Youn H, Kim D, Kim SK, Kwon JT, Cho MH, Lee YS, Jeong JM, Chung JK, Lee DS. Development and *in vivo* imaging of a PET/MRI nanoprobe with enhanced NIR fluorescence by dye encapsulation. Nanomedicine 2012;7:219–229.

[62] Stelter L, Pinkernelle JG, Michel R, Schwartländer R, Raschzok N, Morgul MH, Koch M, Denecke T, Ruf J, Bäumler H, Jordan A, Hamm B, Sauer IM, Teichgräber U. Modification of aminosilanized superparamagnetic nanoparticles: feasibility of multimodal detection using 3T MRI, small animal PET, and fluorescence imaging. Mol Imaging Biol 2010;12:25–34.

[63] Boswell CA, Tesar DB, Mukhyala K, Theil FP, Fielder PJ, Khawli LA. Effects of charge on antibody tissue distribution and pharmacokinetics. Bioconjug Chem 2010;21:2153–2163.

[64] Tinianow JN, Gill HS, Ogasawara A, Flores JE, Vanderbilt AN, Luis E, Vandlen R, Darwish M, Junutula JR, Williams SP, Marik J. Site-specifically 89Zr-labeled monoclonal antibodies for ImmunoPET. Nucl Med Biol 2010;37:289–297.

[65] Shen BQ, Xu K, Liu L, Raab H, Bhakta S, Kenrick M, Parsons-Reponte KL, Tien J, Yu SF, Mai E, Li D, Tibbitts J, Baudys J, Saad OM, Scales SJ, McDonald PJ, Hass PE, Eigenbrot C, Nguyen T, Solis WA, Fuji RN, Flagella KM, Patel D, Spencer SD, Khawli LA, Ebens A, Wong WL, Vandlen R, Kaur S, Sliwkowski MX, Scheller RH, Polakis P, Junutula JR. Conjugation site modulates the *in vivo* stability and therapeutic activity of antibody–drug conjugates. Nat Biotechnol 2012;30:184–189.

[66] Vira S, Mekhedov E, Humphrey G, Blank PS. Fluorescent labeled antibodies—balancing functionality and degree of labeling. Anal Biochem 2010;402:146–150.

[67] Cohen R, Stammes MA, de Roos IH, Stigter-van Walsum M, Visser GW, van Dongen GA. Inert coupling of IRDye800CW to monoclonal antibodies for clinical optical imaging of tumor targets. EJNMMI Res 2011;1:31.

[68] Sampath L, Kwon S, Ke S, Wang W, Schiff R, Mawad ME, Sevick-Muraca EM. Dual-labeled trastuzumab-based imaging agent for the detection of human epidermal growth factor receptor 2 overexpression in breast cancer. J Nucl Med 2007;48:1501–1510.

[69] Khawli LA, Goswami S, Hutchinson R, Kwong ZW, Yang J, Wang X, Yao Z, Sreedhara A, Cano T, Tesar D, Nijem I, Allison DE, Wong PY, Kao YH, Quan C, Joshi A, Harris RJ, Motchnik P. Charge variants in IgG1: isolation, characterization, *in vitro* binding properties and pharmacokinetics in rats. MAbs 2010;2:613–624.

[70] Wakankar A, Chen Y, Gokarn Y, Jacobson FS. Analytical methods for physicochemical characterization of antibody drug conjugates. MAbs 2011;3:161–172.

[71] Vanderheyden JL, Liu G, He J, Patel B, Tait JF, Hnatowich DJ. Evaluation of 99mTc-MAG3-annexin V: influence of the chelate on *in vitro* and *in vivo* properties in mice. Nucl Med Biol 2006;33:135–144.

[72] Dewanjee MK. Methods of radioiodination reactions with several oxidizing agents. In: *Radioiodination: Theory, Practice, and Biomedical Applications*. New York: Springer; 1992. p 129–194.

[73] MacFarland AS. Efficient trace-labeling of proteins with iodine. Nature 1958;782:53.

[74] Batts BD, Gold V. The kinetics of aromatic protio- and deuteriodeiodination. *J Chem Soc* 1964:5753–5762.

[75] Berliner E. Kinetics of aromatic halogenation. V. The iodination of 2,4-dichlorophenol and anisole with iodine monochloride. J Am Chem Soc 1958;80:856 861.

[76] Mohanakrishnan AK, Prakash C, Ramesh N. A simple iodination protocol via *in situ* generated ICl using NaI/FeCl3. Tetrahedron 2006;62:3242–3247.

[77] Taghvaei–Ganjali S, Ghasemian-Dazmiri M, Hosseinzadeh M, Hosseinib A, Khalilzadeh MA. A simple iodination protocol via *in situ* generated ICl using NaIO4/NaCl/Silica sulfuric acid. Iran J Org Chem 2009;3:160–162.

[78] Akhlaghinia B, Rahmani M. Mild and efficient iodination of aromatic compounds with trichloroisocyanuric acid/I_2/wet SiO_2 system. Turk J Chem 2009;33:67–72.

[79] Huguchi T, Hussain A. Mechanism of chlorination of cresol by chloramine-T. Mediation by dichloramine-T. *J Chem Soc* 1967:549–552.

[80] Eckelman WC, Kubota H, Siegel R, Komai T, Rzeszotarski WJ, Reba RC. lodinated bleomycin. An unsatisfactory radiopharmaceutical for tumor localization. J Nucl Med 1976;17: 385–388.

[81] Freychet P, Roth J, Neville DM. Monoiodoinsulin: demonstration of its biological activity and binding to fat cells and liver membranes. Biochem Biophys Res Commun 1971;43: 400–408.

[82] Hornick JL, Khawli LA, Hu P, Epstein AL. Single amino acid substitution in the Fc region of chimeric TNT-3 antibody accelerates clearance and improves immunoscintigraphy of solid tumors. J Nucl Med 2000;41:355–362.

[83] Akanji AG, Muramoto E, Caldeira Filho J, Martinussi Couto RM, Bortoletti de Araújo E. Radiolabeling and biodistribution of monoclonal antibody (MAb) anti-CD20 with iodine-131. Braz Arch Biol Technol 2005;48:69–72.

[84] Markwell MA. A new solid-state reagent to iodinate proteins. I. Conditions for the efficient labeling of antiserum. Anal Biochem 1982;125:427–432.

[85] Chizzonite R, Truitt T, Podlaski FJ, Wolitzky AG, Quinn PM, Nunes P, Stern AS, Gately MK. IL-12: monoclonal antibodies specific for the 40-kDa subunit block receptor binding and biologic activity on activated human lymphoblasts. J Immunol 1991;147:1548–1556.

[86] Regoeczi E. Methods of protein iodination. In: *Iodine Labeled Plasma Proteins*. Volume 1, New York: CRC Press, Inc; 1984. p 35–102.

[87] Youfeng H, Coenen HH, Petzold G, Stocklin G. A comparative study of radioiodination of simple aromatic compounds via N-halosuccinimides and chloramine-T in TFAA. J Labelled Comp Radiopharm 1982;19:807–819.

[88] Khawli LA, Kassis AI. Synthesis of I-125-labeled N-succinimidyl-p-iodobenzoate for use in radiolabeling of antibodies. Nucl Med Biol 1989;16:727–733.

[89] Wilbur DS, Hadley SW, Hylarides MD, Abrams PG, Beaumier PL, Morgan AC, Reno J, Fritzberg AR. Development of a stable radioiodinating reagent to label monoclonal antibodies for radiotherapy of cancer. J Nucl Med 1989;30:216–226.

[90] Mean GE, Feeney RE. Chemical modification of proteins: history and applications. Bioconjug Chem 1990;1:2–12.

[91] Brinkley M. A brief survey of methods for preparing protein conjugates with dyes, haptens, and cross-linking reagents. Bioconjug Chem 1992;3:2–13.

[92] Hadley SW, Wilbur DS, Gray MA, Atcher RW. Astatine-211 labeling of an anti-melanoma antibody and its' Fab fragment using N-succinimidyl p-[211At] astatobenzoate: comparisons *in vivo* with the p-[125I]-iodobenzoyl conjugate. Bioconjug Chem 1991;2:171–179.

[93] Bolton AE, Hunter WM. The labeling of proteins to high specific radioactivities by conjugation to a 125I-containing acylating agent. Biochem J 1973;733:529–539.

[94] Wood FT, Wu MM, Gerhart JC. The radioactive labeling of proteins with an iodinated amidination reagent. Anal Biochem 1975;69:339–349.

[95] Lundblad RL. *Chemical Reagents for Protein Modification*. Volume 1, New York: CRC Press, Inc; 1984. p 127–170.

[96] Regoeczi E. *Iodine-Labeled Plasma Proteins*. Volume 2, *Part B*, New York: CRC Press Inc; 1987. p 43–71.

[97] Kabalka GW, Varma RS. The synthesis of radiolabeled compounds via organometallic intermediates. Tetrahedron 1989;45:6601–6621.

[98] Moerlein SM, Mathis CA, Yano Y. Comparative evaluation of electrophilic aromatic iododemetallation techniques for labeling radiopharmaceuticals with iodine-122. Appl Radiat Isot 1987;38:85–90.

[99] Hylarides MD, Wilbur DS, Hadley SW, Fritzberg AR. Synthesis and iodination of methyl 4-tri-n-butylstannylbenzoate, 4-carbomethoxyphenyl mercuric chloride and 4-carbomethoxyphenyl boronic acid. J Organomet Chem 1989;367: 259–265.

[100] Boswell CA, Ferl GZ, Mundo EE, Schweiger MG, Marik J, Reich MP, Theil FP, Fielder PJ, Khawli LA. Development and evaluation of a novel method for preclinical measurement of tissue vascular volume. Mol Pharm 2010;7:1848–1857.

[101] Khawli LA, Chen F-M, Alauddin MM, Epstein AL. Radioiodinated monoclonal antibody conjugates: synthesis and comparative evaluation. Antibody Immunoconjug Radiopharm J 1991;4:163–182.

[102] Ghosh SS, Kao PM, McCue AW, Chappelle HL. Use of maleimide-thiol coupling chemistry for efficient syntheses of oligonucleotide-enzyme conjugate hybridization probes. Bioconjug Chem 1990;1:71–76.

[103] Birnbaumer ME, Schrader WT, O'Malley BW. Chemical cross-linking of chick oviduct progesterone-receptor subunits by using a reversible bifunctional cross-linking agent. Biochem J 1979;181:201–213.

[104] Das M, Fox CF. Chemical cross-linking in biology. Annu Rev Biophys Bioeng 1979;8:165–193.

[105] Lundblad RL, Noyes CM. The modification of cysteine. In: *Chemical Reagents for Protein Modification*. Boca Raton (FL): CRC Press, Inc; 1984. p 55–93.

[106] Khawli LA, van den Abbeele AD, Kassis I. N-(m-[125I] Iodophenyl)maleimide: an agent for high yield radiolabeling of antibodies. Nucl Med Biol 1992;19:289–295.

[107] Srivastava PC, Buchsbaum DJ, Allred JF, Brubaker PG, Hanna DE, Spicker JK. A new conjugating agent for radioiodination of protein: low *in vivo* deiodination of a radiolabeled antibody in a tumor model. Biotechniques 1990;8: 536–545.

[108] Wilbur DS, Hylarides MD, Hadley SW, Schroeder J, Fritzberg AR. A general approach to radiohalogenation of proteins. Radiohalogenation of organometallic intermediates containing protein reactive substituents. J Label Compd Radiopharm 1989;26:316–318.

[109] Awwad M, Strome PG, Gilman SC, Axelrod HR. Modification of monoclonal antibody carbohydrates by oxidation, conjugation, or deoxymannojirimycin does not interfere with antibody effector functions. Cancer Immunol Immunother 1994;38:23–30.

[110] Thorpe SR, Baynes JW, Chroneos ZC. The design and application of residualizing labels for studies of protein catabolism. FASEB J 1993;7:399–405.

[111] Sugiura G, Kühn H, Sauter M, Haberkorn U, Mier W. Radiolabeling strategies for tumor-targeting proteinaceous drugs. Molecules 2014;19:2135–2165.

[112] Rhodes BA. Direct labeling of proteins with 99mTc. Int J Rad Appl Instru – Part B 1991;18:667–676.

[113] Liu S. Bifunctional coupling agents for radiolabeling of biomolecules and target-specific delivery of metallic radionuclides. Adv Drug Deliv Rev 2008;60:1347–1370.

[114] Dillman RO. Radioimmunotherapy of B-cell lymphoma with radiolabelled anti-CD20 monoclonal antibodies. Clin Exp Med 2006;6:1–12.

[115] Kobayashi H, Wu C, Yoo TM, Sun BF, Drumm D, Pastan I, Paik CH, Gansow OA, Carrasquillo JA, Brechbiel MW. Evaluation of the *in vivo* biodistribution of yttrium-labeled isomers of CHX-DTPA-conjugated monoclonal antibodies. J Nucl Med 1998;39:829–836.

[116] Tolmachev V, Xu H, Wållberg H, Ahlgren S, Hjertman M, Sjöberg A, Sandström M, Abrahmsén L, Brechbiel MW. Evaluation of a maleimido derivative of CH-A″ DTPA for site-specific labeling of Affibody molecules. Bioconjug Chem 2009;19:1579–1587.

[117] Price EW, Orvig C. Matching chelators to radiometals for radiopharmaceuticals. Chem Soc Rev 2014;43:260–290.

[118] Brom M, Joosten L, Oyen WJ, Gotthardt M, Boerman OC. Improved labelling of DTPA- and DOTA-conjugated peptides and antibodies with 111In in HEPES and MES buffer. EJNMMI Res 2012;2:4.

[119] Fichna J, Janecka A. Synthesis of target-specific radiolabeled peptides for diagnostic imaging. Bioconjug Chem 2003;14:3–17.

[120] Pnwar P, Iznaga-Escobar N, Mishra P, Srivastava V, Sharma RK, Chandra R, Mishra AK. Radiolabeling and biological evaluation of DOTA-Ph-Al derivative conjugated to anti-EGFR antibody or egf/r3 for targeted tumor imaging and therapy. Cancer Biol Ther 2005;4:854–860.

[121] Lewis MR, Shively JE. Maleimidocysteineamido-DOTA derivatives: new reagents for radiometal chelate conjugation to antibody sulfhydryl groups undergo pH-dependent cleavage reactions. Bioconjug Chem 2009;9:72–86.

[122] Blower PJ, Lewis JS, Zweit J. Copper radionuclides and radiopharmaceuticals in nuclear medicine. Nucl Med Biol 1996;23:957–980.

[123] Dearling JL, Voss SD, Dunning P, Snay E, Fahey F, Smith SV, Huston JS, Meares CF, Treves ST, Packard AB. Imaging cancer using PET--the effect of the bifunctional chelator on the biodistribution of a (64)Cu-labeled antibody. Nucl Med Biol 2011;38:29–38.

[124] Lewis MR, Boswell CA, Laforest R, Buettner TL, Ye D, Connett JM, Anderson CJ. Conjugation of monoclonal antibodies with TETA using activated esters: biological comparison of 64Cu-TETA-1A3 with 64Cu-BAT-2IT-1A3. Cancer Biother Radiopharm 2001;16:483–494.

[125] Cole WC, DeNardo SJ, Meares CF, McCall MJ, DeNardo GL, Epstein AL, O'Brien HA, Moi MK. Comparative serum stability of radiochelates for antibody radiopharmaceuticals. J Nucl Med 1987;28:83–90.

[126] Anderson C, Dehdashti F. 64Cu-TETA-octreotide as a PET imaging agent for patients with neuroendocrine tumors. J Nucl Med 2001;42:213–221.

[127] Perk LR, Visser OJ, Walsum MSV, Vosjan MJWD, Visser GWM, Zijlstra JM, Huijgens PC, van Dongen GAMS. Preparation and evaluation of Zr-89-Zevalin for monitoring of Y-90-Zevalin biodistribution with positron emission tomography. Eur J Nucl Med Mol Imaging 2006;33: 1337–1345.

[128] Nagengast WB, de Vries EG, Hospers GA, Mulder NH, de Jong JR, Hollema H, Brouwers AH, van Dongen GA, Perk LR, Lub-de Hooge MN. *In vivo* VEGF imaging with radiolabeled bevacizumab in a human ovarian tumor xenograft. J Nucl Med 2007;48:1313–1319.

[129] Hong H, Severin G, Yang Y, Engle J, Zhang Y, Barnhart T, Liu G, Leigh B, Nickles R, Cai W. Positron emission tomography imaging of CD105 expression with 89Zr-Df-TRC105. Eur J Nucl Med Mol Imaging 2012;39:138–148.

[130] Holland JP, Caldas-Lopes E, Divilov V, Longo VA, Taldone T, Zatorska D, Chiosis G, Lewis JS. Measuring the pharmacodynamic effects of a novel Hsp90 inhibitor on HER2/neu expression in mice using 89Zr-DFO-Trastuzumab. PLoS One 2010;5:e8859.

[131] Perk LR, Visser GW, Vosjan MJ, Stigter-van Walsum M, Tijink BM, Leemans CR, van Dongen GAMS. (89)Zr as a PET surrogate radioisotope for scouting biodistribution of the therapeutic radiometals (90)Y and (177)Lu in tumor-bearing nude mice after coupling to the internalizing antibody cetuximab. J Nucl Med 2005;46:1898–1906.

[132] Werthen M, Nygren H. Effect of antibody affinity on the isotherm of antibody binding to surfaceimmobilized antigen. J Immunol Methods 1988;115:71–78.

[133] McCormack T, O'Keeffe G, Mac Craith B, O'Kennedy R. Assessment of the effect of increased fluorophore labelling on the binding ability of an antibody. Anal Lett 1996;29: 953–968.

[134] Haughland RP. Coupling of monoclonal antibodies with fluorophores. Methods Mol Biol 1995;45:205–221.

[135] Der-Balian GP, Kameda N, Rowley GL. Fluorescein labeling of Fab' while preserving single thiol. Anal Biochem 1988;173:59–63.

[136] Rao J, Dragulescu-Andrasi A, Yao H. Fluorescence imaging *in vivo*: recent advances. Curr Opin Biotechnol 2007;18:17–25.

[137] Boswell CA, Bumbaca D, Pastuskovas CV, Mundo EE, Shen BQ, Carano RAD, Marik J, Williams SP, Theil FP, Fielder PJ, van Bruggen N, Khawli LA. *Molecular Imaging Probes for Cancer Research*. Singapore: World Scientific; 2012. p 1015–1039.

[138] Ding D, Wu F. Image guided biodistribution and pharmacokinetic studies of theranostics. Theranostics 2012;2: 1040–1053.

[139] Hoppin J, Orcutt KD, Hesterman JY, Silva MD, Cheng D, Lackas C, Rusckowski M. Assessing antibody pharmacokinetics in mice with *in vivo* imaging. J Pharmacol Exp Ther 2011;337:350–358.

[140] Schneider DW, Heitner T, Alicke B, Light DR, McLean K, Satozawa N, Parry G, Yoo J, Lewis JS, Parry R. *In vivo* biodistribution, PET imaging, and tumor accumulation of 86Y- and 111In-antimindin/RG-1, engineered antibody fragments in LNCaP tumor-bearing nude mice. J Nucl Med 2009; 50:435–443.

[141] Misri R, Saatchi K, Ng SS, Kumar U, Häfeli UO. Evaluation of 111In labeled antibodies for SPECT imaging of mesothelin expressing tumors. Nucl Med Biol 2011;38:885–896.

[142] Khawli LA, Biela BH, Hu P, Epstein AL. Stable, genetically engineered F(ab')(2) fragments of chimeric TNT-3 expressed in mammalian cells. Hybrid Hybridomics 2002;21:11–18.

[143] Khawli LA, Biela B, Hu P, Epstein AL. Comparison of recombinant derivatives of chimeric TNT-3 antibody for the radioimaging of solid tumors. Hybrid Hybridomics 2003;22:1–9.

[144] Dennis MS, Jin H, Dugger D, Yang R, McFarland L, Ogasawara A, Williams S, Cole MJ, Ross S, Schwall R. Imaging tumors with an albumin-binding Fab, a novel tumor-targeting agent. Cancer Res 2007;67:254–261.

[145] Boswell CA, Marik J, Elowson MJ, Reyes NA, Ulufatu S, Bumbaca D, Yip V, Mundo EE, Majidy N, Van Hoy M, Goriparthi S, Trias A, Gill HS, Williams SP, Junutula JR, Fielder PJ, Khawli LA. Enhanced tumor retention of a radiohalogen label for site-specific modification of antibodies. J Med Chem 2013;56:9418–9426.

[146] Cohen R, Vugts DJ, Stigter-van Walsum M, Visser GW, van Dongen GA. Inert coupling of IRDye800CW and zirconium-89 to monoclonal antibodies for single- or dual-mode fluorescence and PET imaging. Nat Protoc 2013;8:1010–1018.

[147] Mortimer JE, Bading JR, Colcher DM, Conti PS, Frankel PH, Carroll MI, Tong S, Poku E, Miles JK, Shively JE, Raubitschek AA. Functional imaging of human epidermal growth factor receptor 2-positive metastatic breast cancer using 64Cu-DOTA-trastuzumab PET. J Nucl Med 2014;55: 23–29.

[148] Tamura K, Kurihara H, Yonemori K, Tsuda H, Suzuki J, Kono Y, Honda N, Kodaira M, Yamamoto H, Yunokawa M, Shimizu C, Hasegawa K, Kanayama Y, Nozaki S, Kinoshita T, Wada Y, Tazawa S, Takahashi K, Watanabe Y, Fujiwara Y. 64Cu-DOTA-trastuzumab PET imaging in patients with HER2-positive breast cancer. J Nucl Med 2013;54:1869–1875.

[149] Herbertson RA, Tebbutt NC, Lee FT, MacFarlane DJ, Chappell B, Micallef N, Lee ST, Saunder T, Hopkins W, Smyth FE, Wyld DK, Bellen J, Sonnichsen DS, Brechbiel MW, Murone C, Scott AM. Phase I biodistribution and pharmacokinetic study of Lewis Y-targeting immunoconjugate CMD-193 in patients with advanced epithelial cancers. Clin Cancer Res 2009;15:6709–6715.

[150] Salsano M, Treglia G. PET imaging using radiolabeled antibodies: future direction in tumor diagnosis and correlate applications. Res Rep Nucl Med 2013;3:9–17.

[151] Weissleder R. Scaling down imaging: molecular mapping of cancer in mice. Nat Rev Cancer 2002;2:11–18.

[152] de Wet JR, Wood K, DeLuca M, Helinski DR, Subramani S. Firefly luciferase gene: structure and expression in mammalian cells. Mol Cell Biol 1987;7:725–737.

[153] Weissleder R, Mahmood U. Molecular imaging. Radiology 2001;219:316–333.

[154] Grimm J, Kirsch DG, Windsor SD, Kim CF, Santiago PM, Ntziachristos V, Jacks T, Weissleder R. Use of gene expression profiling to direct *in vivo* molecular imaging of lung cancer. Proc Natl Acad Sci U S A 2005;102:14404–14409.

[155] Weissleder R, Pittet MJ. Imaging in the era of molecular oncology. Nature 2008;452:580–589.

[156] Leblond F, Davis SC, Valdés PA, Pogue BW. Pre-clinical whole-body fluorescence imaging: review of instruments, methods and applications. J Photochem Photobiol B: Biol 2010;98:77–94.

[157] Archibald SJ. Antibody radiolabeling techniques to optimize cellular retention. J Med Chem 2013;56:9415–9417.

[158] Vaidyanathan G, White BJ, Affleck DJ, Zhao XG, Welsh PC, McDougald D, Choi J, Zalutsky MR. SIB-DOTA: a trifunctional prosthetic group potentially amenable for multi-modal labeling that enhances tumor uptake of internalizing monoclonal antibodies. Bioorg Med Chem 2012;20:6929–6939.

20

KNOWLEDGE OF ADME OF THERAPEUTIC PROTEINS IN ADULTS FACILITATES PEDIATRIC DEVELOPMENT

Omoniyi J. Adedokun and Zhenhua Xu

Janssen Research and Development, LLC, Spring House, PA, USA

20.1 INTRODUCTION

Antibody-based therapeutic proteins are a class of biologics that are used to treat a number of different diseases that affect both adults and children; in some cases, these biologics represent the most advanced form of therapeutic interventions available. For diseases such as rheumatoid arthritis (RA), inflammatory bowel diseases (IBD), psoriasis, asthma, and some other immune-mediated disorders, the use of biologics is established in both adult and pediatric patients. As of September 2014, 14 antibody-based proteins have been approved for the treatment of different pediatric conditions (Table 20.1).

Of these therapeutic proteins, palivizumab, a monoclonal antibody (mAb), targets the human respiratory syncytial virus (RSV) and is exclusively used in children to prevent respiratory disease in infants at high risk of developing complications that could result from the viral infection. More commonly, however, therapeutic proteins are predominantly used in adult populations as most of the experience and research into the safety and efficacy of these agents are conducted in adults with only few studies in pediatric populations. Consequently, data in children are often limited, leading to difficulties in the determination of appropriate pediatric dose regimens. A sufficient knowledge and understanding of the disposition of therapeutic proteins in adults is important in order to facilitate the development of these agents in pediatric patients.

This chapter presents an overview of how the knowledge of the disposition (i.e., absorption, distribution, metabolism, and elimination, ADME) of therapeutic proteins in adults can help facilitate the development of these agents in children. Similarities and differences between pediatric and adult patients with respect to the pharmacokinetics (PK) and immunogenicity of these agents are examined. Furthermore, the use of PK modeling and simulation to assist in the determination of the appropriate pediatric dose regimen is also discussed. The application of these concepts in the development of therapeutic proteins for pediatric indications using knowledge from adult data is illustrated with some examples in literature.

Although therapeutic proteins encompass a wide array of biological products; this chapter will primarily focus on mAbs because this group of therapeutic proteins represents the largest class in current drug development efforts.

With the exception of biologics that are indicated for conditions exclusively occurring in children, the evaluation of therapeutic proteins initially focuses on adult populations (aged 18 and above). The knowledge and experience obtained from clinical trials in adult subjects are then applied to the development of the drug in the target pediatric population. In designing pediatric studies, the investigator must be cognizant of potential differences between adults and children with respect to PK, pharmacodynamics (PD), immunogenicity, safety, and efficacy of the biologic agent. In addition, differences in the etiology, course, and severity of the disease may influence responsiveness to the drug, thus a clear understanding of these factors may be critical to the success of drug development efforts in the pediatric population [13–15]. A comparative assessment of the

ADME and Translational Pharmacokinetics/Pharmacodynamics of Therapeutic Proteins: Applications in Drug Discovery and Development, First Edition. Edited by Honghui Zhou and Frank-Peter Theil.

TABLE 20.1 Antibody-Based Therapeutic Proteins with Approved Pediatric Indications

Generic Name (Trade Name)	Target	Indication	Approved Age Range[a]	References
Abatacept (Orencia®)	Cytotoxic T-lymphocyte antigen	Juvenile idiopathic arthritis	≥6 yr	[1]
Adalimumab (Humira®)	Tumor necrosis factor-α	Juvenile idiopathic arthritis	≥4 yr	[2]
		Crohn's disease	≥6 yr	
Basiliximab (Simulect®)	Interleukin-2 receptor-α	Renal transplantation	1–16 yr	[3]
Canakinumab (Ilaris®)	Interleukin-1 receptor-β	Cryopyrin-associated periodic syndromes	≥4 yr	[4]
Daclizumab (Zenapax®)	Interleukin-2 receptor-α	Renal transplantation	≥11 mo	[5]
Eculizumab (Soliris®)	Complement protein C5	Atypical hemolytic uremic syndrome	No minimum age specified	[6]
Etanercept (Enbrel®)	Tumor necrosis factor-α	Juvenile idiopathic arthritis	≥2 yr	[7]
Infliximab (Remicade®)	Tumor necrosis factor-α	Crohn's disease	≥6 yr	[8]
		Ulcerative colitis	≥6 yr	
Omalizumab (Xolair®)	Immunoglobulin E	Asthma	≥12 yr	[9]
Palivizumab (Synagis®)	Respiratory syncytial virus	Respiratory syncytial virus disease	0–2 years	[10]
Rilonacept (Arcalyst®)	Interleukin-1	Cryopyrin-associated periodic syndromes	≥12 yr	[11]
Tocilizumab (Actemra®)	interleukin-6 receptor	Systemic juvenile idiopathic arthritis	≥2 yr	[12]
		Polyarticular juvenile idiopathic arthritis	≥2 yr	

[a] Pediatric age range up to 17 years unless otherwise indicated.

disposition of therapeutic proteins in adults and pediatric patients is discussed in the following sections.

20.2 COMPARATIVE EVALUATION OF ADME OF THERAPEUTIC PROTEINS BETWEEN ADULTS AND CHILDREN

Drug disposition is generally governed by the processes of ADME. The assessment of these processes is of great importance in the development of therapeutic proteins. Due to the rapid growth and development occurring in children, knowledge of the differences in these processes between pediatric and adult patients is especially important for the determination of pediatric dose regimens.

For chemically synthesized small-molecule therapeutics, there is a large body of knowledge with respect to the characterization and role of metabolizing enzymes in drug disposition. The differential function of drug metabolizing enzymes and renal excretion across the pediatric age spectrum is generally taken into account when designing pediatric studies and serves as the basis for dose adjustments [15, 16]. Therapeutic proteins differ from small-molecule drugs by their source, size, and complexity. As a result, the PK and PD of therapeutic proteins are different and often more complicated than those of small-molecule drugs. In addition, knowledge of the effect of developmental changes on the disposition of therapeutic proteins is sparse, making it difficult to base pediatric dose recommendations on assumptions routinely applied to small-molecule therapeutics. A brief review of potential differences between adults and pediatric patients in the disposition of therapeutic proteins is discussed next.

20.2.1 Absorption

Although oral administration of therapeutic proteins is desirable, and research into developing oral formulations of therapeutic proteins is continuing [17], these agents are rarely given by mouth because they are hydrophilic, subject to enzymatic degradation, and insufficiently absorbed from the gastrointestinal tract. Presently, all approved therapeutic proteins are given by parenteral administration, usually via intravenous (IV), subcutaneous (SC), or less commonly by intramuscular injection. Some biologics are employed for local administration at the site of action, for example, ranibizumab is administered via intravitreal injection in the treatment of wet age-related macular degeneration.

Therapeutic proteins administered via SC or intramuscular route undergo absorption that is known to occur mainly through the lymphatic system and to a lesser extent through capillary diffusion. Given the size of these agents along with the protracted lymphatic flow rate, absorption into the systemic circulation is usually a slow process [18]. After SC administration, therapeutic proteins attain peak systemic concentrations in 2–8 days [19, 20]. Some of the drug undergoes proteolytic degradation at the site of injection or in the process of absorption leading to incomplete absolute bioavailability that generally ranges from 50% to 80% [19, 21].

It is unknown if physiological differences between adults and children in the absorption pathway have a significant impact on the rate or extent of absorption of therapeutic proteins.

Few reports of differences between adults and children in SC absorption and bioavailability parameters of the approved therapeutic proteins are found in the literature. It is of interest that a recent population PK analysis of canakinumab involving subjects with ages ranging from 4 to 74 years showed that the first-order absorption rate constant (ka) following SC administration appeared to decrease 1.5-fold with a doubling of age, thus suggesting that canakinumab absorption might be faster in children [22]. It is not known whether neonatal Fc receptor (FcRn) expression is sufficiently different to impact absorption of mAbs in younger children compared to adults.

20.2.2 Distribution

Typically, after a drug reaches the systemic circulation, it is distributed into various body compartments, the extent of which is determined by the size, polarity, and solubility of the drug. For therapeutic proteins, the distribution occurs mainly by convective transport through capillary pores, as well as through transcytosis into the extracellular space [19]. Not surprisingly, the distribution of large therapeutic proteins is primarily confined to the systemic circulation and, to a less extent, to the extracellular space due to their large molecular weight and hydrophilicity. As a result, the volume of distribution is generally low, often approximating blood volume. The volume of distribution of some therapeutic proteins is influenced by other factors such as the drug's binding affinity, target antigen distribution, antigen–antibody complex formation, and the turnover of the surface receptors to which the drug binds. In some oncology conditions, shedding of target receptors may lead to altered antibody distribution. The presence of solubilized receptors can block antibody-binding sites resulting in diminished antibody binding to the target tissues [23].

In contrast to small-molecule drugs, developmental changes in body composition across various age groups do not appear to significantly influence the biodistribution of therapeutic proteins [24–26]. Nevertheless, differences in the production or degradation rate of the target antigen or ligand, as well as differences in protein or target–tissue binding could exist between adult and pediatric patients. Such differences could result in dissimilar distribution of the therapeutic protein following initial dosing. Accounting for the variation in body size should minimize any clinically meaningful difference in the distribution of therapeutic proteins between adult and pediatric patients.

20.2.3 Metabolism and Elimination

From most reported studies, it does not appear that hepatic metabolizing enzymes, such as the cytochrome P450 family of enzymes, play a significant role in the metabolism of mAbs. Rather, it is believed that therapeutic proteins are catabolized by proteolytic enzymes, which are widely available throughout the body, into peptides and amino acids, which are then recycled or eliminated by physiological processes [27]. In general, the metabolism of therapeutic proteins does not produce active metabolites that may elicit biological effects as sometimes observed with small-molecule drugs.

Although the precise mechanism for the process of elimination of mAbs is unclear, specific and nonspecific elimination pathways have been identified for mAbs [27, 28]. A nonspecific pathway mediated via Fc receptors (Fcγ and FcRn receptors) is commonly shared by endogenous IgG and mAbs that target soluble antigens (e.g., adalimumab, bevacizumab, canakinumab, golimumab, infliximab, and ustekinumab). The FcRn protects both endogenous and therapeutics IgGs from rapid elimination, thus prolonging the half-life of these proteins [29]. Since these nonspecific processes are not known to be age-related, differences in the elimination of mAbs are not expected for adult and pediatric patients via this pathway. This expectation is supported by the fact that the clearance of some of these agents has been found to be comparable between adult and pediatric patients after body size related PK differences are adequately taken into account [26, 30].

Target-mediated elimination that leads to the formation of antibody–antigen complexes with subsequent lysosomal degradation has been observed with some mAbs [27, 28, 31]. In many cases, this process is more common with mAbs that directly interact with their specific targets on the cell surface (e.g., natalizumab, panitumumab, tocilizumab, and trastuzumab). In some cases (e.g., omalizumab and denosumab), however, target-mediated elimination also occurs with mAbs that bind to soluble ligands [32, 33]. In general, target-mediated processes result in saturable or nonlinear disposition of the mAb. The formation of antigen–antibody complexes is dose-dependent, with saturation occurring at sufficiently high mAb doses. As a result, the PK of these mAbs is not usually dose-proportional across the dose range evaluated in pharmacological studies. The clearance of many mAbs involves both nonspecific and target-mediated processes and the disposition of the mAbs will depend on the relative contributions of each process to the overall elimination of the drug. In terms of drug concentrations, nonlinear PK behavior is observed when drug concentrations are low relative to that of the target, while at higher drug concentrations with the target antigen saturated, the nonspecific pathway predominates and the elimination of drug becomes linear. Typically, PK models that simultaneously account for the two elimination pathways result in a more accurate description of the observed data compared to models incorporating only the linear or the nonlinear component.

For drugs with linear PK characteristics, allometric approaches may be employed in the determination of pediatric doses [34, 35], while for those exhibiting nonlinear PK, standard allometric body weight scaling may not be appropriate for estimating pediatric dose regimens. Differences in antigen expression and turnover rates between adults and children may further complicate the nonlinear elimination process, making it more difficult to determine appropriate pediatric dose regimens [36].

Another important and unique component of the clearance process of therapeutic proteins is the development of antibodies to the drug (immunogenicity). In general, the formation of antidrug antibodies (ADAs) leads to increased clearance and a decrease in the systemic level of the drug, with such decrease being proportional to the severity of the immunogenic response. Currently, there are little published data to adequately examine the similarity or dissimilarity in immunogenicity between children and adults. A review of the package of inserts of some approved mAbs shows that higher ADA incidences were reported for adalimumab, abatacept, and infliximab in children with juvenile idiopathic arthritis (JIA)/juvenile rheumatoid arthritis (JRA) when compared with the respective adult populations with RA. On the other hand, the incidences of ADA were similar for etanercept and tocilizumab in children with JIA compared to the respective adult populations with RA, as well as for infliximab in pediatric ulcerative colitis (UC) or Crohn's disease compared to the respective adult patients.

While it is conceivable that children may elicit different immune responses to mAb biotherapeutics, the interpretation of ADA results is often complicated by differences in study designs, patient-related factors (e.g., concomitant immunomodulators and prior biologic exposure), and sample sizes between adult and pediatric studies. For example, some pediatric studies employ a randomized withdrawal design (treatment with placebo after exposure to active drug) that is generally associated with an increased incidence of ADA. If the corresponding adult study employs a placebo-controlled parallel study design, the incidence of ADA could be substantially different between pediatric and adult studies. Nevertheless, given the possibility that a child's immune system may react to "foreign" therapeutic proteins differently from that of an adult, efforts should be made to properly evaluate any significant discrepancy in immunogenicity between pediatric and adult populations. Such an evaluation should take into account any differences that could influence the interpretation of the immunogenicity data.

20.3 EXTRAPOLATION OF EFFICACY FROM ADULTS TO PEDIATRIC PATIENTS

In many instances, the determination of pediatric dose regimens relies to a great degree on the extrapolation of clinical data obtained from studies conducted in adult patients [37–39]. In the context of pediatric drug development, extrapolation may be defined as a framework that allows the use of information available from adult studies to make inferences in a pediatric population in order to reduce the need to generate additional data, which otherwise would be needed to arrive at the right conclusions for the target pediatric population. The primary reason why extrapolation is often advocated in pediatric drug development is the need to avoid unnecessary clinical studies in pediatric patients due to ethical, feasibility, or efficiency considerations.

While extrapolation of data from adults to pediatrics is often proposed as a reasonable approach to enable children to have access to drugs already approved in adults, it is recognized that appropriate application of extrapolation is contingent on meeting certain conditions. A framework that provides guidance on clinical pharmacology considerations for the extrapolation of a drug's effectiveness in pediatric patients was proposed by the FDA in 2003 (Fig. 20.1 [40]).

Inherent in the pediatric study decision tree is the expectation that an investigator demonstrates a reasonable understanding of the similarity or otherwise of the target disease, response to treatment, PK, PD, and exposure–response (E–R) of the drug of interest in the adult and pediatric populations. Importantly, the decision algorithm provides recommendations on the type of clinical pharmacology studies that may be performed to support the safe and effective use of drugs in the pediatric population. These recommended approaches have been described elsewhere as no extrapolation, partial extrapolation, and full or complete extrapolation [39].

20.3.1 No Extrapolation Approach

In the event that the nature or progression of a disease is unique to pediatric patients, or the response to intervention is dissimilar between pediatric and adult patients, then a standard clinical development program (i.e., adequate Phase III pediatric studies) is required to provide the evidence of safety and effectiveness in pediatric patients. A possible exception to this would be in the pediatric oncology setting that may be assessed less rigorously due to the rare nature of the disease and limited availability of effective therapeutic intervention.

20.3.2 Partial Extrapolation Approach

When the disease and response to treatment are similar in pediatric and adult patients but the E–R relationship is either not similar or not adequately defined, then a partial extrapolation approach may be applicable. In this situation, if the E–R relationship in the adult population is well characterized, then an E–R study may be performed in pediatric patients and data from such a study may then be compared with those in adults. Systemic drug concentrations or other PK metrics can represent the exposure while a biomarker or

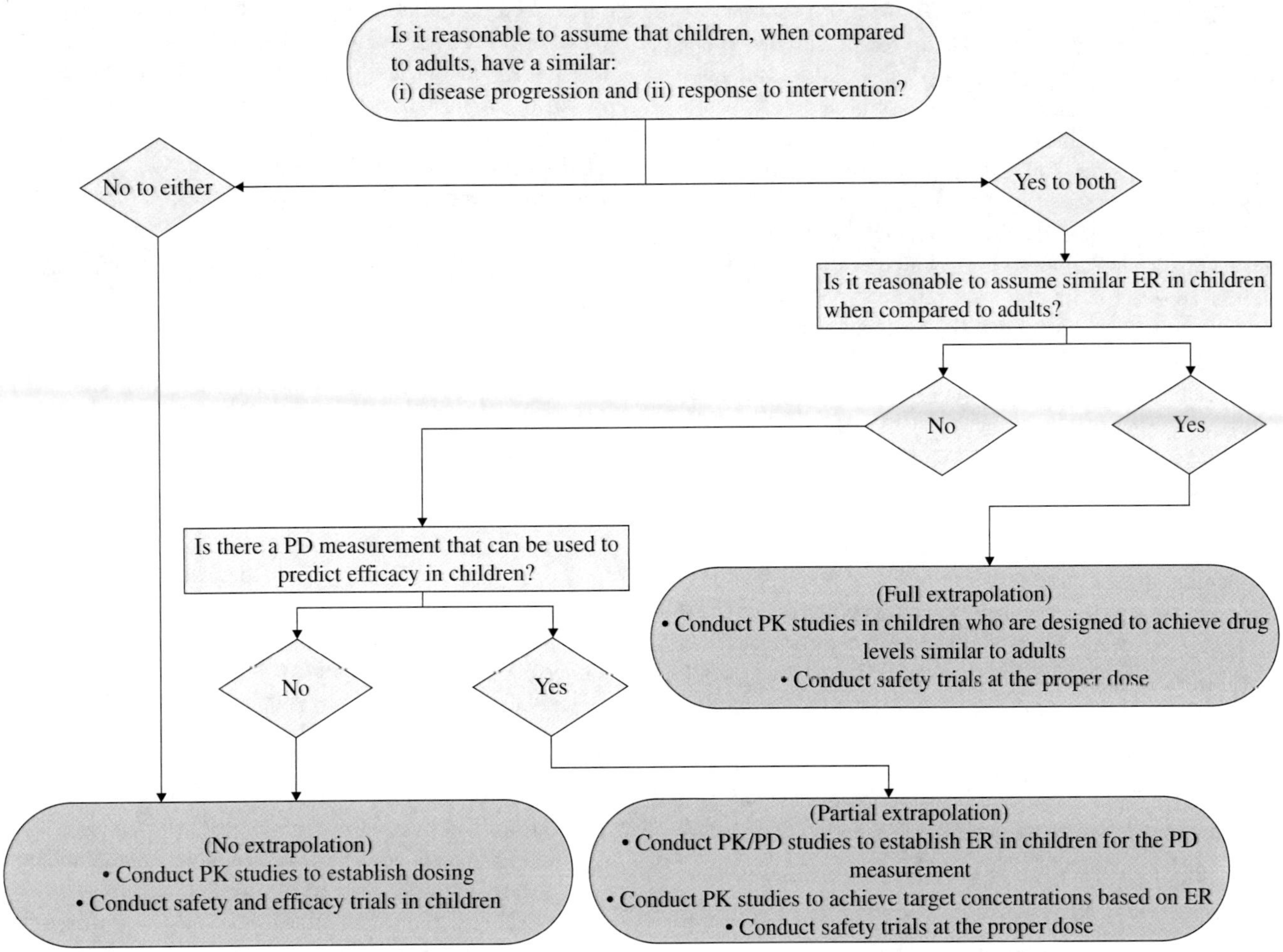

FIGURE 20.1 FDA pediatric study decision tree. **This algorithm can be applied to therapeutic proteins**. When applicable, the pediatric dose and dosage regimen can be estimated from adult and pediatric pharmacokinetic data. ER indicates **exposure response**; PD, pharmacodynamic; PK, pharmacokinetic.

clinical endpoint may be used as the response for this type of analysis.

If the E–R relationship is unclear in adults, then a study evaluating efficacy and safety in addition to PK data may be required to demonstrate effectiveness in pediatric patients. Examples of the partial extrapolation approach in the evaluation of therapeutic proteins may be seen in the approvals of adalimumab and canakinumab in JIA, as well as infliximab in pediatric UC where clinical effectiveness and exposure-matching in comparison to adults were demonstrated [41].

20.3.3 Full Extrapolation Approach

When it can be demonstrated or reasonably assumed that the progression of a disease, its response to treatment, and E–R are similar between adults and pediatric patients, studies focusing on characterizing the PK of the biologic in pediatric patients may be sufficient for extrapolating efficacy. The objectives of the PK study would be to identify and confirm the pediatric dose regimen that matches the systemic exposure obtained in adult patients. Such PK studies are usually performed in children with the disease of interest so that they could derive benefits from the use of the drug.

Currently, no antibody-based therapeutic protein has been approved by the FDA using the approach of full extrapolation. The safety and efficacy of most approved therapeutics were established through adequately controlled Phase III trials in the target pediatric population, for example, adalimumab in JIA. Nevertheless, opportunities exist for full extrapolation for therapeutic proteins where the efficacy and safety of the same therapeutic class have been well established in both adults and children, particularly where clinical data have shown children to have similar or better clinical outcomes to the class of therapeutics. In one such example, infliximab, an anti-TNFα agent is already approved for adult and pediatric UC, while golimumab, another anti-TNFα mAb, is approved for use in adult UC and is currently being investigated for use in the treatment of pediatric UC. Given that both drugs are in the same therapeutic class, along with the similarity of the disease in pediatric and adult patients, the established effectiveness of both drugs in adult UC, and demonstrable E–R relationships, it appears reasonable to consider full extrapolation for golimumab in pediatric UC.

In another example, the PASCAL study (pediatric study of certolizumab pegol) appears to be geared toward a full extrapolation approach with PK as the primary outcome of the study [42]. This is notable as JIA is considered a different or more heterogeneous disease compared to adult RA for which certolizumab pegol is already approved. The established safety and efficacy for the same class of anti-TNFα agents may thus allow for full extrapolation even in circumstances where the pediatric disease is not entirely similar to the adult disease.

In summary, regardless of the extrapolation approach employed, at least two types of studies are expected to be conducted in a pediatric clinical development program. First, a PK study is needed to establish dose recommendations or E–R as applicable. Secondly, a study to establish safety based on the assumption that safety cannot be extrapolated considering the potential differences in safety profile between adult and pediatric patients. In practice, a registry study is generally employed to further assure the safety of the biologic in the target pediatric population, particularly those used for chronic therapy.

20.4 PEDIATRIC DOSE STRATEGIES

The determination of dose regimens to be studied in clinical trials is a critical component of any pediatric drug development plan [43]. Typically, pediatric doses of therapeutic proteins are calculated by scaling established adult doses using a measure of body size such as body weight or body surface area (BSA) (Table 20.2). Although several covariates such as immunogenicity, serum albumin, C-reactive protein, concomitant immunomodulators, and target antigen levels have been shown to influence the PK of therapeutic proteins, the most frequently identified covariate influencing drug clearance is body size. Unlike some small-molecule drugs, age is not usually used to adjust doses for most therapeutic proteins indicated for children. This is because age is typically not a covariate of systemic exposure for therapeutic proteins once body size is accounted for [47]. In line with this observation, PK exposure metrics of therapeutic proteins are generally comparable between adults and children after correcting for the difference in body size.

For convenience, body weight-adjusted dosing is the most commonly applied approach for pediatric dose determination. However, across the spectrum of the pediatric age range, body weight-adjusted dosing is often performed using different approaches including fixed mg/kg, variable mg/kg dosing, body weight tiered-fixed dosing, or a hybrid of these approaches. Ideally, the appropriate dose strategy should be contingent on an understanding of the relationship between body weight and the clearance of the therapeutic protein.

20.4.1 Body Weight-Based (Linear) Dose-Adjustment Approach

Historically, many therapeutic proteins (e.g., infliximab) have been administered using a mg/kg approach. This approach assumes that drug clearance increases proportionally with body weight; however, this is often not the case with many therapeutic proteins [48, 49]. In the event that drug clearance increases with body weight in a less than proportional manner, dose regimens employing a mg/kg approach will tend to lead to lower exposure in patients in the lower body weight range due to non-proportionally higher increased clearance (see Table 20.3). Under this scenario, if the same mg/kg dose is used across the pediatric age continuum, younger children (due to their lower body weights) may achieve lower systemic exposure possibly impacting efficacy. This finding has been demonstrated for infliximab through population PK-based simulations intended to predict exposure in younger children who were not within the age range evaluated in some infliximab pediatric trials [26].

In line with the above findings, it has been suggested that the resulting lower systemic infliximab exposure in some younger children with JRA who received the same 3-mg/kg infliximab dose approved for adults with RA may partly explain the less robust efficacy of this infliximab dose in pediatric patients with JRA [26, 50]. Unlike JRA, pediatric patients with IBD who received the same 5 mg/kg infliximab dose regimen approved in adults demonstrated similar efficacy to that seen in adult IBD patients [51, 52]. The E–R following the 5-mg/kg regimen in pediatric UC was consistent with that seen in adult patients with UC [53, 54]. The difference in the outcomes for the clinical trials of infliximab in JRA and IBD underscores the need to understand the relationship between systemic exposure and efficacy in the context of pediatric dose determination.

Given the above scenario, a variable weight-based dosing approach (where different mg/kg doses are administered to children of different age or body weight groups) is sometimes employed to better correct for the exposure differences that could occur with a fixed mg/kg dosing regimen. An example of this approach is seen with tocilizumab where doses of 12 or 8 mg/kg are given to children with systemic JIA who weigh less than 30 kg, or equal to or greater than 30 kg, respectively [55].

TABLE 20.2 Dose Regimens and Pharmacokinetic Characteristics of Approved Therapeutic Proteins with Pediatric Indications[a]

Therapeutic Protein	Dose Regimen		Pharmacokinetic Characteristics		Notes
	Adult	Pediatrics	Adult	Pediatrics	
Abatacept (Orencia)	Rheumatoid arthritis: <60 kg: 500 mg IV 60–100 kg: 750 mg IV >100 kg: 1000 mg IV (administered as infusions at Weeks 0, 2, 4, and q4w thereafter)	Juvenile idiopathic arthritis (6–17 yr): <75 kg: 10 mg/kg 75–100 kg: 750 mg IV >100 kg: 1000 mg IV (administered as infusions at Weeks 0, 2, 4, and q4w thereafter)	Clearance: 0.22 mL/h/kg Terminal half-life: 13.1 d	Clearance: 0.40 mL/h/kg	Linear PK. CL increases nonlinearly with body weight. No impact of age on CL after accounting for body weight in a population PK model
Adalimumab (Humira)	Rheumatoid arthritis: 40 mg SC q2w	Juvenile idiopathic arthritis (4–17 yr): 10 to <15 kg: 10 mg SC q2w 15 to <30 kg: 20 mg SC q2w ≥30 kg: 40 mg SC q2w	Clearance: 12 mL/h Terminal half-life: 2 weeks (parameters are after single IV infusion)	Derived PK parameters not reported in package insert	Linear PK. BSA-based dosing (24 mg/m² SC q2w up to a maximum of 40 mg) in children with JIA achieved comparable steady-state concentrations to those in adults with RA
	Crohn's disease: 160 mg SC at Week 0, 80 mg SC at Week 2, 40 mg SC q2w thereafter	Pediatric Crohn's disease (4–17 yr): 17 to <40 kg: 80 mg SC at Week 0, 40 mg SC at Week 2, 20 mg SC q2w thereafter ≥40 kg: same as adult CD dose			
Basiliximab (Simulect)	Renal transplantation: 20 mg IV (administered before, and 4 d after transplantation)	Renal transplantation: <35 kg: 10 mg IV ≥35 kg: 20 mg IV (administered before, and 4 d after transplantation)	Clearance: 41 mL/h Terminal half-life: 7.2 d	Children 1–11 yr: Clearance: 17 mL/h Terminal half-life: 9.5 d Children 12–17 yr: Clearance: 31 mL/h Terminal half-life: 9.1 d	Linear PK. Disposition parameters were not affected to a clinically relevant extent by age, body weight, or body surface area
Canakinumab (Ilaris)	Cryopyrin-associated periodic syndromes: ≥15–40 kg: 2 mg/kg SC q8w >40 kg: 150 mg SC q8w	Cryopyrin-associated periodic syndromes (4–17 yr): ≥15–40 kg: 2 mg/kg SC q8w (increase to 3 mg/kg if response is inadequate) >40 kg: Same dose as adults with CAPS Systemic juvenile idiopathic arthritis (2–17 yr): ≥7.5 kg: 4 mg/kg SC (up to a maximum of 300 mg) q4w	Clearance: 0.174 L/d in a 70-kg patient with cryopyrin-associated periodic syndromes Terminal half-life: 26 d	Clearance: 0.11 L/d in a 33-kg patient with systemic juvenile idiopathic arthritis Terminal half-life: 22.9–25.7 d	Linear PK. No age-related PK differences after correction for body weight

(*Continued*)

TABLE 20.2 (Continued)

Therapeutic Protein	Dose Regimen		Pharmacokinetic Characteristics		Notes
	Adult	Pediatrics	Adult	Pediatrics	
Daclizumab (Zenapax)	Renal transplantation: 1 mg/kg IV (five doses with first dose given within 24 h before transplantation and q2w thereafter)	Renal transplantation (11 mo to 17 yr): 1 mg/kg IV (five doses with first dose given within 24 h before transplantation and q2w thereafter)	Clearance: 15 mL/h in a 45-year-old, male Caucasian with a body weight of 80 kg without proteinuria Terminal half-life: 20 d	Clearance: 10 mL/h in a 29.7-kg patient Terminal half-life: 13 d	Influence of body weight on CL supports mg/kg dosing. No dose adjustment based on identified covariates (age, gender, proteinuria, race)
Eculizumab (Soliris)	Atypical hemolytic uremic syndrome: 900 mg IV weekly for the first 2 weeks, followed by 1200 mg for the fifth dose 1 week later, then 1200 mg q2w thereafter	Atypical hemolytic uremic syndrome (<18 yr): 5 to <10 kg: 300 mg IV weekly × 2 doses, then 300 mg IV q3w thereafter 10 to <20 kg: 600 mg IV, then 300 mg IV at Week 2, then 300 mg IV q2w thereafter 20 to <30 kg: 600 mg IV weekly × 2 doses, then 600 mg IV at Week 3; then 600 mg IV q2w thereafter 30 to <40 kg: 600 mg IV × 2 doses, then 900 mg IV at Week 3, then 900 mg IV q2w thereafter ≥40 kg: 900 mg IV × 4 doses, then 1200 mg IV at Week 5, then 1200 mg IV q2w thereafter	Clearance: 14.6 mL/h in a 70-kg patient Terminal half-life: 12.1 d Clearance: 14.6 mL/h	Derived PK parameters not reported in package insert	Age does not influence the PK of eculizumab
Etanercept[b] (Enbrel)	Rheumatoid arthritis: 50 mg SC weekly	Polyarticular juvenile idiopathic arthritis (2–17 yr): <63 kg: 0.8 mg/kg SC weekly ≥63 kg: 50 mg weekly	Clearance: 160 mL/h Terminal half-life: 102 h	Terminal $t_{1/2}$: 70.7 and 94.8 h for females and males, respectively	Limited data suggest etanercept clearance is slightly reduced in children aged 4–8 yr
Infliximab[c] (Remicade)	Crohn's disease: 5 mg/kg at 0, 2, 6 weeks, then q8w. May increase dose to 10 mg/kg on loss of response Ulcerative colitis: 5 mg/kg at 0, 2, 6 weeks, then q8w	Pediatric Crohn's disease (6–17 yr): 5 mg/kg at 0, 2, 6 weeks, then q8w Pediatric ulcerative colitis (6–17 yr): 5 mg/kg at 0, 2, 6 weeks, then q8w	Clearance: 5.39 mL/kg/d Terminal half-life: 12.4 d	Clearance: 5.43 mL/kg/d Terminal half-life: 13.2 d	Linear PK. Population PK analysis showed that median systemic exposure in pediatric patients aged 2–6 yr was about 40% lower than that in adults. No effect of age after correction for body weight

Palivizumab[d] (Synagis)	No adult indication	Respiratory syncytial virus disease (≤24 mo): 15 mg/kg IM monthly (first dose before and remaining doses during the RSV season)	Clearance: 198 mL/d in a 70-kg individual	Clearance: 11 mL/d in a 4.5-kg patient Terminal half-life: 24.5 d	No covariate factor explains the interindividual variability in palivizumab PK
Rilonacept (Arcalyst)	Cryopyrin-associated periodic syndromes: 320 mg SC, then 160 mg weekly	Cryopyrin-associated periodic syndromes (12–17 yr) 4.4 mg/kg SC (up to a maximum of 320 mg), then 2.2 mg/kg (up to a maximum of 160 mg) weekly	No derived PK parameters reported in package insert		Limited data suggest exposure in pediatric patients are comparable to those in adults at recommended doses
Tocilizumab (Actemra)	Rheumatoid arthritis: *IV infusion*	Polyarticular juvenile idiopathic arthritis (2–17 yr) <30 kg: 10 mg/kg IV q4w	Linear clearance: 12.5 mL/h	Linear clearance: 5.8 mL/h (polyarticular juvenile idiopathic arthritis)	Clearance decreased with increased doses. Total clearance is concentration dependent. Concentration-dependent nonlinear clearance plays a major role at low tocilizumab concentrations. Age had no impact on tocilizumab PK but linear clearance increased with body size
	4 mg/kg q4 weeks followed by an increase in weekly doses based on clinical response	≥30 kg: 8 mg/kg IV q4w	Terminal half-life: *IV infusion*	7.1 mL/h (systemic juvenile idiopathic arthritis)	
	SC injection	Systemic juvenile idiopathic arthritis (2–17 yr)	11 d for the 4-mg/kg regimen and 13 d for the 8-mg/kg regimen	Terminal half-life: *IV infusion*	
	<100 kg: 162 mg SC q2w followed by an increase in weekly doses based on clinical response	<30 kg: 12 mg/kg IV q2w	*SC injection*	16 d (polyarticular juvenile idiopathic arthritis regimen)	
	≥100 kg: 162 mg SC weekly	≥30 kg: 8 mg/kg IV q2w (SC presentation not approved in pediatric indications)	13 d for the 162-mg q week regimen, and 5 d for the 162-mg q2w regimen	23 d (systemic juvenile idiopathic arthritis regimen)	

IV, intravenous(ly); PK, pharmacokinetic(s); q2w, every 2 weeks; q3w, every 3 weeks; q4w, every 4 weeks; q8w, every 8 weeks; SC, subcutaneous.

[a] Information in table obtained mainly from the respective product prescribing information. Information or data from other references are noted as appropriate.

[b] Additional information obtained from Yim et al. [44].

[c] Additional information obtained from Fasanmade et al. [45] and Xu et al. [26].

[d] Additional information obtained from Robbie et al. [46].

TABLE 20.3 Relationship between Drug Clearance, Body Weight, and Allometric Coefficient

Body Weight (kg)	Dose (Based on 5 mg/kg) (mg)	Clearance (L/d)			
		Allometric Coefficient (α)			
		1	0.75	0.5	0.25
10	50	0.07	0.12	0.19	0.31
20	100	0.14	0.20	0.27	0.37
30	150	0.21	0.26	0.33	0.40
40	200	0.29	0.33	0.38	0.43
50	250	0.36	0.39	0.42	0.46

Calculation of clearances based on the following hypothetical equation: $CL = 0.5 \times (WT/70)^{\alpha}$, where *CL* is the individual clearance, 0.5 is the population clearance corresponding to a 70-kg individual, *WT* is the individual body weight and α is the allometric power coefficient. As shown, a less than proportional increase in clearance occurs with increasing body weight when $\alpha < 1$.

20.4.2 BSA-Based (Linear) Dose-Adjustment Approach

Therapeutic proteins are sometimes dosed based on BSA. Although this approach is more common for drugs used in oncology, it has also been adopted for some therapeutic proteins in other indications, for example, adalimumab and golimumab were evaluated in clinical trials of children with juvenile arthritis using BSA dose adjustment. The BSA approach emanates from the notion that many physiological processes in mammals are constant with respect to the surface area [56].

When applied to pediatric dose adjustment, BSA can help reduce the variability in the resulting systemic exposure of the biologic particularly in younger children. Notwithstanding, similar to the mg/kg approach, the same BSA-based dose regimen may not lead to consistent exposure across the entire pediatric age range because the assumption of unit exponent in the scaling equation is not always valid over this range. As a result, placing a cap on the amount of drug, or using different mg/m^2 doses for different age ranges may be required to achieve consistent exposure across the pediatric age range of interest for the biologic. An example of this approach is seen in the recommended dose of adalimumab for patients with enthesitis-related arthritis 6 years of age and older, that is, 24 mg/m^2 of BSA up to a maximum single dose of 40 mg adalimumab administered every other week via SC injection [57].

Compared to body weight-based dosing, dose adjustment using BSA is often less desirable because of the potential risk of error in the computation of BSA stemming from the availability of multiple formulae that almost invariably involves a child's height measurement.

20.4.3 Tiered-Fixed Dose-Adjustment Approach

In certain situations, a tiered dosing approach has been employed in order to simplify the dose regimen of therapeutic proteins. This approach is contingent on the fact that the therapeutic window for therapeutic proteins is often wide, thus permitting some degree of variability in systemic exposure therapeutic proteins without a corresponding adverse impact on efficacy or safety. Tiered-fixed dosing is not only convenient but may also reduce the likelihood of dosing errors compared to body size-adjusted dosing in pediatric indications across a wide range of ages. One downside to this approach is the need for multiple presentations of the drug and/or delivery device to cover the range of pediatric doses to be administered.

An example of this approach is seen with basiliximab that is administered as a two-dose regimen, one dose each before, and 4 days after renal transplantation. Pediatric patients who weigh greater than 35 kg are given two IV doses of 10 mg, while those weighing 35 kg and above receive two 20 mg IV doses [3].

20.4.4 Hybrid Dose-Adjustment Approach

In some instances, combinations of the dose-adjustment approaches discussed in the previous sections are often employed to achieve consistent exposure in the pediatric age range. Thus, it is not uncommon to see therapeutic proteins being dosed on a mg/kg or mg/m^2 basis at the lower age or body weight range, and then a fixed dose often equivalent to adult dose in older or higher body weight children. This hybrid approach is exemplified by canakinumab in children with cryopyrin-associated periodic syndromes (CAPS) with SC dose regimens of 2 mg/kg q8w for those with body weights between 15 and 40 kg and 150 mg q8w for those weighing more than 40 kg [4].

20.4.5 Other Dose-Adjustment Approaches

For some therapeutic proteins, the relationship between the systemic drug exposure and clinical or biomarker response is sufficiently understood to be used in the determination of an appropriate dose regimen. Such is the case with omalizumab for which both adults and pediatric patients (12–17 years) with asthma are indicated to receive an SC dose of 150–375 mg q2w or q4w depending on the serum IgE level and body weight [9].

Of note, the presence of target-mediated drug disposition (TMDD) in the kinetics of a therapeutic protein may create additional challenges in the determination of pediatric dose regimens. Clearly, a thorough characterization and understanding of the kinetics and/or dynamics of biologic agents exhibiting TMDD will be helpful in guiding appropriate dose selection for children [58].

20.5 SAMPLE-SIZE DETERMINATION FOR PEDIATRIC STUDIES

A perennial challenge in conducting pediatric trials is the determination of an appropriate sample size that would enable the objectives of the study to be achieved. It is neither ethical to enroll more pediatric patients than necessary as this leads to undue exposure of children, nor is it desirable for a study to enroll a too small number of children from whom reliable inferences about PK, efficacy, or safety of the drug cannot be made. Thus, efforts should be made to determine the optimal sample size needed to achieve the objectives of a pediatric study. Expectedly, historical information from adults or children in previous studies is usually the starting point for reliable sample size estimation for a new pediatric study. Bayesian methods can be used to incorporate prior information into sample-size calculations for the new trial. If the number of subjects needed is not feasible for a particular study design, then alternative designs or extrapolation approaches have to be explored.

With respect to PK or PD studies, approaches to determine the sample size of pediatric studies based on precision of parameter estimates in the adult population have been proposed [59]; however, there is currently no consensus with respect to the appropriateness or universal application of such approaches [60].

The incorporation of prior information obtained from adult populations into the calculation of sample size for pediatric trials may be illustrated with infliximab in the treatment of IBD. In the REACH study, a sample size of 82 subjects was required to achieve an estimating precision within 10% from the true proportion of pediatric subjects in clinical response at Week 10. The sample-size calculation was based on a clinical response rate of 67% at Week 10 using all randomized subjects to the infliximab 5 mg/kg maintenance treatment group in the ACCENT 1 study of adult subjects with Crohn's disease ($n=192$). The sample size was then increased to include approximately 110 subjects to ensure the collection of adequate safety data in pediatric subjects over the course of 1 year (Janssen, unpublished data). Notably, the sample size in pediatric Crohn's disease was adequate to facilitate the comparison between the PK of infliximab in children and adults with Crohn's disease [45].

In a similar manner, the infliximab pediatric UC study enrolled 60 subjects based on achieving an estimating precision of within 12% from the true proportion of pediatric subjects in clinical response at Week 8 using a 95% confidence interval [52]. This sample size calculation assumed a clinical response rate of 67% at Week 8 from the same 5 mg/kg treatment regimen in the clinical trials of adult subjects with UC in the pivotal ACT 1 and ACT 2 trials ($n=242$). Although these 60 pediatric subjects were sufficient to characterize infliximab PK and allow for comparison with the PK in adult subjects, the sample sizes for the maintenance data were not sufficiently robust for the purpose of assessing E–R during maintenance [53]. This was because subjects in clinical response ($n=45$) were randomized into two maintenance treatment groups at Week 8 and were also allowed to escalate their maintenance dose if they lost response. This resulted in fewer subjects remaining in their randomized treatment groups at the end of 1 year. This limitation of the maintenance portion of the infliximab pediatric UC study was, however, obviated by the availability of population PK data from the study that facilitated the comparison with the adult PK maintenance data.

20.6 MODELING AND SIMULATION IN PEDIATRIC DRUG DEVELOPMENT FACILITATED BY EXISTING ADULT MODELS

Modeling and simulation is increasingly being advocated and utilized for planning, decision making, and data analysis in drug development [61]. The ability to analyze sparse data by the population-based approach is particularly appealing in pediatric studies due to the well-documented constraints of limited data and ethical issues associated with clinical trials in children. The use of modeling and simulation is being widely encouraged by regulatory authorities as a critical part of drug development programs for pediatric populations [62]. Guidelines that are currently being developed by regulators emphasize the critical role of modeling and simulation in the clinical pharmacology content of the pediatric drug development plans, especially as a platform to facilitate extrapolation [63].

Modeling and simulation integrate information about dose, systemic exposure and efficacy, or PD outcomes in a mathematical and statistical manner. It takes into account prior knowledge of the target pediatric population and accumulated data on the use of the therapeutic protein of interest in adults. The scope of modeling and simulation in pediatric drug development is broad, addressing potential questions such as the following: What dose regimens should be studied in children? What should be the sample size and the sampling scheme for the pediatric study? What should be the study endpoints? How do we evaluate the efficacy or safety outcomes from the trial? What alternative designs could be used? Which pediatric formulation or device should be commercialized? Modeling techniques such as population PK, physiologically based PK, PK–PD, E–R, disease progression, and clinical trial simulation are usually employed to answer these questions. The ensuing models provide a framework for simulating and predicting the behavior of the biologic, thus contributing to the optimal design of clinical trials and selection of pediatric dose regimens.

20.6.1 Modeling and Simulation Framework for Therapeutic Proteins in Pediatric Drug Development

In a proposed strategy for pediatric dose optimization of antibody-based therapeutic proteins (Fig. 20.2), population PK modeling plays an important and almost ubiquitous role in the determination of dose regimens for pediatric labeling

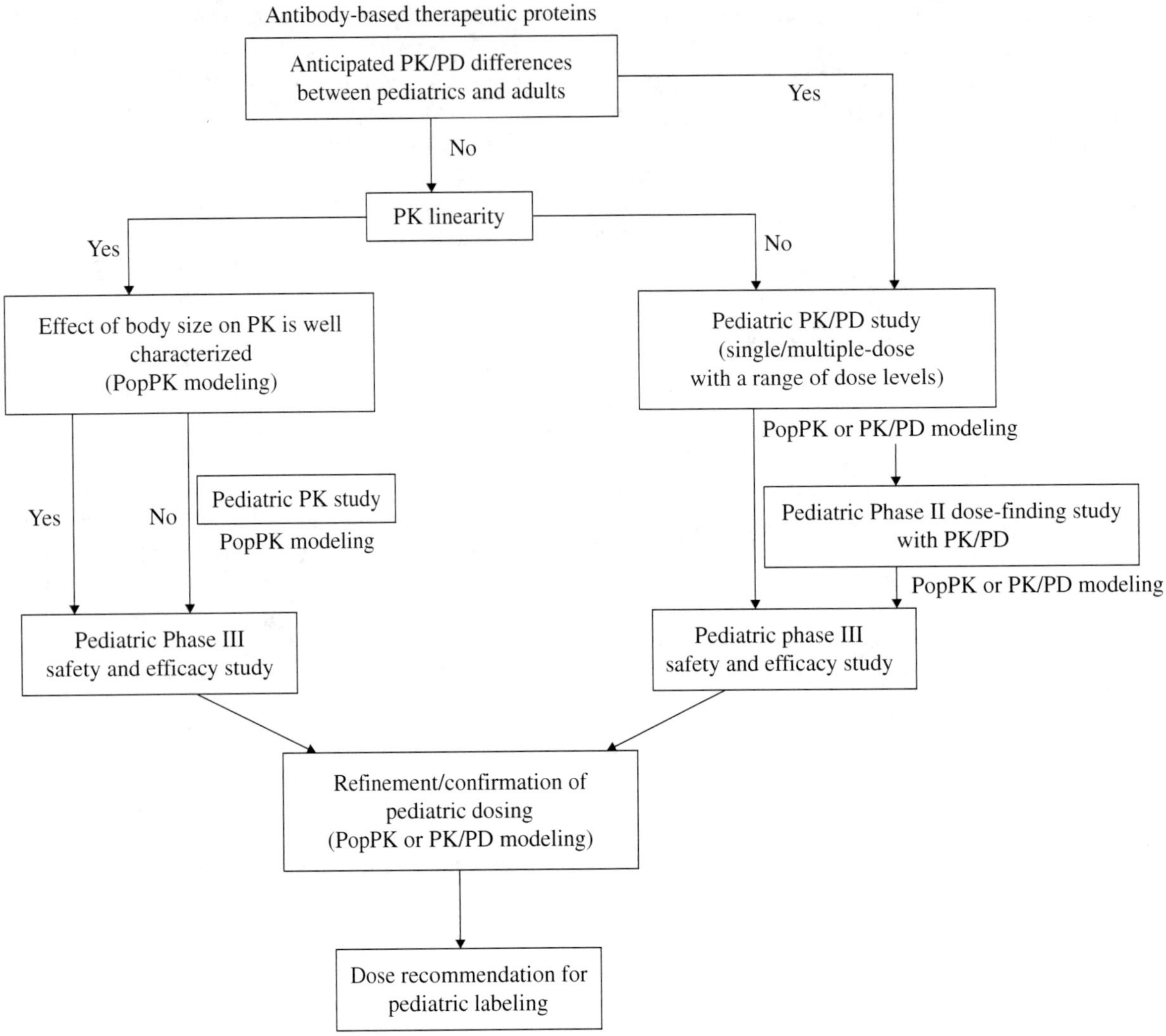

FIGURE 20.2 Proposed strategy for pediatric dose optimization of antibody-based therapeutic proteins.

for these therapeutic proteins [64]. Population PK methodology is widely employed in pediatric studies because of its ability to utilize sparse sampling schemes in the computation of PK parameters and the quantification of the impact of covariates on the disposition of the therapeutic proteins. A typical population PK modeling and simulation procedure for a therapeutic protein already studied in adults may consist of the following steps:

1. Establish a population PK model in adult subjects.
2. Refine the existing population PK model with any available pediatric data, sometimes from other indications where appropriate, and incorporate only covariates with significant impact on exposure (usually body size) in the final model.
3. Identify exposure metrics such as trough concentration (C_{trough}), maximum concentration (C_{max}), and area under the concentration–time curve (AUC), which are either relevant for efficacy or safety, or considered appropriate for comparison of systemic exposure between pediatric and adult subjects.
4. Using clinical trial simulation, simulate PK profiles based on an established distribution of covariates such as body weight, height, gender, and age and obtain the identified systemic exposure metrics. Typically, such covariate data distributions come from previous adult and pediatric clinical trials in the same disease or the Centers for Disease Control and Prevention (CDC) growth chart [65].
5. Compare exposure distributions for the relevant age and body weight groups quantitatively, then determine the appropriate dose regimen for further study or labeling proposal.

The above steps can be conceptually adapted for other types of modeling exercise including PK/PD, E–R, and disease progression models. The use of population PK modeling to leverage adult data in pediatric drug development of therapeutic proteins is illustrated in the following examples for infliximab.

20.6.2 Examples of the Application of Modeling and Simulation in the Development of Therapeutic Proteins in Pediatric Patients

20.6.2.1 Infliximab in Pediatric UC Since its approval for the treatment of pediatric UC, the clinical development program for infliximab in pediatric UC has been cited as an example of the partial extrapolation approach in the development of therapeutics for children [41]. The approval of infliximab for the treatment of pediatric patients with UC was primarily based on a Phase III clinical trial (C0168T72) that enrolled 60 subjects with moderate-to-severe UC. Supportive data were also obtained from two Phase III adult UC studies (ACT 1 and ACT 2 [66]), along with another Phase III trial in pediatric patients with Crohn's disease (REACH [51]). Details about the pediatric UC program for infliximab, including the analysis of efficacy and PK data, are described elsewhere [52, 53].

Importantly, prior knowledge of infliximab in adult UC facilitated the development of this biologic for the pediatric UC indication with substantial contribution from modeling and simulation analysis. Notably, the PK and immunogenicity sampling scheme for the pediatric study was similar to that of the adult study ensuring the feasibility of comparing the infliximab concentrations and the incidence of antibodies to infliximab (ATI) in both populations. Furthermore, a confirmatory population PK analysis approach based on prior knowledge from an existing population PK model in adult subjects with UC was used to evaluate the population PK in the pediatric population. In addition, a conventional population PK analysis served as a sensitivity analysis for assessing the consistency of these two approaches [67]. An E–R model was also developed to examine the relationship between infliximab concentration and clinical efficacy outcomes [53].

The results of the clinical pharmacology assessment showed that infliximab PK in pediatric UC patients, as assessed by peak and trough concentrations, and terminal $t_{1/2}$, were generally comparable to those from adult subjects with UC and pediatrics subjects with Crohn's disease (Table 20.4). Differences in infliximab concentrations between pediatric and adult patients in this evaluation were deemed to be small, considering the variability of the PK of infliximab and other therapeutic proteins.

The results of the population PK analysis of infliximab in pediatric patients with UC supported the assertion of PK comparability with the adult population. In the covariate analysis, body weight was found to be a covariate on the volume of distribution but age did not influence any PK parameter. Further assessment of the PK comparability of infliximab between pediatric and adult patients with UC using simulations based on the population PK models for the respective populations showed that there was substantial overlap of infliximab exposure between pediatric and adult patients (Fig. 20.3). Based on the simulated area under concentration–time curve at steady-state (AUCss), the median systemic infliximab exposure was determined to be 20% lower than that of adults, thus corroborating the observed PK data. In addition, the E–R model (Fig. 20.4) showed that the relationship between infliximab concentration and clinical response was comparable between pediatric and adult patients with UC [53].

With respect to immunogenicity, the incidence of ATI based on the same validated ELISA (enzyme linked immunosorbent assay) assay was low and comparable between pediatric and adults patients with UC (7.7% vs 6.8%, respectively).

TABLE 20.4 Comparison of Infliximab Pharmacokinetics between Pediatric Patients with UC, Pediatric Patients with Crohn's Disease, and Adult Patients with UC

	Pediatric UC	Pediatric Crohn's Disease	Adult UC
Median peak serum concentration during induction (Week 2) (μg/mL)[a]	115.1[b]	108.7[b]	131.6[b]
Median serum concentration at end of induction (Week 8) (μg/mL)[a]	27.5	Not assessed	33.3
Median steady-state trough serum concentration (Week 30) (μg/mL)[a]	1.9	1.8	2.5
Median half-life (d)	10.8[c]	10.7	11.7[d]

Data are presented for
[a] 5 mg/kg q8w treatment group,
[b] at Week 2,
[c] obtained from confirmatory population PK analysis, and
[d] derived from ACT 1 study data [66].

In summary, based on the overall assessment of safety, efficacy, and PK results from both pediatric and adult patients with UC, the same mg/kg dosing regimen in adults was considered appropriate for the treatment of UC in children with 6–17 years of age.

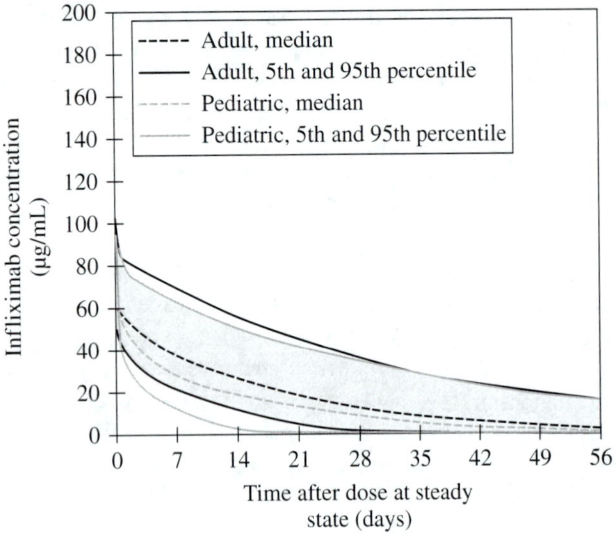

FIGURE 20.3 Comparison of model-based infliximab exposure between pediatric and adult patients with UC following a 5 mg/kg infliximab dose regimen.

20.6.2.2 Use of Infliximab in Younger Children with UC Although infliximab has been studied in a number of adult and pediatric disease populations, the safety and efficacy of infliximab in younger pediatric subjects aged 2 to less than 6 years have not been established for any indications. Clinical trial data in this age group are thus very limited compared to older pediatric subjects (6–17 years). Nevertheless, knowledge of infliximab PK in younger subjects aged 2 to less than 6 years may provide useful information for investigators and physicians who treat patients in this lower age group with infliximab.

To provide a basis for the rational use of infliximab in a younger pediatric population with UC, a modeling analysis was conducted using integrated PK data from 6 infliximab studies across different disease populations to determine whether the PK profiles of infliximab in pediatric patients of all age groups, including subjects aged 2 to less than 6 years, could be predicted by population PK modeling [26]. The integrated analysis included available data from two adult UC trials ($n = 483$) and four pediatric studies ($n = 305$) comprising children with JRA, pediatric UC, pediatric Crohn's disease studies, and an investigator-initiated study in infants and children with Kawasaki disease. Of the pediatric subject, 28 were less than 6 years of age (14 with JRA and 14 with Kawasaki disease). The inclusion of the adult UC data was considered appropriate

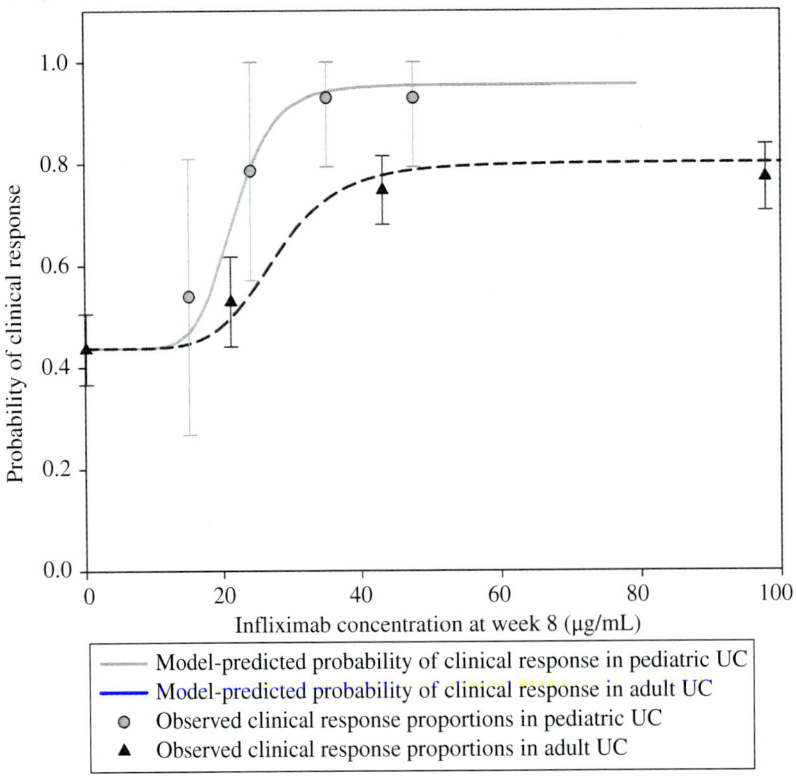

FIGURE 20.4 Comparison of infliximab E–R between pediatric and adult UC patients. Adapted from Janssen Biotech Inc. [68].

because it provides a broader range of body size and included certain covariates shared by adult and pediatric subjects with UC, and also enabled comparison of the pediatric data with adult UC data, which was a key objective of the analysis.

The population PK analysis showed that following administration of 5 mg/kg infliximab q8w, the median steady-state infliximab exposure (AUCss) in pediatric subjects aged 2 to less than 6 years was predicted to be approximately 40% lower than that in adults [26]. The difference in infliximab exposure was attributed to the nonlinear relationship between body weight and infliximab clearance (see Section 20.4.1). Of note, the efficacy and safety of infliximab in pediatric subjects aged 2 to less than 6 years have not been evaluated in adequately designed clinical studies, thus the clinical significance of the lower infliximab exposure predicted by the population PK model in this young age group is unknown. Nevertheless, given that several reports have shown that serum infliximab concentration is associated with efficacy outcomes in IBD [69, 70], it is reasonable to infer that a higher than the approved 5 mg/kg dose regimen may be required for children aged 2 to less than 6 years based on this modeling analysis.

20.7 FUTURE DIRECTIONS

In this chapter, we have shown that the knowledge of the PK properties of therapeutic proteins in adults can facilitate the development of such a drug in children. With the rapid increase in the number of therapeutic proteins at different stages of preclinical and clinical development, there is a need to ensure that the opportunities presented by these drugs in adult patients are extended to pediatric patients in an optimal manner. To this end, further research into the mechanistic understanding of the ADME of therapeutic proteins in pediatric subpopulations (particularly neonates and infants) is needed to gain insight, which may lead to better prediction of pediatric doses. The incorporation of data that quantify the impact of age-related changes of FcRn expression in children could improve the utility of PK/PD models. In addition, novel approaches and strategies such as physiologically based PK models, optimal design, adaptive trials, and sample-size evaluation for precise PK parameter evaluation should be encouraged to bridge the knowledge gaps between adults and pediatrics when these innovative drugs are developed. The ongoing development of new guidelines by regulatory authorities dealing with clinical pharmacology considerations for pediatric drug development including the role of modeling and simulation will help provide clarity to industry and other stakeholders who are engaged in the difficult task of bringing these drugs to the vulnerable population of children who need them.

REFERENCES

[1] Bristol Myers Squibb. 2014. Orencia® US prescribing information. Available at http://packageinserts.bms.com/pi/pi_orencia.pdf. Accessed 2015 Feb 4.

[2] Abbvie. 2005. Humira® US prescribing information. Available at http://www.rxabbvie.com/pdf/humira.pdf. Accessed 2015 Feb 4.

[3] Novartis. 1998. Simulect® US prescribing information. Available at http://www.pharma.us.novartis.com/product/pi/pdf/simulect.pdf. Accessed 2015 Jul 10.

[4] Novartis. 2009. Ilaris® US prescribing information. Available at http://www.pharma.us.novartis.com/product/pi/pdf/ilaris.pdf. Accessed 2015 Feb 4.

[5] Hoffman-La Roche. 2005. Zenapax® US prescribing information. Available at http://www.accessdata.fda.gov/drugsatfda_docs/label/2005/103749s5059lbl.pdf. Accessed 2015 Feb 4.

[6] Alexion. 2014. Soliris® US prescribing information. Available at http://soliris.net/sites/default/files/assets/soliris_pi.pdf. Accessed 2015 Feb 4.

[7] Amgen. 2013. Enbrel® US prescribing information. Available at http://pi.amgen.com/united_states/enbrel/derm/enbrel_pi.pdf. Accessed 2015 Feb 4.

[8] Janssen Biotech Inc. 2015. Remicade® US prescribing information. Available at http://www.remicade.com/shared/product/remicade/prescribing-information.pdf. Accessed 2015 Feb 4.

[9] Genentech. 2003. Xolair® US prescribing information. Available at http://www.gene.com/download/pdf/xolair_prescribing.pdf. Accessed 2015 Jul 10.

[10] MedImmune. 2014. Synagis® US prescribing information. Available at http://www.medimmune.com/docs/default-source/pdfs/prescribing-information-for-synagis.pdf. Accessed 2015 Feb 4.

[11] Regeneron. 2014. Arcalyst® US prescribing information. Available at http://www.arcalyst.com/images/pdf/arcalyst_patpi.pdf. Accessed 2015 Feb 4.

[12] Genentech. 2010. Actemra® US prescribing information. Available at http://www.gene.com/download/pdf/actemra_prescribing.pdf. Accessed 2015 Jul 10.

[13] Gilman JT, Gal P. Pharmacokinetic and pharmacodynamic data collection in children and neonates. A quiet frontier. Clin Pharmacokinet 1992;23:1–9.

[14] Busse W, Banks-Schlegel SP, Larsen GL. Childhood- versus adult-onset asthma. Am J Respir Crit Care Med 1995;151:1635–1639.

[15] Kearns GL, Abdel-Rahman SM, Alander SW, Blowey DL, Leeder JS, Kauffman RE. Developmental pharmacology--drug disposition, action, and therapy in infants and children. N Engl J Med 2003;349:1157–1167.

[16] de Wildt SN, Kearns GL, Leeder JS, van den Anker JN. Glucuronidation in humans. Pharmacogenetic and developmental aspects. Clin Pharmacokinet 1999;36:439–452.

[17] Bhol KC, Tracey DE, Lemos BR, Lyng GD, Erlich EC, Keane DM, Fox BS. AVX-470: a novel oral anti-TNF antibody with therapeutic potential in inflammatory bowel disease. Inflamm Bowel Dis 2013;19 (11):2273–2281.

[18] Porter CJ, Charman SA. 2000. Lymphatic transport of proteins after subcutaneous administration. J Pharm Sci 89, 297–310.

[19] Lobo ED, Hansen RJ, Balthasar JP. Antibody pharmacokinetics and pharmacodynamics. J Pharm Sci 2004;93:2645–2668.

[20] Xu Z, Wang Q, Zhuang Y, Frederick B, Yan H, Bouman-Thio E, Marini JC, Keen M, Snead D, Davis HM, Zhou H. Subcutaneous bioavailability of golimumab at 3 different injection sites in healthy subjects. J Clin Pharmacol 2010;50:276–284.

[21] Keizer RJ, Huitema AD, Schellens JH, Beijnen JH. Clinical pharmacokinetics of therapeutic monoclonal antibodies. Clin Pharmacokinet 2010;49:493–507.

[22] Chakraborty A, Tannenbaum S, Rordorf C, Lowe PJ, Floch D, Gram H, Roy S. Pharmacokinetic and pharmacodynamic properties of canakinumab, a human anti-interleukin-1β monoclonal antibody. Clin Pharmacokinet 2012;51 (6):e1–e18.

[23] Junghans RP, Carrasquillo JA, Waldmann TA. Impact of antigenemia on the bioactivity of infused anti-Tac antibody: implications for dose selection in antibody immunotherapies. Proc Natl Acad Sci U S A 1998;95:1752–1757.

[24] Anderson GD, Lynn AM. Optimizing pediatric dosing: a developmental pharmacologic approach. Pharmacotherapy 2009;29:680–690.

[25] Tabrizi M, Bornstein GG, Suria H. Biodistribution mechanisms of therapeutic monoclonal antibodies in health and disease. AAPS J 2010;12:33–43.

[26] Xu Z, Mould DR, Hu C, Ford J, Keen M, Davis HM, Zhou H. Population pharmacokinetic analysis of infliximab in pediatrics using integrated data from six clinical trials. Clin Pharmacol Drug Dev 2012;1:203.

[27] Tabrizi MA, Tseng CM, Roskos LK. Elimination mechanisms of therapeutic monoclonal antibodies. Drug Discov Today 2006;11:81–88.

[28] Mould DR, Green B. Pharmacokinetics and pharmacodynamics of monoclonal antibodies: concepts and lessons for drug development. BioDrugs 2010;24:23–39.

[29] Junghans RP, Anderson CL. The protection receptor for IgG catabolism is the beta2-microglobulin-containing neonatal intestinal transport receptor. Proc Natl Acad Sci U S A 1996;93:5512–5516.

[30] Burns JC, Best BM, Mejias A, Mahony L, Fixler DE, Jafri HS, Melish ME, Jackson MA, Asmar BI, Lang DJ, Connor JD, Capparelli EV, Keen ML, Mamun K, Keenan GF, Ramilo O. Infliximab treatment of intravenous immunoglobulin-resistant Kawasaki disease. Pediatrics 2008;153:833–838.

[31] Bugelski PJ, Martin PL. Concordance of preclinical and clinical pharmacology and toxicology of therapeutic monoclonal antibodies and fusion proteins: cell surface targets. Br J Pharmacol 2012;166:823–846.

[32] Meno-Tetang GM, Lowe PJ. On the prediction of the human response: a recycled mechanistic pharmacokinetic/pharmacodynamic approach. Basic Clin Pharmacol Toxicol 2005;96:182–192.

[33] Sutjandra L, Rodriguez RD, Doshi S, Ma M, Peterson MC, Jang GR, Chow AT, Pérez-Ruixo JJ. Population pharmacokinetic meta-analysis of denosumab in healthy subjects and postmenopausal women with osteopenia or osteoporosis. Clin Pharmacokinet 2011;50:793–807.

[34] Mahmood I. Prediction of drug clearance in children from adults: a comparison of several allometric methods. Br J Clin Pharmacol 2006;61:545–557.

[35] West GB, Brown JH, Enquist BJ. A general model for the origin of allometric scaling laws in biology. Science 1997;276:122–126.

[36] EMA. 2011. Assessment report for tocilizumab Type II variation EMEA/H/C/955/II/15 for systemic idiopathic juvenile arthritis. Available at http://www.emea.europa.eu/docs/en_GB/document_library/EPAR_-_Assessment_Report_-_Variation/human/000955/WC500111086.pdf. Accessed 2015 Feb 4.

[37] Cucchiara S, Morley-Fletcher A. "New drugs: kids come first": children should be included in trials of new biological treatments. Inflamm Bowel Dis 2007;13:1165–1169.

[38] Gartlehner G, Hansen RA, Jonas BL, Thieda P, Lohr KN. Biologics for the treatment of juvenile idiopathic arthritis: a systematic review and critical analysis of the evidence. Clin Rheumatol 2008;27:67–76.

[39] Dunne J, Rodriguez WJ, Murphy MD, Beasley BN, Burckart GJ, Filie JD, Lewis LL, Sachs HC, Sheridan PH, Starke P, Yao LP. Extrapolation of adult data and other data in pediatric drug-development programs. Pediatrics 2011;128:e1242–e1249.

[40] FDA. *Guidance for Industry: Exposure-Response Relationships: Study Design, Data Analysis, and Regulatory Applications*. Washington (DC): Food and Drug Administration; 2003. . Available at www.fda.gov/downloads/Drugs/GuidanceComplianceRegulatoryInformation/Guidances/UCM072109.pdf. Accessed 2015 Feb 4..

[41] FDA. 2011. Gastrointestinal Advisory Committee for Remicade in Pediatric Ulcerative Colitis development. July 21, 2011. Available at http://www.fda.gov/downloads/AdvisoryCommittees/CommitteesMeetingMaterials/Drugs/GastrointestinalDrugsAdvisoryCommittee/UCM266697.pdf. Accessed 2015 Feb 4.

[42] UCB Biosciences GmbH. Pediatric Arthritis Study of Certolizumab Pegol (PASCAL). In: *ClinicalTrials.gov [Internet]*. Bethesda (MD): National Library of Medicine (US); 2012. . Available from https://clinicaltrials.gov/ct2/show/NCT01550003. Accessed 2015 Feb 4. NLM Identifier: NCT01550003.

[43] Anderson BJ, Holford NH. Mechanism-based concepts of size and maturity in pharmacokinetics. Annu Rev Pharmacol Toxicol 2008;48:303–332.

[44] Yim DS, Zhou H, Buckwalter M, Nestorov I, Peck CC, Lee H. Population pharmacokinetic analysis and simulation of the time-concentration profile of etanercept in pediatric

patients with juvenile rheumatoid arthritis. J Clin Pharmacol 2005;45:246–256.

[45] Fasanmade AA, Adedokun OJ, Blank M, Zhou H, Davis HM. Pharmacokinetic properties of infliximab in children and adults with Crohn's disease: a retrospective analysis of data from 2 Phase III clinical trials. Clin Ther 2011;33:946–964.

[46] Robbie GJ, Zhao L, Mondick J, Losonsky G, Roskos LK. Population pharmacokinetics of palivizumab, a humanized anti-respiratory syncytial virus monoclonal antibody, in adults and children. Antimicrob Agents Chemother 2012;56:4927–4936.

[47] Dirks NL, Meibohm B. Population pharmacokinetics of therapeutic monoclonal antibodies. Clin Pharmacokinet 2010;49:633–659.

[48] Wang DD, Zhang S, Zhao H, Men AY, Parivar K. Fixed dosing versus body size-based dosing of monoclonal antibodies in adult clinical trials. J Clin Pharmacol 2009;49:1012–1024.

[49] Bai S, Jorga K, Xin Y, Jin D, Zheng Y, Damico-Beyer LA, Gupta M, Tang M, Allison DE, Lu D, Zhang Y, Joshi A, Dresser MJ. A guide to rational dosing of monoclonal antibodies. Clin Pharmacokinet 2012;51:119–135.

[50] Ruperto N, Lovell DJ, Cuttica R, Wilkinson N, Woo P, Espada G, Wouters C, Silverman ED, Balogh Z, Henrickson M, Apaz MT, Baildam E, Fasth A, Gerloni V, Lahdenne P, Prieur AM, Ravelli A, Saurenmann RK, Gamir ML, Wulffraat N, Marodi L, Petty RE, Joos R, Zulian F, McCurdy D, Myones BL, Nagy K, Reuman P, Szer I, Travers S, Beutler A, Keenan G, Clark J, Visvanathan S, Fasanmade A, Raychaudhuri A, Mendelsohn A, Martini A, Giannini EH, Paediatric Rheumatology International Trials Organisation, Pediatric Rheumatology Collaborative Study Group. A randomized, placebo-controlled trial of infliximab plus methotrexate for the treatment of polyarticular-course juvenile rheumatoid arthritis. Arthritis Rheum 2007;56:3096–3106.

[51] Hyams J, Crandall W, Kugathasan S, Griffiths A, Olson A, Johanns J, Liu G, Travers S, Heuschkel R, Markowitz J, Cohen S, Winter H, Veereman-Wauters G, Ferry G, Baldassano R, REACH Study Group. Induction and maintenance infliximab therapy for the treatment of moderate-to-severe Crohn's disease in children. Gastroenterology 2007;132:863–873.

[52] Hyams J, Damaraju L, Blank M, Johanns J, Guzzo C, Winter HS, Kugathasan S, Cohen S, Markowitz J, Escher JC, Veereman-Wauters G, Crandall W, Baldassano R, Griffiths A, T72 Study Group. Induction and maintenance therapy with infliximab for children with moderate to severe ulcerative colitis. Clin Gastroenterol Hepatol 2012;10:391–399.

[53] Adedokun OJ, Xu Z, Padgett L, Blank M, Johanns J, Griffiths A, Ford J, Zhou H, Guzzo C, Davis HM, Hyams J. Pharmacokinetics of infliximab in children with moderate-to-severe ulcerative colitis: results from a randomized, multicenter, open-label, phase 3 study. Inflamm Bowel Dis 2013;19:2753–2762.

[54] Mehrotra N, Garnett C, Zhang L, Fang L, Wang YM, Fiorentino RP. Role of exposure-response relationship to guide dose selection in pediatric drug development when extrapolating efficacy from adults. Inflamm Bowel Dis 2011;17:S5.

[55] Genentech Inc. 2013. Actemra® US prescribing information. Available at http://www.gene.com/download/pdf/actemra_prescribing.pdf. Accessed 2015 Feb 4.

[56] Crawford JD, Terry ME, Rourke GM. Simplification of drug dosage calculation by application of the surface area principle. Pediatrics 1950;5:783–790.

[57] Abbvie. 2014. Humira® EU summary of product characteristics. Available at http://www.ema.europa.eu/docs/en_GB/document_library/EPAR_-_Product_Information/human/000481/WC500050870.pdf. Accessed 2015 Feb 4.

[58] Zheng S, Gaitonde P, Andrew MA, Gibbs MA, Lesko LJ, Schmidt S. Model-based assessment of dosing strategies in children for monoclonal antibodies exhibiting target-mediated drug disposition. CPT Pharmacometrics Syst Pharmacol 2014;3:1–10.

[59] Wang Y, Jadhav PR, Lala M, Gobburu JV. Clarification on precision criteria to derive sample size when designing pediatric pharmacokinetic studies. J Clin Pharmacol 2012;52:1601–1606.

[60] Salem F, Ogungbenro K, Vajjah P, Johnson TN, Aarons L, Rostami-Hodjegan A. Precision criteria to derive sample size when designing pediatric pharmacokinetic studies: which measure of variability should be used? J Clin Pharmacol 2014;54:311–317.

[61] Johnson TN. Modeling approaches to dose estimation in children. Br J Clin Pharmacol 2005;59:663–669.

[62] FDA. 2012. Advisory Committee for Pharmaceutical Science and Clinical Pharmacology: clinical pharmacology aspects of pediatric clinical trial design and dosing to optimize pediatric drug development. Available at http://www.fda.gov/AdvisoryCommittees/Calendar/ucm286681.htm. Accessed 2015 Feb 4.

[63] FDA. 2014. General clinical pharmacology considerations for pediatric studies for drugs and biological products. Guidance for industry (draft guidance). Available at http://www.fda.gov/downloads/Drugs/GuidanceCompliance RegulatoryInformation/Guidances/UCM425885.pdf. Accessed 2015 Feb 4.

[64] Xu Z, Davis HD, Zhou H. Rational development and utilization of antibody-based therapeutic proteins in pediatrics. Pharmacol Ther 2013;137 (2):225–247.

[65] Centers for Disease Control. 2009. CDC growth charts. Available at http://www.cdc.gov/growthcharts/percentile_data_files.htm. Accessed 2015 Feb 4.

[66] Rutgeerts P, Sandborn WJ, Feagan BG, Reinisch W, Olson A, Johanns J, Travers S, Rachmilewitz D, Hanauer SB, Lichtenstein GR, de Villiers WJ, Present D, Sands BE, Colombel JF. Infliximab for induction and maintenance therapy for ulcerative colitis. N Engl J Med 2005;353: 2462–2476.

[67] EMA. 2012. Assessment report for infliximab Type II variation EMA/126772/2012 for pediatric ulcerative colitis. Available at http://www.ema.europa.eu/docs/en_GB/document_library/EPAR_-_Assessment_Report_-_Variation/human/000240/WC500124425.pdf. Accessed 2015 Jun 8.

[68] Janssen Biotech Inc. 2011. Briefing Document for Food and Drug Administration Gastrointestinal Drugs Advisory Committee. Available at http://www.fda.gov/downloads/AdvisoryCommittees/CommitteesMeetingMaterials/Drugs/GastrointestinalDrugsAdvisoryCommittee/UCM263459.pdf. Accessed 2015 Feb 4.

[69] Seow CH, Newman A, Irwin SP, Steinhart AH, Silverberg MS, Greenberg GR. Trough serum infliximab: a predictive factor of clinical outcome for infliximab treatment in acute ulcerative colitis. Gut 2010;59:49–54.

[70] Adedokun OJ, Sandborn WJ, Feagan BG, Ruutgeerts P, Xu Z, Marano CW, Johanns J, Zhou H, Davis HM, Cornillie F, Reinisch W. Association between serum concentration of infliximab and efficacy in adult patients with ulcerative colitis. Gastroenterology 2014;147 (6):1296–1307.

21

LC/MS VERSUS IMMUNE-BASED BIOANALYTICAL METHODS IN QUANTITATION OF THERAPEUTIC BIOLOGICS IN BIOLOGICAL MATRICES

Bo An, Ming Zhang and Jun Qu
State University of New York at Buffalo, Buffalo, NY, USA

21.1 INTRODUCTION

Biotherapeutics, especially monoclonal antibodies (mAbs), represent the largest and fastest-growing category of drugs in the market. Accurate, sensitive, and high throughput quantification methods are highly critical for the absorption, distribution, metabolism, and excretion (ADME) studies of mAb and the developments of novel products and biosimilars [1, 2]. Currently, a variety of quantitative methods based on ligand-binding assay (LBA) or liquid chromatography mass spectrometry (LC/MS) are developed and prevalently utilized.

LBA is the most frequently practiced method for protein quantification in past several decades. Enzyme-linked immunosorbent assay (ELISA), the most popular form of LBA, had been first introduced in 1970s [3]. From then on, advances in reagent labeling chemistries, assay configurations, and instrumentations were introduced to improve assay performance (e.g., selectivity, Accuracy, sensitivity) [4]. Particularly, the sandwich ELISA method, which employs two different critical reagents to recognize the target protein by two unique epitopes [5], significantly increased selectivity in a complex matrix. Currently, LBA is considered to provide sufficient sensitivity and throughput for pharmacokinetic (PK), pharmacodynamic (PD), and toxicokinetic (TK) studies [6, 7].

However, LBA methods may fall short in that the quantitative specificity and accuracy may be compromised by the interferences from biomatrices, mAb modification/degradation, and anti-mAb antibody, especially when highly specific critical reagents are not available, for example, at initial discovery stage [8, 9]. Also, these methods are often matrix and species dependent, rendering that methods developed in one matrix/species cannot be easily transferred to another one. Moreover, the background interferences in samples from patients may be different from healthy subjects that are used for method development, and thus additional effort should be made to correct such biases/variations [9]. Most importantly, the development of LBA methods is often time-consuming and costly, which is particularly problematic at the stages of discovery and initial development [10].

Recently, the LC/MS method emerged as an alternative strategy for quantification of proteins of interest in complex biological matrices [11, 12]. Among the available techniques, selected reaction monitoring (SRM) is the method of choice in complex matrix on account of its excellent selectivity/specificity. LC/SRM-MS has superior sensitivity, higher accuracy, and wider dynamic range for targeted protein quantification over other LC/MS techniques, and can be easily multiplexed, for example, quantification of multiple analytes in a single LC/MS analysis by quickly switching among different precursor/product transitions [13]. LC/SRM-MS method has been widely used for small molecule drug quantitative analysis for years, and recently, it has been adapted for targeted quantification of protein

ADME and Translational Pharmacokinetics/Pharmacodynamics of Therapeutic Proteins: Applications in Drug Discovery and Development, First Edition. Edited by Honghui Zhou and Frank-Peter Theil.

based on the analysis of signature peptide (SP) derived from the protein [12–16].

Previous review articles have compared the LC/MS and LBA methods in detail on technical aspects [2, 8, 17–22] (Table 21.1). At different developmental stages of biologics, both LC/MS and LBA methods have their own unique advantages, providing different levels of information. In this chapter, the main characteristics of these two methods are reviewed.

21.2 COMPARISON OF THE CHARACTERISTICS IN METHOD DEVELOPMENT

21.2.1 Method Development Time

Despite of the many formats of LBA assay, the basic elements are the same: capture and detection. For example, the basic principle for ELISA is to specifically extract the analyte from a complex matrix [18], which requires critical reagents with high quality (e.g., structure integrity and purity) [23]. Critical reagent development and evaluation are the most crucial parts for LBA method development, which often take months. Furthermore, as critical reagents are typically produced by biology processes such as these in cell culture, signature features of cells, for example, posttranslational modifications (PTMs), variability in the purity, and potency of critical reagents among lots, may pose significant problems [24]. Apparently, careful management of critical reagents via thorough physicochemical and biological characterizations, such as primary structures, higher-order structures, stability, specification, and potency, are required for each batch of production to maintain high performance of LBA [1, 23, 25].

Unlike LBA method that is dependent on critical reagents, LC/MS method is based on the direct analysis of a specific SP derived from the target protein [2], without using critical reagent. The development process of LC/MS method takes only 1–2 weeks, posing a prominent advantage over LBA in initial development process. Technical details on method development can be found in previous review articles [11, 22, 26]. The critical steps in LC/MS method development are the selection of SP and the optimization of the MS analysis conditions for the SP, which profoundly impact the quantitative reliability and sensitivity. These crucial aspects will be discussed in the following.

21.2.2 Specificity

Based on the specificity of binding between critical reagents and analyte, LBA can be categorized into two classes: specific assay and generic assay [27]. Among the common types of critical reagents, anti-antigen and anti-idiotype antibodies belong to the specific category, and anti-IgG/anti-species antibodies are the generic ones. Assay should be designed, established, and validated based on a "fit-for-purpose" manner depending on the specificity needed, principles for which can be found in previous review articles [28]. For both categories, nonspecific and multispecific immunoglobulins can bind reagent antibodies, resulting in positive biases [8, 29, 30]. The specificity issues will be discussed in Section 21.3.

By comparison, LC/MS method inherently has high specificity and selectivity, which enables sensitive detection of a slight difference between two proteins/peptides [31, 32]. The cooperation of multiple analyzers, which are commonly used in LC/MS analysis, facilitating the selective and sensitive detection of selected specific fragments [13]; higher selectivity can be further achieved by using a higher extent of LC separation.

21.2.3 Characteristics of Method Development

For LBA, antigen capture assays are often recommended for PK study, because such methods may directly quantify the *in vivo* active form of the targeted protein. One of the major challenges is the lot-to-lot variability of the critical reagents [1, 25]. Anti-idiotype assay has also been introduced for PK studies. The anti-idiotype reagent can be generated by multiple attempts to find the optimal reagent with desired traits [28], which enhances the likelihood of success for method development. Furthermore, anti-idiotype can be used as a positive control in antidrug antibody (ADA) assay [33]. However, the specificity/sensitivity of anti-idiotype assay should be evaluated when the project progresses, for example, the animal or chimeric antibody may be gradually humanized, resulting in binding affinity change [28].

Sensitive detection of low abundance proteins in complex matrices can usually be achieved by well-established LBA methods. Nonetheless, high costs, long development cycle, and high failure rates are associated with *de novo* method development for new targets [34]. At the early discovery stage for development of biologics, many candidates need to be investigated. Timely delivery of high quality data is critical for decision making at this stage but it may be cost- and time-prohibitive to develop optimal reagents for each candidate. Moreover, large sample numbers, the need for evolution in different matrices, and limited sample volumes at the discovery stage further increase the difficulties in finding a suitable specific assay [35]. Therefore, generic assays, such as anti-IgG- and anti-species-based capture, which are commercially available, are often adapted for quantification in the discovery stage.

Well-defined generic assays have been reported in the previous literature [36, 37], which can be established in a relatively short period of time, or obtained from commercial sources. Nevertheless, these approaches are vulnerable toward inaccurate results, matrix interference, and high

TABLE 21.1 Ligand-Binding Assay and LC–MS Challenges Encountered When Supporting Regulated Pharmacokinetic Studies

Characteristic	Chromatographic Assays	Ligand-Binding Assays
Assay–Reagent Differences		
Reference standard	Small: <1000 MW Chemically synthesized Homogeneous/high purity Fully characterized with certificate of analysis Stability: chemical Solubility: often hydrophobic Commercially available Not endogenous Usually a solid	Large: >5000 MW Biologically developed Heterogeneous: posttranslational modification, glycosylation, and phosphorylation Less well characterized Stability: chemical, biological, and physical Solubility: often hydrophilic Lot-to-lot variability Not commercially available: obtain from innovator Equivalency to dosed material? May be endogenous Often in solution
Internal standard	Typically used Analog or stable isotope label (MS)	Not typically used Can be used if the antibody used is not recognized by the reference material and a different method is employed
Critical assay reagents	Extraction product, analytical columns, and derivatization agents: lot-to-lot difference uncommon Readily available	Antibody or antibody pairs: derived from biological sources, assay reoptimization may be important for lot changes Not readily available
Assay–Development Differences		
Assay format	LC–MS HPLC GC GC–MS	ELISA RIA Luminex Meso scale discovery Gyros Others
Detection	Direct Based on direct physiochemical properties *m/z*, fluorimetric, electrochemical, and UV	Indirect Based on interaction between macromolecule analyte and capture/detection antibody Fluorimetric, chemiluminescence, radiometric, colorimetric, and electrochemiluminescence
Matrix treatment	Analyte(s) measured following matrix extraction SPE, LLE, PPT, and SLE Organic assay environment at varying pH 2–11 matrix depletion rarely necessary	Analyte(s) measured directly from matrix Aqueous assay environment at physiological pH 6–8 matrix depletion often necessary to remove endogenous analyte(s) Antidrug antibody treatment
Interference selectivity specificity	Depends on extraction, chromatographic, and detector selectivity: endogenous components metabolites: well defined Comeds/OTC Ion suppression Chromatographic reproducibility and adequate separation of interferences	Depends on interactions with other biological molecules: endogenous components, serum binding proteins, hemolysis, lipemia, MRD, autoantibodies, antireagent antibodies, metabolites: not well-defined, soluble target(s)/receptors
DOE	Assay optimization in days to weeks	Assay optimization at many levels over weeks/months
Curve regression	Linear response to concentration relationship Simple regression models Nonlinearity indicative of an assay issue such as detector saturation, matrix effect, ionization competition, dimerization/multimer, and background interference or carryover Assay range usually broad	Nonlinear response to concentration relationship Complex regression equations Hook effects. Dilution of sample to be within the range of the curve may yield a concentration that is inconsistent with the undiluted sample. Sample may contain a concentration above the ULOQ, but due to nonlinear curve many not be detected in the initial assay Anchor points, changes in LLOQ and ULOQ due to endogenous components Assay range limited (2 logs)

C_{max}, maximum concentration; CV, coefficient of variation; DOE, design-of-experiment; ISR, incurred-sample reproducibility; LLE, liquid–liquid extraction; MRD, minimum required dilution; OTC, over the counter; PPT, protein precipitation; RIA, radioimmunoassay; and SLE, supported liquid extraction.

background noise [28]. For example, the responses of the same level of a mAb in different species (e.g., rats vs mice) may not be the same. Moreover, when analyzing the same samples, substantial discrepancy among results using critical reagents from different sources is often observed [38], resulting in high variability among different methods or laboratories.

For LC/MS-based analysis, the selection and optimization of the best SP derived from the target protein is key to achieve a accurate, sensitive, and reliable quantification. Traditionally, *in silico* method, for example, the use of software package such as PeptideAtlas, Skyline, and MRMaid, is commonly used for SP candidate discovery [39–42]. However, this method may not accurately predict the most sensitive proteolytic peptides and the optimal matrix-dependent parameters such as chemical interferences in biological samples [14, 16]. Moreover, in order to discover the best SP, it is often necessary to use synthesized peptides to validate the *in silico* predictions, which may be time-consuming and costly.

Experimental discovery of the optimal SP in the target matrix (e.g., plasma or tissue digest) is the most reliable approach to achieve an accurate discovery and optimization of SP [14, 16]; however, it is challenging to obtain optimal MS conditions for multiple candidates in the digest of biological samples. It is also important to choose stable peptides as the SP [14, 16] to prevent quantitative variation and bias arising from peptide instability, which has been often overlooked. Moreover, it is critical to choose the most sensitive peptide to ensure the detection of low concentration samples. Finally, it is beneficial to use more than one SP for quantification of one protein, because mAb could be truncated biologically outside the SP domain or certain residues within the SP domain could be biologically modified [8], resulting in significant quantification biases that are not perceivable if a lone SP was used.

Our laboratory established a novel experimental strategy for high throughput and accurate method development (Fig. 21.1). Briefly, the pool of SP candidates was generated by a data-dependent peptide discovery using high resolution LC/MS, followed by a stringent filtering step to remove peptides that are not unique to the target, containing labile amino acid, possessing known modification, or possessing miss cleavage sites [44, 45]. The optimal LC/MS conditions of all SP candidates were accurately obtained by a high throughput and on-the-fly orthogonal array optimization (OAO) [14–16], which enables accurate and reproducible optimization of SRM conditions for greater than 100 candidates within a single LC/MS analysis of a spiked biological sample. Using the developed LC/MS conditions, stability and sensitivity of all the SP candidates are evaluated in the matrix digest. Among the stable peptides, two peptides with the highest S/N were selected as the SP. Technical specifics of the OAO method can be found in previous publications [14–16].

Another strategy to enable rapid method development is the use of species-specific SP for certain preclinical applications. For example, Furlong et al. described the use of a "universal surrogate peptide" derived from the constant Fc region of human antibody for quantification of human antibodies in nonhuman animal models [46].

21.3 COMPARISON OF ASSAY PERFORMANCE

If an LBA method for biotherapeutics quantification is well established, it usually provides sufficient sensitivity and selectivity and is considered the method of choice. LC/MS method is often considered a promising alternative to LBA, in that it provides fast method development and highly specific and selective analysis. Both methods have unique pros and cons, and thus should be selected and utilized in a "fit-for-purpose" manner, by balancing the considerations, such as time requirement for method development, reagent availability, requirement of sensitivity and specificity, and possible concentration range [2, 18, 27]. Assay performances by both methods are discussed in a general sense, in order to help researchers to make the correct decision.

21.3.1 Sample Preparation

Development and optimization of sample preparation approaches are critical in method development because an optimal sample preparation procedure lays a solid foundation for excellent quantitative precision, accuracy, and reproducibility. Due to the high complexity of plasma or tissue samples, substantial sample preparation is needed for both methods. For a well-established LBA method where the critical reagent(s) enables selective extraction, a sample dilution is often sufficient; nonetheless, well-trained operators are required to maintain acceptable assay accuracy and reproducibility [47, 48], and inter-/intra-batch variances and inaccuracy are mainly ascribed to dilution [24, 49].

By comparison, the sample preparation for the LC/MS method requires more steps. Protein digestion is the essential part for LC/SRM-MS method, aiming at efficient and reproducible release of SP from the targeted protein. In addition, in order to ensure a sensitive, reproducible, and robust analysis, an effective and quantitative sample cleanup is desirable to remove matrix components that may negatively impact the digestion and LC/MS procedures. A universal and optimal preparation procedure for the quantification of mAb in biological matrices has not yet been established, largely because tissue and plasma samples are highly complex and the structure of a typical mAb renders it resistant to enzymatic digestion [50, 51]. Moreover, due to low drug concentrations in tissues and the lack of an efficient protein extraction procedure that is compatible with LC/SRM-MS analysis, it is challenging to prepare tissue samples for quantification of biotherapeutics [14].

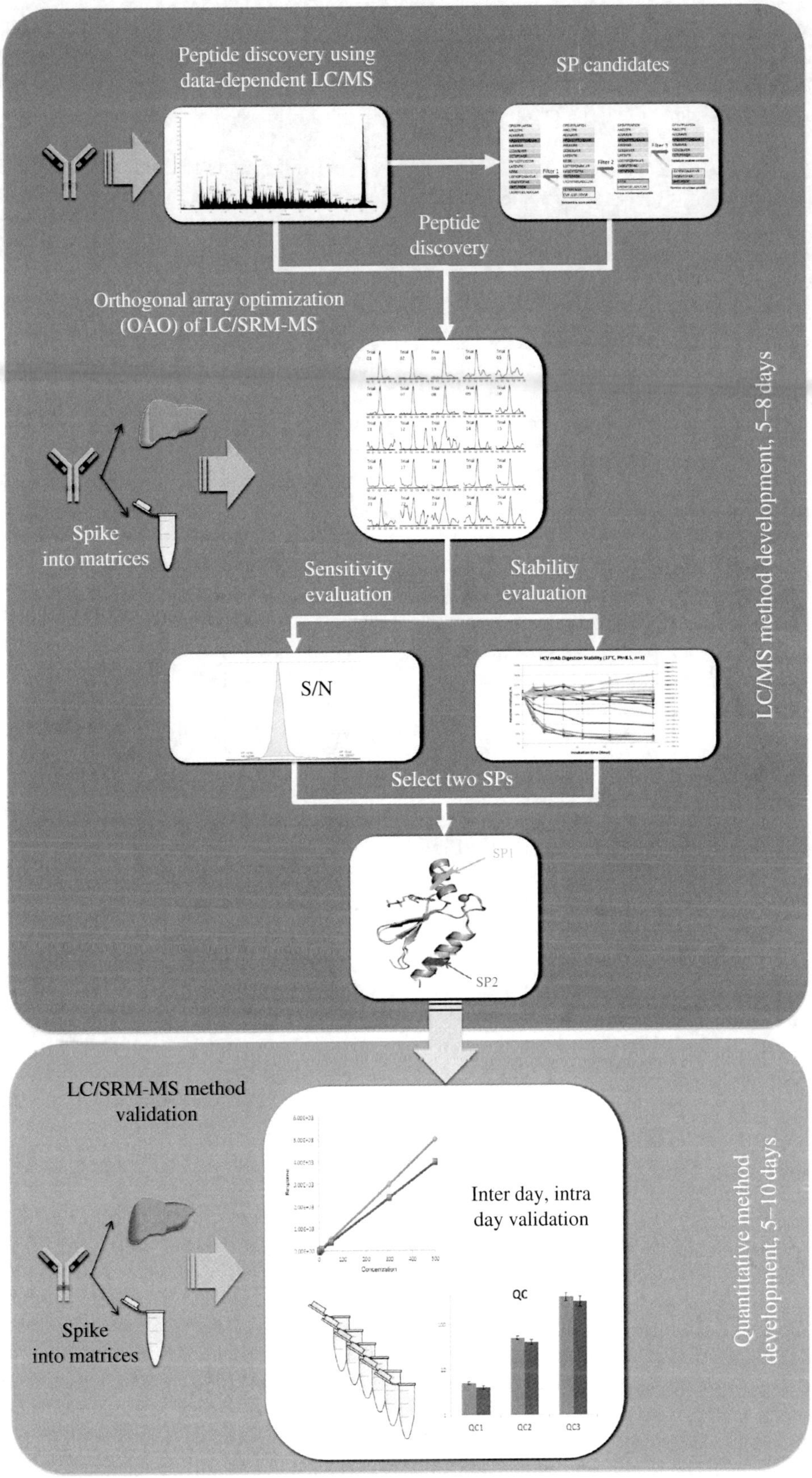

FIGURE 21.1 Flow chart of a novel LC-SRM-MS method development process based on OAO optimization. (An et al. [43]. Reproduced with permission of *Drug Metab Dispos*.)

In-solution digestion method is the most frequently practiced protocol for biotherapeutics quantification [20, 32, 52]. However, as this strategy does not remove nonprotein matrix components, the digestion efficiency may be compromised and an extra postdigestion step is often needed to avoid deterioration of LC/MS analysis [53]. Other methods, such as in-gel digestion method [54] and filter-aided sample preparation method [53, 55, 56], were developed to clean up the protein samples; nonetheless, these methods often suffer from low peptide recovery [57].

Recently, we developed a straightforward, efficient, and reproducible on-pellet digestion method, which provides high and reproducible peptide recovery for both proteomics and targeted protein analysis [14, 15, 58–60], and the advantage of the on-pellet digestion for analysis of protein in plasma had been demonstrated independently by Ouyang et al. [51] and Yuan et al. [50].

Finally, compared with LBA, LC/MS is a much more versatile technique, which may be readily conjugated to a variety of other techniques to improve analytical performance. For example, it is compatible with many matrix cleanup techniques on both protein and peptide levels, such as protein precipitation, solid-phase extraction (SPE), strong cation exchange (SCX), and ultrafiltration [22, 41, 61–63].

21.3.2 Calibration Curve and Linearity Range

As the response versus concentration relationship for LBA is nonlinear, complex regression equations are involved in these methods [2]. Furthermore, the dynamic concentration range in LBA is very narrow, typically within 2 orders of magnitude [17]. Consequently, dilution is needed for samples with concentrations higher than upper limit of quantification (ULOQ); this may introduce quantitative biases or errors, in that the background noise will be reduced after dilution and handling error may occur. Moreover, the samples with high concentrations may have deviated responses due to saturation in some formats of LBA (e.g., sandwich ELISA without the wash step), recognized as hook effect (prozone) [8, 17]. On the other hand, sensitivity of some LBA methods may be insufficient, for example, PK study requires the analysis of samples long after dosing to capture the terminal phase (often five half-lives and more after last or single dosing) or using low doses, producing samples with drug concentrations lower than lower limit of quantification (LLOQ) [18]. The poor performance of LBA for samples beyond LLOQ and ULOQ often leads to bias in ADME research such as false estimation of AUC, clearance, and half-life.

By comparison, for LC/MS analysis, simple linear regression is used to conduct error-free estimation of incurred sample concentration and the quantification dynamic range usually reaches greater than 3 orders of magnitude. Especially, SRM operated on a triple quadrupole MS, which is by far the most commonly utilized technique, exhibits wider dynamic range for targeted protein quantification [13]. Because of the high specificity and selectivity, LC/SRM-MS is less vulnerable to interferences and matrix effect from complex biological samples. Suitable quantification dynamic range can be easily established to meet the demands of wide concentration ranges in typical PK studies.

21.3.3 Applicability

Multiplexed capacity and selectivity are critical features determining the applicability of a bioanalytical method. A method that can be applied in different types of studies in the drug R&D procedure (e.g., applicable in plasma and tissue samples, and suitable for preclinical/clinical investigations and PK, PD, and TK studies) is highly desirable. This section discusses the applicability of LBA and LC/MS.

First, a method with multiplexing capacity (i.e., the ability to quantify multiple targets in a single analysis) is highly valuable for the R&D of biotherapeutics. For instance, quantification of multiple drug candidates in the same sample is necessary for certain administration strategies such as cassette dosing [32, 64]. It is very challenging to develop an LBA method capable of quantifying multiple proteins in one analysis because traditional LBA are designed to detect one target in one analysis. A multiplexed method for quantification of multiple cytokines has been reported, using a bead-array cytometric analyzer [65] or a plate-based proteome array [66]. However, the performance of such strategies remains to be extensively evaluated and several technical challenges remain. For example, interferences between critical reagents may severely impact assay performance, and inter-batch variations are high when the critical reagents are from different manufacturers [67]. Conversely, LC/MS platform, especially an SRM-based method, can be easily multiplexed by quickly switching among different precursor/product transitions [13], which is the method of choice for simultaneous investigation of multiple drugs and proteins [32, 64, 68].

Secondly, the LBA and LC/MS show quite distinct features in terms of selectivity. LBA relies on binding of a specific epitope of the target protein and thereby exhibit limited selectivity, such as the poor ability to distinguish the target protein from degradation products, PTM forms, and some endogenous proteins that interfere with the binding [5, 69]. One advantage of LC/MS method is high specificity at the molecular level, which enables differentiation of modified products from the unmodified targets. Furthermore, a multiplexed LC/SRM-MS approach may enable the simultaneous quantification of both modified and unmodified forms.

In addition, the comprehensive understanding of the forms of mAb in circulation is essential; researchers are often interested in mAb_{free}, most likely the active drug [70].

In theory, if optimal critical reagents were developed, different LBA methods may be developed to, respectively, characterize free, total, or complexed mAb targets [71]. Nonetheless, investigation of different forms of mAb in circulation has been rarely conducted in LBA methods development, largely owing to the challenges in developing specific critical reagents for these forms [9]. Apart from introducing uncertainty to ADME studies, this ambiguity also jeopardizes the comparison between different PK studies that is quantified with different assays [9]. Also, mAb_{total} is drawing increasing interest because it could provide valuable information on on- or off-target effects mAb, for safety evaluation [22, 70]. A typical LC/MS method often quantifies the mAb_{total} with high certainty; yet if a specific separation/enrichment strategy (e.g., affinity capture or fractionation) is employed, it can be utilized to quantify other forms of mAb [70, 72].

Finally, in different matrices or species, the extents of interferences and cross-reactions from matrix components vary considerably [33], rendering it difficulties to transfer an LBA method among matrices (e.g., from plasma to a tissue or among different tissues) [8, 9, 18]. By comparison, as LC/SRM-MS has high specificity and minimizes matrix effects by employing isotope-labeled internal standards (ISs) and sufficient chromatographic separation [13], the methods are often readily transferrable among different matrices. For example, recently, we applied the same LC/SRM-MS method for the quantification of therapeutic mAb in mouse plasma and tissues, such as brain, heart, liver, spleen, kidney, and lung, with rapid and simple revalidation in different matrices [14].

21.3.4 Accuracy

High accuracy ensures credible data for downstream ADME and other studies. LBA often suffers from accuracy issues as there is no IS employed in to correct the variance introduced in analytical process, for example, the pipetting variance will be amplified by the additional pipetting required for dilutions of most protein samples [48], resulting in significant inter-/intra-batch and interlaboratory biases and variations.

One of advantage of LC/MS-based technology is the possibility of utilizing an IS for quantification [73]. Stable isotope-labeled (SIL) IS can be used to effectively correct the variance during sample preparation and LC/MS analysis. For instance, SIL-full-length protein is the ideal IS for protein quantification, which is capable to correct variances in sample extraction, treatment, digestion, loading, ionization, and SRM analysis [5, 22, 73]. Nevertheless, it is costly and time-consuming to produce SIL protein with high isotope purity, and SIL protein is not available for many species [22, 74].

One alternative is protein analog, which is expected to have similar digestion behavior as the target [62, 75]. However, it is very difficult to find an ideal protein analog. Synthesized SP as calibrator and SIL SP as IS is another alternative; the peptides can be easily produced from commercial sources at low cost, and thus this method is the most popular method for protein quantification. As a result, this approach is the most popular for protein quantification. However, this method only corrects the error in LC/MS analysis process, not sample preparation, digestion, and so on [22, 73]. SIL-extended peptide with cleavable sites on both ends of the SP domain was introduced to correct the variability during analysis and digestion, but it only offers partial correction of digestion variation [22, 73].

The choice of calibrators and IS is among the most critical factors governing the reliability and accuracy of the LC/SRM-MS-based quantification. Different calibration approaches require different procedures, as illustrated in Figure 21.2. Recently, we performed a systematic evaluation of prevalent calibration approaches [76]. It was those calibration approaches at peptide and extended-peptide levels resulted in severe negative biases. By comparison, "hybrid calibration" methods that utilize intact protein calibrator and SIL-peptide/extended-peptide IS delivered accurate and reproducible quantification with results comparable to those obtained by protein-level calibration, which is the gold standard for LC/MS-based protein quantification. Therefore, the hybrid calibration approaches provide a cost-effective method for high quality quantification of mAb (Fig. 21.3).

21.3.5 Sensitivity

Well-established LBA methods usually provide adequate sensitivity for quantification analysis of biotherapeutics in preclinical/clinical PK/PD study [6, 7]. However, due to the intrinsic features of detection techniques used in LBA, interference from endogenous antibodies may compromise the selectivity and thus decrease sensitivity [77]; this problem is often more pronounced for newly developed methods. Strategies such as sandwich ELISA and competitive ELISA may mitigate this problem to some extent [78, 79].

By comparison, LC/MS has higher selectivity and sensitivity over ultraviolet (UV) and fluorescence detectors that are used for LBA [73]. However, insufficient sensitivity remains a major concern for quantification of therapeutic mAb, largely due to two reasons: (i) as the signal response of LC/MS is dependent on the molar rather than mass amounts of the analyte, the large molecular weights of mAb pose a disadvantage; and (ii) high protein contents and low concentrations of target mAb in plasma or tissue samples require dilution or enrichment before analysis [15, 16, 50, 60, 80, 81].

In order to improve sensitivity for targeted protein analysis, our laboratory developed a robust nano-flow LC/SRM-MS strategy [14–16], which typically lowers the limit of quantification (LOQ) by approximately 30 to 50-fold compared to a conventional-flow LC/SRM-MS. Another

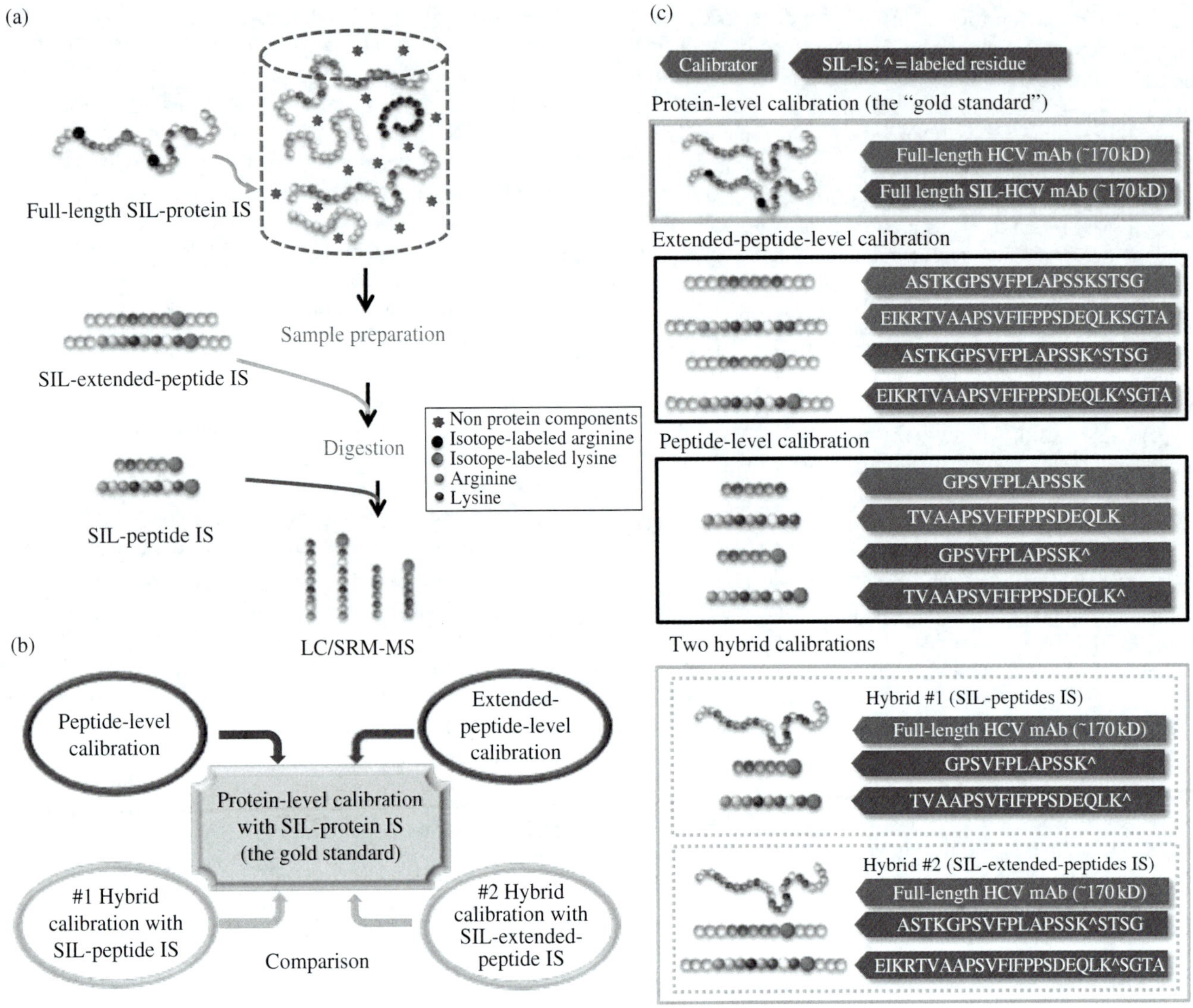

FIGURE 21.2 Illustrations of various calibration approaches for targeted protein quantification and the study design of the comparative study. (a) The three options of internal standard (IS) methods used for targeted protein quantification. Full-length stable isotope-labeled (SIL) protein IS is added to the sample before any preparation; SIL-extended-peptide IS is added right before digestion and SIL peptide IS is spiked into the digest mixture after digestion. (b) Scheme of the comparison. The peptide- and extended-peptide-level calibrations and two hybrid calibrations were compared against the protein-level calibration (the gold standard) for quantitative performances in both protein-spiked QC samples and a full rat PK study. (c) The calibrators and IS employed for each calibration approach for quantification of the anti-HCV mAb in plasma. For each calibration approach, two sets of calibration curves (one for each SP) were independently established and then used for quantification. (Nouri-Nigjeh et al. [76]. Reproduced with permission of *Anal Chem.*)

approach to improve sensitivity is to enrich target proteins or peptides before LC/SRM-MS analysis by affinity capturing. For instance, Dubois et al. achieved an LOQ at 0.02 μg/mL for quantification of a chimeric mAb in human serum samples with an enrichment procedure [77]; Lin et al. utilized immunoprecipitation enrichment before LC/SRM-MS analysis, which achieved an LOQ of 10 ng/mL for mAb analysis [82]. More recently, Neubert et al. developed a series of affinity-based platforms for quantitative enrichment of target proteins and/or SPs in plasma, achieving ultrasensitive quantification of circulating biomarkers in plasma [83–85]. Stable isotope standards and capture by anti-peptide antibodies (SISCAPA) technique was developed to enrich SPs using polyclonal antibodies [86]. Furthermore, besides increase in sensitivity, affinity capture-based method can also enable more specific quantification, for example, for free or total mAb [87]. Figure 21.4 shows the sensitivity of different methods.

A variety of other techniques have been developed to increase the sensitivity for LC/SRM-MS-based targeted

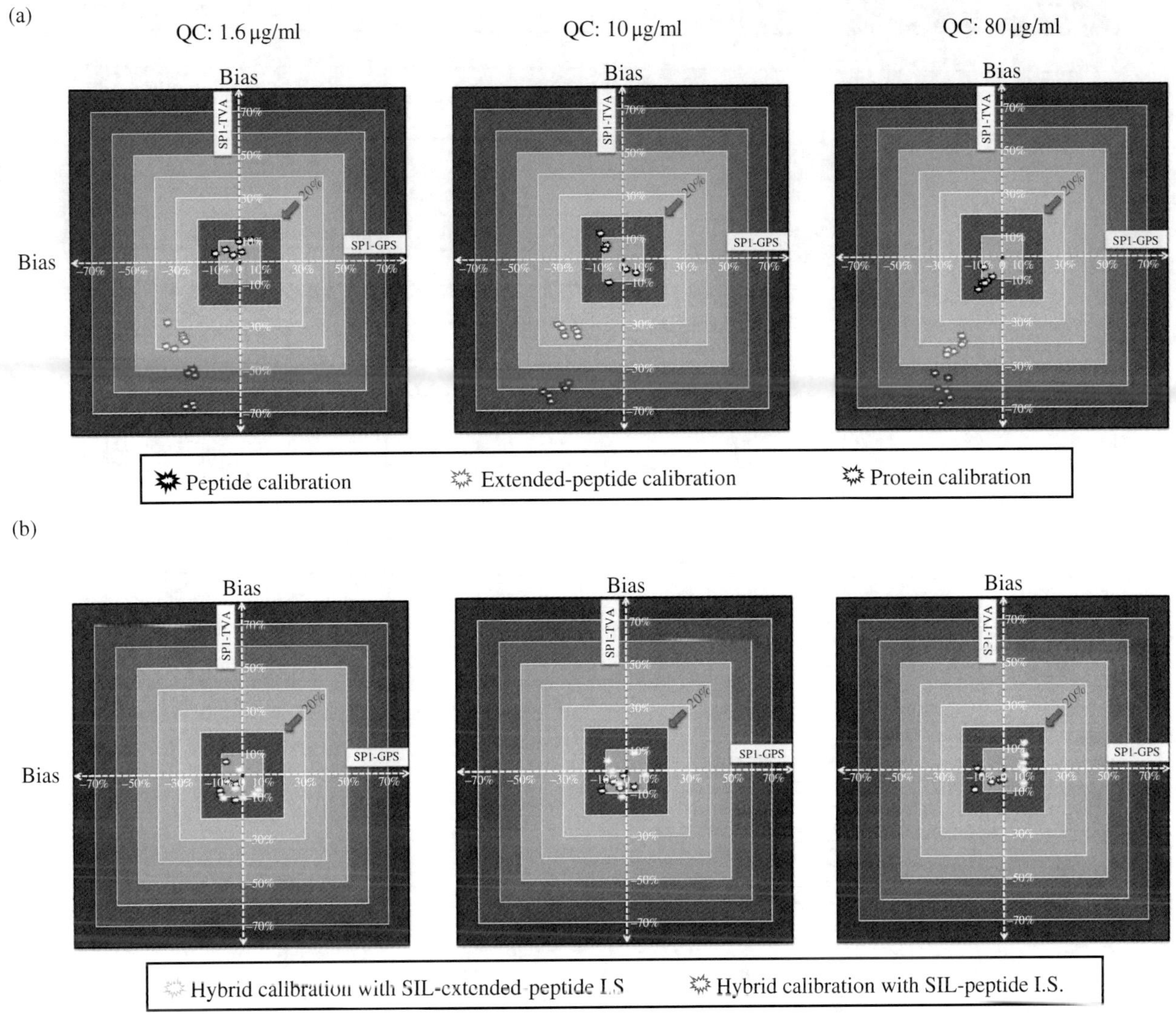

FIGURE 21.3 Two-dimensional representations of the quantitative biases by (a) peptide-, extended-peptide-, and protein-level calibration approaches and (b) the two "hybrid" calibration approaches. For every calibration method, the quantitative values were obtained independently using the two signature peptides (SPs), that is, the GPSVFPLAPSSK (GPS) and TVAAPSVFIFPPSDEQLK (TVA) peptides. The two axes represent the quantitative biases by the two SPs. The red box in the center of each panel denotes the zone of less than 20% bias, while the golden box signifies the zone of <10% bias. (*See insert for color representation of this figure.*) (Nouri-Nigjeh et al. [76]. Reproduced with permission of *Anal Chem.*)

protein quantification, although these have yet been extensively examined in the quantification of biotherapeutics; for example, SCX fractionation [88], high pH fractionation before LC/MS analysis [34], and the use of long columns to obtain high S/N of target peptides [89].

21.3.6 Reproducibility

The critical reagents used for LBA are produced via biological processes, which are inherently prone to variability arising from PTM, variations of reagents, and biological interference [23, 24]. Since LBA does not employ an IS as discussed previously, stringently controlled operations are required to prevent deterioration of assay performance [24, 48]. The critical reagents cross-react with a varying extent of metabolites, for example, the PTM products [41]. Moreover, LBA strategy suffers from various interferences, such as nonspecific interferences (e.g., nonspecific binding with matrix components, assay reagents, and hardware) and specific ones (e.g., high affinity binding proteins, soluble or cell surface targets, and ADA) [18]. Consequently, it is often challenging to maintain high inter-batch/interlaboratory consistency, which is essential to correlate results among batches/studies and to transfer

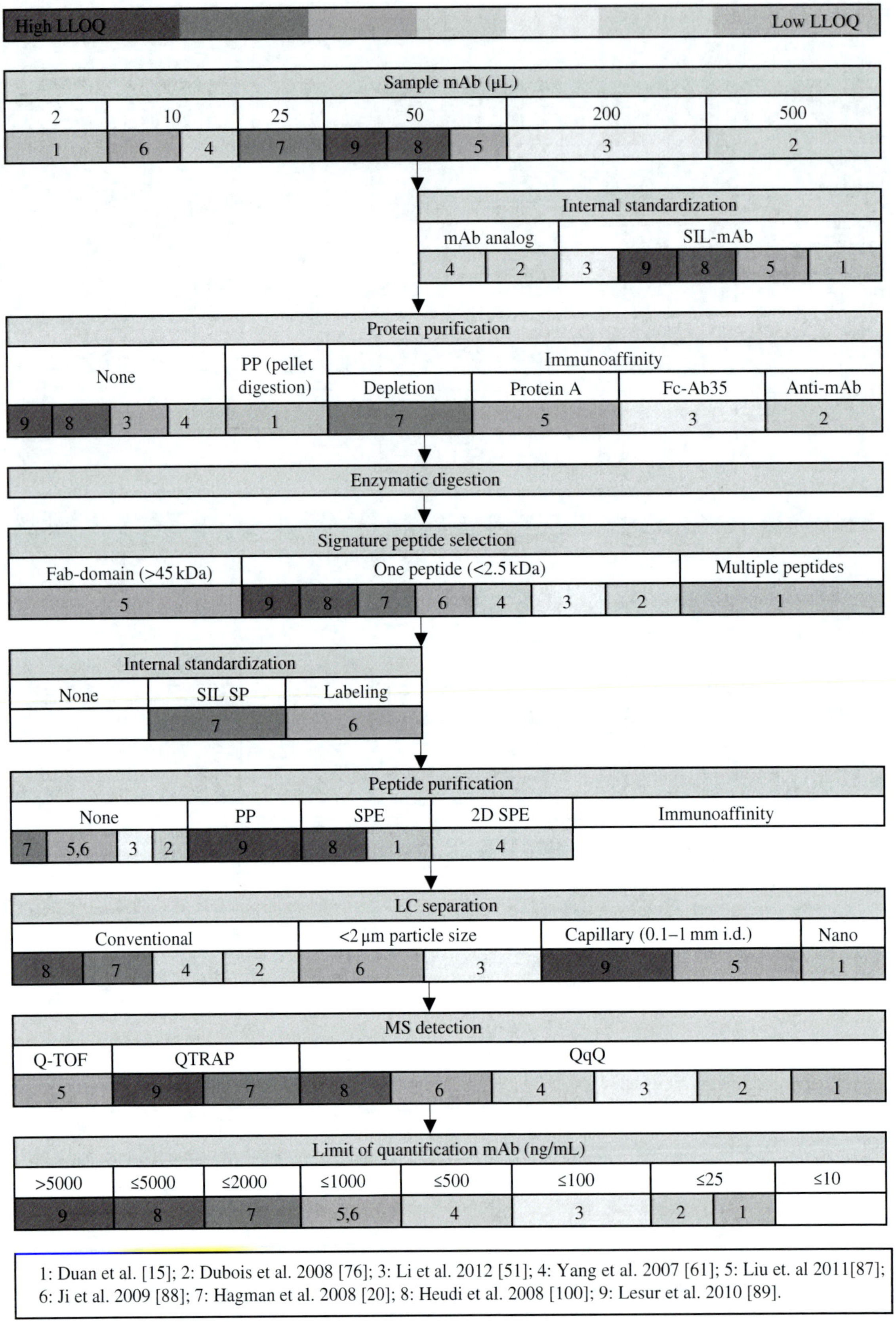

FIGURE 21.4 Sensitivity comparison of different strategies in targeted protein quantification. (*See insert for color representation of this figure.*) (van den Broek et al. [22]. Reproduced with permission of *J Chromatogr.*)

validated methods between laboratories [18]. Furthermore, researchers involved in LBA is confronted by reproducibility problems when, for instance, previous critical reagent supply is discontinued [67] or implementing an interbatch/interlaboratory comparison.

On the contrary, analytical variation is frequently much less a concern for LC/SRM-MS approaches since critical reagents are not required, and isotope-labeled IS is commonly used, which effectively correct analytical variations introduced by LC/MS analysis and matrix effects [11, 22, 73]. In practice, the performance of a developed and validated LC/SRM-MS method is usually quite robust as long as the instrument maintenance and quality control are carried out properly.

21.4 APPLICATION OF LBA AND LC/MS IN THE ANALYSIS OF THERAPEUTIC PROTEINS

21.4.1 Quantification of mAb in Plasma and Tissues

As LBA has been widely used to quantify therapeutic proteins for decades, numerous works in this regard were reported. Here, only several selected, unique works are exemplified. Recently, Joyce et al. established a Gyrolab-based LBA method for mAb quantification and applied it in a study of PK method. Using this strategy, it was proved that mouse serial sampling strategy is more practicable, particularly with limited drug supply or specialized animal models, since there is no significant difference among the results from the PK study by serial sampling and composite sampling from different sampling sites [90]. Bodenlenz et al. utilized an ELISA method to monitor TNF-α (tumor necrosis factor alpha) changes to facilitate a dermal PK/PD study [91]. Many other ADME studies employed LBA as quantitative tools, and a myriad of successful LBA assays were established against CD2 [38], IGF (insulin-like growth factor) receptor [92], CD40 [93], integrin [94], CD33 [95], HER2 [96], and CH1 [97].

Some representative LC/MS-based works for quantification of circulating mAb are also exemplified here. Heudi et al. developed and validated an accurate quantitative method for a candidate mAb in marmoset serum, using postdigestion SPE cleanup and a SIL-full-length protein as IS [98], which was applied to PK analysis in marmosets at dose level of 150 mg/kg. Li et al. developed a universal LC-SRM/MS approach for quantification of a variety of therapeutic mAb in preclinical studies based on the utilization of a full-length SIL mAb as the universal IS [52]. Hagman and coworkers developed an LC/SRM-MS method to quantify a human mAb in the serum of cynomolgus monkey [20], with improved analytical sensitivity by an albumin depletion procedure before LC/MS analysis. Ouyang et al. described the combination of on-pellet digestion with LC/SRM-MS for reproducible analysis of a mAb drug candidate in monkey plasma [51]. Lu and coworkers employed albumin depletion, protein A capture, and antibody capture coupled to LC/SRM-MS for sensitive quantification of a mAb candidate (CNTO736) [99]. Fernandez Ocana et al. established a LC/MS strategy to quantify free and total anti-MadCAM mAb (PF-00547, 659) in human serum [87], which captured free mAb with a biotinylated anti-idiotypic antibody followed by enrichment with streptavidin magnetic beads; total target mAb was enriched by protein G magnetic beads. Our laboratory described a sensitive nano-LC/SRM-MS method for quantification of a chimeric mAb (cT84.66) in mouse serum [15]. Owing to the high sensitivity and selectivity achieved, the method was successfully applied to the preclinical PK study with a subcutaneous dosing at 1 mg/kg.

The determination of the concentrations of biotherapeutics in tissues is critical for PK studies, but such works have rarely been reported with either LC/MS or LBA, due to technical challenges such as the low drug concentrations, lack of proper sample preparation strategy, and interferences of tissue matrices. Although LBA method has been applied in the quantification of protein markers in tissues, for example, human heparanase in mice liver and lung tissues [100], urokinase-type plasminogen activator and its receptor complex in cancerous lung tissue [101], and breast cancer tissue [102, 103], to our knowledge, an LBA-based quantification method for quantification of therapeutic antibody in tissue has not been reported. Using a sensitive nano-LC/SRM-MS, effective sample preparation, and a high throughput method optimization strategy, we described sensitive quantification of two mAbs (8c2 and cT84.66) in seven tissues with LLOQ in the range of 0.156–0.312 μg/g tissue [14]. The work demonstrated that LC/SRM-MS is a promising alternative to radiolabeling strategies for tissue analysis, which may fall short in problems associated to assay accuracy and specificity, degradation of labeled protein, and radiation exposure to investigators and animals [14]. Not long ago, Sleczka et al. established an LC-MS/MS assay for human mAbs quantification in mice tissues, which can be applied for the quantitative analysis of other mAb and Fc-fusion biotherapeutics in a variety of animal tissues, achieving a quantification range of 20–20,000 ng/mL [104].

21.4.2 Application in Multiplexed Analysis

Recently, a novel immunoassay, fluorescent multiplexed bead-based immunoassay (FMIA), had been reported for multiplexed IgG determination in human serum [105], which can facilitate etiological diagnosis and vaccine studies. The multiplexed beads, which have coupling concentrations in the range of 1.6–50 μg/million beads, provide good correlation with singleplexed assay. Other reported multiplexed LBA methods were focusing on biomarker detection [106–108], and none was developed for analysis of biotherapeutics *in vivo*. As already discussed in this chapter, LC/SRM-MS is capable of performing simultaneous quantification of multiple targets, therefore enabling the study of multiple drug candidates in one experiment, for example, the cassette dosing. Although cassette dosing has been prevalently used to screen small molecule candidates [109, 110], it was found to be even more suitable for preliminary investigation of mAb candidates [32] because (i) studies of multiple mAb usually do not carry the risk of drug–drug interaction, and the PK of proteins is not affected by CYP450 and transporters [32, 111], and (ii) it is fairly straightforward to find an optimal, common formulation for multiple mAbs [112, 113]. Jiang et al. developed and validated an LC/SRM-MS method for simultaneous quantitation of two coadministrated mAbs, which showed good sensitivity, reproducibility, and accuracy for both targets. The method was successfully applied to TK study in monkeys [64]. Li et al. reported an analytical method to quantify four mAbs after subcutaneous cassette administration with LOQ of 0.1–0.5 μg/mL in plasma [32]. Recently, Xu et al. developed an affinity capture-aided LC/MS method that can simultaneously quantify two mAbs in human serum for combination therapy study [68].

21.4.3 Characterization of Antibody–Drug Conjugates (ADC)

Antibody–drug conjugate (ADC) development is one of the hottest topics in the field of biotherapeutics. Drug-to-antibody ratios (DARs) and drug load distribution are critical parameters that profoundly determine the *in vivo* efficacy and toxicity of an ADC [114, 115], which is essential for ADC development and their PK/PD studies. LBA such as ELISA cannot provide unbiased total antibody analysis for ADC, because payload distributed on the antibody surface could influence the binding affinity, resulting in quantitative error [116], and it is hard to obtain DAR information by LBA method. LC/MS with high resolution analyzer has become a promising technology for characterization of ADC in the recent years. An affinity capture capillary LC coupled to quadrupole-time-of-flight (TOF) MS has been employed to analyze anti-MUC16 TDC, which obtained both *in vitro* and *in vivo* DAR information [117]. Valliere-Douglass et al. developed a native LC/MS method for the determination of DAR via analysis of intact protein [118]. Recently, a more sensitive method was reported by Chen et al., which employed a native nano-ESI-TOF analysis in conjunction with a limited digestion by cysteine protease to obtain DAR information [119].

21.5 SUMMARY AND FUTURE PERSPECTIVE

LBA is recognized as the standard method for quantification of biotherapeutics. In spite of the drastically increasing role of LC/MS in biotherapeutics analysis, it will not fully replace traditional methods such as LBA in the foreseeable future. First, if a well-developed, specific LBA method is available for the target molecule, LBA may be advantageous over LC/MS in that it could be more sensitive [120] and easy to conduct (does not require LC/MS instruments). Secondly, LBA can be developed to quantify free or total mAb, while LC/MS only detects the total mAb unless specific affinity capture enrichment is employed [70].

LC/MS represents a promising alternative to traditional LBA methods for the analysis of biotherapeutics because LC/MS (i) can be readily adapted to quantification in plasma and tissues and across various species, (ii) provides extraordinary specificity and reproducibility with low sample consumption, and (iii) is capable of simultaneous quantification of multiple proteins (e.g., biotherapeutics and/or their targets, or multiple mAb in cassette-dosing study) in one analysis. Furthermore, the development and validation of an LC/MS method are rapid at a relatively low cost, which is a highly desirable feature that facilitates the rapid development of biotherapeutics.

When choosing the analytical method for ADME study, researcher should take the method development and assay performance factors into consideration, such as development cycle, resources (reagents and instruments) available, samples, and purpose of study. More importantly, a hybrid method, which combines different analysis features together (e.g., coupling sensitivity of LBA and specificity of LC/MS), is feasible to facilitate ADME analysis. For instance, immunocapture/immunodepletion of the samples can greatly enhance the quantitative sensitivity of LC/MS; on the other hand, the specificity and sensitivity requirements of critical reagents used for this purpose are much lower than those for a traditional LBA, owing to the high specificity of LC/MS analysis. Consequently, the method development is much quicker than that for a traditional LBA method. Moreover, immunoprecipitation or native gel separation before LC/MS analysis will provide important information on the target protein, such as binding, aggregation, and degradation states. Metabolism and immunogenicity, and drug–drug interaction studies, which are hard to accomplish by traditional LBA and LC/MS approaches [121], can be investigated by the hybrid method.

REFERENCES

[1] Geist BJ, Egan AC, Yang TY, Dong Y, Shankar G. Characterization of critical reagents in ligand-binding assays: enabling robust bioanalytical methods and lifecycle management. Bioanalysis 2013;5 (2):227–244.

[2] Nowatzke WL, Rogers K, Wells E, Bowsher RR, Ray C, Unger S. Unique challenges of providing bioanalytical support for biological therapeutic pharmacokinetic programs. Bioanalysis 2011;3 (5):509–521.

[3] Engvall E, Perlmann P. Enzyme-linked immunosorbent assay, Elisa. 3. Quantitation of specific antibodies by enzyme-labeled anti-immunoglobulin in antigen-coated tubes. J Immunol 1972;109 (1):129–135.

[4] Yang TY, Uhlinger DJ, Ayers SA, O'Hara DM, Joyce AP. Challenges in selectivity, specificity and quantitation range of ligand-binding assays: case studies using a microfluidics platform. Bioanalysis 2014;6 (8):1049–1057.

[5] Zheng J, Mehl J, Zhu Y, Xin B, Olah T. Application and challenges in using LC-MS assays for absolute quantitative analysis of therapeutic proteins in drug discovery. Bioanalysis 2014;6 (6):859–879.

[6] Urva SR, Yang VC, Balthasar JP. An ELISA for quantification of T84.66, a monoclonal anti-CEA antibody, in mouse plasma. J Immunoassay Immunochem 2010;31 (1):1–9.

[7] Shah DK, Balthasar JP. PK/TD modeling for prediction of the effects of 8C2, an anti-topotecan mAb, on topotecan-induced toxicity in mice. Int J Pharm 2014;465 (1–2):228–238.

[8] Hoofnagle AN, Wener MH. The fundamental flaws of immunoassays and potential solutions using tandem mass spectrometry. J Immunol Methods 2009;347 (1–2):3–11.

[9] Damen CW, Schellens JH, Beijnen JH. Bioanalytical methods for the quantification of therapeutic monoclonal antibodies and their application in clinical pharmacokinetic studies. Hum Antibodies 2009;18 (3):47–73.

[10] Savoie N, Garofolo F, van Amsterdam P, Bansal S, Beaver C, Bedford P, Booth BP, Evans C, Jemal M, Lefebvre M, Lopes de Silva AL, Lowes S, Marini JC, Massé R, Mawer L, Ormsby E, Rocci ML Jr, Viswanathan C, Wakelin-Smith J, Welink J, White JT, Woolf E. 2010 white paper on recent issues in regulated bioanalysis & global harmonization of bioanalytical guidance. Bioanalysis 2010;2 (12):1945–1960.

[11] Pan S, Aebersold R, Chen R, Rush J, Goodlett DR, McIntosh MW, Zhang J, Brentnall TA. Mass spectrometry based targeted protein quantification: methods and applications. J Proteome Res 2009;8 (2):787–797.

[12] Qu J, Jusko WJ, Straubinger RM. Utility of cleavable isotope-coded affinity tagged reagents for quantification of low-copy proteins induced by methylprednisolone using liquid chromatography/tandem mass spectrometry. Anal Chem 2006;78 (13):4543–4552.

[13] Qu J, Straubinger RM. Improved sensitivity for quantification of proteins using triply charged cleavable isotope-coded affinity tag peptides. Rapid Commun Mass Spectrom 2005;19 (19):2857–2864.

[14] Duan X, Abuqayyas L, Dai L, Balthasar JP, Qu J. High-throughput method development for sensitive, accurate, and reproducible quantification of therapeutic monoclonal antibodies in tissues using orthogonal array optimization and nano liquid chromatography/selected reaction monitoring mass spectrometry. Anal Chem 2012;84 (10):4373–4382.

[15] Duan X, Dai L, Chen SC, Balthasar JP, Qu J. Nano-scale liquid chromatography/mass spectrometry and on-the-fly orthogonal array optimization for quantification of therapeutic monoclonal antibodies and the application in preclinical analysis. J Chromatogr A 2012;1251:63–73.

[16] Cao J, Gonzalez-Covarrubias V, Straubinger RM, Wang H, Duan X, Yu H, Qu J, Blanco JG. A rapid, reproducible, on-the-fly orthogonal array optimization method for targeted protein quantification by LC/MS and its application for accurate and sensitive quantification of carbonyl reductases in human liver. Anal Chem 2010;82 (7):2680–2689.

[17] DeSilva B, Smith W, Weiner R, Kelley M, Smolec J, Lee B, Khan M, Tacey R, Hill H, Celniker A. Recommendations for the bioanalytical method validation of ligand-binding assays to support pharmacokinetic assessments of macromolecules. Pharm Res 2003;20 (11):1885–1900.

[18] Ezan E, Bitsch F. Critical comparison of MS and immunoassays for the bioanalysis of therapeutic antibodies. Bioanalysis 2009;1 (8):1375–1388.

[19] Ezan E, Dubois M, Becher F. Bioanalysis of recombinant proteins and antibodies by mass spectrometry. Analyst 2009;134 (5):825–834.

[20] Hagman C, Ricke D, Ewert S, Bek S, Falchetto R, Bitsch F. Absolute quantification of monoclonal antibodies in biofluids by liquid chromatography-tandem mass spectrometry. Anal Chem 2008;80 (4):1290–1296.

[21] Leary BA, Lawrence-Henderson R, Mallozzi C, Fernandez Ocaña M, Duriga N, O'Hara DM, Kavosi M, Qu Q, Joyce AP. Bioanalytical platform comparison using a generic human IgG PK assay format. J Immunol Methods 2013;397 (1–2):28–36.

[22] van den Broek I, Niessen WM, van Dongen WD. Bioanalytical LC-MS/MS of protein-based biopharmaceuticals. J Chromatogr B Analyt Technol Biomed Life Sci 2013;929:161–179.

[23] O'Hara DM, Theobald V, Egan AC, Usansky J, Krishna M, TerWee J, Maia M, Spriggs FP, Kenney J, Safavi A, Keefe J. Ligand binding assays in the 21st century laboratory: recommendations for characterization and supply of critical reagents. AAPS J 2012;14 (2):316–328.

[24] Lee JW, Kelley M. Quality assessment of bioanalytical quantification of monoclonal antibody drugs. Ther Deliv 2011;2 (3):383–396.

[25] King LE, Farley E, Imazato M, Keefe J, Khan M, Ma M, Pihl KS, Sriraman P. Ligand binding assay critical reagents and their stability: recommendations and best practices from the Global Bioanalysis Consortium Harmonization Team. AAPS J 2014;16 (3):504–515.

[26] Liu G, Ji QC, Dodge R, Sun H, Shuster D, Zhao Q, Arnold M. Liquid chromatography coupled with tandem mass spectrometry for the bioanalysis of proteins in drug development: practical considerations in assay development and validation. J Chromatogr A 2013;1284:155–162.

[27] King LE, Leung S, Ray C. Discovery fit-for-purpose ligand-binding PK assays: what's really important? Bioanalysis 2013;5 (12):1463–1466.

[28] Clark TH, King LE. Discovery biotherapeutics bioanalysis: challenges and possible solutions. Bioanalysis 2012;4 (9):1003–1007.

[29] Giovanella L, Ceriani L, Lumastro C, Ghelfo A. False-positive serum chromogranin A assay due to heterophile antibody interference. Clin Chim Acta 2007;379 (1–2):171–172.

[30] Willman JH, Martins TB, Jaskowski TD, Hill HR, Litwin CM. Heterophile antibodies to bovine and caprine proteins

causing false-positive human immunodeficiency virus type 1 and other enzyme-linked immunosorbent assay results. Clin Diagn Lab Immunol 1999;6 (4):615–616.

[31] Geist BJ, Davis D, McIntosh T, Yang TY, Goldberg K, Han C, Pendley C, Davis HM. A novel approach for the simultaneous quantification of a therapeutic monoclonal antibody in serum produced from two distinct host cell lines. MAbs 2013;5 (1):150–161.

[32] Li H, Ortiz R, Tran LT, Salimi-Moosavi H, Malella J, James CA, Lee JW. Simultaneous analysis of multiple monoclonal antibody biotherapeutics by LC-MS/MS method in rat plasma following cassette-dosing. AAPS J 2013;15 (2):337–346.

[33] Pendley C, Shankar G. Bioanalytical interferences in immunoassays for antibody biotherapeutics. Bioanalysis 2011;3 (7):703–706.

[34] Shi T, Fillmore TL, Sun X, Zhao R, Schepmoes AA, Hossain M, Xie F, Wu S, Kim JS, Jones N, Moore RJ, Pasa-Tolić L, Kagan J, Rodland KD, Liu T, Tang K, Camp DG 2nd, Smith RD, Qian WJ. Antibody-free, targeted mass-spectrometric approach for quantification of proteins at low picogram per milliliter levels in human plasma/serum. Proc Natl Acad Sci U S A 2012;109 (38):15395–15400.

[35] Roman J, Qiu J, Dornadula G, Hamuro L, Bakhtiar R, Verch T. Application of miniaturized immunoassays to discovery pharmacokinetic bioanalysis. J Pharmacol Toxicol Methods 2011;63 (3):227–235.

[36] Yang J, Ng C, Lowman H, Chestnut R, Schofield C, Sandlund B, Ernst J, Bennett G, Quarmby V. Quantitative determination of humanized monoclonal antibody rhuMAb2H7 in cynomolgus monkey serum using a Generic Immunoglobulin Pharmacokinetic (GRIP) assay. J Immunol Methods 2008;335 (1–2):8–20.

[37] Stubenrauch K, Wessels U, Lenz H. Evaluation of an immunoassay for human-specific quantitation of therapeutic antibodies in serum samples from non-human primates. J Pharm Biomed Anal 2009;49 (4):1003–1008.

[38] Blasco H, Lalmanach G, Godat E, Maurel MC, Canepa S, Belghazi M, Paintaud G, Degenne D, Chatelut E, Cartron G, Le Guellec C. Evaluation of a peptide ELISA for the detection of rituximab in serum. J Immunol Methods 2007;325 (1–2):127–139.

[39] Cham Mead JA, Bianco L, Bessant C. Free computational resources for designing selected reaction monitoring transitions. Proteomics 2010;10 (6):1106–1126.

[40] Halquist MS, Thomas Karnes H. Quantitative liquid chromatography tandem mass spectrometry analysis of macromolecules using signature peptides in biological fluids. Biomed Chromatogr 2011;25 (1–2):47–58.

[41] Rauh M. LC-MS/MS for protein and peptide quantification in clinical chemistry. J Chromatogr B Analyt Technol Biomed Life Sci 2012;883–884:59–67.

[42] Stergachis AB, MacLean B, Lee K, Stamatoyannopoulos JA, MacCoss MJ. Rapid empirical discovery of optimal peptides for targeted proteomics. Nat Methods 2011;8 (12):1041–1043.

[43] An B, Zhang M, Qu J. Towards sensitive and accurate analysis of antibody biotherapeutics by LC/MS. Drug Metab Dispos 2014;42(11):1858–1866. doi:10.1124/dmd.114.058917.

[44] Keshishian H, Addona T, Burgess M, Kuhn E, Carr SA. Quantitative, multiplexed assays for low abundance proteins in plasma by targeted mass spectrometry and stable isotope dilution. Mol Cell Proteomics 2007;6 (12):2212–2229.

[45] Picotti P, Bodenmiller B, Mueller LN, Domon B, Aebersold R. Full dynamic range proteome analysis of S. cerevisiae by targeted proteomics. Cell 2009;138 (4):795–806.

[46] Furlong MT, Ouyang Z, Wu S, Tamura J, Olah T, Tymiak A, Jemal M. A universal surrogate peptide to enable LC-MS/MS bioanalysis of a diversity of human monoclonal antibody and human Fc-fusion protein drug candidates in pre-clinical animal studies. Biomed Chromatogr 2012;26 (8):1024–1032.

[47] Wild D. *The Immunoassay Handbook*. 3rd ed. Amsterdam; Boston (MA): Elsevier; 2005. p 26, 930 p.

[48] Pandya K, Ray CA, Brunner L, Wang J, Lee JW, DeSilva B. Strategies to minimize variability and bias associated with manual pipetting in ligand binding assays to assure data quality of protein therapeutic quantification. J Pharm Biomed Anal 2010;53 (3):623–630.

[49] Ray CA, Zhou L, Tsoi J, Uy L, Gu J, Malella J, Desimone D, Gunn H, Ma M, Lee J, DeSilva B. A strategy for improving comparability across sites for ligand binding assays measuring therapeutic proteins. J Pharm Biomed Anal 2010;53 (3):729–734.

[50] Yuan L, Arnold ME, Aubry AF, Ji QC. Simple and efficient digestion of a monoclonal antibody in serum using pellet digestion: comparison with traditional digestion methods in LC-MS/MS bioanalysis. Bioanalysis 2012;4 (24):2887–2896.

[51] Ouyang Z, Furlong MT, Wu S, Sleczka B, Tamura J, Wang H, Suchard S, Suri A, Olah T, Tymiak A, Jemal M. Pellet digestion: a simple and efficient sample preparation technique for LC-MS/MS quantification of large therapeutic proteins in plasma. Bioanalysis 2012;4 (1):17–28.

[52] Li H, Ortiz R, Tran L, Hall M, Spahr C, Walker K, Laudemann J, Miller S, Salimi-Moosavi H, Lee JW. General LC-MS/MS method approach to quantify therapeutic monoclonal antibodies using a common whole antibody internal standard with application to preclinical studies. Anal Chem 2012;84 (3):1267–1273.

[53] Manza LL, Stamer SL, Ham AJ, Codreanu SG, Liebler DC. Sample preparation and digestion for proteomic analyses using spin filters. Proteomics 2005;5 (7):1742–1745.

[54] Shevchenko A, Tomas H, Havlis J, Olsen JV, Mann M. In-gel digestion for mass spectrometric characterization of proteins and proteomes. Nat Protoc 2006;1 (6):2856–2860.

[55] Wisniewski JR, Zougman A, Nagaraj N, Mann M. Universal sample preparation method for proteome analysis. Nat Methods 2009;6 (5):359–362.

[56] Wisniewski JR, Ostasiewicz P, Mann M. High recovery FASP applied to the proteomic analysis of microdissected formalin fixed paraffin embedded cancer tissues retrieves known colon cancer markers. J Proteome Res 2011;10 (7):3040–3049.

[57] Hustoft HK, Reubsaet L, Greibrokk T, Lundanes E, Malerod H. Critical assessment of accelerating trypsination methods. J Pharm Biomed Anal 2011;56 (5):1069–1078.

[58] Tu C, Li J, Bu Y, Hangauer D, Qu J. An ion-current-based, comprehensive and reproducible proteomic strategy for comparative characterization of the cellular responses to novel anti-cancer agents in a prostate cell model. J Proteomics 2012;77:187–201.

[59] Tu C, Li J, Jiang X, Sheflin LG, Pfeffer BA, Behringer M, Fliesler SJ, Qu J. Ion-current-based proteomic profiling of the retina in a rat model of Smith-Lemli-Opitz syndrome. Mol Cell Proteomics 2013;12 (12):3583–3598.

[60] Tu C, Li J, Young R, Page BJ, Engler F, Halfon MS, Canty JM Jr, Qu J. Combinatorial peptide ligand library treatment followed by a dual-enzyme, dual-activation approach on a nanoflow liquid chromatography/orbitrap/electron transfer dissociation system for comprehensive analysis of swine plasma proteome. Anal Chem 2011;83 (12):4802–4813.

[61] Ocana MF, Neubert H. An immunoaffinity liquid chromatography-tandem mass spectrometry assay for the quantitation of matrix metalloproteinase 9 in mouse serum. Anal Biochem 2010;399 (2):202–210.

[62] Yang Z, Hayes M, Fang X, Daley MP, Ettenberg S, Tse FL. LC-MS/MS approach for quantification of therapeutic proteins in plasma using a protein internal standard and 2D-solid-phase extraction cleanup. Anal Chem 2007;79 (24):9294–9301.

[63] Callesen AK, Madsen JS, Vach W, Kruse TA, Mogensen O, Jensen ON. Serum protein profiling by solid phase extraction and mass spectrometry: a future diagnostics tool? Proteomics 2009;9 (6):1428–1441.

[64] Jiang H, Zeng J, Titsch C, Voronin K, Akinsanya B, Luo L, Shen H, Desai DD, Allentoff A, Aubry AF, Desilva BS, Arnold ME. Fully validated LC-MS/MS assay for the simultaneous quantitation of coadministered therapeutic antibodies in cynomolgus monkey serum. Anal Chem 2013;85 (20):9859–9867.

[65] Ray CA, Bowsher RR, Smith WC, Devanarayan V, Willey MB, Brandt JT, Dean RA. Development, validation, and implementation of a multiplex immunoassay for the simultaneous determination of five cytokines in human serum. J Pharm Biomed Anal 2005;36 (5):1037–1044.

[66] Toedter G, Hayden K, Wagner C, Brodmerkel C. Simultaneous detection of eight analytes in human serum by two commercially available platforms for multiplex cytokine analysis. Clin Vaccine Immunol 2008;15 (1):42–48.

[67] Yohrling J. Ligand-binding assays: risk of using a platform supported by a single vendor. Bioanalysis 2009;1 (3):629–636.

[68] Xu K, Liu L, Maia M, Li J, Lowe J, Song A, Kaur S. A multiplexed hybrid LC-MS/MS pharmacokinetic assay to measure two co-administered monoclonal antibodies in a clinical study. Bioanalysis 2014;6 (13):1781–1794.

[69] Wang Y, Qu Y, Bellows CL, Ahn JS, Burkey JL, Taylor SW. Simultaneous quantification of davalintide, a novel amylin-mimetic peptide, and its active metabolite in beagle and rat plasma by online SPE and LC-MS/MS. Bioanalysis 2012;4 (17):2141–2152.

[70] Lee JW, Kelley M, King LE, Yang J, Salimi-Moosavi H, Tang MT, Lu JF, Kamerud J, Ahene A, Myler H, Rogers C. Bioanalytical approaches to quantify "total" and "free" therapeutic antibodies and their targets: technical challenges and PK/PD applications over the course of drug development. AAPS J 2011;13 (1):99–110.

[71] Ahene AB. Application and interpretation of free and total drug measurements in the development of biologics. Bioanalysis 2011;3 (11):1287–1295.

[72] Wang SJ, Wu ST, Gokemeijer J, Fura A, Krishna M, Morin P, Chen G, Price K, Wang-Iverson D, Olah T, Weiner R, Tymiak A, Jemal M. Attribution of the discrepancy between ELISA and LC-MS/MS assay results of a PEGylated scaffold protein in post-dose monkey plasma samples due to the presence of anti-drug antibodies. Anal Bioanal Chem 2012;402 (3):1229–1239.

[73] Bronsema KJ, Bischoff R, van de Merbel NC. Internal standards in the quantitative determination of protein biopharmaceuticals using liquid chromatography coupled to mass spectrometry. J Chromatogr B Analyt Technol Biomed Life Sci 2012;893–894:1–14.

[74] Pritchard C, Quaglia M, Ashcroft AE, O'Connor G. Considering the advantages and pitfalls of the use of isotopically labeled protein standards for accurate protein quantification. Bioanalysis 2011;3 (24):2797–2802.

[75] Halquist MS, Karnes HT. Quantification of Alefacept, an immunosuppressive fusion protein in human plasma using a protein analogue internal standard, trypsin cleaved signature peptides and liquid chromatography tandem mass spectrometry. J Chromatogr B Analyt Technol Biomed Life Sci 2011;879 (11–12):789–798.

[76] Nouri-Nigjeh E, Zhang M, Ji T, Yu H, An B, Duan X, Balthasar J, Johnson RW, Qu J. Effects of calibration approaches on the accuracy for LC-MS targeted quantification of therapeutic protein. Anal Chem 2014;86 (7):3575–3584.

[77] Dubois M, Fenaille F, Clement G, Lechmann M, Tabet JC, Ezan E, Becher F. Immunopurification and mass spectrometric quantification of the active form of a chimeric therapeutic antibody in human serum. Anal Chem 2008;80 (5):1737–1745.

[78] Rasmussen M, Dahl M, Buus S, Djurisic S, Ohlsson J, Hviid TV. Evaluation of a competitive enzyme-linked immunosorbent assay for measurements of soluble HLA-G protein. Tissue Antigens 2014;84 (2):206–215.

[79] Kim J, Kim J, Rho TH, Lee JH. Rapid chemiluminescent sandwich enzyme immunoassay capable of consecutively quantifying multiple tumor markers in a sample. Talanta 2014;129:106–112.

[80] Dams R, Huestis MA, Lambert WE, Murphy CM. Matrix effect in bio-analysis of illicit drugs with LC-MS/MS: Influence of ionization type, sample preparation, and biofluid. J Am Soc Mass Spectrom 2003;14 (11):1290–1294.

[81] Chambers E, Wagrowski-Diehl DM, Lu Z, Mazzeo JR. Systematic and comprehensive strategy for reducing matrix effects in LC/MS/MS analyses. J Chromatogr B Analyt Technol Biomed Life Sci 2007;852 (1–2):22–34.

[82] Lin D, Alborn WE, Slebos RJ, Liebler DC. Comparison of protein immunoprecipitation-multiple reaction monitoring with ELISA for assay of biomarker candidates in plasma. J Proteome Res 2013;12 (12):5996–6003.

[83] Neubert H, Muirhead D, Kabir M, Grace C, Cleton A, Arends R. Sequential protein and peptide immunoaffinity capture for mass spectrometry-based quantification of total human beta-nerve growth factor. Anal Chem 2013;85 (3):1719–1726.

[84] Ocaña MF, Neubert H. An immunoaffinity liquid chromatography–tandem mass spectrometry assay for the quantitation of matrix metalloproteinase 9 in mouse serum. Anal Biochem 2010;399 (2):202–210.

[85] Palandra J, Finelli A, Zhu M, Masferrer J, Neubert H. Highly specific and sensitive measurements of human and monkey interleukin 21 using sequential protein and tryptic peptide immunoaffinity LC-MS/MS. Anal Chem 2013;85 (11): 5522–5529.

[86] Anderson NL, Anderson NG, Haines LR, Hardie DB, Olafson RW, Pearson TW. Mass spectrometric quantitation of peptides and proteins using Stable Isotope Standards and Capture by Anti-Peptide Antibodies (SISCAPA). J Proteome Res 2004;3 (2):235–244.

[87] Fernandez Ocana M, James IT, Kabir M, Grace C, Yuan G, Martin SW, Neubert H. Clinical pharmacokinetic assessment of an anti-MAdCAM monoclonal antibody therapeutic by LC-MS/MS. Anal Chem 2012;84 (14):5959–5967.

[88] Keshishian H, Addona T, Burgess M, Mani DR, Shi X, Kuhn E, Sabatine MS, Gerszten RE, Carr SA. Quantification of cardiovascular biomarkers in patient plasma by targeted mass spectrometry and stable isotope dilution. Mol Cell Proteomics 2009;8 (10):2339–2349.

[89] Shi T, Fillmore TL, Gao Y, Zhao R, He J, Schepmoes AA, Nicora CD, Wu C, Chambers JL, Moore RJ, Kagan J, Srivastava S, Liu AY, Rodland KD, Liu T, Camp DG 2nd, Smith RD, Qian WJ. Long-gradient separations coupled with selected reaction monitoring for highly sensitive, large scale targeted protein quantification in a single analysis. Anal Chem 2013;85 (19):9196–9203.

[90] Joyce AP, Wang M, Lawrence-Henderson R, Filliettaz C, Leung SS, Xu X, O'Hara DM. One mouse, one pharmacokinetic profile: quantitative whole blood serial sampling for biotherapeutics. Pharm Res 2014;31 (7):1823–1833.

[91] Bodenlenz M, Höfferer C, Magnes C, Schaller-Ammann R, Schaupp L, Feichtner F, Ratzer M, Pickl K, Sinner F, Wutte A, Korsatko S, Köhler G, Legat FJ, Benfeldt EM, Wright AM, Neddermann D, Jung T, Pieber TR. Dermal PK/PD of a lipophilic topical drug in psoriatic patients by continuous intradermal membrane-free sampling. Eur J Pharm Biopharm 2012;81 (3):635–641.

[92] Cohen BD, Baker DA, Soderstrom C, Tkalcevic G, Rossi AM, Miller PE, Tengowski MW, Wang F, Gualberto A, Beebe JS, Moyer JD. Combination therapy enhances the inhibition of tumor growth with the fully human anti-type 1 insulin-like growth factor receptor monoclonal antibody CP-751,871. Clin Cancer Res 2005;11 (5):2063–2073.

[93] Acerini CL, Clayton KL, Hintz R, Baker B, Watts A, Holly JM, Dunger DB. Serum insulin-like growth factor II levels in normal adolescents and those with insulin dependent diabetes mellitus. Clin Endocrinol (Oxf) 1996;45 (1):13–19.

[94] Ricart AD, Tolcher AW, Liu G, Holen K, Schwartz G, Albertini M, Weiss G, Yazji S, Ng C, Wilding G. Volociximab, a chimeric monoclonal antibody that specifically binds alpha-5beta1 integrin: a phase I, pharmacokinetic, and biological correlative study. Clin Cancer Res 2008;14 (23):7924–7929.

[95] Dowell JA, Korth-Bradley J, Liu H, King SP, Berger MS. Pharmacokinetics of gemtuzumab ozogamicin, an antibody-targeted chemotherapy agent for the treatment of patients with acute myeloid leukemia in first relapse. J Clin Pharmacol 2001;41 (11):1206–1214.

[96] Agus DB, Gordon MS, Taylor C, Natale RB, Karlan B, Mendelson DS, Press MF, Allison DE, Sliwkowski MX, Lieberman G, Kelsey SM, Fyfe G. Phase I clinical study of pertuzumab, a novel HER dimerization inhibitor, in patients with advanced cancer. J Clin Oncol 2005;23 (11):2534–2543.

[97] Stubenrauch K, Wessels U, Essig U, Kowalewsky F, Vogel R, Heinrich J. Characterization of murine anti-human Fab antibodies for use in an immunoassay for generic quantification of human Fab fragments in non-human serum samples including cynomolgus monkey samples. J Pharm Biomed Anal 2013;72:208–215.

[98] Heudi O, Barteau S, Zimmer D, Schmidt J, Bill K, Lehmann N, Bauer C, Kretz O. Towards absolute quantification of therapeutic monoclonal antibody in serum by LC-MS/MS using isotope-labeled antibody standard and protein cleavage isotope dilution mass spectrometry. Anal Chem 2008;80 (11):4200–4207.

[99] Lu Q, Zheng X, McIntosh T, Davis H, Nemeth JF, Pendley C, Wu SL, Hancock WS. Development of different analysis platforms with LC-MS for pharmacokinetic studies of protein drugs. Anal Chem 2009;81 (21):8715–8723.

[100] Shafat I, Zcharia E, Nisman B, Nadir Y, Nakhoul F, Vlodavsky I, Ilan N. An ELISA method for the detection and quantification of human heparanase. Biochem Biophys Res Commun 2006;341 (4):958–963.

[101] de Witte H, Pappot H, Brünner N, Grøndahl-Hansen J, Hoyer-Hansen G, Behrendt N, Guldhammer-Skov B, Sweep F, Benraad T, Danø K. ELISA for complexes between urokinase-type plasminogen activator and its receptor in lung cancer tissue extracts. Int J Cancer 1997;72 (3):416–423.

[102] Sweep CG, Geurts-Moespot J, Grebenschikov N, de Witte JH, Heuvel JJ, Schmitt M, Duffy MJ, Jänicke F, Kramer MD, Foekens JA, Brünner N, Brugal G, Pedersen AN, Benraad TJ. External quality assessment of trans-European multicentre antigen determinations (enzyme-linked immunosorbent assay) of urokinase-type plasminogen activator (uPA) and its type 1 inhibitor (PAI-1) in human breast cancer tissue extracts. Br J Cancer 1998;78 (11):1434–1441.

[103] Pedersen AN, Høyer-Hansen G, Brünner N, Clark GM, Larsen B, Poulsen HS, Danø K, Stephens RW. The complex between urokinase plasminogen activator and its type-1 inhibitor in breast cancer extracts quantitated by ELISA. J Immunol Methods 1997;203 (1):55–65.

[104] Sleczka BG, Mehl JT, Shuster DJ, Lewis KE, Moore R, Vuppugalla R, Rajendran S, D'Arienzo CJ, Olah TV. Quantification of human mAbs in mouse tissues using generic affinity enrichment procedures and LC-MS detection. Bioanalysis 2014;6 (13):1795–1811.

[105] Andrade DC, Borges IC, Laitinen H, Ekström N, Adrian PV, Meinke A, Barral A, Nascimento-Carvalho CM, Käyhty H. A fluorescent multiplexed bead-based immunoassay (FMIA) for quantitation of IgG against Streptococcus pneumoniae, Haemophilus influenzae and Moraxella catarrhalis protein antigens. J Immunol Methods 2014;405:130–143.

[106] Boesch AW, Zhao Y, Landman AS, Garcia MR, Fahey JV, Wira CR, Ackerman ME. A multiplexed assay to detect

antimicrobial peptides in biological fluids and cell secretions. J Immunol Methods 2013;397 (1–2):71–76.

[107] Adamcova M, Ruzickova S, Simko F. Multiplexed immunoassays for simultaneous quantification of cardiovascular biomarkers in the model of H(G)-nitro-L-arginine methylester (L-NAME) hypertensive rat. J Physiol Pharmacol 2013;64 (2):211–217.

[108] Tang D, Hou L, Niessner R, Xu M, Gao Z, Knopp D. Multiplexed electrochemical immunoassay of biomarkers using metal sulfide quantum dot nanolabels and trifunctionalized magnetic beads. Biosens Bioelectron 2013;46:37–43.

[109] Smith NF, Raynaud FI, Workman P. The application of cassette dosing for pharmacokinetic screening in small-molecule cancer drug discovery. Mol Cancer Ther 2007;6 (2):428–440.

[110] White RE, Manitpisitkul P. Pharmacokinetic theory of cassette dosing in drug discovery screening. Drug Metab Dispos 2001;29 (7):957–966.

[111] Zhou H, Mascelli MA. Mechanisms of monoclonal antibody-drug interactions. Annu Rev Pharmacol Toxicol 2011;51:359–372.

[112] Spencer S, Bethea D, Raju TS, Giles-Komar J, Feng Y. Solubility evaluation of murine hybridoma antibodies. MAbs 2012;4 (3):319–325.

[113] Dani B, Platz R, Tzannis ST. High concentration formulation feasibility of human immunoglubulin G for subcutaneous administration. J Pharm Sci 2007;96 (6):1504–1517.

[114] Ducry L. *Antibody-drug conjugates*. In: *Methods in Molecular Biology*. New York: Humana Press; 2013. p 1 online resource (x, 317 pages).

[115] Kaur S, Xu K, Saad OM, Dere RC, Carrasco-Triguero M. Bioanalytical assay strategies for the development of antibody-drug conjugate biotherapeutics. Bioanalysis 2013;5 (2):201–226.

[116] Kozak KR, Tsai SP, Fourie-O'Donohue A, dela Cruz Chuh J, Roth L, Cook R, Chan E, Chan P, Darwish M, Ohri R, Raab H, Zhang C, Lin K, Wong WL. Total antibody quantification for MMAE-conjugated antibody-drug conjugates: impact of assay format and reagents. Bioconjug Chem 2013;24 (5):772–779.

[117] Xu K, Liu L, Saad OM, Baudys J, Williams L, Leipold D, Shen B, Raab H, Junutula JR, Kim A, Kaur S. Characterization of intact antibody-drug conjugates from plasma/serum *in vivo* by affinity capture capillary liquid chromatography-mass spectrometry. Anal Biochem 2011; 412 (1):56–66.

[118] Valliere-Douglass JF, McFee WA, Salas-Solano O. Native intact mass determination of antibodies conjugated with monomethyl Auristatin E and F at interchain cysteine residues. Anal Chem 2012;84 (6):2843–2849.

[119] Chen J, Yin S, Wu Y, Ouyang J. Development of a native nanoelectrospray mass spectrometry method for determination of the drug-to-antibody ratio of antibody-drug conjugates. Anal Chem 2013;85 (3):1699–1704.

[120] de Dios K, Manibusan A, Marsden R, Pinkstaff J. Comparison of bioanalytical methods for the quantitation of PEGylated human insulin. J Immunol Methods 2013;396 (1–2):1–7.

[121] Lee JW. ADME of monoclonal antibody biotherapeutics: knowledge gaps and emerging tools. Bioanalysis 2013;5 (16):2003–2014.

22

BIOSIMILAR DEVELOPMENT: NONCLINICAL AND CLINICAL STRATEGIES AND CHALLENGES WITH A FOCUS ON THE ROLE OF PK/PD ASSESSMENTS

SUSAN HURST[1] AND DONGHUA YIN[2]

[1] *Pfizer Worldwide Research and Development, Groton, CT, USA*

[2] *Pfizer Worldwide Research and Development, San Diego, CA, USA*

22.1 INTRODUCTION

Biosimilars are biotherapeutic products that have been developed to be highly similar, as opposed to identical, to a licensed biotherapeutic product (the reference or innovator product). The European Medicines Agency (EMA), U.S. Food and Drug Administration (FDA), and World Health Organization (WHO) generally have similar requirements for the demonstration of biosimilarity using stepwise approaches to develop the evidence. Biosimilar approval is granted based on the totality of the evidence. Biosimilars do not require demonstration of a role of the target or identification of a potential mechanism of action because these, by definition, are the same as the innovator biotherapeutic agent. A potential biosimilar is filed with an independent application to regulatory authorities by the pharmaceutical company developing the proposed biosimilar that if approved would not become active until the marketed innovator pharmaceutical company's patents and market exclusivity periods have expired for the original marketed product.

This process may seem similar to small molecule generics; however, unlike small molecule generics that are chemically identical to their original marketed counterparts, biotherapeutics are complex products that are produced in living systems such as bacteria, yeast, or mammalian cell lines [1, 2]. There are also multistep manufacturing processes involved in the production of a biotherapeutic product. For a biotherapeutic product produced in a cell line, this would include choice of the specific cell line, fermentation/culture/biosynthesis processes, purification steps, formulation choices, filling procedures, and packaging choices. These multifactorial steps may result in minor differences between the proposed biosimilar and the reference biotherapeutic product (such as glycosylation patterns); however, as long as these difference do not impact the overall quality, safety, and efficacy, the proposed biosimilar upon regulatory authority review may be approved as a biosimilar product [3–5].

In addition, these manufacturing processes for biotherapeutics are usually proprietary and the information around these processes may not be public knowledge and therefore would not be available to other pharmaceutical companies as they set up their manufacturing processes to make a biosimilar. Also, due to the length of time between the original product approval and the loss of market exclusivity, the manufacturing processes and analytical assays for the reference biotherapeutic product may potentially use older methodologies that would no longer be considered state of the art. Since a new potential biosimilar being developed would be required to utilize current technology and meet current standards, this would be an additional source of potential variability.

Other nomenclature used for biosimilars that may differ by country and company includes follow-on biologics, bioanalogs, subsequent-entry biologicals, and similar medicinal

ADME and Translational Pharmacokinetics/Pharmacodynamics of Therapeutic Proteins: Applications in Drug Discovery and Development, First Edition. Edited by Honghui Zhou and Frank-Peter Theil.

biological products. All of these names essentially describe a "copy" of an original innovator's product. This terminology does not include either "noncomparable" versions of existing products that have not been approved as biosimilars (i.e., intended copies) or "bio-better" versions that are deemed "better" or "improved" compared to an existing product.

As patent protection for several key biotherapeutic products has expired and with more set to expire in the future, there has been an increased interest in the development of biosimilar products. It has been estimated that approximately $60 billion worth of patented biotherapeutic products will go off-patent between 2012 and 2019 [6], with additional patents continuing to expire over time. The availability of lower priced quality biosimilars would potentially allow increased global access to biotherapeutic products in key therapeutic areas including cancer and inflammatory conditions, thus enhancing patient care.

The goal of this chapter is to introduce the topic of biosimilarity with a focus on nonclinical and clinical aspects that incorporate pharmacokinetic/pharmacodynamic (PK/PD) and immunogenicity considerations including regulatory requirements and study design considerations.

22.2 ASPECTS OF BIOSIMILARITY

The determination of biosimilarity is based on a head-to-head comparison of the proposed biosimilar to the biotherapeutic reference product in terms of quality, safety, and efficacy. These aspects are determined via a hierarchical stepwise process that includes analytical/biological studies, nonclinical *in vivo* studies, and clinical studies (see Fig. 22.1). The overall assessment of biosimilarity includes a "totality-of-data approach" with the nonclinical *in vivo* and clinical studies designed to support the evaluation of biosimilarity and to answer any residual questions based on the results of the *in vitro* analytical/biological studies.

Analytical/biological studies are the determination of the quality characteristics (also referred to as quality attributes) of the proposed biosimilar and the biotherapeutic reference product that have been deemed critical for the specific reference product to maintain similarity. Examples of analytical structural characterization include the primary structure (amino acid sequence); secondary, tertiary, and quaternary structures; posttranslational modifications (e.g., glycoforms); and potential variants (e.g., aggregation). Biological functional characterizations are also critical for comparing a proposed biosimilar to its reference product and may include *in vitro* potency characterization, target binding, binding to FcRn, and other functional assays such as antibody-dependent cellular cytotoxicity (ADCC), complement-dependent cytotoxicity (CDC), and complement activation. However, the specific functional characterization would need to be optimized to describe the critical characteristics of the specific reference product.

Since reference-product quality characteristics are not often publically available for marketed compounds, the biosimilar manufacturer has to obtain enough varied reference

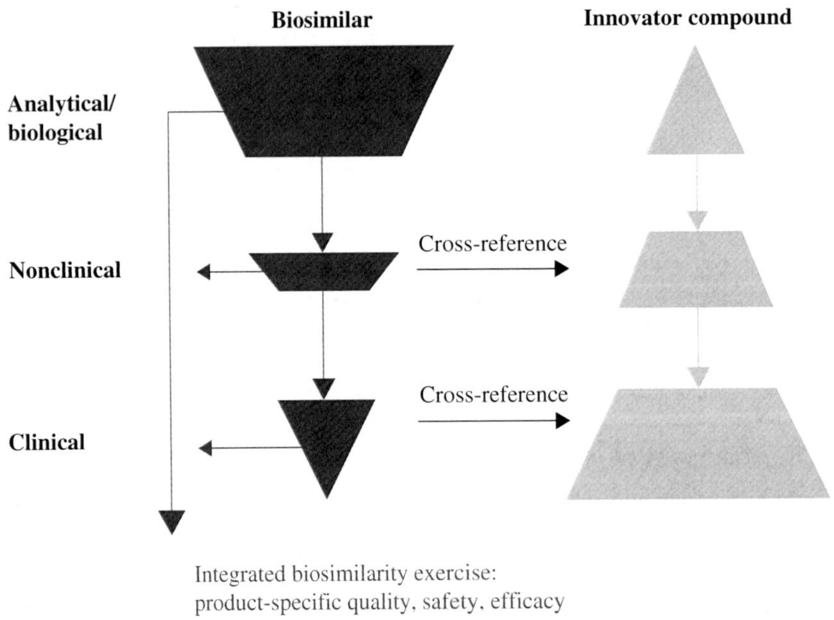

FIGURE 22.1 Biosimilarity hierarchical process.

product lots to determine which specific quality reference product attributes may be critical to maintain for the development of a biosimilar. Once critical quality attributes are identified, then ranges for these attributes across multiple reference product lots can be determined for comparison with the proposed biosimilar [7, 8]. Because reference products can have variations depending on their manufacturer source and time period, it is also important to ensure that the acquired reference product lots are chosen appropriately to evaluate the variation within the marketed product including global considerations. Thus, the overall analytical/biological assessments for a biosimilar usually exceed the analytical/biological assessments that would be conducted for a novel biotherapeutic product. This can be seen in Figure 22.1 where the analytical/biological section is much larger for the biosimilar than for a novel biotherapeutic but where the corresponding nonclinical and clinical sections are smaller for the biosimilar with some of the nonclinical *in vivo* and clinical data cross-referenced back to the original innovator product.

Manufacturing changes may also occur within a reference product (i.e., potential changes to cell line, manufacturing site, formulations, and/or dosage forms) and, for those changes, an analysis of the "comparability" of those lots pre and post change would be conducted by the marketing company. A comparability assessment is similar to biosimilarity in that the overarching goal is to ensure that manufacturing changes do not impact quality, safety, or efficacy; however, in the case of comparability, there is a wealth of data and experience over the entire life span of the reference product that is available for understanding the potential impact of any variations between the pre and post change reference product [9]. Thus, for a comparability assessment, it may be possible to evaluate the pre and post change reference product using only analytical and *in vitro* biological studies without the need for additional *in vivo* nonclinical or clinical assessments.

In contrast, biosimilar assessments would include an *in vivo* component (potentially nonclinical and clinical components) where these studies would be focused on determining the similarity between the proposed biosimilar and a reference product rather than characterizing the full safety and efficacy profile of a novel biotherapeutic, and thus the *in vivo* studies would not necessarily be as extensive as would be required for the development of a novel biotherapeutic.

The extent or requirement of an *in vivo* nonclinical or clinical component for biosimilar evaluations is based on regulatory requirements and optimized for the particular biosimilar based on the characteristics of the particular reference product and the outcomes (any potential identified uncertainties) of the analytical/biological biosimilar assessments.

22.3 BIOSIMILARS' REGULATORY/HISTORICAL PERSPECTIVE

22.3.1 European Union

The European Union (EU) first established a scientific regulatory framework for high quality biosimilars in 2003 and issued its first guideline for the approval of biosimilars in 2005. Since then, numerous guidance documents have been published or are in development from the EMA regarding biosimilars [3, 10]. The EMA guidance documents promote the use of a risk-based stepwise similarity evaluation in regard to quality, safety, and efficacy to determine whether a proposed biosimilar is similar to the approved reference product. These include an overarching biosimilar guidance, nonclinical and clinical development aspects for biosimilars, and specific product class guidelines in the areas of insulins, epoetins, filgrastims, growth hormones, alfa interferons, beta interferons, follitropins, low molecular-weight heparins, and monoclonal antibodies (mAbs) [3, 4, 10–13].

The earliest authorized biotherapeutic product using the biosimilar pathway in the EU was a growth hormone (somatropin) in 2006. Subsequently, more complex biotherapeutic products have been developed as biosimilars with the first mAb biosimilar (infliximab) approved in 2013. A current list of biosimilar products for the EU can be found in Table 22.1 [14].

22.3.2 EMA Nonclinical *In Vivo* Considerations

As was noted earlier regarding the hierarchal approach for biosimilarity, it is recommended in the EMA guidelines [12, 13] that the *in vitro* studies are conducted first and then a decision is made as to the extent of or whether nonclinical *in vivo* studies are necessary. If animal studies are needed, then the studies should be optimized to obtain the maximum amount of information using the fewest animals. The study design should be justified based on the available reference product information including its clinical use. This may include transgenic animals or transplant models if there is a scientific need for additional *in vivo* information.

It is noted that "When the [nonclinical] model allows, the PK and PD of the biosimilar and the reference medicinal product should be quantitatively compared, including, if feasible, a dose–concentration–response assessment including the intended exposure in humans."

Repeat-dose toxicology studies are not recommended to be conducted in nonhuman primates (NHPs) and toxicity studies are not recommended if there is no pharmacologically or toxicologically relevant nonclinical species. In addition, safety pharmacology, reproduction toxicology, and carcinogenicity studies would not be required for biosimilars (unless there is a specific scientific need).

Immunogenicity testing in nonclinical studies (while noted as not being predictive of humans) was recommended

TABLE 22.1 European Medicine Agency: Biosimilars

Active Substance	Common Name	Biotherapeutic Type	Marketing Authorization Holder	Authorization Year
Recombinant human interferon alfa-2a	Interferon alfa-2a	Interferon	BioPartners GmbH	Refused
Somatropin	Somatropin	Growth hormone	Sandoz GmbH	2006
Somatropin	Somatropin	Growth hormone	BioPartners GmbH	2006 (status withdrawn)
Epoetin alfa	Epoetin alfa	Glycoprotein hormone	Medice Arzneimittel Pütter GmbH & Co. KG	2007
Epoetin alfa	Epoetin alfa	Glycoprotein hormone	Sandoz GmbH	2007
Epoetin alfa	Epoetin alfa	Glycoprotein hormone	Hexal AG	2007
Epoetin zeta	Epoetin zeta	Glycoprotein hormone	Hospira UK Limited	2007
Epoetin zeta	Epoetin zeta	Glycoprotein hormone	Stada Arzneimittel AG	2007
Filgrastim	Filgrastim	GCSF	AbZ-Pharma GmbH	2008
Filgrastim	Filgrastim	GCSF	Ratiopharm GmbH	2008
Filgrastim	Filgrastim	GCSF	Teva GmbH	2008
Filgrastim	Filgrastim	GCSF	Hexal AG	2009
Filgrastim	Filgrastim	GCSF	Sandoz GmbH	2009
Filgrastim	Filgrastim	GCSF	Hospira UK Limited	2010
Filgrastim	Filgrastim	GCSF	Apotex Europe BV	2013
Infliximab	Infliximab	Monoclonal antibody	Hospira UK Limited	2013
Follitropin alfa	Follitropin alfa	Follicle stimulating hormone	Teva Pharma B.V.	2013
Infliximab	Infliximab	Monoclonal antibody	Celltrion Healthcare Hungary Kft.	2013
Follitropin alfa	Follitropin alfa	Follicle stimulating hormone	Finox Biotech AG	2014

Table based on data obtained from the EMA website (accessed October 14, 2014) [14]. Biosimilar medicines authorized via the Agency.
GCSF, granulocyte colony-stimulating factor.

if needed to understand the PK/toxicokinetic data within the nonclinical *in vivo* study(s).

22.3.3 EMA Clinical Considerations (Related to PK/PD)

The clinical biosimilar studies are also a stepwise approach utilizing the analytical/biological data and nonclinical data (if conducted) to inform and optimize the clinical study designs.

The clinical biosimilar studies start with a PK comparison and incorporate PD if feasible followed by clinical efficacy and safety trial(s) or potentially confirmatory PK/PD studies [12, 13]. Comparative PK studies are considered an essential part of the biosimilar development program. The biosimilar comparability limits for the PK parameters need to be defined before the initiation of the study. The location and width of the confidence interval for comparison are also important in determining similarity.

It is recommended that PD markers be added to PK studies whenever possible and in certain circumstances where there is confidence in the predictability of the PD markers and their relationship with efficacy/safety, a single and multiple dose exposure-response study at two or more dose levels may be sufficient to waive a clinical efficacy safety study. However, this would need to be discussed in advance and agreed upon by the agency.

Immunogenicity is recommended to be assessed in the conduct of the clinical studies. Immunogenicity assays should be developed using state-of-the-art technology. It is recommended that the immunogenicity analytical assay should detect antidrug antibodies (ADAs) from both the reference product and the proposed biosimilar but at a minimum should be able to detect all antibodies against the proposed biosimilar. In addition, other endpoints of immunogenicity may be assessed including cross-reactivity, target epitopes, and neutralizing activity and interpreted as to their potential impact on the clinical efficacy and safety parameters.

While the proposed biosimilar should be similar to the reference product, it was noted that lower immunogenicity in the biosimilar "would not preclude approval as a biosimilar" [13]. However, it may increase the complexity of understanding the impact of the biosimilar on efficacy in the clinical efficacy study requiring subpopulations (with and without immunogenicity) to be analyzed.

22.3.4 United States

In 2009, the Biologics Price Competition and Innovation (BPCI) Act stated that a biotherapeutic product may be demonstrated to be "biosimilar" if data "among other things" show that the potential biosimilar product is "highly similar" to an already FDA-approved biotherapeutic product [15]. This abbreviated pathway (351(k) application) for licensure was signed into law in 2010 as part of the Patient Protection and Affordable Care Act that amended the Public Health Service Act (PHS) governing the licensure for FDA approval of a novel/innovator biotherapeutic product (351(a) application).

Two FDA draft guidance documents followed in 2012 that focused on Quality Considerations (analytical/biological considerations) and Scientific Considerations (analytical/biological, nonclinical, and clinical considerations) demonstrated biosimilarity to a reference protein product [5, 16]. The FDA guidance documents provide the quality considerations necessary for a biosimilar and discuss the risk-based "totality of evidence" approach that would be used to evaluate and demonstrate biosimilarity. In 2014, a draft guidance regarding clinical pharmacology data used to support the demonstration of biosimilarity was issued, which further elaborates on degrees of similarity as well as provides more detail around clinical study expectations [17].

The first biosimilar application that was accepted for FDA review under the BPCI Act (351(K)) was an application for filgrastim by Sandoz in 2014. This was followed by a filing of the first mAb (infliximab) biosimilar application to the FDA by Celltrion in 2014 [18]. Subsequently, filgrastim (Zarxio®) has been approved by the FDA as the first biosimilar product in the United States [19].

22.3.5 FDA Nonclinical *In Vivo* Considerations

Similar to the EMA, the FDA recommends a stepwise approach to biosimilar assessment, thus the nonclinical studies would be designed to address any remaining uncertainties regarding safety of the proposed biosimilar following the analytical/biological biosimilarity assessment and before initiating clinical studies in humans [16]. In general, the nonclinical *in vivo* study would be comparative in nature using both the proposed biosimilar and the reference product.

Nonclinical *in vivo* toxicity studies would not be considered useful in the case where there is no relevant nonclinical species that provides biological results that do not correspond to the human response. However, it is noted that it may be useful to use a nonrelevant nonclinical species (including rodents) to test a biosimilar product that has not been previously tested in human subjects by evaluating comparative PK and systemic tolerability.

The scope and extent of an animal toxicity study would depend on the available information regarding the reference product and if there were any differences observed between the reference product and the proposed biosimilar in the analytical/biological studies. If an animal toxicity study is conducted, the design should provide meaningful toxicological comparison between the proposed biosimilar and the reference product. The relatively small sample size and potential intraspecies variations should be kept in mind when interpreting the study results.

Single-dose PK/PD studies in a relevant nonclinical species may add to the overall biosimilar assessment. PK/PD may also be incorporated into any toxicity study, where appropriate. However, a nonclinical PK/PD study would not replace the need to conduct clinical PK and PD studies.

Nonclinical immunogenicity assessments, while not predictive of humans, are recognized as a possible marker for potential differences between the proposed biosimilar and the reference product if there are marked differences in the immunogenicity. These data may be useful to assist in the design of the clinical immunogenicity assessment.

22.3.6 FDA Clinical Considerations (Related to PK/PD)

A biosimilar clinical program must include a clinical study(s) that is sufficient to demonstrate that the proposed biosimilar is similar to the reference product in terms of safety, purity, and potency in one or more conditions for which the reference product is licensed and intended for use and that the proposed biosimilar is seeking a license. The clinical program must include a comparative PK or PD assessment, as well as assessments of effectiveness, safety, and immunogenicity. The scope and size of the clinical program will be dependent on the residual uncertainties around biosimilarity remaining after the analytical/biological studies (and possibly the nonclinical studies) and based on the available information on the reference product in regard to any safety risks or other safety and efficacy concerns [16].

If a relevant PD marker is available, it would be expected to be included in the biosimilarity assessment unless the biosimilar sponsor can scientifically justify that it is not necessary. The FDA recommends the selection of PD markers that are relevant to clinical outcomes that can be measured in a timely manner with appropriate precision, and has the sensitivity to detect clinically meaningful differences between the proposed biosimilar and the reference product. The biosimilar sponsor should also provide a justification for the selection of the PK and PD study population and should predefine and justify the criteria for PK and PD parameters utilized to establish biosimilarity. If a proposed biosimilar demonstrates similar effect compared to the reference product in regard to a PD measure that is known to be clinically related to safety and effectiveness, this information can be utilized to optimize the clinical safety/efficacy studies. It was noted "in certain circumstances, human PK and PD data may provide sufficient clinical data to support a demonstration of biosimilarity" [16].

The draft FDA guidance for Industry entitled Clinical Pharmacology Data to Support a Demonstration of Biosimilarity to a Reference Product was issued in May 2014 and contains detailed information on the recommendations for the design of clinical pharmacology studies [17], which will be discussed in Section 22.5. In addition to overall critical considerations for biosimilarity and study design considerations, the guidance also discusses how to define the appropriate PD-time profile, recommendations for statistical comparison of PK and PD results, and the utility of simulation tools in study design and data analysis.

22.3.7 The WHO and Other Global Markets

The WHO published a guidance in 2010 that shared similar scientific principles to the earlier issued guidance by the EMA [20]. Most recently, the WHO has issued draft guidance in December 2014 regarding Scientific Principles for Regulatory Risk Assessment of Biotherapeutic Products [21].

Multiple countries have guidance documents for biosimilars that have either been issued or are in draft form; these include but may not be limited to the EU, Turkey, Malaysia, Taiwan, S. Korea, Japan, Singapore, Canada, S. Africa, Brazil, Argentina, Mexico, Cuba, Ireland, India, Peru, the United States, Columbia, Jordan, Thailand, and China [22–33].

As can be expected, given the number of available guidance documents on biosimilarity, there is an overarching consistency in requirements for high quality well-designed studies; however, the details across the individual country guidelines may vary. In addition, specific requirements may depend on the characteristics of the reference product and available data on the particular biosimilar. In fact, some countries have issued specific requirement guidelines for individual product types (e.g., the EMA [10]) and even guidelines for specific biosimilars (e.g., Mexico (COFEPRIS http://www.cofepris.gob.mx/Paginas/Inicio.aspx) with specific draft guidance documents including those for etanercept, rituximab, and infliximab).

Thus, the regulatory environment for biosimilars is currently evolving and must be constantly monitored by the biosimilar manufacturers as they work to develop the proposed biosimilar for the global market. A cross-country comparison of regulatory and clinical considerations for biosimilar oncology drugs was recently published by Bennett et al. in December 2014 [34]. This included a comparison summary for recommendations from the EU, the United States, India, Latin America, South Korea, Australia, Canada, and Japan.

22.4 NONCLINICAL ASSESSMENTS IN THE DEVELOPMENT OF BIOSIMILARS

22.4.1 Biosimilars Nonclinical Development

Nonclinical *in vivo* studies are the bridge between the *in vitro* analytical/biological biosimilar studies and the clinical biosimilar studies. These studies will be the first time a proposed biosimilar has been dosed to an intact organism. An important purpose of the nonclinical *in vivo* studies is to evaluate any potential remaining uncertainty resulting from the *in vitro* analytical/biological biosimilar studies before advancing the proposed biosimilar into humans. In fact, the EMA has stated "if the biosimilar comparability exercise for the physicochemical and biological characteristics and the nonclinical *in vitro* studies are considered satisfactory and no issues are identified which would block direct entrance into humans, an *in vivo* animal study is usually not considered necessary" [12]. However, *in vivo* nonclinical studies are often considered necessary as part of a complete biosimilarity assessment by other global regulatory authorities.

Thus, the first step in understanding the potential requirements for *in vivo* nonclinical studies is to understand both the proposed clinical plan including timing for executing the subsequent clinical studies and the proposed global registration plan. Since global regulatory authorities are not always aligned on requirements for *in vivo* nonclinical studies, it is important to ensure that the regulatory requirements for the key global markets for the proposed biosimilar are fully understood so that the necessary nonclinical data are available when it is required. Also, since clinical data may be able to be submitted in place of nonclinical data dependent on agreement of the engaged regulatory authority, understanding when that data will be available for regulatory discussions is essential.

The next step after understanding the regulatory requirements for the class of the proposed biosimilar is to fully understand the specific characteristics and potential issues associated with the reference product. This would include any publically available data on the reference product (such as the U.S. Summary Basis of Approval, European Public Assessment Report (EPAR), the Canadian Product Monographs, and any published research papers or reviews) as well as the results of the biosimilar biopharmaceutical company's analytical/biological biosimilarity evaluation around quality attributes. This then extends into understanding any potential differences between the proposed biosimilar and the reference product in terms of its critical quality attributes and assessing whether any differences if present could potentially impact safety or efficacy. The FDA has stated that "the scope and extent of any animal toxicity studies will depend on the body of information available on the reference product, the proposed product, and the extent of known similarities or differences between the two" [16].

If conducted, *in vivo* toxicity studies should be designed to demonstrate that the biosimilar does not exhibit any biologically meaningful differences from the reference product and if relevant for the specific reference product that the biosimilar does not exhibit any new or unexpected toxicities. "The selection of dose, regimen, duration, and test species for these studies should provide a meaningful toxicological comparison between the two products" [16].

22.4.2 Designing the Nonclinical *In Vivo* Study

The goal of the nonclinical study design is to select the species, dose, regimen, length of study, and potential biomarker evaluation to provide the most meaningful comparison between the proposed biosimilar and the reference product while being animal sparing.

Current guidance documents do allow for situations where nonclinical studies are not necessary if the *in vitro* analytical/biological data result in no residual uncertainty when planning for the clinical study. However, a single-dose PK study in a single sex may be conducted to gather additional assurance that in an *in vivo* system no relevant differences are observed between the proposed biosimilar and the reference product especially if there were no relevant *in vivo* changes reported in the innovators toxicity studies with the reference product.

If a toxicity study is needed, a single-dose level may be appropriate, and the dose chosen should match the original innovator studies and/or an approved clinical dose. A saturating dose should be avoided to avoid saturating a response and thus potentially obscuring any potential differences between the biosimilar and the reference product. Since the goal of the study is to provide comparative data, a single-dose study may be sufficient and recovery data may not be needed or (complete resolution if a recovery phase is included). There are also some circumstances where a toxicity study may be solely conducted with the proposed biosimilar if scientifically justified with the appropriate regulatory authorities. For example, a safety study may be conducted solely with the proposed biosimilar if the analytical/biological data for the proposed biosimilar and reference product were deemed sufficient evidence of biosimilarity to continue to the clinical comparison; however, a small safety study with just the proposed biosimilar may be requested to provide assurance of safety.

Species selection is usually based on the findings observed and the species choice for the original reference product studies conducted by the innovator company. The species chosen should be one in which the reference product has either an established toxicological or PD profile. Also for critical targets or endpoints, the relevant animal species should ensure that the nonclinical findings are reflective of humans (i.e., if the biotherapeutic product is given to an appropriate animal species, the product should bind to the animal target with similar potency as it binds to the human target).

However, in some circumstances, there is no available animal species that can provide pharmacologically relevant data for the biotherapeutic product. In this situation, rodents (rats or mice) may be an option since if the studies are to be conducted in nonpharmacology-relevant species, the goal would be to reduce the use of NHP if possible. In regard to mAbs, rodents are considered an appropriate species because the rat neonatal Fc receptor can recognize the Fc fragment of human IgG [35].

Given the focus on optimizing the nonclinical studies while balancing the ethical considerations to minimize animal usage, the nonclinical studies are usually not powered for statistical significance. It is important to understand the inherent variability that may be expected based on the reference product data and whether immunogenicity may impact the ability to evaluate the collected data.

22.4.3 Designing the Nonclinical Study: Immunogenicity/Bioanalytical

The nonclinical immunogenicity assays for ADA and the bioanalytical drug assays should both be based on current state-of-the-art technology and validated using established industry standards under GLP conditions. The additional considerations for the biosimilar bioanalytical drug assay are presented in Section 22.5.

While immunogenicity in animals is not considered to be predictive for humans, the understanding of expected immunogenicity within the nonclinical study assists in both the study design (appropriate species, number of animals needed, and potential length of the study) and the evaluation of the data. Usually, the proposed biosimilar and the reference product are compared within the concentration–time window where there are no large impacts on the variability of the data due to immunogenicity that would potentially mask any actual difference between the two products. So, understanding the immunogenicity would include not only the extent of the induction of the immune response and thus the potential subsequent decrease in product concentrations but the timing (time of onset), potential transience, and the extent of the immunogenicity with the measured titer values and its impact on PK and/or PD. Since the intended purpose of immunogenicity testing in nonclinical studies is to assist in the interpretation of the nonclinical study, mechanistic immunogenicity assays or neutralizing assays would not be conducted unless scientifically justified.

Also, while the nonclinical immunogenicity is not considered predictive of the clinical immunogenicity, if there is marked difference between the proposed biosimilar product and the reference product in a nonclinical study, this difference would need to be understood in regard to any particular quality attributes that could be impacting this differentiation. If any quality attributes are identified that could be causing the differences in immunogenicity nonclinically, then the potential impact of those differences in the quality attributes on the clinical situation would need to be considered. It is also important to realize that immunogenicity data may differ from the historical innovator data even for the reference compound due to potential improvements in immunogenicity assay technology since the original innovator data were collected.

22.4.4 Designing the Nonclinical *In Vivo* Study—PK and PD Focus

Nonclinical *in vivo* studies can be utilized as part of the overall biosimilar assessment and they can also be used to specifically evaluate the importance of quality attribute differences. In a study evaluating a proposed biosimilar for recombinant human chorionic gonadotropin (rhCG; DA-3803), additional isoform fractions were also tested that evaluated the impact of the sialic acid content on both PK and PD [36].

The PK were evaluated following a single-dose 50-µg/kg (~1300 IU/kg) subcutaneous injection of either Ovidrel® (the reference product), DA-3803, the acidic isoform, or the basic isoform to male Sprague–Dawley rats. The mean sialic acid content of the reference product, DA-3803, the acidic isoform, and the basic isoform were 13.86, 13.79, 14.40, and 8.13 mol/mol rhCG, respectively. DA-3803 PK appeared similar to the reference product with mean parameter values of 13830.8 and 12472.7 mIU·h/mL for area under the concentration-time curve (AUC) from time zero extrapolated to infinite time ($AUC_{0-\infty}$), 689.4 and 713.8 mIU/mL for maximum drug concentration (C_{max}), 9.0 and 9.0 h for time to maximum concentration (T_{max}), and 6.1 and 5.3 h for time to half-life ($t_{½}$), respectively.

In contrast, the corresponding acidic isoforms PK parameter values were higher ($AUC_{0-\infty}$, C_{max}, T_{max}, and $t_{½}$) and basic isoform PK parameter values were lower ($AUC_{0-\infty}$, C_{max}, and T_{max} but $t_{½}$ was higher) than those observed for DA-3803 and the reference product. The differences were also marked in the comparison of the PD marker (seminal vesicle weight gain assay) based on mean potency (%) with mean potency (%) values for the reference product, DA-3803, the acidic isoform, and the basic isoform of 99.4, 98.9, 120.0, and 15.7, respectively. These results add to the similarity comparison of the proposed biosimilar DA-3803 to the reference product. These results also show sialic acid content has the potential to impact both PK and PD an *in vivo* species, thus indicating that sialic acid content is a critical quality attribute to be measured and controlled for during the manufacturing process.

PK and PD evaluations have also been conducted for mAbs in the evaluation of potential biosimilars for rituximab where immunogenicity had to be considered during the data analysis. For GP2013 (a proposed rituximab biosimilar), comparative single-dose and repeat-dose studies were conducted in cynomolgus monkeys with PK, PD, and immunogenicity endpoints evaluated [7]. The PK analyses for both the single-dose (5 mg/kg) and the repeat-dose (20 and 100 mg/kg) studies following intravenous (IV) administration were limited due to the development of ADA for both GP2013 and MabThera® (rituximab) between Days 9 and 14 (single dose) and starting Day 14 (repeat dose). For the single-dose study, the concentration–time profiles over the first 9 days were similar for the two molecules and the AUC interval was noted to meet bioequivalence standards within 0.80–1.25 at 90% confidence interval. The C_{max} of GP2013 was approximately 13% lower than the reference product but this was thought to be due to heterogeneity in the early sample collection. For the repeat-dose study, the proposed biosimilar and reference product were comparable over the 14-day observation period and the corresponding AUC intervals met the bioequivalence acceptance criteria.

B-cell depletion was used as a PD biomarker (B-cell specific markers such as CD19, CD20, and CD40 (which indicates a defined subset of cellular surface receptors where CD stands for cluster of differentiation) in the two studies. For the single-dose studies, the depletion was comparable in the CD20 low subset up to Day 7 with a slightly lower 9% depletion for GP2013 when compared to MabThera. This small difference was lessened when the comparison was made using the CD20 high subset. For both subsets, the PD biomarker fell within the bioequivalence acceptance criteria. It should also be noted that mouse xenograft models were also evaluated as part of the nonclinical *in vivo* biosimilarity assessment [7].

Another proposed rituximab biosimilar (PF-05280586) was also compared to MabThera in single-dose and repeat-dose studies in male and female cynomolgus monkeys [37]. The single-dose (0, 2, 10, or 20 mg/kg) and repeat-dose (0 and 20 mg/kg) studies were given via IV administration and were evaluated for PK, PD, immunogenicity, and clinical assessments of tolerability. The repeat-dose study also included anatomical pathology. The animals were observed for 92 days in the single-dose study and 121 days in the repeat-dose study.

In the single-dose study, all animals (both PF-05280586 and the MabThera) were positive for ADAs by Day 29 (the first day tested postdose) and remained positive until Day 85 (the last time point evaluated). In the repeat-dose study, some animals for both PF-05280586 and the reference product were detected positive for ADAs by Day 22 (the first day tested post-dose). During the dosing phase test, circulating PF-05280586 or MabThera levels may have interfered with the ability to detect ADAs. During the recovery phase, ADAs were detected in three of six (50%) animals in each 20-mg/kg group (log_{10} titers of 1.87–6.13 and 2.01–5.70 for animals administered PF-05280586 and MabThera, respectively). The PK parameters (C_{max}, T_{max}, and AUC_{0-168h}) and concentration–time profiles were similar for both PF-05280586 and the reference product. In both studies, the B-cell depletion–repletion profiles of PF-05280586 were similar to those of the reference product via evaluation of multiple B-cell markers including CD19, CD20, and CD40. Both molecules were well tolerated at all doses, and in all the endpoints evaluated, PF-05280586 exhibited similarity to MabThera.

The two rituximab-proposed biosimilar case studies above emphasize the importance of monitoring immunogenicity in the nonclinical studies. The studies also highlight that, for a given reference product and based on the reference product characteristics and the regulatory guidance documents for biosimilars, different biosimilar programs will have common study design characteristics. However, there may be some differences based on the individual pharmaceutical company's interpretation of the reference product data, the biosimilar guidance documents, their interactions with the regulatory authorities, and potentially the outcomes of the respective analytical/biological data. Still the results from both case studies supported the continued development of the respective proposed biosimilars (GP2013 and PF-05280586) and added to their overall biosimilarity assessments.

22.4.5 Designing the Nonclinical *In Vivo* Study—No Relevant Nonclinical Species

PF-05280014 is being developed as a potential biosimilar to trastuzumab (Herceptin®), a humanized mAb that binds to the HER2 protein [38]. Nonclinical studies were designed to evaluate the similarity of PF-05280014 to trastuzumab-US (trastuzumab products sourced from the United States) and trastuzumab-EU (trastuzumab products sourced from the European Union) using *in vivo* PK, immunogenicity, and tolerability assessments. For trastuzumab, there was no pharmacologically relevant nonclinical species since trastuzumab does not recognize the mouse counterpart of human HER2, and unlike humans, NHP species do not overexpress p185HER2 or produce shed antigen [39]. Thus, for the trastuzumab nonclinical biosimilar study, mice were used to evaluate the nontarget-mediated and FcRn-dependent clearance of PF-05280014, trastuzumab-US, and trastuzumab-EU [40] and the overall tolerability profile.

CD-1 male mice were administered a single dose (0, 1, 10, or 100 mg/kg) of PF-05280014, trastuzumab-US, or trastuzumab-EU. The mice were monitored for drug concentrations, ADAs, and clinical signs and body weight changes over a 4-month period indicative of tolerability. Composite profiles for each mAb (five animals/group/time point) were generated for PK evaluation.

PF-05280014, trastuzumab-US, and trastuzumab-EU showed similar PK profiles over the first 24 h and the entire study time course (see Fig. 22.2). The overall incidence of ADA in mice was low, specifically, 8/74 (10.8%), 6/75 (8%), and 8/75 (10.7%) of the animals tested positive for the induction of antibodies against PF-05280014, trastuzumab-US, and trastuzumab-EU, respectively. The animals with

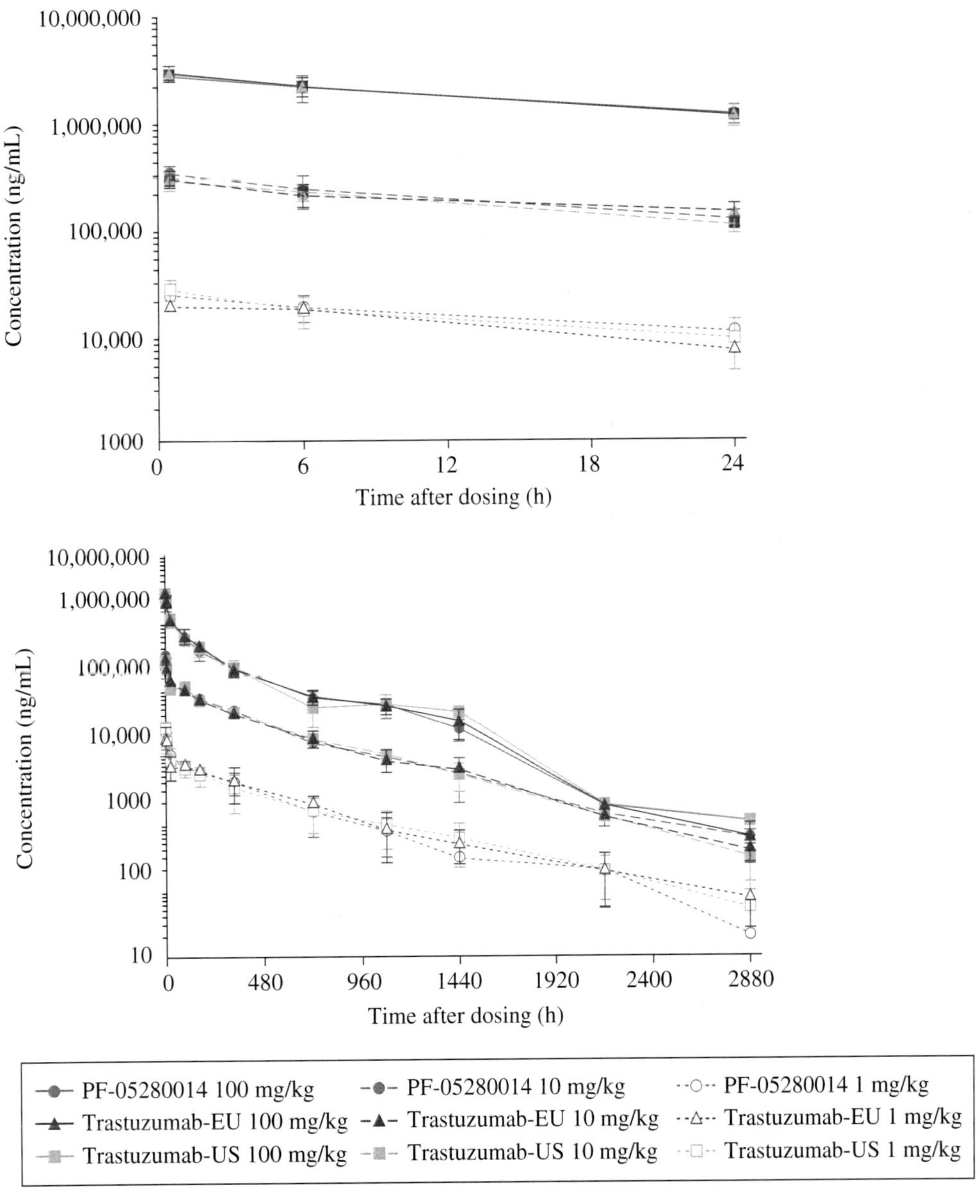

FIGURE 22.2 Mean (±SD) concentrations of PF-05280014, trastuzumab-US, and trastuzumab-EU for (a) up to 24 h and (b) over time for the whole study duration up to 2880 h after a single bolus intravenous injection of 1, 10, and 100 mg/kg in male CD-1 mice.

ADA in general had lower exposures than those without ADA; however, there was no impact on the overall PK profile evaluation because of the overall low incidence of ADA within each dose group.

For all animals (and the subset of ADA negative animals), the C_{max} and $AUC_{0-2880h}$ values in mice were similar across all three tested mAbs and increased with increasing dose. The *CL*, *V*ss, and $t_{½}$ values were similar across all three mAbs. The PK parameters for all the animals are shown in Table 22.2. The PK parameter ratios for all animals and ADA negative animals are illustrated in Table 22.3. PF-05280014, trastuzumab-US, and trastuzumab-EU were well tolerated during the 4-month observation period following a single dose of up to 100 mg/kg.

The results of these studies support the continued development of PF-05280014 as a proposed biosimilar for trastuzumab. For cases where there are no relevant nonclinical models, the type of study and the extent of the studies (i.e., single-dose PK/tolerability vs repeat-dose toxicity) are dependent on the discussions and feedback of the involved regulatory authorities.

22.5 CLINICAL PK AND PD ASSESSMENTS IN THE DEVELOPMENT OF BIOSIMILARS

22.5.1 Biosimilars Clinical Development

Clinical development of a proposed biosimilar product needs to address the residual uncertainty in biosimilarity determination remaining after extensive analytical, functional, and

TABLE 22.2 PF-05280014, Trastuzumab-US, and Trastuzumab-EU—Nonclinical PK Parameters (All Mice)

PK Parameters	Dose Level (mg/kg)	PF-05280014	Trastuzumab-EU	Trastuzumab-US
C_{max} ± SD (μg/mL)	1	22.8 ± 1.90	18.6 ± 8.55	26.3 ± 5.83
	10	318 ± 49.4[a]	281 ± 49.9	269 ± 52.7
	100	2,520 ± 219	2,700 ± 450	2,620 ± 332
$AUC_{0-2880h}$[b] ± SEM (μg h/mL)	1	4,200 ± 287	4,590 ± 337	4,050 ± 334
	10	51,400 ± 1560[a]	51,500 ± 1370	49,800 ± 1430
	100	285,000 ± 11,000	298,000 ± 10,300	289,000 ± 12,500
CL (mL/h/kg)	1	0.237	0.215	0.245
	10	0.193[a]	0.193	0.200
	100	0.350	0.335	0.346
V_{ss} (mL/kg)	1	104	113	129
	10	84.9[a]	85.4	86.7
	100	120	116	130
$t_{1/2}$ (h)	1	380	536	416
	10	440[a]	392	352
	100	309	280	320

SD, standard deviation and SEM, standard error of the mean.
[a] Nominal dose (actual dose 11.4 mg/kg).
[b] The percent extrapolation from the $AUC_{0-2880h}$ to $AUC_{0-\infty}$ ranged from 0.081% to 1.28%.

TABLE 22.3 Nonclinical Comparison of Pharmacokinetic Exposure Parameters between Test and Reference Products

			Ratio % (All Mice)		Ratio % (ADA Negative Mice)	
Test	Reference	Dose (mg/kg)	C_{max}	$AUC_{0-2880h}$	C_{max}	$AUC_{0-2880h}$
PF-05280014	Trastuzumab-EU	1	123	91.5	123	98.0
		10[a]	113	100	113	101
		100	93.0	96.0	93.0	96.0
PF-05280014	Trastuzumab-US	1	87.0	104	87.0	108
		10[a]	118	103	118	104
		100	96.0	99.0	96.0	99.0
Trastuzumab-US	Trastuzumab-EU	1	141	88.0	141	91.0
		10	96.0	97.0	96.0	97.0
		100	97.0	97.0	97.0	97.0

[a] PF-05280014 nominal dose (actual dose 11.4 mg/kg).

animal studies. The clinical program is typically composed of comparative PK or PD studies in healthy subjects or a patient population and one or more effectiveness and safety study in target patient population(s). Collectively, these clinical studies should provide data demonstrating that there are no clinically meaningful differences in PK, PD, effectiveness, safety, and immunogenicity between the proposed biosimilar product and a reference product [3, 12, 13, 16, 17].

The biosimilar clinical development normally starts with a clinical pharmacology study comparing the PK of the proposed biosimilar product to a reference product. The primary objective of this study is to demonstrate PK similarity between the proposed biosimilar product and the reference product using key PK exposure parameters and predefined acceptance criteria. Secondary objectives of the study include assessments of safety and immunogenicity of the proposed biosimilar product. When clinically relevant PD markers related to the mechanism of action or therapeutic outcome are available and appropriate for measurement in the chosen study population, a comparative PD assessment can also be integrated into the study [12, 13, 17].

Establishing PK and PD similarity in the early clinical pharmacology study can form the basis to design subsequent clinical studies in a selected and targeted manner. In certain circumstances, demonstration of PK and PD similarity may be accepted as sufficient clinical evidence to support biosimilarity determination [12, 13]. For example, clinical evidence to support the authorization of Zarzio®, a biosimilar filgrastim, by EMA in 2009 was based on four comparative PK/PD studies in a total of 146 healthy subjects, which demonstrated PK similarity as well as similar increases in blood neutrophil counts between the proposed biosimilar product and the originator product, Neupogen® [41]. In many cases, established PD markers with clinical relevance are not available as surrogates for effectiveness comparison; further comparative assessments of effectiveness and safety in at least one patient population become necessary.

22.5.2 Bioanalytical Assays for Biosimilars PK and PD Investigations

Drug concentration assays to be used in biosimilars PK studies should be designed in a way to allow sensitive detection of potential differences in PK between the proposed biosimilar and the originator products. Ligand-binding assays based on the interaction of drug molecules with the drug target or an antibody are most commonly used. The drug concentration assays should be validated according to relevant regulatory guidance documents and industry best practices [42–45].

A unique challenge in developing a drug concentration assay for biosimilar PK studies is that the assay must demonstrate comparable performance characteristics in analyzing both the proposed biosimilar and the originator drug molecules [46–48]. It is generally accepted that, whenever possible, a single assay with the use of a common calibration standard should be developed for quantifying both the proposed biosimilar and the originator drug concentrations. In the absence of an international public reference standard, the drug concentration assay may use the originator product or the proposed biosimilar product as the common calibration standard. The bioanalytical equivalence of the assay for the proposed biosimilar and the originator molecules may be established in multiple steps. In the assay development stage, standard curves prepared using both the proposed biosimilar and the originator products can be compared over the full calibration range. In the assay validation stage, the accuracy, precision, and selectivity (matrix interference) of the assay may be evaluated for full sets of quality control samples prepared from both the proposed biosimilar and originator products, using the common calibration standard. A systematic difference of the assay in quantifying the proposed biosimilar and the originator products should be of concern, as the bioanalytical difference could affect the capability to compare intrinsic PK properties between the products in a PK study.

While the assay platform may vary depending on the specific PD marker of interest, development of a PD marker assay should follow similar general considerations for drug concentration assays, in that the PD marker assay needs to be accurate, precise, specific, and reproducible. Assay-related variability should be within an acceptance limit to sensitively discern the difference in PD marker responses by the proposed biosimilar and the originator products.

22.5.3 Design Considerations for Phase I PK and PD Similarity Studies

The PK and PD similarity studies should be designed based on prior knowledge of the PK and PD properties, as well as safety and immunogenicity profiles of the originator product, with the ultimate goal to sensitively and efficiently detect any difference in PK and PD between the products.

For therapeutic proteins with a relatively long disposition half-life, a parallel group design is usually needed. A crossover design may be considered for products with a relatively short half-life and minimal concern for increased immune response following repeated-dose administration. An adequate wash-out of the drug molecules is needed for the crossover design.

The study population should be selected after considering prior clinical experience and safety information, sensitivity to detect the difference, and intersubject variability in candidate populations. When safety allows, healthy subjects are generally the preferred population since this population is associated with less complicating factors affecting PK (e.g., disease status and concomitant medications) and thus less variability in comparison to patient populations. Also, a healthy subject study can be conducted as a single-dose study, whereas a study in patients typically involves repeated therapy and thus is longer in duration. It should also be noted that healthy subjects may differ from a patient population in immunogenicity response after treatment due to potential differences in host immune status and the common use of immune-modulating concomitant medications in patients. When it is anticipated that the formation of ADA can alter PK, it may be desirable to conduct the PK similarity study in a less immunogenic population to minimize the interference of ADA on PK assessment. Within the study population, subjects with prior exposure to the same or similar biologic product need to be excluded from the study, since a prior exposure could alter the immune response to a repeated exposure and cause complications in PK and safety assessments.

The dose and route of administration used for the PK and PD similarity studies should be selected to allow the most sensitive detection of potential PK and PD differences. For PK studies, a low dose level may be more sensitive than a high dose for detecting difference in the target-mediated disposition that is involved in the distribution and elimination mechanisms of some products. For products that can be administered both IV and subcutaneously, PK studies using the subcutaneous route would allow comparing both absorption and elimination characteristics. The dose for the PD studies, on the other hand, should be at the steep part of the dose–response curve to ensure sensitive detection of potential difference in the PD response.

For a single-dose PK similarity study, key parameters for the PK similarity assessment should include the C_{max}, AUC from zero to the last time point with measurable concentration (AUC_{0-t}), and $AUC_{0-\infty}$. If the PK similarity is to be assessed using steady-state PK after repeated dosing, the key parameters should include C_{max}, AUC within one dose interval (AUC_{tau}), and the steady-state trough concentration (C_{trough}). PK similarity of the proposed biosimilar to the reference product is established through bioequivalence testing of the key PK parameters, using predefined acceptance criteria. The commonly accepted criteria are that the 90% confidence intervals of the test-to-reference ratio for the key exposure parameters need to be within 80–125%. When the acceptance criteria of PK similarity are not met, investigations will be needed to determine whether the observed difference is due to study design limitations or due to difference in intrinsic PK properties. Further assessment will be needed to address the residual uncertainty in PK and to determine the clinical meaningfulness of the observed difference.

22.5.4 PK Similarity Study of PF-05280014, a Proposed Biosimilar to Trastuzumab: An Example

PF-05280014 is a potential biosimilar of trastuzumab that is currently under development. As the first clinical study of PF-05280014, a randomized, three-arm PK study was conducted in healthy subjects to compare the PK of PF-05280014 and the originator trastuzumab-EU and trastuzumab-US [49]. The comparison between trastuzumab-EU and trastuzumab-US was to provide the PK bridging data so that a single reference product could be used in further Phase III studies to support the registration in both the EU and U.S. regions. The study enrolled a total of 105 subjects who were randomized to receive a single 6-mg/kg dose of PF-05280014, trastuzumab-EU, or trastuzumab-US via IV infusion. The PK, safety, and immunogenicity were assessed over a 10-week period after dose administration.

PK results showed that the individual subject serum concentration–time profiles of the three products were similar (Fig. 22.3) and that the median and mean profiles were superimposable. For the PK parameters, the 90% confidence intervals of the test-to-reference ratios of C_{max}, AUC_{0-t}, and $AUC_{0-\infty}$ were within the acceptance criteria of 80–125% for all three pair-wise comparisons of PF-05280014 to either trastuzumab-EU or trastuzumab-US and trastuzumab-EU to trastuzumab-US (Tables 22.4 and 22.5). The safety and immunogenicity profiles observed for the three products in this study were consistent with previous reports for trastuzumab.

22.5.5 Extrapolation of Clinical Data

The ability to extrapolate data from one clinical study to others for different therapeutics areas is a key issue in the development of a biosimilar product.

A biosimilar may apply for one or more therapeutic indications that have been previously granted to the innovator for the reference product. In some situations, a biosimilar may apply for therapeutic indications granted to the reference product without biosimilar clinical studies specifically designed for those therapeutic indications. In these cases, an extrapolation from the available biosimilar data to the additional therapeutic indication may be made based on comparative PK/PD data to bridge two or more indications. It may also be possible to extrapolate clinical data to other indications where it can be fully justified based on mechanism(s) of action; pathophysiological mechanism(s) of the disease(s) or conditions involved; safety profile in the respective conditions and/or populations; and clinical experience with the reference product. mAbs, for examples, are large, complex, and multifunctional, sometimes would render demonstration of biosimilarity without clinical testing beyond technical capabilities. There are substantial reasons (e.g., different immune status,

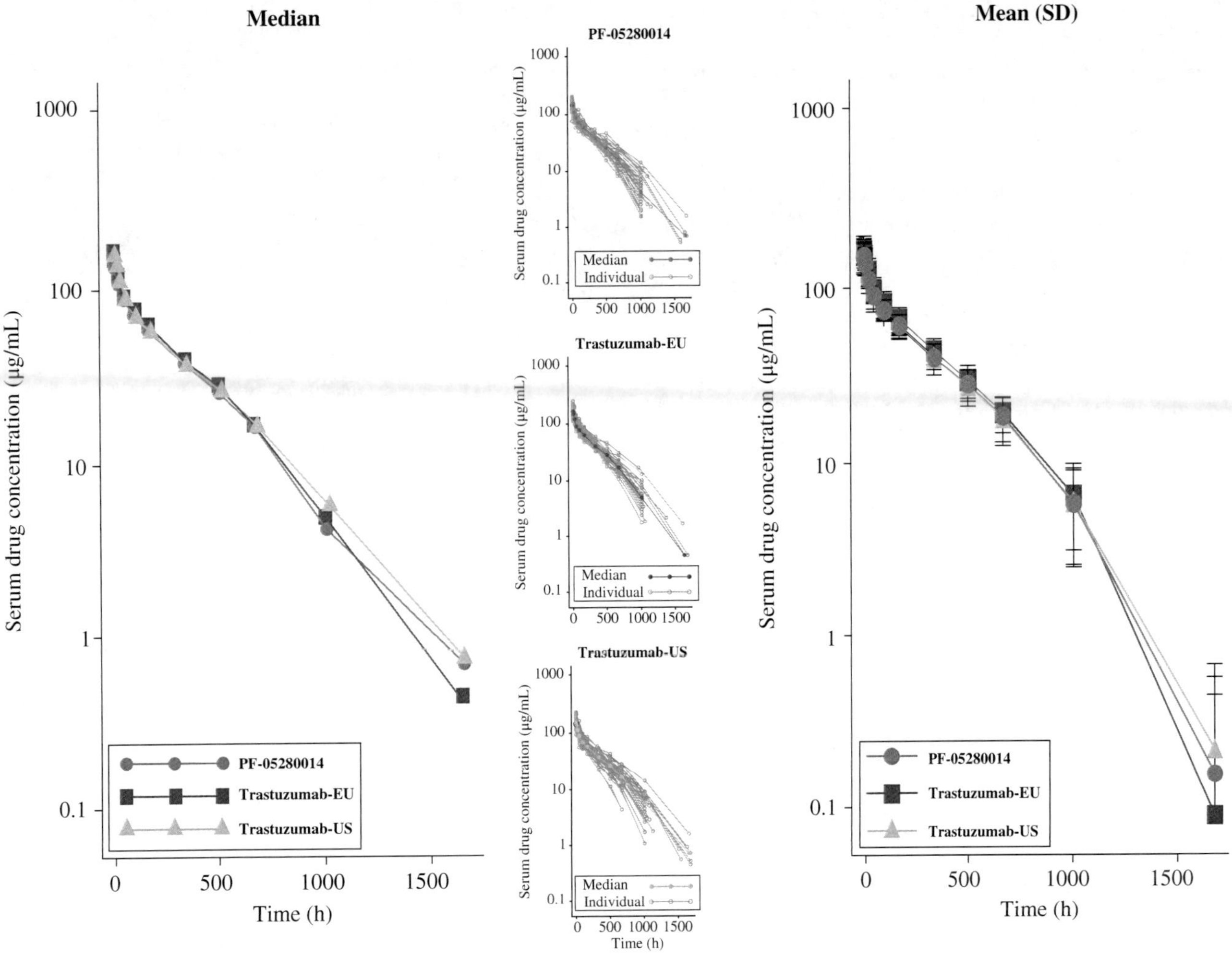

FIGURE 22.3 Median (with embedded individual profiles) and mean ± SD serum concentration–time profiles of PF-05280014, trastuzumab-EU, and trastuzumab-US following a single dose of 6 mg/kg in healthy subjects.

TABLE 22.4 Clinical PK Parameters (Geometric Mean)

	PK Parameter[a]		
Reference	C_{max} (μg/mL)	AUC_{0-t} (μg·h/mL)	$AUC_{0-\infty}$ (μg·h/mL)
PF-05280014	157	35,210	36,650
Trastuzumab-EU	171	38,000	39,770
Trastuzumab-US	161	35,230	36,710

[a] Geometric mean.

TABLE 22.5 Clinical Comparison of Pharmacokinetic Exposure Parameters between Test and Reference Products

PK Parameter	Test:Reference	Ratio (%)	90% CI (%)
C_{max} (μg/mL)	PF:EU	91.49	85.32, 98.09
	PF:US	97.41	90.71, 104.62
	EU:US	106.48	99.20, 114.30
$AUC_{0-2880h}$ (μg·h/mL)	PF:EU	92.66	86.44, 99.34
	PF:US	99.94	93.08, 107.31
	EU:US	107.85	100.50, 115.75
$AUC_{0-\infty}$ (μg·h/mL)	PF:EU	92.15	86.03, 98.69
	PF:US	99.83	93.06, 107.09
	EU:US	108.34	101.05, 116.16

EU, Trastuzumab-EU; PF, PF-05280014; and US, Trastuzumab-US.

different concomitant medications, and different sites of action in different disease populations) to believe that clinical studies conducted in some patient populations are not sensitive to certain potential differences between an innovator product and a biosimilar product, and that these differences might manifest as clinically significant in other indications. Potential safety issues in different subpopulations would therefore need to be adequately addressed.

A detailed scientific rationale addressing the benefits and risks of such a proposal to extrapolate across therapeutic areas would need to be provided to and agreed upon by the appropriate regulatory authorities.

22.6 CONCLUDING REMARKS

In developing a biosimilar, it is critical to understand the regulatory requirements in the regions of interest, the characteristics of the innovator reference product, and to design optimized studies starting with analytical/biological studies, then subsequently design nonclinical and clinical studies to address any remaining uncertainty and to further characterize the biosimilar as needed.

Comparative PK and/or PD assessments are important components of this biosimilarity determination. To compare the PK and/or PD properties between a proposed biosimilar product and the originator's product, the study(s) should be designed in a way that any difference in the properties could be sensitively detected. As part of the stepwise approach of biosimilarity determination, demonstration of PK and PD similarity could reduce the residual uncertainty, thus allowing a selective and targeted approach in study design.

ACKNOWLEDGMENTS

The authors thank Dr. Ira Jacobs and Dr. Greg Finch for reviewing and providing comments to help improve this chapter. Graphic design of the figures and data verification was provided by Engage Scientific Solutions and was funded by Pfizer. The authors also thank Dr. Honghui Zhou and Dr. Frank-Peter Theil for their excellent editorial assistance for this chapter.

REFERENCES

[1] Hu WS, Aunins JG. Large-scale mammalian cell culture. Curr Opin Biotechnol 1997;8:148–153.

[2] Li F, Vijayasankaran N, Shen AY, Kiss R, Amanullah A. Cell culture processes for monoclonal antibody production. MAbs 2010;2:466–479.

[3] EMA. 2015. Similar biological medicinal products. April 2015 ed. Available at http://www.ema.europa.eu/docs/en_GB/document_library/Scientific_guideline/2014/10/WC500176768.pdf. Accessed 2015 Jun 8.

[4] EMA. 2014. Similar biological medicinal products containing biotechnology-derived proteins as active substance: quality issues. June 2014 ed. Available at http://www.ema.europa.eu/docs/en_GB/document_library/Scientific_guideline/2014/06/WC500167838.pdf. Accessed 2015 Jun 8.

[5] FDA. 2015. Quality considerations in demonstrating biosimilarity to a reference protein product: Guidance for industry. Available at http://www.fda.gov/downloads/drugs/guidance complianceregulatoryinformation/guidances/ucm291134.pdf. accessed 2015 Aug 17.

[6] Narayanamoni M. 2013. Bio-dynamism – insights into the biosimilars market: an overall perspective. Available at http://www.grantthornton.in/healthcare-life-sciences/bio-dynamism-insights-into-the-biosimilars-market/ Accessed 2015 Jul 21.

[7] da Silva A, Kronthaler U, Koppenburg V, Fink M, Meyer I, Papandrikopoulou A, Hofmann M, Stangler T, Visser J. Target-directed development and preclinical characterization of the proposed biosimilar rituximab GP2013. Leuk Lymphoma 2014;55:1609–1617.

[8] Schiestl M, Stangler T, Torella C, Cepeljnik T, Toll H, Grau R. Acceptable changes in quality attributes of glycosylated biopharmaceuticals. Nat Biotechnol 2011;29:310–312.

[9] ICH. 2004. Q5E – comparability of biotechnological/biological products subject to changes in their manufacturing process. November 2004 ed. Available at http://www.ich.org/products/guidelines/quality/quality-single/article/comparability-of-biotechnologicalbiological-products-subject-to-changes-in-their-manufacturing-proc.html. Accessed 2015 Jun 8.

[10] EMA. European Medicines Agency's scientific guidelines on biosimilar medicines. Available at http://www.ema.europa.eu/ema/index.jsp?curl=pages/regulation/general/general_content_000408.jsp&mid=WC0b01ac058002958c.. Accessed 2015 Jun 8.

[11] EMA. 2006. Similar biological medicinal products containing biotechnology-derived proteins as active substance: non-clinical and clinical issues. 2006 ed. Available at http://www.ema.europa.eu/docs/en_GB/document_library/Scientific_guideline/2009/09/WC500003920.pdf. Accessed 2015 Jun 8.

[12] EMA. 2015. Similar biological medicinal products containing biotechnology-derived proteins as active substance: non-clinical and clinical issues. July 2015 ed. Available at http://www.ema.europa.eu/docs/en_GB/document_library/Scientific_guideline/2015/01/WC500180219.pdf. Accessed 2015 Jun 8.

[13] EMA. 2012. Similar biological medicinal products containing monoclonal antibodies: non-clinical and clinical issues. June 2012 ed. Available at http://www.ema.europa.eu/docs/en_GB/document_library/Scientific_guideline/2012/06/WC500128686.pdf. Accessed 2015 Jun 8.

[14] EMA. European public assessment reports: table of biosimilar products. Available at http://www.ema.europa.eu/ema/index.jsp?curl=pages%2Fmedicines%2Flanding%2Fepar_search.jsp&mid=WC0b01ac058001d124&searchTab=searchByAuthType&alreadyLoaded=true&isNewQuery=true&status=Authorised&status=Withdrawn&status=Suspended&status=Refused&keyword=Enter+keywords&searchType=name&taxonomyPath=&treeNumber=&searchGenericType=biosimilars. Accessed 2015 Jun 8.

[15] FDA. Information for industry (Biosimilars). April 18, 2015 update. Available at http://www.fda.gov/Drugs/DevelopmentApprovalProcess/HowDrugsareDevelopedandApproved/ApprovalApplications/TherapeuticBiologicApplications/Biosimilars/ucm241720.htm. Accessed 2015 Aug 17.

[16] FDA. 2012. Guidance for industry: scientific considerations in demonstrating biosimilarity to a reference product (draft guidance). February 2012 ed. Available at http://www.fda.gov/downloads/Drugs/GuidanceComplianceRegulatoryInformation/Guidances/UCM291128.pdf. Accessed 2015 Jun 8.

[17] FDA. 2014. Guidance for industry: clinical pharmacology data to support a demonstration of biosimilarity to a reference product (draft guidance). May 2014 ed. Available at http://www.fda.gov/downloads/Drugs/GuidanceCompliance RegulatoryInformation/Guidances/UCM397017.pdf. Accessed 2015 Jun 8.

[18] Celltrion. 2014. Celltrion files for US FDA approval of Remsima®. [Media release: August 11, 2014]. Available at http://www.celltrion.com/en/company/notice_view.asp?idx=456&code=ennews&intNowPage=1&menu_num=&align_year=all. Accessed 2015 Jun 8.

[19] FDA. 2015. FDA approves first biosimilar product Zarxio. FDA News Release. March 06, 2015 ed. Available at http://www.fda.gov/NewsEvents/Newsroom/PressAnnouncements/ucm436648.htm. Accessed 2015 Jun 8.

[20] WHO. 2009. Guidelines on evaluation of similar biotherapeutic products (SBPs). April 2010 ed. Available at http://www.who.int/biologicals/areas/biological_therapeutics/BIOTHERAPEUTICS_FOR_WEB_22APRIL2010.pdf?ua=1. Accessed 2015 Jun 8.

[21] WHO. 2014. Guidance on scientific principles for regulatory risk assessment of biotherapeutic products (Draft). Available at http://www.who.int/biologicals/WHO_Risk_Assessment_for_BTP_2nd_PC_10_Dec_2014.pdf. Accessed 2015 Jun 8.

[22] Wang J, Chow SC. On the regulatory approval pathway of biosimilar products. Pharmaceuticals (Basel) 2012;5:353–368.

[23] Abas A. Regulatory guidelines for biosimilars in Malaysia. Biologicals 2011;39:339–342.

[24] Korean Food and Drug Administration (KFDA). 2009. Evaluation guidelines for biosimilars. Available at http://www.biosimilars.ca/docs/Evalutation_Guidelines_for_Biosimilars.pdf. Accessed 2015 Jun 8.

[25] Suh SK, Park Y. Regulatory guideline for biosimilar products in Korea. Biologicals 2011;39:336–338.

[26] Health Sciences Authority. 2009. Guidance on registration of similar biological products in Singapore. Available at http://www.biosimilars.ca/docs/Guidance_Registration_Similar_Biological Products_Singapore.pdf. Accessed 2015 Jun 8.

[27] Poh J, Tam KT. Registration of similar biological products–Singapore's approach. Biologicals 2011;39:343–345.

[28] Health Canada. 2010. Guidance for sponsors: information and submission requirements for subsequent entry biologics (SEBs). Available at http://www.hc-sc.gc.ca/dhp-mps/brgtherap/applic-demande/guides/seb-pbu/seb-pbu_2010-eng.php. Accessed 2015 Jun 8.

[29] Kay J, Feagan BG, Guirguis MS, Keystone EC, Klein AV, Lubiniecki AS, Mould DR, Nyarko KA, Ridgway AA, Trudeau ME, Wang J. Health Canada/BIOTECanada Summit on regulatory and clinical topics related to subsequent entry biologics (biosimilars), Ottawa, Canada, 14 May 2012. Biologicals 2012;40:517–527.

[30] Hechavarria Nunez Y, Perez Massipe RO, Orta Hernandez SD, Munoz LM, Jacobo Casanueva OL, Pérez Rodríguez V, Domínguez Morales RB, Pérez Cristiá RB. The regulatory framework for similar biotherapeutic products in Cuba. Biologicals 2011;39:317–320.

[31] Thanaphollert P, Tungsanga K. Towards regulation of similar biotherapeutic products: Thailand's perspective. Biologicals 2011;39:346–347.

[32] Liang C, Wang J. China's perspective on similar biotherapeutic products. Biologicals 2011;39:312–316.

[33] Malaysia Ministry of Health. 2008. National pharmaceutical control bureau, guidance document and guidelines for registration of biosimilars in Malaysia. Available at http://portal.bpfk.gov.my/view_file.cfm?fileid=302. Accessed 2015 Jun 8.

[34] Bennett CL, Chen B, Hermanson T, Wyatt MD, Schulz RM, Georgantopoulos P, Kessler S, Raisch DW, Qureshi ZP, Lu ZK, Love BL, Noxon V, Bobolts L, Armitage M, Bian J, Ray P, Ablin RJ, Hrushesky WJ, Macdougall IC, Sartor O, Armitage JO. Regulatory and clinical considerations for biosimilar oncology drugs. Lancet Oncol 2014;15: e594–e605.

[35] Wallace KH, Rees AR. Studies on the immunoglobulin-G Fc-fragment receptor from neonatal rat small intestine. Biochem J 1980;188:9–16.

[36] Seo KS, Yoon JW, Na KH, Bae EJ, Woo JG, Lee SH, Kang SH, Yang JM. Evaluation of process efficiency and bioequivalence of biosimilar recombinant human chorionic gonadotropin (rhCG). BioDrugs 2011;25:115–127.

[37] Ryan AM, Sokolowski SA, Ng CK, Shirai N, Collinge M, Shen AC, Arrington J, Radi Z, Cummings TR, Ploch SA, Stephenson SA, Tripathi NK, Hurst SI, Finch GL, Leach MW. Comparative nonclinical assessments of the proposed biosimilar PF-05280586 and rituximab (MabThera(R)). Toxicol Pathol 2014;42:1069–1081.

[38] Hurst S, Ryan AM, Ng CK, McNally JM, Lorello LG, Finch GL, Leach MW, Ploch SA, Fohey JA, Smolarek TA. Comparative nonclinical assessments of the proposed biosimilar PF-05280014 and trastuzumab (Herceptin((R)). BioDrugs 2014;28:451–459.

[39] EMA. 2005. EPAR product information – herceptin. September 1, 2014 ed. Available at http://www.ema.europa.eu/ema/index.jsp?curl=pages/medicines/human/medicines/000278/human_med_000818.jsp&mid=WC0b01ac058001d124. Accessed 2015 Aug 17.

[40] Wang W, Prueksaritanont T. Prediction of human clearance of therapeutic proteins: simple allometric scaling method revisited. Biopharm Drug Dispos 2010;31:253–263.

[41] EMA. 2009. Zarzio (filgrastim) EPAR summary for the public. February 02, 2009 ed. Available at http://www.ema.europa.eu/docs/en_GB/document_library/EPAR_-_Summary_for_the_public/human/000917/WC500046526.pdf. Accessed 2015 Jun 8.

[42] FDA. 2001. Guidance for industry: bioanalytical method validation. May 2001 ed. Available at http://www.fda.gov/downloads/Drugs/Guidances/ucm070107.pdf. Accessed 2015 Jun 8.

[43] DeSilva B, Smith W, Weiner R, Kelley M, Smolec J, Lee B, Khan M, Tacey R, Hill H, Celniker A. Recommendations for the bioanalytical method validation of ligand-binding assays to support pharmacokinetic assessments of macromolecules. Pharm Res 2003;20:1885–1900.

[44] Findlay JW, Smith WC, Lee JW, Nordblom GD, Das I, DeSilva BS, Khan MN, Bowsher RR. Validation of immunoassays for bioanalysis: a pharmaceutical industry perspective. J Pharm Biomed Anal 2000;21:1249–1273.

[45] Viswanathan CT, Bansal S, Booth B, DeStefano AJ, Rose MJ, Sailstad J, Shah VP, Skelly JP, Swann PG, Weiner R. Quantitative bioanalytical methods validation and implementation: best practices for chromatographic and ligand binding assays. Pharm Res 2007;24:1962–1973.

[46] Cai XY, Wake A, Gouty D. Analytical and bioanalytical assay challenges to support comparability studies for biosimilar drug development. Bioanalysis 2013;5:517–520.

[47] Islam R. Bioanalytical challenges of biosimilars. Bioanalysis 2014;6:349–356.

[48] Cai XY, Gouty D, Baughman S, Ramakrishnan M, Cullen C. Recommendations and requirements for the design of bioanalytical testing used in comparability studies for biosimilar drug development. Bioanalysis 2011;3:535–540.

[49] Yin D, Barker KB, Li R, Meng X, Reich SD, Ricart AD, Rudin D, Taylor CT, Zacharchuk CM, Hansson AG. A randomized phase 1 pharmacokinetic trial comparing the potential biosimilar PF-05280014 with trastuzumab in healthy volunteers (REFLECTIONS B327-01). Br J Clin Pharmacol 2014;78:1281–1290.

23

ADME PROCESSES IN VACCINES AND PK/PD APPROACHES FOR VACCINATION OPTIMIZATION

José David Gómez-Mantilla, Iñaki F. Trocóniz and María J. Garrido
University of Navarra, Pamplona, Spain

23.1 INTRODUCTION

From a biopharmaceutic and pharmacodynamic point of view, vaccines are quite unique; no other type of therapeutics is expected to confer a life-time protection with only one or few minimal doses. This high expectation on vaccines contrasts with the very limited knowledge about their mechanisms of action or about their absorption, distribution, metabolism, and excretion (ADME) processes in the human body. Vaccines are also unique regarding their ADME; contrary to most drugs, antigens (Ag) from vaccines are expected to be absorbed and distributed in the lymphatic system, where the immune response is started. Interestingly, vaccine metabolism and excretion are consistently ignored in the literature. In fact, experimental pharmacokinetic (PK) studies are not required from regulatory agencies for vaccine approval.

In general, ADME processes are routinely studied to characterize the PK of a therapeutic agent and to investigate the changes in formulation, administration, or dosage schedules (biopharmaceutic optimization), which could lead to some desired effects, such as stronger, faster, localized, or prolonged response or minimized toxicity or drug interactions. This is possible if we know where and how fast a drug should be available to produce a desired effect. The more we know about mechanism(s) of action of a given drug, the more useful is this approach. In the case of vaccines, biopharmaceutic optimization is hindered by the fact that neither their mechanisms of action nor their ADME are well known. The knowledge about vaccine ADME is very limited because quantifying Ag concentration in the lymphatic system at different time points encounters many technical difficulties; to date, there are no studies that quantitatively describe the absorption or distribution of vaccine.

Nevertheless, knowledge on vaccine ADME, and its impact on immunization efficiency, could also lead to a more rational vaccine design, regarding formulation, administration route, and dosage schedules. Mathematical models may also be used to optimize vaccination. These tools can predict particular outputs, such as vaccination efficacy in an individual, using a series of input variables, like dosage schedules. Although, the utility of these tools is limited by the incomplete knowledge of immunization mechanisms, these models have been useful, in certain situations, to predict certain responses after immunization. Most importantly, as the knowledge of immunization mechanism and kinetics improves, mathematical models are expected to become more useful and accurate.

In order to present the current knowledge on vaccine ADME and vaccine modeling, it is convenient to briefly summarize how vaccines are developed, what type of vaccines are available, and how much it is known about their mechanisms of action.

23.1.1 Vaccine Development

During preclinical phases in vaccine development, animal models are studied to evaluate certain vaccine features, such as seroconversion rates, geometric mean antibody titers, or cellular immunity. In some cases, if a vaccine consists of more than one Ag, the response to each Ag must be evaluated in areas such as the characterization of antibody class, avidity,

ADME and Translational Pharmacokinetics/Pharmacodynamics of Therapeutic Proteins: Applications in Drug Discovery and Development,
First Edition. Edited by Honghui Zhou and Frank-Peter Theil.

affinity, half-life, memory, and potential induction of cellular immunity. However, preclinical data should be interpreted with caution because immune response and vaccine ADME vary across species [1]. Yet, studies in animals can provide valuable though not definite information to support subsequent clinical studies, regarding the Ag selection, dose, schedule, and routes of administration or formulation. Dose projection from mice and rabbits to human may be misleading as a predictor of immune response in clinical trials, suggesting in some cases the use of primates as a more informative model.

23.1.1.1 Clinical Programs Phase I studies include a naive population with few healthy subjects aiming to evaluate the vaccine safety [2]. In this phase, besides safety and tolerability assessment, evaluation of PK and pharmacodynamics represent an essential component in the development of nonvaccine therapeutic biologic products. However, because vaccine Ag is administered in discrete, widely spaced, and relatively small amounts, the vaccine ADME is rarely described, while local or general toxicity and types of immune responses are usually characterized in detail.

Dose optimization may also be evaluated in dose–response studies, from the minimal dose that elicits response or antibody titers until increment in dose no longer induces an increment in titers, that is, reaching a plateau of the immune response. These trials are conducted in blinded design with a control group and last for 6 months after the last injection, although the duration can also be considered for several years.

Phase II clinical studies involve larger numbers of subjects and are intended to provide preliminary information about a vaccine's ability to produce its desired effect (usually immunogenicity) in the target population and its general safety.

Phase III clinical studies are pivotal by nature on which the decision on whether to grant the license is based. Sufficient data must be obtained to demonstrate that a vaccine is both safe and effective, while the proof of efficacy may take years—because a control group exposure to the infectious pathogen is not allowed for ethical reasons, efficacy can be evaluated only as a protective effect between groups of vaccinated and unvaccinated individuals observed in long period of time.

23.1.2 Types of Vaccines

Different vaccine types have different chemical properties and therefore different ADME properties, formulation challenges, and immunogenicity. Vaccines are classified according to the Ag used for their preparation in:

1. **Live Attenuated Vaccine** (LAV).

   This type of vaccine contains the live pathogen responsible for the disease, but attenuated in the laboratory. It cannot cause disease, but can elicit strong cellular and humoral response after one or two doses.

   Examples of vaccines included in this group are tuberculosis (bacille Calmette–Guérin (BCG)), oral polio vaccine (OPV), measles, rotavirus, and yellow fever.

2. **Killed Antigen:**

   Vaccines are prepared with pathogens that are inactivated by using some chemical treatment, heat, or radiation. These vaccines are more stable and safer than LAV. The main disadvantage of this class of vaccines is the need for several doses or booster shots to maintain the immunity.

   Examples of vaccines included in this group are whole-cell pertussis (wP) and inactivated polio virus (IPV).

3. **Subunit (Purified Antigen):**

   Vaccines are prepared with specific parts or subunits of the pathogens that are able to stimulate the immune system. Due to their low immunogenicity, these vaccines are normally conjugated with adjuvants.

   Examples of vaccines included in this group are acellular pertussis (aP), haemophilus influenzae type B (Hib), pneumococcal (PCV-7, PCV-10, PCV-13), and hepatitis B (HepB).

4. **Toxoid (Inactivated Toxins):**

   These vaccines include bacterial toxins relevant for the infectious process or its symptoms. Toxins are inactivated by chemical treatment using products such as formaldehyde. Such detoxified toxins are known as toxoids. They are safe and can produce antibodies that block the toxin.

   Some examples of vaccines included in this group are tetanus toxoid (TT) and diphtheria toxoid.

5. **DNA Vaccines:**

   These types of vaccines are DNA sequences (normally plasmids) designed to replicate inside the cells of vaccinated individual, where the Ag of interest is expressed and further presented to the recipient's immune system. Many of DNA vaccines have been tested in clinical trials, including HIV, influenza, HPV (human papillomavirus), cancer (metastatic breast, B-cell lymphoma, prostate, colorectal, and melanoma), hepatitis B, hepatitis C, and malaria I/II 7 [3].

6. **Dendritic Cell (DC) Vaccines:**

   Isolated DC loaded with Ag *ex vivo* and administered as a cellular vaccine have been found to induce protective and therapeutic immunity [4]. This strategy has been explored specially for cancer vaccination [5] and HIV [6].

23.1.3 Basic Immunological Mechanism of Vaccine Development

The main goal of vaccination is to provide protection against diseases caused by a microorganism through the activation of the immune mechanisms, specially the formation of

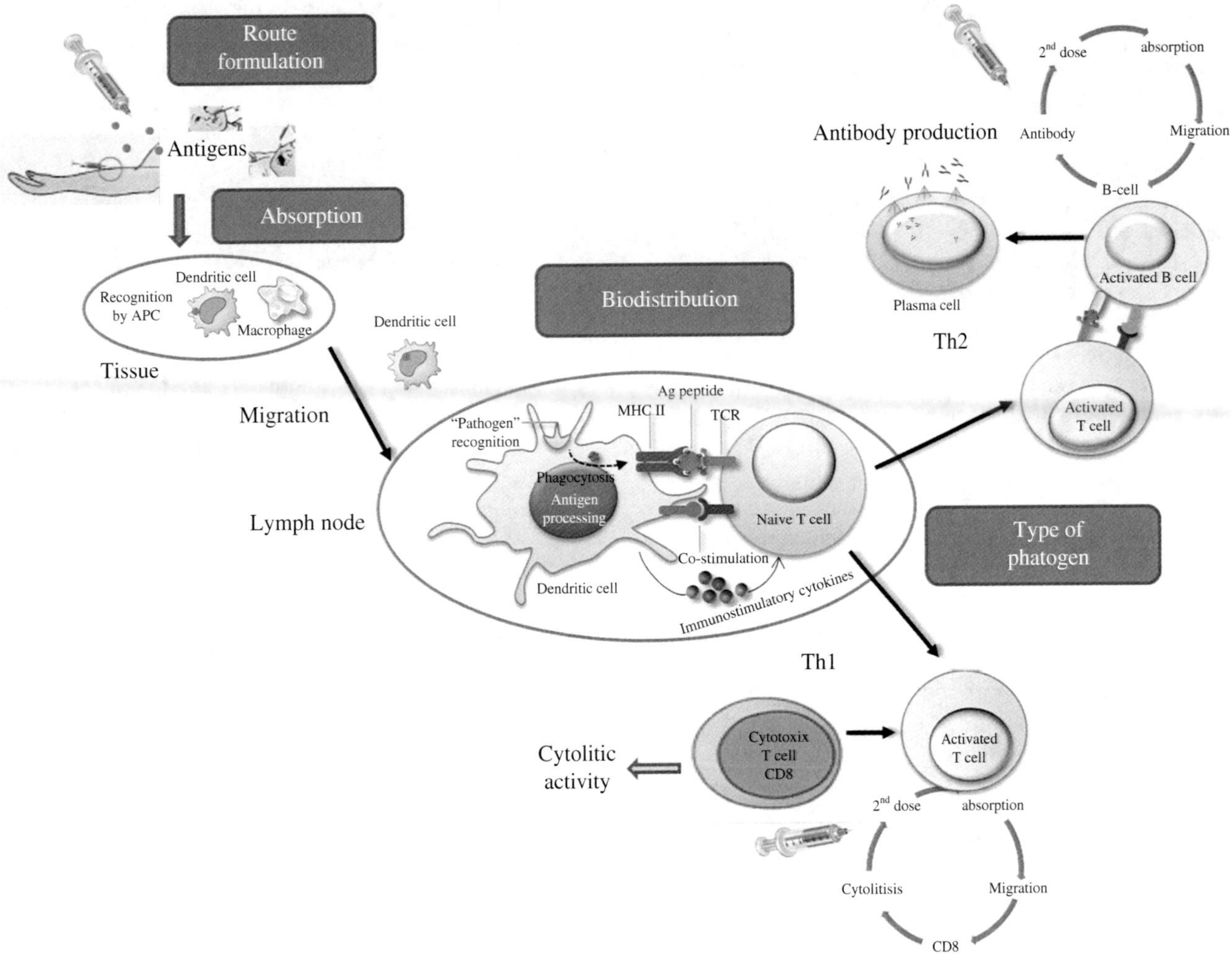

FIGURE 23.1 Schematic representation of vaccine mechanism of action depending on the pathogen type and the main vaccine ADME processes involved in the triggering of the immune response. The vaccine formulation and route of administration determine the antigen absorption, vaccine encounters dendritic cells (DC) in tissues neighboring the administration site, and after antigen processing by DC in tissue, these cells migrate to the nearest LN where the antigen is presented to the naive T-cells. Depending on the antigen type, either Th2 or Th1 response is activated. Alternatively, vaccine formulation or site of administration may achieve antigen delivery directly into the LN without the transport by DC from tissue.

immune memory. Figure 23.1 briefly depicts the responses triggered after vaccination, which are dependent on the type of vaccine pathogen. The immune response could be humoral (mediated by B-cells producing antibodies) or cellular (T-cells) immunity. For certain infections, protective immunity by antibodies is sufficient; however, for others such as HIV, *Mycobacterium tuberculosis*, and malaria, both humoral and cellular seem to be relevant to induce a long-term protective response.

Innate immune response, which includes phagocytic cells, antimicrobial peptides, and complement, is the primary barrier involved in the defense against infection. This response also plays an important role in regulating the adaptive response. Innate immune recognition of infectious pathogens is mediated by germline-encoded receptors that recognize a limited number of conserved microbial structures or pathogen-associated molecular patterns (PAMPs). These PAMPs interact with pattern-recognition receptors, called toll-like receptors that are expressed on various cells of the innate immune response, including the major antigen-presenting cells (APCs). Immunogenic stimuli such as lipopolysaccharides, mycobacterial Ag, or specific DNA sequences bind to toll-like receptors and stimulate the secretion of cytokines and chemokines that increase the activity of APCs, which are responsible for T-cell activation; this represents the starting point of the adaptive immune response. In this way, both immunity responses, innate and adaptive, can be activated by vaccines.

23.1.3.1 Antigen Presentation Adaptive cellular immunity: The primary immune response is initiated by the presentation of Ag to naïve CD4+ T-cells that are associated with the major histocompatibility complex (MHC) class II expressed on specific APCs (mainly, DC, macrophages, and B-cells). DC are found in lymphoid tissues and are strategically located in potential pathogen entry site, such as mucosal tissue and skin.

Following activation, naive CD4+ T-cells can be split into T helper 1 (Th1) and T helper 2 (Th2) types of cells. Th1 cells produce IL-2 and IFN-γ and trigger a direct cytolytic mechanism by CD8+ T-cells known as cytotoxic T lymphocytes (CTLs). Activated CTL induces a cascade of signals that lead to apoptosis of target cells, controlling viral and nonviral intracellular pathogens (fungi, protozoa, and certain bacteria) by lysis of infected target cells and/or by cytokine-induced inhibition of pathogen replication. Th2 cells produce IL-4, IL-5, IL-10, and IL-13 and activate the B-cells to produce Ag-specific antibodies that can bind to the Ag and promote its elimination by forming complexes with the immune complement. In both cases, Ag-specific CTL and B-cells can remain in the body after the vaccine is eliminated, creating a pool of memory cells that can respond faster after a new encounter with the Ag, and subsequent Ag administrations increase the pool of memory cells and can promote the formation of new ones.

23.2 BIOPHARMACEUTIC CONSIDERATIONS ON VACCINE ADME PROCESSES

Despite the great success of vaccination in virtually eradicating several infectious diseases, little is known about the impact of biopharmaceutics on the safety and efficacy of immunization agents. As a result, choosing of vaccine formulation, route, and dosing during development has been driven mostly by trial and error. When a vaccine system fails to induce a protective response, it is unknown whether the Ag is incapable of triggering the formation of memory cells, or whether the Ag does not reach the right tissues, in the right amount and timing, in order to be properly presented by APCs to the T-cells.

Regardless of the entrance route of a pathogen, effective Ag presentation able to elicit the production of memory cells occurs in the lymph node (LN) [7], where the high density of lymphocytes allows a proper immune response. Under infection, APCs, especially DC, migrate from the site of pathogen encounter to the LN, where Ag presentation to T-cells occurs and triggers the immune response. Therefore, two main biopharmaceutic aspects must be considered for effective vaccination:

1. The exposure of Ag to DC in pathogen-relevant tissues, which is one of the most decisive factors in vaccination success. Ag exposure to DC *in vivo* is normally not measured after immunization, the closest indicator of this process that has been assessed for several immunization procedures is the presence of Ag in the LN and lymph-related organs as spleen and thymus. Although the presence of Ag in the LN does not necessarily indicate an appropriate Ag presentation by the APCs, the presence of Ag in these sites has been linked to effective immunization [7].
2. The exposure time of Ag presentation is also crucial for immunization. Nonantigenic substances could elicit an immune response over repeated exposures as occurs in some allergies [8]; while chronic exposure of antigenic substances is also required to reach tolerance, which explains why in normal subjects there is no inflammatory or cytotoxic response against gut microbiota. Some Ag require a single administration to produce a protective immune response, while others may require several doses; some Ag elicit a stronger immune response by a fast exposure and clearance [9], while others require a persistent exposure that can last several days, even weeks [10]. What must be considered in vaccine design is that the immune response does not depend on the Ag exclusively and that the formulation, route of administration, and dosing also play a crucial role in immunization success. For a better application of biopharmaceutic concepts in vaccine development, deeper knowledge on immunogenicity dynamics is required, especially on the complex immune mechanisms to elicit effective responses, or generate tolerance against heterologous substances.

Immunization through vaccination can be severely affected by aspects such as the formulation, type of vaccine, and the route of administration. However, most vaccines are administered through subcutaneous (SC) or intramuscular (IM) injection. Failure of promising immunizing agents using these routes has created a new field of research in which it has been shown that a formulation can trigger a greater immune response only by changing the site of vaccine administration [11]. The purpose of next section is to summarize the current knowledge on the impact of administration routes, dosing schedules, and formulation on vaccine ADME processes, focusing only on the vaccine properties that can be altered without modifying the Ag.

23.3 VACCINES AND ADME PROCESSES

Most drug formulations attempt to increase the bioavailability of an administered therapeutic compound, expecting that higher and more persistent drug levels in blood are proportional to the desired therapeutic effect (as long as toxicity is acceptable or not present), even if the drug site of action is located in a peripheral tissue, concentrations in blood are assumed to be proportional to drug concentrations in these tissues and, therefore, proportional to the therapeutic effect. As a result, most of literature in drug absorption is concerned about transport of drugs from the administration site to blood flow.

In the case of vaccines, concentration of Ag in blood is normally not correlated with the vaccine immunoprotective efficacy. In fact, the entrance of Ag in blood is usually counterproductive because it increases Ag metabolism and hinders Ag distribution into the lymphatic system, where the immune response is triggered. Ag absorption from the administration site into lymphatic system is much more relevant in vaccine biopharmaceutics than Ag kinetics in blood. Kinetics of Ag levels in lymphoid organs or lymphatic system is not easy to evaluate because (i) the distribution in the lymphatic system is not uniform as in blood and (ii) quantification of Ag in these tissues cannot be performed *in vivo*, but only after animals are killed. These restrictions prevent the study of Ag absorption kinetics into the lymphatic system and explain the lack of published works in this field. Despite these limitations, some aspects of Ag absorption can be inferred from the available studies on Ag biodistribution after vaccination. That is why absorption and distribution must be considered simultaneously in this section.

Once a pathogen enters the body, DC are responsible for pathogen recognition, processing, and Ag transportation to LN. After vaccination, Ag can be present at the LN by two paths: vaccine delivery into the lymphatic circulation due to an efficient biopharmaceutics and transportation from tissue by DC. According to the studies published in this field, formulation and administration route contribute more to the presence of Ag in LN after vaccination than Ag transport by DC. However, the presence of Ag in nodes due to migration of Ag-activated DC could be more relevant for protective immunization because it mimics more the natural response to a pathogen. It may be also possible that vaccine formulation or administration route promotes the DC uptake in a pathogen-related tissue, but this effect is more difficult to prove. The limited numbers of studies that explore Ag distribution after vaccination are more focused on distribution into LN than into Ag DC uptake from pathogen-related tissues.

Table 23.1 [12–33] summarizes the available qualitative data for Ag biodistribution. The most listed formulations are able to deliver Ag to the LN; however, none of them reports to Ag delivery exclusively to nodes or a targeted tissue. Independent of Ag type, a wide biodistribution is described for most vaccines, especially for intravenous (IV) formulations. Because data come from different animal models and doses, a quantitative comparison of the impact of route of administration or formulation on Ag absorption and distribution is not possible between these studies, especially, considering that biodistribution of the same Ag may vary across species [1]. Another limitation is that not all the reviewed studies measured the presence of Ag in the same tissues; therefore, this limits the evaluation of what factors promote distribution into particular tissues. Fortunately, several studies compare absorption and biodistribution into lymph organs between formulations and administration routes. In the following section, we discuss the current knowledge of the impact of formulation and route of administration on Ag absorption and distribution. Explanations for the lack of studies on vaccine metabolism and excretion are also commented.

23.3.1 Effect of Vaccine Formulation on ADME

First vaccine formulations consisted of solutions or suspensions of the immunizing agent. The main objective of these formulations was to achieve vaccine stability, safety, and suitability for administration. These types of formulations normally do not impact the Ag exposure in LN or DC uptake, but seemed to suffice in the initial attempts of vaccine development. As unsuccessful vaccination cases started to appear, more attention was directed to formulation design, aiming to improve immunogenicity. Initially, vaccine design was focused on finding immunogenic adjuvants that, by activating the immune system, can indirectly trigger a stronger immunogenic response against the vaccination Ag by enhancing the density of APCs and excitability of T-cells. Some of these adjuvants have also shown to improve DC uptake of the vaccine Ag and Ag absorption and biodistribution into the lymphatic system. Other vaccine systems aim to protect the Ag- and target-specific cells or tissues, prolonging Ag exposure in relevant tissues and increasing the Ag uptake by DC.

23.3.1.1 Effect of Adjuvants in Vaccine Absorption and Distribution Several hundred natural and synthetic compounds have been identified as adjuvants, yet aluminum-based compounds are the most used adjuvants for approved human vaccines [34]. Originally, it was believed that these aluminum compounds boost the immune systems by producing a depot effect, prolonging Ag exposure [35, 36]. Recently, it has been stated that Ag adsorbed in aluminum particles are recognized and processed by phagocytes, promoting lymphocyte recruitment, upregulation of Ag-presenting related molecules in APCs, and production of proinflammatory cytokines [37–39].

Apart from boosting the immune response, the role of adjuvants in facilitating a prolonged Ag exposure in relevant tissues has proven to be crucial for effective immunization.

Several adjuvants have shown to improve retention time in LN, promoting DC uptake, and generation of Ag-specific Ab titers [7, 40], even in weak immunogenic vaccines as DNA vaccines [27, 41]. Lipid-based adjuvants in particular have shown to increase the Ag exposition in LN and other lymph organs [21, 42].

TABLE 23.1 Summary of Some Published Studies on Vaccine Biodistribution in Different Organs and Tissues

Vaccine	Antigen	Type	Formulation	Route	Dosing	LN	Sp	Th	Bl	Br	Hr	Lv	Lu	Kd	It	AT	Gd	Mc	Sk	References
Melanoma	Tumor antigens	DC vaccine	Solution	IV	Two doses	++	++	n.a.	++	n.a.	n.a.	++	++	++	n.a.	n.a.	n.a.	–	–	[12]
				SC		++	++		+			+	+	+				++	++	
				IP		++	++		+			+	+	+				++	+	
Coccidioides	Ag2/FRA cDNA	DC-vaccine	Solution	IN mice	Single dose	++	++	++	++	–	–	++	++	–	++	–	–	–	–	[13]
Model	DT	Toxoid	Immuno PLGA_MP	SC mice	Single dose	++	++	n.a.	n.a.	n.a.	n.a.	n.a.	n.a.	n.a.	n.a.	n.a.	n.a.	n.a.	++	[14]
Model	Spermine-alginate	Peptide	NP	IN mice	Single dose	++	++	n.a.	n.a.	n.a.	n.a.	++	++	n.a.	n.a.	n.a.	n.a.	n.a.	n.a.	[15]
Model	OVA	Peptide	NP	ID mice	Single dose	++	++	n.a.	n.a.	n.a.	n.a.	n.a.	n.a.	n.a.	n.a.	n.a.	n.a.	n.a.	n.a.	[16]
Brucellosis	Subcellular antigen HS (*Brucella ovis*)	Subunit	NP	IO mice	Single dose	n.a.	+	+	+	n.a.	n.a.	+	+	+	+	n.a.	n.a.	++	++	[17]
Model	OVA	Peptide	PEG-Lp	SC mice	Single dose	++	+	n.a.	+	n.a.	+	+	+	+	n.a.	n.a.	n.a.	n.a.	n.a.	[18]
Model	Model	—	PG-NP	SC rabbit	Single dose	++	+	n.a.	+	–	+	++	+	+	++	++	–	++	+	[19]
Model	Streptavidin	Peptide	Influenza Virosomes	IM mice	Single dose	n.a.	++	n.a.	+	n.a.	+	++	+	++	+	n.a.	n.a.	++	n.a.	[20]
Tuberculosis	BCG	LAV	BCG/lipid-based adjuvants	SC mice	Single dose	++	++	n.a.	n.a.	n.a.	n.a.	+	++	n.a.	n.a.	n.a.	n.a.	n.a.	n.a.	[21]
HIV	Multigenic (fusion protein)	DNA vaccine	Attenuated genetically engineered Measles virus (MV) Schwarz	IM	Three doses	++	++	–	–	n.a.	n.a.	–	–	–	++	n.a.	n.a.	n.a.	–	[22]
HIV	Multigenes	DNA vaccine	Naked DNA	IV/ID/IM rat	Single dose	+	+	n.a.	++	–	–	–	+	–	n.a.	n.a.	–	–	–	[23]
						++	–		–	–	–	–	–	–			–	++	++	
						++	–		–	–	–	–	–	–			–	–	–	
Model plasmid	—	DNA vaccine	Naked DNA	IN mice	Single dose	++	++	n.a.	++	++	++	++	++	++	n.a.	n.a.	n.a.	++	n.a.	[24]
Pseudorabies virus	GP B, C, D	DNA vaccine	Naked DNA	IM swine	Single dose	+	+	n.a.	n.a.	n.a.	n.a.	n.a.	+	+	n.a.	n.a.	+	++	n.a.	[1]
HIV	Multigenes	DNA vaccine	GP 120 proteins	IM/ID rabbit	Single dose	–	–	–	–	–	–	–	–	–	–	–	–	+	++	[25]
						–	–	–	–	–	–	–	–	–	–	–	–	++		
Schistosomiasis	Sj23	DNA vaccine	Naked DNA	IM mice	Single dose	n.a.	++	n.a.	n.a.	++	++	++	++	++	n.a.	n.a.	n.a.	++	n.a.	[26]
Model Th1 response	HBsAg	DNA vaccine	Adjuvant SAINT-18	ID rat	Single dose	++	++	n.a.	++	++	++	++	++	++	n.a.	n.a.	++	++	++	[27]
HIV	Multigenes	DNA vaccine	Naked DNA	EP rabbit	Three doses	+	–	–	–	–	–	–	+	–	n.a.	n.a.	–	++	++	[28]
H5N1 influenza	VCL-IIPT1	DNA vaccine	Adjuvant Vaxfectin	IM rabbit	Single dose	+	++	n.a.	++	++	+	++	++	+	n.a.	n.a.	+	++	++	[29]
HIV	Multigenes	DNA vaccine	Naked DNA	ID-EP mice	Single dose	n.a.	++	n.a.	n.a.	–	–	–	++	–	–	n.a.	++	++	++	[30]
Hantan virus	S segment hantavirus	DNA vaccine	Naked DNA	IM mice	Single dose	n.a.	++	n.a.	n.a.	++	n.a.	++	n.a.	++	n.a.	n.a.	n.a.	++	n.a.	[31]
Dental caries	pGJA-P/VAX1	DNA vaccine	Naked DNA	IN mice	Two doses	++	++	++	++	++	++	++	++	++	++	n.a.	n.a.	++	n.a.	[32]
Measles	Measles proteins	DNA vaccine	Naked DNA	ID rabbit	Single dose (4 injection)	++	–	n.a.	+	–	–	–	–	–	n.a.	n.a.	n.a.	++	++	[33]

+, found in tissue; –, not detected in tissue;
++, repeated findings of the Ag at different time points (days) after immunization; AT, adipose tissue; Bl, blood; Br, brain; EP, electroporation; Gd, gonads; GPs, glycoproteins; Hr, heart; IM, intramuscular; IO, intraocular; It, intestines; IP, intraperitoneal; IN, intranasal; ID, intradermal; IV, intravenous; Kd, kidney; LN, lymph node; Lu, lungs; Lv, liver; Mc, muscle; MV, measles virus; n.a., not available; NPs, nanoparticles; PEG-Lps, pegylated liposomes; PG-NPs, polyglucosamine nanoparticles; PLGA-MP, polylactide-coglycolide microspheres; SC, subcutaneous; Sk, skin; Sp, spleen; and Th, thymus.

23.3.1.2 Effect of Carriers in Vaccine Absorption and Distribution New vaccine formulation systems have also shown to increase the Ag DC uptake, by improving biodistribution or targeting specific tissues or components of the immune system such as DC, M cells, or the complement. Polymeric and lipidic nanoparticles (NPs) have been extensively studied in the last two decades as carrier systems for drug delivery in order to (i) achieve local targeting, (ii) enhanced permeability, (iii) achieve controlled released, (iv) improve the stability of pharmacological substances in biological fluids, and (v) explore the use of alternative routes of administration. Unfortunately, this intensive research in NPs systems has not yet been translated into effective systems for human vaccination, although some findings in this field are promising and the lessons from available NPs studies in vaccination are valuable.

At preclinical stages, NPs systems have also exhibited targeting, enhanced absorption, improved distribution, controlled release, and higher stability for vaccines [43–46]. In addition, these systems allow simultaneous administration of Ag from different pathogens, potentially conferring protection against more than one disease. Besides, it has been argued that conjugation of Ag into NPs mimics more realistically the Ag presentation by the pathogen, therefore triggering a similar response [34].

The effect of NPs systems on Ag absorption and distribution depends on several factors as follows.

Particle Size: In general, vaccine formulation tends to induce more DC uptake and biodistribution in LN in a size-dependent manner, achieving better performance at smaller vaccine particle sizes. NPs of 50 nm particle size have shown to increase Ag uptake and distribution into LN or pathogen-related mucosa/tissue from (i) intradermal [16, 47], (ii) SC [19, 48], (iii) IV [49], (iv) oral [14, 15], (v) intranasal [50], (vi) intraocular [17], and (vii) interstitial [51] vaccine administration. In some instances, even a 30-fold increase in DC uptake can be achieved by using NPs compared to the soluble Ag alone [52–54]. This increment in the Ag uptake or exposure in LN was also correlated with an improved immune response measured by the production of antibody titers or the number of CD8+ T-cells. Besides increasing DC uptake, polylactide-coglycolide (PLGA) NPs also can raise the priming of Ag-specific CD8+ T-cells inducing a balanced TH1/TH2-type antibody response, compared to bigger particles or sustained release from a local depot [55].

Controlled Release: The prolonged exposure achieved by slower Ag released from NPs seemed to play a crucial role in the higher uptake and increased immunogenicity of these formulations [56, 57]. The chemical and physical properties of these NPs also play an important role in vaccination. It has been observed that changes in the composition and manufacturing process of NPs can modify the immune response profile [47, 48, 51].

Targeting: Some NP systems can increment Ag exposure by targeting and activating components of the complement system, in which the vaccine complement complex is processed by phagocytes [58]. M cells, located in the follicle-associated epithelium of Peyer's patch as well as in bronchus-associated lymphoid tissue (BALT), which are considered an important route of entrance for pathogens, have also been targeted with NPs for vaccination [59].

Type of Polymers: There are different types of polymers, synthetic polymers such as PLG (polyglycolide), PLA (polylactide), and PCL (poly(ε-caprolactone)), or natural polymers such as alginate, gelatin, and chitosan. All of them are highly biocompatible and biodegradable material for *in vivo* applications, although chitosan is the most valued due to its mucoadhesive properties and its ability to induce mucosal and systemic immunization [56, 60, 61]. For liposomes, low levels of pegylation (1%) at the surface have been correlated with prolonging Ag retention in LN and enhance DC uptake [18]. However, high levels of pegylation induce a faster drainage to the LN, reducing the total time of Ag exposure and even increasing the blood circulation, which is not desired in vaccination [62, 63].

Zeta Potential: Cationic charge plays a relevant role in Ag exposure and distribution, compared to neutral vesicles. Cationic vesicles as Ag carriers can exhibit more Ag retention in the immune-related tissues, facilitating long-term (14 days) Ag presentation to circulating Ag-specific T-cells and inducing a clear Th1 response [64].

Despite all the mentioned advantages of NPs, these systems do not have a longstanding safety profile in human use [34]. Therefore, further research in NPs toxicity is necessary to fully address the real use of these systems. Other delivery systems such as influenza virosomes (proteoliposomes composed of influenza surface glycoproteins and a mixture of natural and synthetic phospholipids) or endogenous retrovirus envelope-coated baculovirus vector also induce a depot effect after IM vaccination in mice, leading to a longer Ag exposition and reduced clearance [20, 65].

23.3.2 Effect of Route of Administration

Keeping in mind that Ag presentation occurs mainly in the LN, vaccine administration should be planned to deliver Ag to these organs. However, most of the commercial vaccines are injected in the arm, the scapular region deltoid muscle [11]. In addition, DC at different LN react differently even to the same pathogen and can trigger different immune responses under the same stimulus [66]. Therefore, apart from the importance of Ag delivery to LN, targeting the right node could make a difference between successful and unsuccessful immunization. If it is desired to target a specific node to optimize vaccination, its selection must consider the LN that the pathogen first encounters in an infection process.

After vaccination, DC near an administration site can capture the Ag, migrate to a node, and then present themselves to T-cells. Alternatively, vaccines can directly deliver the Ag into the nodes, where there is a high density of DC, providing the Ag presentation by these DC. Then, the migration of these cells is not required. Ag presentation in a particular node would be potentiated if vaccine administration occurs near the targeted LN. Consequently, site of administration should be located into tissues rich in DC or near to the nodes responsible for the first response after pathogen infection. Interestingly, tissues around traditional routes of administration, such as muscles and the SC layers, possess a limited population of DC [67].

LN under the armpits play a key role orchestrating responses against intracellular infections. The particular type of DC and the high density of T lymphocytes in these nodes can send strong Th1 signals in response to Ag exposure, suggesting that parenteral or topical administration in this area may be ideal sites for vaccine administration, especially for the protection against intracellular pathogens. Unfortunately, this particular route has not been used so far. Complementarily, the DC present in the abdominal LN tend to promote humoral response after Ag exposure. Therefore, routes and sites of vaccine administration that would facilitate Ag exposure in these nodes would be ideal for protection against mucosal infections. In the case of blood infections, the spleen is responsible to a great extent for the response against such pathogens. Again, despite its potential as a site/route for vaccine administration for protection against blood pathogens as malaria, research in this field is very limited.

In summary, not only the route but also the site of vaccine administration is crucial for effective protection against infectious diseases. The type of pathogen (intracellular, mucosal, blood-borne, etc.) and the desired type of immune response, humoral or cellular, should be considered in vaccine development in order to plan which lymph organs must be targeted, and accordingly choose the appropriate route and site of vaccine administration and the formulation that is able to promote sufficient absorption and distribution into this lymph organs. In the following section, the current knowledge about the impact of site and route of vaccine administration on the absorption and biodistribution of Ag is summarized and discussed, with special emphasis on distribution into LN or DC.

23.3.2.1 Oral Route Ag exposure in abdominal LN is especially relevant for inducing humoral response, which is crucial for mucosal infections, making oral administration the easiest way to reach this area. Ag absorption after oral vaccination occurs in the M cells [68], which are epithelial cells highly specialized in (i) the phagocytosis of gut lumen macromolecules, particulate Ag and pathogenic, or commensal microorganisms and (ii) the transport of those substances across epithelium to the mucosa-associated lymphoid tissue (MALT) (Peyer's patches). M cells transfer Ag to DC in Peyer's patches, where the Ag presentation occurs directly [69].

Despite the aggressive environment of the gastrointestinal tract that compromises Ag stability, oral vaccination induces a strong immune response able to confer protection against several pathogens. This type of vaccines have been successful in human against poliomyelitis (LAV), cholera (toxin B subunit and inactivated and LAV), typhoid (Ty21a LAV), and rotavirus (LAV and pentavalent live vaccine).

Nevertheless, degradation of the Ag is still an obstacle for immunization following oral administration. In addition, poor oral absorption, interaction with other nonspecific host factors, and preexisting immunity contribute negatively to oral immunization [70]. To overcome those limitations, currently there are two main strategies to improve efficiently the immunity response after oral vaccination:

1. **Ag-Replicating Delivery Systems**: Vaccines containing viral or bacterial vectors, which have been genetically engineered to proliferate in host tissues, replicate the Ag after immunization.
2. **Use of Carriers**: Liposomes, proteasomes, and polymeric microparticles and NPs can protect Ag in the gastro intestinal tract (GIT), while their small size allows absorption through the M cells into Peyer's patches.

The main problem for the Ag-replicating delivery systems is the preexisting immunity to the carrier organism that limits its replication in the host, thus reducing the efficacy of repeated administrations. Furthermore, these systems are generally less immunogenic and normally require the use of adjuvants.

In addition to their protective effect, carrier systems are popular for their ability to release Ag in a controlled rate, increase the time of Ag exposure [56, 57], and to achieve an effect by adhesion of the carrier components to the GIT mucosa [71, 72]. Components of these carrier systems can also exhibit immunogenic properties and serve simultaneously as adjuvants and carriers [61, 73]. Absorption and distribution to LN can be improved using carriers that are able to target receptors of epithelial cells [74], or M cells [75, 76].

23.3.2.2 Intranasal Intranasal vaccine administration is optimal for Ag distribution into the nasal-associated lymphoid tissue (NALT) [15], which contains high levels of DC and T-cells. NALT is especially relevant for immune response against airborne pathogens and to a lesser extent to mucosal infections, due to its predominant polarization to humoral response. This route is characterized by a rapid and direct systemic absorption. Nasal mucosa is highly vascularized and supplied with nerves, glands, and immune cells. The epithelium contains microvilli and M cells that increase the absorption of bacteria, pathogens, and xenobiotics and transport through the pharynx. Intranasal administration has shown to produce greater Ag C_{max} (maximum concentration) and AUC (area under curve of the pathogen or molecule administered as vaccine) compared to IM administration and promote a higher exposure of the Ag in the pathogen-associated mucosae and LN [32]. This route has been able to deliver Ag to different LN including distal ones, demonstrating absorption and distribution into the lymphatic system [24]. Delivery of DC vaccines by this route has shown also to distribute into the lung, spleen, and proximal LN where the administered DC effectively activate naive T-cells leading to the formation of memory cells [13].

However, enzymatic degradation, poor permeability, and reduced exposure time of Ag by this route hinder immunization by intranasal vaccination. This has opened a new field of research to overcome these obstacles by the use of LAVs, adjuvants, mucoadhesives, particulate delivery systems, and virus-like particles [77]. At the moment, there is only one intranasal vaccine approved for human use, a spray formulation against influenza containing LAV subtypes A and B [78], although new systems are under development. For example, chitosan NPs have widely been tested for intranasal formulations in animal studies [57, 61, 73, 79]. This polymer seems to interact with the protein kinase C system and opens the tight junctions between epithelial cells, enhancing the transport of drugs across membranes. Furthermore, chitosan may also protect the drug/agent from enzymatic metabolism, sustain drug release, and prolong its effect. Therefore, this strategy, by delaying Ag release during its residence in the nasal cavity (15–20 min), may be applicable to humans. Chitosan has been tested with three different vaccines, influenza, pertussis, and diphtheria in various animal models and in humans. Another degradable polymer NP (50 nm) based on polypropylene sulfide has also been used for intranasal vaccine delivery showing the penetration of nasal mucosae, transit via M cells, and uptake by APCs in the NALT [50].

Liposomes have been tested for intranasal vaccination, but they show some limitations in terms of sensitivity to host enzymes, instability in storage, and high cost of manufacture. Those caveats lead to the modified systems such as ISCOM (immune-stimulating complexes), virosomes, proteosomes, proteoliposomes, and cochleates. Intranasal evaluation of virosomal influenza vaccines in a Phase I study showed that the use of coadjuvant was necessary to obtain a humoral response. Initial clinical study in healthy subjects has shown intranasally administered proteosome-based influenza and Shigella vaccines to be effective and well tolerated [24]. Unfortunately, Ag absorption and distribution to LN from these studies has not been published.

Intranasal delivery of vaccines into nostrils is an attractive mode of immunization because the lymph nodules in nostrils are very important in the response against pathogens ingested with food [80], but there are risks of Ag transport into the brain through olfactory nerves and could induce important side effects. This route has been used by Oh and coworkers [24] to immunized mice with 50 μg of a DNA vaccine plasmid into the nostrils. The plasmid was found in the systemic circulation at 5 min following intranasal administration with a T_{max} (time-to-reach maximum concentration) of 90 min, while at 15 min the highest distribution was observed in the liver followed by the kidney, heart, lung, brain, and the spleen. Regarding immune-related organs, the highest Ag levels were found in cervical LN followed by mesenteric and iliac. The authors explained the high brain Ag levels by the administration into nostrils [24].

23.3.2.3 ***Intradermal or Transcutaneous*** Skin tissues are populated with different types of very effective APCs that are able to induce proper immune responses against pathogens. Langerhans cells are found in epidermis and DC in dermis. DC in skin preferably trigger a humoral immune response, while Langerhans cells activate the CTLs more efficiently [81], therefore, covering a wide range of infections for which immunization can be achieved by these routes. DNA vaccination using intradermal administration is also associated with a higher number of Ag at the injection site compared to IM route, prolonging the Ag exposure time [25]. In addition, most LN are accessible from these routes if the administration site is close to the targeted nodes. However, the stratum corneum, the top layer of the skin, is difficult to overcome due to its function as an effective barrier for protection against heterologous substances including vaccines. To enhance the Ag penetration across this barrier, the following strategies have been used [82, 83]:

Patch Formulated in Hydrogels: This patch is able to generate a high concentration gradient to force Ag penetration. This system facilitates the global distribution due to the wide region available for its administration. NPs formulated with polymers such as PLGA, PLA, and chitosan seem to be promising carriers. Some of them can penetrate and distribute through hair follicles reaching the mesenteric LN and the spleen, thereby enhancing the immune response [14].

Metal Nanoparticles (Au-NP): Systems with a particle size smaller than 10 nm seem to be skin permeable allowing for the coadministration of antigen-Au-NP. Barfoed et al. [84] showed that this route was more efficient for immunization than the combined

IM/intradermal injection of a DNA vaccine against reproductive and respiratory syndrome virus in pigs. This system has also been examined and shown promising results in preclinical and clinical studies employing a hepatitis B DNA vaccine [85].

Other Nanosystems: Other nanosystems such as modified ISCOM particles or liposomes, especially elastic liposomes, can be mixed with stratum corneum lipids to enhance Ag deposition in skin. There are several studies for hepatitis B and tetanus toxoid using these types of formulations [86–88]. Niosomes, transfersomes, or microemulsions represent the systems under development for transcutaneous vaccination. In addition, physical techniques, for example, iontophoresis, microneedles, electroporation, and ultrasound, have been investigated with encouraging results [82, 83]. Electroporation in particular has allowed DNA vaccines to reach LN to enhance the cellular immune response in mice [28].

23.3.2.4 *Intravenous/Intramuscular/Subcutaneous*

Popularity of these administration (i.e., IV, IM, or SC) routes contrast with the low density of DC and other APCs in these tissues (blood, muscle, and fat). Furthermore, proximity to LN or potential to reach the lymphatic system is normally not considered in the selection of vaccine site of administration by these routes.

Since late 1970s, it has been shown that the route of administration impacts the Ag biodistribution and the type of immune response elicited. Gerbrandy and Bienenstock in 1976 [89] studied the kinetics and the localization of immunoglobulin E (IgE) tetanus antibody in mice after intratracheal (ITR), intraperitoneal (IP), or SC immunization with tetanus toxoid and *Bordetella pertussis* organisms. The authors found that ITR and IP immunization produced similar patterns of response with higher serum PCA (passive cutaneous anaphylaxis) titers in the bronchial LN, spleen, and the serum. On the other hand, SC immunization led to a low primary response and Ag distribution into the lung and draining LN with a higher concentration or recruitment of IgE precursors.

Currently, SC and IM immunization are the preferred routes of administration of licensed vaccines and the same vaccine can be administered by either route (SC or IM) depending on the country of administration [69]. In some cases, IM and SC routes trigger similar results as demonstrated for Zostavax, a commercial herpes zoster LAV for adults aged greater than 50 years. Zostavax is administered by SC route in United States while in Europe the common route is IM. In order to establish the comparison between both routes, a clinical trial conducted in Germany and Spain was published recently. The study found no significant difference in the production of antibodies between the SC and IM routes. Regarding safety, the main difference observed was fewer injection-site reactions after IM than after SC. This observation was in agreement with another trial in children comparing two varicella vaccines, Varivax and ProQuad, containing the same Ag. In the case of pneumococcal and influenza vaccines in adults, immunogenicity and systemic safety were comparable for IM and SC administration, although, in general, the immunogenicity was stronger with IM than with SC [90]. SC administration in DC vaccines has proven to be more effective compared to IV administration, and this effect is associated with better absorption and distribution of the administered DC into the lymphatic system [12].

Following IM administration of 10 μg (1012–1013 copies) of the plasmid pDNAX (pVAX-Hsp60TM814) vaccine against *Trichophyton mentagrophytes*, naked plasmid can persist in muscle tissue for more than 1 year [91]. In the case of pSO2C1 DNA vaccine harboring the *Bacillus thuringiensis* cry11Bb gene, it was reported that plasmid DNA persisted in the muscle of mice for up to 2 years after IM administration of a dose of 5 μg [92]. Needleless delivery systems have also shown to increase the humoral response in rabbits after IM vaccination presumably due to greater plasmid deposition in skin [29].

IV administration of DNA vaccines as naked DNA plasmids normally leads to a rapid blood and tissue degradation of the vectors, while after IM administration their persistence in muscle tissues has been shown to vary depending on the DNA vaccine dose, vector, and use of adjuvants [93]. For example, for *Mycobacterium leprae* 65-kDa heat-shock protein, plasmid DNA was detected in muscle for up to 180 days at a dose of 100 μg/mouse, and at a dose of 20 μg/mouse, plasmid DNA was undetectable in muscle after 30 days.

Tissue targeting has also been reported for DNA vaccines and it depends mainly on the employed vector. For example, adenovirus type 5 (Ad5)-vectored DNA vaccine was found to be distributed only to the spleen and the liver by binding to coxsackievirus and adenovirus receptors [94]. Recombinant vesicular stomatitis virus expressing HIV-1 Gag showed greater persistence in LN compared to other tissues after IM administrations [95].

23.3.2.5 *Other Routes*

Several other routes, and in particular sites of vaccine administration, have a lot of potential to target mucosal and systemic protection. MALT, BALT, conjunctiva-associated lymphoid tissue (CALT), and larynx-associated lymphoid tissues are aggregations of lymph cells with great capacity to trigger protective immune responses. Considering that most harmful pathogens enter the body through mucosal surfaces by ingestion, inhalation, or sexual contact, the MALT is an ideal candidate site for vaccination. However, despite its potential, research in this field is very incipient [96]. Vaccine administered to BALT was found to be more effective than parenteral or nasal administration [97]. Similar results were found after pulmonary immunization against influenza [98], but the use of this route is still in initial stages.

Ophthalmic administration is optimal for reaching CALT and confers protection against not only ocular infections but also mucosal and systemic pathogens [99, 100]. Biodistribution studies after ophthalmic administration of mannosylated NPs loaded with a *Brucella ovis* Ag in mice demonstrated persistent distribution in the nasal and ocular mucosa as well as gastrointestinal tract, making this an effective system of Ag release accompanied by an elevated immune response [17]. Other mucosal routes for vaccination have been attempted including vaginal, pulmonary, and rectal; however, no relevant results have been reported to date [101].

23.3.3 Metabolism and Excretion

The metabolism and excretion processes are not well studied for vaccines because PK studies are not required for vaccine approval, and also because they are assumed to be irrelevant regarding vaccine efficacy or interaction with other drugs. Considering the low and expected few doses administered in vaccination, these assumptions seem reasonable. However, as chronic Ag exposure is associated with tolerance development, complete Ag elimination should be guaranteed in order to avoid chronic exposure leading to a decrease in vaccine efficacy. Regarding DNA vaccines, one of the main concerns is the plasmid integration of vaccine into host DNA. This integration depends mainly on the nature of the foreign plasmid and DNA, but it must be considered that a very low elimination (which can take years) increases the chances of plasmid integration. Therefore, demonstration of complete elimination of these vaccines may become relevant to assure safety or to avoid interaction of vaccine DNA with other pathogens or microorganisms.

23.3.3.1 Effect of Dose Level and Dosing Schedule It is well established that under repeated Ag administration, antibody affinity increases and can reach affinities even 100-fold higher than after first administration [102]. To a lesser extent, this effect is also observed in CTL affinity to Ag-expressing cells. That affinity increase after several Ag exposures explains the greater immunization success upon repeated vaccine doses [103]. Although chronic exposure to Ag may lead to tolerance development, greater vaccination efficacy caused by increasing the number of doses has been reported for hepatitis B [104], rotavirus [105], and pneumonia [106], but the increase in effectivity is not always cost-effective [107]. Nevertheless, not only the number of doses but also the number of sites of administration seems to affect the vaccination efficacy [108].

Dosing schedule also impacts vaccination efficacy. Thus, Hepburn et al. [109] compared the immune response of two regimens with the same number of doses and the same initial and final points of immunization for the UK anthrax vaccine: normal schedules (0, 3, 6, and 32 weeks) and extended (0, 10, 13, and 32 weeks) schedules. Results showed that concentrations of protective antibodies were higher ($p<0.05$) among recipients of the extended versus regular schedule.

Although protective effect of vaccines is dose dependent, the maximum effect-type behavior has also been reported for immunization [108]. The optimum DNA-dosing regimen using particle mediated epidermal delivery (PMED), a DNA vaccine against hepatitis B, in humans was determined in several clinical trials. A minimum of 1.0 μg and two delivery sites were required to induce a detectable response in most subjects, while higher responses and a better response rate could be achieved by increasing the dose to 4.0 μg DNA administered into four delivery sites [110]. A subsequent study in preimmune subjects showed that increasing the DNA dose up to 8.0 μg (four delivery sites) provided no additional benefit, indicating that a 4.0-μg DNA dose was sufficient using PMED in humans [108].

Similar results were reported by Song et al. [111] when comparing the long-term immunogenicity of intradermal influenza vaccination at one-fifth and one-half of the conventional IM Ag dose. They found that the immunogenicity was comparable for both formulations at 1 month postvaccination [111].

23.3.4 PK Considerations

The influence of biopharmaceutics properties on vaccine PK processes such as absorption and distribution has been commented in detail in the previous section of this chapter. Because experimental PK studies are not currently required for vaccines' regulatory registration process across the world [42, 112–114], there are not much data in terms of PK parameters associated with the apparent volume of distribution or total clearance. This section illustrates with some studies from the literature, the most common PK parameters reported during vaccine development, and the factors that may affect them.

PK studies have been carried out to characterize the kinetics of adjuvants or vehicles coadministered in vaccination or even to characterize the kinetics of excipients, as illustrated with formaldehyde used in certain vaccines to inactivate viruses and to detoxify bacterial toxins [42, 115]. Because Ag is administered in a relatively small amount and widely spaced, processes such as distribution, accumulation, and elimination are very difficult to quantify, even in animal models. Nevertheless, for some vaccines, Ag kinetic studies have been conducted in preclinical and clinical studies and certain PK parameters have been reported. These parameters are represented by C_{max}, T_{max}, AUC, and half-life, which is in general longer than for small molecules. Table 23.2 lists some examples reported in the literature showing the PK parameters studied during the first phases of vaccine development.

TABLE 23.2 Pharmacokinetic Parameters Reported in the Literature for Some Studies Developed in Preclinical and Clinical Stages

Vaccine	Antigen	Type	Formulation	Route/Species	Dosing	Dose	Model	K_{el}	$T_{1/2}$	*AUC*	C_{MAX}	T_{MAX}	References
Luteinizing hormone-releasing hormone (LHRH)	Synthetic molecule (LHRH)	—	Emulsion Water-in-oil adjuvant	IM Rats	Single dose	25 μg	Noncompart	—	158.67 h	(0–28 d 4.5 h%ID/g	0.026%ID/g	24 h	[116]
Human papilloma virus (HPV)	HPV16L1	DNA vaccine	PBS solution (naked)	IM BALB/c Mice	Single dose	100 μg	Noncompart	—	87.66 h	14.02 g/L*h	2.5 g/L	0.08 h	[65]
Human papilloma virus (HPV)	HPV16L1	DNA vaccine	PBS solution (naked)	IV BALB/c Mice	Single dose	100 μg	Noncompart	—	99.27 h	4839.3 g/L*h	19033.28 g/L	—	[65]
Human papilloma virus (HPV)	HPV16L1	DNA vaccine	Baculovirus	IM BALB/c Mice	Single dose	1.55 ng	Noncompart	—	289.13 h	25.63 g/L*h	0.75 g/L	1 h	[65]
Human papilloma virus (HPV)	HPV16L1	DNA vaccine	Baculovirus	IV BALB/c Mice	Single dose	1.55 ng	Noncompart	—	174.27 h	46.31 g/L*h	6.49 g/L	—	[65]
Tetanus	Human tetanus Ig (P-HTIG)	—	Solution	IM Healthy volunteers	Single dose	500 IU	Noncompart	—	—	0–42 d 6.11 IU*d/mL	0.313 IU/mL	4.46 d	[117]
Tetanus	Human tetanus Ig (P-HTIG)	—	Solution	IM Healthy volunteers	Single dose	500 IU	One-compart	0.0246/d	28.17 d	—	—	—	[117]
Anticaries	pGJA-P/ VAX1	DNA vaccine	Solution Coadministered with bupivacaine	Nostril (intra nasal) BALB/c Mice	Two doses 0–14 d	100 μg	Noncompart	—	—	(0–24 h) 50.89 ng*h/mL	57.254 ng/mL	30 min	[32]

ID, percentage of the injected dose and PBS, phosphate buffer solution, IU, international units.

Chang et al. [116] have reported the PK of UBITh-LHRH a synthetic luteinizing hormone-releasing hormone peptide-based immunotherapeutic vaccine for the treatment of androgen-responsive prostate cancer. Currently, this type of cancer is treated with an luteinizing hormone-releasing hormone (LHRH) antagonist, cetrorelix, which is administered subcutaneously. Despite the structural similarity between the two molecules, there are significant differences in their PK characteristics. The half-life for UBITh was much longer than that for cetrorelix, 168.57 versus 35.6 h, as well as the parameter associated with absorption, T_{max}, 24 h compared to 2 h, respectively. The strategy used for UBITh, based on the association of peptides with a low ability to induce immunogenicity with an adjuvant, clearly improved the immune response through a modification of PK properties.

There are much more PK studies for DNA vaccines than for other types of vaccines. The ability to detect either the plasmid or its expression might provide an important advantage for PK characterization of DNA vaccines. However, the PK characterization of naked DNA has a serious limitation because DNA is very unstable in blood, which would hamper its access to target organ in order to induce a proper response. To overcome that limitation, DNA in these vaccines is normally incorporated in a vector that totally alters its PK.

HPV is an infection preventable by prophylactic vaccination with virus-like particles, similar to the case of Gardasil and Cervarix vaccines. However, the vaccination efficacy is low (~30%), therefore, significant efforts need to be made to improve the immunogenic effect. For example, Cho and coworkers [65] have reported the PK of a new HPV vaccine formulated on a nonreplicating baculovirus vector to deliver HPV16L1-encoding genes. This vaccine was able to trigger a greater cellular immune response compared to Gardasil [118]. The impact of the route of administrations, IM and IV, was also evaluated by estimating the bioavailability of this vaccine together with other PK parameters such as C_{max}, T_{max}, and *AUC*. The absolute bioavailability (F), calculated by the ratio between AUC_{IM} and AUC_{IV}, was 55% for baculovirus formulation, while 0.03% for naked DNA, showing the clear benefit of the encapsulation strategy for this type of vaccines. In addition, it is also known that IM administration exhibits a longer half-life than IV. But even independently of the route, baculovirus formulation plays an important role in establishing a depot effect, which can also protect the gene from serum nucleases. However, this persistence of foreign DNA in muscle leads to exploring whether there is an integration of genes into chromosomes. In fact, as has been previously commented, the FDA guidance require this type of study when the copy numbers of DNA vaccine are greater than 30,000/μg of host DNA 90 days after administration. In the study carried out by Cho et al., the dose was inferior than 1100 copies/μg of DNA 30 days after IM administration.

Another example that shows the relevance of PK studies is presented in Table 23.2 for tetanus. It is a Phase I one-center, double-blind, randomized trial in healthy subjects to characterize the safety and PK of a new pasteurized, human tetanus immunoglobulin (P-HTIG) vaccine. Although the descriptive PK parameters correspond to the antibody titers produced after vaccination, some inferences about the Ag PK can be extracted. In that study, two groups received two different batches of the P-HTIG alone and two others received the P-HTIG combined with tetanus–diphtheria (Td) vaccine. Local reactions and tetanus antibody titers at Day 0 (D0) or baseline and until Day 42 (D42) postinjection were evaluated by noncompartmental and compartmental model analysis.

For groups treated with P-HTIG, the time profiles of tetanus titers were described by the Bateman function, using a one-compartment model, where the first-order absorption rate constant, k_a, was estimated in 15 per day, the time of latency (T_{lag}) was 0.865 days, and the first-order elimination rate constant was 0.0246 per day. With this approach, the authors were able to quantify the input and output functions, although this is not a common analysis. In fact, to compare the tetanus vaccination and tetanus plus tetanus–diptheria, a noncompartmental analysis was performed. The geometric mean titers for Day 0 and Day 42 for each individual, AUC calculated by trapezoidal method from 0 to 42 days, and the parameters, T_{max} and C_{max}, from graphical representation were obtained. In that analysis, C_{max} and T_{max} were 0.313 ± 2.49 IU/mL and 4.46 ± 1.92 days, respectively, for P-HTIG plus placebo group, while 15.2 ± 2.42 IU/mL and 18.80 ± 1.40 days for P-HTIG plus Td vaccine group. Despite the differences between both groups, the times to reach the concentration threshold (0.01 or 0.1 IU/mL) were similar for both groups, although the times to reach the threshold of 0.01 IU/mL (minimum seroprotective level) were delayed for the group combined with Td vaccines (1.47 ± 2.13 days) compared to the P-HTIG plus placebo group (1.64 ± 1.62 days).

PK of antibody titers has also been used to evaluate the equivalence of two rabies vaccines after administration in humans. Although bioequivalence based on *AUC* and C_{max} was not demonstrated, similar absorption and equal clinical seroconversion between two rabies vaccines with the same Ag were shown [119].

It is clear that different types of vaccines, together with the formulation or routes, can influence the outcome *in vivo*. Several peptide-based vaccines have been included in various clinical trials to induce protective immunogenicity against a variety of infections and tumors. However, some of them have limited peptide-specific immune response with undetectable therapeutic effects. For example, peptides (HPV)16 E7 and (Ad5) E1A that are recognized by CTL lead to very different effect when used for HPV vaccination. (HPV)16 E7 peptide is able to induce a specific CTL response that prevents the outgrowth of HPV16 E7-expressing tumors while the peptide encoding an Ad5 E1A CTL epitope results in CTL tolerance and enhanced growth of an Ad5 E1A-expressing tumor. In order to explain that difference, the PK of both types of tritiated peptides was studied in mice after SC vaccination. Results showed that the tolerizing peptide spread through the body 16 times faster than the activating peptide and was cleared at least 2 times faster. In addition, there was clear evidence that T-cell activation or tolerization was determined by Ag concentration over time and Ag persistence. The HPV16 E7 peptide kinetics correlated with the kinetics of HPV16 E7-specific CTL induction, while Ad5 E1A peptide resulted in physical deletion of preexisting Ad5 E1A-specific CTLs within 24 h after injection. The ability of adjuvants to prevent tolerance is believed to be a result of slow Ag release from the adjuvant, providing a depot effect that allows APCs to enter the vaccine site and collect the peptide. These Ag-loaded APCs then migrate to the LN and initiate T-cell activation. This study demonstrates that the activating HPV16 E7 peptide persists in all organs longer than the tolerizing Ad5 E1A peptide as a result of a slow release of the HPV16 E7 peptide from the vaccine site. Differences in kinetics may be the consequence of a combination of intrinsic peptide characteristics, that is, hydrophobicity, charge, structure, and MHC class I binding affinities of the peptide. Ad5 E1A peptide is more hydrophilic than the HPV16 E7 peptide and negatively charged compared with a positive charge of HPV16 E7. These variables may influence the PK of the peptide-based vaccines and therefore its immunogenicity. In consequence, prediction of the effects that peptides may have on the immune system using PK profiles may optimize the safety and efficacy of peptide-based vaccinations in humans [10].

23.4 MATHEMATICAL MODELING FOR VACCINE OPTIMIZATION IN CANCER TREATMENT

Several authors have made efforts to develop mathematical models in order to describe and predict the fate of an immunological therapy based on vaccine, generating additional knowledge about the kinetics of the immune system response, the effects of treatments, and the dynamics of the disease. Most of the examples reported in the literature are focused on cancer therapy. In this particular case, vaccination requires the activation and expansion of tumor-specific T-cytotoxic cells to allow their migration into the tumor tissue in order to induce the antitumor effect. One of the main benefits of mathematical models is the opportunity to simulate *in silico* several dosing scenarios and predict the outcome in preclinical models, and therefore use this information to drive the drug development into Phase I studies.

Mathematical models in this field are basically divided into two categories: agent-based modeling (ABM) and models based on differential equations (DEs). ABM gives useful information at the cellular level for the design of optimal schedules considering some biological hypotheses previously proved in simulation experiments. ABM can describe complex systems in a flexible way, including the handling of entity heterogeneity and physical space. ABMs aim at recreating and predicting the cellular interactions through simulating the behavior and the interactions of autonomous entities (cells and molecules). The dynamics of agents can be described as a function of time, position, and an internal state that includes most important properties of the agent [120].

DE models include different components of the immune system, cells, pathogens, and treatments from a population and try to estimate the model that better describe the population behavior as well as the individual data. This approach is less accurate in describing specific immunological processes, but allows the extraction of fundamental properties about the parameter's space via sensitivity analysis.

Pappalardo et al. in 2014 [121] developed a DE mathematical model to evaluate the number and frequency of vaccine boosts required to reach a long-lasting and protecting memory T-cell response from immunization with a DC-based vaccine. Other examples, such as those reported by Wilson and Levy [122] and Parra-Guillen et al. [123], were focused on the antitumor effect as the final response of vaccination, considering the effect of other endogenous molecules involved in the immune systems, and describing the main features implicated in the dynamics of the tumor growth influenced by the time at which the vaccine was administered.

This approach allowed the development of a semimechanistic model to describe the efficacy induced by a single IV dose of CyaA-E7, a vaccine able to target Ag to DC, in mice bearing tumor cells. In this mathematical model, the antitumor effect induced by CyaA-E7 was delayed in time for treatment-sensitive animals and was also associated with a resistance effect that was observed in some mice with a certain tumor size. This structure, shown in Figure 23.2, included the relapse observed in those mice sensitive to the vaccine, but with a regrowth over time. Therefore, this model allows the simultaneous description of the two types of populations, responders, and nonresponders for treatment together with control or nontreated animals. Note that PK data were not available in the study assuming an instantaneous input followed by an exponential elimination.

The authors explored the applicability of this model using model-based computing simulations to evaluate which components had higher impact on the efficacy in order to optimize the dosing schedule or the combination therapy. The model represents a useful framework to maximize the information obtained in animals to explore clinical designs for immunotherapies with a similar mechanism of action.

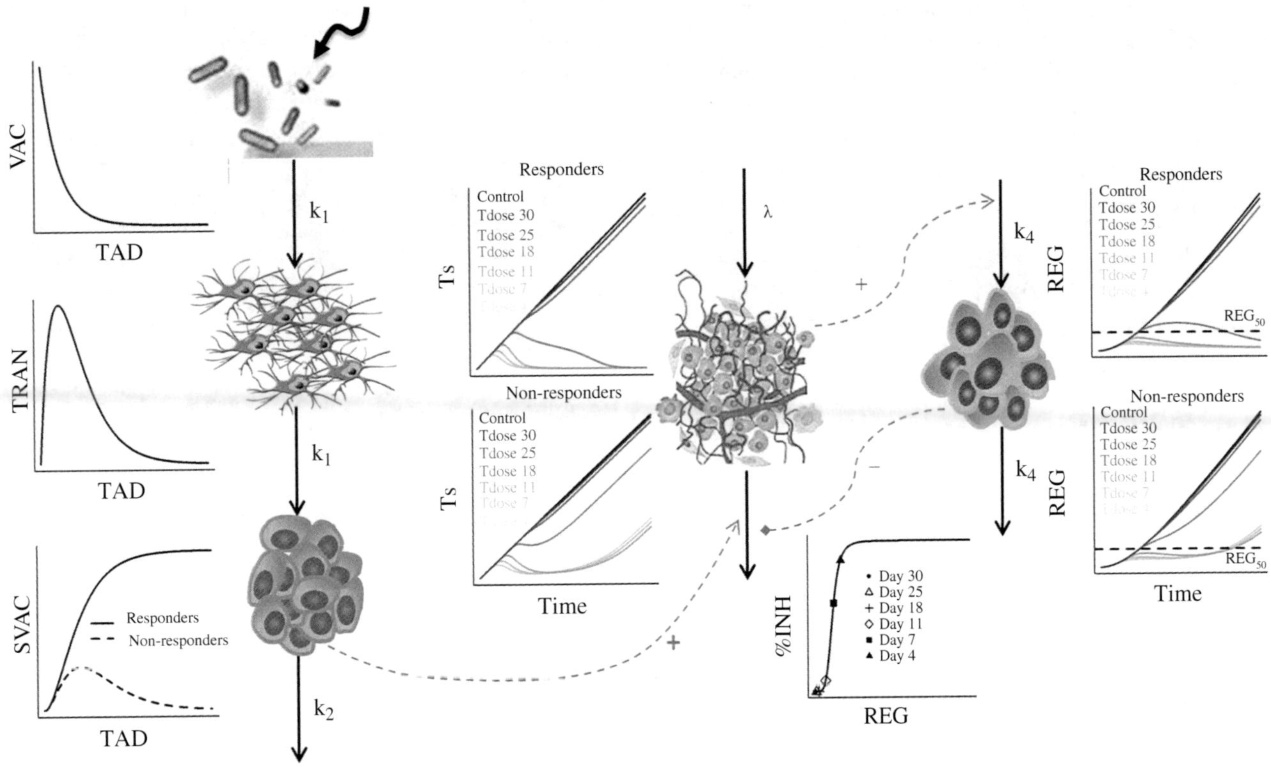

FIGURE 23.2 Schematic representation of the mechanism proposed for the vaccine, CyaA-E7, in tumor-bearing mice and the predictions from the mathematical model. After vaccine (VAC) administration and through a transit compartment (TRAN), the vaccine triggers a signal (SVAC) able to decrease tumor size (Ts). Two different populations (responder and nonresponders) at the level of SVAC elimination were identified. An inhibition of vaccine efficacy induced by a regulator compartment (REG) controlled by tumor size was detected. The behavior of the different model compartments over time for both populations, along with the percentage of inhibition (%INH) versus the amount in the regulator compartment under no vaccine administration and highlighting the REG amount at relevant time points, is presented. k_1: first-order rate constant controlling vaccine elimination and transit between compartments; k_2: first-order rate constant accounting for SVAC degradation; λ: zero-order rate constant of tumor growth; k_3: the vaccine efficacy second order rate constant; k_4: the first-order rate constant controlling the regulator compartment dynamics; REG_{50}: amount in the regulator compartment needed to inhibit vaccine activity by a half; γ: the shape of that inhibitory process. TAD: time after dose; T_{dose}: day of vaccine administration, represented by different shades of gray.

Despite the acute efficacy of CyaA-E7, reaching a long-lasting immunogenic response is normally not achieved. As a consequence, this vaccine was also administered in combination with other therapeutic agents. Parra-Guillen and coworkers proposed an extension of the previous model by incorporating certain biological mechanisms promoted by the application of CyaA-E7 with CpG (a TLR9 ligand), a coadjuvant, and/or cyclophosphamide (CTX) [124]. CpG is a TLR9 ligand able to increase the innate immune system, while CTX induces an inhibition of the immunosuppression elicited by the tumor, through depletion of the regulator T cells (Treg) cells. The effect of CPG led to an amplification of the immune response of the vaccine and a reduction of time delay required to start the response, while CTX was able to decrease the tumor-induced inhibition of vaccine efficacy over time, along with a delayed induction of tumor cell death. The model provided favorable results regarding the percentage of cured individuals and faster tumor shrinkage for the tritherapy compared to biotherapy; a good agreement between model predictions and experimental data was obtained and the benefit of combination therapies with different mechanisms of action was highlighted. In this work, model simulations based on the preclinical data predicted a higher probability of cure for CpG and CTX compared to monotherapy; these predictions were confirmed by later clinical studies with these agents. This example confirmed the applicability of the model in the optimization of clinical trials. Although this model does not reproduce the complexities of the immune system response after vaccination in detail, this framework exploits the main known features of the combined therapy while providing the flexibility to incorporate these features in a semimechanistic framework.

Applying mathematical models for a better understanding of this type of complex systems represents a useful strategy to identify mechanisms that are able to describe the interaction between the different immune cell populations and tumors [125, 126].

23.5 SYSTEMS VACCINOLOGY: APPLICATION OF SYSTEMS BIOLOGY IN PERSONALIZED VACCINATION

Unsuccessful vaccination attempts after decades of research have highlighted the need for a better understanding of the pathogen–host interactions, and the types of immune responses necessary to confer protective immunity. In the last years, computational biology approaches are gaining importance in guiding vaccine design and vaccine optimization and personalization, and subsequently systems vaccinology has emerged as a field able to tackle the main challenges of vaccination.

Systems vaccinology aims to understand what are the factors that determine a successful immunization, highlighting that the focus of vaccine development should not be on the vaccine alone but also on the subjects whom to be immunized. Some vaccines have been effective in some individuals and not in others; therefore a deep understanding of the immune response against a pathogen, and the knowledge of the immune state of a particular individual can predict, not only how effective a vaccine could be in a particular individual, but also how to boost vaccination to immunize a subject with low chances of getting protective immunization after normal vaccination [127].

Recent advances in nanotechnology, robotics, optics, and high throughput techniques allow laboratories to assess entire genomes, sets of transcripts (transcriptome), proteins (proteome), and metabolites (metabolome) of cells and tissues. Systems biology utilizes and integrates the large amount of data generated by these techniques in order to describe the complex interactions between all parts of a biological system, with the ultimate goal of predicting the behavior of the system [128–130].

Researchers can now make system-wide measurements of immune responses, and use computational approaches to identify molecular signatures (e.g., patterns of gene expression induced after vaccination) that correlate with and predict subsequent adaptive immune responses [127, 131–133]. The first successful example of systems vaccinology came from the analysis of immunization with the yellow fever LAV YF-17D, which was selected as a gold standard because of its high efficacy [134]. A single immunization with YF-17D stimulated robust Ag-specific CD8+ T-cells and neutralized antibody responses persisting for several decades [135]. Querec et al. and Sékaly et al. [131, 136] performed independently transcriptomic analysis of peripheral blood mononuclear cells isolated on the days after vaccination with YF-17D. They identified signatures of gene expression, induced 3 or 7 days after vaccination, which correlated with the magnitude of the later Ag-specific CD8+ T-cell and neutralizing antibody responses. By subsequent use of machine-learning techniques, they validated the predictive capacity of such signatures by assessing their ability to predict the magnitude of the CD8+ T-cell and neutralizing antibody response in an independent clinical study with subjects vaccinated with YF-17D [127].

However, it was later proved that gene expression signature that predicted effective immunization after vaccination with YF-17D did not apply to other pathogens. As different pathogens activate different immune responses, gene expression signatures are different for each pathogen and even for each vaccine. Different gene expression signatures have been described for different vaccines against the same pathogen, as was reported in the studies with the trivalent inactivated influenza vaccine (TIV) and the live-attenuated influenza vaccine (LAIV) [132]. Similar results were described for studies with the carbohydrate meningococcal vaccine (Menimmune) and the conjugate meningococcal vaccine (Menectra) [137].

Genetic and environmental factors can affect the individual response to a vaccine. Several studies have estimated the influence of genetics on vaccine induced immunity, finding heritability for antibody response to range from 39% to 89% [138–141]. In addition, individual immune response is also affected by the presence of obesity [142] or undernutrition [143], and the composition of the local microbiome [144]. Apart from keeping the search of early predictors of effective immunization through vaccination, the next goal of systems vaccinology is to find optimal and personalized procedures to boost vaccination in individuals, whose genetic features or metabolic states prevent them to trigger a proper immune response that is able to confer protection against pathogens after vaccination.

The major challenge that systems vaccinology faces is the integration of enormous amount of information from different origins (proteomics, genomics, transcriptomics, etc.). The analysis of these enormous multivariate data sets, which normally include tens of thousands of variables measured at different time points, is a complex statistical task with no unique answer. There are different methodologies to build gene expression-based predictive models and analysis of the same data from different teams could lead to different models [145]. In addition, normal "omics" determinations are measured from blood samples, and even when this information has been used to successfully find gene expressions signatures of effective vaccination, a much more relevant data should be obtained from "omics" studies from immune-related tissues samples, but this arises many technical difficulties.

Systems vaccinology has also contributed to the understanding of the immune system as this approach is able to identify the immune function of molecules, which are considered as not immune relevant, and to identify interactions between immune components that are unknown [131, 132]. It is expected that systems vaccinology to be applied to other aspects of vaccination. For example, the effect of ADME processes or biopharmaceutics in vaccination has not been evaluated with systems vaccinology; evidently this approach would be of great help in vaccine development to find optimal vaccine adjuvants, and to compare between vaccine formulations or different routes and sites of vaccine administration. We expect the following years to bring many exciting findings in the fusion of systems vaccinology with vaccinology biopharmaceutics.

23.6 CONCLUDING REMARKS

Whereas there has been a significant advance in the knowledge of vaccine biopharmaceutics and its impact on immune response, including the effect of formulation and route of administration, the lack of information of the disposition characteristics represented by the ADME processes, and the vaccine mechanisms of actions represents a major limitation to optimize vaccine efficacy, to establish personalized dosing schedule, and to identify patients associated with nonprotective response. Therefore, there is an urgent need to perform studies based on PK, pharmacodynamics, and systems biology principles, to get the longitudinal information required, that once analyzed under the model-based framework (as it represents the current practice for a vast array of therapeutic strategies in other fields of drug development), can be linked to the biopharmaceutical characteristics of the vaccine formulation.

REFERENCES

[1] Gravier R, Dory D, Laurentie M, Bougeard S, Cariolet R, Jestin A. *In vivo* tissue distribution and kinetics of a pseudorabies virus plasmid DNA vaccine after intramuscular injection in swine. Vaccine 2007;25:6930–6938.

[2] Bloom BR, Lambert P-H. *The Vaccine Book*. Academic Press; 2002.

[3] Ferraro B, Morrow MP, Hutnick NA, Shin TH, Lucke CE, Weiner DB. Clinical applications of DNA vaccines: current progress. Clin Infect Dis 2011;53:296–302.

[4] Timmerman JM, Levy R. Dendritic cell vaccines for cancer immunotherapy. Annu Rev Med 1999;50:507–529.

[5] Radford KJ, Caminschi I. New generation of dendritic cell vaccines. Hum Vaccin Immunother 2013;9:259–264.

[6] García F, Plana M, Climent N, León A, Gatell JM, Gallart T. Dendritic cell based vaccines for HIV infection: the way ahead. Hum Vaccin Immunother 2013;9:2445–2452.

[7] Cantisani R, Pezzicoli A, Cioncada R, Malzone C, De Gregorio E, D'Oro U, Piccioli D. Vaccine adjuvant MF59 promotes retention of unprocessed antigen on lymph node macrophage compartments and follicular dendritic cells. J Immunol. 2015;194:1717–1725. DOI: 10.4049/jimmunol.1400623.

[8] Genuis SJ. Chemical sensitivity: pathophysiology or pathopsychology? Clin Ther 2013;35:572–577.

[9] Silvestrini B. Immunokinetics: a new approach to vaccines. A working hypothesis. Ann Ist Super Sanita 2013;49:306–308.

[10] Weijzen S, Meredith SC, Velders MP, Elmishad AG, Schreiber H, Kast WM. Pharmacokinetic differences between a T cell-tolerizing and a T cell-activating peptide. J Immunol 2001; 166:7151–7157.

[11] Malik B, Rath G, Goyal AK. Are the anatomical sites for vaccine administration selected judiciously? Int Immunopharmacol 2014;19:17–26.

[12] Eggert AA, Schreurs MW, Boerman OC, Oyen WJ, de Boer AJ, Punt CJ, Figdor CG, Adema GJ. Biodistribution and vaccine efficiency of murine dendritic cells are dependent on the route of administration. Cancer Res 1999;59:3340–3345.

[13] Vilekar P, Awasthi V, Lagisetty P, King C, Shankar N, Awasthi S. *In vivo* trafficking and immunostimulatory potential of an intranasally-administered primary dendritic cell-based vaccine. BMC Immunol 2010;11:60.

[14] Peyre M, Fleck R, Hockley D, Gander B, Sesardic D. *In vivo* uptake of an experimental microencapsulated diphtheria vaccine following sub-cutaneous immunisation. Vaccine 2004;22:2430–2437.

[15] Eyles JE, Bramwell VW, Williamson ED, Alpar HO. Microsphere translocation and immunopotentiation in systemic tissues following intranasal administration. Vaccine 2001; 19:4732–4742.

[16] Eby JK, Dane KY, O'Neil CP, Hirosue S, Swartz MA, Hubbell JA. Polymer micelles with pyridyl disulfide-coupled antigen travel through lymphatics and show enhanced cellular responses following immunization. Acta Biomater 2012;8:3210–3217.

[17] Sánchez-Martínez M, da Costa MR, Quincoces G, Gamazo C, Caicedo C, Irache JM, Peñuelas I. Radiolabeling and biodistribution studies of polymeric nanoparticles as adjuvants for ocular vaccination against brucellosis. Rev Esp Med Nucl Imagen Mol 2013;32:92–97.

[18] Zhuang Y, Ma Y, Wang C, Hai L, Yan C, Zhang Y, Liu F, Cai L. PEGylated cationic liposomes robustly augment vaccine-induced immune responses: role of lymphatic trafficking and biodistribution. J Control Release 2012;159:135–142.

[19] Vicente S, Goins BA, Sanchez A, Alonso MJ, Phillips WT. Biodistribution and lymph node retention of polysaccharide-based immunostimulating nanocapsules. Vaccine 2014;32: 1685–1692.

[20] Zurbriggen R, Novak-Hofer I, Seelig A, Glück R. IRIV-adjuvanted hepatitis A vaccine: *in vivo* absorption and biophysical characterization. Prog Lipid Res 2000;39:3–18.

[21] Derrick SC, Yang A, Parra M, Kolibab K, Morris SL. Effect of cationic liposomes on BCG trafficking and vaccine-induced immune responses following a subcutaneous immunization in mice. Vaccine 2015;33:126–132.

[22] Lorin C, Segal L, Mols J, Morelle D, Bourguignon P, Rovira O, Mettens P, Silvano J, Dumey N, Le Goff F, Koutsoukos M, Voss G, Tangy F. Toxicology, biodistribution and shedding profile of a recombinant measles vaccine vector expressing HIV-1 antigens, in cynomolgus macaques. Naunyn Schmiedebergs Arch Pharmacol 2012;385:1211–1225.

[23] Tuomela M, Malm M, Wallen M, Stanescu I, Krohn K, Peterson P. Biodistribution and general safety of a naked DNA plasmid, GTU®-MultiHIV, in a rat, using a quantitative PCR method. Vaccine 2005;23:890–896.

[24] Oh YK, Kim JP, Hwang TS, Ko JJ, Kim JM, Yang JS, Kim CK. Nasal absorption and biodistribution of plasmid DNA: an alternative route of DNA vaccine delivery. Vaccine 2001;19: 4519–4525.

[25] Pal R, Yu Q, Wang S, Kalyanaraman VS, Nair BC, Hudacik L, Whitney S, Keen T, Hung CL, Hocker L, Kennedy JS, Markham P, Lu S. Definitive toxicology and biodistribution study of a polyvalent DNA prime/protein boost human immunodeficiency virus type 1 (HIV-1) vaccine in rabbits. Vaccine 2006;24:1225–1234.

[26] Liu H-F, Li W, Lu M-B, Yu L-J. Pharmacokinetics and risk evaluation of DNA vaccine against Schistosoma japonicum. Parasitol Res 2013;112:59–67.

[27] Endmann A, Oswald D, Riede O, Talman EG, Vos RE, Schroff M, Kleuss C, Ruiters MH, Juhls C. Combination of MIDGE-Th1 DNA vaccines with the cationic lipid SAINT-18: studies on formulation, biodistribution and vector clearance. Vaccine 2014;32:3460–3467.

[28] Dolter KE, Evans CF, Ellefsen B, Song J, Boente-Carrera M, Vittorino R, Rosenberg TJ, Hannaman D, Vasan S. Immunogenicity, safety, biodistribution and persistence of ADVAX, a prophylactic DNA vaccine for HIV-1, delivered by *in vivo* electroporation. Vaccine 2011;29:795–803.

[29] Doukas J, Morrow J, Bellinger D, Hilgert T, Martin T, Jones D, Mahajan R, Rusalov D, Sullivan S, Rolland A. Nonclinical biodistribution, integration, and toxicology evaluations of an H5N1 pandemic influenza plasmid DNA vaccine formulated with Vaxfectin®. Vaccine 2011;29:5443–5452.

[30] Bråve A, Gudmundsdotter L, Sandström E, Haller BK, Hallengärd D, Maltais AK, King AD, Stout RR, Blomberg P, Höglund U, Hejdeman B, Biberfeld G, Wahren B. Biodistribution, persistence and lack of integration of a multigene HIV vaccine delivered by needle-free intradermal injection and electroporation. Vaccine 2010;28:8203–8209.

[31] Wang S, Nie Q, Zheng L-Y, Hu J, Luo E-J. *In vivo* kinetics and biodistribution of a Hantaan virus DNA vaccine after intramuscular injection in mice. Virol Sin 2010;25: 177–182.

[32] Liu C, Fan M, Xu Q, Li Y. Biodistribution and expression of targeted fusion anti-caries DNA vaccine pGJA-P/VAX in mice. J Gene Med 2008;10:298–305.

[33] Ramirez K, Barry EM, Ulmer J, Stout R, Szabo J, Manetz S, Levine MM, Pasetti MF. Preclinical safety and biodistribution of sindbis virus measles DNA vaccines administered as a single dose or followed by live attenuated measles vaccine in a heterologous prime–boost regimen. Hum Gene Ther 2008;19:522–531.

[34] Gregory AE, Titball R, Williamson D. Vaccine delivery using nanoparticles. Front Cell Infect Microbiol 2013;3:13.

[35] Glenny AT, Buttle GAH, Stevens MF. Rate of disappearance of diphtheria toxoid injected into rabbits and guinea – pigs: toxoid precipitated with alum. J Pathol 1931;34:267–275.

[36] Harrison WT. Some observations on the use of alum precipitated diphtheria toxoid. Am J Public Health Nations Health 1935;25:298–300.

[37] Devarajan P, Chen Z. Autoimmune effector memory T cells: the bad and the good. Immunol Res 2013;57:12–22.

[38] Li H, Nookala S, Re F. Aluminum hydroxide adjuvants activate caspase-1 and induce IL-1beta and IL-18 release. J Immunol 2007;178:5271–5276.

[39] McKee AS, Munks MW, MacLeod MKL, Fleenor CJ, Van Rooijen N, Kappler JW, Marrack P. Alum induces innate immune responses through macrophage and mast cell sensors, but these sensors are not required for alum to act as an adjuvant for specific immunity. J Immunol 2009;183: 4403–4414.

[40] Didierlaurent AM, Collignon C, Bourguignon P, Wouters S, Fierens K, Fochesato M, Dendouga N, Langlet C, Malissen B, Lambrecht BN, Garçon N, Van Mechelen M, Morel S. Enhancement of adaptive immunity by the human vaccine adjuvant AS01 depends on activated dendritic cells. J Immunol 2014;193:1920–1930.

[41] Vilalta A, Shlapobersky M, Wei Q, Planchon R, Rolland A, Sullivan S. Analysis of biomarkers after intramuscular injection of Vaxfectin-formulated hCMV gB plasmid DNA. Vaccine 2009;27:7409–7417.

[42] Tegenge MA, Mitkus RJ. A physiologically-based pharmacokinetic (PBPK) model of squalene-containing adjuvant in human vaccines. J Pharmacokinet Pharmacodyn 2013;40: 545–556.

[43] Saluja SS, Hanlon DJ, Sharp FA, Hong E, Khalil D, Robinson E, Tigelaar R, Fahmy TM, Edelson RL. Targeting human dendritic cells via DEC-205 using PLGA nanoparticles leads to enhanced cross-presentation of a melanoma-associated antigen. Int J Nanomedicine 2014;9:5231–5246.

[44] Salem AK. Nanoparticles in vaccine delivery. AAPS J 2015;17:289–291. DOI: 10.1208/s12248-015-9720-1.

[45] Ghaffar KA, Giddam AK, Zaman M, Skwarczynski M, Toth I. Liposomes as nanovaccine delivery systems. Curr Top Med Chem 2014;14:1194–1208.

[46] Zhao L, Seth A, Wibowo N, Zhao C-X, Mitter N, Yu C, Middelberg AP. Nanoparticle vaccines. Vaccine 2014; 32:327–337.

[47] Hirosue S, Kourtis IC, van der Vlies AJ, Hubbell JA, Swartz MA. Antigen delivery to dendritic cells by poly(propylene sulfide) nanoparticles with disulfide conjugated peptides: cross-presentation and T cell activation. Vaccine 2010;28: 7897–7906.

[48] Reddy ST, Swartz MA, Hubbell JA. Targeting dendritic cells with biomaterials: developing the next generation of vaccines. Trends Immunol 2006;27:573–579.

[49] Demento SL, Cui W, Criscione JM, Stern E, Tulipan J, Kaech SM, Fahmy TM. Role of sustained antigen release

from nanoparticle vaccines in shaping the T cell memory phenotype. Biomaterials 2012;33:4957–4964.

[50] Stano A, van der Vlies AJ, Martino MM, Swartz MA, Hubbell JA, Simeoni E. PPS nanoparticles as versatile delivery system to induce systemic and broad mucosal immunity after intranasal administration. Vaccine 2011;29:804–812.

[51] Reddy ST, Rehor A, Schmoekel HG, Hubbell JA, Swartz MA. *In vivo* targeting of dendritic cells in lymph nodes with poly(propylene sulfide) nanoparticles. J Control Release 2006;112:26–34.

[52] Uto T, Wang X, Sato K, Haraguchi M. Targeting of antigen to dendritic cells with poly (γ-glutamic acid) nanoparticles induces antigen-specific humoral and cellular immunity. J Immunol 2007;178:2979–2986. . Available at http://www.jimmunol.org/content/178/5/2979.short. Accessed 2015 Jun 8.

[53] Uto T, Akagi T, Toyama M, Nishi Y, Shima F, Akashi M, Baba M. Comparative activity of biodegradable nanoparticles with aluminum adjuvants: antigen uptake by dendritic cells and induction of immune response in mice. Immunol Lett 2011;140:36–43.

[54] Akagi T, Baba M, Akashi M. Biodegradable nanoparticles as vaccine adjuvants and delivery systems: regulation of immune responses by nanoparticle-based vaccine. In: *Polymers in Nanomedicine*. Berlin, Heidelberg: Springer; 2012. Advances in Polymer Science; p 31–64.

[55] Silva AL, Rosalia RA, Varypataki E, Sibuea S, Ossendorp F, Jiskoot W. Poly-(lactic-co-glycolic-acid)-based particulate vaccines: particle uptake by dendritic cells is a key parameter for immune activation. Vaccine 2015;33:847–854.

[56] Des Rieux A, Fievez V, Garinot M, Schneider Y-J, Préat V. Nanoparticles as potential oral delivery systems of proteins and vaccines: a mechanistic approach. J Control Release 2006;116:1–27.

[57] Islam MA, Firdous J, Choi Y-J, Yun C-H, Cho C-S. Design and application of chitosan microspheres as oral and nasal vaccine carriers: an updated review. Int J Nanomedicine 2012;7:6077–6093.

[58] Thomas SN, van der Vlies AJ, O'Neil CP, Reddy ST, Yu SS, Giorgio TD, Swartz MA, Hubbell JA. Engineering complement activation on polypropylene sulfide vaccine nanoparticles. Biomaterials 2011;32:2194–2203.

[59] Song X, Zhao X, Zhou Y, Li S, Ma Q. Pharmacokinetics and disposition of various drug loaded biodegradable poly(lactide-co-glycolide) (PLGA) nanoparticles. Curr Drug Metab 2010;11:859–869.

[60] Gurunathan S, Klinman DM, Seder RA. DNA vaccines: immunology, application, and optimization*. Annu Rev Immunol 2000;18:927–974.

[61] Van der Lubben IM, Verhoef JC, Borchard G, Junginger HE. Chitosan for mucosal vaccination. Adv Drug Deliv Rev 2001;52:139–144.

[62] Kaur R, Henriksen-Lacey M, Wilkhu J, Devitt A, Christensen D, Perrie Y. Effect of incorporating cholesterol into DDA:TDB liposomal adjuvants on bilayer properties, biodistribution, and immune responses. Mol Pharm 2014;11:197–207.

[63] Kaur R, Bramwell VW, Kirby DJ, Perrie Y. Pegylation of DDA:TDB liposomal adjuvants reduces the vaccine depot effect and alters the Th1/Th2 immune responses. J Control Release 2012;158:72–77.

[64] Henriksen-Lacey M, Christensen D, Bramwell VW, Lindenstrøm T, Agger EM, Andersen P, Perrie Y. Liposomal cationic charge and antigen adsorption are important properties for the efficient deposition of antigen at the injection site and ability of the vaccine to induce a CMI response. J Control Release 2010;145:102–108.

[65] Cho H-J, Lee S, Im S, Kim M-G, Lee J, Lee HJ, Lee KH, Kim S, Kim YB, Oh YK. Preclinical pharmacokinetics and biodistribution of human papillomavirus DNA vaccine delivered in human endogenous retrovirus envelope-coated baculovirus vector. Pharm Res 2012;29:585–593.

[66] Joffre O, Nolte MA, Spörri R, Reis e Sousa C. Inflammatory signals in dendritic cell activation and the induction of adaptive immunity. Immunol Rev 2009;227:234–247.

[67] Combadière B, Vogt A, Mahé B, Costagliola D, Hadam S, Bonduelle O, Sterry W, Staszewski S, Schaefer H, van der Werf S, Katlama C, Autran B, Blume-Peytavi U. Preferential amplification of CD8 effector-T cells after transcutaneous application of an inactivated influenza vaccine: a randomized phase I trial. PLoS One 2010;5:e10818.

[68] Mabbott NA, Donaldson DS, Ohno H, Williams IR, Mahajan A. Microfold (M) cells: important immunosurveillance posts in the intestinal epithelium. Mucosal Immunol 2013;6:666–677.

[69] Mowat AM. Anatomical basis of tolerance and immunity to intestinal antigens. Nat Rev Immunol 2003;3:331–341.

[70] O'Hagan DT. Oral delivery of vaccines. Formulation and clinical pharmacokinetic considerations. Clin Pharmacokinet 1992;22:1–10.

[71] Kammona O, Kiparissides C. Recent advances in nanocarrier-based mucosal delivery of biomolecules. J Control Release 2012;161:781–794.

[72] Rosales-Mendoza S, Salazar-González JA. Immunological aspects of using plant cells as delivery vehicles for oral vaccines. Expert Rev Vaccines 2014;13:737–749.

[73] Jabbal-Gill I, Watts P, Smith A. Chitosan based delivery systems for mucosal vaccines. Expert Opin Drug Deliv 2012;9:1051–1067.

[74] Devriendt B, De Geest BG, Goddeeris BM, Cox E. Crossing the barrier: targeting epithelial receptors for enhanced oral vaccine delivery. J Control Release 2012;160:431–439.

[75] Ma T, Wang L, Yang T, Ma G, Wang S. M-cell targeted polymeric lipid nanoparticles containing a Toll-like receptor agonist to boost oral immunity. Int J Pharm 2014;473:296–303.

[76] Kim S-H, Jang Y-S. Antigen targeting to M cells for enhancing the efficacy of mucosal vaccines. Exp Mol Med 2014;46:e85.

[77] Jabbal-Gill I. Nasal vaccine innovation. J Drug Target 2010;18:771–786.

[78] Fortuna A, Alves G, Serralheiro A, Sousa J, Falcão A. Intranasal delivery of systemic-acting drugs: small-molecules and biomacromolecules. Eur J Pharm Biopharm 2014; 88:8–27.

[79] Kang ML, Cho CS, Yoo HS. Application of chitosan microspheres for nasal delivery of vaccines. Biotechnol Adv 2009;27:857–865.

[80] Payne JM, Derbyshire JB. Portals of entry for bacterial infection in calves and piglets with particular reference to the tonsil. J Pathol Bacteriol 1963;85:171–178.

[81] Banchereau J, Thompson-Snipes L, Zurawski S, Blanck J-P, Cao Y, Clayton S, Gorvel JP, Zurawski G, Klechevsky E. The differential production of cytokines by human Langerhans cells and dermal CD14(+) DCs controls CTL priming. Blood 2012;119:5742–5749.

[82] Li N, Peng L-H, Chen X, Nakagawa S, Gao J-Q. Transcutaneous vaccines: novel advances in technology and delivery for overcoming the barriers. Vaccine 2011;29:6179–6190.

[83] Bal SM, Ding Z, van Riet E, Jiskoot W, Bouwstra JA. Advances in transcutaneous vaccine delivery: do all ways lead to Rome? J Control Release 2010;148:266–282.

[84] Barfoed AM, Rodriguez F, Therrien D, Borrego B, Sobrino F, Kamstrup S. DNA immunization with 2C FMDV nonstructural protein reveals the presence of an immunodominant CD8+, CTL epitope for Balb/c mice. Antiviral Res 2006; 72:178–189.

[85] Fuller DH, Shipley T, Allen TM, Fuller JT, Wu MS, Horton H, Wilson N, Widera G, Watkins DI. Immunogenicity of hybrid DNA vaccines expressing hepatitis B core particles carrying human and simian immunodeficiency virus epitopes in mice and rhesus macaques. Virology 2007;364:245–255.

[86] Pandey RS, Dixit VK. Evaluation of ISCOM vaccines for mucosal immunization against hepatitis B. J Drug Target 2010;18:282–291.

[87] Pandey RS, Dixit VK. Evaluation of ISCOMs for immunization against hepatitis B. Curr Pharm Biotechnol 2009;10:709–716.

[88] Heldens JGM, Pouwels HGW, Derks CGG, Van de Zande SMA, Hoeijmakers MJH. Duration of immunity induced by an equine influenza and tetanus combination vaccine formulation adjuvanted with ISCOM-Matrix. Vaccine 2010;28: 6989–6996.

[89] Gerbrandy JL, Bienenstock J. Kinetics and localization of IgE tetanus antibody response in mice immunized by the intratracheal, intraperitoneal and subcutaneous routes. Immunology 1976;31:913–919.

[90] Diez-Domingo J, Weinke T, Garcia de Lomas J, Meyer CU, Bertrand I, Eymin C, Thomas S, Sadorge C. Comparison of intramuscular and subcutaneous administration of a herpes zoster live-attenuated vaccine in adults aged ≥50 years: a randomised non-inferiority clinical trial. Vaccine 2014;33 (6):789–795. DOI: 10.1016/j.vaccine.2014.12.024.

[91] Orság P, Kvardová V, Raska M, Miller AD, Ledvina M, Turánek J. Quantitative real-time PCR study on persistence of pDNA vaccine pVax-Hsp60 TM814 in beef muscles. Genet Vaccines Ther 2008;6:11.

[92] Armengol G, Ruiz LM, Orduz S. The injection of plasmid DNA in mouse muscle results in lifelong persistence of DNA, gene expression, and humoral response. Mol Biotechnol 2004;27:109–118.

[93] Coelho-Castelo AAM, Trombone AP, Rosada RS, Santos RR Jr, Bonato VLD, Sartori A, Silva CL. Tissue distribution of a plasmid DNA encoding Hsp65 gene is dependent on the dose administered through intramuscular delivery. Genet Vaccines Ther 2006;4:1.

[94] Sheets RL, Stein J, Bailer RT, Koup RA, Andrews C, Nason M, He B, Koo E, Trotter H, Duffy C, Manetz TS, Gomez P. Biodistribution and toxicological safety of adenovirus type 5 and type 35 vectored vaccines against human immunodeficiency virus-1 (HIV-1), Ebola, or Marburg are similar despite differing adenovirus serotype vector, manufacturer's construct, or gene inserts. J Immunotoxicol 2008;5:315–335.

[95] Johnson JE, Coleman JW, Kalyan NK, Calderon P, Wright KJ, Obregon J, Ogin-Wilson E, Natuk RJ, Clarke DK, Udem SA, Cooper D, Hendry RM. *In vivo* biodistribution of a highly attenuated recombinant vesicular stomatitis virus expressing HIV-1 Gag following intramuscular, intranasal, or intravenous inoculation. Vaccine 2009;27:2930–2939.

[96] Lamichhane A, Azegamia T, Kiyonoa H. The mucosal immune system for vaccine development. Vaccine 2014; 32:6711–6723.

[97] Smith DJ, Bot S, Dellamary L, Bot A. Evaluation of novel aerosol formulations designed for mucosal vaccination against influenza virus. Vaccine 2003;21:2805–2812.

[98] Amorij J-P, Saluja V, Petersen AH, Hinrichs WLJ, Huckriede A, Frijlink HW. Pulmonary delivery of an inulin-stabilized influenza subunit vaccine prepared by spray-freeze drying induces systemic, mucosal humoral as well as cell-mediated immune responses in BALB/c mice. Vaccine 2007;25: 8707–8717.

[99] Seo KY, Han SJ, Cha H-R, Seo S-U, Song J-H, Chung SH, Kweon MN. Eye mucosa: an efficient vaccine delivery route for inducing protective immunity. J Immunol 2010;185: 3610–3619.

[100] Phillips TE, Sharp J, Rodgers K, Liu H. M cell–targeted ocular immunization: effect on immunoglobulins in tears, feces, and serum. Invest Ophthalmol Vis Sci 2010;51: 1533–1539.

[101] Pavot V, Rochereau N, Genin C, Verrier B, Paul S. New insights in mucosal vaccine development. Vaccine 2012;30: 142–154.

[102] Gómez-Mantilla JD, Trocóniz IF, Parra-Guillén Z, Garrido MJ. Review on modeling anti-antibody responses to monoclonal antibodies. J Pharmacokinet Pharmacodyn 2014;41: 523–536.

[103] Mooney M, McWeeney S, Canderan G, Sékaly R-P. A systems framework for vaccine design. Curr Opin Immunol 2013;25:551–555.

[104] Hipgrave DB, Maynard JE, Biggs B-A. Improving birth dose coverage of hepatitis B vaccine. Bull World Health Organ 2006;84:65–71.

[105] Ali SA, Kazi AM, Cortese MM, Fleming JA, Parashar UD, Jiang B, McNeal MM, Steele D, Bhutta Z, Zaidi A. Impact of different dosing schedules on the immunogenicity of the human rotavirus vaccine in infants in Pakistan: a randomized trial. J Infect Dis 2014;210:1772–1779.

[106] Loo JD, Conklin L, Fleming-Dutra KE, Deloria Knoll M, Park DE, Kirk J, Goldblatt D, O'Brien KL, Whitney CG. Systematic review of the effect of pneumococcal conjugate vaccine dosing schedules on prevention of pneumonia. Pediatr Infect Dis J 2014;33 (Suppl 2):S140–S151.

[107] Standaert BA, Curran D, Postma MJ. Budget constraint and vaccine dosing: a mathematical modelling exercise. Cost Eff Resour Alloc 2014;12:3.

[108] Fuller DH, Loudon P, Schmaljohn C. Preclinical and clinical progress of particle-mediated DNA vaccines for infectious diseases. Methods 2006;40:86–97.

[109] Hepburn MJ, Hugh Dyson E, Simpson AJH, Brenneman KE, Bailey N, Wilkinson L, Hornby R, Mateczun AJ, Bell MG, Baillie LW. Immune response to two different dosing schedules of the anthrax vaccine precipitated (AVP) vaccine. Vaccine 2007;25:6089–6097.

[110] Tacket CO, Roy MJ, Widera G, Swain WF, Broome S, Edelman R. Phase 1 safety and immune response studies of a DNA vaccine encoding hepatitis B surface antigen delivered by a gene delivery device. Vaccine 1999;17: 2826–2829.

[111] Song JY, Cheong HJ, Noh JY, Yang TU, Seo YB, Hong KW, Kim IS, Choi WS, Kim WJ. Long-term immunogenicity of the influenza vaccine at reduced intradermal and full intramuscular doses among healthy young adults. Clin Exp Vaccine Res 2013;2:115–119.

[112] Sun Y, Gruber M, Matsumoto M. Overview of global regulatory toxicology requirements for vaccines and adjuvants. J Pharmacol Toxicol Methods 2012;65:49–57.

[113] Wolf JJ, Kaplanski CV, Lebron JA. Nonclinical safety assessment of vaccines and adjuvants. Methods Mol Biol 2010;626:29–40.

[114] WHO. WHO guidelines on nonclinical evaluation of vaccines. WHO Technical Report Series No. 927, Annex 1; 2005.

[115] Mitkus RJ, Hess MA, Schwartz SL. Pharmacokinetic modeling as an approach to assessing the safety of residual formaldehyde in infant vaccines. Vaccine 2013;31: 2738–2743.

[116] Chang CH, Hsu WC, Wang CY, Jan ML, Tsai TH, Lee TW, Lynn SG, Yeh CH, Chang TJ. Longitudinal microSPECT/CT imaging and pharmacokinetics of synthetic luteinizing hormone-releasing hormone (LHRH) vaccine in rats. Anticancer Res 2007;27:3251–3257.

[117] Forrat R, Dumas R, Seiberling M, Merz M, Lutsch C, Lang J. Evaluation of the safety and pharmacokinetic profile of a new, pasteurized, human tetanus immunoglobulin administered as sham, postexposure prophylaxis of tetanus. Antimicrob Agents Chemother 1998;42:298–305.

[118] Lee H-J, Park N, Cho H-J, Yoon JK, Van ND, Oh YK, Kim YB. Development of a novel viral DNA vaccine against human papillomavirus: AcHERV-HP16L1. Vaccine 2010; 28:1613–1619.

[119] Lang J, Attanath P, Quiambao B, Singhasivanon V, Chanthavanich P, Montalban C, Lutsch C, Pepin-Covatta S, Le Mener V, Miranda M, Sabchareon A. Evaluation of the safety, immunogenicity, and pharmacokinetic profile of a new, highly purified, heat-treated equine rabies immunoglobulin, administered either alone or in association with a purified, Vero-cell rabies vaccine. Acta Trop 1998;70: 317–333.

[120] Pappalardo F, Flower D, Russo G, Pennisi M, Motta S. Computational modelling approaches to vaccinology. Pharmacol Res 2014;92:40–45. DOI: 10.1016/j.phrs.2014.08.006.

[121] Pappalardo F, Pennisi M, Ricupito A, Topputo F, Bellone M. Induction of T-cell memory by a dendritic cell vaccine: a computational model. Bioinformatics 2014;30:1884–1891.

[122] Wilson S, Levy D. A mathematical model of the enhancement of tumor vaccine efficacy by immunotherapy. Bull Math Biol 2012;74:1485–1500.

[123] Parra-Guillen ZP, Berraondo P, Grenier E, Ribba B, Troconiz IF. Mathematical model approach to describe tumour response in mice after vaccine administration and its applicability to immune-stimulatory cytokine-based strategies. AAPS J 2013;15:797–807.

[124] Parra-Guillen ZP, Berraondo P, Ribba B, Trocóniz IF. Modeling tumor response after combined administration of different immune-stimulatory agents. J Pharmacol Exp Ther 2013;346:432–442.

[125] Cappuccio A, Elishmereni M, Agur Z. Cancer immunotherapy by interleukin-21: potential treatment strategies evaluated in a mathematical model. Cancer Res 2006;66: 7293–7300.

[126] Agur Z, Vuk-Pavlović S. Mathematical modeling in immunotherapy of cancer: personalizing clinical trials. Mol Ther 2012;20:1–2.

[127] Pulendran B. Systems vaccinology: probing humanity's diverse immune systems with vaccines. Proc Natl Acad Sci U S A 2014;111:12300–12306.

[128] Ideker T, Galitski T, Hood L. A new approach to decoding life: systems biology. Annu Rev Genomics Hum Genet 2001; 2:343–372.

[129] Kitano H. Computational systems biology. Nature 2002; 420:206–210.

[130] Nakaya HI, Li S, Pulendran B. Systems vaccinology: learning to compute the behavior of vaccine induced immunity. Wiley Interdiscip Rev Syst Biol Med 2012;4:193–205.

[131] Querec TD, Akondy RS, Lee EK, Cao W, Nakaya HI, Teuwen D, Pirani A, Gernert K, Deng J, Marzolf B, Kennedy K, Wu H, Bennouna S, Oluoch H, Miller J, Vencio RZ, Mulligan M, Aderem A, Ahmed R, Pulendran B. Systems biology approach predicts immunogenicity of the yellow fever vaccine in humans. Nat Immunol 2009;10:116–125.

[132] Nakaya HI, Wrammert J, Lee EK, Racioppi L, Marie-Kunze S, Haining WN, Means AR, Kasturi SP, Khan N, Li GM, McCausland M, Kanchan V, Kokko KE, Li S, Elbein R, Mehta AK, Aderem A, Subbarao K, Ahmed R, Pulendran B. Systems biology of vaccination for seasonal influenza in humans. Nat Immunol 2011;12:786–795.

[133] Vahey MT, Wang Z, Kester KE, Cummings J, Heppner DG Jr, Nau ME, Ofori-Anyinam O, Cohen J, Coche T, Ballou WR, Ockenhouse CF. Expression of genes associated with

immunoproteasome processing of major histocompatibility complex peptides is indicative of protection with adjuvanted RTS, S malaria vaccine. J Infect Dis 2010;201: 580–589.

[134] Pulendran B. Learning immunology from the yellow fever vaccine: innate immunity to systems vaccinology. Nat Rev Immunol 2009;9:741–747.

[135] Pulendran B, Oh JZ, Nakaya HI, Ravindran R, Kazmin DA. Immunity to viruses: learning from successful human vaccines. Immunol Rev 2013;255:243–255.

[136] Gaucher D, Therrien R, Kettaf N, Angermann BR, Boucher G, Filali-Mouhim A, Moser JM, Mehta RS, Drake DR III, Castro E, Akondy R, Rinfret A, Yassine-Diab B, Said EA, Chouikh Y, Cameron MJ, Clum R, Kelvin D, Somogyi R, Greller LD, Balderas RS, Wilkinson P, Pantaleo G, Tartaglia J, Haddad EK, Sékaly RP. Yellow fever vaccine induces integrated multilineage and polyfunctional immune responses. J Exp Med 2008;205:3119–3131.

[137] Li S, Rouphael N, Duraisingham S, Romero-Steiner S, Presnell S, Davis C, Schmidt DS, Johnson SE, Milton A, Rajam G, Kasturi S, Carlone GM, Quinn C, Chaussabel D, Palucka AK, Mulligan MJ, Ahmed R, Stephens DS, Nakaya HI, Pulendran B. Molecular signatures of antibody responses derived from a systems biology study of five human vaccines. Nat Immunol 2014;15:195–204.

[138] Höhler T, Reuss E, Evers N, Dietrich E, Rittner C, Freitag CM, Vollmar J, Schneider PM, Fimmers R. Differential genetic determination of immune responsiveness to hepatitis B surface antigen and to hepatitis A virus: a vaccination study in twins. Lancet 2002;360:991–995.

[139] Lee YC, Newport MJ, Goetghebuer T, Siegrist CA, Weiss HA, Pollard AJ, Marchant A, MRC Twin Study Group. Influence of genetic and environmental factors on the immunogenicity of Hib vaccine in Gambian twins. Vaccine 2006;24:5335–5340.

[140] Konradsen HB, Henrichsen J, Wachmann H, Holm N. The influence of genetic factors on the immune response as judged by pneumococcal vaccination of mono- and dizygotic Caucasian twins. Clin Exp Immunol 1993;92: 532–536.

[141] O'Connor D, Pollard AJ. Characterizing vaccine responses using host genomic and transcriptomic analysis. Clin Infect Dis 2013;57:860–869.

[142] Bandaru P, Rajkumar H, Nappanveettil G. The impact of obesity on immune response to infection and vaccine: an insight into plausible mechanisms. Endocrinol Metab Synd 2013;2 (2):113. DOI: 10.4172/2161-1017.1000113.

[143] Haque R, Snider C, Liu Y, Ma JZ, Liu L, Nayak U, Mychaleckyj JC, Korpe P, Mondal D, Kabir M, Alam M, Pallansch M, Oberste MS, Weldon W, Kirkpatrick BD, Petri WA Jr. Oral polio vaccine response in breast fed infants with malnutrition and diarrhea. Vaccine 2014;32:478–482.

[144] Kau AL, Ahern PP, Griffin NW, Goodman AL, Gordon JI. Human nutrition, the gut microbiome and the immune system. Nature 2011;474:327–336.

[145] Shi L, Campbell G, Jones WD, Campagne F, Wen Z, Walker SJ, Su Z, Chu TM, Goodsaid FM, Pusztai L, Shaughnessy JD Jr, Oberthuer A, Thomas RS, Paules RS, Fielden M, Barlogie B, Chen W, Du P, Fischer M, Furlanello C, Gallas BD, Ge X, Megherbi DB, Symmans WF, Wang MD, Zhang J, Bitter H, Brors B, Bushel PR, Bylesjo M, Chen M, Cheng J, Cheng J, Chou J, Davison TS, Delorenzi M, Deng Y, Devanarayan V, Dix DJ, Dopazo J, Dorff KC, Elloumi F, Fan J, Fan S, Fan X, Fang H, Gonzaludo N, Hess KR, Hong H, Huan J, Irizarry RA, Judson R, Juraeva D, Lababidi S, Lambert CG, Li L, Li Y, Li Z, Lin SM, Liu G, Lobenhofer EK, Luo J, Luo W, McCall MN, Nikolsky Y, Pennello GA, Perkins RG, Philip R, Popovici V, Price ND, Qian F, Scherer A, Shi T, Shi W, Sung J, Thierry-Mieg D, Thierry-Mieg J, Thodima V, Trygg J, Vishnuvajjala L, Wang SJ, Wu J, Wu Y, Xie Q, Yousef WA, Zhang L, Zhang X, Zhong S, Zhou Y, Zhu S, Arasappan D, Bao W, Lucas AB, Berthold F, Brennan RJ, Buness A, Catalano JG, Chang C, Chen R, Cheng Y, Cui J, Czika W, Demichelis F, Deng X, Dosymbekov D, Eils R, Feng Y, Fostel J, Fulmer-Smentek S, Fuscoe JC, Gatto L, Ge W, Goldstein DR, Guo L, Halbert DN, Han J, Harris SC, Hatzis C, Herman D, Huang J, Jensen RV, Jiang R, Johnson CD, Jurman G, Kahlert Y, Khuder SA, Kohl M, Li J, Li L, Li M, Li QZ, Li S, Li Z, Liu J, Liu Y, Liu Z, Meng L, Madera M, Martinez-Murillo F, Medina I, Meehan J, Miclaus K, Moffitt RA, Montaner D, Mukherjee P, Mulligan GJ, Neville P, Nikolskaya T, Ning B, Page GP, Parker J, Parry RM, Peng X, Peterson RL, Phan JH, Quanz B, Ren Y, Riccadonna S, Roter AH, Samuelson FW, Schumacher MM, Shambaugh JD, Shi Q, Shippy R, Si S, Smalter A, Sotiriou C, Soukup M, Staedtler F, Steiner G, Stokes TH, Sun Q, Tan PY, Tang R, Tezak Z, Thorn B, Tsyganova M, Turpaz Y, Vega SC, Visintainer R, von Frese J, Wang C, Wang E, Wang J, Wang W, Westermann F, Willey JC, Woods M, Wu S, Xiao N, Xu J, Xu L, Yang L, Zeng X, Zhang J, Zhang L, Zhang M, Zhao C, Puri RK, Scherf U, Tong W, Wolfinger RD, MAQC Consortium. The MicroArray Quality Control (MAQC)-II study of common practices for the development and validation of microarray-based predictive models. Nat Biotechnol 2010;28:827–838.

24

DRUG DEVELOPMENT STRATEGIES FOR THERAPEUTIC BIOLOGICS: INDUSTRY PERSPECTIVES

Theresa Yuraszeck and Megan Gibbs
Amgen Inc., Thousand Oaks, CA, USA

24.1 INTRODUCTION

Insulin was the first recombinant biologic brought to market, and its approval by the U.S. FDA in 1982 initiated a wave of investment in biologics research and development. While diabetic patients could be treated with insulin isolated from bovine and porcine pancreas before this advance, production was expensive, availability was limited, and immunogenicity was an issue [1]. Recombinantly produced human insulin solved these problems. Three decades later, biologics comprise a significant and increasing share of the worldwide pharmaceutical market. Between 1980 and 1994, investigational new drug (IND) applications for biologics increased 10-fold, and by 2010 there were over 200 commercially available biopharmaceuticals, including biosimilars [2, 3]. Of the top 20 biopharmaceuticals based on worldwide sales in 2012, 40% were biologics [4]. In the United States, sales of biologics increased approximately 20% from 2011 to 2012 and accounted for approximately $64 billion in revenue (Fig. 24.1) [5]. Moreover, over 900 biologics are currently in preclinical and clinical development [6]. Clearly, biologics will play an increasingly important role in healthcare for the foreseeable future.

Development costs for biologics and small molecules are similar, but the success rate for biologics has been substantially higher than that for small molecules for a variety of reasons that include better target specificity and reduced off-target toxicity and activity [7, 8]. Moreover, biologics comprise an increasing proportion of the pharmaceutical market [3]. Still, the success rate of biologics is declining, and consideration of absorption, disposition, metabolism, and excretion (ADME) properties can help maintain success rates by facilitating the selection of the correct dose and dosing regimen, an approach that has already proven effective for small molecule development. In 1991, poor pharmacokinetic (PK) properties were accountable for 40% of all failed drugs. Ten years later, the failure rate due to poor PK properties was reduced to just 8% as a result of greater attention to PK properties earlier in drug discovery [9, 10]. The applicability of this approach to biologics development is demonstrated by the success of drugs like Lantus®, a form of long-acting insulin, and Aranesp® (darbepoetin alfa), a hyperglycosylated analog of recombinant human erythropoietin (EPO); although these drugs were initially successful, the improvement in their PK properties and in particular the longer half-lives of Lantus® and Aranesp® compared to their predecessors offered substantial benefits to patients through more convenient dosing regimens. Improved understanding of the relationship between PK, pharmacodynamics (PD), and safety signals can reduce attrition rates and development costs further; 16% of failed first-time applications can be attributed to uncertainties in the dose [11].

While general principles from small molecule development are certainly relevant to biologics, there are a variety of considerations unique to biologics development owing to the inherent differences in their properties and characteristics. These differences dictate different approaches to both preclinical and clinical development, and many of the studies conducted for small molecules are not necessary for biologics. Likewise, some studies are conducted only for

ADME and Translational Pharmacokinetics/Pharmacodynamics of Therapeutic Proteins: Applications in Drug Discovery and Development,
First Edition. Edited by Honghui Zhou and Frank-Peter Theil.

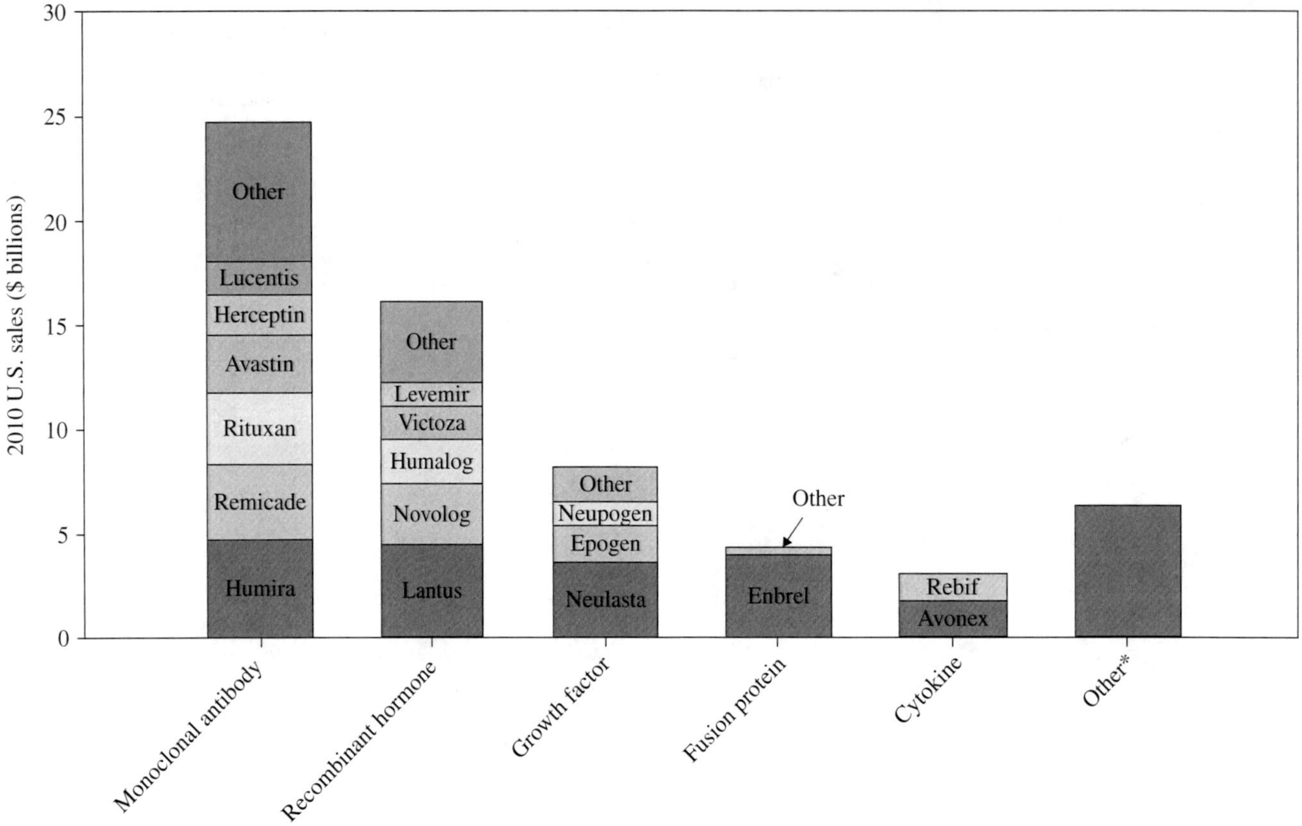

FIGURE 24.1 Sales of biologics drugs in the United States—2012 [5]. Includes therapeutic enzymes, blood factors, recombinant vaccines, and anticoagulants. Blockbuster biologics, or biologics whose sales exceed $1 billion annually, are highlighted. (*See insert for color representation of this figure.*)

biologics. We will discuss the properties that differentiate biologics from small molecules and the impact they have on the preclinical and clinical development strategy for biologics.

24.1.1 Biologics Properties and Classification

Biologics range in size from 1 to 150 kDa (small molecules are <0.5 kDa) and can be divided into several groups with distinctive properties: (i) monoclonal antibodies (mAbs), including bispecific antibodies and antibody–drug conjugates (ADCs); (ii) recombinant proteins, including hormones, growth factors, blood factors, anticoagulants, and cytokines; (iii) peptides and fusion proteins; (iv) nucleic acid molecules; (v) gene therapies; (vi) cell-based therapies; or (vii) vaccines [12]. The biologics market is currently dominated by mAbs, with sales exceeding $24 billion in 2012. Targets are concentrated in the cancer and anti-inflammatory therapeutic areas, with a smaller but still significant presence in neuroscience, cardiovascular diseases, metabolic disorders, ophthalmology, and infectious diseases (Table 24.1) [5, 13–22]. The introduction of insulin analogs with extended half-lives and glucagon-like peptide 1 agonists has also driven growth in the recombinant protein sector, which has the highest sales after mAbs [3, 5, 23].

Monoclonal antibodies and Fc-fusion proteins are protected from degradation by FcRn (neonatal Fc receptor)-mediated recycling and can have long half-lives in man of 3–4 and 1–2 weeks, respectively [24]. Recombinant proteins and peptides are smaller and tend to have shorter half-lives ranging from minutes to hours that require frequent dosing or constant infusions to maintain concentrations in an efficacious range. Techniques such as pegylation, glycoengineering, and the addition of fusion protein (to a peptide, for example) can be used to extend the half-life of such biologics and allow for less frequent dosing [25, 26]. Half-life extension can also result in efficacy and safety benefits. The utility of pegylation is demonstrated by its application to the interferons, which facilitated once-weekly instead of thrice-weekly dosing and prevented the emergence of drug-resistant virus by maintaining stable plasma concentrations for a longer duration (interferons are typically used to treat hepatitis C or other viral infections). Darbepoetin alfa, a hyperglycosylated analog of EPO, is an example of a successful application of glycoengineering. Its extended half-life resulted in a more convenient dosing regimen for patients

TABLE 24.1 A Selected List of Biologics for Indications in Growing Therapeutic Areas

Therapeutic Area	Indication	Drug	Status
Cardiovascular diseases and metabolic disorders	Hypercholesterolemia	Evolucumab	Ph III
		Alirocumab	Ph III
		Bococizumab	Ph III
Neuroscience	Alzheimer's disease	Solanezumab (Aβ)	Ph III
		Crenezumab (Aβ)	Ph II
		Gantanerumab (Aβ)	Ph III
		AADvac1 (tau)	Ph I
Ophthalmology	Macular degeneration	Bevacizumab	Approved (not for this indication, but widely prescribed)
	Macular degeneration	Ranibizumab	Approved
Infectious diseases	Respiratory syncytial virus	Palivizumab	Approved
	HIV	Ibalizumab	Ph III

while providing similar efficacy/safety profiles [27–29]. Development of molecules in the latter three categories has special considerations that will not be detailed here as these compounds make up a small share of the biologics market and pipeline.

Biologics can also be classified according to their mechanism of action, which can be particularly useful when developing mechanistic PK/PD models and selecting appropriate animal species in which to conduct preclinical experiments. According to this paradigm, biologics can (i) replace a deficient or abnormal protein (insulin, growth factors, and coagulation factors), (ii) augment an endogenous pathway (EPO, filgrastim, and interferon), (iii) provide a novel function or activity, (iv) interfere with a molecule or organism (oncology agents such as bevacizumab and trastuzumab and anti-inflammatory agents such as adalimumab and etanercept), (v) deliver other compounds, such as cytotoxic small molecules, to targets (as is the case with ADCs), and (vi) act as vaccines [1]. There has been tremendous interest recently in leveraging and stimulating patients' endogenous immune system to treat their own cancer. Cancer immunotherapies span a variety of biologics platforms and include cell-based therapies, such as CAR (chimeric antigen receptor) T-cell therapies, antibody-based therapies such as BiTEs (bispecific T-cell engagers) molecules, and cytokine-based therapies in which endogenously present molecules such as interferon-alpha are administered. A comprehensive understanding of the underlying biology of the indication of a molecule is being developed to guide which of the aforementioned paradigms is chosen.

Compared to small molecules, biologics have a lower potential for off-target toxicity due to the high selectivity and specificity with which they bind their targets (a consequence of their size and interaction with a larger area of the target), but they cannot be dosed orally, do not penetrate tissues well, and cannot easily access intracellular targets or CNS-based targets. While the lack of off-target toxicity differentiates biologics from small molecules, differentiation among biologics on the basis of safety profiles may be difficult, since all molecules hitting a particular target would be expected to have similar safety profiles. There is, however, opportunity for differentiation between biologics according to the type of target; mAbs and recombinant proteins typically target receptor–ligand interactions, while peptides have greater tissue penetration and better access to intracellular targets.

It is well established that many biologics can evoke an immune response, resulting in the development of antidrug antibodies (ADAs) that can affect exposure, efficacy, and safety [30]. A variety of patient-, disease-, and drug-specific factors influence the development of immunogenicity, including the patient's genetic background, age, concomitant medications or exposure to similar drugs, the immune function of the patient, protein structure and formulation, and the dose and duration of treatment. Immunogenicity is a particular problem for biologics of nonhuman origin such as murine and chimeric antibodies. Humanized and fully human antibodies are more difficult to produce, but they are generally associated with a lower risk for immunogenicity, although the risk is not completely mitigated [31]. Adalimumab, for example, is a fully humanized anti-tumor necrosis factor (TNF) drug for which the development of ADAs has been observed [32, 33]. In a long-term study of approximately 300 patients with rheumatoid arthritis, anti-adalimumab antibodies were detected in 28% of patients, and these patients had correspondingly lower adalimumab trough concentrations and a poorer clinical outcome [32]. Most patients who developed ADAs did so within a few months of beginning treatment [32]. The mechanism of immunogenicity of fully human antibodies is not completely understood, and while the immunological risk may never be eliminated, their tendency to be less immunogenic than their murine and chimeric counterparts is driving their increasing share of the development pipeline [31, 34]. Small proteins and peptides are less likely to cause immunogenicity [23, 24, 26, 35–37].

24.1.2 Assay Development and Validation

The analysis and interpretation of PK and toxicokinetic (TK) data depend on the underlying reliability of the assays used to generate that data. Accuracy, precision, selectivity, sensitivity, reproducibility, and stability are key aspects of assay development for both biologics and small molecules [38]. Small molecules are typically quantified using chromatographic methods and the assays are relatively straightforward and quick to develop. Immunoassays, such as ligand-binding assays (LBAs), are more commonly used to quantify biologics and can take much longer to develop owing to the complex nature of the molecules [38].

The general challenges faced during assay development are similar for all biologics. First, appropriate capture or detection reagents must be prepared, as the ligands (antibodies, binding proteins, receptors, oligonucleotides, or peptide fragments) that are the foundation of LBAs are usually not commercially available. Since the reagents themselves are often biologics, their production can be as challenging as the production of the drug. Concerns about stability, lot-to-lot variability, and heterogeneity must be addressed; such factors can affect binding kinetics and therefore bias measurement of drug levels. Failure to consider the potential for cross-reactivity with endogenous proteins in the biological matrix can also result in biased measurements. The dynamic range of LBAs is typically narrow, spanning 2 orders of magnitude, in contrast to small molecule assays that can span 4–5 orders of magnitude. While high concentration samples can be diluted into the assay range, dilutions can impact the equilibrium between the protein and its ligand and bias the results [39]. Nonlinear calibration curves can also bias the results. Assays for small molecules detect the actual physicochemical properties of the molecule and therefore the relationship between the assay measurement and concentration is usually direct and linear. In contrast, levels of biologics are measured indirectly through the interaction of the drug with the assay ligand, and the relationship between response and concentration is often nonlinear with heteroscedastic variability [40].

Whether the assay will measure free or total drug must also be considered. Ideally, the pharmacologically active free forms, including unbound and monovalently bound forms, would be measured, but it can be challenging to develop such assays and, when the drug concentrations are in excess of their target, total drug may be an acceptable surrogate for free drug [39]. Special considerations related to average drug load and drug load distribution must also be addressed during assay development for the quantitation of ADCs, a type of biologic in which an antibody is linked to a cytotoxic drug [41, 42].

In addition to the LBA to quantitate drug levels, an assay to detect ADAs must also be developed, since the immunogenicity potential of biologic agents is well established. Such assays operate by comparing the inhibition of pharmacological activity of a test sample with the activity of a sample with a known drug concentration. Many of the issues already discussed as challenges for LBAs also apply to immunogenicity assays, such as reagent availability, although there are several unique obstacles to be overcome for assays to detect ADAs [43]. Because ADAs are inherently heterogeneous, it is difficult to establish a positive control or reference standard, and therefore ADA testing starts with screening assays that simply confirm the presence of ADAs in a sample [43]. Platform selection is another important question for ADA assay development. The FDA recommends cell-based assays to be used for ADA detection, although non-cell-based receptor and target-binding assays are acceptable for soluble targets or when appropriate cell lines cannot be identified [44, 45].

24.2 PRECLINICAL DEVELOPMENT

The primary objectives of a preclinical development program for a new biologic are to (i) assess its safety profile, (ii) characterize its ADME properties and PK profile, (iii) establish the relationship between PK and PD in an appropriate animal model, and (iv) translate the preclinical findings to the clinic. A host of *in vitro* and *in vivo* experiments are carried out in support of these objectives. Studies conducted to address ADME questions include single- and multiple-dose PK studies, TK studies, and PD studies in appropriate animal species. These studies provide a basis for projecting the PK profile in humans, estimating efficacious doses, and calculating exposure margins to support the starting-dose selection.

Species selection is an important factor for preclinical PK, PD, and TK studies and can impact the fidelity with which the clinical efficacy and safety outcomes are predicted by the preclinical data. Most importantly, the compound should be pharmacologically active in the selected species to improve the predictive capability of the toxicology model, as most toxicity associated with biologics is on-target and can manifest as exaggerated pharmacology, although this is not necessarily the case for ADCs that deliver cytotoxic small molecules to targets and can therefore be associated with off-target toxicities. The importance of species selection is demonstrated by the case of TGN1412, an antibody that induced a life-threatening phenomenon called cytokine release syndrome in all six healthy subjects who received it during the first-in-human (FIH) trial. The preclinical data failed to predict this outcome due to a lack of understanding of species differences [46–49].

Several other factors must be considered when selecting an appropriate species for preclinical studies. Expression and distribution of the relevant receptor or isotope in the selected species should be similar to humans, lest the PK data be under- or overestimated, and the PD effects be unpredictable from the preclinical data. Immunogenicity should not limit study duration due to the formation of ADAs that reduce drug exposure because it is important that high exposures be achieved in toxicology studies to establish the therapeutic window for clinical development. These requirements usually limit the PK, PD, and TK studies to a single species. The FDA will accept toxicology studies from a single species if the biologic is not active in multiple species.

Nonhuman primates (NHPs) are often chosen for preclinical studies due to their genetic similarity to humans, although species selection should be driven by consideration of the relevant differences in cross-selectivity to the target. Healthy normal animals are typically used in these studies unless there is a clear biological rationale for selecting an animal model of disease [50, 51]. An animal model of disease may be appropriate for biologics that replace or augment a deficient protein, for example, or biologics that reduce excess protein levels because toxicity effects can be misleading in a normal individual. In cases where pharmacological activity cannot be identified in any preclinical species, homologous proteins or transgenic mice may be used for preclinical safety assessments [50–52].

Fewer preclinical studies are conducted to assess preclinical ADME properties of biologics compared to small molecules. In both cases, PK is assessed in single- and multiple-dose settings and within the toxicology studies, although usually assessed in multiple species for small molecules. Biologics are administered parenterally, so bioavailability is assessed for compounds dosed subcutaneously (SC) or intramuscularly; factors such as the size of the compound and the administration site can affect bioavailability [53]. However, metabolism and distribution studies are not common for biologics. Large molecule catabolism generally mitigates the need for the biotransformation and mass balance studies that are routinely performed during small molecule development. Moreover, the amino acid products of catabolism are not considered safety risks [50]. Such studies may be conducted for some biologics, such as peptides, whose relatively smaller size may make them subject to renal elimination [54]. Distribution studies are not typically conducted because the relatively large size of biologics limits tissue penetration and are not a regulatory requirement [55, 56]. However, they may be conducted in cases where biodistribution is critical to understanding the drug's mechanism of action. For example, some next-generation anti-TNF compounds are designed to be of a lower molecular weight to increase biodistribution and potential efficacy, and preclinical distribution studies might be conducted to assess this approach. Such studies might also be conducted for drugs targeting solid tumors, where tissue penetration is critical [57]. Physiologically based pharmacokinetic (PBPK) models can be employed to assess tissue distribution but validation and translation may be challenging and this approach has greater predictive value when combined with experimental biodistribution data [58]. *In vitro* assessments of drug–drug interaction (DDI) potential are also uncommon because most of the therapeutic protein–drug interactions are indirect and mediated via disease status or target physiology, therefore the *in vitro* assessment of interaction potential typically performed for small molecule drugs (which are based on direct interaction with cytochrome P450 enzymes) are not predictive [59].

Immunogenicity studies, on the other hand, are commonly conducted for biologics. It is well accepted that even fully humanized agents are likely to evoke an immune response, and that the resulting ADAs may inhibit the effectiveness of the therapeutic agent or cause toxic side effects. The objective of immunogenicity studies is to aid in the interpretation of the preclinical PK data because, unfortunately, immunogenicity in humans is not predictable from animal studies [60].

Preclinical PK, PD, and TK studies, and metabolism and distribution studies, and immunogenicity studies facilitate the characterization of ADME properties and their relationship with efficacy and toxicity. Several other studies are routinely performed in preclinical development for which ADME properties are less relevant but are still nevertheless critical. A brief review of those studies is warranted, as well as other studies that are standard for small molecule development but unnecessary for biologics.

The initiation of reproductive performance and developmental toxicity studies depends on the product, indication, and intended patient population. If information is available for other compounds in the same class and the only relevant species is NHP, these studies may not be necessary, although in practice such studies are conducted regardless [50]. Given the route of administration, local tolerance must also be evaluated. Safety pharmacology studies are conducted, although *in vitro* assessments of cardiovascular risk are not very useful and should again be undertaken on a case-by-case basis driven by the biology; cardiovascular risk is more appropriately assessed within the context of the repeat-dose toxicology studies and followed by an appropriate clinical study if a signal is observed [50, 61, 62]. There is usually little potential for cardiovascular toxicity, as the large size of most biologics prevents access to the inner pore of the hERG channel (encoded by the human ether-a-go-go-related gene), the binding site for most compounds that interact with hERG [61]. Cardiovascular risk for biologics tends to be associated with the biological mechanism by which the drug acts (in cases where the target is expressed in the cardiovascular system, for example) and that should be considered when interpreting the preclinical data. Consider the case of trastuzumab (Herceptin®), a mAb associated with myocardial toxicity via its mechanism of action as an erB2 inhibitor, which disrupts myocardial function in cardiomyocytes. The preclinical data did not predict cardiovascular toxicity, but its mechanism of action clearly suggested a cardiovascular risk, which was assessed clinically [63, 64]. But in general, in the absence of a mechanism-related risk, cardiovascular toxicity assessments are usually limited to preclinical repeat-dose toxicology studies and electrocardiogram (ECG) assessments in the Phase I trials.

While required to support the advancement of small molecules to the clinic, genotoxicity and carcinogenicity studies are conducted on a case-by-case basis for biologics. In general, they are not needed unless the mechanism of action suggests a risk. For example, carcinogenicity studies for growth factors are appropriate and assessments would begin with an *in vitro* study to evaluate cell proliferation upon drug exposure [50].

24.2.1 FIH Starting Dose

The translation of preclinical PK and PD data to predict human PK/PD is a critical objective of any preclinical development program, and one in which modeling and simulation play an important role. Translation of the preclinical data provides an estimate of the efficacious dose and regimen in humans and a foundation for the design of the FIH trial. After the incident with TGN1412, the European Medicines Agency (EMA) introduced the concept of the MABEL (minimum anticipated biological effect level) as a guide for selecting a safe starting dose for FIH trials. Had this approach been used in the development of TGN1412, the FIH dose would have been 200-fold lower than what was actually administered [48]. While the FDA guidance bases the maximum recommended starting dose (MRSD) on the no adverse event level (NOAEL) established in preclinical toxicology studies, in practice the NOAEL-based MRSD is used in conjunction with the PK/PD predictions to select a safe starting dose with minimal pharmacological activity.

Translation of the preclinical PK/PD data first involves predicting human PK, then using human-predicted exposure to predict the PD response, accounting for species differences that may influence PK and PD [65]. For small molecules, allometric scaling using data from three species is typically used to predict the human PK, although emerging evidence suggests that single-species allometric scaling may be equally accurate [66]. Single-species, fixed exponent allometric scaling based on noncompartmental PK data from NHPs is typically used for biologics, as it is generally difficult to identify more than one relevant species for preclinical PK studies. Several studies have shown that this method works well for biologics that exhibit linear PK [55, 67–69]. Predicting human PK for compounds exhibiting nonlinear PK is more challenging. In this case, a population PK model of target-mediated disposition is typically employed to describe the preclinical PK data [70]. The target-mediated drug disposition (TMDD) model directly accounts for species differences in target expression, affinity of the compound for the target, and target turnover, all of which impact PK in the nonlinear region. It has been demonstrated that such differences can impact PK [71]. The model can suffer from identifiability problems, and several simplified versions have been proposed as an alternative; one commonly used version takes the form of a Michaelis–Menten equation [72]. Following the estimation of the model parameters using a nonlinear mixed-effect modeling approach, allometric scaling is applied to the clearance, volume, and intercompartmental transport parameters while assuming that bioavailability and absorption rate (for SC administered therapeutics) are species independent. TMDD-related parameters (receptor density, receptor turnover rate, and binding kinetics) are estimated from *in vitro* data [70, 73, 74]; V_{max} is scaled as clearance and the translation of K_m is informed by *in vitro* binding characteristics for the simplified Michaelis–Menten version of the TMDD model. Predictions based on the Michaelis–Menten model are more reliable in the dose range where the nonlinear clearance pathway was saturated, and may not adequately describe the data at lower doses [55]. The lack of mechanistic detail may also hinder translatability by not accounting for species differences in target-mediated disposition.

PD translations can be very difficult and failure to translate PK/PD is a major source of Phase II attrition [75]. Challenges involved in the translation of biologics PK/PD include species differences in target recognition and expression, immunogenicity, and clearance mechanisms [76]. Disease state can also affect pharmacological assessments and this can be another source of error in the translation of preclinical data to the clinic. The impact of pharmacological effects on PK, a phenomenon not normally seen with small molecules, further complicates the translation.

Mechanistic PK/PD models and systems pharmacology models may address these challenges and show greater predictive power than the empirical models that have traditionally been used. These models do require extensive data to avoid parameter identifiability issues and scaling mechanistic PD parameters between species is not straightforward. For these reasons, such models are still not widely used to support FIH dose selection, and there are many reports of successful translations based on traditional models including those for TMDD [73, 77].

24.3 CLINICAL DEVELOPMENT

The clinical development of biologics typically takes more than 7 years, with an additional year required for regulatory review and approval. Development costs have been trending upwards for years and are estimated to exceed $1 billion [7, 78]. Phase I starts with single and multiple ascending dose studies in which assessments of safety, tolerability, PK, and immunogenicity are the typical objectives. These studies are usually performed in healthy subjects but, in certain cases, FIH trials should be conducted in patients [79]. The safety risk profile of anticancer therapies, for example, may preclude testing in healthy subjects. Patient-centric FIH trials should also be considered when the target is absent in healthy subjects, when the agent invokes an immune response, or when the agent is replacing a deficient or absent protein that is present at normal levels in healthy subjects. Phase I studies also involve an assessment of the PD effect, although often a biomarker that is a surrogate for the actual clinical endpoint is used to assess these effects [80]. The relationship between exposure and response is then used to select the dose for Phase II, with adjustments for differences in the biomarker response and the clinical endpoint where relevant and where such data are available [81]. Phase II starts with a proof of concept study and then a larger efficacy, safety, and tolerability trial. Phase III studies are the pivotal studies upon which FDA approval rests and they are large and costly; drugs for serious or life-threatening illnesses with few treatment options may be candidates for accelerated approval based on Phase II data, as has been the case for many HIV drugs and the cancer drug Gleevec® (imatinib mesylate). Population PK/PD modeling is again typically employed to analyze the Phase II data and select the Phase III dose, and may even be used to support a request to waive one of the two required Phase III trials [82, 83]. Model-based meta-analyses may also be employed throughout development to leverage external data and competitive intelligence when making program decisions [84–86]. Interest in systems pharmacology modeling is increasing, but thus far its use has been limited due to the complexity of developing and validating such models [87, 88].

A key part of any biologic license application (BLA) is the characterization of the exposure–response and exposure–toxicity relationships. These relationships are based on the Phase I and Phase II data, and data from a variety of other Phase I and Phase II studies that are conducted to evaluate the impact of factors that can impact exposure and potentially alter efficacy and safety profiles. Such factors include disease state, demographic characteristics, and concomitant medications. Modeling and simulation play a crucial role in characterizing these relationships and is used to (i) inform the FIH starting dose, (ii) inform the dose selection for Phase II, (iii) justify the marketed dose and dosing regimen, (iv) determine whether dosing adjustments are needed for special populations, pediatric subjects, subjects with particular demographic characteristics, and subjects taking concomitant medications, (v) evaluate immunogenicity, and (vi) inform clinical trial design [89, 90].

24.3.1 Intrinsic and Extrinsic Factors

The influence of intrinsic factors such as age, race, body weight, and sex and extrinsic factors such as diet and use of concomitant medications on ADME properties, exposure levels, and pharmacological effect must be characterized so that the appropriate dose adjustments may be made to ensure that efficacy is maintained and toxicity risks are not increased. These factors also influence the selection of the final dose and dosing regimen, with reduction of intersubject variability as a key consideration. Often, PK exposure is used as the anchor for investigating these effects and full safety and efficacy studies are not conducted. In fact, age and sex effects are commonly explored through population PK analyses, although studies dedicated to assessing these effects are typically conducted for small molecules [91].

Most biologics are dosed based on body size, even in adults, and the body-size effects discussed in the context of pediatric development also apply to the development of biologics in adults. However, population PK and PK/PD analyses are more easily applied due to the relatively narrow weight range in the adult population compared to the pediatric population and lack of size-based physiological or pathophysiological differences. Body weight is the most commonly identified covariate in population PK models and simulations based on these models can be used to justify fixed weight or body-size adjusted dosing, and determine the appropriate body-size correction [62, 92, 93]

Pharmacogenomic information is another intrinsic factor that may be assessed for its influence on efficacy, safety, or DDIs; such information may be used to stratify patient populations and identify those most likely to respond, or those most likely to experience an adverse event [94]. Consider trastuzumab, a mAb targeting HER-2-positive breast cancers. Before the introduction of trastuzumab, HER-2 overexpression was associated with a poor prognosis for patients, but trastuzumab offered a treatment option in which the best responses were associated with the highest levels of HER-2 overexpression. In fact, the use of trastuzumab is limited to patients overexpressing HER-2 and its approval was accompanied by a diagnostic test to identify such patients. Cetuximab is another mAb that is most effective in the subpopulation of cancer patients in which epidermal growth factor receptor (EGFR) overexpression is present. Although there are relatively few such examples of patient stratification on the basis of pharmacogenomics for approved biologics, interest in this area continues to grow [94].

Extrinsic factors such as diet, tobacco use, and use of concomitant medications may also alter ADME properties. Diet is mainly associated with altered absorption, a phenomenon known as the "food effect," and tobacco use is associated with changes in elimination due to the induction of the cytochrome P450 (CYP) enzymes CYP1A1, CYP1A2, and CYP2E1. Since biologics typically cannot be dosed orally, nor are they metabolized by the CYP enzymes, these factors are expected to have little impact on their PK. Biologics DDI, on the other hand, are the subject of increasing scrutiny [95–97]. Historically, few DDI studies were conducted with biologics because DDI are generally caused by interference with the CYP pathway. Because biologics do not undergo metabolism by CYP enzymes, the risk of DDI has been considered very low. Indeed, there are currently few restrictions on the concomitant use of biologics and other biologics or small molecule drugs and no dose adjustments have been required for any biologic because of DDI. However, recent data show that therapeutic proteins can perpetrate DDIs by inducing CYP enzymes, which would be expected to decrease the victim drug's exposure and compromise efficacy, and might require dose adjustments or label warnings [96]. Biologics-mediated DDI is likely a consequence of coadministration with another biologic or small molecule that has a similar mechanism of action, resulting in alterations to the biology of the target like its expression level, or a disease–drug interaction, in which CYP enzymes downregulated by the disease state of the subject are normalized by the administration of the biologic. Other potential mechanisms for DDI include changes immunogenicity by the concomitant medication, modulation of FcRn receptor expression, and displacement from binding proteins. Indeed, there are established examples of the latter two mechanisms of DDI, which are demonstrated by the effect of methotrexate on adalimumab exposure and heparin on palifermin exposure, respectively [98]. There is even some evidence that mAbs can alter the behavior of intestinal P-gp and cause DDI with transporter-mediated small molecule drugs [99]. These data, increasing interest in polypharmacy, and development of more complex biologic molecules have triggered renewed interest in DDI of biologics, and consequently the FDA has issued guidance to help innovators determine when biologics DDI studies should be conducted; such a study would typically be conducted in patients in a Phase Ib setting or in a PK subgroup of a Phase II/III study [62, 95, 98, 100]. Nevertheless, DDI studies with biologics are still uncommon and biologics labels are usually devoid of information on DDI potential [97].

DDI studies involving biologics are logistically challenging as they often need to be conducted in patient populations and long half-lives preclude the use of crossover designs. However, population PK approaches that pool data across a variety of studies are a viable means of assessing the DDI potential of biologics with a host of advantages, provided that accurate information about concomitant medication is captured, including PK samples if the biologic is being assessed for its perpetrator potential; this approach has been accepted by the FDA [89]. In 2013, the population PK TPDI Working Group, comprised of pharmaceutical industry and FDA representatives, issued a series of recommendations for employing the population PK approach to DDI assessment to meet regulatory expectations [101]. PBPK models can also be useful tools to evaluate DDI potential of biologics. This approach was used to mitigate the need for a DDI study to assess the effect of transient (IL-6) (interleukin-6) elevation after blinatumomab administration on the CYP 450 enzymes [102].

24.3.2 Special Populations: Renal and Hepatic Impairment

For small molecules that are extensively metabolized by the liver, exposures can be significantly higher in hepatically impaired subjects and these subjects may require dose adjustments to prevent unexpected toxicities. But for biologics that are mainly cleared through catabolism, hepatic impairment is not expected to influence exposure and therefore studies in hepatically impaired subjects are not generally recommended [103]. Likewise, the size of most biologics prevents their clearance through renal elimination, precluding the need for a renal impairment study. A renal impairment study may be indicated for smaller biologics (<69 kDa) that can undergo renal elimination [104]. Such studies are also necessarily conducted in cases where the intended patient population has renal insufficiency, as was the case for darbepoetin alfa administered to patients with chronic kidney disease (CKD), and population modeling can also be employed to understand the effect of hepatic or renal insufficiency on PK and PD [105]. Label considerations and competitive positioning may also influence the decision to conduct renal and hepatic impairment studies.

24.3.3 Special Populations: Pediatrics

Pediatric studies were not routinely conducted until the late 1990s when the Pediatric Research Equity Act and the Best Pharmaceuticals for Children Act were passed, which offered 6 months of additional patent protection in exchange for conducting the now required pediatric studies. Pediatric studies are much more routinely conducted as a result, and only waived in special cases, such as for indications that are rarely or never diagnosed in children [106]. A typical pediatric investigation plan includes single- and multiple-dose Phase I studies followed by a multiple-dose Phase II safety and efficacy trial to understand the differences in PK, PD, and immunogenicity in children and justify the dose selection for this population. Dose-finding studies are not performed, partly due to the logistical limitations of conducting such studies, but a Phase III trial is required unless there is evidence that the exposure–response relationship is similar to adults [107]. Pediatric studies are typically initiated once sufficient data to characterize the exposure–response relationship in adults are collected, which is used to inform the starting-dose selection.

The design of studies for pediatric subjects presents a number of logistical challenges not necessarily encountered with adults. First, studies often have to be conducted in infants, young children, and adolescents. Enrollment in these studies is challenging despite high participation rates by pediatric patients in clinical trials and minimization of the number of study subjects required because the total population size is small. It can take years to enroll a pediatric trial, particularly for studies in infants or very young children. Then the data themselves are sparse, as the small size of pediatric patients limits blood volume. Furthermore, the differences in PK, PD, and immunogenicity between adults and children are not well understood, which complicates the starting-dose selection.

Selection of the starting dose in pediatric subjects is based on the relationship between exposure and response in adults and differences in body size between adults and children. The commonly employed strategies for body-size adjusted dosing in children are body weight-based dosing, body surface area (BSA) based dosing, a tiered fixed dosing strategy, or a hybrid approach [108, 109]. The tiered approach offers the easiest and most convenient dosing paradigm, while also managing differences in PK/PD in children compared to adults. Body weight-based dosing and tiered fixed strategies were most commonly employed for the 12 antibody-based therapeutics approved for use in pediatric populations by 2012 [108].

Given the limitations associated with conducting pediatric trials, population PK and PD modeling approaches play a critical role in selecting and justifying the dose selection for children, and for designing informative clinical trials in spite of the small sample sizes and sparse PK sampling. The pediatric dose selection for etanercept, which is approved for the treatment of juvenile rheumatoid arthritis, demonstrates the utility of modeling and simulation in pediatric development. The population PK model was based on sparse PK data from 69 pediatric subjects who received 0.4 mg/kg SC twice weekly for 12 weeks [110]. A more convenient once-weekly dose of 0.8 mg/kg SC was investigated through simulation. The results showed that the PK of the two regimens was largely overlapping, and no difference in patient outcome was expected with the simulated dosing regimen. On the basis of these modeling and simulation results, the FDA approved the use of the once-weekly 0.8-mg/kg SC dose for 12 weeks in children [109].

24.4 BIOSIMILARS

Generic small molecule drugs account for a large share of the U.S. prescription drug market and offer expanded access with reduced cost to patients [111]. Because the physicochemical properties of small molecules can be completely characterized and they are manufactured using reproducible chemical processes, generic versions of small molecules are expected to be structurally identical to the reference product. The development of small molecule generics is generally straightforward, only requiring demonstration of bioequivalence for approval. Biologics, on the other hand, are heterogeneous, complex molecules produced by living cells; the posttranslational modifications and protein folding associated with the production of biologics are difficult factors to control. Thus, it is not expected that biosimilars, follow-on versions of original biologic products, will be structurally identical to the reference product. Because the effect of minor differences in structure is unpredictable, comparability studies must be conducted to demonstrate that there are no clinically meaningful differences between the biosimilar and the innovator product, although the EMA may accept evidence that the PK/PD relationship for a biomarker is similar to that of the reference compound. As of 2013, there were no biosimilars approved by the FDA, although many are in development. In the European Union, 14 biosimilar medicines have been approved [110].

The FDA has given guidance that it will generally evaluate the totality of evidence in reviewing biosimilars for approval, since they will not be exactly identical to the reference product, and suggest a step-wise approach to the development of biosimilars [112]. First, the structure and function of the biosimilars should be characterized, followed by animal toxicology studies. Clinically, human PK studies are required, as well as PD studies if a relevant biomarker exists, to show comparable exposure and response to the reference product; immunogenicity should also be evaluated in these studies. Safety and efficacy trials with a direct comparison to the reference product may be required to complete the approval package; the FDA has offered an abbreviated pathway to approval, where a comparative analytical characterization showing that the biosimilar product is highly similar or highly similar with fingerprint-like similarity may be the basis for a more "selected and targeted" approach to development [113]. Modeling and simulation approaches can streamline this process, as the exposure–response data from the reference compound can be used to design targeted clinical studies and evaluate the probability of a successful outcome [114]. Because of the limited data supporting biosimilars' approval compared to the reference product, additional pharmacovigilance (postapproval safety monitoring) may be required for biosimilars [112, 115]. The biosimilar development approach recommended by the FDA is fundamentally equivalent to that recommended by the EMA [116].

24.5 EMERGING MARKETS

Clinical development has traditionally been concentrated in the United States and western Europe, but recently an increasing number of clinical trials have been conducted in the emerging markets of Asia, South America, and eastern Europe [117]. This increase has been driven partly by the significant expansion of the markets in these countries, with sales in Asia expected to exceed sales in Europe by 2020 [118]. Clinical trial globalization may speed the launch of new medicines in places such as China and India, and may reduce development times by speeding up recruitment and enrollment. Phase III studies are now routinely conducted on a global scale. Some countries, such as Taiwan and South Korea, accept results from global multinational studies to support registration, provided those studies contain a sufficient number of subjects from the local population. The Japanese Pharmaceutical and Medical Devices Agency, on the other hand, will typically require a full clinical development program in the local population, although they have indicated they may accept data from a global Phase III trial if there are safety data in Japanese subjects. In general, biologics are not expected to be ethnically sensitive; dosing adjustments have not been required for the 12 mAbs approved in both the United States and Japan [119].

24.6 CONCLUSIONS

From an ADME perspective, biologics suffer from poor oral bioavailability, limited tissue distribution, lack of access to targets in the central nervous system, and nonlinear, time-, dose-, and disease-dependent PK that potentially translates to variability in safety and efficacy outcomes. But these challenges are far outweighed by the benefits offered by

TABLE 24.2 Clinical Development Programs for Biologics and Small Molecules

Small Molecule Development	Biologics Development
FIH starting dose: allometric scaling based on three species or population PK/PD model (in some cases one or two species may be used, but three species is preferred)	*FIH starting dose:* single-species allometric scaling or PK/PD model
Phase I: single and multiple ascending dose studies to evaluate safety, tolerability, and PK; <100 subjects	*Phase I:* single and multiple ascending dose studies to evaluate safety, tolerability, PK, and immunogenicity; <100 subjects
Phase II: dose-ranging studies to expand safety data and obtain preliminary efficacy data; <500 subjects	*Phase II:* dose-ranging studies to expand safety data and obtain preliminary efficacy data; <500 subjects
Phase III: Large-scale studies in heterogenous patient populations; 100s–1000s of subjects	*Phase III:* Large-scale studies in homogeneous or heterogeneous patient populations, depending on the indication and mechanism of action (some biologics have been approved in highly selected and often "molecularly" specified study populations, e.g., trastuzumab, which was studied in women with HER2 overexpressing metastatic breast cancer); 100–1000s of subjects
Hepatic impairment: studies routinely conducted	*Hepatic impairment:* studies generally not required due to lack of liver metabolism
Renal impairment: routinely conducted	*Renal impairment:* routinely conducted for molecules <69 Da, but otherwise such studies are considered on a case-by-case basis
Intrinsic factors: studies to evaluate the effects of age and gender are routinely conducted; body weight is typically assessed with population PK models; studies to evaluate the effect of race are determined on a case-by-case basis	*Intrinsic factors:* age, gender, and body weight are typically assessed with population PK models; studies to evaluate the effect of race are determined on a case-by-case basis
Extrinsic factors: studies to evaluate the effect of food, concomitant medications, and lifestyle factors (e.g., nicotine use) are routinely conducted. DDI risk is associated with CYP metabolism/ inhibition or induction of CYP enzymes and transporters	*Extrinsic factors:* parenteral administration mitigates need for food effect studies; lack of CYP metabolism mitigates the need to study lifestyle factors (e.g., nicotine use); DDI studies should be considered on a case-by-case basis
ADME: studies to assess routes and rates of excretion (including mass balance) and characterize the metabolite profile are required	*ADME:* studies generally not required due to large molecule catabolism and lack of metabolites
Pediatric studies: generally required	*Pediatric studies:* generally required
Cardiovascular risk assessment: thorough QT/QT_c study is routinely conducted	*Cardiovascular risk assessment:* usually limited to ECG assessments in Phase I, unless preclinical data or MOA (mechanism of action) suggest a risk
Generics: requires demonstration of bioequivalence	*Biosimilars:* "totality-of-evidence" approach to approval; requires comparison of PK, safety, and efficacy to reference compound

biologics, which are highly specific molecules with limited off-target toxicity that offer long half-lives with prolonged efficacy and more convenient, comfortable dosing regimens for patients. Indeed, the ability to rationally design highly specific molecules with appropriate PK properties through techniques such as pegylation and glycoengineering has already resulted in major benefits for patients. Biologics also have a lower risk of DDI compared with small molecules, an increasingly important property given the interest in combination therapies as a means to improve response, particularly in oncology, and the tendency of elderly patients to be prescribed multiple medications simultaneously [120–122]. These properties lead to some differences in the development of biologics compared to small molecules (Table 24.2), although in both cases the development goals are the same: identify an efficacious, safe, and tolerable dose.

Continued optimization of ADME properties will help to reduce attrition rates through the development process. In the preclinical space, this optimization should include improved methods for translating the PK, PD, and TK data, reducing the opportunity costs incurred when promising compounds are eliminated from consideration or inappropriately taken into development, and the direct costs are incurred when the starting dose is too low or too high, resulting in increased development times and unnecessarily large clinical trials or unexpected toxicities that derail development. Modeling and simulation will be equally important during clinical development, to support the characterization and interpretation of ADME properties and the rational design of molecules with optimal PK, ensuring that efficacious, safe, and convenient dosing regimens are available to patients.

REFERENCES

[1] Leader B, Baca QJ, Golan DE. Protein therapeutics: a summary and pharmacological classification. Nat Rev Drug Discov 2008;7:21–39.

[2] Dimasi JA. New drug development in the United States from 1963 to 1999. Clin Pharmacol Ther 2001;69:286–296.

[3] Walsh G. Biopharmaceutical benchmarks 2010. Nat Biotechnol 2010;28:917–924.

[4] Carroll J. 2012. The 15 best-selling drugs of 2012. Available at www.fiercepharma.com. Accessed 2015 Jun 5.

[5] Aggarwal RS. What's fueling the biotech engine-2012 to 2013. Nat Biotechnol 2014;32:32–39.

[6] Evens RP, Kaitin KI. The biotechnology innovation machine: a source of intelligent biopharmaceuticals for the pharma industry–mapping biotechnology's success. Clin Pharmacol Ther 2014;95:528–532.

[7] DiMasi JA, Grabowski HG. The cost of biopharmaceutical R&D: is biotech different? Manage Decis Econ 2007;28:469–479.

[8] DiMasi JA, Feldman L, Seckler A, Wilson A. Trends in risks associated with new drug development: success rates for investigational drugs. Clin Pharmacol Ther 2010;87:272–277.

[9] Prueksaritanont T, Tang C. ADME of biologics-what have we learned from small molecules? AAPS J 2012;14:410–419.

[10] Kola I, Landis J. Can the pharmaceutical industry reduce attrition rates? Nat Rev Drug Discov 2004;3:711–715.

[11] Sacks LV, Shamsuddin HH, Yasinskaya YI, Bouri K, Lanthier ML, Sherman RE. Scientific and regulatory reasons for delay and denial of FDA approval of initial applications for new drugs, 2000–2012. JAMA 2014;311:378–384.

[12] Lobo ED, Hansen RJ, Balthasar JP. Antibody pharmacokinetics and pharmacodynamics. J Pharm Sci 2004;93: 2645–2668.

[13] Elvin JG, Couston RG, van der Walle CF. Therapeutic antibodies: market considerations, disease targets and bioprocessing. Int J Pharm 2013;440:83–98.

[14] Lambracht-Washington D, Rosenberg RN. Anti-amyloid beta to tau – based immunization: developments in immunotherapy for Alzheimer disease. Immunotargets Ther 2013;2013: 105–114.

[15] Lobello K, Ryan JM, Liu E, Rippon G, Black R. Targeting Beta amyloid: a clinical review of immunotherapeutic approaches in Alzheimer's disease. Int J Alzheimers Dis 2012;2012:628070.

[16] Both L, Banyard AC, van Dolleweerd C, Wright E, Ma JK, Fooks AR. Monoclonal antibodies for prophylactic and therapeutic use against viral infections. Vaccine 2013;31: 1553–1559.

[17] Kontsekova E, Zilka N, Kovacech B, Novak P, Novak M. First-in-man tau vaccine targeting structural determinants essential for pathological tau-tau interaction reduces tau oligomerisation and neurofibrillary degeneration in an Alzheimer's disease model. Alzheimers Res Ther 2014;6:44.

[18] Lannfelt L, Moller C, Basun H, Osswald G, Sehlin D, Satlin A, Logovinsky V, Gellerfors P. Perspectives on future Alzheimer therapies: amyloid-beta protofibrils – a new target for immunotherapy with BAN2401 in Alzheimer's disease. Alzheimers Res Ther 2014;6:16.

[19] Martin DF, Maguire MG, Ying GS, Grunwald JE, Fine SL, Jaffe GJ, CATT Research Group. Ranibizumab and bevacizumab for neovascular age-related macular degeneration. N Engl J Med 2011;364:1897–1908.

[20] Magdelaine-Beuzelin C, Pinault C, Paintaud G, Watier H. Therapeutic antibodies in ophthalmology: old is new again. MAbs 2010;2:176–180.

[21] Catapano AL, Papadopoulos N. The safety of therapeutic monoclonal antibodies: implications for cardiovascular disease and targeting the PCSK9 pathway. Atherosclerosis 2013;228:18–28.

[22] Saylor C, Dadachova E, Casadevall A. Monoclonal antibody-based therapies for microbial diseases. Vaccine 2009;27 (6 Suppl):G38–G46.

[23] Strohl WR, Knight DM. Discovery and development of biopharmaceuticals: current issues. Curr Opin Biotechnol 2009;20:668–672.

[24] Czajkowsky DM, Hu J, Shao Z, Pleass RJ. Fc-fusion proteins: new developments and future perspectives. EMBO Mol Med 2012;4:1015–1028.

[25] Werle M, Bernkop-Schnurch A. Strategies to improve plasma half life time of peptide and protein drugs. Amino Acids 2006;30:351–367.

[26] Veronese FM, Mero A. The impact of PEGylation on biological therapies. BioDrugs 2008;22:315–329.

[27] Macdougall IC, Gray SJ, Elston O, Breen C, Jenkins B, Browne J, Egrie J. Pharmacokinetics of novel erythropoiesis stimulating protein compared with epoetin alfa in dialysis patients. J Am Soc Nephrol 1999;10:2392–2395.

[28] Egrie JC, Dwyer E, Browne JK, Hitz A, Lykos MA. Darbepoetin alfa has a longer circulating half-life and greater *in vivo* potency than recombinant human erythropoietin. Exp Hematol 2003;31:290–299.

[29] Glaspy JA, Jadeja JS, Justice G, Kessler J, Richards D, Schwartzberg L, Tchekmedyian NS, Armstrong S, O'Byrne J, Rossi G, Colowick AB. Darbepoetin alfa given every 1 or 2 weeks alleviates anaemia associated with cancer chemotherapy. Br J Cancer 2002;87:268–276.

[30] Shankar G, Shores E, Wagner C, Mire-Sluis A. Scientific and regulatory considerations on the immunogenicity of biologics. Trends Biotechnol 2006;24:274–280.

[31] Harding FA, Stickler MM, Razo J, DuBridge RB. The immunogenicity of humanized and fully human antibodies: residual immunogenicity resides in the CDR regions. MAbs 2010;2:256–265.

[32] Bartelds GM, Krieckaert CL, Nurmohamed MT, van Schouwenburg PA, Lems WF, Twisk JW, Dijkmans BA, Aarden L, Wolbink GJ. Development of antidrug antibodies against adalimumab and association with disease activity and treatment failure during long-term follow-up. JAMA 2011;305:1460–1468.

[33] Billioud V, Sandborn WJ, Peyrin-Biroulet L. Loss of response and need for adalimumab dose intensification in Crohn's disease: a systematic review. Am J Gastroenterol 2011;106:674–684.

[34] Nelson AL, Dhimolea E, Reichert JM. Development trends for human monoclonal antibody therapeutics. Nat Rev Drug Discov 2010;9:767–774.

[35] Xin Y, Bai S, Damico-Beyer LA, Jin D, Liang WC, Wu Y, Theil FP, Joshi A, Lu Y, Lowe J, Maia M, Brachmann RK, Xiang H. Anti-neuropilin-1 (MNRP1685A): unexpected pharmacokinetic differences across species, from preclinical models to humans. Pharm Res 2012;29:2512–2521.

[36] Craik DJ, Fairlie DP, Liras S, Price D. The future of peptide-based drugs. Chem Biol Drug Des 2013;81:136–147.

[37] Beck A, Reichert JM. Therapeutic Fc-fusion proteins and peptides as successful alternatives to antibodies. MAbs 2011;3:415–416.

[38] U.S. Department of Health and Human Services, Food and Drug Administration, Center for Drug Evaluation and Research, Center for Veterinary Medicine. 2013. Guidance for industry: bioanalytical method validation. Rockville (MD).

[39] Lee JW, Kelley M, King LE, Yang J, Salimi-Moosavi H, Tang MT, Lu JF, Kamerud J, Ahene A, Myler H, Rogers C. Bioanalytical approaches to quantify "total" and "free" therapeutic antibodies and their targets: technical challenges and PK/PD applications over the course of drug development. AAPS J 2011;13:99–110.

[40] Findlay JW. Some important considerations for validation of ligand-binding assays. J Chromatogr B Analyt Technol Biomed Life Sci 2009;877:2191–2197.

[41] Kelley M, DeSilva B. Key elements of bioanalytical method validation for macromolecules. AAPS J 2007;9:E156–E163.

[42] Alley SC, Anderson KE. Analytical and bioanalytical technologies for characterizing antibody–drug conjugates. Curr Opin Chem Biol 2013;17:406–411.

[43] Dodge R, Daus C, Yaskanin D. Challenges in developing antidrug antibody screening assays. Bioanalysis 2009;1: 699–704.

[44] U.S. Department of Health and Human Services, Food and Drug Administration, Center for Drug Evaluation and Research, Center for Biologics Evaluation and Research. 2009. Guidance for industry: assay development for immunogenicity testing of therapeutic proteins. Rockville (MD).

[45] Civoli F, Kroenke MA, Reynhardt K, Zhuang Y, Kaliyaperumal A, Gupta S. Development and optimization of neutralizing antibody assays to monitor clinical immunogenicity. Bioanalysis 2012;4:2725–2735.

[46] Eastwood D, Findlay L, Poole S, Bird C, Wadhwa M, Moore M, Burns C, Thorpe R, Stebbings R. Monoclonal antibody TGN1412 trial failure explained by species differences in CD28 expression on CD4+ effector memory T-cells. Br J Pharmacol 2010;161:512–526.

[47] Suntharalingam G, Perry MR, Ward S, Brett SJ, Castello-Cortes A, Brunner MD, Panoskaltsis N. Cytokine storm in a phase 1 trial of the anti-CD28 monoclonal antibody TGN1412. N Engl J Med 2006;355:1018–1028.

[48] Hunig T. The storm has cleared: lessons from the CD28 superagonist TGN1412 trial. Nat Rev Immunol 2012;12: 317–318.

[49] Stebbings R, Findlay L, Edwards C, Eastwood D, Bird C, North D, Mistry Y, Dilger P, Liefooghe E, Cludts I, Fox B, Tarrant G, Robinson J, Meager T, Dolman C, Thorpe SJ, Bristow A, Wadhwa M, Thorpe R, Poole S. "Cytokine storm" in the phase I trial of monoclonal antibody TGN1412: better understanding the causes to improve preclinical testing of immunotherapeutics. J Immunol 2007;179: 3325–3331.

[50] U.S. Department of Health and Human Services, Food and Drug Administration, Center for Drug Evaluation and Research, Center for Biologics Evaluation and Research. 1997. Guidance for industry: S6 preclinical safety evaluation of biotechnology derived pharmaceuticals. Rockville (MD).

[51] Bussiere JL. Species selection considerations for preclinical toxicology studies for biotherapeutics. Expert Opin Drug Metab Toxicol 2008;4:871–877.

[52] Chapman K, Pullen N, Graham M, Ragan I. Preclinical safety testing of monoclonal antibodies: the significance of species relevance. Nat Rev Drug Discov 2007;6:120–126.

[53] Baumann A. Early development of therapeutic biologics–pharmacokinetics. Curr Drug Metab 2006;7:15–21.

[54] Wu B, Johnson J, Soto M, Ponce M, Calamba D, Sun YN. Investigation of the mechanism of clearance of AMG 386, a selective angiopoietin-1/2 neutralizing peptibody, in

splenectomized, nephrectomized, and FcRn knockout rodent models. Pharm Res 2012;29:1057–1065.

[55] Dong JQ, Salinger DH, Endres CJ, Gibbs JP, Hsu CP, Stouch BJ, Hurh E, Gibbs MA. Quantitative prediction of human pharmacokinetics for monoclonal antibodies: retrospective analysis of monkey as a single species for first-in-human prediction. Clin Pharmacokinet 2011;50:131–142.

[56] Vugmeyster Y, Xu X, Theil FP, Khawli LA, Leach MW. Pharmacokinetics and toxicology of therapeutic proteins: advances and challenges. World J Biol Chem 2012; 3:73–92.

[57] Tabrizi M, Bornstein GG, Suria H. Biodistribution mechanisms of therapeutic monoclonal antibodies in health and disease. AAPS J 2010;12:33–43.

[58] Luu KT, Kraynov E, Kuang B, Vicini P, Zhong WZ. Modeling, simulation, and translation framework for the preclinical development of monoclonal antibodies. AAPS J 2013;15:551–558.

[59] Evers R, Dallas S, Dickmann LJ, Fahmi OA, Kenny JR, Kraynov E, Nguyen T, Patel AH, Slatter JG, Zhang L. Critical review of preclinical approaches to investigate cytochrome p450-mediated therapeutic protein drug-drug interactions and recommendations for best practices: a white paper. Drug Metab Dispos 2013;41:1598–1609.

[60] van Meer PJ, Kooijman M, Brinks V, Gispen-de Wied CC, Silva-Lima B, Moors EH, Schellekens H. Immunogenicity of mAbs in non-human primates during nonclinical safety assessment. MAbs 2013;5:810–816.

[61] Vargas HM, Bass AS, Breidenbach A, Feldman HS, Gintant GA, Harmer AR, Heath B, Hoffmann P, Lagrutta A, Leishman D, McMahon N, Mittelstadt S, Polonchuk L, Pugsley MK, Salata JJ, Valentin JP. Scientific review and recommendations on preclinical cardiovascular safety evaluation of biologics. J Pharmacol Toxicol Methods 2008;58: 72–76.

[62] Zhao L, Ren TH, Wang DD. Clinical pharmacology considerations in biologics development. Acta Pharmacol Sin 2012;33:1339–1347.

[63] Grazette LP, Boecker W, Matsui T, Semigran M, Force TL, Hajjar RJ, Rosenzweig A. Inhibition of ErbB2 causes mitochondrial dysfunction in cardiomyocytes: implications for herceptin-induced cardiomyopathy. J Am Coll Cardiol 2004;44:2231–2238.

[64] Speyer J. Cardiac dysfunction in the trastuzumab clinical experience. J Clin Oncol 2002;20:1156–1157.

[65] Gibbs JP. Prediction of exposure-response relationships to support first-in-human study design. AAPS J 2010;12:750–758.

[66] Tang H, Hussain A, Leal M, Mayersohn M, Fluhler E. Interspecies prediction of human drug clearance based on scaling data from one or two animal species. Drug Metab Dispos 2007;35:1886–1893.

[67] Oitate M, Masubuchi N, Ito T, Yabe Y, Karibe T, Aoki T, Murayama N, Kurihara A, Okudaira N, Izumi T. Prediction of human pharmacokinetics of therapeutic monoclonal antibodies from simple allometry of monkey data. Drug Metab Pharmacokinet 2011;26:423–430.

[68] Deng R, Iyer S, Theil FP, Mortensen DL, Fielder PJ, Prabhu S. Projecting human pharmacokinetics of therapeutic antibodies from nonclinical data: what have we learned? MAbs 2011;3:61–66.

[69] Ling J, Zhou H, Jiao Q, Davis HM. Interspecies scaling of therapeutic monoclonal antibodies: initial look. J Clin Pharmacol 2009;49:1382–1402.

[70] Mager DE, Jusko WJ. General pharmacokinetic model for drugs exhibiting target-mediated drug disposition. J Pharmacokinet Pharmacodyn 2001;28:507–532.

[71] Vugmeyster Y, Szklut P, Wensel D, Ross J, Xu X, Awwad M, Gill D, Tchistiakov L, Warner G. Complex pharmacokinetics of a humanized antibody against human amyloid beta peptide, anti-abeta Ab2, in nonclinical species. Pharm Res 2011;28:1696–1706.

[72] Gibiansky L, Gibiansky E. Target-mediated drug disposition model: approximations, identifiability of model parameters and applications to the population pharmacokinetic-pharmacodynamic modeling of biologics. Expert Opin Drug Metab Toxicol 2009;5:803–812.

[73] Luu KT, Bergqvist S, Chen E, Hu-Lowe D, Kraynov E. A model-based approach to predicting the human pharmacokinetics of a monoclonal antibody exhibiting target-mediated drug disposition. J Pharmacol Exp Ther 2012;341:702–708.

[74] Mager DE, Neuteboom B, Efthymiopoulos C, Munafo A, Jusko WJ. Receptor-mediated pharmacokinetics and pharmacodynamics of interferon-beta1a in monkeys. J Pharmacol Exp Ther 2003;306:262–270.

[75] Agoram BM, Martin SW, van der Graaf PH. The role of mechanism-based pharmacokinetic-pharmacodynamic (PK-PD) modelling in translational research of biologics. Drug Discov Today 2007;12:1018–1024.

[76] Wu B, Sun YN. Pharmacokinetics of peptide-Fc fusion proteins. J Pharm Sci 2014;103:53–64.

[77] Betts AM, Clark TH, Yang J, Treadway JL, Li M, Giovanelli MA, Abdiche Y, Stone DM, Paralkar VM. The application of target information and preclinical pharmacokinetic/pharmacodynamic modeling in predicting clinical doses of a Dickkopf-1 antibody for osteoporosis. J Pharmacol Exp Ther 2010;333:2–13.

[78] Reichert JM. Metrics for antibody therapeutics development. MAbs 2010;2:695–700.

[79] Dresser R. First-in-human trial participants: not a vulnerable population, but vulnerable nonetheless. J Law Med Ethics 2009;37:38–50.

[80] Frank R, Hargreaves R. Clinical biomarkers in drug discovery and development. Nat Rev Drug Discov 2003;2:566–580.

[81] Salinger DH, Endres CJ, Martin DA, Gibbs MA. A semi-mechanistic model to characterize the pharmacokinetics and pharmacodynamics of brodalumab in healthy volunteers and subjects with psoriasis in a first-in-human single ascending dose study. Clin Pharmacol Drug Dev 2014;3:276–283.

[82] Peterson MC, Jang G, Kim W, Gurrola E, Kinsey A, Dansey R. Selection of a phase 3 dose regimen for denosumab based on pharmacokinetic (PK), pharmacodynamic (PD), and

safety data from multiple subcutaneous (SC) dosing regimens in breast cancer patients (pts) with bone metastases (BM). J Clin Oncol 2006;24: Abstract no. 3086. Available at http://meeting.ascopubs.org/cgi/content/short/24/18_suppl/3086. Accessed 2014 Jun 16.

[83] Lee H, Yim DS, Zhou H, Peck CC. Evidence of effectiveness: how much can we extrapolate from existing studies? AAPS J 2005;7:E467–E474.

[84] Mandema JW, Salinger DH, Baumgartner SW, Gibbs MA. A dose-response meta-analysis for quantifying relative efficacy of biologics in rheumatoid arthritis. Clin Pharmacol Ther 2011;90:828–835.

[85] Mandema JW, Gibbs M, Boyd RA, Wada DR, Pfister M. Model-based meta-analysis for comparative efficacy and safety: application in drug development and beyond. Clin Pharmacol Ther 2011;90:766–769.

[86] Dodds MG, Salinger DH, Mandema J, Gibbs JP, Gibbs MA. Clinical trial simulation to inform phase 2: comparison of concentrated vs. distributed first-in-patient study designs in psoriasis. CPT Pharmacometrics Syst Pharmacol 2013;2:e58.

[87] Vicini P, van der Graaf PH. Systems pharmacology for drug discovery and development: paradigm shift or flash in the pan? Clin Pharmacol Ther 2013;93:379–381.

[88] Sorger PK, Allerheiligen SRB, Abernethy DR, Altman RB, Brouwer KLR, Califano A, D'Argenio DZ, Iyengar R, Jusko WJ, Lalonde R, Lauffenberger DA, Shoichet B, Stevens JL, Subramanian S, Van der Graaf P, Vicini P. Quantitative and systems pharmacology in the post-genomic era: new approaches to discovering drugs and understanding therapeutic mechanisms. An NIH white paper by the *QSP Workshop Group* 2011; October 2011.

[89] Sutjandra L, Rodriguez RD, Doshi S, Ma M, Peterson MC, Jang GR, Chow AT, Pérez-Ruixo JJ. Population pharmacokinetic meta-analysis of denosumab in healthy subjects and postmenopausal women with osteopenia or osteoporosis. Clin Pharmacokinet 2011;50:793–807.

[90] Doshi S, Sutjandra L, Zheng J, Sohn W, Peterson M, Jang G, Chow AT, Pérez-Ruixo JJ. Denosumab dose selection for patients with bone metastases from solid tumors. Clin Cancer Res 2012;18:2648–2657.

[91] U.S. Department of Health and Human Services, Food and Drug Administration, Center for Drug Evaluation and Research, Center for Biologics Evaluation and Research. 2003. Guidance for industry: exposure-response relationships – study design, data analysis, and regulatory applications. Rockville (MD).

[92] Wang DD, Zhang S, Zhao H, Men AY, Parivar K. Fixed dosing versus body size-based dosing of monoclonal antibodies in adult clinical trials. J Clin Pharmacol 2009;49: 1012–1024.

[93] Zhang S, Shi R, Li C, Parivar K, Wang DD. Fixed dosing versus body size-based dosing of therapeutic peptides and proteins in adults. J Clin Pharmacol 2012;52:18–28.

[94] Lacana E, Amur S, Mummanneni P, Zhao H, Frueh FW. The emerging role of pharmacogenomics in biologics. Clin Pharmacol Ther 2007;82:466–471.

[95] Girish S, Martin SW, Peterson MC, Zhang LK, Zhao H, Balthasar J, Evers R, Zhou H, Zhu M, Klunk L, Han C, Berglund EG, Huang SM, Joshi A. AAPS workshop report: strategies to address therapeutic protein-drug interactions during clinical development. AAPS J 2011;13:405–416.

[96] Huang SM, Zhao H, Lee JI, Reynolds K, Zhang L, Temple R, Lesko LJ. Therapeutic protein-drug interactions and implications for drug development. Clin Pharmacol Ther 2010;87: 497–503.

[97] Lee JI, Zhang L, Men AY, Kenna LA, Huang SM. CYP-mediated therapeutic protein-drug interactions: clinical findings, proposed mechanisms and regulatory implications. Clin Pharmacokinet 2010;49:295–310.

[98] Kraynov E, Martin SW, Hurst S, Fahmi OA, Dowty M, Cronenberger C, Loi CM, Kuang B, Fields O, Fountain S, Awwad M, Wang D. How current understanding of clearance mechanisms and pharmacodynamics of therapeutic proteins can be applied for evaluation of their drug-drug interaction potential. Drug Metab Dispos 2011;39:1779–1783.

[99] Mahmood I, Green MD. Drug interaction studies of therapeutic proteins or monoclonal antibodies. J Clin Pharmacol 2007;47:1540–1554.

[100] U.S. Department of Health and Human Services, Food and Drug Administration, Center for Drug Evaluation and Research. 2012. Guidance for industry: drug interaction studies – study design, data analysis, implications for dosing, and labeling recommendations. Rockville (MD).

[101] Chow AT, Earp JC, Gupta M, Hanley W, Hu C, Wang DD, Zajic S, Zhu M, Population PK TPDI Working Group. Utility of population pharmacokinetic modeling in the assessment of therapeutic protein-drug interactions. J Clin Pharmacol 2013;54 (5):593–601.

[102] Xu Y, Hijazi Y, Wolf A, Wu B, Sun YN, Zhu M. *Evaluation of the Effect of Blinatumomab-Mediated Cytokine Elevations on Cytochrome P450 Enzymes Using a Physiologically Based Pharmacokinetic (PBPK) Model.* Atlanta (GA): American Society for Clinical Pharmacology and Therapeutics; 2014.

[103] Yang J, Shord S, Zhao H, Men Y, Rahman A. Are hepatic impairment studies necessary for therapeutic proteins? Clin Ther 2013;35:1444–1451.

[104] Meibohm B, Zhou H. Characterizing the impact of renal impairment on the clinical pharmacology of biologics. J Clin Pharmacol 2012;52:54S–62S.

[105] Doshi S, Chow A, Perez Ruixo JJ. Exposure-response modeling of darbepoetin alfa in anemic patients with chronic kidney disease not receiving dialysis. J Clin Pharmacol 2010;50:75S–90S.

[106] Field MJ, Ellinger LK, Boat TF. IOM review of FDA–approved biologics labeled or studied for pediatric use. Pediatrics 2013;131:328–335.

[107] U.S. Department of Health and Human Services, Food and Drug Administration, Center for Drug Evaluation and Research, Center for Biologics Evaluation and Research.

2000. Guidance for industry: E11 clinical investigation of medicinal products in the pediatric population. Rockville (MD).

[108] Xu Z, Davis HM, Zhou H. Rational development and utilization of antibody-based therapeutic proteins in pediatrics. Pharmacol Ther 2013;137:225–247.

[109] Yim DS, Zhou H, Buckwalter M, Nestorov I, Peck CC, Lee H. Population pharmacokinetic analysis and simulation of the time-concentration profile of etanercept in pediatric patients with juvenile rheumatoid arthritis. J Clin Pharmacol 2005;45:246–256.

[110] Lovell DJ, Giannini EH, Reiff A, Cawkwell GD, Silverman ED, Nocton JJ, Stein LD, Gedalia A, Ilowite NT, Wallace CA, Whitmore J, Finck BK. Etanercept in children with polyarticular juvenile rheumatoid arthritis. Pediatric Rheumatology Collaborative Study Group. N Engl J Med 2000;342:763–769.

[111] Kozlowski S, Woodcock J, Midthun K, Sherman RB. Developing the nation's biosimilars program. N Engl J Med 2011;365:385–388.

[112] U.S. Department of Health and Human Services, Food and Drug Administration, Center for Drug Evaluation and Research, Center for Biologics Evaluation and Research. 2012. Guidance for industry: scientific considerations in demonstrating biosimilarity to a reference product. Rockville (MD).

[113] U.S. Department of Health and Human Services, Food and Drug Administration, Center for Drug Evaluation and Research, Center for Biologics Evaluation and Research. 2014. Guidance for industry: clinical pharmacology data to support a demonstration of biosimilarity to a reference product. Rockville (MD).

[114] Dodds M, Chow V, Markus R, Perez-Ruixo JJ, Shen D, Gibbs M. The use of pharmacometrics to optimize biosimilar development. J Pharm Sci 2013;102:3908–3914.

[115] Casadevall N, Edwards IR, Felix T, Graze PR, Litten JB, Strober BE, Warnock DG. Pharmacovigilance and biosimilars: considerations, needs and challenges. Expert Opin Biol Ther 2013;13:1039–1047.

[116] European Medicines Agency. *Guideline on Similar Biological Medicinal Products Containing Biotechnology-Derived Proteins as Active Substance: Non-Clinical and Clinical Issues*. European Medicines Agency Committee for Medicinal Products for Human Use; 2013.

[117] Thiers FA, Sinskey AJ, Berndt ER. Trends in the globalization of clinical trials. Nat Rev Drug Discov 2008; 7:13–14.

[118] Hughes B. Evolving R&D for emerging markets. Nat Rev Drug Discov 2010;9:417–420.

[119] Zhou H, Tsukamoto Y, Davis HM. Should clinical pharmacokinetic bridging studies between Caucasian and Asian populations be required for approval of monoclonal antibodies? J Clin Pharmacol 2012;52:1273–1276.

[120] Loya AM, Gonzalez-Stuart A, Rivera JO. Prevalence of polypharmacy, polyherbacy, nutritional supplement use and potential product interactions among older adults living on the United States-Mexico border: a descriptive, questionnaire-based study. Drugs Aging 2009;26:423–436.

[121] Li F, Zhao C, Wang L. Molecular-targeted agents combination therapy for cancer: developments and potentials. Int J Cancer 2014;134:1257–1269.

[122] Yap TA, Omlin A, de Bono JS. Development of therapeutic combinations targeting major cancer signaling pathways. J Clin Oncol 2013;31:1592–1605.

25

REVIEW: THE CRITICAL ROLE OF CLINICAL PHARMACOLOGY IN THE DEVELOPMENT OF BIOLOGICS

LIANG ZHAO[1], DIANE WANG[2], PING ZHAO[1], ELIZABETH Y. SHANG[1], YANING WANG[1] AND VIKRAM SINHA[1]

[1] *US Food and Drug Administration, Silver Spring, MD, USA*

[2] *Pfizer Oncology, San Diego, CA, USA*

25.1 INTRODUCTION

Compared with small-molecule drugs (SMDs), the development of biologics has become one of the fastest growing areas in research and drug development owing to advancement in biotechnology sciences such as recombinant protein technology. Biologics have high potency and specificity resulting in the potential for relatively cleaner safety profiles. Research in biologics has opened doors for new drug therapies, including recombinant insulin products for diabetes, enzyme replacement therapy for Gaucher disease, and tumor necrosis factor-alpha (TNF-α) products for rheumatoid arthritis. However, challenges remain in biological product development because of the larger molecular size causing poor oral bioavailability, less membrane permeability, and low stability.

Many biologics are large complex molecules/mixtures that are not easily identified or characterized as biologics and are defined as a virus, therapeutic serum, toxin, antitoxin, vaccine, blood product, blood component or derivative, allergenic product (or any other analogous product), or an arsphenamine or derivative of arsphenamine (or any other trivalent organic arsenic compound) applicable to the prevention, treatment, or cure of a disease/condition of human beings. Refer to the U.S. Public Health Services Act 42 U.S.C. §262(i).

In this chapter, we will focus on the clinical pharmacology aspects of biologics, including their basic pharmacokinetic (PK)/pharmacodynamic (PD) characteristics, first-in-human (FIH) dose selection, model-based dose determination, and trial design. This chapter discusses monoclonal antibodies (mAbs), proteins, and peptides. Vaccines and stem cell therapies will not be covered given their different development pathways.

25.2 PK AND PD OF BIOLOGICS

Compared to SMDs, biologics have unique characteristics in absorption, distribution, metabolism, and elimination (ADME), which play a role in their use as therapeutic interventions and lead to significant differences in their development.

25.2.1 Structural Difference between SMDs and Biological Products

The primary distinctions between biologics and SMDs are size, how they are produced, and structural complexity. Compared to the typical molecular weight (MW) of a few kilodaltons to 1000 kDa (i.e., IgM mAbs), the MW of an SMD is typically less than 1 kDa (20–100 atoms). Biologics

ADME and Translational Pharmacokinetics/Pharmacodynamics of Therapeutic Proteins: Applications in Drug Discovery and Development, First Edition. Edited by Honghui Zhou and Frank-Peter Theil.

are commonly generated from bacteria, yeast, insects, plants, or mammalian cells engineered with the gene of interest, but they can also be purified from natural sources, while SMDs are usually synthesized. Besides the primary structure, therapeutic proteins have secondary, tertiary, and quaternary structures. Many structural properties influence the biological functionality of biologics, such as protein folding, denaturation, amino acid substitution, deamidation, N- and C-terminal modifications, protein aggregation, oxidation, O/N-linked glycosylation, truncation, phosphorylation, sulfation, PEGylation, carbamylation/carboxylation/acetylation, multimer dissociation, mismatched S—S bonds, truncation, fatty acylation, gamma-carboxy glutamylation, formylation, and methylation. Structure modifications, for example, glycosylation and pegylation, may change the PK characteristics.

25.2.2 Route of Administration and Absorption

The oral bioavailability of therapeutic proteins is negligible because of their size, polarity, and enzymatic degradation in the gastrointestinal (GI) tract. Thus, oral administration is an infeasible route for biologics that need to reach a certain level of systemic exposure for efficacy. However, oral administration can be a useful route for treating GI diseases. Linaclotide, a peptide with 14 amino acids (aa) (MW: 1.5 kDa) is an oral drug for irritable bowel syndrome with constipation and chronic idiopathic constipation [1]. Linaclotide and its active metabolite activate guanylate cyclase-C and act locally on the luminal surface of the intestinal epithelium.

Due to poor oral bioavailability, biologics are usually delivered by parenteral administration, such as intravenous (IV), subcutaneous (SC), or intramuscular (IM) injections. Following SC or IM injection, the absorption is slow with a longer time-to-peak concentration (T_{max}) compared to the SMD. Based on the public data for approved Biologics License Applications (BLAs) by the U.S. Food and Drug Administration (FDA), the T_{max} of 12 approved mAbs or fusion proteins administered by the SC route ranges from 2 to 14 days. The corresponding bioavailability ranges from 50% to 80%. The slow absorption is partially explained by the slow lymphatic uptake determined by the limited lymph flow rate with the maximum flow rate of 1–2 mL/kg/h in the thoracic duct [2]. In comparison, oral dosing is the most common dosing route for SMDs with T_{max} in the order of hours.

25.2.3 Distribution

MW is the key determinant of distribution of biologics. Compared with mAbs, smaller proteins have faster absorption rate and more extensive distribution into tissues. For instance, the steady-state volume of distribution (V_{ss}) of adalimumab (Humira, MW: ~148 kDa) and abatacept (Orencia, MW: ~92 kDa) are 4.7–6.0 L and approximately 6.3 L, respectively. In comparison, the V_{ss} value of anakinra (MW: 17 kDa) is approximately 10 L [3]. This value is greater than blood volume suggesting tissue penetration. Generally, biologics have smaller V_{ss} than SMDs. Most of the SMDs can readily penetrate cell membrane by passive permeation or carrier-mediated mechanisms. In addition, SMDs with lipophilic property are usually not confined to the vascular space either. They exhibit a much larger apparent volume of distribution. For example, theophylline and ethanol have volumes of distributions equivalent to approximately 30 L total body water.

25.2.4 Metabolism and Elimination

SMDs are primarily cleared either through hepatic metabolism or through renal/biliary excretion. Similarly, the biologics with an MW of less than 69 kDa are mainly cleared by renal excretion. Therefore, like renally cleared SMDs, the clearance of these biologics can be compromised in patients with renal impairment [4].

25.2.4.1 Therapeutic mAb mAbs are distinguished by their large molecular size (MW: ~150 kDa), thousands of amino acids, and complex structure. They bind both specifically and nonspecifically to the targets. It is known that mAbs use specific binding domains located in the variable region to bind a target (e.g., soluble antigen/ligand or receptor). They can also bind nonspecifically to neonatal Fc receptor (FcRn) and Fcγ receptors through binding domains located in the Fc region. These binding characteristics lead to unique distribution and elimination behavior of mAb as opposed to SMDs or peptides.

25.2.5 mAb Distribution

mAbs have limited distribution, owing to their large size, surface charge, and target binding. Typical central compartment volume of distribution (V_c) of a mAb derived from a two-compartment PK model is approximately 2–4 L, which is similar to the total plasma volume.

Even though the volumes of distribution of mAbs are relatively small, the mAbs are able to penetrate into peripheral tissues by paracellular and/or transcellular movement [5]. The paracellular movement of mAb is via convective transport instead of passive diffusion often seen for SMDs [6]. The rate of distribution by convection is determined by the rate of fluid movement from blood to tissue, the size and morphology of the paracellular pores in the vascular endothelium, and the size, shape, and charge of the mAb. A unique way of mAb penetration into cellular space is via transcellular trafficking. It consists of three types of translocations: (i) fluid-phase pinocytosis (i.e., mAb taken up from the surrounding fluid space in the vessel wall by endothelial cells), (ii) receptor-mediated endocytosis (mainly via Fcγ receptor binding or through binding to cell surface antigens), and (iii) phagocytosis. Fluid-phase pinocytosis is the main pathway for mAbs entering endothelial cells. After endothelial cell entry, the FcRn recycling pathway participates in the transcytosis step, in which mAbs can be bidirectionally transported to either the interstitial spaces or the vascular space [7–9]. The tissue distribution of mAbs may be heterogeneous because the tight binding between the mAb and its target prohibits deeper penetration, which is typically called antigen barrier [10–13].

25.2.6 Catabolism and Elimination

mAbs are eliminated mainly via intracellular lysosomal proteolytic degradation that occurs throughout the entire body. One exception is IgA-based antibodies, which are mainly eliminated by biliary secretion [14]. More than half of the mAbs on the market exhibit nonlinear PK. The apparent linear elimination component is mainly attributable to Fc-receptor-mediated clearance while the nonlinear elimination is attributable to the target-mediated drug disposition (TMDD). Fc-mediated elimination is a nonspecific elimination pathway for both endogenous IgGs and exogenous therapeutic IgG mAbs involving either FcRn or Fcγ receptors.

In the absence of TMDD, most of the IgG-based mAbs (i.e., IgG1, IgG2, and IgG4) exhibit long half-lives (~3–4 weeks in human). This is primarily due to FcRn-mediated antibody recycling, which salvages the IgG from proteolytic degradation. Under acidic conditions, IgGs entering the endosome via fluid-phase pinocytosis bind to the FcRn receptor and will not be transferred for lysosomal degradation, unlike the unbound antibodies [15, 16]. Fcγ receptors are responsible for clearing soluble mAb-antigen immune complexes or cells opsonized by the mAb [17]. However, the exact mechanism of action of Fcγ receptors in antibody clearance is not fully understood.

25.2.7 Other Biologics

25.2.7.1 Peptides Unlike mAbs that are among the largest biologics, peptides have smaller sizes. There are a few different definitions of peptides. Sato et al. [18] and Latham [19] define peptide as polymers of less than 50 aa with an MW of less than 10 kDa. U.S. FDA guidance defines peptide as less than 40 aa [3]. Because the size of the peptides range from small (<1000 Da) to large (between 1000 and 10,000 Da), they have diverse PK characteristics as a class [20]. Oral absorption of peptides is usually low, although it can be improved by various methods such as modifying the aa backbone and the use of permeation enhancers. The distribution is often limited to the extracellular space with a V_{ss} of 15 L. Similar to mAbs, renal or biliary excretion is usually negligible for peptides. Although a peptide can be filtered through glomeruli, it goes through rapid proteolytic hydrolysis in the proximal tubule cells resulting in a short half-life (<10 min). Hepatic metabolism in general is a minor route of elimination with the exception of some small peptides. Peptides are subject to TMDD.

25.3 CRITICAL ROLE OF CLINICAL PHARMACOLOGY AND RELATED REGULATORY GUIDANCE FOR BIOLOGICS DEVELOPMENT

25.3.1 First-in-Human (FIH) Dose Determination and Study Design

Based on both the FDA guidance and the European Medicines Agency (EMA) guidelines, the starting dose for an FIH study is a dose that should not result in pharmacological or toxicological effects. As a result, the traditional method in determining the FIH dose for new molecular entities is the no observed adverse effect level (NOAEL) approach and to a lesser extent the no observed effect level (NOEL) approach. The NOAEL approach normally includes three steps that are the same for SMDs and biologics. The first step is the determination of a NOAEL dose in the most sensitive or clinically relevant nonclinical species. The second step is to convert the animal NOAEL dose to a human equivalent dose (HED) by applying a body surface area conversion factor. The third step is to determine the maximum recommended starting dose (MRSD).

For SMDs, MRSD is calculated by adjusting the HED with a safety factor ($MRSD = HED \times$ safety factor). The default value for the safety factor is 1/10th the HED but it may be adjusted upward or downward with sound justification. Unlike SMDs, the NOAEL-to-HED conversion step is not needed to calculate MRSD for biologics since usually the only relevant nonclinical animal species is nonhuman primates. To calculate MRSD for biologics, the safety factor can be directly applied to the body weight (BW) -normalized NOAEL dose.

Additional considerations for determining an MRSD are sometimes taken into account under special circumstances. For example, when cross-reactivity between human and animal models cannot be established for the biologics in development, especially for mAbs, a surrogate antibody may be used for animal study since a toxicology study in chimpanzees, the most relevant animal model, is prohibited. Furthermore, an MRSD derived from an animal NOAEL approach may not be reliable because of potential differences in target/antigen binding, distribution, and capacity between human and animals. Therefore, all potential factors leading to changes in PK exposure, efficacy, and safety responses should be taken into account when extrapolating a dose from animals to human.

The widely used technique for human PK predictions is allometric scaling. TMDD modeling and physiologically based pharmacokinetic (PBPK) modeling have also been used for such purpose, but to a lesser extent especially for regulatory submissions.

The EMA has proposed the minimum anticipated biological effect level (MABEL) approach for FIH dose selection. The MABEL approach is mainly for high risk biologics with agonist-like effects, such as biologics with cellular targets that activate downstream intracellular pathways and trigger cytokine release. The determination of a human MABEL dose depends on PK/PD relationships. Such relationships can be based on *in vitro* and/or *in vivo* studies involving human cells or animal species. The differences between animals and humans in systemic exposure, target-binding affinity, and PD potency should all be taken into consideration [21]. One of the examples of using the MABEL approach for FIH dose determination is for an antibody that targets a blood cell surface receptor [22]. Generally speaking, a dose derived from MABEL is more conservative than that from NOAEL and is normally one or more than 1 order of magnitude lower. If necessary, MRSDs determined from the NOEL, the pharmacologically active dose (PAD), and/or the MABEL approaches can be evaluated against the MRSD dose determined from the NOAEL to make the final FIH recommendation.

25.3.2 Critical Considerations from a Standpoint of Clinical Pharmacology in Biologics Development

The U.S. FDA and the EMA have developed several guidelines to regulate biologics development [23, 24]. The current development strategies and regulatory guidance are evolving and subject to change with advancements in both technology and regulatory sciences. During the clinical development of biologics, general aspects from a clinical pharmacology perspective as shown by Table 25.1 should be considered. Assay development for PK and immunogenicity assays is out of the scope of this chapter.

Clinical pharmacology considerations for biologics are depicted in Figure 25.1. Ultimately, it is the size of the biologics that distinguishes the clinical development plan for biologics and SMDs. The inherent characteristics of mAbs, such as specific functions associated with their FcRn, FcγR, and C1Q domains, differentiate them from other therapeutic proteins in many respects, such as PK, PD, and pharmacological functions. Consequently, special considerations should be given to each individual biological product based on its own characteristics.

25.3.2.1 Evaluation of Drug–Drug Interactions (DDI) for Biologics Like SMDs, biological products can have potentials of drug–drug interaction (DDI). For instance, a DDI between a biological product modulating cytokines and an SMD has been reported. However, the reported magnitudes of interaction are mostly mild and less common than those for SMDs, owing to the difference in drug elimination mechanisms. Cytokine-mediated changes in drug-metabolizing enzymes are the most well-documented therapeutic DDI mechanisms for biologics. Because of a lack of predictive *in vitro* and preclinical animal models for DDIs, clinical study is usually the approach for biologic DDI assessment. Clinical investigations on biologics as a victim drug include the impact of altered target protein levels by the concomitant medication on the clearance of therapeutic proteins, the displacement of therapeutic proteins from binding proteins, and the modulation of Fcγ receptor expression. When designing a DDI study for biologics, one should consider factors such as patient population, disease status, concomitant medications, elimination mechanisms of a therapeutic protein, and effect of biologics on cytochrome P450 (CYP) activities. While the crossover study design is the most often used approach for SMD DDI assessment, it may not be a feasible approach for most biologics due to their long half-lives and/or propensity of developing immune response. Even for evaluating the effect of biologics on PK of SMDs, a sequential study design (an SMD administered in Period 1 or Lead-in phase of a

TABLE 25.1 General Considerations of Clinical Pharmacology Development

General Clinical Pharmacology
1. Single (and multiple) dose PK and dose proportionality, identify potential PD endpoints, characterize immunogenicity, and capture large cardiac safety signals.
2. The analytical method validation used to determine the concentrations.
3. Evaluation of impact of immunogenicity on PK, PD, safety, and efficacy of the biological product.
4. Assay validations that will be used for the detection of antiproduct antibodies (APAs). The qualification results should include data demonstrating that the assay is specific, sensitive, and reproducible and should include information on the sensitivity of the assay to product interference. The validated assay should be capable of sensitively detecting APA responses in the presence of biologic levels that are expected to be present at the time of patient sampling. An assay should also be developed that is able to delineate neutralizing APA responses. Store patient samples under appropriate storage conditions until an assay(s) is been developed.
5. The potential for a pharmacokinetic interaction between the biologics and the approved drug/combination therapy will need to be evaluated during the development of this combination therapy.
Pharmacometrics
1. Conduct population PK analysis to evaluate the effect of intrinsic and extrinsic factors on the PK of the biological product in humans.
2. Explore the exposure–response relationships for the proposed product for measures of both effectiveness and safety.

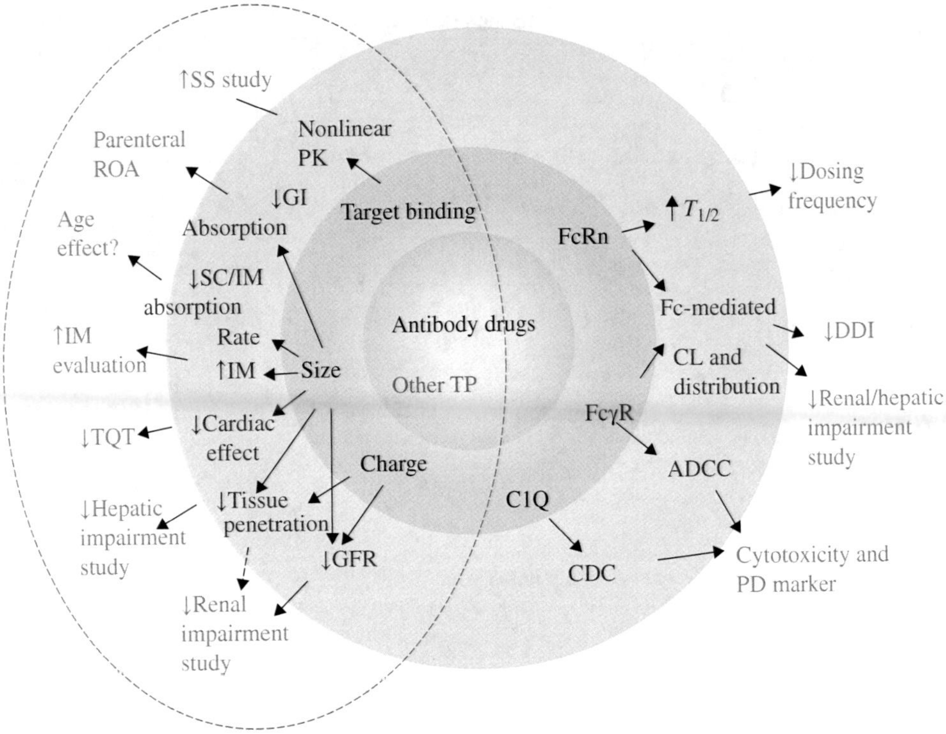

FIGURE 25.1 Grand view of clinical pharmacology aspects for biologics development and their origins. Common clinical pharmacology considerations are shown in the unshaded area. Typical PK, PD, physiological, and pharmacological properties of biologics are shown in the third layer of the shaded area. Biochemical structural properties are shown in the second shaded layer. The two common classes of biologics, antibody drugs and other therapeutic proteins (TPs), are shown in the inner layer. The broken circle highlights the features common to all TPs. TP, therapeutic protein; DDI, drug–drug interaction; CDC, complement dependent cytotoxicity; ADCC, antibody dependent cellular cytotoxicity; SS, steady state; TQT, thorough QT prolongation; GFR, glomerular; GI, gastrointestinal; ROA, route of administration; IM, intramuscular; and SC, subcutaneous. (Adapted from Zhao et al. [25].)

Phase II or III study, SMD + biologics administered in Period 2 or Day 1 of a Phase II or III study) is often used to avoid long washout period for biologics. In addition, DDI assessment for biologics is often conducted in patients instead of healthy subjects. This is mainly due to: (i) potential difference in PK and PD between patients and healthy subjects; (ii) DDI may not be direct and target biology and disease status often served as mediator for DDIs of biologics; (iii) toxicities of the biologics and SMDs especially for oncology compounds that prohibit evaluation of DDI in healthy subjects; and (iv) immunogenicity issues. All these factors make it difficult to conduct dedicated DDI assessment for biologics. Alternatively, population PK method can be used for confirmatory DDI assessment [26]. Population PK approach allows less intensive PK sampling in patients, incorporating DDI assessment in Phase II/III trials, and integrating data generated across multiple studies during different development phases. DDI findings identified by population PK approach have already been exemplified in current labels (e.g., pregabalin, pramipexole, tocilizumab (TCZ), sildenafil, and cilostazol). It should be noted that biologics and SMDs share the same principles in terms of the DDI data analysis method and labeling language regarding dosing and DDI potential [27]. Figure 25.2 shows the decision tree and steps involved in DDI evaluations of biologics.

25.3.2.2 QT$_c$ Prolongation by Biologics The QT interval measures the time from the start of the Q wave to the end of the T wave in a heart's electrical cycle. QT$_c$ represents the QT interval corrected for heart rate (RR interval) because QT interval is heart rate dependent. In general, ICH-14 guidance currently only recommends a thorough QT (TQT) study be conducted for SMDs with systemic bioavailability. In oncology, alternative proposals to the "TQT" study may be appropriate with adequate safety monitoring. Based on the FDA's in-house, retrospective review on QT evaluation on 14 mAbs, it appears that mAbs are not likely to cause QT prolongation at doses being studied (unpublished data). In the published literature domain, there are many publications on TQT study results for mAbs. For TCZ, the TQT results following single-dose administration at therapeutic and supratherapeutic doses in healthy subjects showed no QT$_c$ interval change from time-matched baseline at all the

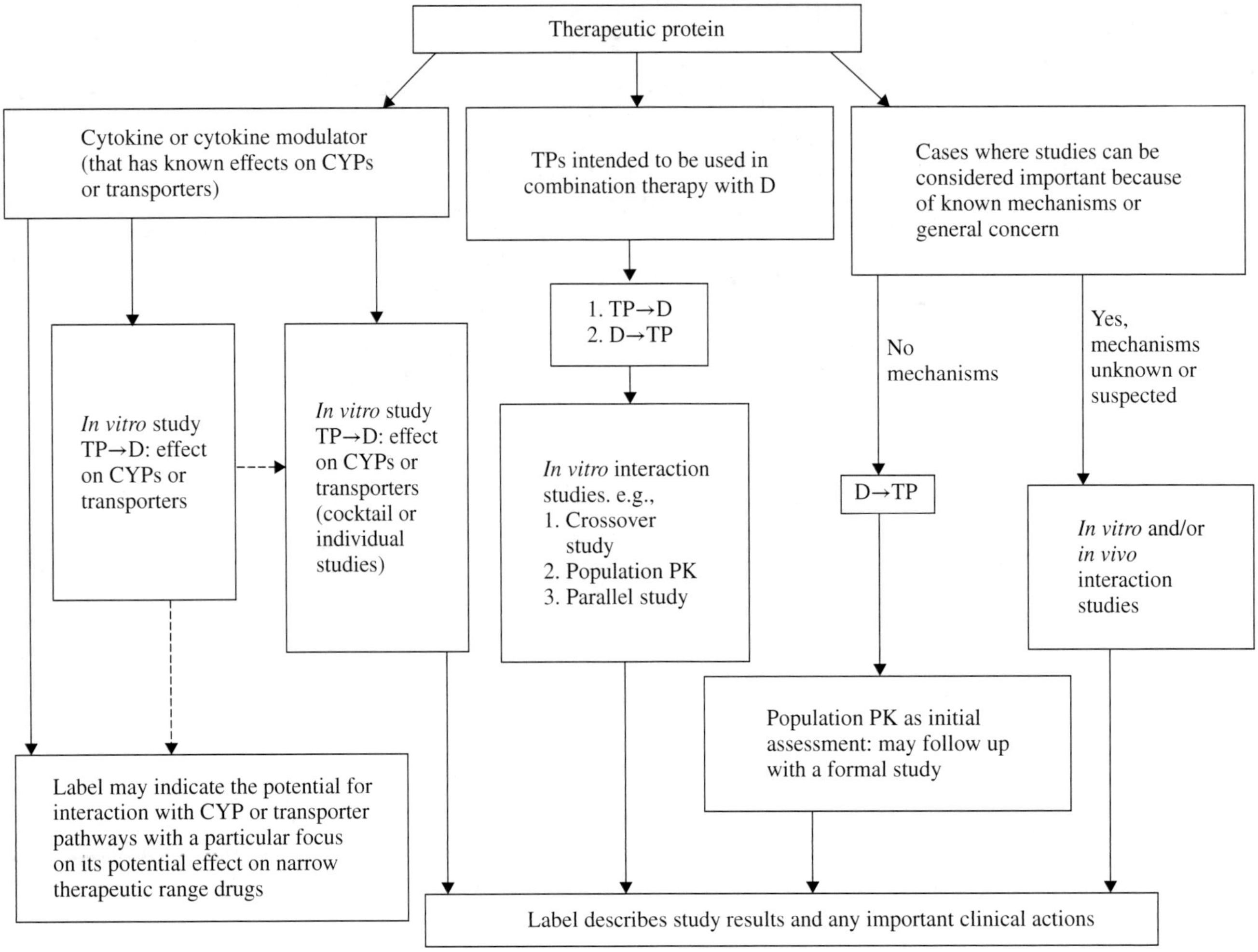

FIGURE 25.2 Decision tree for DDI study for biologics proposed by Huang et al. TP, therapeutic protein; D, small-molecule drug; TP → D, an evaluation of the effect of TP on D; D → TP, an evaluation of the effect of D on TP; broken lines, the limited use of *in vitro* studies for informing *in vivo* study design or labeling; CYP, cytochrome P450. (Chirmule et al. [28]. Reproduced with permission of AAPS J.)

time points and the associated upper bound of the one-sided 95% confidence limit was below threshold (10 ms) [29]. Since mAbs have shown low potential for QT prolongation, a TQT study is not required. However, extensive ECG monitoring (e.g., triplicate ECGs, frequent measures around T_{max}, and predose) should be undertaken in the early phase of clinical development to monitor cardiac safety and to confirm the need of a TQT study because off-target cardiac effects remain in question. Exposure–response analysis for QT_c change can be conducted early in the development phase to estimate the degree of QT prolongation [30]. Particular attention should be paid to smaller size of the biologics (<5 kDa), biologics with heart and/or vasculature targets or targets with the same nature, antibody–drug conjugates (ADCs), and biologics with positive nonclinical cardiovascular safety signals. A TQT study may be needed on a case-by-case basis, depending on the information gained from the ongoing clinical development.

25.3.2.3 Immunogenicity Assessment and Risk Mitigation Another distinction between SMDs and biologics is that biologics can be immunogenic. Owing to the size of biologics and the human body's mechanisms for protecting against foreign invasion, biologics have the potential for immunogenicity by inducing antiproduct antibodies (APAs). The formation of APA against the corresponding biologic or immune complexes that trigger proteolytic elimination in the reticuloendothelial system (RES) will cause increased clearance of a biologic. Infliximab, a chimeric IgG1κ mAb (composed of human constant and murine variable regions) specific for human TNF-α, was cleared more rapidly in patients who developed HACAs (human antichimeric antibodies) [31]. A decreased trough concentration with increasing immunogenicity was also reported for golimumab, a human IgG1κ mAb specific for human TNF-α. In contrast to the notion that immunogenicity always increases drug clearance, changes in the elimination rate due to immunogenicity may be bidirectional. For example, an immune complex that does not trigger proteolytic elimination in the RES may slow down drug elimination by serving as a depot for the therapeutic proteins. This has been observed quite often for cytokines and hormones [32].

The degree of humanization, route of administration, duration of therapy, and dose level can also impact immunogenicity. The incidence of immunogenicity is negatively correlated with the degree of humanization of mAbs [33]. In the case of tositumomab, a murine antibody, the incidence of developing HAMA (human antimouse antibody) seropositivity was 70% in patients with low grade non-Hodgkin's lymphoma [34]. In comparison, the reported incidence of immunogenicity for fully human, humanized, and chimeric mAbs ranged from less than 1% to approximately 10% [6]. The incidence of immunogenicity is positively correlated with the duration of therapy and the reintroduction of drugs [34, 35]. It has also been reported that immunogenicity is negatively related to the dose [6]. However, a conclusion regarding the relationship between immunogenicity and dose could not be reached. The potential to develop APA could be higher following SC or IM administration than IV administration because phagocytes and natural killer (NK) cells, which are responsible for the initial, innate immune response, are found under the skin and in the mucosal epithelia [17].

The major consequences of immunogenicity include loss of efficacy and serious adverse events. The loss of efficacy may be a result of neutralizing APA blocking the effect of a biological product on its intended target. The safety events can originate from an intensified general immune response or cross-reactivity of APA with endogenous substances that are critical for maintaining physiological function. APA may alter PK profile of a biological product and result in change in its exposure–response relationships for efficacy and safety. Alteration in PK of biologics is one of the early indications of APA formation and a surrogate for later changes in efficacy and safety.

Measuring and predicting the effect of immunogenicity of a specific product is challenging. It depends on how the diseases change the immune response, characteristics of the product, and individual patient's genetic background and immune status. Challenges in developing a validated bioanalytical method for measuring APA add another layer of difficulties in quantifying the magnitude of immunogenicity. Therefore, it is important to look at the totality of the data from PK, APA, PD, and efficacy/toxicity study when evaluating the impact of immunogenicity [36]. When analyzing the impact of immunogenicity on PK/PD, one should consider the different dosing schedules and sampling schedules, bioanalytical methods used for detecting APA, differentiating APA responses by concentrations, and assessing PK profiles through cross-sectional analysis of PK data [28].

In the past few years, population PK approach has been applied to quantitative evaluation on the effect of APA on PK. Statistical and mathematical PK/APA models are developed with the hypothesis that observed changes in PK of biologics reflect the time course and magnitude of APA generation [37, 38]. By comparing PK data in the presence or absence/very low incidence of APA, the effect of immunogenicity can be confirmed.

During the development of a biological product, a risk mitigation strategy for potential immunogenicity-related adverse events should be developed and implemented in the early phases. The risk mitigation strategy may be allowed to evolve at different stages of drug development as the understanding of the risk factors for immunogenicity and the degree of its impact on drug safety and efficacy improves. A typical mitigation plan is formulated by considering the protein structure (e.g., its similarity to endogenous substances), the manufacturing process, and the target population. It normally includes a bioanalytical assay strategy for both nonclinical and clinical samples, a titer testing approach to identify the types of APA responses, a medical treatment plan for potential immunogenicity-related safety events (e.g., coadministration of methotrexate with the biologics presenting significant immunogenicity issues [31]), and appropriate study-stopping criteria.

25.3.2.4 Specific Populations

Pediatrics The clinical program in the pediatric population is the same for both biologics and SMDs. The principle that underlies current regulatory guidance for pediatric studies is to minimize the trial burden in the pediatric population [39]. For example, the relative bioavailability of pediatric oral formulations compared to adult oral formulations should be evaluated in the adult population for an SMD.

To conduct a pediatric PK study, the use of population PKs and sparse sampling based on the optimal sampling theory is recommended. For biologics that exhibit nonlinear PKs in adults, such as a result of TMDD, a steady-state study in the pediatric population is normally needed. Based on similarities in disease progression and the exposure–response between children and adults, a different clinical development plan is needed. In the event of a different disease progression pattern for pediatrics, it is expected that both safety and efficacy trials will be carried out in the pediatric population (Figure 25.3). The volume of blood drawn should be minimized [41]. If a pediatric study is needed, current recommendations include the use of the patient population instead of healthy subjects, body-size-based dosing (i.e., BW- or body surface area (BSA)-based dosing even if fixed dosing is used for adults), conservative dosing in anticipation of potential differences in PK parameters, and the use of a formulation and vial strength suitable for the pediatric population [42]. One example is the

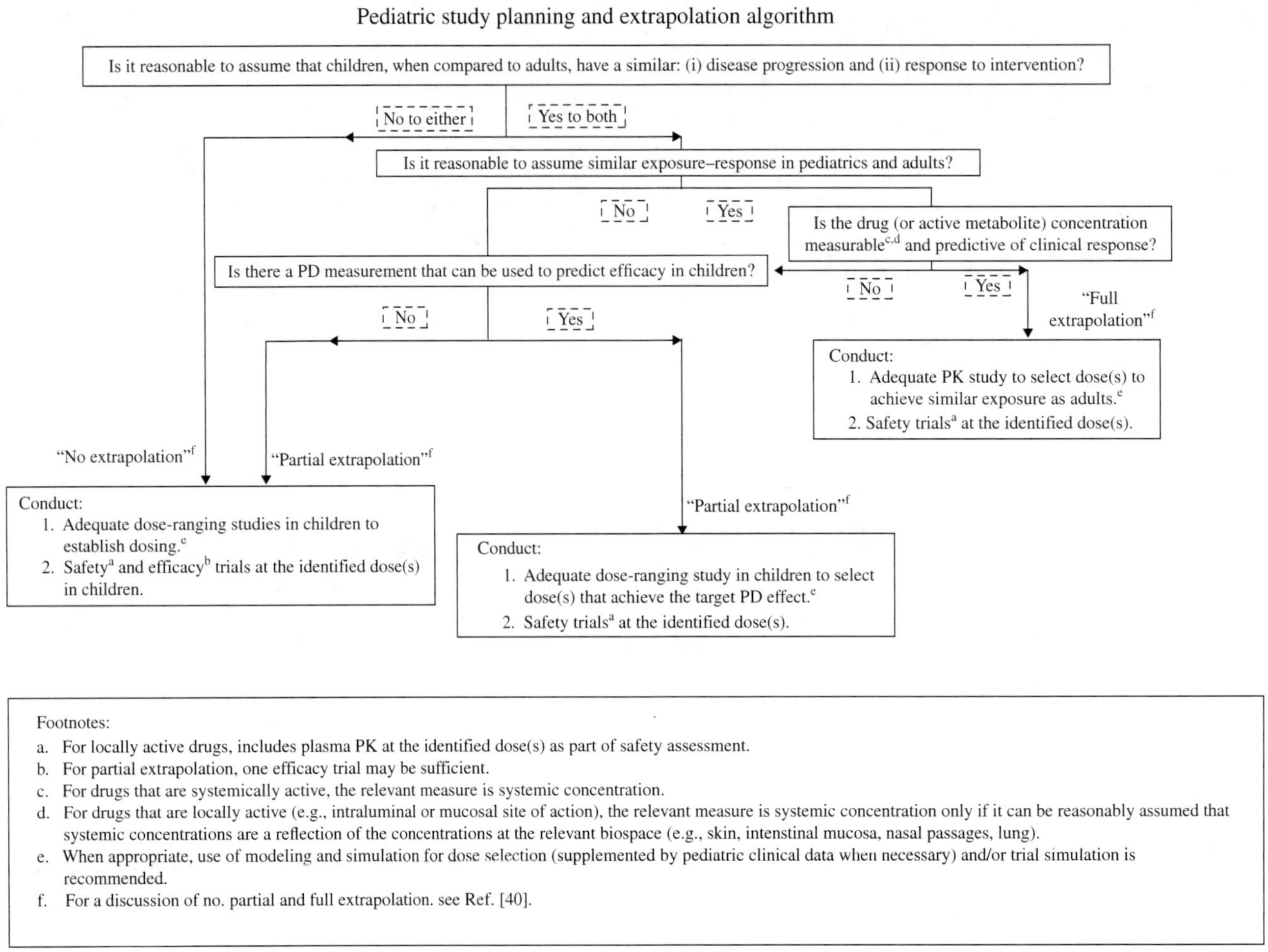

FIGURE 25.3 Pediatric study decision tree. (Bonate et al. [37]. Reproduced with permission of J Pharmacokinet Pharmacodyn.)

development of TCZ for systemic juvenile idiopathic arthritis (sJIA). TCZ 4 mg/kg IV every 4 weeks followed by an increase to 8 mg/kg every 4 weeks was first approved for adult patients with rheumatoid arthritis [43]. The dose finding for patients with sJIA employed population PK modeling and simulation [44, 45]. TCZ 8 mg/kg IV every 2 weeks for sJIA patients was first tested in Japanese patients. The results showed that patients with lower BW had lower systemic exposures. The proportion of patients reaching American College of Rheumatology (ACR) 50 and ACR 70 responses was also lower in these patients. In order to achieve uniform exposure across the entire BW range, a population PK modeling and simulation was performed using pooled PK data from the Japanese patients. The *post hoc* estimates of systemic exposures of TCZ using final population PK model predicted that the 12-mg/kg dose in patients less than 30 kg would yield systemic exposures of TCZ similar to the patients with a BW of ≥30 kg receiving 8 mg/kg IV. The prediction was then confirmed in a pivotal study. TCZ 12 mg/kg IV for patients less than 30 kg and 8 mg/kg IV for patients ≥30 kg every 2 weeks were therefore approved for the treatment of sJIA in patients 2 years or older.

Renal Impairment Dedicated renal impairment study is not required for mAb drugs and can be evaluated in population PK analysis when a sufficient number of patients are included in the dataset. However, a dedicated study may not be waived for biologics with smaller size. The effect of renal impairment on the exposure of therapeutic proteins with an MW of less than 69 kDa is inconsistent across different proteins. Renal impairment has been shown to potentially increase the exposure to the cytokines or cytokine modulators with an MW of less than 69 kDa. The examples include anakinra and oprelvekin. A dose reduction is required for these therapeutic proteins in patients with severe renal impairment.

For Kineret [46], a nonglycosylated form of the human interleukin-1 receptor antagonist with an MW of 17 kDa, a renal impairment study revealed that its plasma clearance was incrementally reduced in patients with mild, moderate, or severe renal impairment. For PEGINTRON [47], a PEGylated interferon alfa-2b with an MW of 31 kDa, a renal impairment study indicated that its clearance decreased by 17% and 44% in patients with moderate or severe renal impairment, respectively. Thus, dose reductions were recommended for both Kineret and PEGINTRON in patients with renal impairment. However, the degree of exposure changes for biologics with an MW of less than 69 kDa in patients with renal impairment may not lead to dose adjustment. For example, for ranibizumab, a recombinant humanized IgG1 kappa isotype mAb fragment with an MW of 48 kDa, the exposure change caused by renal impairment is not considered clinically significant [48].

Dedicated renal impairment study to evaluate the impact of renal impairment on PK is recommended for small proteins (<69 kDa) with presumed or known renal clearance as the dominant excretion pathway, similar to the approach undertaken for SMDs.

Hepatic Impairment Similar to the strategy for assessing the impact of renal impairment on PK, hepatic impairment study is not required for mAb drugs and can be evaluated in population PK analysis if a sufficient number of patients are included in the dataset. The effect of hepatic impairment on the exposure to biologics can be minimal. There have been very few reports of specific studies investigating the impact of hepatic impairment on the PK of biologics. Although a dedicated study to evaluate hepatic clearance in patients with impaired hepatic function is normally not required for biologics, the impact of hepatic impairment may be evaluated on a case-by-case basis. For smaller size of the biologics, for example, small peptide drugs, hepatic metabolism becomes more important. Dipeptide bortezomib (MW: 384.24 Da) undergoes extensive hepatic metabolism mainly by CYP 3A4, 2C19, and 1A2 [49]. In the case of teduglutide (MW: 3752 Da, 33 aa), a 15% difference in AUClast was observed between the patients with moderate hepatic impairment and healthy subjects in a dedicated hepatic impairment study [50]. For ADCs, the drug (small molecule) released from the antibody is subjected to hepatic metabolism. Mylotarg is a cytotoxic antitumor antibiotic (calicheamicin) linked to a recombinant humanized IgG4 kappa antibody for targeted delivery. The metabolism of Mylotarg has been investigated in human liver microsomes, human liver cytosol, and human leukemia cells. After *in vitro* incubation with Mylotarg, a total of 11 metabolites were found because of the metabolism of calicheamicin. As a result, Mylotarg product label notes that "extra caution should be taken when administering Mylotarg in patients with hepatic impairment" [51].

Geriatrics Reports on clinical pharmacology studies of biologics in elderly subjects have been scarce. However, there are examples where age has an effect on PK parameters of certain biologics. For instance, canakinumab, an IgG1-based antibody drug, has been found to have a slightly reduced absorption rate, but not a reduced overall drug exposure, in the elderly [52]. Levemir, a recombinant long-acting basal insulin with an MW of approximately 6 kDa, shows a higher exposure in the elderly than in younger subjects but no difference in overall safety or effectiveness [1]. Under most circumstances, age effect is evaluated as a covariate (e.g., for panitumumab) in their corresponding population PK models.

Race and Sex The effect of race or sex on the exposure of biologics is often insignificant after the difference in BW has been taken into account. Race and sex effects have often been investigated as covariates on PK parameters using population PK modeling. For example, age was found not to be a significant covariate on PK parameters for panitumumab and natalizumab. Definitive statements regarding the effect of race or sex on drug exposure are rarely found in biological product labels. However, pharmaceutical sponsors may wish to facilitate quick entry to other ethnic regions and cross the regulatory barriers early by conducting bridging studies to investigate potential differences in drug exposure in different ethnic groups.

25.4 MODEL-BASED DRUG DEVELOPMENT FOR BIOLOGICS

Population PK and PK/PD-clinical response models play a central role in dosing regimen determination. There are different types of models, such as mechanism-based models, physiologically based models, empirical models, semiempirical models, and meta-analysis for biologics of similar molecular structure and the same target. All of these models can be used to analyze different types of relationships, such as exposure–biomarker, exposure–response, and biomarker–response relationships. Empirical exposure–response models have gained popularity mainly because of their practicality and convenience, as many of the downstream actions after a drug binds to its target remain unknown. In this regard, empirical exposure–response models used for SMDs can be directly applied to characterize the drug effects of biologics. It should be noted that predictive empirical models, as well as all other types of models, rely on the availability of sufficient high quality data, which calls for well-designed studies with prospectively defined PK, PD, and clinical response endpoints. Another modeling tool that has been increasingly used is meta-analysis. Meta-analysis synthesizes reported clinical data from drugs in the same class to enrich the information for the dose/exposure response of the drug candidate in development [53–56] and also provides a benchmark for comparison purposes. Simultaneous modeling of exposure–PD response, PD response–clinical response, and exposure/PD response–clinical response has been considered an ideal

TABLE 25.2 Selected Therapeutic Peptides and Proteins and Their Dosing Approaches for Adult Patients

Generic Name	Brand Name	Approval Date	MW (Da)	Type	Target	Dosing Approach
Abatacept	Orencia	2005	92,300	Fusion protein	CD80/CD86	mg/kg
Daptomycin	Cubicin	2004	1,620	Peptide	LTA synthesis	mg/kg
Darbepoetin alfa	Aranesp	2001	37,100	Protein	EpoR	μg/kg
Degarelix	Firmagon	2008	1,632	Peptide	GnRHR	mg
Emfilermin		Discontinued	22,007	Protein	LIFR	μg/kg
Enfuvirtide	Fuzeon	2003	4,492	Peptide	gp41	mg
Erythropoietin alpha	EPOGEN	1989	30,400	Protein	EpoR	Units/kg
Erythropoietin beta	NeoRecormon	1993	30,000	Protein	EpoR	μg/kg
Etanercept	Enbrel	1999	150,000	Fusion protein	TNF	mg
Hematide		In development	NR[a]	Pegylated peptide	EpoR	mg/kg
Lanreotide autogel	Somatuline	2007	1,096	Peptide	IGF-1	mg
Octreotide acetate	Sandostatin	1988	1,019	Peptide	SSTR2/5	μg
Onercept		Discontinued	18,000	Fusion protein	TNFR	mg/kg
PEG interferon alpha-2b	PEG-Intron A	2001	19,271	Protein	IFNAR1/2	μg/kg
Plitidepsin	Aplidin	2004	1,110	Peptide	EGFR	mg/m^2
Recombinant Factor VIIa	NovoSeven	1999	50,000	Protein	TF	μg/kg
rhGH[a]	Norditropin	1987	22,000	Protein	GH receptor	mg/kg
u-hFSH[a]	Metrodin HP	Discontinued	30,000	Protein	FSH receptor	IU

[a] EGFR, epidermal growth factor receptor; FSH, follicle-stimulating hormone; GH, growth hormone; IGF, insulin-like growth factor; LTA, lipoteichoic acid; NR, not reported; PEG, pegylation; rhGH, recombinant human growth hormone; TNFR, tumor necrosis factor receptor; and u-hFSH, urinary human follicle stimulating hormone.

Chow et al. [26]. Reproduced with permission of *J Clin Pharmacol*.

approach for dosing regimen justification. However, the data requirements, the difficulty in identifying measurable biomarkers, and the lack of a relationship between PD response and clinical response have limited the application of PK/PD-clinical response models in certain therapeutic areas.

With an exposure–response analysis for efficacy identified for biologics, careful examinations should be undertaken before moving forward with the results. It has been found in a few cases that mAb exposure was confounded with disease severity, which can be potentially explained by varying target capacity and TMDD. Therefore, patients who have low drug exposure may at the same time have high disease severity and consequently less efficacy response. Under this circumstance, case–control analysis can be used to have a relatively more objective assessment of efficacy in the subgroup populations based on drug exposure.

25.4.1 Fixed Dosing versus Body Size-Adjusted Dosing

Fixed dosing is the most common dosing approach for SMDs in adult patients. Fixed dosing regimen is favored for its ease of use. However, biologic products are often dosed based on body size, such as BW and BSA. Whether a drug should be administered based on a patient's body size mainly depends on the effect of the body size on the PK, PD, and the therapeutic index of the drug (Table 25.2). A good dosing strategy is to provide reduced interpatient variability in PK, sometimes in PD, while achieving optimized therapeutic outcomes. PK/PD or E-R modeling plays a critical role in this matter.

Two retrospective studies evaluated the potential benefits of fixed dosing and body-size-based dosing by comparing the ability of each of the two approaches to reduce PK and/or PD variability in adults for 30 biologics with published population PK and/or PD models [57, 58]. Of these 30 biologics, 12 were mAbs [57] and 18 were not mAbs (these included therapeutic proteins and peptides) [58]. At the population level, the intersubject variability in exposure (area under the curve (AUC) and C_{max}) was examined in 1000 subjects for both dosing approaches. At the individual level, the difference between the exposure of patients with extreme body sizes and the typical exposure following both approaches was compared. The results, as illustrated by a representative plot (Figure 25.4), show that the two dosing approaches perform similarly across the biologics investigated, with fixed dosing being better for some biologics and body-size-based dosing being better for the others. Based on these findings, fixed dosing can be used for FIH adult studies because it offers advantages including the ease of preparation, lesser costs, and a reduced chance of dosing errors. When sufficient data become available, a full assessment of the body-size effect on PK/PD should be conducted to determine the dosing approach for Phase 3 trials.

25.4.2 Mechanism- and Physiologically Based Models for mAbs

Mechanism-based models aim to mathematically characterize the underlying biological and pharmacological processes. This type of model has been applied to describe

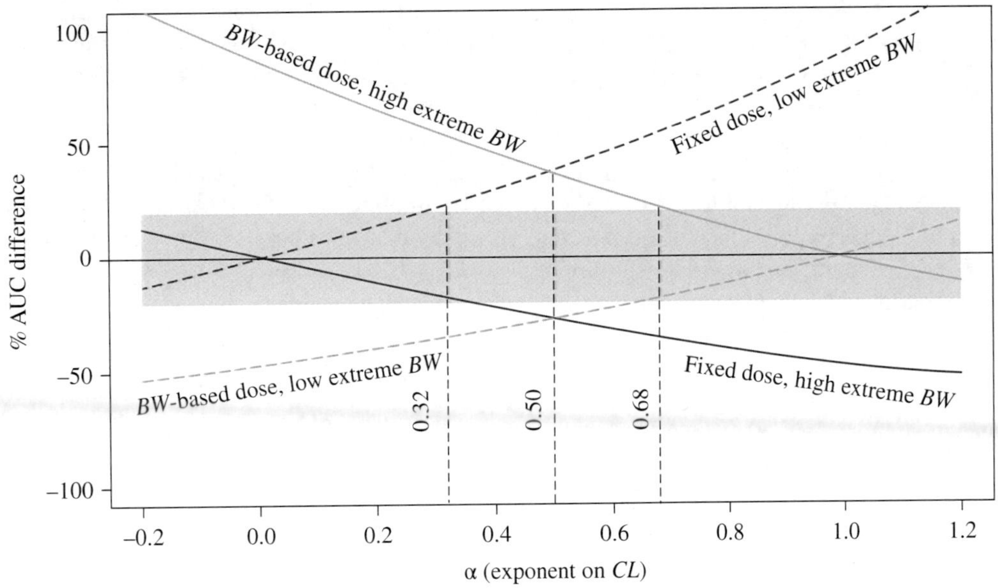

FIGURE 25.4 The % difference of AUC for patients with extremely low body weight (BW) (40 kg, broken lines) and extremely high BW (140 kg, solid lines) from those for patients with a median BW of 75 kg as a function of the α values following a fixed (dark grey) and a BW-based (light grey) dose, assuming $CL = CL_{\text{typical}} \cdot \left(\frac{BW}{BW_{\text{typical}}}\right)^{\alpha}$. The shaded area represents AUC values within 100 ± 20% of typical AUC. (Chow et al. [26]. Reproduced with permission of J Clin Pharmacol.)

antibody drug disposition by incorporating target-mediated binding and disposition, also known as target suppression or neutralization. Of note, receptor occupancy has been used as a measure of target neutralization for cell surface targets. The degree of target suppression or receptor occupancy has been used to estimate a clinically efficacious dose.

A PBPK model can also be used to understand the distribution of biologics in organs/tissues of interest in order to better establish the exposure/response relationship, with the notion that drug exposure in the blood stream may not reflect the exposure in the targeted tissues. With a good understanding of human body dynamics and the associated physiological parameters, PBPK models can be developed by using a series of differential equations to describe drug kinetics in different organs. PBPK has been applied in drug development and regulation for SMDs [59–61]. Science in the application of the PBPK modeling for biologics is emerging but its clinical pharmacology application in the development of biologics is scarce.

A typical whole-body PBPK approach models the whole human body as a closed circulatory system with interconnecting compartments representing organs or specific tissues, with organ- and tissue-specific blood flow as the corresponding volume input to and output from the compartment. The arterial and venous blood connects most organs, whereas the flow from the GI tract, spleen, and pancreas goes to the liver via the portal vein before it reaches the venous side. All of the blood that flows from various organs will converge at the lungs and then return to the blood compartment to complete a cycle. In a PBPK model, a flow balance is maintained by ensuring that the sum of input flow (i.e., blood + lymphatic flow) is equal to the sum of output flows for each compartment. A schematic chart of a PBPK model can be found in many publications [14, 62, 63]. Major hurdles exist for the broader use of PBPK in biologic drug development, including consideration regarding FcRn binding, TMDD, and lymph physiology when building a PBPK model for biologics.[64, 65]. Refer Chapter 12 (by Yanguang Cao and William Jusko) in the current book for more insight about PBPK models in the development of therapeutic mAbs.

25.4.3 Utility of Meta-Analysis

The meta-analyses of approved drug dose (exposure)–responses use available clinical information of approved mAbs to assist in the development of new mAbs with the same target. Parametric models can be built across all relevant trials to study disease progression, biomarker profile, magnitudes of efficacy, and safety responses. Therefore, a virtual head-to-head comparison of drug efficacy potencies or other profiles can be made to guide treatment regimen optimization, evaluate the presence of correlation between biomarker and clinical endpoints, and offer insight in designing better trials. Prominent meta-analyses for SMDs in the past included modeling the effects of age, dosage, and duration of parathyroid hormone treatment on bone mineral density changes (15 trials) [66], evaluating potential correlations between low density lipoprotein cholesterol, non-high density lipoprotein cholesterol, and apolipoprotein

B levels with the risk of cardiovascular events among patients treated with statin therapy (8 trials) [67], and modeling anticoagulants effect for venous thromboembolism prevention after hip/knee replacement (89 trials) [53]. Typically, the published data sources provided only study-level data that have been put into the public domain. However, constructing high quality data for meta-analyses faces the challenge of publication bias, incomplete description of trial design, potential auto-correlation among response data within each study, and the appropriate incorporation of patient-level data if available [68].

The latest meta-analysis performed for biologics used in treating rheumatoid arthritis was mainly based on the percentage of patients attaining ACR 20, 50, and 70 responses [56]. In this analysis, data were extracted from 50 randomized controlled trials encompassing 21,500 patients and 9 biologics with 5 mechanisms of action. The analysis revealed that all TNF inhibitors (anti-TNFs) (i.e., golimumab, infliximab, adalimumab, etanercept, and certolizumab) would have similar dose–response relationships for ACR 20/50/70 if differences in potency were accounted by normalizing the corresponding doses by IC_{50} values. However, dissimilar dose–response relationships were found between anti-TNFs and other biologics, which may indicate their differences in efficacy that require different strategies for dose titration [56, 68–70]. Similar analyses have been conducted for anti-TNFs (i.e., infliximab, adalimumab, and etanercept) along with the IL-1 inhibitor anakinra for the treatment of rheumatoid arthritis. A mixed-effects logistic regression model that adjusted the log odds ratio for study-level prognostic factors was used to compare both efficacy and safety [56, 68–70]. The analysis stated that the anti-TNFs as a class were not different from each other in treatment effects and the apparent differences in the randomized trials among TNF-α antagonists came as a result of difference in prognostic factors. A network meta-analysis of randomized controlled trials of biologics (i.e., abatacept, adalimumab, anakinra, etanercept, infliximab, and rituximab) for rheumatoid arthritis based on Cochrane reviews was also conducted and an indirect comparison of the treatment effects among these biologics was made based on mixed-effects logistic regression [56, 68–70]. Because of differences in study population characteristics among the trials, the findings must be interpreted with caution and longer comparative studies to provide data about the relative and absolute benefit and safety of biologics during various stages of rheumatoid arthritis (early, established, and late) were still needed to draw a definite conclusion.

The common considerations used to develop metadata parametric models are the differences in patient demographics, baseline disease status/biomarker quantity, study-level covariates, concomitant medications, trial design characteristics, and variability structures. A logistic E_{max} model has been used for binary clinical outcomes.

Other forms of dose–response relationship can be used to describe a different mechanism of action. Similar models can be assumed for drugs with the same target or with the same mechanism of action. For sigmoidal E_{max} models, ED_{50} is the dose required to achieve 50% of E_{max} and is drug specific. Linear models can be used on a case-by-case basis. Appropriate layers of error structures should be added if more than one clinical endpoint are considered for the same drug [6, 62].

25.4.4 Utility of Case–Control Analysis in Biologics Development

The utility of case–control analysis, which has been used to adjust for measured confounding factors in observational studies, has recently been used in exposure–response analysis [71]. A case–control comparison was used to evaluate whether the proposed dosing regimen for trastuzumab in patients with metastatic gastric cancer (mGC) was optimal. The main idea was to retrospectively create a virtually randomized study with respect to identified risk factors that on *post hoc* analysis for a subgroup population of interest.

For antibody drugs, exposure–response analysis can potentially overpredict the underlying exposure–efficacy response relationship. For anticancer therapies, it has caught the attention of pharmacometricians at FDA who were reviewing intriguing cases that patients on placebo treatment responded better than patients with low drug exposures on active treatment. Furthermore, it was observed that the exposure–response analysis showed unrealistic efficacy benefit between low and high exposure groups. These phenomena have been exemplified in two recent cases.

In the case of ado-trastuzumab emtansine (Kadcyla), a human epidermal growth factor 2-directed ADC consisting of trastuzumab linked to the small-molecule emtansine [72], the current approved dose of 3.6 mg/kg IV every 3 weeks that was determined as the maximum tolerated dose (MTD) in the Phase 1 stage, was approved for patients with HER2-positive metastatic breast cancer. In the analysis of its Phase 3 study results [73], Kaplan–Meier survival analysis stratified by the trough concentrations of Kadcyla revealed that patients on control treatment (lapatinib plus capecitabine) had longer median overall survival than patients with exposures in the lowest quartile (Q1) on Kadcyla add-on treatment by 9 months (25.1 vs 16.1 months). The analysis also showed that, with increasing drug exposure, the median survival times for patients in Q1, Q2, and Q3 were 16.1, 26.5, and 34.1 months, respectively, whereas the median survival time had not been reached for patients in Q4 at the time of reporting. An increase of 18 months in median survival time from Q1 to Q3 group could reflect the potential for overprediction of drug efficacy if other confounding risk factors to drug exposure were not taken into account. Cox proportional hazard model incorporating all the identified baseline risk factors for efficacy, such as HER2 extracellular domain, Eastern Cooperative Oncology Group performance score, measurable disease, and tumor burden, plus Kadcyla PK exposure suggested that there may be an opportunity to optimize Kadcyla dose in the patient subgroup with low T-DM1 exposure for improved efficacy with acceptable tolerability. Consequently, a postmarketing commitment study to further characterize the exposure–response for both efficacy and safety with respect to Kadcyla trough concentrations was requested [74].

Similarly, exposure–response analysis was applied to the data of trastuzumab for mGC. Trastuzumab was approved for metastatic HER2-overexpressing gastric cancer with a fixed dosing regimen: 8 mg/kg IV as initial dose and 6 mg/kg every 3 weeks as maintenance dose in combination with fluoropyrimidine and cisplatin. Based on the corresponding Phase III study results for approval, the median survival time for patients in the lowest exposure quartile group (referred to as Q1, T+FC) (FC, fragment crystallizable) was 7.7 months, which is approximately 4 months shorter than patients in the active control treatment arm and 8 months shorter than the median survival time for the rest of the patients with higher trastuzumab exposures. Further analysis showed that certain baseline risk factors identified by Cox proportional hazard model, such as Eastern Cooperative Oncology Group performance score, prior surgery, number of metastatic sites, and percentage of the non-Asian population, were not evenly distributed between Q1 and Q2–4 patients groups.

Case–control analysis was therefore applied to the trastuzumab case to evaluate whether the identified risk factors can explain the better efficacy response for patients following active control treatment than patients in the Q1 group following the trastuzumab add-on treatment. After exploring various propensity score-matching methods and the weighted Mahalanobis metric method, Yang et al. [71] employed the weighted Mahalanobis metric method and patients receiving active control treatment were matched 1:1 to Q1 patients for five significant potential risk factors identified in a stepwise Cox model. After case matching, differences in patient baseline risk factors became balanced between active control and Q1 groups and the median survival times for these two patient groups became comparable, whereas it still showed a survival advantage for other exposure groups overactive control treatment after similar case-matching procedures. Therefore, a lack of survival benefit in the low exposure subgroup indicated that the trastuzumab in the proposed add-on dose did not generate additional benefit and was considered a safety concern. Subsequently, a postmarketing requirement for a clinical study to evaluate the dose optimization for the subpopulation with lower trough concentrations following Cycle 1 treatment and improve the survival benefit was issued [75].

Currently, the utility of case–control analysis has been applied to other cases in situations where drug exposure can be confounded by other risk factors. However, it may be arguable that some potential risk factors still remain elusive to the scientific community and thus not monitored in trials, and case–control analysis cannot substitute the actual randomized studies.

25.5 CONCLUSIONS

Biologics, including mAbs and other therapeutic proteins such as cytokines and growth hormones, have unique characteristics compared to SMDs. Their unique properties define their unique development pathways in terms of the determination of a starting dose for FIH studies, DDIs, QT_c prolongation, immunogenicity, and studies in specific populations, as summarized.

Quantitative methods, such as mechanism-based PK, PBPK, PK/PD, and exposure–response models, offer great benefit for biologics development. With the advancement of new quantitative tools [76, 77], drug developers and researchers can gain powerful insight into designing the most effective therapeutic regimens.

25.6 DISCLAIMER

The views expressed in this chapter are those of the authors and do not reflect the official policy of the FDA. No official support or endorsement by the FDA is intended or should be inferred.

REFERENCES

[1] Novo Nordisk. 2013. Levemir (package insert).

[2] O'Driscoll C. Anatomy and physiology of the lymphatics. In: Charman W, Stella V, editors. *Lymphatic Transport of Drugs*. Boca Raton (FL): CRC Press; 1992. p 1–36.

[3] US Food and Drug Administration Draft Guidance for Industry Biosimilars. 2012. Questions and answers regarding implementation of the biologics price competition and innovation act of 2009.

[4] US Food and Drug Administration Guidance for Industry. 2010. Pharmacokinetics in patients with impaired renal function – study design, data analysis, and impact on dosing and labeling (Draft).

[5] Lobo ED, Hansen RJ, Balthasar JP. Antibody pharmacokinetics and pharmacodynamics. J Pharm Sci 2004;93: 2645–2668.

[6] Wang W, Wang EQ, Balthasar JP. Monoclonal antibody pharmacokinetics and pharmacodynamics. Clin Pharmacol Ther 2008;84:548–558.

[7] Antohe F, Radulescu L, Gafencu A, Ghetie V, Simionescu M. Expression of functionally active FcRn and the differentiated bidirectional transport of IgG in human placental endothelial cells. Hum Immunol 2001;62:93–105.

[8] Dickinson BL, Badizadegan K, Wu Z, Ahouse JC, Zhu X, Simister NE, Blumberg RS, Lencer WI. Bidirectional FcRn-dependent IgG transport in a polarized human intestinal epithelial cell line. J Clin Invest 1999;104:903–911.

[9] McCarthy KM, Yoong Y, Simister NE. Bidirectional transcytosis of IgG by the rat neonatal Fc receptor expressed in a rat kidney cell line: a system to study protein transport across epithelia. J Cell Sci 2000;113 (Pt 7):1277–1285.

[10] Bill A. Plasma protein dynamics: albumin and IgG capillary permeability, extravascular movement and regional blood flow in unanesthetized rabbits. Acta Physiol Scand 1977; 101:28–42.

[11] Hedger MP, Hettiarachchi S. Measurement of immunoglobulin G levels in adult rat testicular interstitial fluid and serum. J Androl 1994;15:583–590.

[12] Juweid M, Strauss HW, Yaoita H, Rubin RH, Fischman AJ. Accumulation of immunoglobulin G at focal sites of inflammation. Eur J Nucl Med 1992;19:159–165.

[13] Weinstein JN, Eger RR, Covell DG, Black CD, Mulshine J, Carrasquillo JA, Larson SM, Keenan AM. The pharmacology of monoclonal antibodies. Ann NY Acad Sci 1987;507: 199–210.

[14] Garg A, Balthasar JP. Physiologically-based pharmacokinetic (PBPK) model to predict IgG tissue kinetics in wild-type and FcRn-knockout mice. J Pharmacokinet Pharmacodyn 2007; 34:687–709.

[15] Dall'Acqua WF, Woods RM, Ward ES, Palaszynski SR, Patel NK, Brewah YA, Wu H, Kiener PA, Langermann S. Increasing the affinity of a human IgG1 for the neonatal Fc receptor: biological consequences. J Immunol 2002;169: 5171–5180.

[16] Ghetie V, Popov S, Borvak J, Radu C, Matesoi D, Medesan C, Ober RJ, Ward ES. Increasing the serum persistence of an IgG fragment by random mutagenesis. Nat Biotechnol 1997;15: 637–640.

[17] Murphy K, Travers P, Walport M. *Janeway's Immunobiology*. Garland Science; 2007.

[18] Sato AK, Viswanathan M, Kent RB, Wood CR. Therapeutic peptides: technological advances driving peptides into development. Curr Opin Biotechnol 2006;17:638–642.

[19] Latham PW. Therapeutic peptides revisited. Nat Biotechnol 1999;17:755–757.

[20] Diao L, Meibohm B. Pharmacokinetics and pharmacokinetic-pharmacodynamic correlations of therapeutic peptides. Clin Pharmacokinet 2013;52:855–868.

[21] Muller PY, Milton M, Lloyd P, Sims J, Brennan FR. The minimum anticipated biological effect level (MABEL) for selection of first human dose in clinical trials with monoclonal antibodies. Curr Opin Biotechnol 2009;20: 722–729.

[22] Yu J, Karcher H, Feire AL, Lowe PJ. From target selection to the minimum acceptable biological effect level for human study: use of mechanism-based PK/PD modeling to design safe and efficacious biologics. AAPS J 2011;13: 169–178.

[23] US Food and Drug Administration. Drug guidance, compliance and regulatory information. Available at http://www.fda.gov/Drugs/GuidanceComplianceRegulatoryInformation/Guidances/. Accessed 2015 Jul 17.

[24] European Medicines Agency. (2007) Clinical investigation of the pharmacokinetics of therapeutic proteins, July 2007.

[25] Zhao L, Ren TH, Wang DD. Clinical pharmacology considerations in biologics development. Acta Pharmacol Sin 2012; 33:1339–1347.

[26] Chow AT, Earp JC, Gupta M, Hanley W, Hu C, Wang DD, Zajic S, Zhu M, Population PK TPDI Working Group. Utility of population pharmacokinetic modeling in the assessment of therapeutic protein-drug interactions. J Clin Pharmacol 2014; 54:593–601.

[27] Huang SM, Zhao H, Lee JI, Reynolds K, Zhang L, Temple R, Lesko LJ. Therapeutic protein-drug interactions and implications for drug development. Clin Pharmacol Ther 2010;87: 497–503.

[28] Chirmule N, Jawa V, Meibohm B. Immunogenicity to therapeutic proteins: impact on PK/PD and efficacy. AAPS J 2012;14:296–302.

[29] Grange S, Schmitt C, Banken L, Kuhn B, Zhang X. Thorough QT/QTc study of tocilizumab after single-dose administration at therapeutic and supratherapeutic doses in healthy subjects. Int J Clin Pharmacol Ther 2011;49:648–655.

[30] Garnett CE, Beasley N, Bhattaram VA, Jadhav PR, Madabushi R, Stockbridge N, Tornøe CW, Wang Y, Zhu H, Gobburu JV. Concentration-QT relationships play a key role in the evaluation of proarrhythmic risk during regulatory review. J Clin Pharmacol 2008;48:13–18.

[31] de Vries MK, Brouwer E, van der Horst-Bruinsma IE, Spoorenberg A, van Denderen JC, Jamnitski A, Nurmohamed MT, Dijkmans BA, Aarden LA, Wolbink GJ. Decreased clinical response to adalimumab in ankylosing spondylitis is associated with antibody formation. Ann Rheum Dis 2009; 68:1787–1788.

[32] Meibohm B, Braeckman R. Pharmacokinetics and pharmacodynamics of peptide and protein drugs. In: Crommelin D, Sindelar R, Meibohm B, editors. *Pharmaceutical Biotechnology: Fundamentals and Applications*. New York: Informa Healthcare; 2007. p 95–124.

[33] Schellekens H. Immunogenicity of therapeutic proteins: clinical implications and future prospects. Clin Ther 2002; 24:1720–1740. ; discussion 1719.

[34] GlaxoSmithKline LLC. 2011. Bexxar (package insert).

[35] Janssen Biotech, Inc. 2011. Remicade (package insert).

[36] Sailstad JM, Amaravadi L, Clements-Egan A, Gorovits B, Myler HA, Pillutla RC, Pursuhothama S, Putman M, Rose MK, Sonehara K, Tang L, Wustner JT, Global Bioanalysis Consortium. A white paper--consensus and recommendations of a global harmonization team on assessing the impact of immunogenicity on pharmacokinetic measurements. AAPS J 2014;16:488–498.

[37] Bonate PL, Sung C, Welch K, Richards S. Conditional modeling of antibody titers using a zero-inflated poisson random effects model: application to Fabrazyme. J Pharmacokinet Pharmacodyn 2009;36:443–459.

[38] Chen X, Hickling T, Kraynov E, Kuang B, Parng C, Vicini P. A mathematical model of the effect of immunogenicity on therapeutic protein pharmacokinetics. AAPS J 2013;15: 1141–1154.

[39] US Food and Drug Administration Guidance for Industry. 2014. General clinical pharmacology considerations for pediatric studies for drugs and biological products (PDF - 375KB), 2014 Dec 8.

[40] Dunne J, Rodriguez WJ, Murphy MD, Beasley BN, Burckart GJ, Filie JD, Lewis LL, Sachs HC, Sheridan PH, Starke P, Yao LP. Extrapolation of adult data and other data in pediatric drug-development programs. Pediatrics 2011;128 (5):e1242–e1249.

[41] US Food and Drug Administration Guidance for Industry. 2000. E11 clinical investigation of medicinal products in the pediatric population.

[42] US Food and Drug Administration Guidance for Industry. 2003. Exposure–response relationships—study design, data analysis, and regulatory applications.

[43] Genentech, Inc. 2013. Actemra (package insert).

[44] Zhang X, Morcos PN, Saito T, Terao K. Clinical pharmacology of tocilizumab for the treatment of systemic juvenile idiopathic arthritis. Expert Rev Clin Pharmacol 2013;6: 123–137.

[45] Roy P. 2012. Clinical pharmacology review: actemra.

[46] Swedish Orphan Biovitrum AB. 2011. Kineret (package insert).

[47] Schering Corporation, a subsidiary of Merck & Co., Inc. 2011. Pegintron (package insert).

[48] Genetech. 2014. Ranibizumab (package insert).

[49] Millennium Pharmaceuticals, Inc. 2012. Bortezomib (package insert).

[50] Fang Y. 2012. Clinical pharmacology review: teduglutide.

[51] Wyeth. 2005. Mylotarg (package insert).

[52] EM Agency. 2009. European medicines agency CHMP assessment report for Ilaris.

[53] Mandema JW, Boyd RA, DiCarlo LA. Therapeutic index of anticoagulants for prevention of venous thromboembolism following orthopedic surgery: a dose-response meta-analysis. Clin Pharmacol Ther 2011;90:820–827.

[54] Mandema JW, Cox E, Alderman J. Therapeutic benefit of eletriptan compared to sumatriptan for the acute relief of migraine pain--results of a model-based meta-analysis that accounts for encapsulation. Cephalalgia 2005;25:715–725.

[55] Mandema JW, Gibbs M, Boyd RA, Wada DR, Pfister M. Model-based meta-analysis for comparative efficacy and safety: application in drug development and beyond. Clin Pharmacol Ther 2011;90:766–769.

[56] Mandema JW, Salinger DH, Baumgartner SW, Gibbs MA. A dose-response meta-analysis for quantifying relative efficacy of biologics in rheumatoid arthritis. Clin Pharmacol Ther 2011;90:828–835.

[57] Wang DD, Zhang S, Zhao H, Men AY, Parivar K. Fixed dosing versus body size-based dosing of monoclonal antibodies in adult clinical trials. J Clin Pharmacol 2009;49: 1012–1024.

[58] Zhang S, Shi R, Li C, Parivar K, Wang DD. Fixed dosing versus body size-based dosing of therapeutic peptides and proteins in adults. J Clin Pharmacol 2012;52:18–28.

[59] Rowland M, Peck C, Tucker G. Physiologically-based pharmacokinetics in drug development and regulatory science. Annu Rev Pharmacol Toxicol 2011;51:45–73.

[60] Zhao P, Rowland M, Huang SM. Best practice in the use of physiologically based pharmacokinetic modeling and simulation to address clinical pharmacology regulatory questions. Clin Pharmacol Ther 2012;92:17–20.

[61] Sinha V, Zhao P, Huang SM, Zineh I. Physiologically based pharmacokinetic modeling: from regulatory science to regulatory policy. Clin Pharmacol Ther 2014;95:478–480.

[62] Thygesen P, Macheras P, Van Peer A. Physiologically-based PK/PD modelling of therapeutic macromolecules. Pharm Res 2009;26:2543–2550.

[63] Baxter LT, Zhu H, Mackensen DG, Butler WF, Jain RK. Biodistribution of monoclonal antibodies: scale-up from mouse to human using a physiologically based pharmacokinetic model. Cancer Res 1995;55:4611–4622.

[64] Zhao L, Ji P, Li Z, Roy P, Sahajwalla CG. The antibody drug absorption following subcutaneous or intramuscular administration and its mathematical description by coupling physiologically based absorption process with the conventional compartment pharmacokinetic model. J Clin Pharmacol 2013;53:314–325.

[65] Dostalek M, Gardner I, Gurbaxani BM, Rose RH, Chetty M. Pharmacokinetics, pharmacodynamics and physiologically-based pharmacokinetic modelling of monoclonal antibodies. Clin Pharmacokinet 2013;52:83–124.

[66] Schwarz P, Jorgensen NR, Mosekilde L, Vestergaard P. Effects of increasing age, dosage, and duration of PTH treatment on BMD increase–a meta-analysis. Calcif Tissue Int 2012;90: 165–173.

[67] Boekholdt SM, Arsenault BJ, Mora S, Pedersen TR, LaRosa JC, Nestel PJ, Simes RJ, Durrington P, Hitman GA, Welch KM, DeMicco DA, Zwinderman AH, Clearfield MB, Downs JR, Tonkin AM, Colhoun HM, Gotto AM Jr, Ridker PM, Kastelein JJ. Association of LDL cholesterol, non-HDL cholesterol, and apolipoprotein B levels with risk of cardiovascular events among patients treated with statins: a meta-analysis. J Am Med Assoc 2012;307: 1302–1309.

[68] Lalonde RL, Kowalski KG, Hutmacher MM, Ewy W, Nichols DJ, Milligan PA, Corrigan BW, Lockwood PA, Marshall SA, Benincosa LJ, Tensfeldt TG, Parivar K, Amantea M, Glue P, Koide H, Miller R. Model-based drug development. Clin Pharmacol Ther 2007;82:21–32.

[69] Nixon RM, Bansback N, Brennan A. Using mixed treatment comparisons and meta-regression to perform indirect comparisons to estimate the efficacy of biologic treatments in rheumatoid arthritis. Stat Med 2007;26:1237–1254.

[70] Singh JA, Christensen R, Wells GA, Suarez-Almazor ME, Buchbinder R, Lopez-Olivo MA, Ghogomu ET, Tugwell P. A network meta-analysis of randomized controlled trials of biologics for rheumatoid arthritis: a Cochrane overview. Can Med Assoc J 2009;181:787–796.

[71] Yang J, Zhao H, Garnett C, Rahman A, Gobburu JV, Pierce W, Schechter G, Summers J, Keegan P, Booth B, Wang Y. The combination of exposure-response and case-control analyses in regulatory decision making. J Clin Pharmacol 2013;53: 160–166.

[72] Genetech, Inc. 2014. Kadcyla (ado-trastuzumab emtansine) injection, powder, lyophilized, for solution (package insert).

[73] Wang J, Song P, Schrieber S, Liu Q, Xu Q, Blumenthal G, Amiri Kordestani L, Cortazar P, Ibrahim A, Justice R, Wang Y, Tang S, Booth B, Mehrotra N, Rahman A. Exposure-response relationship of T-DM1: insight into dose optimization for patients with HER2-positive metastatic breast cancer. Clin Pharmacol Ther 2014;95:558–564.

[74] US Food and Drug Administration. 2013. Approval Letter Kadcyla, original BLA.

[75] US Food and Drug Administration. 2010. Approval Letter for Herceptin (Trastuzumab), supplemental Biologics License Application (sBLA), dated April 19.

[76] Zhao L. New developments in using stochastic recipe for multi-compartment model: inter-compartment traveling route, residence time, and exponential convolution expansion. Math Biosci Eng 2009;6:663–682.

[77] Zhao L, Li N, Yang H. A new stochastic approach to multi-compartment pharmacokinetic models: probability of traveling route and distribution of residence time in linear and nonlinear systems. J Pharmacokinet Pharmacodyn 2011;38: 83–104.

26

INVESTIGATING THE NONCLINICAL ADME AND PK/PD OF AN ANTIBODY–DRUG CONJUGATE: A CASE STUDY OF ADO-TRASTUZUMAB EMTANSINE (T-DM1)

JAY TIBBITTS
UCB Celltech, Slough, UK

26.1 INTRODUCTION

The treatment of cancer has seen substantial advances in the past decades, with the development of monoclonal antibodies (mAbs) and other targeted therapies improving efficacy and reducing the often severe toxicities seen with traditional chemotherapeutic agents. Despite these positive developments, cancer remains a deadly disease and even these new treatments often just prolong the time to disease advancement and death. Breast cancer is one of the leading causes of death among women. In approximately 20% of breast cancers, the cell surface protein human epidermal growth factor receptor 2 (HER2) is overexpressed, and this characteristic is associated with shortened survival. A humanized monoclonal immunoglobulin G1 (IgG1) antibody to HER2 (trastuzumab, Herceptin®) has shown substantial efficacy in treating patients with HER2-positive breast cancer and is now a mainstay of therapy in this disease. However, not all patients treated with trastuzumab will respond, and in a substantial fraction of those who do, their disease will eventually progress. In addition, while trastuzumab is generally well tolerated, to be maximally efficacious it is commonly combined with chemotherapy, which can cause substantial toxicity. As such, new approaches for the treatment of HER2+ breast cancer and other cancers are needed.

Antibody–drug conjugates (ADCs) are a growing class of anticancer agents combining the targetting ability of an antibody with the potency of a cytotoxic agent. ADCs are frequently described as highly complex owing to their having characteristics of both large and small molecule drugs [1]. This is certainly true and applies to not only the understanding of the behavior of these fascinating and promising molecules, but also to their development. Ado-trastuzumab emtansine (Kadcyla®, or Trastuzumab-DM1 (T-DM1)) is currently approved for the treatment of HER2+ metastatic breast cancer [2] and in development for other HER2+ indications (Clinical Trials.gov, http://clinicaltrials.gov/show/NCT01641939). T-DM1 is composed of trastuzumab conjugated via a stable and noncleavable linker ((N-maleimidomethyl)-cyclohexane-1-carboxylate, MCC) to the tubulin-binding cytotoxic agent DM1 (Fig. 26.1). The development of T-DM1 was initiated at a time when little precedent existed in the understanding of the pharmacokinetic/pharmacodynamic (PK/PD) and absorption, distribution, metabolism, and excretion (ADME) behavior of ADCs and where the regulatory environment was (and to some degree still is) unclear. Over the course of several years of research and development, a large body of knowledge was created describing the PK/PD and ADME characteristics of T-DM1 and other ADCs, which have formed the basis for a much stronger understanding of this class of molecules, and has guided drug developers and regulators on the key characteristics and behaviors necessary for the development of safe and effective molecules. In this case study, we will explore the PK/PD and ADME strategy and data for T-DM1.

At the initiation of any drug development program, it is important to take into consideration the mechanism of action of the drug and optimize drug candidates with these

ADME and Translational Pharmacokinetics/Pharmacodynamics of Therapeutic Proteins: Applications in Drug Discovery and Development, First Edition. Edited by Honghui Zhou and Frank-Peter Theil.

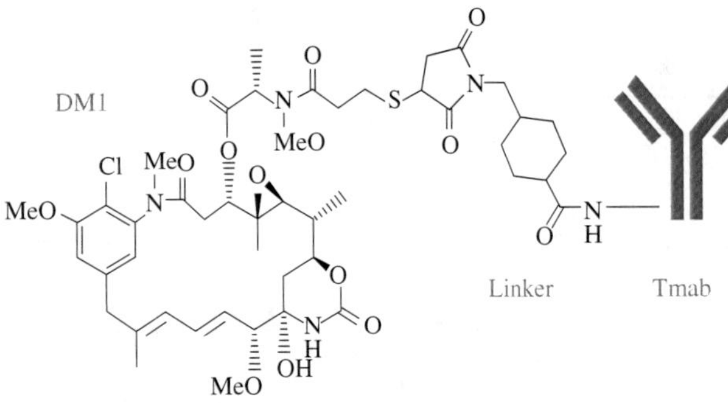

FIGURE 26.1 Trastuzumab-DM1.

principles in mind, and ADCs are no exception. However, ADCs possess a comparatively complicated mechanism of action that is dependent on each of its components: antibody, linker, and cytotoxic drug. The mechanism of action of ADCs has been described in several excellent reviews [3–5]. Briefly, the ADC binds to the target on cancer cells via the highly specific complementarity-determining regions of the antibody, which facilitates its internalization into the cell and trafficking to the endosome. The ADC can be recycled to the cell surface with its receptor or subsequently be degraded in the lysosome [6–9]. In the course of this degradation process, the cytotoxic drug can be released either by cleavage of the linker attaching the drug to the antibody or by complete degradation of the antibody releasing the drug–linker–amino acid complex [6, 9, 10]. The release of the cytotoxic moiety inside the cell allows binding of this molecule to its cellular target, thus causing cell death. With some considerations of this process, it becomes apparent that the ideal ADC is one that possesses the following features: (i) it is composed of an antibody against a highly abundant and rapidly internalized target on tumor cells, (ii) it bears a highly potent cytotoxic drug to facilitate effective killing of the target cell, and (iii) it possesses a sufficiently stable linker that is either efficiently cleaved inside the lysosome or endosome or whose attachment to the cytotoxic drug does not affect the its potency. For T-DM1, the first criterion was well validated. The clinical success of trastuzumab in HER2+ cancers had clearly demonstrated that this target is present on tumor cells in certain cancers, and extensive previous work has shown that HER2 is highly overexpressed in these tumors and exhibits a rapid internalization rate [11–13]. The cytotoxic drug conjugated to trastuzumab to form T-DM1, DM1, is a derivative of maytansine, an ansamitocin product of the microorganism Actinosynnema [14]. Maytansines are potent tubulin-binding agents that act as antimitotics by causing cells to arrest in the G2/M phase of the cell cycle [14]. In *in vitro* assays, DM1 is a highly potent cytotoxin against tumor cells with an IC_{50} of 1–5 nM [13], and when conjugated to trastuzumab showed potent antitumor activity *in vitro* (IC_{50}s of <5 nM in HER2-expressing cells) and *in vivo* mouse models of HER2+ breast cancer [13, 15]. Conjugation of the drug to the antibody is typically achieved using chemical reactions between side chains of antibody amino acids and components of the linker. In the case of T- DM1, this is achieved by first conjugating *N*-succinimidyl-4-(*N*-maleimidomethyl)-cyclohexane-1-carboxylate (SMCC) to side-chain amines of lysine residues on trastuzumab, followed by the addition of DM1 forming a succinimide thioether bond connecting these two components [16]. By controlling the chemical reaction process, only a fraction of more than 80 lysine residues on trastuzumab is conjugated. The average number of drugs per antibody (drug-to-antibody ratio, DAR) is typically 3.5. However, there is considerable heterogeneity with the final product in both the number and location of conjugated DM1 molecules, with individual T-DM1 molecules in the product having a DAR of 0–8. This mixture can be described by a Poisson distribution, with the highest fraction of DARs in the 3–5 range [16, 17]. As it is conceivable that the number of cytotoxic drugs conjugated to T-DM1 could affect its behavior, this heterogeneity can complicate many aspects of the characterization and understanding of T-DM1, sometimes making interpretation of assays, PK, and efficacy difficult. A full discussion of this topic is outside the scope of this chapter, and interested readers are encouraged to see excellent publications on this topic [18–21]. The choice of linker (and drug) for T-DM1 was the result of a systematic evaluation of several options comparing PK, efficacy, and toxicity. *In vitro*, there was little difference in potency in HER2+ tumor cells between T-DM1 and other trastuzumab–maytansinoid conjugates [13]. However, comparisons of tolerability of T-DM1, which uses the stable, nonreducible, thioether-based linker, SMCC, with T-SPP-DM1, which used the less stable, reducible, disulfide linker SPP (*N*-succinimidyl-4-(2-pyridyldithio)pentanoate), clearly showed superior tolerability with the more stable SMCC linker, an observation attributed to less release of the cytotoxic drug in the systemic circulation and subsequently less toxicity. This hypothesis was supported by PK data for both conjugates showing that the PK of the antibody components of the ADCs did not differ while the clearance of the ADC (a measure of the sum of the antibody clearance and loss of cytotoxic drug) clearly showed slower release of DM1 from T-DM1 compared with T-SPP-DM1. In summary, T-DM1 satisfies all the critical criteria for an effective ADC, combining the clinically validated antibody trastuzumab with the potent cytotoxic agent DM1 via the stable thioether linker SMCC.

26.2 IMPORTANCE OF ADME FOR ADCs

The ADME characteristics of xenobiotic agents are important aspects of their understanding and development. For much of the recent history of small molecule development, the ADME characteristics of drug candidates have been thoroughly studied and form an important body of work in any regulatory submission. Such work provides guidance on the rate and extent of absorption, which can guide drug selection, dosing regimens, and formulation development. Investigations into distribution help determine tissue uptake kinetics, assess potential for tissue accumulation, and can guide understanding of efficacy and toxicity. Extensive investigation of the metabolism and excretion of small molecule drugs is conducted to provide critical information on the mechanisms and rates of formation and the identity of metabolites. Knowledge derived from these studies can, among other things, guide the understanding of drug toxicity related to metabolic products, inform the risk of interactions with other drugs, and help in the extrapolation of human PK. Over the years, a comprehensive set of regulatory guidelines have been developed to guide drug developers of new chemical entities (NCEs) in their ADME studies, with a number of specific requirements in place for both nonclinical and clinical ADME investigations during drug development. In addition, a large array of *in vitro* and *in vivo* methods and technologies has been developed to support the investigation of the ADME of small molecule drugs.

For biotherapeutic drugs such as antibodies, the current understanding of their ADME properties is less well-developed. In many instances, the biotherapeutic is either identical to (e.g., erythropoietin or insulin) or is highly similar to (e.g., mAbs) an endogenous protein. As such, it has been assumed that these molecules will have ADME properties similar to their endogenous counterparts, in effect catabolized to their amino acid constituents by physiologic processes that are generally nonsaturable and not likely to be subject to competition with other substrates, generally avoiding relevant PK interactions with other drugs. As such, the regulatory framework surrounding the ADME of biotherapeutics is very limited with only brief mentions of the topic but little specific guidance [22, 23]. Regarding ADCs, in the International Conference for Harmonization guidance S6(R1), which outlines the preclinical evaluation of biotechnology-derived pharmaceutics, it is inferred but not explicitly stated that the development of ADCs should be consistent with these guidelines. The only specific mention of ADCs is with respect to species selection for preclinical studies, where it is recommended that the strategy applied be consistent with that used for the unconjugated molecule. In the Guideline on the Clinical Investigation of the Pharmacokinetics of Therapeutic Proteins [24], there is no specific mention of ADCs, but the principles described for bioanalysis and assessment of ADME can be applied to ADCs. Perhaps as a result, the technology and strategies necessary for investigating biotherapeutic ADME have also lagged behind those for small molecules.

With respect to ADME, ADCs present a complicated scenario. The ADME characteristics of a molecule composed primarily of an antibody are expected to be dominated by the biotherapeutic component of the molecule, with distribution and excretion processes similar to that of an antibody. However, the presence of a potent small molecule cytotoxin conjugated to the antibody and capable of its own ADME properties suggests that additional considerations are required. The understanding and regulatory guidance associated with such molecules is more well established, and this topic is discussed later in this chapter.

In the course of the development of T-DM1, a substantial body of information was generated describing the ADME characteristics of the ADC and its component parts. The strategy was to develop a comprehensive understanding of the behavior of T-DM1, including its PK/PD, distribution, *in vivo* and *ex vivo* stability, catabolic fate, and excretion.

An important characteristic of any therapeutic agent is its PK. An understanding of the rate of elimination and extent of distribution can tell an investigator about the behavior of that agent, can help guide dose and dose regimen determination, and also inform the extrapolation of PK from nonclinical species to humans. For ADCs such as T-DM1, a well-designed PK and bioanalytical strategy can also provide vital information about linker stability and the effects of drug conjugation on antibody behavior. For T-DM1, nonclinical PK studies were conducted in mice, rats, and cynomolgus monkeys. The objectives of these studies were to characterize the PK of T-DM1 to guide dosing and interpretation of efficacy and toxicity studies [15, 25]. The antibody component of T-DM1, trastuzumab, binds only to HER2 in primates. As such, PK and ADME studies conducted in rodents could be used to explore the target independent (nonspecific) behavior of T-DM1 while studies in primates were used to better understand its target-dependent and -independent behavior.

26.3 T-DM1 BIOANALYTICAL STRATEGY AND METHODS

The foundation upon which PK and ADME evaluation is built is thorough and robust bioanalytical methods capable of determining the presence of key components of the molecule of interest. For a complex molecule such as T-DM1 with two pharmacologically active moieties, the antibody and the cytotoxic drug, it was important to employ a bioanalytical method for T-DM1 that was capable of ensuring that both entities were present. As such, the PK assay for T-DM1 was an immunoassay that required the presence on the detected molecule of both trastuzumab (via its ability to bind to HER2) and DM1 (via an anti-DM1 antibody) [1, 25] and was capable of detecting T-DM1 with one or more DM1 molecules attached (Fig. 26.2a). While this assay can provide information on the PK of T-DM1, it alone is not able to provide information on other important

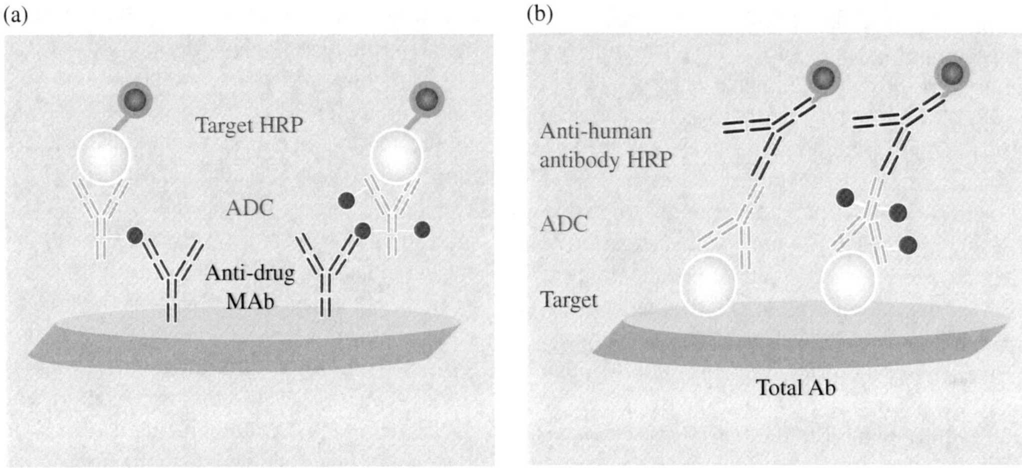

FIGURE 26.2 Generic schematic of an immunoassay to measure concentrations of ADC (a) and total antibody (b).

aspects of T-DM1 behavior. The stability of the SMCC linker, and the perceived risk of the premature release of DM1 into the circulation, can only be evaluated by employing additional assays. As can be inferred from the design of the T-DM1 assay, decreases in T-DM1 concentrations could be attributed to either elimination of the entire molecule or loss of all DM1 molecules. To better differentiate these two processes, assays measuring total antibody and free (or unconjugated) DM1 were developed. The total antibody assay was an adaptation of the immunoassay used for trastuzumab and was capable of measuring the concentration of intact antibody, regardless of the presence or absence of DM1 bound to T-DM1 (Fig. 26.2b). Concentrations of unconjugated DM1 were determined by liquid chromatography coupled to mass spectrometry (LC/MS), which was specific for DM1 bearing the free thiol (S–H) that in T-DM1 is attached to the linker via a thioether bond. If DM1 is released from the linker, this chemically active free thiol will likely react to form a disulfide bond with endogenous molecules bearing free thiols (e.g., albumin, cysteine, and glutathione), or form DM1 dimers. To account for this and to ensure an accurate determination of total amount of DM1 in the sample that is not conjugated to T-DM1, a reduction step was included in the sample preparation to convert any disulfide-bound DM1 to the free thiol form to facilitate analysis. As such, the DM1 assay as employed determined the concentration (and PK) of unconjugated DM1 existing in many forms in plasma.

26.4 *EX VIVO* LINKER STABILITY

A critical characteristic of an ADC is the stability of the linker between the antibody and its potent cytotoxic cargo. Rapid release of the cytotoxic drug following administration could have potentially severe toxicologic effects, in addition to rendering the ADC less potent. To understand the degree of stability conferred by the SMCC linker, evidence can be obtained from both *ex vivo* and *in vivo* studies. *Ex vivo* plasma stability studies are commonly conducted by incubating an ADC in the plasma of humans and other relevant animal species for several days at 37 °C and determining the concentration of the total antibody and ADC in plasma samples over time using assays described above. The data from such studies can provide valuable information regarding the stability of the ADC linker and also allow cross-species comparisons providing at least qualitative assurance of acceptable linker stability in humans before starting clinical trials. Such studies performed with T-DM1 confirmed that the SMCC linker is stable in plasma with a slow rate of DM1-related product release, showing that approximately 80% of T-DM1 has at least one conjugated DM1 after 96 h of incubation [8, 26].

26.5 PLASMA PK

The plasma (or serum) PK of T-DM1, as determined by the assay shown in Figure 26.2, is a composite of the elimination of the antibody and the loss of the DM1 or DM1-related products from the antibody. As such, the PK of T-DM1, while similar to that of an antibody (low volume of distribution, slow clearance, long half-life), exhibits an approximately two to three times faster elimination in all species than its parent antibody [7, 8, 27–29]. The relative contributions of the two processes driving T-DM1 PK can be distinguished by comparing T-DM1 and total antibody PK. Doing this confirmed that this difference in PK between T-DM1 and trastuzumab can be explained by the loss of DM1 (or other DM1-related products) from T-DM1 after administration [8, 29] and that the PK of the antibody component (i.e., total antibody) of T-DM1 differs little from that of trastuzumab.

Consideration of these clearance rates, with some understanding of the assays used to provide them, allows one to estimate what is called the deconjugation clearance. The assay used to measure concentrations of T-DM1 (and, typically, other ADCs) is an immunoassay that relies on the presence of antibody and at least one cytotoxic drug. When either the ADC (or unconjugated antibody) is eliminated or the ADC has lost all of its DM1-related products, the signal in the assay is lost. The assay used to measure total antibody relies only on the presence of intact trastuzumab. Thus, by taking the difference between the clearance of T-DM1 and total trastuzumab, one can estimate the deconjugation clearance, a value analogous to the rate of loss of cytotoxic drug from trastuzumab. Indeed, a model-based analysis of this process was conducted, using data from both animals and humans. In rats and cynomolgus monkeys, the deconjugation clearance (CL) was estimated to be approximately 9 and 6 mL/day/kg, while in humans the CL was 5.75 mL/day/kg, values similar to total antibody clearance indicating that both the proteolytic clearance of trastuzumab and the loss of DM1-related products occur at a similar, slow rate. In addition, the data indicate that the rate of loss of DM1-related products is similar across species, consistent with the *ex vivo* plasma stability findings [8, 29]. Also, by comparing the proteolytic degradation clearance of T-DM1 with the clearance of trastuzumab, it was confirmed that the conjugation of DM1 to trastuzumab has little impact on the underlying PK of the antibody. Bender et al. also used a more informative analytical method to explore the mechanistic aspects of the deconjugation of T-DM1 in animals [21]. By using affinity capture LC/MS, a method by which paramagnetic beads coated with HER2 are used to capture trastuzumab-containing molecules in plasma followed by LC/MS determination of the relative abundance of different DAR species in the sample, it was possible to estimate the concentrations of individual DAR species and to assess their change in concentration over time following dosing in animals. From these data, a population PK analysis suggested that the deconjugation rate of DM1-related products was dependent on the DAR; such that higher DAR species had a higher rate of deconjugation than lower DAR species, with the rate-limiting step the deconjugation from a DAR of 1 to 0. The cause of this observation is not known, but has been postulated to be related to the higher probability of deconjugation with a higher DAR, or that the more chemically labile DM1-related products are lost more rapidly followed by more stably linked products.

The typical decrease in clearance with increasing body weight and similarity in distribution volume (per kg) across species (Table 26.1) [8, 15, 25]. Also observed is dose-related nonlinearity in PK in primates and humans due to binding to HER2, a finding consistent with that of trastuzumab [27]. Interestingly, the difference in T-DM1 CL between cynomolgus monkey and human is less than the typical twofold commonly described for mAbs [31], which may be related to the somewhat slow clearance of total trastuzumab in cynomolgus monkeys (~5.5 mL/day/kg) [8] rather than differences in the rate of deconjugation of DM1.

It is conceivable that the stability of the linker could also be determined by evaluating the concentrations of free cytotoxic drug in systemic circulation. If the linker is labile and the cytotoxic drug has a sufficiently low clearance and distribution volume, the levels of cytotoxic drug could become appreciable. Integrating this information with independently determined cytotoxic drug PK could allow deconvolution of the rate of deconjugation. For T-DM1, it was observed that the concentrations of free DM1 in plasma relative to T-DM1 were low, with a molar ratio approximately 50-fold less than T-DM1 concentrations, and a plasma concentration–time profile similar to T-DM1 [30, 32]. This is consistent with formation rate-limited PK of DM1, and a higher clearance and volume relative to T-DM1. Indeed, PK data generated in the rat confirm this with DM1 exhibiting a CL of approximately 30–35 mL/min/kg and a large volume of distribution (approximately 2–4 L/kg) [25, 33]. The rapid clearance of DM1, coupled with its slow formation rate, results in very low concentrations that minimize the risk of toxicity while

TABLE 26.1 Mean Pharmacokinetic Parameters for T-DM1 and Total Trastuzumab Following Dosing with T-DM1

Parameter	Analyte	Clearance (mL/day/kg)	Terminal Half-life (days)	Volume of Distribution (mL/kg)
Mouse[a]	T-DM1	18.1	5.61	56.8
Rat[b]	T-DM1	18.4	4.77	42.8
	Total trastuzumab	9.6	8.84	
Cyno[b]	T-DM1	11.6	5.21	42.8
	Total trastuzumab	4.8	10.5	
Human[b,c]	T-DM1	10	3.94	48.1
	Total trastuzumab	4.3	~9	

[a] 15 mg/kg.
[b] At 3.6 mg/kg.
[c] CL and V (central volume) were estimated based on rat, cyno, and human body weights of 0.25, 3.6, or 70 kg, respectively [8, 15, 29, 30].

maintaining a high fraction of DM1 conjugated to T-DM1 for delivery to HER2-expressing tumors. What is less expected from careful inspection of the free DM1 data is the observation that the peak free DM1 concentrations are observed immediately after T-DM1 dosing [25, 30, 34], a phenomenon atypical for a catabolite. The cause for this observation is not known, but may be related to a combination of factors. As described earlier, the assay used to measure free DM1 relies on the reduction of the plasma sample to cleave disulfide bonds between free DM1 with itself or with matrix components (e.g., glutathione, cysteine, and albumin) [25]. Trastuzumab is known to have a small number of free cysteines in its sequence (data not shown) that can allow the conjugation of DM1 via a disulfide bond. Thus, these disulfide-conjugated DM1 molecules could be released from T-DM1 during the sample processing used in the free DM1 assay, thus providing a source of free DM1 in the sample that is generally proportional to the amount of T-DM1 in the sample. Such a phenomenon would explain the observed free DM1 plasma concentrations, and suggest that measured free DM1 concentrations in plasma may be overestimated.

Another factor that complicates the use of DM1 as a determinant of the stability of the SMCC linker used in T-DM1 is that substantial evidence exists that DM1 is not the only catabolic product of T-DM1. Data from Shen et al. show that lys–MCC–DM1 and MCC–DM1 exist in plasma in rats at levels comparable to DM1 following dosing with T-DM1, consistent with observations in humans [32]. The source of lys–MCC–DM1 is certainly antibody catabolism, thus not a product of deconjugation, but the source of MCC–DM1 is less clear and could be related to deconjugation. Nevertheless, without more information regarding the identities and PK properties of all the deconjugation products of T-DM1, it is not possible to use this information to assess linker stability.

26.6 DISTRIBUTION OF T-DM1

The tissue distribution of antibodies has been thoroughly investigated, providing a generally strong understanding of the rate and extent of penetration into most tissues [35, 36]. The distribution of antibodies into tissues is limited primarily by their size, which prevents the free movement between plasma and extracellular fluid, resulting in extracellular fluid concentrations that are a fraction of plasma. The conjugation of an antibody with a cytotoxic drug, while contributing in a very small way to the overall size of the ADC (typically <2–3% of the overall mass), can have profound effects on its distribution. This was clearly demonstrated by studies of tissue distribution in rodents where conjugation of an antibody with MMAE (monomethylauristatin E) increased distribution to the liver and a study in humans showing conjugation with calicheamicin having profound effects on uptake in the tumor and other tissues [19, 37, 38]. While the exact causes for the change in distribution in these cases is not known, effects of conjugation on antibody hydrophobicity or aggregation have been postulated. The presence of the cytotoxic drug on an ADC increases the importance of understanding its distribution, confirming that conjugation is not substantively impacting the typical distribution of the antibody. For T-DM1, a tissue distribution study in rats with the ADC-bearing radiolabeled DM1 was conducted [32]. Because rats do not express HER2, the distribution of T-DM1 observed in this study is independent of target binding. The use of radiolabeled DM1 allowed the determination of the distribution of the cytotoxic agent, both bound and released from the ADC. The resulting data showed that the distribution of DM1 was consistent with that of an antibody that does not bind to antigen, demonstrating a slow decrease in tissue concentrations that is proportional to changes in plasma, and tissue-to-plasma ratios less than 1.

26.7 T-DM1 CATABOLISM AND ELIMINATION

The systemic elimination of an antibody can occur via several mechanistic processes. For antibodies that target cell surface antigens, binding to target can result in internalization of the antibody–antigen complex with subsequent proteolytic degradation [39]. Alternatively, antibodies can be internalized by cells via nonspecific processes including pinocytosis and Fcγ-receptor-mediated binding. Pinocytosis is a phenomenon exhibited by nearly all cells, while Fcγ receptor binding is limited to cells primarily associated with immune function. The consequence of cellular uptake of antibodies is often proteolytic degradation. The proteolytic process produces small peptides and amino acids, which are subsequently reused by the cells and are likely to have no pharmacologic activity. The conjugation of a cytotoxin to an antibody should have no impact on the underlying elimination processes of the antibody, as the cytotoxin is typically too small to affect pinocytosis, should not interfere with binding to target or Fc receptors, and should not impact cellular trafficking and proteolysis. However, the catabolic products of ADCs and the pharmacologic activity of those products have the potential to be quite different from unconjugated antibodies, thus indicating the need for thorough characterization of their identity, fate, and pharmacologic activity. In the course of the development of T-DM1, an extensive series of studies was conducted to explore its catabolic processing and fate.

At this point, it is probably appropriate to clearly explain the use of the terms linker cleavage and catabolism. In the context applied here, linker cleavage describes the release of DM1 or DM1-related products from the antibody resulting from cleavage of the MCC linker. Theoretically, this can occur both within and outside cells (e.g., in plasma).

Catabolism specifically describes the intracellular degradation of the antibody resulting in the release of DM1-related products. On first principles, the use of the "stable" MCC linker suggested that linker cleavage should be minimal, and the rate and extent of this process was discussed earlier. Indeed, plasma incubation studies and *in vivo* PK demonstrated the presence of slow release of drug from the ADC [8, 25, 26, 29] without catabolism (i.e., linker cleavage). The mechanism and products of the linker cleavage of T-DM1 have been described by two groups. Fishkin et al. determined that the thioether bond linking DM1 to its linker was susceptible to slow oxidation under unbuffered plasma incubation conditions or other oxidizing conditions, particularly if the pH increases above the typical physiologic range (e.g., pH 8.5). This oxidation of the thioether can result in the release of a DM1-sulfonate product that is susceptible to reduction in the presence of a reducing agent such as tris-2-carboxyethyl phosphine (TCEP), a reagent used in the sample preparation for analysis of DM1 concentrations in plasma [40]. Under the conditions used in this study, no evidence of the release of other DM1-related products was observed and these data suggest that the product of the slow loss of DM1-related products from T-DM1 both *ex vivo* and *in vivo* is DM1-sulfonate, which would appear as free DM1 if the serum or plasma sample was treated with TCEP. As a pharmacologic agent, DM1-sulfonate should have low potency due to its poor cell penetration, thus minimizing the risk for free cytotoxin mediated toxicity. In contrast to these *ex vivo* observations, *in vivo* studies of T-DM1 and other mAb–MCC–DM1 ADCs have shown no evidence of the generation of DM1-sulfonate in the plasma, the liver, or other body fluids (bile, urine), even when matrix samples were not treated with TCEP [32, 41]. The cause of this discrepancy is unclear, but may be related to differences in the physiologic processes occurring *ex vivo* versus *in vivo*. Investigators at Genentech have postulated that the linker cleavage can occur via a reverse Michael reaction of the thioether bond linking the thiol of DM1 to the maleimide ring of MCC, thus resulting in the slow release of free DM1 [26]. This hypothesis is based on *ex vivo* plasma stability data for T-DM1, which shows increasing concentrations of free DM1 in the incubation mixture; however, a thorough mechanistic investigation was not provided to support this hypothesis. Despite these studies, there remain significant gaps in our understanding of the mechanisms and products of linker cleavage of T-DM1.

The catabolic fate of T-DM1 was extensively evaluated in rats using T-DM1 conjugated with ^{3}H-DM1 [32]. Employing this strategy allowed the identification of DM1-related products, which could be pharmacologically active, and provided high sensitivity. The rat was chosen as a suitable species based on the assumption that: (i) the DM1-related products (linker cleavage, proteolytic degradation) should be qualitatively similar between species, (ii) doing such studies in rats was more practical than in primates, (iii) and a substantial body of toxicologic data with both DM1 and T-DM1 was available in rats, which could facilitate the integration of catabolic product data with toxicologic data. Following a dose of radiolabeled T-DM1, greater than 95% of radioactivity in plasma was found to be associated with the ADC, consistent with the previous data indicating the slow release of DM1-related products and low DM1 concentrations in plasma. In plasma, the main DM1-related products were DM1, MCC–DM1, and lys–MCC–DM1, all at similar concentrations and in total constituting less than 5% of the total plasma radioactivity. Comparison of the plasma catabolites identified in rats with those observed in the plasma of patients treated with T-DM1 showed that the catabolites identified in rats were also present in human plasma, but no data were reported regarding the presence of human-specific catabolites. Direct comparison of the concentrations and relative abundances of the catabolites is difficult due to differences in the dose, and lack of information about sampling times in the rat study. In humans, mean (SD) concentrations of MCC–DM1 at 1 h post-T-DM1 dosing were substantially higher than the other catabolites (34.4 (24.30) ng/mL for MCC–DM1 vs. 5.32 (1.48) and 1.35 (0.133) ng/mL for DM1 and lys–MCC–DM1, respectively). By Day 8, the concentrations of all catabolites were substantially lower, with low concentrations of lys–MCC–DM1 being detectable (~1.2 ng/mL) and the others not detectable (<2 ng/mL). Considering the likely large volume of distribution of MCC–DM1 and DM1, these concentrations appear inordinately high, and if correct could constitute a substantial fraction of the total DM1 in the body at the time of sample. Indeed, an estimate based on the central compartment volume (V/F) of DM1 in humans being similar to that reported in rats (~10,000 mL/kg) [25] and the human plasma concentration of free DM1 at 1 h after dosing (5.32 ng/mL), more than 80% of the administered DM1 would need to be free (unconjugated) in plasma to account for this concentration. This is clearly inconsistent with other data indicating that the vast majority of DM1 remains conjugated to T-DM1 suggesting that other explanations for the free DM1 and MCC–DM1 concentrations beyond deconjugation or catabolism are necessary. In the absence of more robust sampling, and PK information on each catabolite, it is difficult to provide a more comprehensive assessment of the available human catabolism data.

Clearly, the mechanism by which some of these DM1-related products forms is not entirely clear. The best mechanistic understanding is for the formation of lys–MCC–DM1. Previous *in vitro* and *in vivo* data demonstrated that mAb–MCC–DM1 conjugates are subject to extensive cellular catabolism to a molecule consisting of linker, drug, and the amino acid to which it is conjugated on the antibody (lysine) [42, 43]. Thus, the presence of this catabolite in plasma can be interpreted as a reflection of the proteolytic degradation of T-DM1 in tissue with cellular release of the product. It is interesting to note that lys–MCC–DM1 is a large, polar zwitterion at neutral pH, and is thus highly cell impermeable, and should have a high binding affinity for tubulin [44]. Thus, the mechanism by which it leaves the cell is unclear but may require either cell death or active transport. Indeed, some evidence exists that lys–MCC–DM1 may be a substrate for efflux transporters such as the multidrug-resistance protein 1 (MDR1) [45]. Lys–MCC–DM1, due to its physicochemical nature, is expected to be considerably less potent than DM1 when applied extracellularly, and this has been shown to be the case with lys–MCC–DM1 having an *in vitro* IC_{50} against cancer cells several log units higher than maytansine or DM1 [41, 46]. In addition to its low potency, lys–MCC–DM1 is rapidly eliminated from plasma, consistent with its low concentration being the result of slow formation, a relatively large distribution volume, and rapid elimination. These factors suggest that this catabolite possesses a low risk for causing substantial toxicity following T-DM1 administration.

The mechanism of formation of the other DM1-related products observed in plasma is more difficult to explain. As discussed earlier, potential explanations have been proposed for the formation of DM1 in plasma but these do not extend to processes occurring within cells. *In vitro* studies of mAb–MCC–DM1 conjugates have not observed the formation of DM1, making it unlikely that this is a product of cellular degradation of the ADC. The presence of MCC–DM1 in plasma is also difficult to explain. No plausible mechanistic explanation for the formation of this product from a mAb–MCC–DM1 conjugate has been reported. There have been some observations of MCC–DM1 in *in vitro* studies of cellular catabolism of mAb–MCC–DM1 ADCs, but this has been attributed to contamination of the ADC starting material with trace amounts of MCC–DM1 [41]; however, most similar studies of such ADCs (including T-DM1) have shown no evidence of this DM1-related product [42, 43, 47]. Regardless of its source, MCC–DM1 has a substantially lower potency than DM1 or maytansine [41], and has a large distribution volume and systemic clearance in rats (~200 mL/min/kg) and was not shown to be substantially metabolized to DM1 [33], suggesting that there is a low risk that this product contributes substantially to the risk of toxicity.

Overall, substantial gaps remain in the mechanistic understanding of the linker cleavage and catabolism of T-DM1. However, the low plasma concentrations of these measured products, coupled with their low cytotoxic potency, suggest that there is a very low risk associated with these molecules following administration of T-DM1. Importantly, analysis of DM1-related product concentrations in humans following dosing with T-DM1 identified similar products.

The excretion of the DM1-related products of T-DM1 following dosing with T-DM1 has also been investigated. Following dosing with T-^{3}H-DM1 in rats, the majority of the excreted products was in bile with 80% of the radioactivity being recovered in feces over 14 days. A smaller fraction was excreted in urine. Lys–MCC–DM1 was the predominant product found in excreta by 7 days, constituting approximately 70% of the excreted radioactivity in bile and urine, with MCC–DM1 accounting for approximately 10% and DM1 only about 3%. This suggests that cellular catabolism to lys–MCC–DM1 is the predominant elimination process for T-DM1 in rats over this period.

As described above, following the administration of T-DM1, small quantities of small molecule DM1-related products are found in plasma. Beyond their cytotoxic potential, these products may be perpetrators or victims of metabolic drug–drug interactions with coadministered drugs. The potential for such interactions is a major focus of investigation during the development of NCEs, with clearly defined regulatory expectations and a well-developed arsenal of methodologies for determining the nature and consequences of such interactions. For ADCs, the situation is less clear. Certainly, conducting studies of antibody-based ADCs in *in vitro* systems that assess cytoplasmic or organellar phenomena (CYPs or hepatocytes) is not appropriate; however, well-reasoned investigations of the small molecule products of linker cleavage or ADC catabolism may be warranted. For T-DM1, such studies have been conducted using DM1. At the time of publication, data are available only in abstract form and are thus somewhat limited. *In vitro* studies using the major human cytochromes P450 enzymes (CYP450) demonstrated that DM1 appears to be metabolized primarily by CYP3A4 to oxidative and hydrolytic metabolites. DM1 was found to be neither an inducer nor inhibitor of major CYP isoforms. When tested in MDCKII-MDR1 cells, DM1 was a substrate but not an inhibitor of P-glycoprotein [48]. Further studies of the *in vivo* disposition and metabolism of DM1 were conducted in rats. DM1 rapidly distributed to multiple tissues, consistent with its somewhat hydrophobic nature [45, 49]. DM1 was extensively metabolized to multiple metabolites, with the predominant route of elimination being via the bile. This is consistent with the data generated for T-DM1, which also demonstrated that the DM1 present in plasma following T-DM1 administration was excreted primarily in bile. It is not known whether additional studies with the other measured DM1-related products observed in plasma following T-DM1 administration have been conducted.

26.8 T-DM1 NONCLINICAL PK/PD

A thorough understanding of the PK/PD relationship of a drug can provide valuable insights into optimized dosing, interspecies translation, and enhanced mechanistic understanding. The complexity of T-DM1 related to its heterogeneity, its complicated mechanism of action, and its multiple active components provides both challenges and opportunities for the application of PK/PD. Published studies of the PK/PD of T-DM1 have focused primarily on its nonclinical behavior. An important application of PK/PD modeling is to guide the determination of a potentially efficacious dose and dose regimen in humans. Mouse efficacy studies using tumor models that are relevant representations of human disease can provide useful data for such translations. Jumbe et al. conducted PK/PD studies in mice bearing orthotopically implanted breast tumors that overexpressed HER2 and were insensitive to trastuzumab, mimicking the clinical situation in early trials of T-DM1 [15]. Different doses and dose regimens of T-DM1 were tested, and a semimechanistic PK/PD model developed. The analysis determined that the tumor response to T-DM1 was independent of dosing regimen and was able to provide an estimate of a plasma T-DM1 concentration required for tumor stasis, a value that could be used in the clinic to guide dose and dose regimen. The results from this study were supplemented by data from additional mouse tumor models of HER2+ cancer, and clinical PK and response data for T-DM1 in HER2-positive breast cancer patients who had previously progressed on trastuzumab [50]. Integrating the nonclinical tumor response data with human PK data provided estimates of human efficacious doses that were consistent with those observed in patients, thus confirming both the utility of the nonclinical tumor models and the PK/PD analysis. Both of these studies were somewhat empirical in incorporating the complexities of T-DM1, using plasma T-DM1 concentrations as the driver for drug effect, ignoring the detailed mechanistic behavior of the ADC. This aspect of T-DM1 activity was explored using T-DM1 conjugated with ^{3}H-DM1. Mice bearing HER2+ tumors were dosed with T-^{3}H-DM1, and the concentrations of T-DM1 and its active tumor catabolite (lys–MCC–DM1) were determined. Integrating this information with tumor response and plasma PK information, it was possible to explore the mechanistic processes that determine tumor response, including tumor uptake, cellular catabolism, and cellular efflux of active catabolic products. The results confirmed many previously held assumptions about the behavior of T-DM1. The results also provided a quantitative framework for describing the *in vivo* mechanism of action of T-DM1. This allowed the investigators to conduct simulations exploring how changing the rate of tumor cell catabolism of T-DM1 or the rate of tumor catabolite elimination from the cell affects tumor catabolite concentrations and tumor response. The simulations suggested that both increased T-DM1 catabolism within tumor cells and decreased elimination of these active products would result in improved tumor response, but that decreased catabolite elimination had a far greater impact. Such information could be useful in the design of improvements to T-DM1.

26.9 CONCLUSIONS

ADCs are rapidly becoming a potent weapon in the treatment of multiple forms of cancer [5]. The combination of a highly specific targeting component and a highly potent cytotoxic agent has the potential to improve response and reduce systemic toxicity. However, our understanding of the design and behavior of these complex molecules is still incomplete. Also, this complexity necessitates a rational and comprehensive drug development strategy to ensure that they can be safely and effectively administered. Understanding the ADME and PK/PD properties of ADCs is an important component of this strategy and such investigations, as described herein for T-DM1, go well beyond those conducted for unconjugated biotherapeutics. This started with the rational development of a bioanalysis strategy focused on the critical components of the T-DM1, facilitating a thorough understanding of not only its PK but also some elements of its disposition. Distribution studies confirmed similar distribution to the unconjugated antibody, but also assessed the distribution of the cytotoxic drug. Also, an exhaustive evaluation of T-DM1 catabolism and excretion addressed the fate of the conjugate and its catabolic products. This work provided a rich array of information to guide patients, physicians, and regulators in the use of the important drug, and can also inform scientists and other investigators in their efforts to understand the intricacies of these molecules and to develop new ADCs that can improve the lives of patients.

REFERENCES

[1] Kaur S, Xu K, Saad OM, Dere RC, Carrasco-Triguero M. Bioanalytical assay strategies for the development of antibody-drug conjugate biotherapeutics. Bioanalysis 2013;5 (2):201–226.

[2] Amiri-Kordestani L, Blumenthal GM, Xu QC, Zhang L, Tang SW, Ha L, Weinberg WC, Chi B, Candau-Chacon R, Hughes P. FDA approval: ado-trastuzumab emtansine for the treatment of patients with HER2-positive metastatic breast cancer. Clin Cancer Res 2014;20:4436.

[3] Alley SC, Okeley NM, Senter PD. Antibody–drug conjugates: targeted drug delivery for cancer. Curr Opin Chem Biol 2010;14 (4):529–537.

[4] Casi G, Neri D. Antibody–drug conjugates: basic concepts, examples and future perspectives. J Control Release 2012;161 (2):422–428.

[5] Sievers EL, Senter PD. Antibody-drug conjugates in cancer therapy. Annu Rev Med 2013;64:15–29.

[6] Austin CD, Wen X, Gazzard L, Nelson C, Scheller RH, Scales SJ. Oxidizing potential of endosomes and lysosomes limits intracellular cleavage of disulfide-based antibody–drug conjugates. Proc Natl Acad Sci U S A 2005;102 (50):17987–17992.

[7] Bender BC, Schaedeli-Stark F, Koch R, Joshi A, Chu Y-W, Rugo H, Krop IE, Girish S, Friberg LE, Gupta M. A population

pharmacokinetic/pharmacodynamic model of thrombocytopenia characterizing the effect of trastuzumab emtansine (T-DM1) on platelet counts in patients with HER2-positive metastatic breast cancer. Cancer Chemother Pharmacol 2012;70 (4):591–601.

[8] Bender B, Leipold DD, Xu K, Shen B-Q, Tibbitts J, Friberg LE. A mechanistic pharmacokinetic model elucidating the disposition of trastuzumab emtansine (T-DM1), an antibody–drug conjugate (ADC) for treatment of metastatic breast cancer. AAPS J 2014;16 (5):994–1008.

[9] Alley SC, Zhang X, Okeley NM, Anderson M, Law C-L, Senter PD, Benjamin DR. The pharmacologic basis for antibody-auristatin conjugate activity. J Pharmacol Exp Ther 2009;330 (3):932–938.

[10] Erickson HK, Phillips GDL, Leipold DD, Provenzano CA, Mai E, Johnson HA, Gunter B, Audette CA, Gupta M, Pinkas J. The effect of different linkers on target cell catabolism and pharmacokinetics/pharmacodynamics of trastuzumab maytansinoid conjugates. Mol Cancer Ther 2012;11 (5):1133–1142.

[11] Koeppen H, Wright B, Burt A, Quirke P, McNicol A, Dybdal N, Sliwkowski M, Hillan K. Overexpression of HER2/neu in solid tumours: an immunohistochemical survey. Histopathology 2001;38 (2):96–104.

[12] Austin CD, De Mazière AM, Pisacane PI, van Dijk SM, Eigenbrot C, Sliwkowski MX, Klumperman J, Scheller RH. Endocytosis and sorting of ErbB2 and the site of action of cancer therapeutics trastuzumab and geldanamycin. Mol Biol Cell 2004;15 (12):5268–5282.

[13] Phillips GDL, Li G, Dugger DL, Crocker LM, Parsons KL, Mai E, Blättler WA, Lambert JM, Chari RV, Lutz RJ. Targeting HER2-positive breast cancer with trastuzumab-DM1, an antibody–cytotoxic drug conjugate. Cancer Res 2008;68 (22):9280–9290.

[14] Lambert JM. Drug-conjugated monoclonal antibodies for the treatment of cancer. Curr Opin Pharmacol 2005;5 (5):543–549.

[15] Jumbe NL, Xin Y, Leipold DD, Crocker L, Dugger D, Mai E, Sliwkowski MX, Fielder PJ, Tibbitts J. Modeling the efficacy of trastuzumab-DM1, an antibody drug conjugate, in mice. J Pharmacokinet Pharmacodyn 2010;37 (3):221–242.

[16] Wakankar AA, Feeney MB, Rivera J, Chen Y, Kim M, Sharma VK, Wang YJ. Physicochemical stability of the antibody – drug conjugate trastuzumab-DM1: changes due to modification and conjugation processes. Bioconjug Chem 2010;21 (9):1588–1595.

[17] Kim MT-J, Chen Y, Marhoul J, Jacobson F. Statistical modeling of the drug load distribution on trastuzumab emtansine (Kadcyla®), a lysine-linked antibody drug conjugate. Bioconjug Chem 2014;25:1223–1232.

[18] Kaur S, Xu K, Saad O, Liu L, Slattery T, Dere R. Mass spectrometry of antibody–drug conjugates in plasma and tissue in drug development. In: *Characterization of Protein Therapeutics Using Mass Spectrometry*. Springer; 2013. p 279–304.

[19] Boswell CA, Mundo EE, Zhang C, Bumbaca D, Valle NR, Kozak KR, Fourie A, Chuh J, Koppada N, Saad O, Gill H, Shen BQ, Rubinfeld B, Tibbitts J, Kaur S, Theil FP, Fielder PJ, Khawli LA, Lin K. Impact of drug conjugation on pharmacokinetics and tissue distribution of anti-STEAP1 antibody-drug conjugates in rats. Bioconjug Chem 2011;22 (10):1994–2004.

[20] Dere R, Yi J-H, Lei C, Saad OM, Huang C, Li Y, Baudys J, Kaur S. PK assays for antibody-drug conjugates: case study with ado-trastuzumab emtansine. Bioanalysis 2013;5 (9):1025–1040.

[21] Xu K, Liu L, Dere R, Mai E, Erickson R, Hendricks A, Lin K, Junutula JR, Kaur S. Characterization of the drug-to-antibody ratio distribution for antibody-drug conjugates in plasma/serum. Bioanalysis 2013;5 (9):1057–1071.

[22] ICH Guideline. 1997. Preclinical safety evaluation of biotechnology-derived pharmaceuticals s6 (r1).

[23] EM Agency. 2007. Guideline on the clinical investigation of the pharmacokinetics of therapeutic proteins. London.

[24] Committee for Medicinal Products for Human Use. 2007. Guideline on the clinical investigation of the pharmacokinetics of therapeutic proteins.

[25] Poon KA, Flagella K, Beyer J, Tibbitts J, Kaur S, Saad O, Yi J-H, Girish S, Dybdal N, Reynolds T. Preclinical safety profile of trastuzumab emtansine (T-DM1): mechanism of action of its cytotoxic component retained with improved tolerability. Toxicol Appl Pharmacol 2013;273 (2):298–313.

[26] Pillow TH, Tien J, Parsons-Reponte KL, Bhakta S, Li H, Staben LR, Li G, Chuh J, Fourie-O'Donohue A, Darwish M. Site-specific trastuzumab maytansinoid antibody–drug conjugates with improved therapeutic activity through linker and antibody engineering. J Med Chem 2014;57 (19): 7890–7899.

[27] Bruno R, Washington CB, Lu J-F, Lieberman G, Banken L, Klein P. Population pharmacokinetics of trastuzumab in patients with HER2+ metastatic breast cancer. Cancer Chemother Pharmacol 2005;56 (4):361–369.

[28] Gupta M, Wang B, Carrothers T, Joshi A, LoRusso PM, Chu W, Shih T, Loecke D, Girish S. Exposure-response analysis in patients with HER2-positive (HER2+) metastatic breast cancer (MBC) to assess the effect of T-DM1 on QTc prolongation. Clin Pharmacol Ther 2011;89 (Suppl 1):S58.

[29] Lu D, Joshi A, Wang B, Olsen S, Yi J-H, Krop IE, Burris HA, Girish S. An integrated multiple-analyte pharmacokinetic model to characterize trastuzumab emtansine (T-DM1) clearance pathways and to evaluate reduced pharmacokinetic sampling in patients with HER2-positive metastatic breast cancer. Clin Pharmacokinet 2013;52 (8):657–672.

[30] Girish S, Gupta M, Wang B, Lu D, Krop IE, Vogel CL, Burris HA III, LoRusso PM, Yi J-H, Saad O. Clinical pharmacology of trastuzumab emtansine (T-DM1): an antibody–drug conjugate in development for the treatment of HER2-positive cancer. Cancer Chemother Pharmacol 2012;69 (5):1229–1240.

[31] Deng R, Loyet KM, Lien S, Iyer S, DeForge LE, Theil FP, Lowman HB, Fielder PJ, Prabhu S. Pharmacokinetics of humanized monoclonal anti-tumor necrosis factor-{alpha} antibody and its neonatal Fc receptor variants in mice and cynomolgus monkeys. Drug Metab Dispos 2010;38 (4):600–605.

[32] Shen B-Q, Bumbaca D, Saad O, Yue Q, Pastuskovas CV, Cyrus Khojasteh S, Tibbitts J, Kaur S, Wang B, Chu Y-W. Catabolic fate and pharmacokinetic characterization of trastuzumab emtansine (T-DM1): an emphasis on preclinical and clinical catabolism. Curr Drug Metab 2012;13 (7):901–910.

[33] Leipold D, Moore H, Jumbe S, Baudys J, Saad O, Mai E, Wong WL, Tibbitts J. Development of a pharmacokinetic model examining trastuzumab-MCC-DM1, MCC-DM1 and DM1 in normal rats. Mol Cancer Ther Proceedings of the American Association for Cancer Research 2007 (Molecular Targets Meeting) 2007;6:C155.

[34] Krop IE, Beeram M, Modi S, Jones SF, Holden SN, Yu W, Girish S, Tibbitts J, Yi J-H, Sliwkowski MX. Phase I study of trastuzumab-DM1, an HER2 antibody-drug conjugate, given every 3 weeks to patients with HER2-positive metastatic breast cancer. J Clin Oncol 2010;28 (16):2698–2704.

[35] Shah DK, Betts AM. Towards a platform PBPK model to characterize the plasma and tissue disposition of monoclonal antibodies in preclinical species and human. J Pharmacokinet Pharmacodyn 2012;39 (1):67–86.

[36] Xu X, Vugmeyster Y. Challenges and opportunities in absorption, distribution, metabolism, and excretion studies of therapeutic biologics. AAPS J 2012;14 (4):781–791.

[37] Scott AM, Tebbutt N, Lee F-T, Cavicchiolo T, Liu Z, Gill S, Poon AM, Hopkins W, Smyth FE, Murone C. A phase I biodistribution and pharmacokinetic trial of humanized monoclonal antibody Hu3s193 in patients with advanced epithelial cancers that express the Lewis-Y antigen. Clin Cancer Res 2007;13 (11):3286–3292.

[38] Herbertson RA, Tebbutt NC, Lee FT, MacFarlane DJ, Chappell B, Micallef N, Lee ST, Saunder T, Hopkins W, Smyth FE, Wyld DK, Bellen J, Sonnichsen DS, Brechbiel MW, Murone C, Scott AM. Phase I biodistribution and pharmacokinetic study of Lewis Y-targeting immunoconjugate CMD 193 in patients with advanced epithelial cancers. Clin Cancer Res 2009;15 (21):6709–6715.

[39] Lobo ED, Hansen RJ, Balthasar JP. Antibody pharmacokinetics and pharmacodynamics. J Pharm Sci 2004;93 (11):2645–2668.

[40] Fishkin N, Maloney EK, Chari RV, Singh R. A novel pathway for maytansinoid release from thioether linked antibody–drug conjugates (ADCs) under oxidative conditions. Chem Commun 2011;47 (38):10752–10754.

[41] Sun X, Erickson H. Studies on the metabolism of antibody–drug conjugates. In: *Antibody-Drug Conjugates and Immunotoxins*. Springer; 2013. p 297–316.

[42] Erickson HK, Park PU, Widdison WC, Kovtun YV, Garrett LM, Hoffman K, Lutz RJ, Goldmacher VS, Blättler WA. Antibody-maytansinoid conjugates are activated in targeted cancer cells by lysosomal degradation and linker-dependent intracellular processing. Cancer Res 2006;66 (8):4426–4433.

[43] Erickson HK, Widdison WC, Mayo MF, Whiteman K, Audette C, Wilhelm SD, Singh R. Tumor delivery and *In vivo* processing of disulfide-linked and thioether-linked antibody–maytansinoid conjugates. Bioconjug Chem 2009;21 (1):84–92.

[44] Lopus M, Oroudjev E, Wilson L, Wilhelm S, Widdison W, Chari R, Jordan MA. Maytansine and cellular metabolites of antibody-maytansinoid conjugates strongly suppress microtubule dynamics by binding to microtubules. Mol Cancer Ther 2010;9 (10):2689–2699.

[45] Kovtun YV, Audette CA, Mayo MF, Jones GE, Doherty H, Maloney EK, Erickson HK, Sun X, Wilhelm S, Ab O. Antibody-maytansinoid conjugates designed to bypass multidrug resistance. Cancer Res 2010;70 (6):2528–2537.

[46] Junttila TT, Li G, Parsons K, Phillips GL, Sliwkowski MX. Trastuzumab-DM1 (T-DM1) retains all the mechanisms of action of trastuzumab and efficiently inhibits growth of lapatinib insensitive breast cancer. Breast Cancer Res Treat 2011;128 (2):347–356.

[47] Erickson HK, Lambert JM. ADME of antibody-maytansinoid conjugates. AAPS J 2012;14 (4):799–805.

[48] Wong S, Bumbaca D, Yue Q, Halladay J, Kenny JR, Salphati L, Saad O, Tibbitts J, Khojasteh C, Girish S. Abstract A136: nonclinical disposition, metabolism, and *in vitro* drug-drug interaction assessment of DM1, a component of trastuzumab emtansine (T-DM1). Mol Cancer Ther 2011;10 (Suppl 11):A136–A136.

[49] Kovtun YV, Audette CA, Ye Y, Xie H, Ruberti MF, Phinney SJ, Leece BA, Chittenden T, Blättler WA, Goldmacher VS. Antibody-drug conjugates designed to eradicate tumors with homogeneous and heterogeneous expression of the target antigen. Cancer Res 2006;66 (6):3214–3221.

[50] Shah DK, Haddish-Berhane N, Betts A. Bench to bedside translation of antibody drug conjugates using a multiscale mechanistic PK/PD model: a case study with brentuximab-vedotin. J Pharmacokinet Pharmacodyn 2012;39 (6):643–659.

27

USE OF PK/PD KNOWLEDGE IN GUIDING BISPECIFIC BIOLOGICS RESEARCH AND DEVELOPMENT

ANDREAS BAUMANN[1], SAILETA PRABHU[2] AND JITENDRA KANODIA[2]

[1] *Bayer Pharma Aktiengesellschaft, Berlin, Germany*

[2] *Genentech Research and Early Development, South San Francisco, CA, USA*

27.1 INTRODUCTION

Bispecific antibodies (bsAbs) have a greater functionality as compared to established monoclonal antibodies (mAbs) because they bind to two different targets or, potentially, two epitopes on the same target (dual targeting (DT)). This may result in enhanced binding avidity with preferential binding to only cells that express both targets, or binding to targets on different cells. As early as the 1980s, it was already possible to generate these moieties by simultaneous availability of the mAb, hybrid-hybridoma (or quadroma), and recombinant DNA technologies [1, 2]. However, there have been failures, mainly due to poor clinical safety and efficacy as well as manufacturing problems [3].

Recently, there is a notable revival of interest in this format driven by improved technologies, and by the high unmet medical need in indications such as cancer and inflammation. These complex diseases are often multifactorial, involving a redundancy of disease-mediating ligands and receptors, as well as crosstalk between signaling cascades. The development of resistance to therapy with a single mAb is often associated with upregulation of alternative receptors as well as a switch in the receptor-mediated pathways [4]. The use of combination therapies with mAbs targeting different receptors or different epitopes on the same target have shown improved efficacy, but require more investments for the development of two or more mAbs.

An alternative to the development of the combination strategy and also for the development of therapeutics with increased functionality are the so-called noncanonical mAbs, such as antibody–drug conjugates, engineered antibodies, and antibody fragments and/or domains and bispecific formats. From a scientific and regulatory point of view, bsAbs are individual drug development programs and do not need parallel development activities as compared to combination therapies. In addition, therapy with a single DT drug rather than combinations usually implies better patient compliance. Finally, bispecific formats may also provide a competitive edge to companies. Bispecific formats comprised about 6% of all biologics formats in development for cancer indications in 2012 [5]. Most of the bispecifics are still in preclinical development, and around 20 are in clinical development. Only two bispecifics are approved so far—blinatumomab was recently approved in December, 2014, to treat patients with Philadelphia chromosome-negative precursor B-cell acute lymphoblastic leukemia (B-cell ALL); and catumaxomab (Removab®) is in the market since 2009 necessitating an understanding of this therapeutic platform in greater detail (see Table 27.1).

In the following sections, we describe the different types of bispecific formats as well as the principles used to generate them, before elaborating on the biochemical and pharmacological properties including affinity, avidity, and pharmacokinetics (PK). In addition, assay strategies

ADME and Translational Pharmacokinetics/Pharmacodynamics of Therapeutic Proteins: Applications in Drug Discovery and Development, First Edition. Edited by Honghui Zhou and Frank-Peter Theil.

TABLE 27.1 BsAbs in the Clinic

Sl. No.	Drug	Companies	Target Antigen	Technology	Indication
1.	ALX-0061 [6, 7]	Ablynx N.V., Abbvie Inc.	IL-6R (CD126) × serum albumin	Nanobody	Rheumatoid arthritis
2.	Pasotuxizumab [8]	Bayer	CD3 × PSMA	BiTE	Prostate neoplasms
3.	Blinatumomab/MT103/MEDI-538/AMG103 [9]	Amgen	CD3 × CD19	BiTE	ALL, DLBCL, NHL
4.	Catumaxomab (Removab) [10]	Fresenius Biotech GmbH, Trion Pharma	EpCAM × CD3	Triomab	Malignant ascites, gastric cancer, ovarian cancer
5.	CD20Bi/FBT-A05/Lymphomun [11, 12]	Fresenius Biotech, Trion Pharma	CD3 × CD20	Triomab	B-cell malignancies
6.	Duligotumab/MEHD7945A/RG7597 [13]	Genentech Inc.	EGFR × HER3	Two-in-one Dual action Fab (DAF)	Metastatic colorectal cancer, Recurrent/metastatic squamous cell carcinoma of the head and neck (SCCHN)
7.	Ertumaxomab/Rexomun [14]	Trion Pharma GmBh with Fresenius SE & Co, KGaA	(HER2; ErbB2; neu) × CD3	Triomab	Breast cancer
8.	MM-111 [15]	Merrimack Pharmaceuticals	HER2 × HER3	scFv	Gastric cancer, Advanced refractory HER2 amplified Heregulin-positive cancers, HER2-positive solid tumors
9.	MM-141 [16]	Merrimack Pharmaceuticals	HER3 × IGF-1R	Tetravalent bispecific antibody	Advanced solid tumors
10.	Ozoralizumab (ATN-103) [17]	Originally developed by Pfizer, now under Ablynx	TNF-α × serum albumin	Trivalent bispecific nanobody	Rheumatoid arthritis
11.	SAR156597 [18]	Sanofi	IL-4 × IL-13	Tetravalent bispecific tandem Ig (TBTI)	Idiopathic pulmonary fibrosis
12.	Solitomab (AMG110, MT110) [19, 20]	Amgen	EpCAM × CD3	BiTE	Solid tumors

to measure these compounds and the use of pharmacokinetic/pharmacodynamic (PK/PD) modeling in the design and development of these complex molecules are also covered. These topics are illustrated, where applicable, with examples. Finally, three case studies have been described in detail to highlight some of the challenges in the research and development of bsAbs. We conclude with some open questions that provide an outlook for future developments.

27.2 STRUCTURAL FORMATS AND GENERATION OF BISPECIFIC BIOLOGICS

BsAbs can be classified either through their mechanism of action (targeting) or biochemically by their structural format. DT bsAbs can either (i) act directly on target structures such as cell surface receptors or soluble factors or (ii) act indirectly using DT to recruit a therapeutically active moiety such as effector molecules and effector cells [2], or (iii) function by using the target as a drug delivery platform (see Fig. 27.1). Direct and indirect actions can be also combined within one molecule to further improve efficacy.

More than 45 different bispecific formats have been established in the last 20 years [3]. Baeuerle has grouped 37 bsAb formats into seven distinct classes [1].

1. Asymmetric immunoglobulin G (IgG)-like bsAbs (e.g., Triomab®, the first and only bispecific trifunctional on the market from Trion/Fresenius, knobs-into-holes from Genentech [21], Fab-exchanged by Genmab).
2. Symmetric IgG-like bispecifics (e.g., DT Ig, by GlaxoSmithKline, Two-in-one, by Genentech [22] and mAb2 by F-star).
3. IgG-like bispecifics with fused antibody fragments (e.g., dual variable domain Ig from Abbott, Ts2Ab from Medimmune, TvAb from Roche).
4. Bispecific constructs based on the Fcɣ fragment (e.g., Dual affinity retargeting technology (Fc DART) from MacroGenics, SCORPION from Bristol–Myers Squibb).
5. Bispecifics based on Fab fragments (e.g., F(ab)2 from Medarex/Amgen, Fab-Fv from UCB).
6. Bispecifics based on diabodies or single-chain antibodies (e.g., BiTE®) from Amgen [23, 24], Tandem diabody from Affimed, human serum albumin single-chain variable fragment (scFv) fusion from Merrimack.
7. Bifunctional fusions of antibodies or fragments with other proteins (e.g., Immunocytokines from EMD Serono, Immune Mobilizing mTCR against cancer (ImmTAC) from Immunocore).

Different bispecific formats are included in Table 27.2.

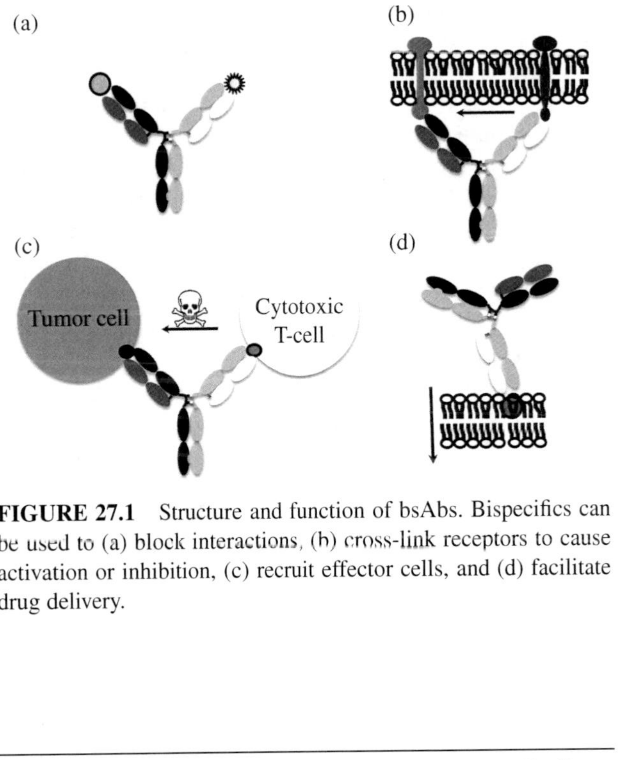

FIGURE 27.1 Structure and function of bsAbs. Bispecifics can be used to (a) block interactions, (b) cross-link receptors to cause activation or inhibition, (c) recruit effector cells, and (d) facilitate drug delivery.

TABLE 27.2 Bispecific Approaches

Intact Ig Hetero-oligomerization Domains	V_H and V_L Based	Intact Ig Homo-oligomerization domains	Single Domain Antibodies (Alternative Scaffolds)
dAb-Ig (Domantis/GSK)	Micromet/Amgen—BiTE	DVD-Ig (Abbott)	Adnectins (Adnexis/BMS)
Knobs-into-hole/2-in-1 (Genentech)	Macrogenics—DART (Fc-DART)	CovX (Pfizer)	Ablynx—Nanobodies
Duobody (Genmab)	Affimed—TandAb (Fc-T and Ab)	Peptibodies (Amgen)	Molecular Partners—DARPin
	LeadArtis—Trimerbodies	SVD-Ig (Imclone/Lilly)	Pieris—Duocalins
		IgG-svFv (Biogen Idec)	Covagen-Fynomers
		scFv2-Fc (MedImmune)	
		CODV (Sanofi)	

27.3 BIOCHEMISTRY AND PHARMACOLOGY OF BISPECIFICS

The following section highlights some of the key properties of bsAbs that may be considered or leveraged while developing these novel and complex therapeutics.

27.3.1 Affinity

Affinity can be defined as the strength of the binding interaction between a single antigen and the single region of the mAb. The affinity between a mAb and its target antigens has important implications in distribution, tumor penetration, or internalization and catabolism that can be naturally extended even to bsAbs. Optimal affinity for the tumor antigen is a major factor influencing antibody localization and efficacy. One explanation for how exceedingly high affinity for the tumor antigens may cause poor penetration of the mAbs is provided by the binding-site barrier model that analyzed different parameters for tumor distribution and penetration for the full mAb and mAb fragments [25]. McCall et al. demonstrated that scFv molecules with very high affinity for Her2/neu have poor tumor penetration from the vasculature than the corresponding lower affinity variants [26]. However, the determination of the optimal affinity for each arm of the bsAb can be complex and influenced by several factors such as the mechanism of pharmacological activity of each arm, site of action, and indication (also see section on PK/PD model informed design of BsAbs). For example, Bortoletto et al. [27] demonstrated that bispecific single-chain antibodies directed against epithelial cell adhesion molecule (EpCAM) and CD3 showed lower capacity to target EpCAM positive tumor with the higher affinity CD3 variant and vice versa. High affinity binding to CD3 may reduce the efficiency of T-cell stimulation and target *the* bispecific molecule to T-cells rather than to tumor cells *in vivo* [26, 27]. Similarly, Yu et al. [28] developed a bsAb with low affinity against transferrin receptor (TfR) to enable transport across the blood–brain barrier (see Section 27.6.1).

27.3.2 Avidity

BsAbs are expected to have an advantage over combination antibodies in targeting multiple pathways because of what is termed "avidity hypothesis." Specifically, the avidity hypothesis predicts that if the increased avidity arising from the binding of two receptors on a target cell leads to greater efficacy, then the DT agent will show greater efficacy than the combination of agents that bind two receptors individually, each binding only a single target receptor [4, 29]. BsAbs may demonstrate an enhanced avidity depending on the properties of the antigen and the binding epitope (s) targeted by the bsAb. BsAb directed against two separate, non-overlapping epitopes on the same target molecule may therefore most likely possess enhanced avidity due to its bivalent paratopic binding. Similarly, a tetravalent IgG-like bsAb with bivalent binding for each antigen is most likely to have a very slow dissociation rate because two dissociation events must occur simultaneously to free the bsAb molecule [30]. Thus, a bsAb molecule (EI-04), constructed with a stability-engineered scFv against IGF-1R attached to the carboxyl terminus of an IgG against epidermal growth factor receptor (EGFR), exhibited high avidity binding to BxPC3 tumor cells coexpressing EGFR and IGF-1R likely due to its tetravalent bispecific format, and consequently improved potency at inhibiting IGF-driven cell growth over the mAb combination [4]. However, the avidity hypothesis is yet to be confirmed *in vivo* in clinical trials.

27.4 PHARMACOKINETICS

Most of the bsAbs in development are still in the nonclinical phase and there is only limited published information on their PK behavior. There are no specific PK properties/mechanisms of bsAbs, which have not been described for other biologics. The overall PK behavior of the bsAb is strongly dependent on the format (e.g., IgG containing or not). In addition, the target-mediated clearance processes through two binding sites with similar affinities (e.g., to two target antigens) can be complex depending on the type of target, its expression profile, as well as its turnover rate.

BsAbs based on IgG may be preferred due to the favorable PK properties (e.g., long serum half-life) as well as the option to modulate the effector functions. In addition, data from canonical mAbs suggest that human PK can be translated from monkeys [31]. There is limited scaling data for bsAbs so far. The nonspecific clearance of MEHD7945A, a dual action Fab (DAF), was successfully projected using PK data from cynomolgus monkeys [32]. BsAbs with lower molecular weight, for example, in the range of 50 kDa and not containing the Fc, may penetrate tissues better than mAbs, and/or may have improved binding to hidden epitopes in cavities due to its size and shape. Biodistribution studies with tumor-targeting bsAbs reveal selective accumulation at the tumor site [33]. However, no final conclusion on the preference of any class of bsAbs can be made due to limited published quantitative biodistribution data. On the other hand, increased clearance of lower molecular weight bsAbs can balance or even outweigh the higher penetration potential by hampering the therapeutic efficacy, as demonstrated with a series of anti-CEA/anti-DOTA bsAbs [34] and single intravenous (IV) administration of blinatumomab [35].

As an alternative to the continuous infusion of a high clearance bsAb, half-life extension strategies can be used for low molecular weight bsAbs to lower their clearance as compared to IgG-containing moieties. These include reduction of the endogenous degradation of the drug, slow release/depot formulations, increasing the hydrodynamic volume by large bulky or linear polymers (e.g., polyethylene glycol), or the addition or binding capability of a large protein that can take up the FcRn-mediated salvage mechanism (recycling) [36, 37]. Association, conjugation, or fusion of low molecular weight bsAbs to albumin or albumin binders is one possibility through implementation of the FcRn-mediated half-life prolongation [38].

There are other ways to overcome the limitation of a low molecular weight bsAb (not containing the Fc-fragment) such as by using adopted administration regimen and routes. Blinatumomab, a bispecific T-cell engager (BiTE) antibody comprising two fused single-chain antibodies against CD19 and CD3, has a short elimination half-life of 2 h in humans after single IV administration. After continuous infusion of blinatumomab in adult patients with relapsed non-Hodgkin lymphoma, the compound demonstrated dose linearity and predictable drug levels throughout, making 24-h infusion the preferred administration schedule [39, 40]. On the other hand, the subcutaneous (SC) administration regimen may offer not only improved convenience for patients, but also less frequent administrations potentially due to flip-flop PK, when the drug is released from the SC site at a slower rate than the actual systemic clearance of the protein therapeutic. However, immediate release (IV bolus, short-term, or long-term infusion) into the blood stream provides typically a terminal concentration–time profile where the concentration decay is determined by the systemic clearance of the therapeutic protein.

27.4.1 PK Assay Strategies Employed for the Development of bsAbs

When developing a PK assay strategy, the needs of the nonclinical and clinical development program have to be considered, such as sensitivity of the assay (based on dose estimations), planned combination treatments, estimated target levels, and PK/PD-modeling approaches [41].

A review of existing literature suggests that the PK assay strategy is primarily governed by the format of the bsAbs and that ligand-based assay technologies including the typical enzyme-linked immunosorbent assay (ELISA) constitute the most commonly used methodology. The choice of capture and detection reagents is guided by the target specificity [16, 42], the bispecific format [4, 43–45], and the objective(s) of the assay to measure either the free or total drug or any other specific species.

The Triomab® catumaxomab (CD3 × EpCAM) was measured in plasma and ascites fluid in patients using a validated two-site ELISA wherein an antirat IgG λ light chain-specific antibody was used to capture catumaxomab and the bound bsAb was detected by an antimouse IgG2a-specific biotin-labeled detection antibody [43]. The assay format was used to measure the total drug by utilizing the chimeric composition of catumaxomab (rat–mouse hybrid mAb) rather than antigen specificity and can be extended to the PK assay of other Triomab antibodies that bind to different tumor-associated antigens. This holds significance for bsAb development in terms of minimizing cost and complexity for developing different bsAb molecules. Similar to the measurement of total antibody by Ruf et al. [43], Sampei et al. used human IgG-specific ELISA to determine plasma concentrations of total anti-FIXA/FX bsAbs in mice and monkeys [44]. The rationale could be once again on selecting a format that can comply with PK assessments for differing bispecificities in different *in vivo* models and thus promise broader applications for bsAb development.

Vugmeyster et al. [42] developed two assays to measure the concentrations of anti-IL17A and anti-IL22 peptide-antibody bispecific genetic fusions (IL-17A binding peptide genetically fused to N-termini of anti-IL22 human IgG1 through either the heavy chains only or through both heavy and light chains). The peptide–antibody fusions were captured by human his-IL17A and detected with goat antihuman IgG antibody horseradish peroxidase conjugate and provided a measure of anti-IL22 antibody that had at least one IL-17A binding peptide attached to it. The other assay provided the total human IgG concentrations, that is, anti-IL22 antibody either with the IL-17A binding peptide(s) attached or without any binding peptide. Thus, the two PK assays together target assessment of the functional production of the engineered peptide–antibody genetic fusion. On a similar note, Dong et al. [4] used ELISA to determine BIIB4-5scFv and 5scFv-BIIB4 (mAb with scFv; both mAb and scFv target IGF-1R but at different epitopes) serum concentrations and additionally tested the ability of the sera to block the binding activity to hIGF-1R to ensure that the scFv was present and functional [4].

Alternative as well as improvised PK assay formats have also been explored in bsAb development. An electrochemiluminescence-based assay was used to determine serum concentrations of BAY2010112 in mice [8]. A polyclonal goat anti-AMG 212/BAY2010112 serum was coated to a high binding plate (Meso Scale Discovery) for binding to the bsAb in the study samples. The detection was performed by anti-penta his-biotin-labeled detection antibody. Another assay format, a semihomogenous assay, was used to measure the dual-antigen binding activity of a bsAb. It was elegantly optimized to negate the "hook" effect of the bsAb at elevated concentrations. The semihomogeneous assay format is a significant improvement over the conventional ELISA method, with increased efficiency through fewer assay steps, good quantitation, and specificity for the bsAb without requirements of new reagents and instruments. With the semihomogeneous assay format, the authors also demonstrated that a DAF antibody indeed could bind two different antigens simultaneously *in vitro* [46].

While a majority of the PK assays for bsAbs is based on the principles of ELISA, there are examples of exceptions. The human serum concentration of the anticancer drug blinatumomab (CD19×CD3) was determined by fluorescence-activated cell sorting (FACS) analysis [35]. The assay is based on upregulation of CD69 on the surface of activated T-cells subject to dual binding of the bsAb to the T-cells as well as lymphoma cells. The activation of CD69 was concentration dependent, which could successfully be monitored by FACS. Given the low serum concentrations of blinatumomab following dosing, the activity-based assay likely provides an advantage over ELISA in terms of detecting the presence of the drug at low concentrations.

Overall, the PK assay strategy used for the mAb-based formats appears to provide a good estimate of the drug concentrations. However, as molecules get complicated in design, for example, anti-IL17A and anti-IL22 peptide-antibody bispecific genetic fusions, it will be necessary to develop assay strategies in order to measure the appropriate species for efficacy and safety evaluations.

27.4.2 Immunogenicity Strategies Employed for the Development of bsAbs

Product-related factors affecting the rate of occurrence of immunogenicity hold true for bsAbs, similar to other types of biologics. There is an increasing risk with the presence of B-cell/T-cell epitopes, novel epitopes (e.g., in fusion proteins), amino acid substitutions, and finally foreign sequences [47]. All these factors have to be considered early during bsAb design despite the fact that there is no validated model to predict the immunogenicity in patients from *in silico*, *in vitro*, or animal data. Importantly, assay strategies for monitoring immunogenicity in the preclinical species have to be developed in time and based on our knowledge from other biologics. They are primarily governed by the bsAb format and disease.

27.5 PHARMACOKINETIC–PHARMACODYNAMIC MODEL-INFORMED DESIGN OF bsAbs

Conceptually, *in vitro* and *in vivo* experiments have demonstrated that targeting multiple targets can be beneficial. However, an understanding of the cases in which bispecific molecules would be advantageous and the corresponding design for the bispecific molecule is unclear. Such an exploration can be accelerated by combining mathematical models with experimental data and thus inform therapeutic design.

The physiologically based pharmacokinetic (PBPK) model by Friedrich et al. for antibody-directed effector cell therapy of tumors provides an example for modeling of quantitative experiments to make therapeutically meaningful predictions [48]. The authors combined experimental data by Bakacs et al. with their PBPK model to explore how antibodies targeting receptors on both lymphocytes and tumor cells can influence the biodistribution of lymphocytes within the body, specifically focusing on redistribution of lymphocytes to tumors [48, 49]. The model systematically investigates a range of antibody properties to predict optimal bsAb binding constants for tumor antigen and lymphocyte binding epitopes to attain maximal efficacy.

Multiple models that focus on binding of multispecific molecules to their targets and subsequent effects have been published [16, 50–52]. These models focus on stoichiometric binding of multispecific antibodies to their targets and subsequent PD variables such as fractional receptor binding. Harms et al. [52] presented a generalized framework and case studies for quantitative investigation of multiarm antibodies based on systematic assessment of interplay between multiple relevant parameters such as affinity, avidity, antibody valency, and target expression. One of the most therapeutically relevant observations from their model is that bispecific molecules should not be treated as "magic bullets" and that a combination of monospecific antibodies can be better than bispecific molecules under certain conditions. Optimal design of bispecific therapies would require careful matching of antibody properties such as affinity and avidity as well as expected concentration profile at the site of action with specific target properties such as relative expression levels and desired pathway to be inhibited.

Fitzgerald et al. [16] investigated the activity of a bsAb (MM141) a step further by predicting cellular response to antibodies in terms of cell-behavior regulators such as ERK and AKT. The authors presented a mechanistic model for PI3K/AKT/mTOR cascade activation by IGF-1R and ErbB3. The model was calibrated to specific antibody (MM141) properties and was used to investigate two therapeutic strategies: combination of two monospecific antibodies versus a bsAb. In accordance with *in vitro* and *in vivo* results, the bsAb was predicted to be a stronger inhibitor than combination of monospecific antibodies due to high avidity inhibition of bispecific molecule to both pathways simultaneously. The approach can be easily generalized to evaluate the effect of antibodies that target other signaling pathways.

A combination of the aforementioned mechanistic PBPK and PD models mimicking the complexity of the biologic system might help inform IND-enabling (IND, investigational new drug) studies and clinical trial designs, although those systems models require more experimental data to ensure the predictive performance. Given the complexity of those models, formal model validation efforts are usually not provided and therefore those models have to be used with caution. Typically for IND-enabling predictions of human PK, scaling techniques such as different types of allometry including species-invariant time method are used commonly to estimate expected human PK and the corresponding exposure based on preclinical animal data, and thus enable efficient design of clinical trial parameters such as dosing strategy, minimum efficacious dose, and maximal efficacious or tolerated dose among others.

27.6 APPLICATION OF PK/PD IN THE RESEARCH AND DEVELOPMENT OF BISPECIFIC BIOLOGICS: CASE EXAMPLES

27.6.1 Anti-TfR/BACE1 to Improve Therapeutic Antibody Transport across the Blood–Brain Barrier

A conceptually different approach to deliver the active protein therapeutic or to improve the uptake of antibodies to target tissues represents employing an antigen that can transcytose across the cell membrane to transport or deliver the therapeutic to the site of action. BsAbs can be leveraged in this regard as a delivery platform where one arm of the mAb binds to a "transporter" antigen while the other arm binds to the "therapeutic" antigen to modulate its expression and/or biological activity. This approach has been used to improve the uptake of antibodies across the blood–brain barrier for the development of biotherapeutics for neurodegenerative diseases such as Alzheimer's disease. Yu et al. [28] engineered a bsAb, anti-TfR/BACE1 (β-site amyloid precursor protein cleaving enzyme 1), using the "knob-into-hole" technology. The "transporter" arm is against the TfR and the other arm recognizes BACE1 (Fig. 27.2). TfR imports iron into cells via receptor-mediated endocytosis of the transferrin–iron complex. TfR is ubiquitously expressed in several tissues including the endothelial cells of the blood–brain barrier. BACE1 initiates the process of cleaving amyloid precursor protein into amyloid beta (Aβ) and has been pursued as a therapeutic target for Alzheimer's disease [54, 55]. The BACE1 arm of anti-TfR/BACE1 was incorporated from the anti-BACE1 monospecific mAb [56]. It binds to BACE1 with high specificity and affinity ($K_d = 5$ nM). Our initial experiments with high affinity anti-TfR mAbs suggested that these antibodies bound to the vasculature with minimal uptake into the brain. In addition, these antibodies were cleared rapidly from the circulation due to TfR-mediated clearance. Hence the affinity was reduced, which improved the uptake into the brain (Fig. 27.2) as well as resulted in a lower systemic clearance [57]. The optimized bsAb showed 10-fold higher brain concentrations in mice compared to the canonical anti-BACE1 mAb. More importantly, it lowered endogenous Aβ levels in the brain to a greater extent than the

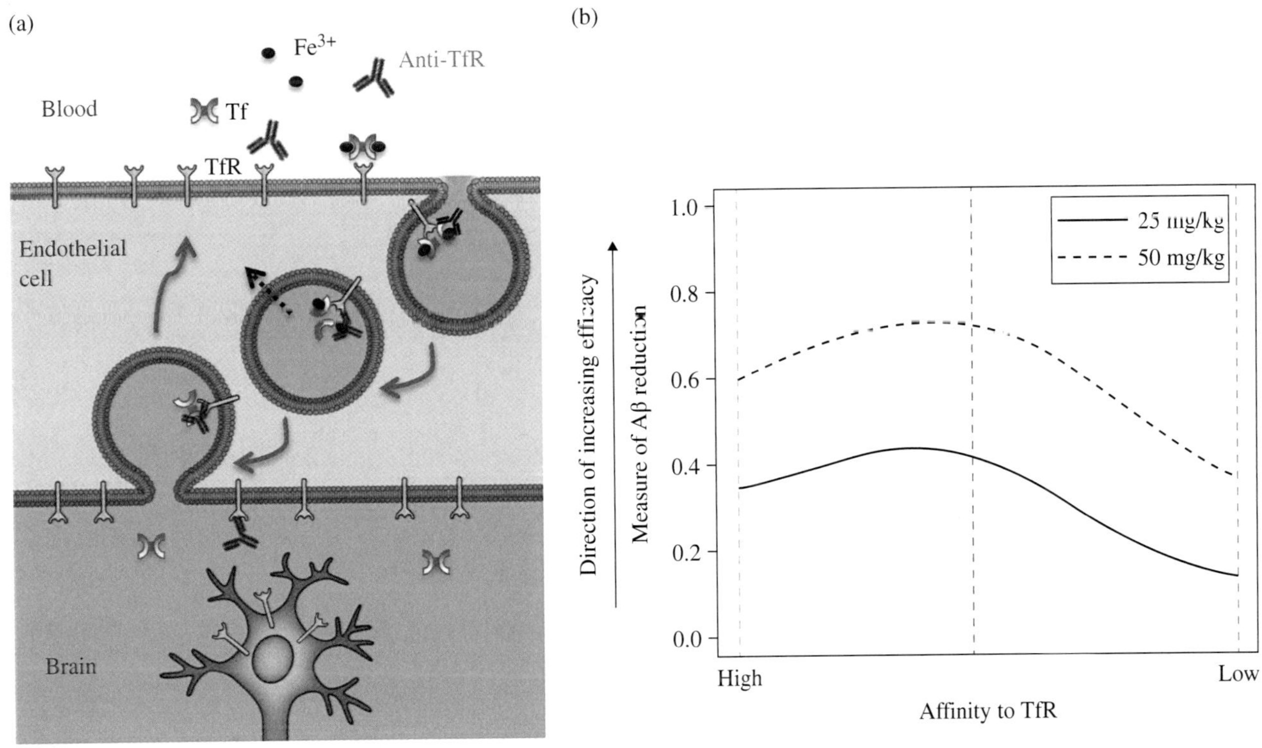

FIGURE 27.2 (a) Schematic for mechanism of TfR-mediated transcytosis. (b) Mechanism-based PK/PD model to support the synthesis of optimal bispecific anti-TfR/BACE1 candidate. Key mechanisms in the mathematical model include target-mediated clearance in periphery, TfR affinity driven transcytosis across the blood–brain barrier and turnover of Aβ in the brain and impact of anti-BACE1 on Aβ production. Adapted from Bumbaca et al. [53]. (*See insert for color representation of this figure.*)

monospecific anti-BACE1. A mechanism-based PK/PD model was developed to predict successfully the impact of the affinity on Aβ levels (Fig. 27.2) [53]. The model accounts for target-mediated drug clearance through TfR binding in systemic compartment, nonmonotonic trends for brain uptake of antibodies with different affinities and the dynamics of inhibition of Aβ production by BACE1. In addition, the model also captures the toxic effect of TfR binding on reticulocytes and thus helps quantitatively define the therapeutic window for the antibody. The bsAb was also designed to remove the Fc effector functions because it was associated with safety findings [57]. This strategy can work to deliver mAbs across the blood–brain barrier and might be broadly applicable to numerous neurological disorders. Although such entities are simple in principle, developing them is a major challenge because of suboptimal PK, modulating the affinity as a function of PK/PD as well as activity and its translation across species, changes in binding affinity as a bispecific (monovalent) compared to monospecific (bivalent), lack of suitable transporter antigens that can efficiently transcytose across the blood–brain barrier [58], need of appropriate PD markers, and potential safety issues. Consequently, the discovery and development strategy requires a rigorous optimization strategy to obtain the structural characteristics of the bsAb, which supports the therapeutic intent.

27.6.2 PK Characterization to Optimize bsAb Molecule Design and Selection for Ophthalmology

BsAbs are also being employed to develop therapeutics for ophthalmic diseases such as age-related macular degeneration [59]. One of the requirements for this class of protein therapeutics is an optimal residence in the intraocular compartment with minimal systemic exposure. To this end, a comprehensive set of molecules including a mAb, a F(ab′)2 with a double-disulfide bond linking the two Fabs (F(ab′)2 double disulfide), and a F(ab′)2 with a single-disulfide bond (F(ab′)2 single disulfide) was generated to investigate the ocular PK (Fig. 27.3); a Fab and a Fc were also generated to evaluate the impact of size and the role of Fc. All molecules except the Fc targeted viral glycoprotein D, a nontherapeutic antigen. The PK was determined in New Zealand white rabbits following intravitreal (IVT) administration. There were no apparent differences in the PK of these molecules (Fig. 27.3). The concentrations of the different formats decreased at comparable rates with vitreous half-lives ranging from 3.1 to 4.6 days. This suggested that proteolytic

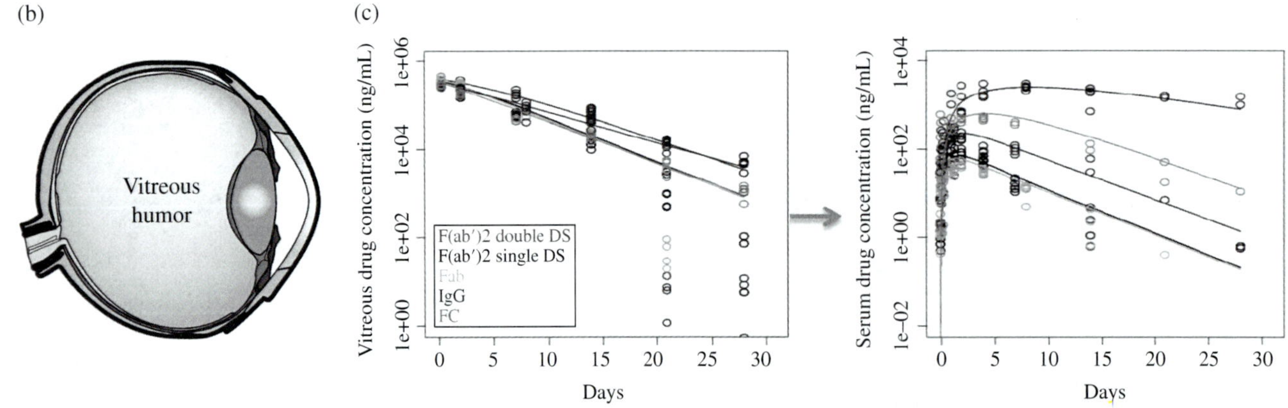

FIGURE 27.3 Pharmacokinetics of different protein formats in rabbit following IVT administration. (a) Structures of different protein formats used in the study. (b) Site of administration in the eye-vitreous humor. (c) Concentration–time profiles for the vitreous humor and serum after bilateral 0.5 mg/eye IVT administration (n=2 animals or four eyes/time point). Serum and vitreous samples were collected over 28 days post dose. Each eye was analyzed independently. Observed data = circles, model results = lines (mean concentration). Concentrations after Day 14 are impacted by antitherapeutic antibodies. Adapted from Le Couter et al. [59]. (*See insert for color representation of this figure.*)

catabolism and FcRn-mediated recycling probably have a limited role in the eye. In contrast, notable differences were seen in the PK of the different formats in the serum. The systemic exposure of mAb was 30-fold higher than F(ab′)2 double disulfide and greater than 100-fold higher than F(ab′)2 single disulfide or Fab. Thus, these data indicated that the use of mAb would clearly result in an unwanted increase in systemic exposure.

When compared between the F(ab′)2 formats, the F(ab′)2 double-disulfide molecules demonstrated slightly longer vitreous half-life than the F(ab′)2 single-disulfide and Fab molecules indicating a stabilizing effect with the second disulfide bond. The better stability of the F(ab′)2 double-disulfide molecules coupled with rapid systemic elimination may translate to superior efficacy and safety when compared to the other molecules. Similar results were obtained with another bsAb targeting vascular endothelial growth factor A and an undisclosed antigen [59].

27.6.3 Pharmacokinetic Studies during Development of a Bispecific T-Cell Engager

Bispecific T-cell engagers, commonly referred to as BiTEs, comprised two different flexibly linked single-chain antibodies, one directed against a tumor antigen and one targeting CD3. BITEs can transiently link tumor cells with resting polyclonal T-cells for induction of a surface target antigen-dependent redirected lysis of tumor cells, closely mimicking a natural cytotoxic T-cell response. *In vitro*, BiTE antibodies activate T-cells in a highly conditional manner that is dependent on the presence of target cells. First-generation BiTE antibodies cross-react only with respective antigens from chimpanzees [23]. To facilitate *in vivo* safety testing, surrogate BiTE antibodies were generated that are cross-reactive with murine antigens. The pharmacological characterization of BiTE antibodies includes in-depth analysis of their effects on tumor as well as on T-cells. Various xenograft models are available for *in vivo* efficacy testing. The second-generation BiTE antibodies are fully human in sequence and cross-react with nonhuman primates [24]. BAY2010112, in development for the treatment of patients with prostate cancer, is bispecific for prostate-specific membrane antigen (PSMA) and the CD3 epsilon subunit of the T-cell receptor complex. BAY2010112 binds PSMA and CD3 of human and macaque origin allowing for the assessment of safety, PD, and PK in a relevant animal species [8]. Cynomolgus monkey PK/toxicokinetic studies were performed with single and repeated SC as well as IV administration of BAY2010112. BAY2010112 was rapidly eliminated from serum with apparent half-lives of between 1 and 3 h as determined in the interval up to 8 h after IV administration. After reaching a plateau between 4 and 8 h, BAY2010112 was eliminated from the serum with a mean half-life of 7 h (4.1–11.2 h) and in between 6/10 and 24 h after SC administration of BAY2010112. The half-life reflects a continuing absorption from the site of injection rather than elimination ("flip-flop phenomenon"). Single species-based allometric scaling was used to estimate the human exposure at first-in-human doses. PK/PD data were generated as well in non-obese diabetic/severe combined immunodeficiency (NOD/SCID) mice that had been injected SC with human PSMA cells after γ irradiation [8]. Additional PK information was generated in BALB/c mice after single IV and SC administration of BAY2010112 indicating an absolute bioavailability of 18% (monkey: 66%).

Due to the short elimination half-life of BAY2010112 in monkeys, distribution into the tumor tissue was investigated by using the ^{14}C labeled compound. It could be demonstrated that ^{14}C-labeled BAY2010112 administered to mice by tail vein injection accumulates in SC implanted LNCaP PCa tumors [60] (Fig. 27.4). Distribution studies with biologics are not required as IND enabling, but they can generate a mechanistic understanding to aid internal decision making. Such studies can provide information on the major tissue distribution compartments and underlying mechanisms of disposition kinetics, elucidate on- and off-target binding kinetics in tissues of interest, and quantify the drug entity and/or its relevant parts.

27.7 OUTLOOK

There are no common design specifications of bsAbs yet. They need to be structurally optimized for each target, target organ, and other biological properties of the disease to achieve the wanted balance between efficacy and safety.

Several key questions have to be addressed in this exercise: (i) Are bsAbs better than the alternative of combining individual antibodies directed toward the targets? (ii) Is the optimum affinity of the two binding domains same as that for optimum individual antibodies? (iii) Does the increased avidity effect due to multiple binding domains pose an advantage by lowering dose or a limitation in the case of presence of decoy receptors? (iv) What are the optimum conditions (e.g., receptor densities) where such bsAbs could provide optimum benefits? (v) Are bivalent bsAbs (one or two target-binding sites) superior to monovalent binding? (vi) Do we want (or avoid) Fc-mediated functions? (vii) How do we align bsAbs format and the required half-life (clearance)?

In order to fully answer these questions and optimally exploit the technology of bsAbs, a thorough fundamental knowledge of their binding properties is critical. In addition, a thorough understanding of the PK/PD relationships and clinical data from the ongoing studies are needed to verify whether bsAbs will play a major role in future therapeutics.

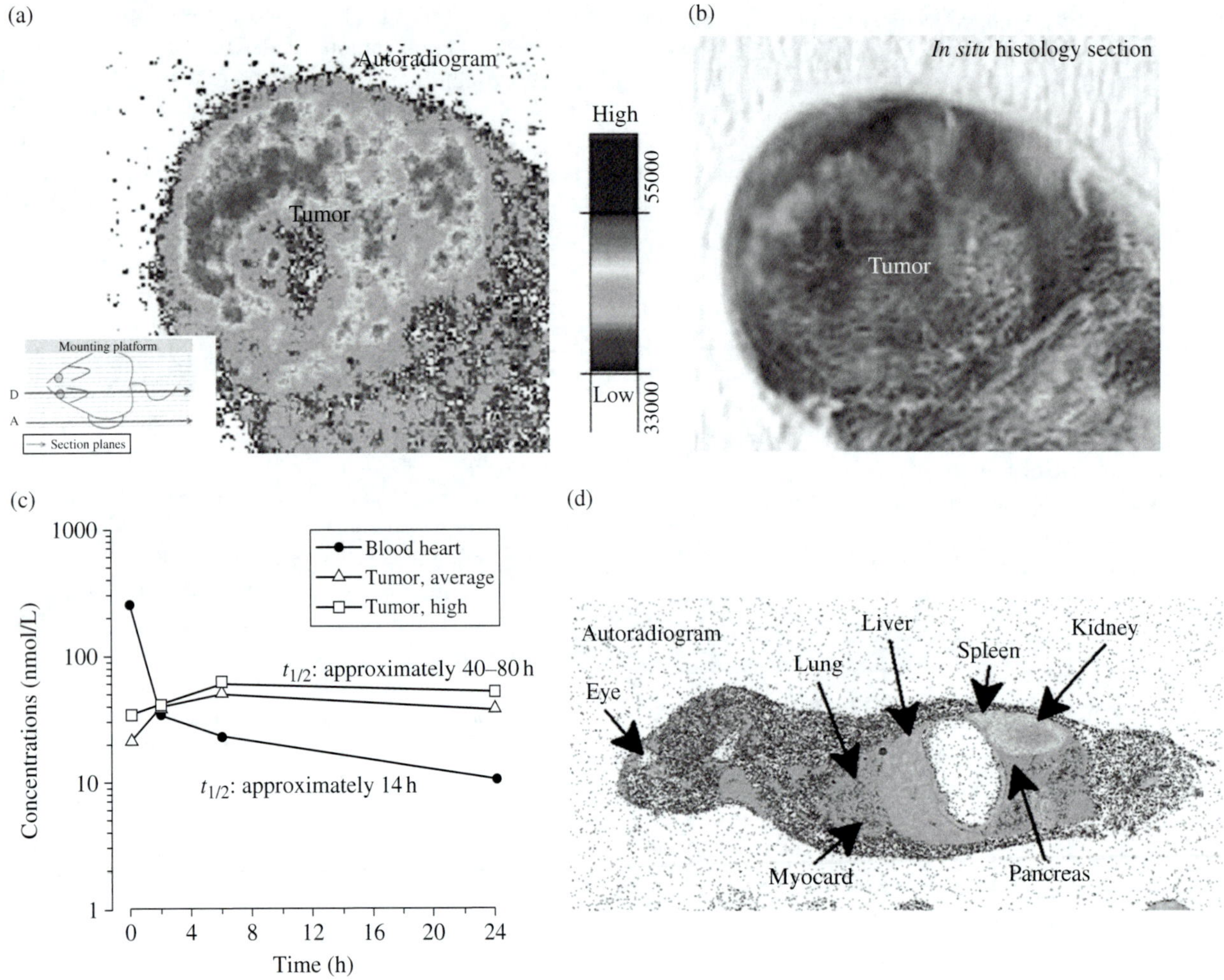

FIGURE 27.4 (a) Autoradiogram of a 50-μm frozen section of a subcutaneous human LNCaP PCa xenograft grown in male NOD/SCID mice 24 h after IV injection of 17 μCi 14C labeled BAY2010112 (910 mCi/mmol). The plane of the section is shown in the schematic diagram of the inset. The distribution of the labeled BiTE antibody was detected by radioluminography and exposure for 4 days. Calibration was performed by simultaneous exposure of nine blood-based internal standards. (b) *In situ* histology of the same native section as used for A. Brown color is attributed to hemoglobin. (c) Pharmacokinetic analysis of radiolabeled PSMA BiTE concentration. In this pilot study, tumor-bearing mice were killed 5 min, 2 h, 6 h, and 24 h after 14C BAY2010112 injection, and its concentration and half-life in the blood and in the tumor were determined based on defined regions of interest. Tumor "high" represents the highest radioactive signal (orange to red color) and tumor "average" includes all tumor areas. Tumor tissue show clear accumulation compared to blood with late t_{max} after 6 h and significantly slower elimination. (d) Whole-body autoradiogram of a section (see inset in (a)) 24 h after injection of the radiolabeled BiTE antibody. Major organs are labeled. Modified based on Friedrich et al. [60]. (*See insert for color representation of this figure.*)

REFERENCES

[1] Baeuerle P. Resurgence of bispecific antibodies. In: Schmidt S, editor. *Fusion Protein Technologies for Biopharmaceuticals: Applications and Challenges*. New York: Wiley; 2013. p 529–543.

[2] Kontermann RE. Dual targeting strategies with bispecific antibodies. MAbs 2012;4 (2):182–197.

[3] Muller D, Kontermann RE. Bispecific antibodies for cancer immunotherapy: current perspectives. BioDrugs 2010;24 (2): 89–98.

[4] Dong J, Sereno A, Snyder WB, Miller BR, Tamraz S, Doern A, Favis M, Wu X, Tran H, Langley E, Joseph I, Boccia A, Kelly R, Wortham K, Wang Q, Berquist L, Huang F, Gao SX, Zhang Y, Lugovskoy A, Martin S, Gouvis H, Berkowitz S, Chiang G, Reff M, Glaser SM, Hariharan K, Demarest SJ. Stable IgG-like bispecific antibodies directed toward the type I insulin-like growth factor receptor demonstrate enhanced ligand blockade and anti-tumor activity. J Biol Chem 2011;286 (6):4703–4717.

[5] Reichert JM, Dhimolea E. The future of antibodies as cancer drugs. Drug Discov Today 2012;17 (17–18):954–963.

[6] Van Beneden K, Verschueren K, Willems W, Wouters H, D'Artois J, De Swert K, Arold G, De Bruyn S. Impact of clinical remission on physical function in patients with rheumatoid arthritis treated with ALX-0061: post-hoc analysis

of Phase I/II data. (The European League Against Rheumatism 2014; 2014; Paris). Ann Rheum Dis 2014;73 (Suppl 2):506. DOI: 10.1136/annrheumdis-2014-eular.2875.

[7] Van Roy M, Van de Sompel A, De Smet K, Jacobs J, Denayer T, Ulrichts H, Baumeister J, Holz JB. ALX-0061, an anti-IL-6R nanobody® for therapeutic use in rheumatoid arthritis, demonstrates *in vitro* a differential biological activity profile as compared to tocilizumab. (The European League Against Rheumatism 2013; 2013; Madrid, Spain). Ann Rheum Dis 2013;72 (Suppl 3):A375. DOI: 10.1136/annrheumdis-2013-eular.1149.

[8] Friedrich M, Raum T, Lutterbuese R, Voelkel M, Deegen P, Rau D, Kischel R, Hoffmann P, Brandl C, Schuhmacher J, Mueller P, Finnern R, Fuergut M, Zopf D, Slootstra JW, Baeuerle PA, Rattel B, Kufer P. Regression of human prostate cancer xenografts in mice by AMG 212/BAY2010112, a novel PSMA/CD3-bispecific BiTE antibody cross-reactive with non-human primate antigens. Mol Cancer Ther 2012;11 (12):2664–2673.

[9] Zimmerman Z, Maniar T, Nagorsen D. Unleashing the clinical power of T cells: CD19/CD3 Bi-specific T cell engager (BiTE®) antibody construct blinatumomab as a potential therapy. Int Immunol 2015;27:31–37.

[10] Seimetz D. Novel monoclonal antibodies for cancer treatment: the trifunctional antibody catumaxomab (removab). J Cancer Educ 2011;2:309–316.

[11] Boehrer S, Schroeder P, Mueller T, Atz J, Chow KU. Cytotoxic effects of the trifunctional bispecific antibody FBTA05 in ex-vivo cells of chronic lymphocytic leukaemia depend on immune-mediated mechanism. Anticancer Drugs 2011;22 (6):519–530.

[12] Buhmann R, Simoes B, Stanglmaier M, Yang T, Faltin M, Bund D, Lindhofer H, Kolb HJ. Immunotherapy of recurrent B-cell malignancies after allo-SCT with Bi20 (FBTA05), a trifunctional anti-CD3 x anti-CD20 antibody and donor lymphocyte infusion. Bone Marrow Transplant 2009;43 (5):383–397.

[13] Schaefer G, Haber L, Crocker LM, Shia S, Shao L, Dowbenko D, Totpal K, Wong A, Lee CV, Stawicki S, Clark R, Fields C, Lewis Phillips GD, Prell RA, Danilenko DM, Franke Y, Stephan JP, Hwang J, Wu Y, Bostrom J, Sliwkowski MX, Fuh G, Eigenbrot C. A two in one antibody against HER3 and EGFR has superior inhibitory activity compared with monospecific antibodies. Cancer Cell 2011;20 (4):472–486.

[14] Kiewe P, Thiel E. Ertumaxomab: a trifunctional antibody for breast cancer treatment. Expert Opin Investig Drugs 2008;17 (10):1553–1558.

[15] McDonagh CF, Huhalov A, Harms BD, Adams S, Paragas V, Oyama S, Zhang B, Luus L, Overland R, Nguyen S, Gu J, Kohli N, Wallace M, Feldhaus MJ, Kudla AJ, Schoeberl B, Nielsen UB. Antitumor activity of a novel bispecific antibody that targets the ErbB2/ErbB3 oncogenic unit and inhibits heregulin-induced activation of ErbB3. Mol Cancer Ther 2012;11 (3):582–593.

[16] Fitzgerald JB, Johnson BW, Baum J, Adams S, Iadevaia S, Tang J, Rimkunas V, Xu L, Kohli N, Rennard R, Razlog M, Jiao Y, Harms BD, Olivier KJ Jr, Schoeberl B, Nielsen UB, Lugovskoy AA. MM-141, an IGF-IR- and ErbB3-directed bispecific antibody, overcomes network adaptations that limit activity of IGF-IR inhibitors. Mol Cancer Ther 2014;13 (2):410–425.

[17] Fleischmann R, Nayiager S, Louw I, Rojkovich B, Fu C, Udata C, Fardipour PA. Multiple ascending dose/proof of concept study of ATN-103 (Ozoralizumab) in rheumatoid arthritis subjects on a background of methotrexate. (American College of Rheumatology/Association of Rheumatology Health Professionals Annual Scientific Meeting 2011: 2011; Chicago, Illinois). Arthritis Rheum 2011;63 (Suppl 10):2630.

[18] Rao E, Li D, Underwood S, Mikol V, Davison M, Kruip J. Generation of a tetravalent bispecific antibody against IL4 and IL13 for the treatment of idiopathic pulmonary fibrosis. (World Bispecific Antibody Summit, September 27–28, 2011). mAbs 2011;4 (1):14–16.

[19] Brischwein K, Schlereth B, Guller B, Steiger C, Wolf A, Lutterbuese R, Offner S, Locher M, Urbig T, Raum T, Kleindienst P, Wimberger P, Kimmig R, Fichtner I, Kufer P, Hofmeister R, da Silva AJ, Baeuerle PA. MT110: a novel bispecific single-chain antibody construct with high efficacy in eradicating established tumors. Mol Immunol 2006;43 (8):1129–1143.

[20] English DP, Bellone S, Schwab CL, Roque DM, Lopez S, Bortolomai I, Cocco E, Bonazzoli E, Chatterjee S, Ratner E, Silasi DA, Azodi M, Schwartz PE, Rutherford TJ, Santin AD. Solitomab, an epithelial cell adhesion molecule/CD3 bispecific antibody (BiTE), is highly active against primary chemotherapy-resistant ovarian cancer cell lines *in vitro* and fresh tumor cells *ex vivo*. Cancer 2015;121 (3):403–412. DOI: 10.1002/cncr.29062. [Epub 2014 Sep 23].

[21] Spiess C, Merchant M, Huang A, Zheng Z, Yang NY, Peng J, Ellerman D, Shatz W, Reilly D, Yansura DG, Scheer JM. Bispecific antibodies with natural architecture produced by co-culture of bacteria expressing two distinct half-antibodies. Nat Biotechnol 2013;31 (8):753–758.

[22] Bostrom J, Yu SF, Kan D, Appleton BA, Lee CV, Billeci K, Man W, Peale F, Ross S, Wiesmann C, Fuh G. Variants of the antibody herceptin that interact with HER2 and VEGF at the antigen binding site. Science 2009;323 (5921):1610–1614.

[23] Baeuerle PA, Kufer P, Bargou R. BiTE: teaching antibodies to engage T-cells for cancer therapy. Curr Opin Mol Ther 2009;11 (1):22–30.

[24] Nagorsen D, Kufer P, Baeuerle PA, Bargou R. Blinatumomab: a historical perspective. Pharmacol Ther 2012;136 (3):334–342.

[25] Rudnick SI, Adams GP. Affinity and avidity in antibody-based tumor targeting. Cancer Biother Radiopharm 2009;24 (2):155–161.

[26] McCall AM, Shahied L, Amoroso AR, Horak EM, Simmons HH, Nielson U, Adams GP, Schier R, Marks JD, Weiner LM. Increasing the affinity for tumor antigen enhances bispecific antibody cytotoxicity. J Immunol 2001;166 (10):6112–6117.

[27] Bortoletto N, Scotet E, Myamoto Y, D'Oro U, Lanzavecchia A. Optimizing anti-CD3 affinity for effective T cell targeting against tumor cells. Eur J Immunol 2002;32 (11):3102–3107.

[28] Yu YJ, Zhang Y, Kenrick M, Hoyte K, Luk W, Lu Y, Atwal J, Elliott JM, Prabhu S, Watts RJ, Dennis MS. Boosting brain uptake of a therapeutic antibody by reducing its affinity for a transcytosis target. Sci Transl Med 2011;3 (84):84ra44.

[29] Croasdale R, Wartha K, Schanzer JM, Kuenkele KP, Ries C, Mayer K, Gassner C, Wagner M, Dimoudis N, Herter S, Jaeger C, Ferrara C, Hoffmann E, Kling L, Lau W, Staack

RF, Heinrich J, Scheuer W, Stracke J, Gerdes C, Brinkmann U, Umana P, Klein C. Development of tetravalent IgG1 dual targeting IGF-1R-EGFR antibodies with potent tumor inhibition. Arch Biochem Biophys 2012;526 (2):206–218.

[30] Jin P, Zhu Z. The design and engineering of IgG-like bispecific antibodies. In: Kontermann RE, editor. *Bispecific Antibodies*. New York: Springer; 2011. p 151–165.

[31] Deng R, Iyer S, Theil FP, Mortensen DL, Fielder PJ, Prabhu S. Projecting human pharmacokinetics of therapeutic antibodies from nonclinical data: what have we learned? MAbs 2011;3 (1):61–66.

[32] Kamath AV, Lu D, Gupta P, Jin D, Xiang H, Wong A, Leddy C, Crocker L, Schaefer G, Sliwkowski MX, Damico-Beyer LA. Preclinical pharmacokinetics of MEHD7945A, a novel EGFR/HER3 dual-action antibody, and prediction of its human pharmacokinetics and efficacious clinical dose. Cancer Chemother Pharmacol 2012;69 (4):1063–1069.

[33] List T, Neri D. Biodistribution studies with tumor-targeting bispecific antibodies reveal selective accumulation at the tumor site. MAbs 2012;4 (6):775–783.

[34] Yazaki PJ, Lee B, Channappa D, Cheung CW, Crow D, Chea J, Poku E, Li L, Andersen JT, Sandlie I, Orcutt KD, Wittrup KD, Shively JE, Raubitschek A, Colcher D. A series of anti-CEA/anti-DOTA bispecific antibody formats evaluated for pre-targeting: comparison of tumor uptake and blood clearance. Protein Eng Des Sel 2013;26 (3):187–193.

[35] Bargou R, Leo E, Zugmaier G, Klinger M, Goebeler M, Knop S, Noppeney R, Viardot A, Hess G, Schuler M, Einsele H, Brandl C, Wolf A, Kirchinger P, Klappers P, Schmidt M, Riethmüller G, Reinhardt C, Baeuerle PA, Kufer P. Tumor regression in cancer patients by very low doses of a T cell-engaging antibody. Science 2008;321 (5891):974–977.

[36] Kontermann RE. Strategies for extended serum half-life of protein therapeutics. Curr Opin Biotechnol 2011;22 (6):868–876.

[37] Muller D, Karle A, Meissburger B, Hofig I, Stork R, Kontermann RE. Improved pharmacokinetics of recombinant bispecific antibody molecules by fusion to human serum albumin. J Biol Chem 2007;282 (17):12650–12660.

[38] Sleep D, Cameron J, Evans LR. Albumin as a versatile platform for drug half-life extension. Biochim Biophys Acta 2013;1830 (12):5526–5534.

[39] Hijazi, Y., Klinger, M., Schub, A., Wu, B., Zhu, M., Kufer, P., Wolf, A., Nagorsen, D. Blinatumomab exposure and pharmacodynamic response in patients with non-Hodgkin lymphoma (NHL). (2013 ASCO Annual Meeting. vol. 31, 2013; Chicago, IL): J Clin Oncol; 2013, 31, 3051 May 20 Supplement: 3051.

[40] Hoffman LM, Gore L. Blinatumomab, a bi-specific anti-CD19/CD3 BiTE((R)) antibody for the treatment of acute lymphoblastic leukemia: perspectives and current pediatric applications. Front Oncol 2014;4:63.

[41] Baumann A. Nonclinical development of biopharmaceuticals. Drug Discov Today 2009;14 (23–24):1112–1122.

[42] Vugmeyster Y, Zhang YE, Zhong X, Wright J, Leung SS. Pharmacokinetics of anti-IL17A and anti-IL22 peptide-antibody bispecific genetic fusions in mice. Int Immunopharmacol 2014;18 (2):225–227.

[43] Ruf P, Kluge M, Jager M, Burges A, Volovat C, Heiss MM, Hess J, Wimberger P, Brandt B, Lindhofer H. Pharmacokinetics, immunogenicity and bioactivity of the therapeutic antibody catumaxomab intraperitoneally administered to cancer patients. Br J Clin Pharmacol 2010;69 (6):617–625.

[44] Sampei Z, Igawa T, Soeda T, Okuyama-Nishida Y, Moriyama C, Wakabayashi T, Tanaka E, Muto A, Kojima T, Kitazawa T, Yoshihashi K, Harada A, Funaki M, Haraya K, Tachibana T, Suzuki S, Esaki K, Nabuchi Y, Hattori K. Identification and multidimensional optimization of an asymmetric bispecific IgG antibody mimicking the function of factor VIII cofactor activity. PLoS One 2013;8 (2):e57479.

[45] Stork R, Campigna E, Robert B, Muller D, Kontermann RE. Biodistribution of a bispecific single-chain diabody and its half-life extended derivatives. J Biol Chem 2009;284 (38):25612–25619.

[46] Jiang G, Lee CW, Wong PY, Gazzano-Santoro H. Evaluation of semi-homogeneous assay formats for dual-specificity antibodies. J Immunol Methods 2013;387 (1–2):51–56.

[47] Tovey MG, Lallemand C. Immunogenicity and other problems associated with the use of biopharmaceuticals. Ther Adv Drug Saf 2011;2 (3):113–128.

[48] Friedrich SW, Lin SC, Stoll BR, Baxter LT, Munn LL, Jain RK. Antibody-directed effector cell therapy of tumors: analysis and optimization using a physiologically based pharmacokinetic model. Neoplasia 2002;4 (5):449–463.

[49] Bakacs T, Lee J, Moreno MB, Zacharchuk CM, Cole MS, Tso JY, Paik CH, Ward JM, Segal DM. A bispecific antibody prolongs survival in mice bearing lung metastases of syngeneic mammary adenocarcinoma. Int Immunol 1995;7 (6):947–955.

[50] Dmitriev DA, Massino YS, Segal OL, Smirnova MB, Pavlova EV, Gurevich KG, Gnedenko OV, Ivanov YD, Kolyaskina GI, Archakov AI, Osipov AP, Dmitriev AD, Egorov AM. Analysis of the binding of bispecific monoclonal antibodies with immobilized antigens (human IgG and horseradish peroxidase) using a resonant mirror biosensor. J Immunol Methods 2002;261 (1–2):103–118.

[51] Poon GM. Quantitative analysis of affinity enhancement by noncovalently oligomeric ligands. Anal Biochem 2013;433 (1):19–27.

[52] Harms BD, Kearns JD, Iadevaia S, Lugovskoy AA. Understanding the role of cross-arm binding efficiency in the activity of monoclonal and multispecific therapeutic antibodies. Methods 2014;65 (1):95–104.

[53] Bumbaca D, Yu J, Couch J, Kenrick M, Atwal J, Dennis M, Joseph S, Ramanujan S, Prabhu S, Watts RJ, Gadkar, K. 2014. Mathematical PKPD and safety model of bispecific TfR/BACE1 antibodies for the optimization of antibody uptake in brain. American Association of Pharmaceutical Scientists. Available at abstracts.aaps.org/Verify/NBC14/PosterSubmissions/T2052.pdf. Accessed 2015 Jun 5.

[54] Vassar R, Kuhn PH, Haass C, Kennedy ME, Rajendran L, Wong PC, Lichtenthaler SF. Function, therapeutic potential and cell biology of BACE proteases: current status and future prospects. J Neurochem 2014;130 (1):4–28.

[55] Yan R, Vassar R. Targeting the beta secretase BACE1 for Alzheimer's disease therapy. Lancet Neurol 2014;13 (3):319–329.

[56] Atwal JK, Chen Y, Chiu C, Mortensen DL, Meilandt WJ, Liu Y, Heise CE, Hoyte K, Luk W, Lu Y, Peng K, Wu P, Rouge L, Zhang Y, Lazarus RA, Scearce-Levie K, Wang W, Wu Y, Tessier-Lavigne M, Watts RJ. A therapeutic antibody targeting BACE1 inhibits amyloid-beta production *in vivo*. Sci Transl Med 2011;3 (84):84ra43.

[57] Couch JA, Yu YJ, Zhang Y, Tarrant JM, Fuji RN, Meilandt WJ, Solanoy H, Tong RK, Hoyte K, Luk W, Lu Y, Gadkar K, Prabhu S, Ordonia BA, Nguyen Q, Lin Y, Lin Z, Balazs M, Scearce-Levie K, Ernst JA, Dennis MS, Watts RJ. Addressing safety liabilities of TfR bispecific antibodies that cross the blood-brain barrier. Sci Transl Med 2013;5 (183):183ra157, 1–12.

[58] Stutz CC, Zhang X, Shusta EV. Combinatorial approaches for the identification of brain drug delivery targets. Curr Pharm Des 2014;20 (10):1564–1576.

[59] Le Couter J, Gadkar K, Elliott JM, Lee T, Meng YG, Zhang L, Kenrick M, Prabhu S, Scheer J. Fragment crystallizable (Fc) region results in an increased systemic exposure with no significant difference in intra-ocular pharmacokinetics. J Invest Ophthalmol Vis Sci 2013;54 (6):1967.

[60] Friedrich M, Deegen P, Voelkel M, Wahl J, Schuhmacher J, Steinke W, Zopf D, Harris RB, Baeuerle PA, Kufer P, Rattel B. Abstract 3526: Subcutaneous administration of PSMA/CD3-bispecific BiTE antibody MT112/BAY 2010112 leads to complete remission of human prostate cancer xenografts in mice. (Proceedings of the 103rd Annual Meeting of the American Association for Cancer Research; 2012 Mar 31-Apr 4; Chicago, IL. Philadelphia, PA: AACR). Cancer Res 2012;72(8 Suppl), Abstract nr : 3526. DOI: 1538-7445. AM2012-3526.

INDEX

ADME and Translational Pharmacokinetics/Pharmacodynamics of Therapeutic Proteins: Applications in Drug Discovery and Development, First Edition. Edited by Honghui Zhou and Frank-Peter Theil.

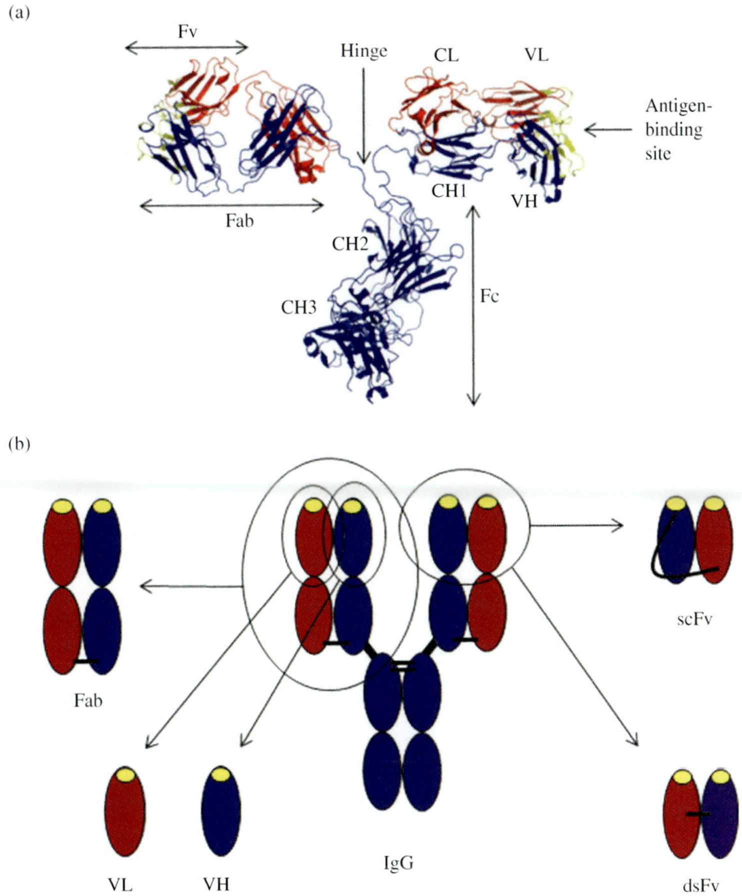

FIGURE 2.1 See text page 15 for figure caption.

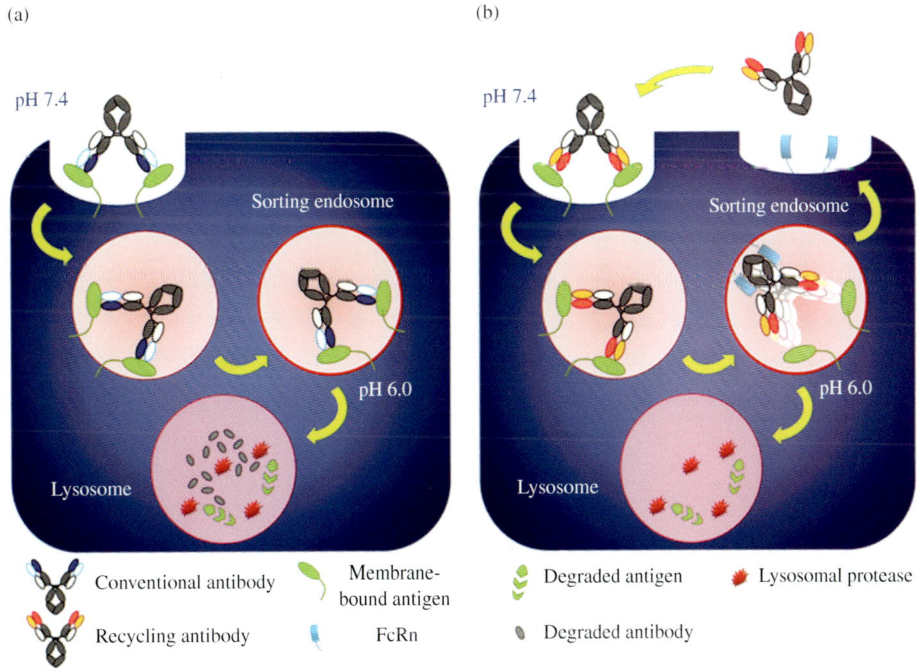

FIGURE 3.7 See text page 30 for figure caption.

ADME and Translational Pharmacokinetics / Pharmacodynamics of Therapeutic Proteins: Applications in Drug Discovery and Development,
First Edition. Edited by Honghui Zhou and Frank-Peter Theil.

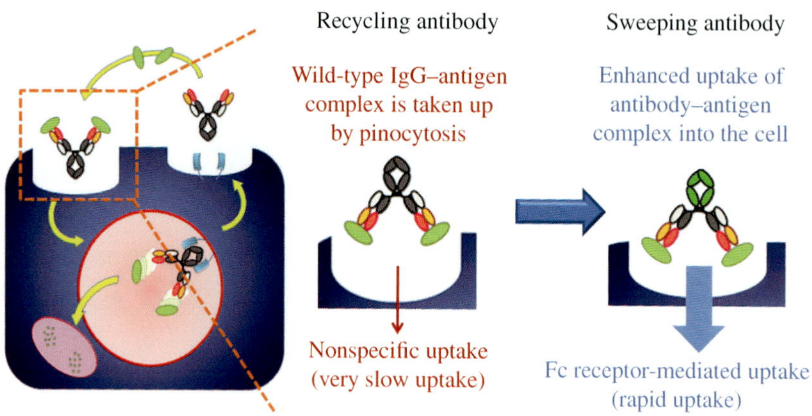

FIGURE 3.8 See text page 31 for figure caption.

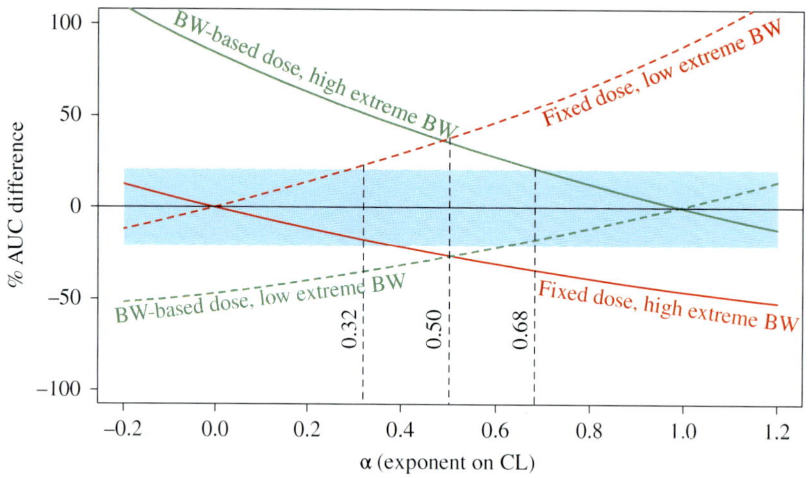

FIGURE 9.1 See text page 109 for figure caption.

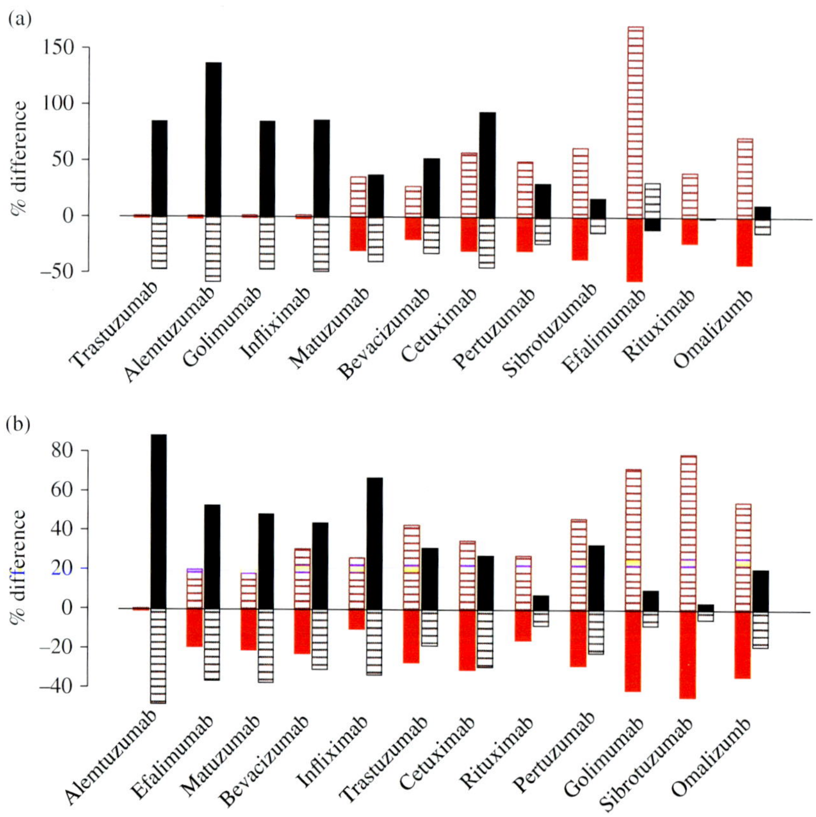

FIGURE 9.3 See text page 115 for figure caption.

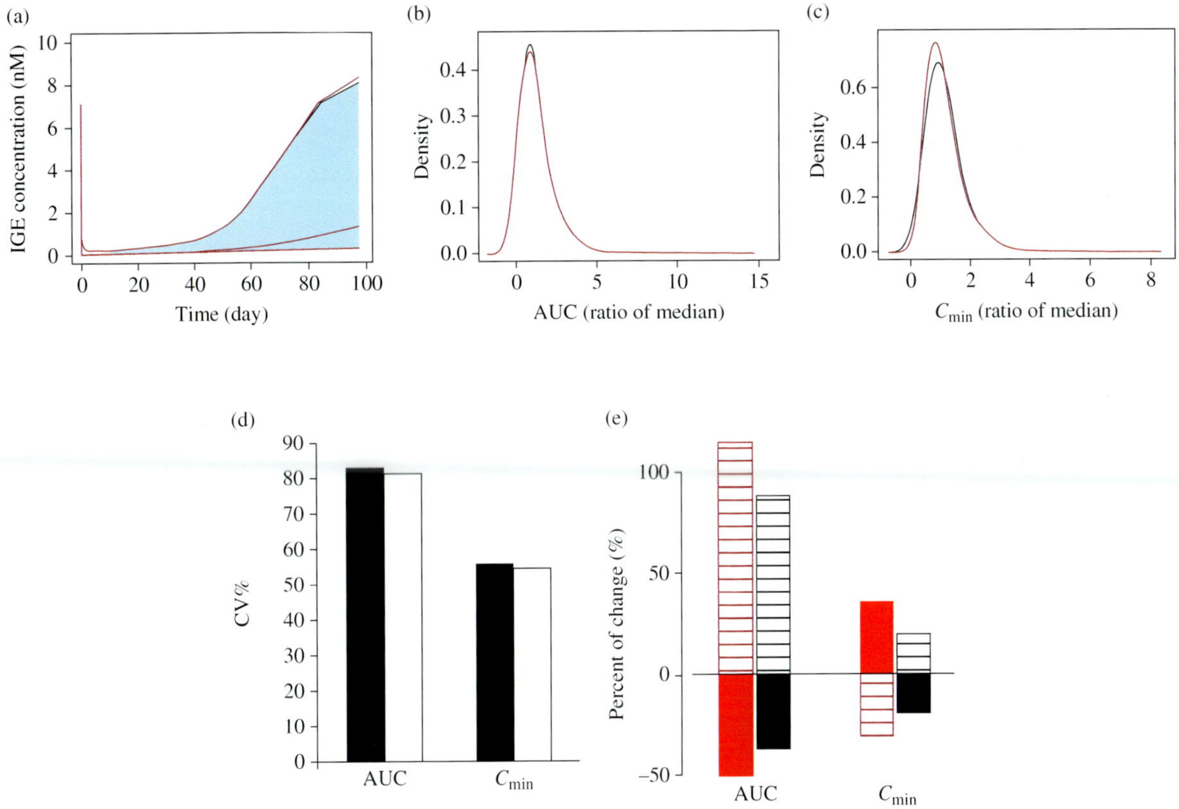

FIGURE 9.4 See text page 116 for figure caption.

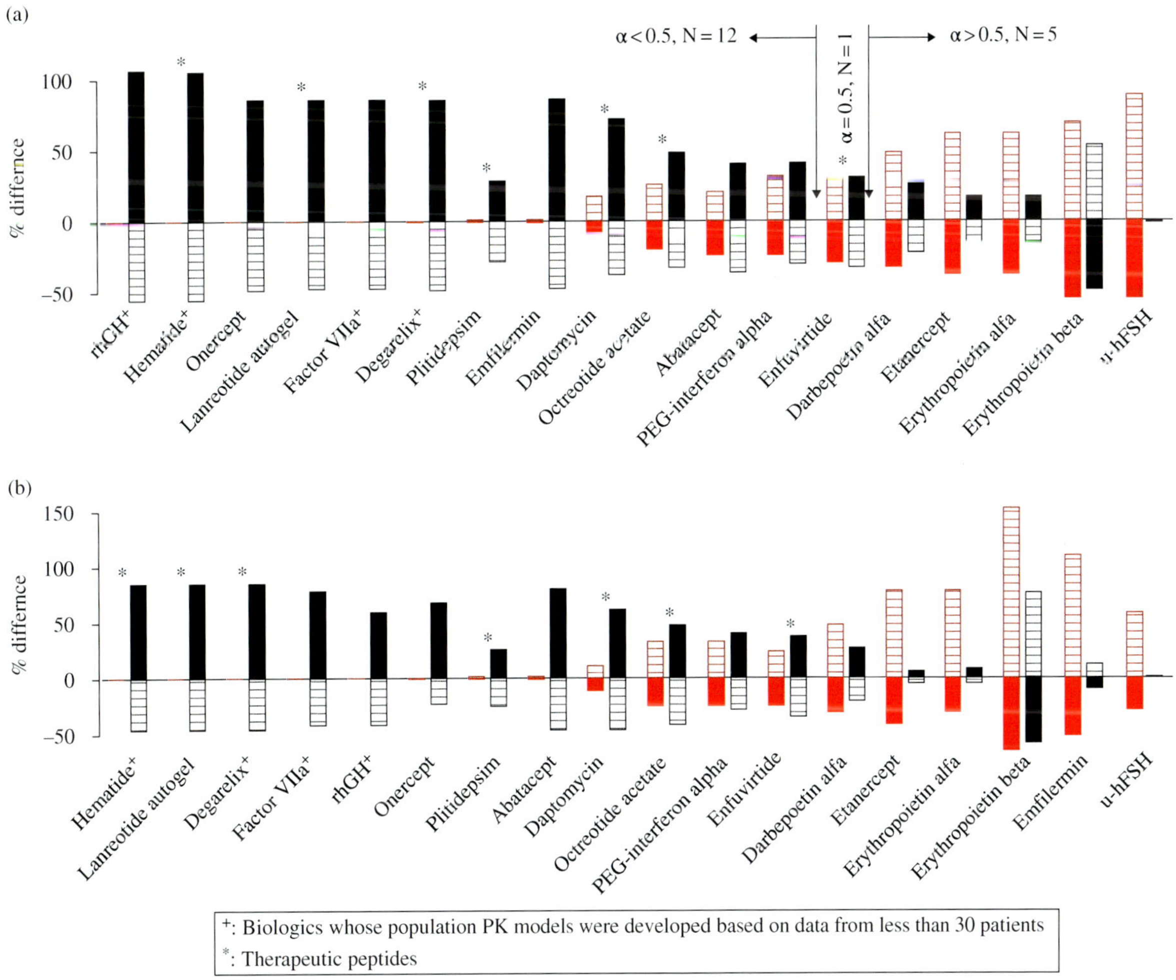

FIGURE 9.6 See text page 119 for figure caption.

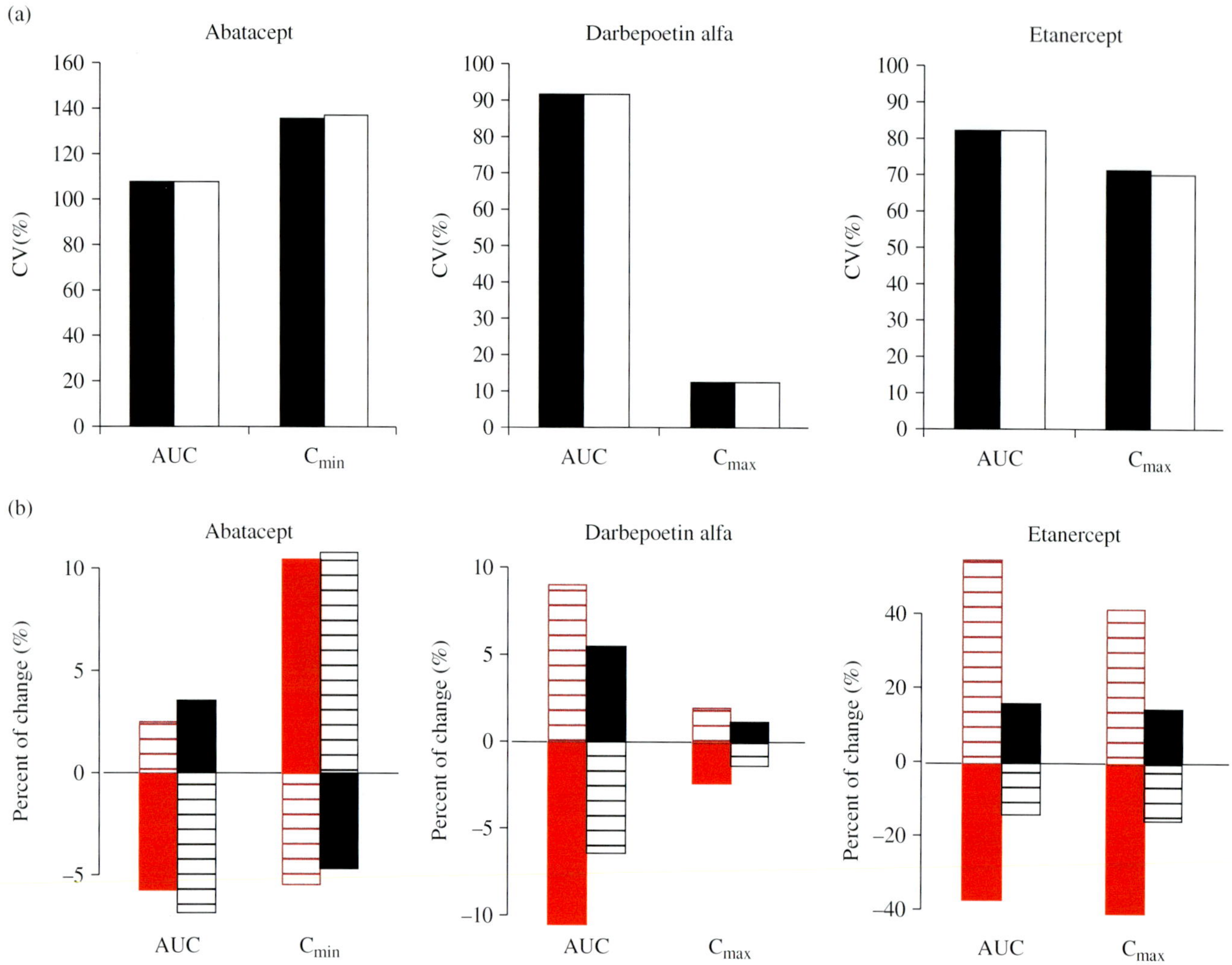

FIGURE 9.7 See text page 121 for figure caption.

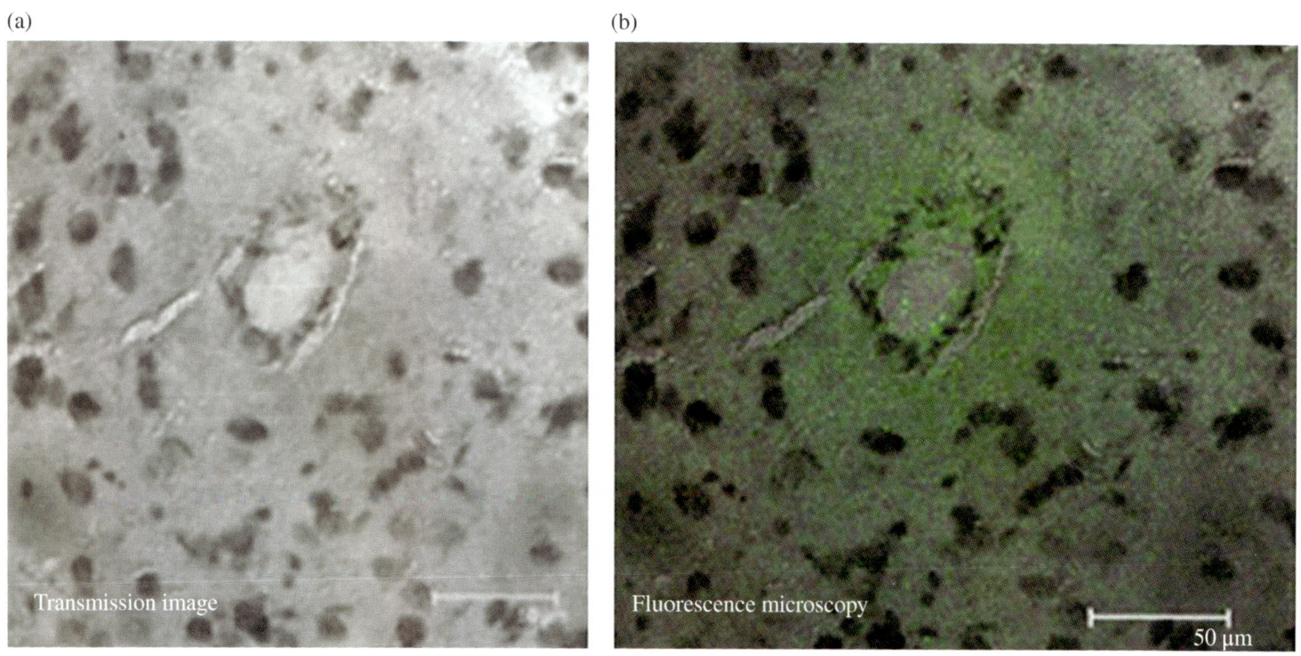

FIGURE 17.3 See text page 250 for figure caption.

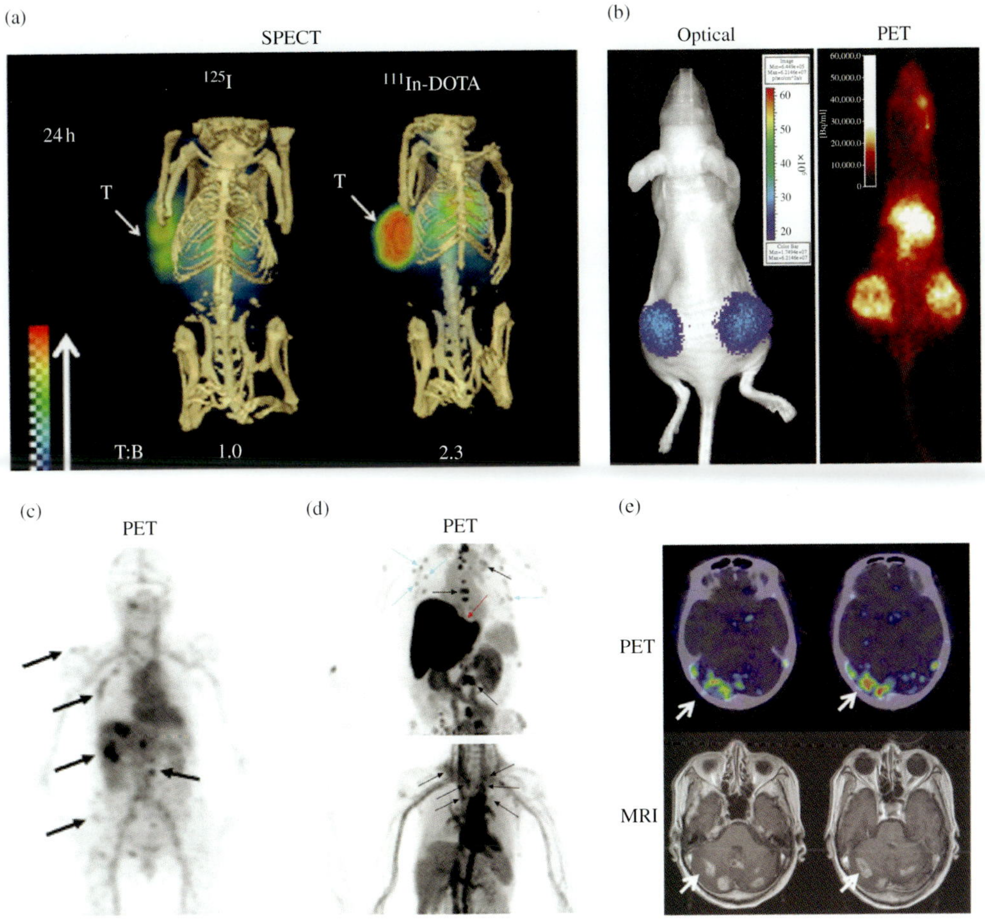

FIGURE 19.5 See text page 287 for figure caption.

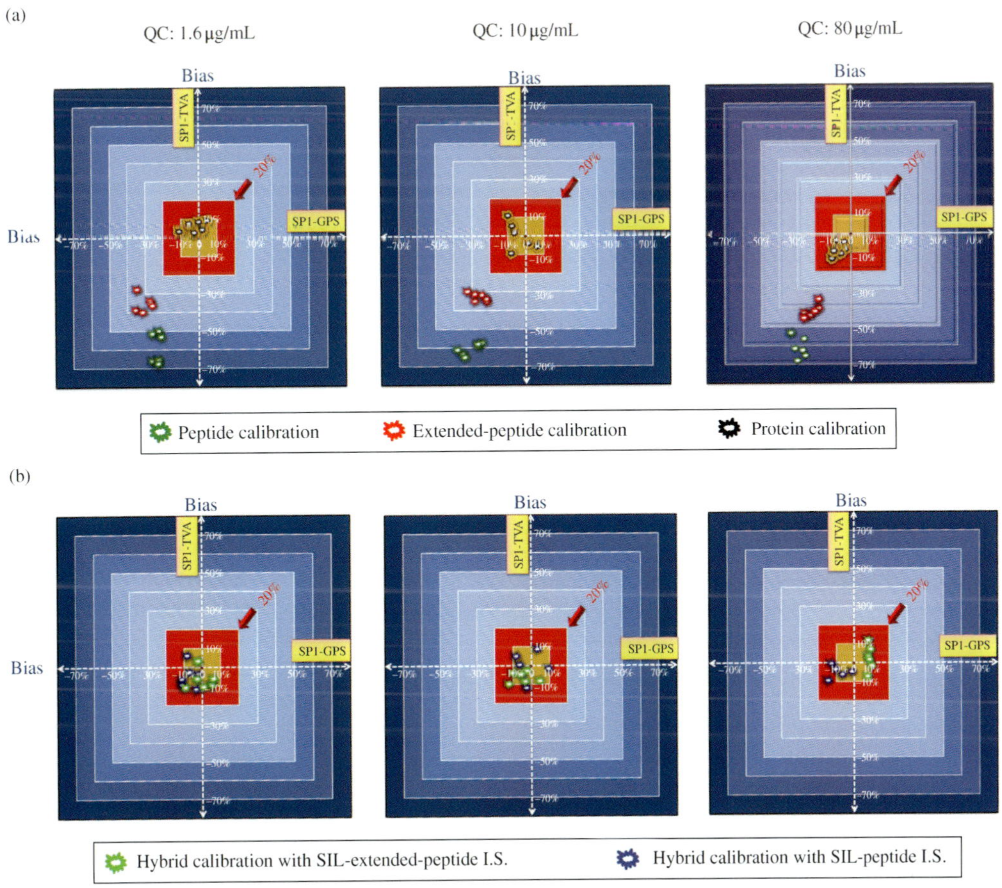

FIGURE 21.3 See text page 321 for figure caption.

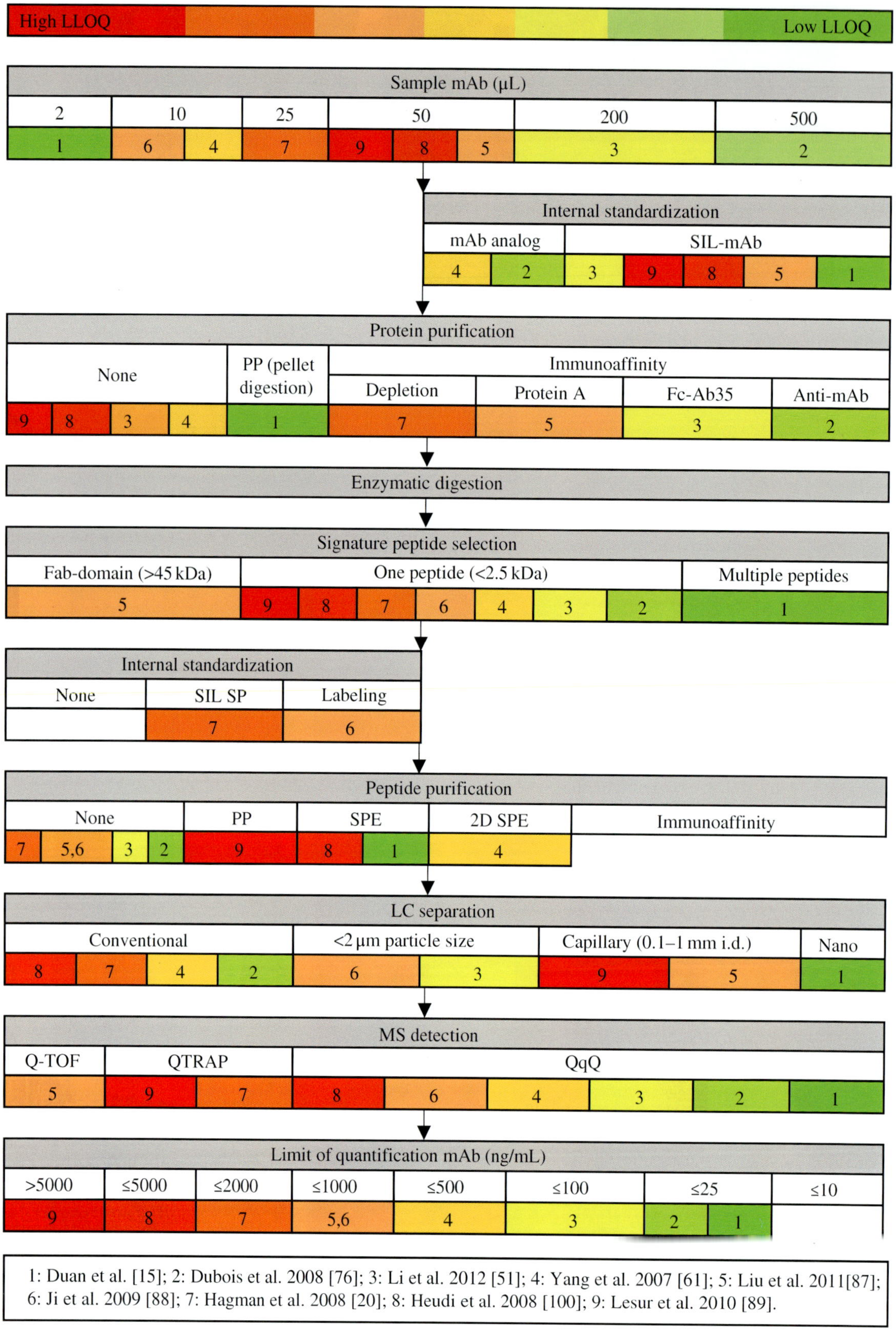

FIGURE 21.4 See text page 322 for figure caption.

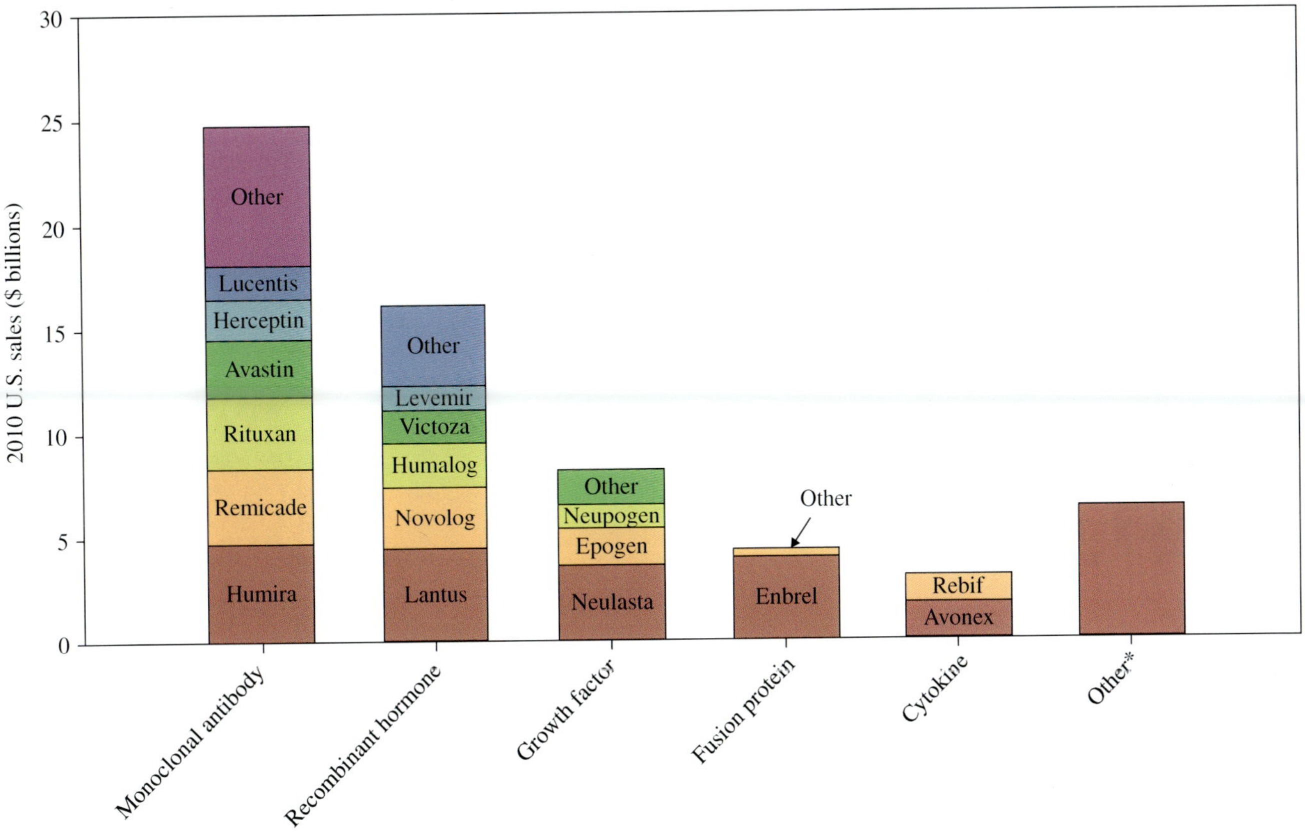

FIGURE 24.1 See text page 370 for figure caption.

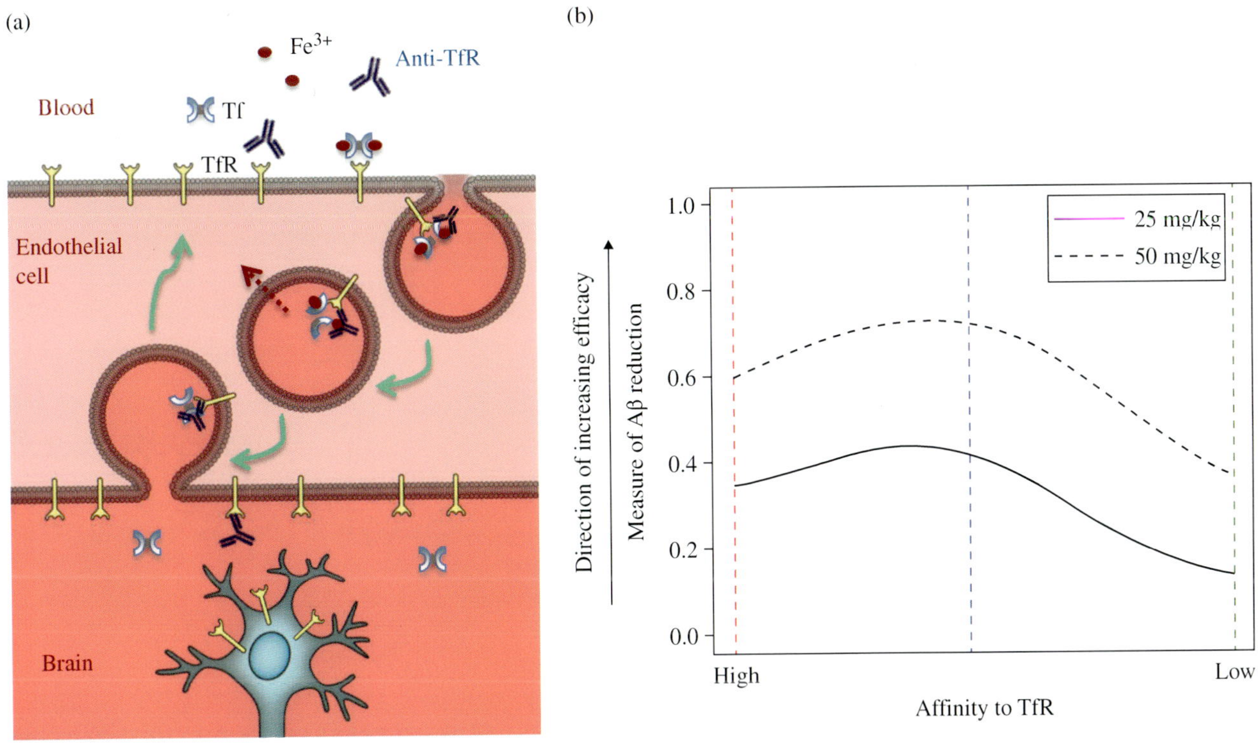

FIGURE 27.2 See text page 419 for figure caption.

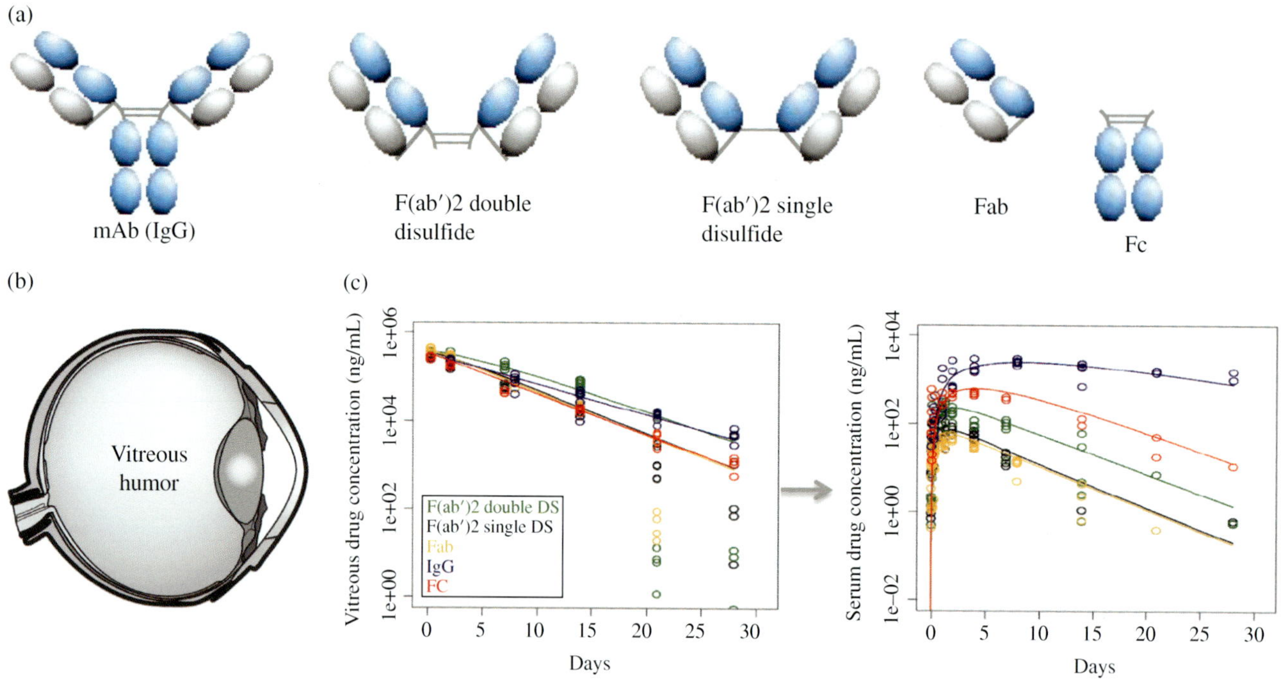

FIGURE 27.3 See text page 420 for figure caption.

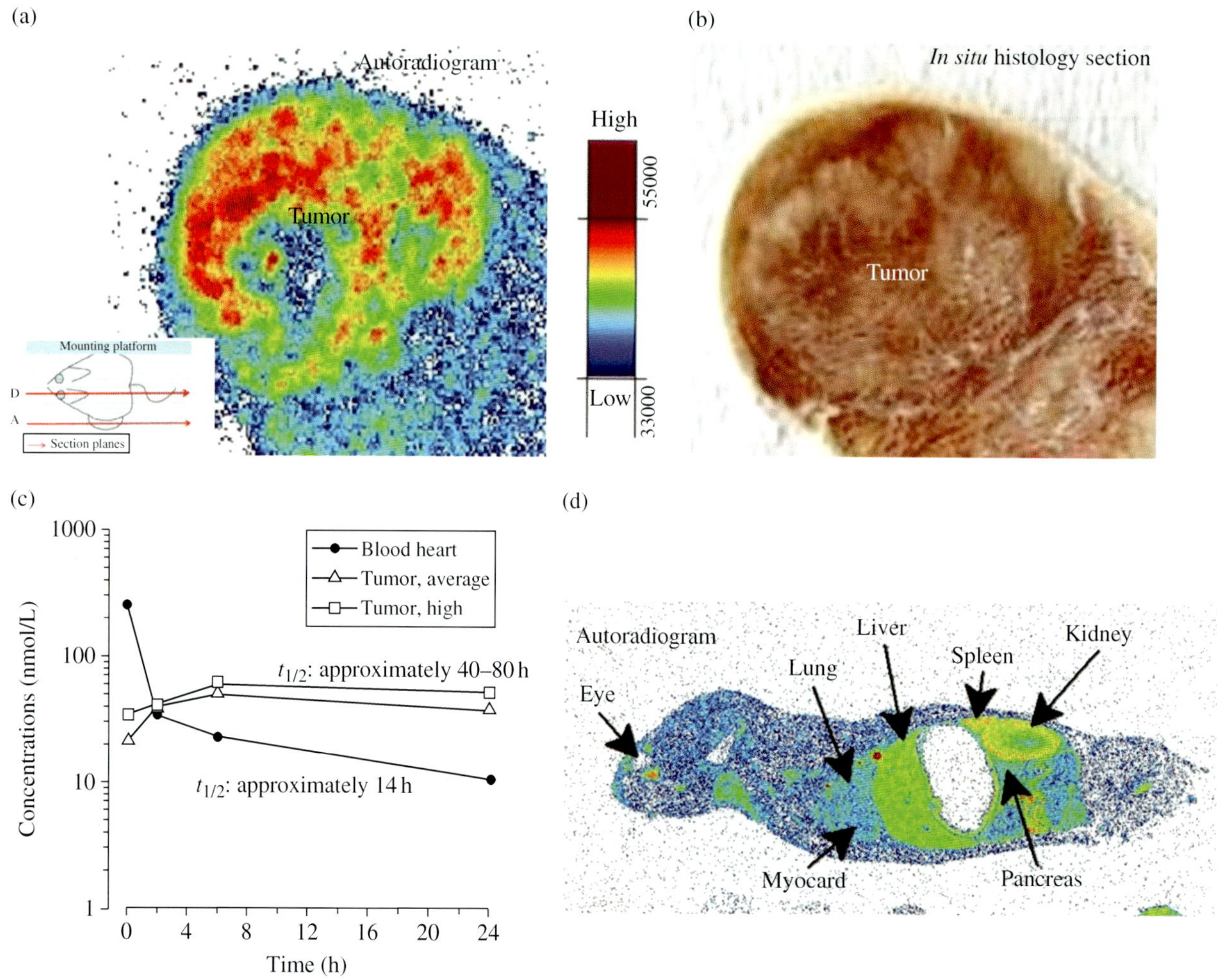

FIGURE 27.4 See text page 422 for figure caption.

Printed in the USA/Agawam, MA
September 21, 2026

910661.039